Biopolymers

Edited by A. Steinbüchel

Volume 7
Polyamides and Complex Proteinaceous Materials I

Edited by S. R. Fahnestock and A. Steinbüchel

Biopolymers

Edited by A. Steinbüchel

Volume 7
Polyamides and Complex Proteinaceous Materials I

Edited by
S. R. Fahnestock and A. Steinbüchel

Editors:

Prof. Dr. Alexander Steinbüchel
Institut für Mikrobiologie
Westfälische Wilhelms-Universität
Corrensstrasse 3
D-48149 Münster
Germany

Dr. Stephen R. Fahnestock
E.I. DuPont de Nemours & Co.
Experimental Station
Wilmington, DE 19880-0328
USA

Library of Congress Card No: applied for

British Library Cataloguing-in-Publication Data: A catalogue record for this book is available from the British Library

Bibliographic information published by Die Deutsche Bibliothek
Die Deutsche Bibliothek lists this publication in the Deutsche Nationalbibliografie; detailed bibliographic data is available in the Internet at <http://dnb.ddb.de>.

Printed in the Federal Republic of Germany
Printed on acid-free paper.

Composition, Printing, and Bookbinding: Konrad Triltsch
Print und digitale Medien GmbH,
Ochsenfurt-Hohestadt

ISBN 3-527-30222-0

Preface

Biopolymers and their derivatives are diverse, abundant, important for life, they exhibit fascinating properties and are of increasing importance for various applications. Living matter is able to synthesize an overwhelming variety of polymers, which can be divided into eight major classes according to their chemical structure: (1) nucleic acids such as ribonucleic acids and deoxyribonucleic acids, (2) polyamides such as proteins and poly(amino acids), (3) polysaccharides such as cellulose, starch and xanthan, (4) organic polyoxoesters such as poly(hydroxyalkanoic acids), poly(malic acid) and cutin, (5) polythioesters, which were reported only recently, (6) inorganic polyesters with polyphosphate as the only example, (7) polyisoprenoids such as natural rubber or Gutta Percha and (8) polyphenols such as lignin or humic acids.

Biopolymers occur in any organism, and in most organisms they contribute to the by far major fraction of the cellular dry matter. Biopolymers possess a wide range of different essential or beneficial functions for the organisms: conservation and expression of genetic information, catalysis of reactions, storage of carbon, nitrogen, phosphorus and other nutrients and of energy, defense and protection against the attack of other cells or hazardous environmental or intrinsic factors, sensors of biotic and abiotic factors, communication with the environment and other organisms, mediators of adhesion to surfaces of other organisms or of non-living matter and many more. In addition, many biopolymers are structural components of cells, tissues, and whole organisms.

To fulfil all these different functions, biopolymers must exhibit rather diverse properties. They must very specifically interact with a large variety of different substances, components and materials, and often they must have extraordinarily high affinities to them. Finally, many of them must have a high strength. Some of these properties are utilized directly or indirectly for various applications. This and the possibility to produce them from renewable resources, as living matter mostly does, make biopolymers interesting candidates to industry.

Basic and applied research have already revealed much knowledge on the enzyme systems catalyzing biosynthesis, degradation and modification of biopolymers as well as on the properties of biopolymers. This has also resulted in an increased interest in biopolymers for various applications in industry, medicine, pharmacy, agriculture, electronics and various other areas. However, considering the developments during the last two decades and reviewing the literature shows that our knowledge is still scarce. The genes for the biosynthesis pathways of many biopolymers are still not available or were identified only recently, many new biopolymers have just been described, and from only a minor fraction of biopolymers the biological, chemical, physical and material properties have been investigated. Often promising biopolymers are not available in sufficient amounts. Nevertheless, polymer chemists, engineers and material scientists in academia and industry have discovered biopolymers as chemicals and materials for many new applications, or they consider biopolymers as models to design novel synthetic polymers.

The first edition of this multivolume handbook comprehensively reviews and compiles information on biopolymers in 10 volumes covering (a) occurrence, synthesis, isolation and production, (b) properties and applications, (c) biodegradation and modification not only of natural but also of synthetic polymers, and (e) the relevant analysis methods to reveal the structures and properties. Volumes 1-8 are structured according to the chemical classes of biopolymers, whereas Volume 9 focusses on aspects of the biodegradation of synthetic polymers and Volume 10 deals with general aspects related to biopolymers.

This book series will hopefully be helpful to many scientists, physicians, pharmaceutics, engineers and other experts in a wide variety of different disciplines, in academia and in industry. It may not only support research and development but may be also suitable for teaching.

Publishing of this book series was achieved by chosing volume editors and authors of the individual volumes and chapters for their recognized expertise and for their excellent contributions to the various fields of research. I am very grateful to these scientists for their willingness to contribute to this reference work and for their engagement. Without them and without their comitment and enthusiasm it would have not been possible to compile such a book series.

I am also very grateful to the publisher WILEY-VCH for recognizing the demand for such a book series, for taking the risk to start such a big new project and for realizing the publication of *Biopolymers* in excellent quality. Special thanks are due to Karin Dembowsky and many of her WILEY-VCH colleagues, especially from production and marketing, for their constant effort, their helpful suggestions, constructive criticism, and wonderful ideas.

Last but not least I would like to thank my family for their patience, and I have to excuse for the many hours the preparation of this book series kept me away from them.

Münster, February 2001 Alexander Steinbüchel

Introduction

Proteins and polyamides are essential components of every biological system. They are catalysts of diverse capability, agents of molecular recognition, directors of self-assembly, transducers of energy and information, media of communication, producers of directed motion, and librarians of the genetic program. Proteinaceous materials provided by nature have been exploited by human technology for millennia. Silks, furs, leather, bone, horn, feather, all have been and many still remain, in spite of modern chemistry, essential materials for all human cultures.

The second half of the twentieth century saw vast advances in our understanding of proteins and our ability to exploit them, given momentum by the insights of the central dogma of molecular biology in the 1950s and ever increasing velocity by the biotechnology revolution it has spawned. Yet as historic as this progress has been, one cannot escape the impression that the biotechnology of protein-based materials is still in its infancy a big bang of creation just beginning.

In this volume and its companion volume 8 we hope to transmit a view of a field in which progress has been broad and deep, but in which change is dizzyingly rapid and promises revolutions to come. The first chapter in this volume surveys the recent advances in the biochemistry of protein synthesis by ribosomes translating the genetic code, then chapter 2 describes how the repertoire of amino acid monomers can be extended by biochemistry. The next four chapters discuss polyamides synthesized by non-ribosomal processes (chapters 3–6). Chapter 7 deals with polyaspartic acid, which is so far only available by synthetic processes; however there is a strong interest to make this biodegradable polymer also available through a biotechnological process. Chapters 8 and 9 introduce membrane-associated proteins and the targeting of proteins to and through membranes. Self-assembling protein systems, likely to play key roles in the coming nanotechnology revolution, are the subjects of the next three chapters (chapters 10–12). Chapter 13 discusses a small but important subset of enzymes, those of commercial utility, including several proteases. Physiological protein degradation is the subject of chapter 14, and finally, chapter 15 describes how the properties of proteins can be altered enzymatically or chemically.

We consider it a strength of this collection that the individual chapters are diverse in style and in purpose. Some paint an area in broad, conceptual strokes; others in fine technical detail. Some present information; others arguments or interpretation. Some summarize past accomplishments; others point to future possibilities. Each is, we trust, in some way appropriate to its topic. As vast as this field is, there is of course no hope of completeness. We have attempted to sample broadly, but key omissions are inevitable and we readily acknowledge them. We will have succeeded if some of these chapters stimulate interest in specific problems, or make the field more accessible to newcomers, or document the state of the art at the beginning of the 21st century.

Whatever is accomplished is of course the accomplishment of the authors. We are most grateful to all of them for devoting so much of their valuable time to this endeavor and for sharing their knowledge and insights so generously. We would also like to thank the publisher WILEY-VCH for publishing Biopolymers with their customary professional excellence, and for their expert assistance throughout the conception and development of this volume. Special thanks are due to Karin Dembowsky, who initiated this book series, and to Andreas Sendtko, who continued and further developed it, and to their colleagues, whose constant efforts made this volume possible.

Wilmington and Münster
November 2002

S. R. Fahnestock
A. Steinbüchel

Contents

1
Ribosomal Protein Synthesis

Prof. Dr. Wolfgang Wintermeyer[1], **Prof. Dr. Marina V. Rodnina**[2]

[1] Institut für Molekularbiologie, Universität Witten/Herdecke, Stockumer Straße 10, 58448 Witten, Germany; Tel.: +49-2302-669140; Fax: +49-2302-669117; E-mail: winterme@uni-wh.de

[2] Institut für Physikalische Biochemie, Universität Witten/Herdecke, Stockumer Straße 10, 58448 Witten, Germany; Tel.: +49-2302-669205; Fax: +49-2302-669117; E-mail: rodnina@uni-wh.de

Aa-tRNA	aminoacyl-tRNA
EF	bacterial elongation factor
eEF	eukaryotic elongation factor
IF	bacterial inititiation factor
eIF	eukaryotic initiation factor
IRES	internal ribosome entry site
mRNA	messenger RNA
Pept-tRNA	peptidyl-tRNA
PT	peptidyl transferase
RF	bacterial release factor
eRF	eukaryotic release factor
rRNA	ribosomal RNA
tRNA	transfer RNA
UTR	untranslated region

1 Introduction

Information for the amino acid sequence of proteins is encoded in the nucleotide sequence of the DNA. Gene expression, i.e., the synthesis of a protein encoded in a gene comprises two main phases: transcription of one DNA strand into the complementary RNA copy (messenger RNA, mRNA) and translation of the mRNA into protein by polymerizing amino acids in the sequence specified by the nucleotide sequence of the mRNA. Three nucleotides out of four specify one of the 20 amino acids incorporated into protein. The substrates of protein synthesis are aminoacylated tRNAs (Aa-tRNAs) that at their 3′ end carry specific amino acids in an energy-rich ester bond. Aa-tRNAs are produced by the action of aminoacyl-tRNA synthetases, which each specifically recognize one amino acid, and the tRNA that is specific for this amino acid, as it has the anticodon triplet matching the respective codon triplet. Thus, the tRNA serves as an adaptor between the amino acid and the codon on the mRNA. Protein synthesis takes place on ribosomes that are large macromolecular assemblies consisting of several RNAs (ribosomal RNA, rRNA) and numerous proteins. The ribosome presents the mRNA such that Aa-tRNAs can bind to their respective codon and take part in peptide bond formation catalyzed by the ribosome. In their functions, ribosomes are assisted by a number of accessory enzymes (translation factors).

In prokaryotes, transcription and translation are coupled spatially and functionally, such that ribosomes initiate translation on mRNA while it is being synthesized. In eukaryotes, the two processes are separated spatially, in that transcription and processing of primary transcripts takes place in the nucleus, and mature mRNAs are exported from the nucleus to the cytosol for translation.

Translational control, i.e., the regulation of the efficiency of translation of a given mRNA, is an important means of regulation by which the cell can adapt to changes of its physiological status or to external stimuli. Compared with the control of transcription and posttranscriptional processing, translational control can elicit a more rapid response. It is particularly important in eukaryotes, and in most cases the regulated step is initiation.

Various components of the translational apparatus, including the ribosome and translation factors, are targeted by naturally occurring inhibitors, such as many of the classic antibiotics, that inhibit specific steps of translation. The development of new inhibitors of translation that can be used for therapeutic purposes is an important challenge for research on translation.

2
Historical Outline

Our present knowledge about ribosomal protein synthesis is the result of five decades of research. Soon after DNA had been identified as the molecule that, in its base sequence, encodes the information for the amino acid sequence of proteins, it became clear that DNA does not serve as a template for protein synthesis directly. In the early 1950s, ribosomes were identified as the site of amino acid incorporation into protein, taking place in the cytosol of the cell. Soon thereafter, protein synthesis on ribosomes was demonstrated in bacteria as well. In the late 1950s, the function of transfer RNA (tRNA) in carrying amino acids to the ribosome was elucidated, and the enzymes that attach the amino acids to tRNAs at the expense of ATP (aminoacyl-tRNA synthetases) were characterized. As ribosomes contain RNA, it was assumed for some time that it was ribosomal RNA (rRNA) that served as a template and determined the sequence of the protein to be synthesized. It took several years until, in 1960, the template was identified as messenger RNA (mRNA) that, in eukaryotic cells, was transcribed from DNA in the nucleus and translated in the cytosol. Thus, by the end of the first decade of research on protein synthesis, the important role of different RNAs (rRNA, tRNA, mRNA) in translation had been recognized and the central function of the ribosome was established.

The genetic code, i.e., the meaning of each of the 64 trinucleotide codons of the mRNA, was deciphered in the first half of the 1960s. The first sequence of a nucleic acid, the 76-nucleotide sequence of alanine-specific tRNA, was determined in 1965, and it was soon followed by the determination of several more tRNA sequences. In these years it was also found that, apart from RNA, protein synthesis required soluble protein factors, called translation factors, in order to proceed at a significant rate. The first translation factors to be characterized were the factors that promote partial reactions of the elongation cycle of protein synthesis. The identification of factors for initiation and termination followed, and by the end of the second decade most of the components of the translation apparatus were known. However, the knowledge of the biochemical mechanisms was still in its infancy, and in subsequent years a large body of biochemical information was gathered as a result of the massive efforts of many groups.

One major development during the 1970s was the determination of the primary structures of ribosomal proteins and ribosomal RNA, such that by 1980 the complete sequence of the ribosome from *Escherichia coli* was known and secondary structure models of ribosomal RNAs were developed. Another important milestone in structure determination was reached when the structure of a tRNA molecule was solved in 1974 – the first tertiary structure of a nucleic acid at atomic resolution. At the same time, the architecture of ribosomes at low resolution was visualized by electron microscopy. The topography of functional centers, such as tRNA binding sites, decoding sites, peptidyl transferase centers, and factor binding sites, as well as of individual ribosomal proteins, was studied by using cross-linking and immuno-electron microscopy techniques. The basal biochemical mechanisms of protein synthesis were worked out, showing similarities and differences in prokaryotic and eukaryotic systems.

As a consequence of the advent, in the early 1980s, of catalytic RNA, and of the difficulty in assigning defined biochemical functions to ribosomal proteins, there was a shift from the protein paradigm, which assumed that major ribosome functions were carried out by ribosomal proteins, to the RNA paradigm, attributing both structural and functional roles to ribosomal RNA. Structural studies on the ribosome employed cross-linking and scattering techniques in order to determine interactions between different parts of ribosomal RNAs and between RNA and proteins, as well as the locations and interactions of proteins. These studies provided fairly well-defined, low-resolution structural models of the ribosome. Atomic structures of aminoacyl-tRNA synthetases and their tRNA complexes became available. Improved preparation procedures and the application of recombinant DNA technology led to an ever-increasing sophistication of mechanistic understanding of ribosome function. Many aspects of translational regulation in eukaryotes, of the initiation step in particular, were worked out and linked translation to general regulatory circuits in the cell.

The last decade was dominated by structure, although major steps in unraveling detailed kinetic mechanisms of ribosome function were made in parallel. Atomic structures of translation factors provided hints as to their function on the ribosome. Cryo-electron microscopy provided three-dimensional models of ribosomes and ribosome complexes with various ligands of ever-increasing detail. The structural work culminated in the recent determination, at atomic resolution, of the structure of ribosomal subunits and, at lower resolution, of whole ribosomes with tRNAs bound to their respective sites. These structures, together with advanced enzymology, form the basis for future work toward molecular mechanisms of translation. It has become clear by now that the ribosome has a dynamic structure that changes in response to substrate binding and the action of translation factors and takes an active part in all steps of translation. As to the enzymology of the ribosome, the structures show that the ribosome is a ribozyme whose active sites consist of RNA and whose assembly and functions are assisted by ribosomal proteins.

3
The Genetic Code

The genetic code provides the nucleic-acid alphabet that is used to encode the sequence of proteins in their respective genes. The structure of the code and the mechanism of decoding are determined by the fact that there are 4 nucleic acid bases and at least 20 amino acids to be coded for. Alterations of the coding sequence by mutations frequently lead to changes of the amino acid sequence of the encoded protein. Recoding events during translation can lead to changes in the amino acid incorporated in response to certain codons or to changes in the frame in which a particular mRNA is translated.

3.1 The Four Letters of the Genetic Code

The bases of the mRNA (guanine, G; adenine, A; cytosine, C; uracil, U) are the four letters of the genetic code, of which three each form one word that specifies an amino acid. From the total of 64 (4^3) code words, 61 stand for the 20 standard amino acids to be incorporated into proteins during translation (sense codons); the remaining three serve as termination signals that specify the end of the coding sequence (nonsense codons) (Table 1). Because the number of codons is much larger than the number of amino acids to be coded for (degenerated code), most amino acids (except methionine and tryptophan) are encoded by two or more codons. One codon, AUG, stands for methionine; in a particular mRNA context, it specifies the start of the coding sequence, implying that the first amino acid of newly synthesized proteins is methionine. The codon triplets of the mRNA are decoded by the formation of three complementary base pairs with the anticodons of tRNAs. Elongator tRNAs decode internal codons. Initiation codons are decoded by a particular methionine-specific tRNA, initiator tRNA, which differs from elongator $tRNA^{Met}$ used for decoding internal methionine codons. The three termination codons, UAA, UAG, UGA, stand for "Stop"; these codons are recognized by proteins, the termination (or release) factors, rather than by tRNAs. In a particular mRNA context, UGA also may be decoded as selenocysteine (see Section 3.3), which is therefore the 21st amino acid incorporated into protein during ribosomal protein synthesis. The genetic code is the same in all organisms. Slight deviations, i.e., a few codons with different meanings, were found in mitochondria and ciliates.

3.2 Mutations

Deletions or insertions of bases in DNA/RNA sequences coding for protein nearly

Tab. 1 Genetic code

1st Base	*2nd Base*								*3rd Base*
	U		*C*		*A*		*G*		
U	UUU	Phe	UCU	Ser	UAU	Tyr	UGU	Cys	U
	UUC	Phe	UCC	Ser	UAC	Tyr	UGC	Cys	C
	UUA	Leu	UCA	Ser	UAA	Stop	UGA	Stop	A
	UUG	Leu	UCG	Ser	UAG	Stop	UGG	Trp	G
C	CUU	Leu	CCU	Pro	CAU	His	CGU	Arg	U
	CUC	Leu	CCC	Pro	CAC	His	CGC	Arg	C
	CUA	Leu	CCA	Pro	CAA	Gln	CGA	Arg	A
	CUG	Leu	CCG	Pro	CAG	Gln	CGG	Arg	G
A	AUU	Ile	ACU	Thr	AAU	Asn	AGU	Ser	U
	AUC	Ile	ACC	Thr	AAC	Asn	AGC	Ser	C
	AUA	Ile	ACA	Thr	AAA	Lys	AGA	Arg	A
	AUG	Met	ACG	Thr	AAG	Lys	AGG	Arg	G
G	GUU	Val	GCU	Ala	GAU	Asp	GGU	Gly	U
	GUC	Val	GCC	Ala	GAC	Asp	GGC	Gly	C
	GUA	Val	GCA	Ala	GAA	Glu	GGA	Gly	A
	GUG	Val	GCG	Ala	GAG	Glu	GGG	Gly	G

always (except when three bases are deleted or inserted) lead to a totally different amino acid sequence, as downstream of the mutated position the reading frame is shifted (frameshift mutation).

Single amino acid exchanges are the result of point mutations in which one base in a codon is replaced with another. Base exchanges in the first and second codon positions frequently result in an exchange of the amino acid incorporated at the mutated codon. Base exchanges that lead to the incorporation of chemically related amino acids are referred to as neutral mutations. There is no change in the amino acid sequence when a base exchange, mostly in the third codon position, creates a codon that specifies the same amino acid (silent mutation).

Single-base substitutions that change a sense codon coding for an amino acid into a termination codon (nonsense mutation), for instance UGG (tryptophan) to UGA (stop), are deleterious, as at such mutated positions the synthesis of the protein is terminated prematurely and full-length, functional protein is not made. Truncated proteins formed as a consequence of nonsense mutations may impair cell functions. In eukaryotic cells, mRNAs containing premature stop codons are subject to rapid, preferential nucleolytic degradation (nonsense-mediated decay).

3.3 Recoding

Recoding is a programmed event (not an error) during translation that leads to a protein that differs in sequence from the sequence predicted from the mRNA on the basis of the standard genetic code and that must take place to obtain functional protein. This can be redefinition of a single codon or a shift of the reading frame (mostly by one base in +1 or −1 direction). Recoding events depend on the presence of particular structural elements in the mRNA which constitute recoding signals of varying efficiency. Frameshift signals are frequently provided by stable hairpins or pseudoknots in the mRNA that lead to ribosome pausing on the sequence on which the frameshift takes place. The nature of the shift sequence is such that it provides sufficiently stable base pairing between codon and anticodon after the anticodon has shifted by one base.

One important example of codon redefinition is the incorporation of selenocysteine at UGA codons, which normally function as termination codons. Selenocysteine is found in the active site of a number of oxidoreductases, such as glutathione peroxidase or deiodinases for thyroxine and triiodothyronine. It is formed from serine by enzymatic transformation of a particular tRNA charged with serine to form Sec-tRNA^{Sec} (Sec, selenocysteine). With the help of a specialized translation factor (SelB in bacteria), Sec-tRNA^{Sec} binds to the ribosome and recognizes UGA when a particular structural element of the mRNA, the selenocysteine insertion sequence, which recruits SelB, is present either within the coding sequence downstream of the UGA (bacteria) or further away in the 3′-untranslated region (3′-UTR) of the mRNA (eukaryotes).

4 The Translational Apparatus

Translation of the nucleotide sequence of mRNAs into protein requires a complex apparatus that consists of a large variety of proteins and RNAs. Transfer RNA (tRNA) functions as an adapter between the codon of the mRNA and the respective amino acid specified by the codon. tRNAs are aminoacylated by aminoacyl-tRNA synthetases.

Decoding and peptide synthesis take place on ribosomes, large ribonucleoprotein complexes consisting of three or four RNAs and 50 to 80 proteins, depending on organism and organelle. Ribosomes are assembled in the nucleus and carry out protein synthesis in the cytosol. All phases of protein synthesis, i.e., initiation, elongation, and termination, require the action of translation factors that interact with the ribosome at defined stages of translation.

4.1 Transfer RNA

tRNAs are small RNAs (75–85 nucleotides) that have very similar secondary and tertiary structure (Figure 1) and contain a large number of modified nucleosides. The functional centers of the tRNA molecule are the anticodon and the 3′ terminus to which the amino acid is attached by an energy-rich ester bond.

Decoding of mRNA codons on the ribosome entails the formation of base pairs between three bases of the codon and complementary bases of the anticodon. Most amino acids are specified by more than one codon, in some cases by up to six (Table 1). In such cases, there are two or more tRNAs that are charged with the same amino acid (isoacceptors) and have different sequences, including different anticodons.

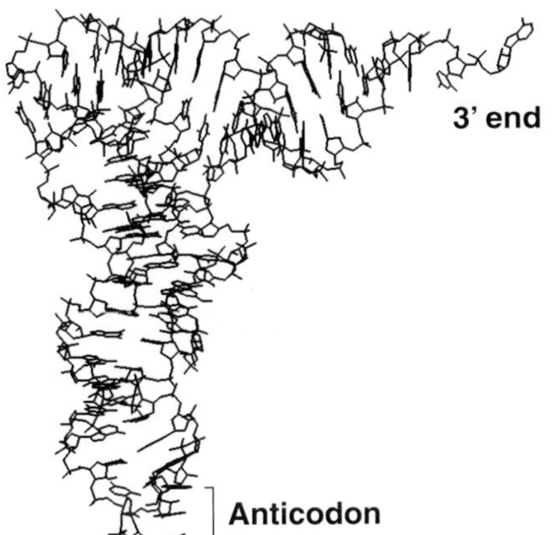

Fig. 1 Tertiary structure of yeast $tRNA^{Phe}$: The three anticodon bases recognize the codon triplet on the ribosome, and during aminoacylation the amino acid is attached to the 3′-terminal adenosine.

In the first and second codon position, base pairing with the anticodon strictly follows the Watson-Crick complementarity rules, i.e., G pairs with C and A pairs with U. Base pairing at the third codon position is less restrictive because also "wobble" pairs (G:U, U:G, or I:U,C,A; I stands for inosine) are "allowed", i.e., energetically favorable. As a consequence, there are tRNAs that can pair with two or more codons differing in the third base, which reduces the total number of different tRNAs required to decode all 61 sense codons. Most organisms have around 50 different tRNAs, and a set of 22 tRNAs suffices for protein synthesis in mitochondria of mammalian cells.

Attachment of the amino acid to the 3′ end of the tRNA (aminoacylation) is catalyzed by aminoacyl-tRNA (Aa-tRNA) synthetases (amino acid-tRNA ligases). Aa-tRNA synthetases are specific for their respective amino acid and tRNA (or group of isoaccepting tRNAs). Thus, every organism contains 20 different Aa-tRNA synthetases, one for every amino acid incorporated. (An interesting exception is an Aa-tRNA synthetase with specificity for two amino acids [Pro, Cys] found in a thermophilic archaeon.)

Aminoacylation takes place in two steps that are both catalyzed by the same Aa-tRNA synthetase:

1) Amino acid + ATP = Aa-AMP + PP_i
2) Aa-AMP + tRNA = Aa-tRNA + AMP

The first step is the activation of the amino acid with ATP by formation of a mixed anhydride formed between the carboxyl group of the amino acid and the phosphoryl group of AMP. The second reaction is the transfer of the aminoacyl residue from Aa-AMP to tRNA to form an energy-rich ester bond with a hydroxyl group of the 3′-terminal ribose.

Aa-tRNA synthetases vary in size and subunit composition; there are monomeric enzymes with molecular masses around 50 kDa and $(\alpha\beta)_2$ heterotetramers with over 200 kDa. Two classes of Aa-tRNA synthetases can be distinguished that differ in the structure of the catalytic center and transfer the amino acid to either the 2′or the 3′ hydroxyl group of the terminal ribose of the tRNA. By spontaneous migration of the aminoacyl residue, the 2′ derivative isomerizes to the 3′ Aa-tRNA, which is the substrate for subsequent steps of translation.

The aminoacylation reaction is highly accurate, i.e., the frequency by which an incorrect amino acid is attached to a tRNA is low (about 10^{-4}), matching the overall error frequency of translation. Thus, Aa-tRNA synthetases must distinguish correct and incorrect amino acids and tRNAs with high accuracy. Most amino acids can be discriminated with sufficient accuracy in a single binding step on the basis of structure, size, or chemical character. In the few cases where these criteria do not suffice for high-accuracy discrimination (isosteric or chemically similar amino acids, such as threonine and isoleucine or isoleucine and valine, respectively), the required low error level is attained by hydrolytically discarding incorrectly formed products (editing). Editing synthetases have a second catalytic site where incorrectly formed Aa-AMP or Aa-tRNA is hydrolyzed.

The recognition of the tRNA substrate is brought about by specific interactions between the Aa-tRNA synthetase and structural elements at various positions of the tRNA molecule (identity elements), frequently including residues in the acceptor arm and the anticodon. Isoaccepting tRNAs that are aminoacylated by the same synthetase possess the same pattern of identity elements.

4.2 Ribosomes

Ribosomes are large (diameter, 20 nm) ribonucleoprotein particles (RNPs) that consist of two subunits of different sizes. They contain several molecules of rRNA and many proteins that are mostly small (10–20 kDa) and basic (Table 2). Ribosomes from bacteria and eukaryotes have similar archi-

Tab. 2 Composition of ribosomes

Organism	*Ribosome*[a]	*Small subunit*[a]	*Large subunit*[a]
Eukaryotes (Mammalia)			
Size	80S (4.2 MDa)	40S (1.4 MDa)	60S (2.8 MDa)
rRNAs		18S rRNA (1874 Nt)	28S rRNA (4718 Nt) 5.8S rRNA (160 Nt) 5S rRNA (120 Nt)
Proteins		33 proteins	49 proteins
Prokaryotes (*E. coli*)			
Size	70S (2.5 MDa)	30S (0.9 MDa)	50S (1.6 MDa)
rRNAs		16S rRNA (1542 Nt)	23S rRNA (2904 Nt) 5S rRNA (120 Nt)
Proteins		21 proteins	34 proteins

[a] Abbreviations: S = sedimentation constant; MDa = megadalton; Nt = nucleotide.

tecture, although eukaryotic ribosomes are larger and have higher molecular mass because they have larger rRNAs and more proteins. Mitochondrial ribosomes resemble bacterial ribosomes but contain smaller rRNAs and many more proteins.

The structures of 30S and 50S ribosomal subunits are known at atomic resolution. In both subunits, the structure is determined by the tertiary structure of the RNA to which the ribosomal proteins are attached. The three domains of 16S rRNA are represented in the body, the platform, and the head domain of the 30S particle (Figure 2A). The decoding center is built of parts of helices 44, 18, and 34 of 16S rRNA, and the only protein that is close to the decoding center is S12, which influences fidelity. 16S rRNA forms contacts with the codon–anticodon complex that are essential for the accuracy of decoding. Monitoring the quality of codon–anticodon interaction is performed by helix 44 through A1493 and A1492 that contact the 1st and 2nd position, respectively, in the minor groove of the codon–anticodon duplex; the 3rd position is monitored less stringently by a contact from G530 (Figure 2B).

Major landmarks of the 50S subunit (Figure 3A) are the L1 region; the central

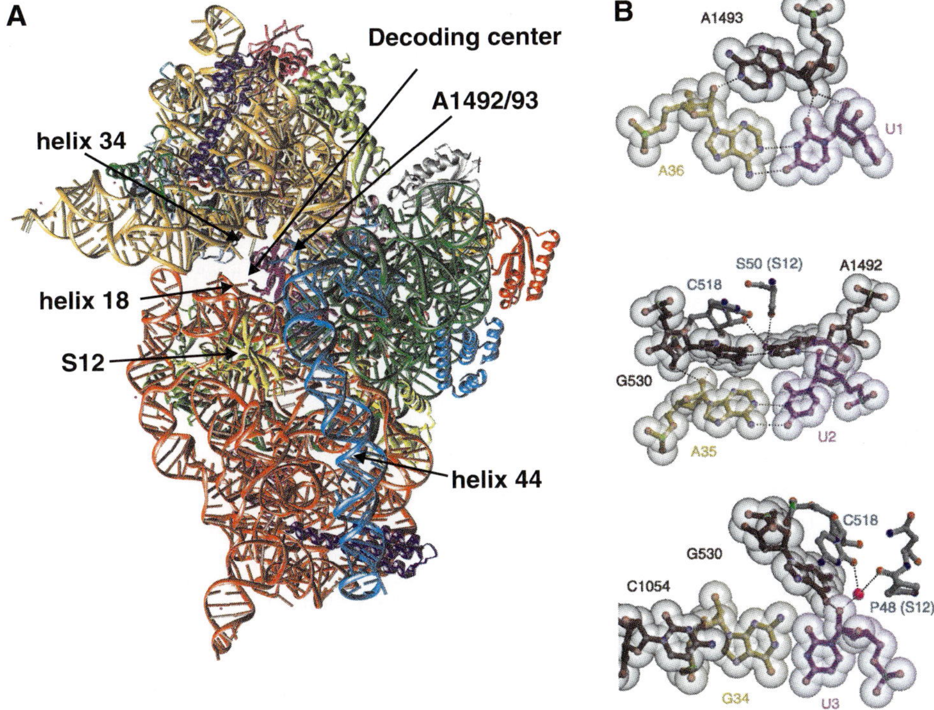

Fig. 2 Atomic structure of the 30S ribosomal subunit of ribosomes: (A) Overview of the 30S structure from *Thermus thermophilus* (ribbon representation). The three major domains of 16S rRNA are colored differently, and proteins are shown in various colors. (B) Recognition of codon–anticodon base pairs by ribosomal residues of the decoding center. Shown are the three anticodon bases (G34A35A36) bound to a codon triplet (U1U2U3) in the 30S crystal. Reprinted from *Science* 292 (2001) 897, with permission, Copyright 2001 American Association for the Advancement of Science.

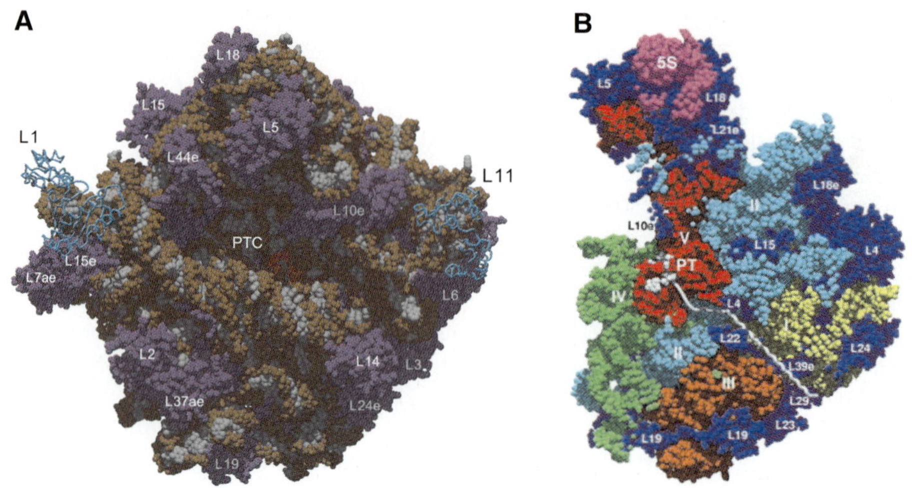

Fig. 3 Atomic structure of the 50S ribosomal subunit: (A) Subunit interface view of 50S subunits from *Haloarcula marismortui*. The space-filling model shows RNA in gray/brown and proteins in blue. The peptidyl transferase center (PTC) is indicated by an inhibitor (red) mimicking the tetrahedral intermediate of the PT reaction. (B) Side view cut to show the peptide exit tunnel. Domains I-V of 23S rRNA are shown in different colors, proteins in blue. The tunnel is indicated by a modeled peptide extending from the PT center to the exit. Reprinted from *Science* 289 (2000) 920, with permission, Copyright 2000 American Association for the Advancement of Science.

protuberance that contains 5S rRNA; and the stalk region where proteins L11, L10, and L7/12 (not seen in the structure) form an important site for elongation factor binding. The peptidyl transferase center, as localized by an analogue mimicking the tetrahedral intermediate of the peptidyl transferase reaction, is made up of RNA only, suggesting that the ribosome is a ribozyme. The exit tunnel for the growing peptide extends from the peptidyl transferase center through the body to the back of the subunit where the peptide emerges (Figure 3B).

The structure of 70S ribosomes with three tRNA molecules bound in the binding sites for Aa-tRNA (A site), Pept-tRNA (P site), and the exit site for deacylated tRNA (E site) was determined by X-ray crystallography at somewhat lower resolution (Figure 4). The arrangement of the tRNAs in A and P sites with their anticodons and CCA ends, respectively, coming close together is clearly seen in the structure. There are a number of connections between the subunits, most of them made up of RNA. One prominent connection (bridge 2a) is formed of helix 69 of 23S rRNA and connects to the 30S decoding site (helix 44), providing the basis for transmitting conformational signals between subunits. Some bridges contact the A-site tRNA, e.g., helix 38 (bridge 1a, or "A-site finger") from above and helix 69 from below; the latter also contacts the P-site tRNA. These interactions stabilize the tRNAs in their binding positions and presumably have to be released to allow tRNA movement during translocation.

Structural and functional analyses suggest that rRNA is involved in the major functions of the ribosome, i.e., decoding, peptide bond formation, and translocation. This is consistent with the fact that the respective functional centers of rRNA are highly conserved. However, ribosomal proteins

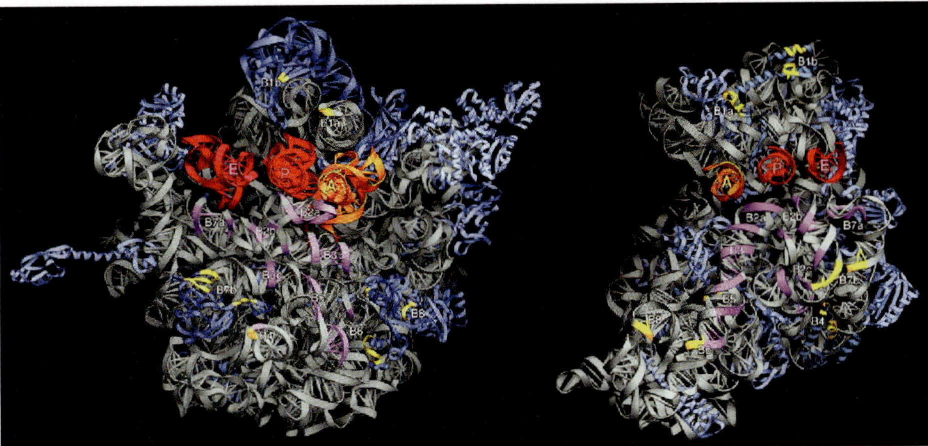

Fig. 4 Crystal structure of 70S ribosomes from *Thermus thermophilus* with tRNAs bound to A, P, and E sites: Shown are the interface sides of 50S (left panel) and 30S (right panel) subunits. 23S and 16S rRNAs are shown in gray, 5S rRNA in blue at the top of the 50S subunit; proteins are in blue. tRNAs are in gold (A site), orange (P site), and red (E site). Connections between subunits are indicated in purple (RNA–RNA bridges) or yellow (RNA–protein bridges). Reprinted from *Science* 292 (2001) 883, with permission, Copyright 2001 American Association for the Advancement of Science.

appear to have an important role as well. First of all, ribosomal proteins are probably required for rapid and accurate assembly of ribosomal particles *in vivo* and are probably essential in stabilizing certain structural elements of rRNA. For several ribosomal proteins a direct involvement in ribosome function has been demonstrated. Mutations in proteins S4, S5, S12 and others have strong effects on the fidelity of Aa-tRNA selection, suggesting an involvement of those proteins in decoding. In fact, all these proteins are located within or close to the decoding center. Proteins L7/12 that form the stalk of the 50S subunit of the bacterial ribosome, and the corresponding P1/P2 proteins of eukaryotic ribosomes, have an important role in the function of translation factors on the ribosome, including an involvement in the stimulation of the GTPase activity of those factors.

4.3 Translation Factors

All phases of ribosomal protein synthesis, except peptide bond formation itself, require that protein factors interact with the ribosome in order to proceed at physiologically relevant rates. There are factors for initiation, elongation, and termination (Table 3).

Tab. 3 Protein factors in translation

	Prokaryotes	*Eukaryotes*
Initiation	IF1	~ 12 eIFs
	IF2	
	IF3	
Elongation	EF-Tu	EF1α
	EF-Ts	EF1β,γ,δ
	EF-G	EF2
		EF3 (fungi)
Termination	RF1	eRF1
	RF2	–
	RF3	eRF3
Ribosome recycling	RRF	–

Several translation factors are GTPases that hydrolyze GTP at one point during their functional cycles. Atomic structures of a number of translation factors are known. The functions of individual factors are discussed below in the context of the respective partial reaction of protein synthesis.

4.4
Messenger RNA

Prokaryotic messenger RNAs (mRNAs) have a 5′-triphosphate, usually an untranslated leader sequence preceding the start codon, and an untranslated 3′ sequence following the termination signal; there is no poly(A) tail. Bacterial mRNAs frequently possess a conserved purin-rich sequence (Shine-Dalgarno sequence), located about 10 nucleotides upstream of the start codon, that is complementary to a sequence near the 3′ end of 16S rRNA. Base-pairing between these complementary sequences promotes the binding of the 30S subunit to the initiation site of the mRNA. Many bacterial mRNAs are polycistronic, i.e., they contain several translatable sequences coding for enzymes that catalyze consecutive or otherwise related reactions in metabolism, e.g., enzymes in biosynthetic or catabolic pathways. Translation of different coding regions of polycistronic mRNAs is usually initiated independently, although there are examples of preferential re-initiation at consecutive coding sequences.

Eukaryotic mRNAs differ from prokaryotic ones in a number of features. They are monocistronic, i.e., they contain only one coding sequence each. Most eukaryotic mRNAs carry at their 5′ end the cap structure $m^7G(5')ppp(5')N$ (N usually being A), which is added during mRNA maturation in the nucleus. The 5′-untranslated region (5′-UTR) precedes the AUG initiation codon; it can comprise 100 nucleotides or more. The coding sequence is followed by the 3′-UTR, and at the 3′ terminus, most eukaryotic mRNAs have a poly(A) sequence that is added posttranscriptionally and, depending on organism and functional state, can comprise up to several hundred adenylyl residues. 5′- and 3′-UTRs and the poly(A) tail have important functions in the regulation of the stability and translation activity of eukaryotic mRNA (see Section 6).

5
Protein Synthesis

Ribosomal protein synthesis proceeds in three phases: initiation, elongation, and termination. During initiation, the ribosome is positioned at the beginning of the coding sequence of the mRNA to be translated, and initiator tRNA (Met-$tRNA_i^{Met}$ in eukaryotes, N-formylated fMet-$tRNA_f^{Met}$ in bacteria) is bound to the AUG start codon in the P site of the ribosome. The elongation phase that follows is a cyclic process of three basic steps. During the first step, Aa-tRNA with an anticodon complementary to the codon presented in the decoding site binds to the A site. In the second step, A site-bound Aa-tRNA reacts with P site-bound initiator tRNA, or peptidyl-tRNA (Pept-tRNA) in subsequent elongation cycles, to form a peptide bond, resulting in deacylated tRNA in the P site and Pept-tRNA, which is one amino acid longer in the A site. The elongation cycle is completed by the translocation of Pept-tRNA from the A site to the P site; during the movement, the mRNA is carried along with the tRNA, performing a movement by one codon triplet, and deacylated tRNA moves out of the P site into the exit (E) site from where it dissociates from the ribosome. The elongation cycle is repeated until the entire coding sequence of the mRNA is translated and a termination codon appears in the decoding site. This is the signal for termination, during which the completed protein is released from Pept-tRNA hydrolytically. Finally, deacylated tRNA dissociates and the ribosome is recycled to enter another round of initiation.

5.1
Initiation of Translation in Bacteria

Initiation of translation in bacteria takes place in two main phases following the dissociation of 70S ribosomes into subunits induced by binding of IF3 (Figure 5). In the first phase, the 30S initiation complex is formed by binding initiator tRNA (N-formylmethionyl-$tRNA_f^{Met}$, or fMet-$tRNA_f^{Met}$) and the mRNA to the 30S ribosomal subunit. Usually, the initiation site is found by base-pairing between a short sequence in 16S rRNA and a partially or fully complementary sequence of the mRNA that is located about 10 nucleotides upstream of the initiation codon (Shine-Dalgarno sequence). The initiation codon is recognized by the anticodon of the initiator tRNA bound to the 30S P site.

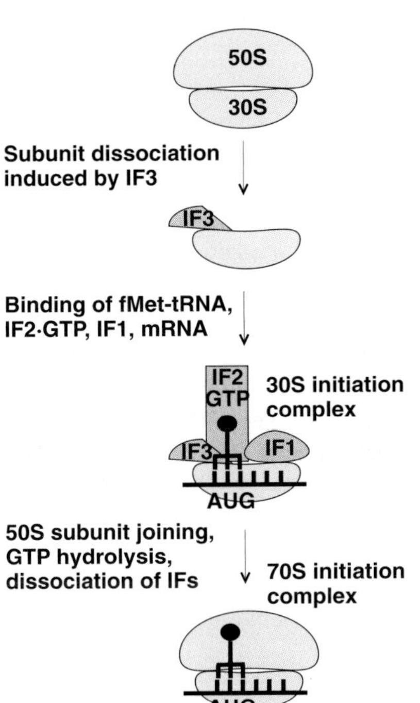

Fig. 5 Initiation of translation in prokaryotes.

30S initiation complex formation in bacteria requires the function of three initiation factors, IF1, IF2, and IF3. IF3 promotes the dissociation of 70S ribosomes into subunits by binding to the 30S subunit and stimulates the binding of both initiator tRNA and mRNA. IF1 stimulates the function of the other two factors; it is bound to the 30S A site, where it may help in directing initiator tRNA to the P site. IF2, a GTPase, is essential for the binding of initiator tRNA.

In the second step of initiation, the 30S initiation complex is joined by the 50S ribosomal subunit. Subunit association triggers GTP hydrolysis by IF2, and subsequently all three initiation factors dissociate from the ribosome. The resulting 70S initiation complex contains initiator tRNA bound to the AUG start codon of the mRNA in the P site, and the codon that follows is presented in the A site. Binding of the respective Aa-tRNA to the 70S initiation complex initiates the first round of elongation.

5.2
Initiation of Translation in Eukaryotes

The initiation of translation of a eukaryotic mRNA requires a large number of initiation factors ($>$ 12), comprising close to 30 polypeptides. In the usual cap-dependent initiation, the first phase of initiation consists in the assembly at the 5′ cap of a complex containing a number of initiation factors and the 40S ribosomal subunit with the initiator Met-$tRNA_i^{Met}$ (Figure 6). The formation of the (cytosolic) cap-binding complex is initiated by the cap-binding

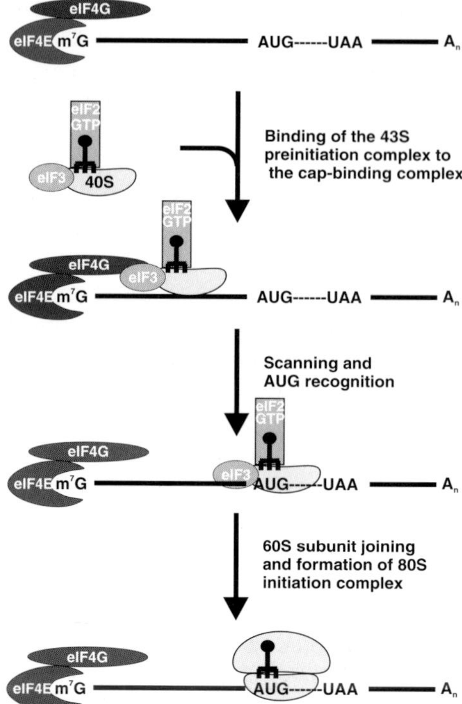

Fig. 6 Cap-dependent initiation of translation in eukaryotes.

protein, eIF4E, which then recruits other factors, including eIF4G, which is a central constituent of the initiation complex. By binding to eIF3 – a large (> 500 kDa) factor consisting of up to 10 subunits, which in turn is bound to the 40S subunit – eIF4G mediates the binding of the 43S initiation complex (40S subunit, Met-tRNA$_i^{Met}$, eIF2-GTP, eIF3) to the cap-binding complex to form the 48S initiation complex. eIF4G additionally binds the poly(A)-binding protein, Pabp1; this interaction is the basis for the strong effects on translation initiation exerted from the polyadenylated 3′ terminus of the mRNA. The second phase of initiation comprises the movement of the 40S subunit with a number of factors bound to it along the 5′-UTR until it reaches an AUG codon (usually the first) and stops there (scanning). Scanning requires ATP hydrolysis by eIF4A, which belongs to the DEAD/H box family of RNA helicases. The next step is the binding of the 60S subunit, which is promoted by eIF5B, another GTPase. 60S binding leads to GTP hydrolysis by both eIF2 and eIF5B and the dissociation of all initiation factors. The resulting 80S initiation complex is ready to enter elongation.

Besides cap-dependent initiation, which is the initiation mechanism for the majority of cellular mRNAs, there is a mechanism for cap-independent initiation on mRNAs lacking the 5′ cap. These mRNAs have, in their 5′-UTR immediately preceding the coding region, particular structures comprising about 300 nucleotides that function as an internal ribosome entry site (IRES), where an initiation complex consisting of 40S subunit, initiator tRNA, and a number of initiation factors and other cellular factors can assemble without prior scanning. The IRES-dependent mechanism of translation initiation is used by small RNA viruses (picorna viruses, such as polio virus or encephalomyocarditis virus), which have uncapped mRNA with IRES elements preceding the coding region. By proteolytically cleaving off the domain of eIF4G that is required for the interaction with the cap-binding complex, these viruses abolish cap-dependent initiation of the host cell. Cap-independent initiation is not restricted to viral mRNAs; a considerable number of uncapped cellular mRNAs containing IRES elements have been found. Translation of these mRNAs is not affected by signals that regulate cap-dependent initiation and may, therefore, continue in situations where cellular translation on the whole is inhibited (see Section 6).

5.3 The Elongation Cycle

The elongation of the peptide chain follows similar mechanisms in both prokaryotes and eukaryotes and is therefore described for both systems together (Figure 7). In the first step, bacterial EF-Tu (or eukaryotic eEF1α) catalyzes the binding of Aa-tRNA to the A site of the ribosome. These factors are GTPases that in their GTP-bound conformation form a high-affinity ternary complex with Aa-tRNA, which, in turn, binds to the ribosome and, after GTP hydrolysis, releases Aa-tRNA to enter the peptidyl transferase center. Peptide bond formation between Aa-tRNA in the A site and Pept-tRNA in the P site is catalyzed by the ribosome. Finally, translocation is catalyzed by EF-G (eEF2 in eukaryotes), another GTPase.

5.3.1 Aa-tRNA Binding

EF-Tu-dependent A-site binding of Aa-tRNA comprises a number of steps, as revealed by biochemical and rapid kinetic analysis (Figure 8). The first step is the codon-independent formation of an unstable initial binding complex. During initial complex formation, ternary complexes bind to the ribosome in a stochastic fashion and dissociate until one with the anticodon matching the A-site codon binds to form the codon-recognition complex (Figure 9). Codon recognition has two important consequences. One is the stabilization of the complex by the formation of base pairs between codon and anticodon as well as by ribosome interactions. The other is stimulation of the GTPase activity of EF-Tu by more than five orders of magnitude (GTPase activation). GTP hydrolysis and P_i release lead to an extensive structural change of EF-Tu from the GTP- to the GDP-bound form, by which it loses the affinity for Aa-tRNA. The aminoacyl end of Aa-tRNA set free from EF-Tu-GDP moves into the peptidyl transferase center (accommodation) and takes part in peptide bond formation (see Section 5.3.3). EF-Tu-GDP dissociates from the ribosome and is reactivated by EF-Ts that acts as a guanine nucleotide exchange factor and catalyzes the exchange of GDP for GTP. The functional cycle of eukaryotic eEF1α is probably similar, the exchange factor being eEF1β,γ. Because of the high cellular concentrations of EF-Tu or eEF1α, practically all Aa-tRNA is bound in the ternary complex and immediately available for entering the ribosome.

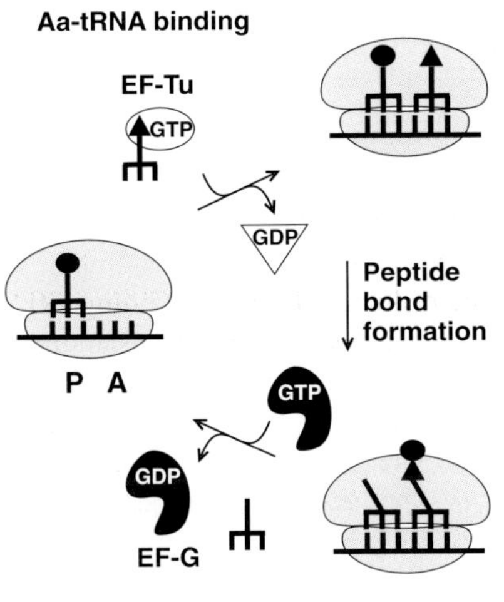

Fig. 7 Schematic of the elongation cycle.

5.3.2 Aa-tRNA Selection

The ribosome discriminates between correct and incorrect Aa-tRNAs, or their complexes with EF-Tu-GTP, according to the match between anticodon and codon in the A site. The average frequency of amino acid misincorporation *in vivo* is about 10^{-3}. This low level of missense errors is difficult to reconcile with the difference in the free energy of binding, ΔG, between codon–

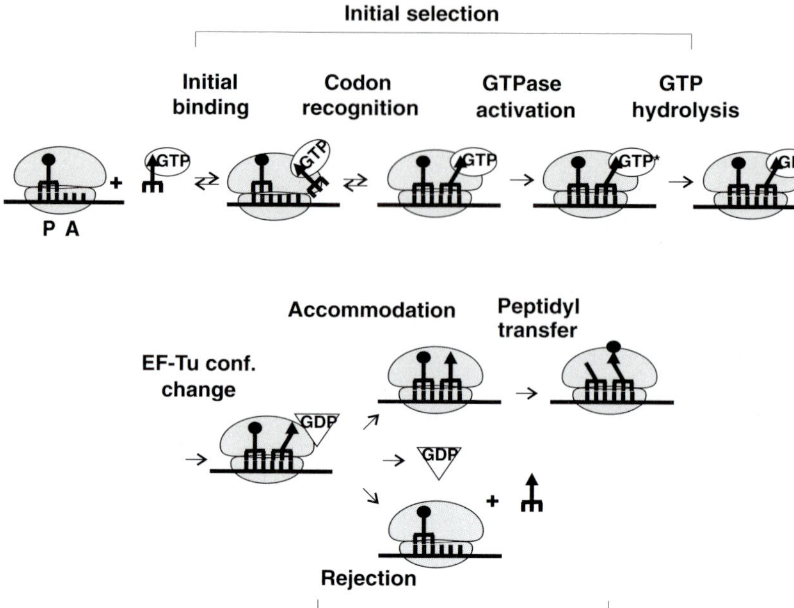

Fig. 8 Kinetic scheme of Aa-tRNA binding to the ribosome.

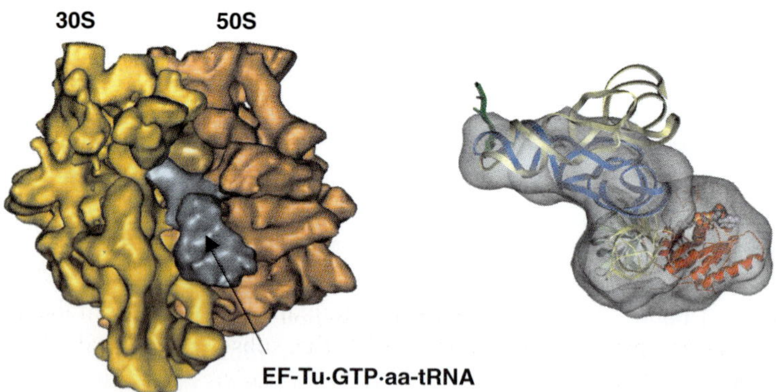

Fig. 9 Codon-recognition complex: The three-dimensional reconstruction was obtained from cyro-electron microscopic images at 13 A resolution. (A) Overview of the complex. Ribosome density is colored yellow for the 30S and brown for the 50S subunit. The ternary complex EF-Tu-GDP-Aa-tRNA complex stalled by binding the antibiotic kirromycin to EF-Tu is shown in blue, with EF-Tu oriented towards the viewer and the tRMNA below, the anticodon arm reaching into the 30S decoding center. (B) Density resulting from the ternary complex is shown with the crystal structure of the ternary complex fitted into the density by adjustments of the position of the tRNA (blue) and domains 2 (yellow) and 3 (green) of EF-Tu. Also shown is the position of tRNA accommodated in the A site (gold) and and an mRNA codon (green).

anticodon pairs that are correct, i.e., fully complementary, called cognates, and those that are nearly correct, having one mismatch, called near-cognates. These differences can be quite small ($\leq$ 10 kJ/mol), predicting error frequencies of up to 1 out of 100 amino acids incorporated. Moreover, even this discrimination potential is unlikely to be used in full because the high rates of the step(s) following codon recognition preclude that the codon-reading step reaches equilibrium. Thus, in order to reconcile an error frequency of at least 10^{-2} predicted for a single-step selection mechanism and the much lower error frequencies actually observed *in vivo*, the ribosome must either augment the precision of a single recognition step or use the same discriminatory interactions more than once in consecutive rejection steps (proofreading). In order to be effective, rejection steps have to be separated by an irreversible, energy-dissipating step.

In fact, Aa-tRNA selection on the ribosome takes place at two stages: prior to GTP hydrolysis (initial selection) and after GTP hydrolysis, but before peptide bond formation (proofreading). In cases where there is no match between anticodon and codon (non-cognate), the respective ternary complexes are rejected in the initial selection phase. By contrast, in near-cognate situations, rejection in initial selection is not sufficient because of rapid GTP hydrolysis. In such cases, final discrimination is accomplished by rejection of Aa-tRNA during the proofreading phase.

In both selection phases, part of the discrimination stems from different dissociation rates that are due to different stabilities of the respective codon–anticodon complexes. However, an important additional contribution to discrimination is provided by induced fit. The quality of codon–anticodon interaction determines the rates of two rearrangement steps (GTPase activation, accommodation) that precede, and limit the rates of, the irreversible chemical steps of GTP hydrolysis and peptide bond formation. Both rearrangements and, with that, both chemical steps are much faster for cognate than for near-cognate codon recognition, thus providing a substantial kinetic contribution to discrimination.

The recent elucidation of ribosome structures at atomic resolution and mutational analyses suggest which residues of the decoding center monitor the codon–anticodon duplex (see Section 4.2, Figure 2). By forming interactions with the correct codon–anticodon duplex, these residues undergo conformational changes that presumably are transmitted to the functional centers of the ribosome that are involved in GTPase activation and peptide bond formation. In utilizing induced fit for substrate discrimination, the ribosome resembles other nucleic acid-programmed polymerases, such as DNA or RNA polymerases.

5.3.3
Peptide Bond Formation

Catalysis of peptide bond formation is the enzymatic activity of the ribosome. The rate of the uncatalyzed reaction is extremely low; one may estimate that catalysis by the ribosome accelerates the reaction by five orders of magnitude or more. The reaction proceeds via nucleophilic attack of the α-amino group of Aa-tRNA (A site) on the carbonyl carbon of the ester bond of Pept-tRNA (P site) and leads to a tetrahedral intermediate that subsequently breaks down to form deacylated tRNA (P site) and a new, one-amino-acid longer Pept-tRNA (A site) (Figure 10). The peptidyl transferase (PT) center is located on the 50S subunit of the ribosome. 50S subunits largely depleted of protein retained some PT activity, suggesting that the activity might reside in 23S rRNA, although attempts to obtain protein-free 23S

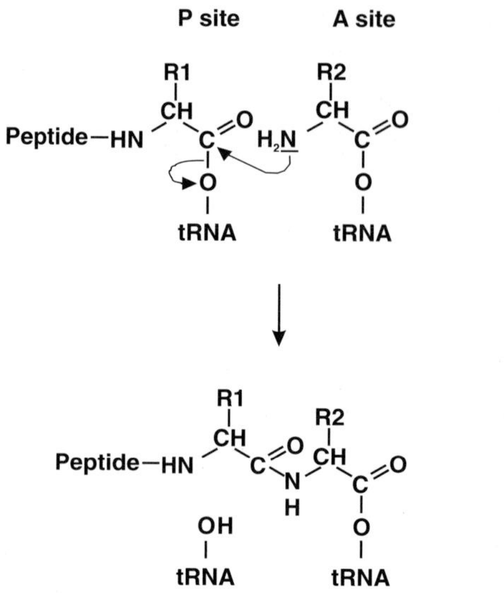

Fig. 10 Schematic of the peptidyl transferase reaction between P site-bound Pept-tRNA and A site-bound Aa-tRNA.

rRNA that was active in the PT reaction were unsuccessful. The atomic structure of the large ribosomal subunit from the halophilic archaeon *Haloarcula marismortui* shows that the PT center is composed of RNA exclusively (Figure 3), suggesting that the reaction in fact is catalyzed by RNA.

The mechanism of catalysis of peptide bond formation is not known in detail. Positioning of the 3′ CCA ends of both Pept-tRNA and Aa-tRNA by forming base pairs with 23S rRNA in the PT center has an important role, as shown by mutational analysis. Several such base pairs between CCA sequences of substrate analogues and bases in the P and A loops of 23S rRNA in the PT center were revealed by the crystal structures. There may be an additional catalytic contribution by general acid–base catalysis, as protonation of a single group of the ribosome with a pK_a around 7.5 slows down the reaction, which proceeds at 50 s^{-1} at pH $>$ 7.8, about 100-fold. The ribosomal residue that may serve the function of the catalytic base is not known with certainty. It may be an adenine base, the pK_a of which is shifted from around 3.5 (unperturbed) to 7.5 by interactions in the catalytic site.

5.3.4
Translocation

The last step of the elongation cycle is translocation. It entails the synchronous movement of two tRNAs and the mRNA on the ribosome: deacylated tRNA moves from the P to the E site and Pept-tRNA moves from the A to the P site. The mRNA is moved along with the tRNAs by exactly one codon triplet. Following the movement, deacylated tRNA spontaneously dissociates from the E site, such that the ribosome in the posttranslocation state has Pept-tRNA in the P site and an empty A site presenting the codon to be decoded in the next round of elongation.

Translocation is catalyzed by elongation factor G in bacteria (eEF2 in eukaryotes), which hydrolyzes one molecule of GTP during the reaction. Rapid kinetic analysis has revealed the reaction sequence shown in Figure 11. Binding of EF-G-GTP to the pretranslocation complex triggers rapid GTP hydrolysis (unbound EF-G has no GTPase activity). Ribosomal residues involved in GTPase activation include proteins L7/L12 and possibly also elements of 23S rRNA, such as the sarcin–ricin loop. GTP hydrolysis and/or release of P_i appear to induce a conformational change of the factor that is coupled to a rearrangement of the ribosome that leads to rapid tRNA movement. The sequence is completed by further conformational changes and, finally, dissociation of EF-G-GDP and decylated tRNA. When GTP hydrolysis is avoided by replacing GTP with non-hydrolyzable GTP analogues, translocation is about 50 times slower but is still considerably faster than

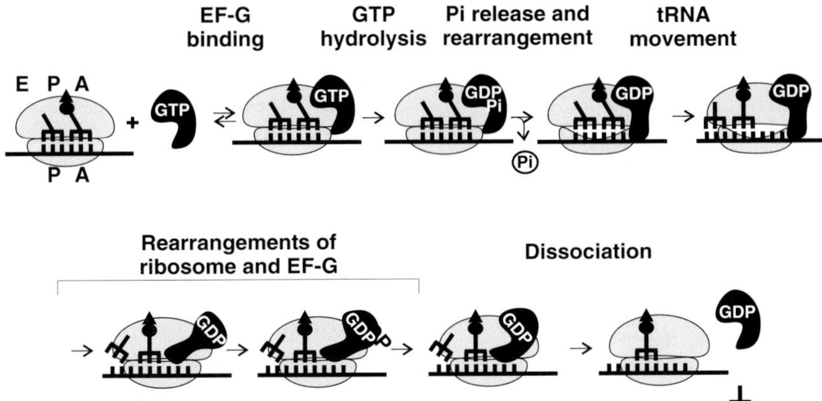

Fig. 11 Kinetic scheme of translocation.

reaction in the absence of EF-G, which is extremely slow. This suggests that EF-G, at least in part, has a mechanochemical function on the ribosome. Exactly how the free energy of GTP hydrolysis is coupled with movement on the ribosome is not known.

5.3.5
Rate of Elongation

One round of the elongation cycle takes 0.05–0.1 s. Thus, a single ribosome needs up to one minute to translate an mRNA coding for a protein of 500 amino acids. High overall rates of protein synthesis in the cell result from the presence of many copies of each mRNA, the fact that one mRNA can be translated by several ribosomes at the same time (polysomes), and the large number of ribosomes, up to 20,000 in bacteria and one million in liver cells, that actively synthesize protein.

5.4
Termination

The elongation cycle is repeated until a termination codon appears in the decoding site. This is the signal for termination, during which the completed protein is released from Pept-tRNA hydrolytically (Figure 12). Termination is catalyzed by termination or release factors (RF). In bacteria, RF1 recognizes the termination codons UAG and UAA, while RF2 is specific for UGA and UAA; in eukaryotes, eRF1, recognizes all three termination codons. The specificity of the release factors in distinguishing termination codons from sense codons is extremely high (up to 10^5), and the frequency of premature termination on sense codons is very low. Binding of RF1/2 to a termination codon in the A site leads to hydrolysis of Pept-tRNA. Apparently, the factor binds to the termination codon in the decoding site and, at the same time, contacts the peptidyl transferase center and induces the hydrolytic activity, possibly by taking part in the activation or positioning of the hydrolytic water molecule. After release of the peptide, RF3 (or eRF3), a GTPase, promotes the dissociation of RF1/2 from the ribosome in a process that requires GTP hydrolysis.

The posttermination complex releases the deacylated tRNA from the P site and dissociates into subunits, assisted by initiation factor IF3 (eIF3), which binds to the small subunit and stabilizes it in the un-

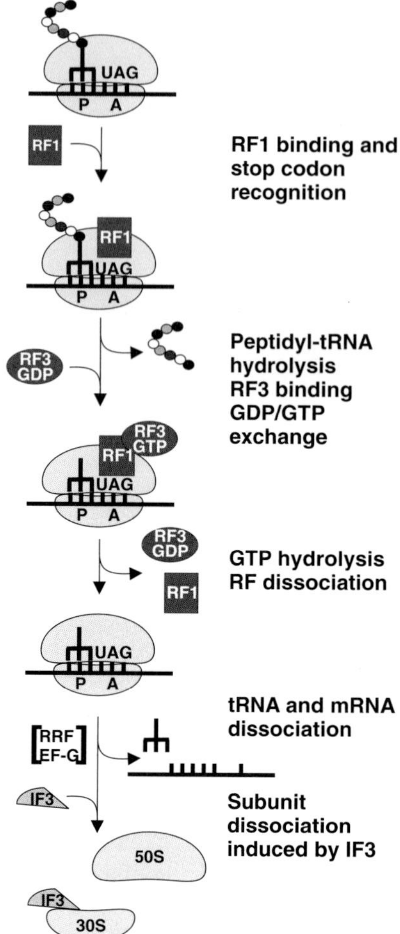

Fig. 12 Termination of prokaryotic translation.

bound form. In bacteria, ribosome recycling requires another factor, ribosome recycling factor (RRF), that together with EF-G promotes the dissociation of the posttermination complex and the formation of free ribosomal subunits.

6 Regulation of Translation in Eukaryotes

In addition to transcriptional and posttranscriptional RNA processing, translation is another important level where gene expression is regulated. By controlling the amount of certain proteins to be synthesized, the cell is able to adequately react according to its own and the organism's physiological condition. Regulatory signals can be transmitted by protein kinases, thereby coupling translation with signal transduction processes in general, or by concentration variations of important metabolites.

Frequently, the regulated reaction is initiation, particularly the formation of the pre-initiation complex at the 5′ cap of the mRNA. Cap-dependent initiation begins with the association of the cap-binding protein, eIF4E, to the 5′ cap, which then recruits eIF4G and the 40S ribosomal subunit with eIF3, eIF2, and initiator tRNA. The formation of the pre-initiation complex is favored by phosphorylation of eIF4E and eIF4G by protein kinases that are coupled with the mitogen activated protein (MAP) kinase signaling pathway and, thereby, to mitogenic stimuli.

An example of negative regulation is the inhibition of initiation by phosphorylation of eIF2. Phosphorylated eIF2-GDP forms a stable complex with the nucleotide exchange factor eIF2B, with the consequence that active eIF2-GTP is not regenerated and initiation ceases. eIF2 is phosphorylated by a group of protein kinases that are regulated by a variety of different signals. In erythropoietic cells that produce large amounts of globin for hemoglobin, the translation of globin mRNA is adapted to the amount of available heme by a heme-regulated eIF2 kinase, heme-regulated inhibitor (HRI). When there is no free heme, the kinase is active and represses globin synthesis by phosphorylating eIF2. Free heme inhibits the kinase, thus allowing more globin to be synthesized. Phosphorylation of eIF2 is also involved in interferon action. Interferon induces the expression of another eIF2 kinase, PKR, which is activated by binding double-stranded RNA (i.e., a replicating viral RNA). Activated PKR phosphorylates eIF2 and thereby inhibits virus replication by inhibiting translation. Other eIF2 kinases are activated by conditions of stress, such as shortage of amino acids or the accumulation of aberrantly folded proteins in the endoplasmic reticulum.

Other mechanisms of translational regulation make use of structural elements located in the 5′- and 3′-UTR of the mRNA. Binding of regulatory proteins to these elements may either inhibit or stimulate initiation. One well-studied example is the regulation of translation of several mRNAs related to iron metabolism. Free iron (Fe^{2+}) in the cell is stored in ferritin complexes. The translation of ferritin mRNA is stimulated when the concentration of free Fe^{2+} ions in the cell is high, such that more Fe^{2+} is stored, and repressed when it is low, resulting in Fe^{2+} release from ferritin stores. The inhibition of translation is brought about by a regulatory protein (iron response element binding protein, IRP) that binds to a hairpin structure in the 5′-UTR (iron response element, IRE) near the 5′ cap and thereby inhibits the formation of the pre-initiation complex (Figure 13). The affinity of IRP for binding to the mRNA is regulated by Fe^{2+}. At low concentrations of Fe^{2+}, an iron-sulfur cluster (Fe_4S_4) bound to IRP releases one Fe^{2+} ion. The resulting conformational change increases the RNA-binding affinity of IRP. At high intracellular concentration of free Fe^{2+}, the cluster is present in the Fe_4S_4 form, and IRP has low affinity for binding to the IRE; therefore, translation of ferritin mRNA is not inhibited and ferritin can be

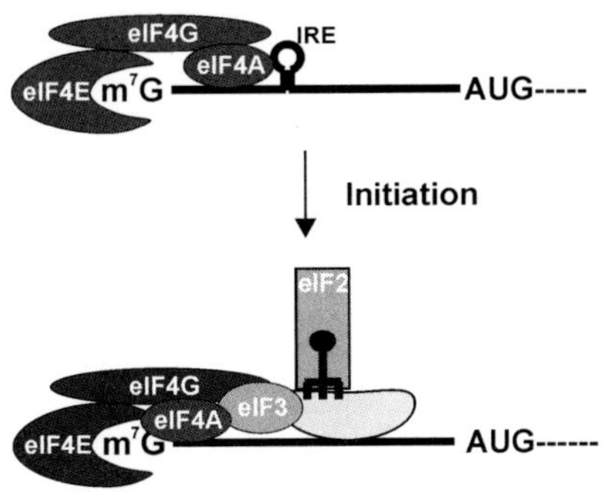

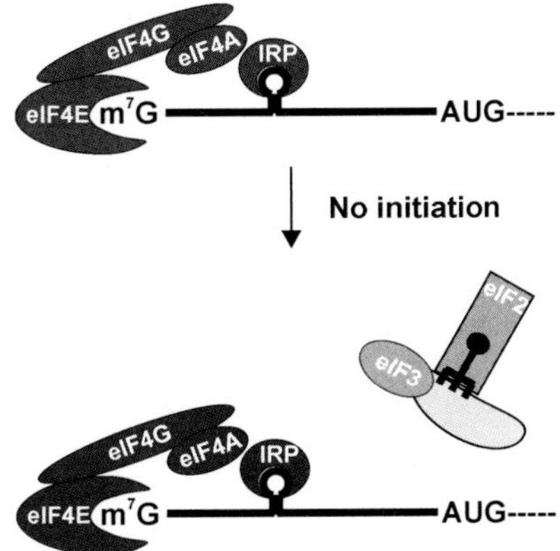

Fig. 13 Regulation of translational initiation on ferritin mRNA by the Fe^{2+} concentration: Binding of iron response element binding protein (IRP) stabilizes the structure of the IRE against helicase eIF4A action. The IRE–IRP structure inhibits the binding of the 43S initiation complex to the cap binding complex.

made. Other proteins involved in iron metabolism are regulated by IRP as well. One is the transferrin receptor that binds the extracellular transporter for Fe^{2+}, transferrin, and mediates the uptake of the transferrin–Fe^{2+} complex by endocytosis. The translation of transferrin receptor mRNA is

Tab. 4 Selected antibiotics affecting prokaryotic translation

Antibiotic	*Action*
Tetracyclin	Inhibits Aa-tRNA binding
Neomycin	Induces misreading
Streptomycin	
Chloramphenicol	Inhibits peptidyl transferase
Fusidic acid	Inhibits translocation
Thiostrepton	
Erythromycin	Inhibits peptide elongation
Puromycin	Peptide release

regulated by the binding of IRP to several IREs in the 3′-UTR. This favors pre-initiation complex formation at the 5′ cap and protects the mRNA against nucleolytic degradation. In a similar fashion, the concentration of free Fe^{2+} regulates the synthesis of porphyrin, the precursor of heme, by regulating the expression of the key enzyme, δ-amino levulinate synthase.

The 3′-UTRs of many mRNAs contain binding sites for translational repressors, i.e., proteins that block translation by binding to the mRNA. This way of regulating translation is found, for instance, with mRNAs that are to be translated only in certain phases of embryogenesis. Also in the 3′-UTR there are elements (e.g., short AU-rich sequences) that, by binding certain proteins, influence the stability of the mRNA against nucleolytic degradation.

These examples demonstrate the functional importance of the 3′-UTR of many mRNAs. There are a number of documented cases where changes in the 3′-UTR are directly related to diseases. For example, in patients suffering from myotonic dystrophy, the 3′-UTR of the gene coding for a cAMP-dependent protein kinase contains a much larger number of CTG triplet repeats ($>$ 1000) than in unaffected persons (5–30 repeats), suggesting that the translation of the kinase mRNA may be deregulated with consequences for signaling by phosphorylation. Finally, the lack of AU-rich elements in 3′-truncated mRNAs, which inhibits mRNA degradation and thereby enhances translation, seems to be related to higher rates of cell proliferation in certain forms of cancer.

7
Inhibitors, Antibiotics, and Toxins

The translation apparatus is the target for many natural inhibitors, antibiotics, and toxins that affect certain steps of ribosomal protein synthesis. A number of antibiotics inhibit bacterial translation with sufficient selectivity as to be useful for antibacterial therapy (Table 4). The use of many of these antibiotics in the meantime is restricted because resistant pathogens have evolved. Much effort is spent, therefore, to develop new antibiotics that act on translation or other targets. The extensive knowledge, in terms of both structure and mechanism of action, about many antibiotics acting on translation is particularly useful here.

Many antibiotics inhibit ribosome function by binding to ribosomal RNA. For instance, certain aminoglycosides (neomycin, paromomycin, gentamicin) bind to 16S rRNA directly in the decoding site (Figure 14) and increase amino acid misincorporation by decreasing the accuracy of decoding. Streptomycin, belonging to another group of aminoglycosides, binds to another site in 16S rRNA but has the same effect. Tetracycline also binds in this region and interferes with productive binding of Aa-tRNA. There are many inhibitors of the peptidyl transferase reaction (e.g., chloramphenicol) that bind to the peptidyl transferase center of 23S rRNA. The binding site of erythromycin is located in the peptide exit tunnel of the 50S subunit, and binding of the antibiotic precludes peptide elongation.

Thiostrepton binds to a single site in 23S rRNA and inhibits translocation by blocking conformational transitions of the ribosome that are important for the reaction. An example of an antibiotic that inhibits translation by binding to a translation factor is fusidic acid; it binds to EF-G after translocation and inhibits its dissociation from the ribosome, thereby blocking further rounds of elongation.

Translocation in eukaryotic cells is the target for the diphtheria toxin of *Corynebacterium diphtheriae*. The A subunit of the toxin enters the cell and inactivates eEF2 by ADP-ribosylation of a single, essential residue, diphthamide (posttranslationally modified histidine). There are other toxic proteins, produced by plants, that have glycosidase activity and cleave off a single adenine residue from 28S rRNA in the so-called sarcin–ricin stem-loop, which is important for the function of translation factors, thereby completely inactivating the ribosome (ribosome inactivating proteins, RIPs). Examples are ricin, contained in the seeds of the ricinus plant, or viscumin from mistletoe, in which the B subunit, a lectin, binds to glycosylated proteins of the cell membrane and mediates the entrance into the cell of the A subunit, which has the N-glycosidase activity. The fungal toxin α-sarcin functions similarly; after entering the cell, it inactivates ribosomes by cleaving a single phophodiester bond in 28S rRNA, again in the sarcin–ricin stem-loop. Some bacteria produce toxins; one of them, colicin E3 from *E. coli*, inactivates ribosomes in other bacteria by specifically cleaving 16S rRNA in the vicinity of the decoding center.

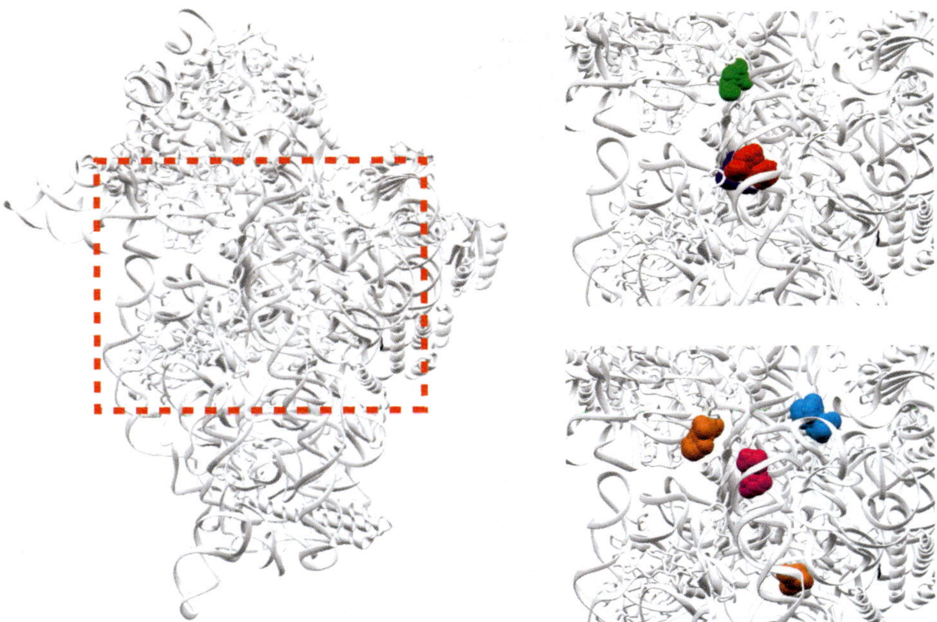

Fig. 14 Antibiotic binding to the 30S ribosomal subunit: The antibiotics streptomycin (blue), spectinomycin (green), paromomycin (red), tetracycline (orange), pactamycin (cyan), and hygromycin (magenta) bind in the central region at or near the decoding center. Figure courtesy of V. Ramakrishnan.

8
Further Reading

Ban N., Nissen P., Hansen J., Moore P. B., Steitz T. A. (2000) The complete atomic structure of the large ribosomal subunit at 2.4 A resolution, *Science* **289**, 905–920.

Dever T. E. (1999) Translation initiation: adept at adapting, *Trends Biochem. Sci.* **24**, 398–403.

Garrett R. A., Douthwaite S. D., Liljas A., Matheson A. T., Moore P. B., Noller H. F., eds. (2000) *The Ribosome: Structure, Function, Antibiotics, and Cellular Interactions.* Washington, DC: ASM Press.

Gesteland R. F., Atkins J. F. (1996) Recoding: dynamic reprogramming of translation, *Annu. Rev. Biochem.* **65**, 741–768.

Green R., Noller H. F. (1997) Ribosomes and translation, *Annu. Rev. Biochem.* **66**, 679–716.

Harms J., Schluenzen F., Zarivach R., Bashan A., Gat S., Agmon I., Bartels H., Franceschi F., Yonath A. (2001) High resolution structure of the large ribosomal subunit from a mesophilic eubacterium, *Cell* **107**, 679–688.

Hellen C. U., Sarnow P. (2001) Internal ribosome entry sites in eukaryotic mRNA molecules, *Genes Dev.* **15**, 1593–1612.

Hershey J. W. B., Mathews M. B., Sonenberg N., eds. (1996) *Translational Control.* Cold Spring Harbor NY: Cold Spring Harbor Laboratory Press.

Kozak M. (1999) Initiation of translation in prokaryotes and eukaryotes, *Gene* **234**, 187–208.

Ogle J. M., Brodersen D. E., Clemons W. M., Tarry M. J., Carter A. P., Ramakrishnan V. (2001) Recognition of cognate transfer RNA by the 30S ribosomal subunit, *Science* **292**, 897–902.

Rodnina M. V., Savelsbergh A., Katunin V. I., Wintermeyer W. (1997) Hydrolysis of GTP by elongation factor G drives tRNA movement on the ribosome, *Nature* **385**, 37–41.

Rodnina M. V., Savelsbergh A., Wintermeyer W. (1999) Dynamics of translation on the ribosome: molecular mechanics of translocation, *FEMS Microbiology Reviews* **23**, 317–333.

Rodnina, M. V., Wintermeyer, W. (2001) Fidelity of aminoacyl-tRNA selection on the ribosome: kinetic and structural mechanisms, *Annu. Rev. Biochem.* **70**, 415–435.

Rodnina, M. V., Wintermeyer, W. (2001) Ribosome fidelity: tRNA discrimination, proofreading and induced fit. *Trends Biochem. Sci.* **26**, 124–130.

Sachs A., Sarnow P., Hentze M. W. (1997) Starting at the beginning, middle, and the end: translational initiation in eukaryotes, *Cell* **89**, 831–838.

Stadtman T. C. (1996) Selenocysteine, *Annu. Rev. Biochem.* **65**, 83–100.

Stark H., Rodnina M. V., Wieden H. J., van Heel M., Wintermeyer W. (2000) Large-scale movement of elongation factor G and extensive conformational change of the ribosome during translocation, *Cell* **100**, 301–309.

Stillman B, ed. (2000) The ribosome, in:. Cold Spring Harbor Symposia on Quantitative Biology LXVI, *Cold Spring Harbor, NY: Cold Spring Harbor Laboratory Press.*

Wilson K. S., Noller H. F. (1998) Molecular movement inside the translational engine, *Cell* **92**, 337–349.

Yusupov M. M., Yusupova G. Z., Baucom A., Lieberman K., Earnest T. N., Cate J. H., Noller H. F. (2001) Crystal structure of the ribosome at 5.5 A resolution, *Science* **292**, 883–896.

2
Proteins Containing Nonnatural Amino Acids

Prof. Dr. Masahiko Sisido
Department of Bioscience and Biotechnology, Okayama University, 3-1-1 Tsushimanaka, Okayama 700-8530, Japan; Tel.: +81-86-251-8018; Fax: +81-86-251-8219; E-mail: sisido@cc.okayama-u.ac.jp

AnqAla	2-anthraquinonylalanine
ARS	aminoacyl tRNA synthetase
AtnDap	β-anthraniloyl-L-α,β-diaminopropionic acid
EDTA	ethylenediamine tetraacetic acid
ET	electron transfer
GFP	green fluorescent protein
HATU	*o*-(7-azabenzotriazol-1-yl)-1,1,3,3-tetramethyluronium hexafluorophosphate
HPLC	high-performance liquid chromatography
Lys(NBD)	ε-(7-nitrobenz-2-oxa-1,3-diazol-4-yl)-L-lysine
PCR	polymerase chain reaction
RID mutagenesis	random insertion/deletion mutagenesis
TOF mass	time-of-flight mass spectroscopy

1 Introduction

Protein engineering has been extended to include various nonnatural amino acids that may be incorporated into specific positions along a polypeptide chain (Bain et al., 1989; Noren et al., 1989). Depending on the type of nonnatural amino acid incorporated, the nonnatural mutants show various functions that cannot be achieved by any sequences and combinations of the twenty naturally occurring amino acids. Nonnatural mutagenesis has been made possible by breakthroughs in several key steps, including: (1) the synthesis of nonnatural amino acids that carry specialty side groups; (2) chemical aminoacylation of nonnatural amino acids to tRNAs; (3) finding of special codon/anticodon pairs that may be assigned to nonnatural amino acids, independently from the naturally occurring ones; and (4) expression of the nonnatural mutant in in-vitro protein-synthesizing systems. Recent progress in the above key steps are described in this review, and some examples of the applications of nonnatural mutagenesis will be detailed in the final section.

In adding artificial components into biochemical systems *in vitro* and in vivo, the concept of orthogonality becomes important. An orthogonal molecule in a biochemical system is defined as the molecule that functions as a new member in the existing system but behaves independently from other members. For example, an orthogonal nonnatural amino acid must be free from the enzymatic action of endogenous aminoacyl tRNA synthetases (ARSs), but must be incorporated into proteins when they were charged onto tRNAs. An orthogonal tRNA/ARS pair must be free from aminoacylation by endogenous ARSs, but must be aminoacylated with a specific nonnatural amino acid by the corresponding ARS. Further, an orthogonal codon/anticodon pair must function independently of the existing three-base codons in order to transport a nonnatural amino acid into the ribosomes.

2 Historical Outline

Nonnatural mutagenesis is based on an earlier finding by Chapeville et al. in 1962, that the codons on mRNA do not recognize amino acids on the 3′-end of tRNA. These authors chemically reduced a cysteine unit of $tRNA^{Cys}$-Cys to give an a $tRNA^{Cys}$-Ala. When the misaminoacylated tRNA was added to an in-vitro biosynthesizing system, Ala units were incorporated in place of Cys units in the polypeptide chain. Similarly, Fahnestock and Rich (1971) chemically converted the amino group of Phe-tRNA to a hydroxy group, giving phelac-tRNA. When the latter misacylated tRNA was added to an in-vitro protein-synthesizing system with poly(U), a polyester of phenyllactic acid was produced (Fahnestock and Rich, 1971) These experiments suggested that, if a nonnatural amino acid were linked to a tRNA by any means, then the amino acid would be introduced into proteins at the position directed by the anticodon of the tRNA.

In nature, a nonstandard amino acid, selenocysteine, has been incorporated into some redox enzymes during translation. (Böck et al., 1991) An UGA stop codon is assigned to selenocysteine, and a $tRNA^{Sec}{}_{UCA}$ is the adapter. The $tRNA^{Sec}{}_{UCA}$ was first aminoacylated with serine, and the latter was converted to selenocysteinyl-$tRNA^{Sec}{}_{UCA}$ by selenocysteine tRNA synthetase. Through this process, selenocysteine was incorporated, for example, into *Escherichia coli* formate dehydrogenase and mammalian tetraiodothyronine deiodinase type I. This mechanism exemplifies that a stop codon can be used to direct a nonnatural amino acid, whilst a suppressor tRNA that carries the UCA anticodon can carry the amino acid into the ribosome.

These examples suggest that when a nonnatural amino acid is charged onto a tRNA, it will be incorporated into a protein at the position of a unique codon on a mRNA. The nonnatural mutagenesis described in the following sections is based on these historical and important experimental results.

3 Extension of Amino Acids

Because the primary purpose of nonnatural mutagenesis is to introduce novel functions into proteins, the synthesis of nonnatural amino acids that carry specialty side group, as well as their screening for efficient incorporation into proteins, are essential. In recent years, a wide variety of nonnatural amino acids have become available commercially, mainly due to the prevalence of peptide library techniques. Very often, however, amino acid that will carry desired functional groups cannot be purchased directly, and must be synthesized in the laboratory. Examples of such currently non-available amino acids include those with a large fluorescent group, those with electron donors/acceptors, those with photochromic groups, and those with nucleic acid bases. Fortunately, several synthetic routes have now been established by which optically pure amino acids with specialty side groups can be produced, and in many cases the synthesis is not difficult (Williams, 1989). For those biochemists to whom facilities for organic synthesis are not available, custom synthesis offers a convenient alternative.

A range of nonnatural amino acids that have been effectively incorporated into proteins through an *E. coli* in-vitro synthesizing system is shown in Figure 1 (Sisido and Hohsaka, 2001). This group includes only amino acids with specialty side groups that may serve to expand protein functions. Amino acids with fluorescent groups, phosphorescent groups, electron donors/acceptors, sugar groups, phosphorylated groups, a biotin, a photochromic group, a spin probe, metal ligands, and a nucleobase have all been found to be incorporated into proteins, indicating that a wide variety of chemical functions is available on various proteins. It is somewhat surprising that amino acids with considerably large side groups can be incorporated into proteins through *E. coli* ribosomal systems. As has been reported previously, whilst aromatic amino acids with laterally extended side groups cannot enter the A-site of the ribosome, those with linearly extended groups can do so (Hohsaka et al., 1999a) For example, whilst 9-anthrylalanine that was charged onto a tRNA could not be incorporated into proteins, 2-anthrylalanine could be incorporated with reasonable efficiency. Similarly, 1-pyrenylalanine was strictly rejected by an *E. coli* ribosome, yet 2-pyrenylalanine was able to enter the A-site, albeit at low efficiency. This selectivity may be attributed to a matching of the amino acid structure with the open space of the A-site, since a similar tendency has been demonstrated for the selectivity of puromycin analogs that carry similar set of nonnatural amino acids (Hohsaka et al., 1993).

It is also interesting to note that some amino acids with very long side groups have been incorporated with reasonable efficiencies. For example, an amino acid with a biotin unit connected with a very long spacer has been successfully incorporated (Gallivan et al., 1997). As the long side group is unlikely to tunnel inside a ribosome from the A-site to the P-site, its successful incorporation indicates that the ribosome has a cleft or a special path along which the very long side group may move. In any case, the successful incorporation of amino acids with very long side groups opened the way to introducing a wide variety of specialty functions into proteins.

As described above, the selectivity of amino acids is determined mainly by their affinity for the ribosome A-site, which suggests that selectivity might also depend on the type of ribosomes. However, in the case of *E. coli* and rabbit reticulocyte ribosomes, amino acid selectivity did not change significantly despite the rabbit ribosome being somewhat more tolerant to large aromatic side groups (Hohsaka et al., 1999a).

Interestingly, some types of nonnatural amino acids have been synthesized enzymatically. For example, pyrazolylalanine has been synthesized from pyrazole and serine by a recombinant bacterium that co-expresses serineacetyl transferase and O-acetylsulfhydrylase. This enzymatic synthesis should lead the way to creating a novel microorganism that synthesizes and utilizes the twenty-first amino acid (Mino et al., 2000).

4
Chemical Aminoacylation

The first step in introducing nonnatural amino acids into proteins is to link them to the 3′-OH group of the terminal A unit of a specific tRNA. So far, this has been achieved by chemical aminoacylation, a technique first developed by Hecht and colleagues

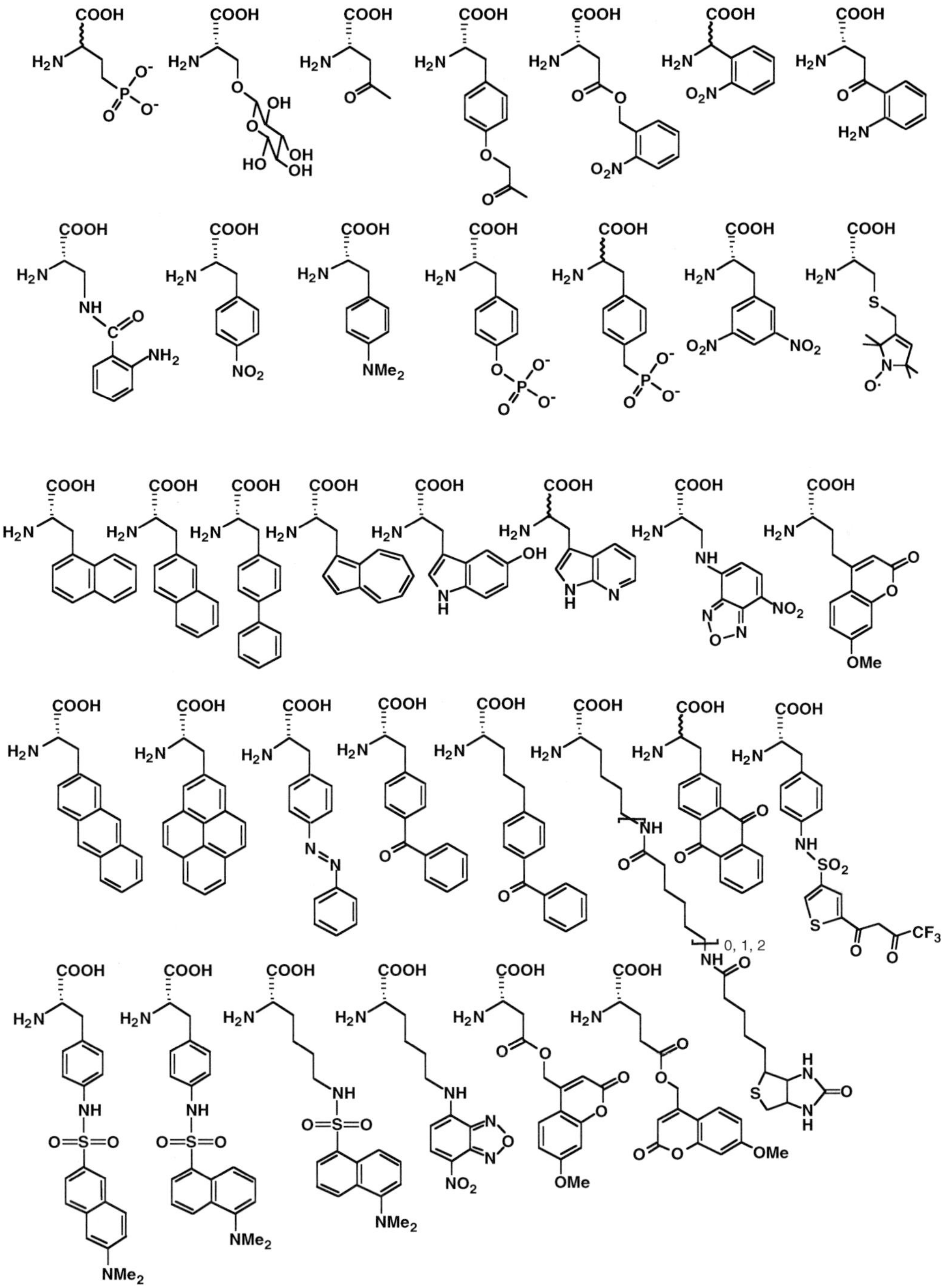

Fig. 1 Nonnatural amino acids that have been incorporated into proteins through in-vitro translation. Only those with specialty side groups are shown. (From Sisido and Hohsaka, 2001a.)

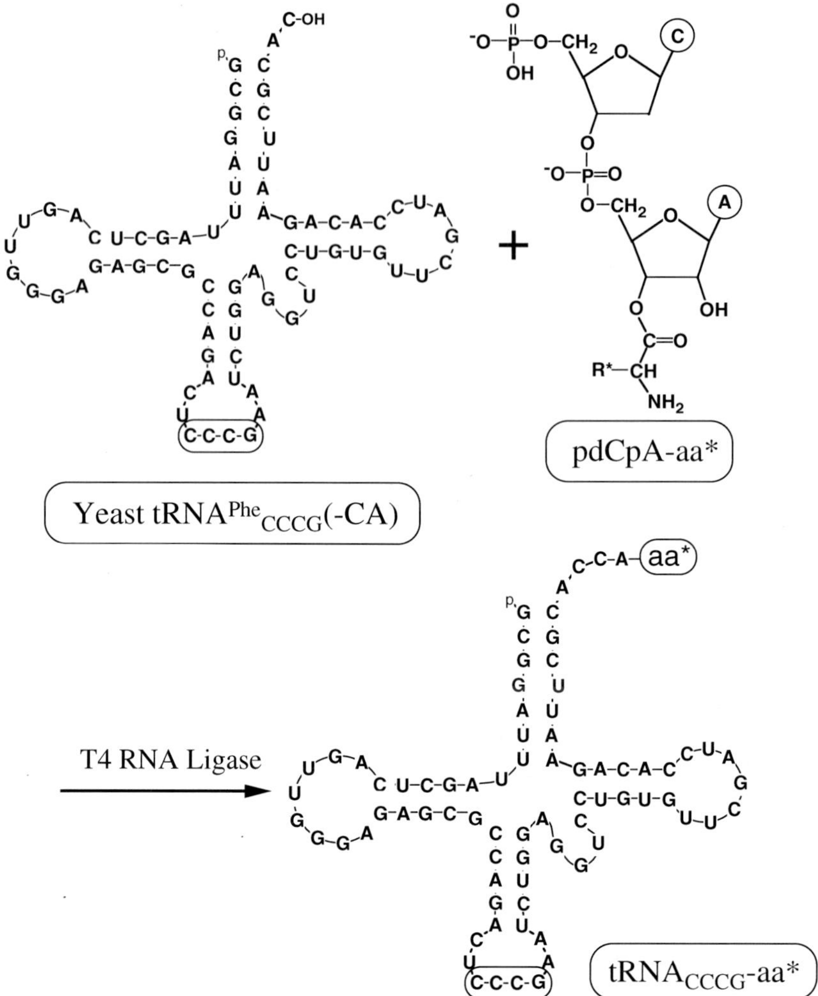

Fig. 2 Procedure of chemical aminoacylation. A tRNA that lacks the CA dinucleotide unit at the 3′ end is enzymatically ligated with a pdCpA mixed dinucleotide that is aminoacylated with a nonnatural amino acid. The pdCpA-aa* is obtained through organic synthesis.

(Heckler et al., 1983; Lodder et al., 1997). The procedure is shown diagrammatically in Figure 2. First, a mixed dinucleotide pdCpA must be synthesized, and this is then aminoacylated with an N-protected amino acid cyanomethyl ester in an organic solvent. The latter reaction occurs specifically at the 2′ or 3′ hydroxyl group of the adenosine unit. The pdCpA-aa* obtained is linked enzymatically to a tRNA that lacks a CA dinucleotide unit at the 3′ end, giving a full-length tRNA that is aminoacylated with any desired nonnatural amino acid (tRNA-aa*). This technique is widely applicable to any type of nonnatural amino acids and has been employed by most laboratories in this field. However, this chemical aminoacylation cannot be employed in living cells where a number of different tRNA molecules co-exist.

Recently, two biochemical techniques have been reported for the nonnatural aminoacylation of tRNAs. Szostak and Suga employed ribozymes that are screened to bind a specific tRNA and, at the same time, a specific amino acid-activated ester (Lohse and Szostak, 1996; Suga et al., 1998; Lee et al., 2000) According to their report, these authors screened and minimized RNA sequences and finally obtained a 57-nucleotide RNA that can charge *N*-biotinyl-L-phenylalanine onto an engineered tRNA with high efficiency and high specificity. The ribozyme approach is very attractive because of its high tRNA-recognition potentiality, and is likely to lead to a ribozyme-assisted nonnatural aminoacylation in the living cells. These authors also showed that the ribozyme could be immobilized onto a resin in order to facilitate nonnatural aminoacylation in the in-vitro system (Saito et al., 2001; Murakami et al., 2002a).

Schultz's group has also identified several orthogonal tRNA/ARS pairs that function independently from other tRNA/ARS pairs in the living *E. coli* system (Liu et al., 1997; Liu and Schultz, 1999; Pastrnak et al., 2000; Wang et al., 2001, 2002). The orthogonal ARSs have been mutated and screened not to accept any naturally occurring amino acids, but to accept a specific nonnatural amino acid. Thus, the orthogonal tRNA/ARS pairs may transfer a specific nonnatural amino acid to a specific tRNA and, subsequently, the aminoacylated tRNA may bring the nonnatural amino acid into the ribosomes. By using the orthogonal pair that is mutated to accept a nonnatural amino acid, these authors succeeded in creating a novel type of *E. coli* that utilized either *O*-methyltyrosine (Wang et al., 2001) or 2-naphthylalanine (Wang et al., 2002) as the twenty-first amino acid. The selection of orthogonal tRNA/ARS pairs that function in the *E. coli* system, as well as the screening of engineered tRNA to accept nonnatural amino acids, have also been attempted by Nishikawa's group (Ohno et al., 2001). It will be of interest to note whether these approaches can be further extended to include nonnatural amino acids with larger and functionally more interesting side groups.

Nonnatural aminoacylation is currently the most complex and crucial step that limits wide application of nonnatural mutagenesis using in-vitro protein synthesis. Aminoacylation also acts as a bottle-neck stage towards the realization of the in-vivo synthesis of nonnatural mutants, and in this respect a facile, dependable and widely applicable method of nonnatural aminoacylation must be developed.

5
Extension of Codons (Four-base Codons)

Since most triplet codons have been allocated to naturally occurring amino acids, it is necessary to determine a new class of codons for directing nonnatural amino acids into particular positions on a protein. Schultz, Chamberlin, and other groups have been employing the amber stop codon (UAG) for this purpose. The UAG codon in mRNA was successfully translated by a suppressor tRNA that carries a CUA anticodon. Although the nonsense suppressor method has advantages of simplicity and compatibility to the existing codon/anticodon system, one disadvantage is that its competition with a termination factor often limits the incorporation efficiency of nonnatural amino acids at a low level. Another drawback is that, practically, only one of the three stop codons may be assigned to a nonnatural amino acid, which means that only one type of nonnatural amino acid can be incorporated into a single protein.

Codons may be extended in different ways. It has been found that several four-base sequences may function as four-base codons, without disturbing the existing three-base codon system to any great degree (Hohsaka et al., 1996, 2001a; Murakami et al., 1998). The principle of the four-base codon is illustrated in Figure 3. When a $tRNA_{CCCG}$ that carries a CCCG four-base anticodon successfully reads a CGGG sequence as the four-base codon, the protein synthesis continues to the end and a full-length protein containing a nonnatural amino acid will be obtained. If the CGGG sequence is read, however, as the CGG three-base codon by the endogenous $tRNA_{CCG}^{Arg}$, the subsequent reading frame is shifted backward by one base, resulting in an encounter with the stop codon, UAA. Thus, the undesired translation as the three-base codon results in the truncation of protein synthesis. In this way, the four-base codon CGGG is practically orthogonal to the existing three-base codon system, except for CGG that is scarcely used in *E. coli*. In practical applications, the CGG codon in the target gene must be replaced by CGU or other codons that direct arginine. In fact, when a CGGG codon was inserted in place of three-base codons at different positions of streptavidin, both the full-length streptavidin and the truncated protein were detected in a mixture of the in-vitro translation system that contained $tRNA_{CCCG}$ aminoacy-

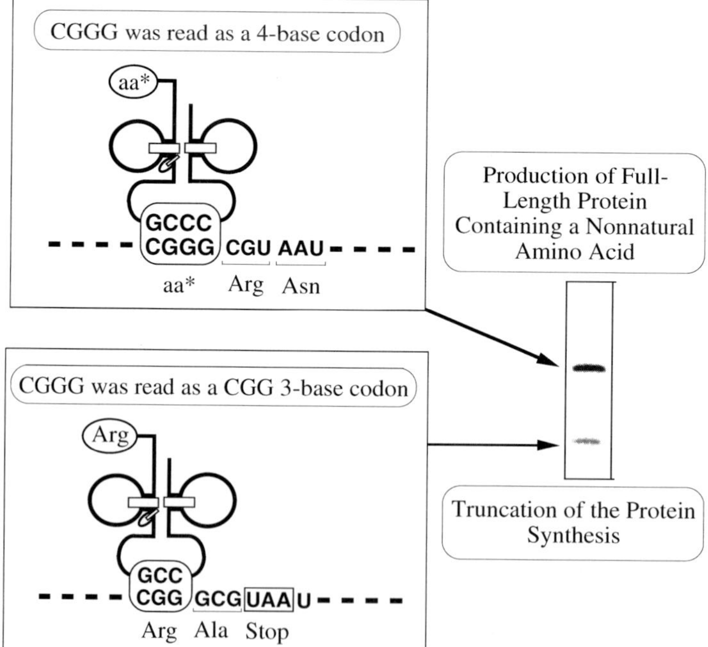

Fig. 3 Principle of the four-base codon strategy. When a CGGG sequence is translated by a Xaa-tRNACCCG, a full-length protein that contains a nonnatural amino acid Xaa, will be synthesized. When the CGGG sequence is read by a Arg-tRNACCG, the reading frame will be shifted by one base backward that leads to a truncation of the protein synthesis at the UAA stop codon. (From Sisido and Hohsaka, 2001a.)

lated with *p*-nitrophenylalanine. The yields of the full-length protein were not very sensitive to the position of the four-base codons, indicating that the translation efficiency of the CGGG four-base codon does not depend on the gene context (Murakami et al., 1998).

Other four-base codon-anticodon pairs have been examined for their translation efficiencies, and the following five four-base codons were found to be effective: AGGU, CGGU, CCCU, CUCU, and GGGU (Hohsaka et al., 2001a). These four-base codons were also orthogonal to each other and to the existing three-base codons. The translation efficiency depends on the type of the four-base codons; the highest efficiency has been observed with GGGU (about 80% with respect to the wild-type protein), while that of the others ranged from 40 to 60%. The high efficiency may be due primarily to the scarceness of tRNAs that are cognate to the original minor codons, AGG, CGG, CCC, CUC, and GGG. An extended codon table that includes orthogonal four-base codons is shown in Figure 4.

As reported from Schultz's group, four-base codons also function in vivo, despite possible interference from the corresponding three-base codons in the endogenous mRNAs (Magliery et al., 2001).

6 Adaptability of Four-base Codons in the Ribosome System and Extension to Five-base Codons

It is somewhat surprising to find that the tRNA with an extended anticodon loop–from seven unpaired bases to eight–is able to function in the ribosome. A plausible explanation is provided by the X-ray crystallographic structure of the ribosome from *Thermus thermophilus*, where tRNAs are bound to the A-site, P-site, and E-site (Yusupov et al., 2001). As shown in Figure 5, configurations of the mRNA and tRNAs in the ribosome are compatible with the four-base codons. The mRNA is bent and twisted at the junction of the two codons, while the two tRNAs are almost parallel-

	U	C	A	G
U	UUU Phe UUC UUA Leu UUG	UCU UCC Ser UCA UCG	UAU Tyr UAC UAA Stop UAG	UGU Cys UGC UGA Stop UGG Trp
C	CUCU CUCA CUU CUC Leu CUA CUG	CCCU CCU CCC Pro CCA CCG	CAU His CAC CAA Gln CAG	CGU CGC Arg CGA CGG CGGU CGGG
A	AUU AUC Ile AUA AUG Met	ACU ACC Thr ACA ACG	AAU Asn AAC AAA Lys AAG	AGU Ser AGC AGA Arg AGG AGGU
G	GUU GUC Val GUA GUG	GCU GCC Ala GCA GCG	GAU Asp GAC GAA Glu GAG	GGU GGC Gly GGA GGG GGGU

Fig. 4 Extended codon table that includes five additional four-base codons that can be assigned to different nonnatural amino acids. (From Sisido and Hohsaka, 2001a.)

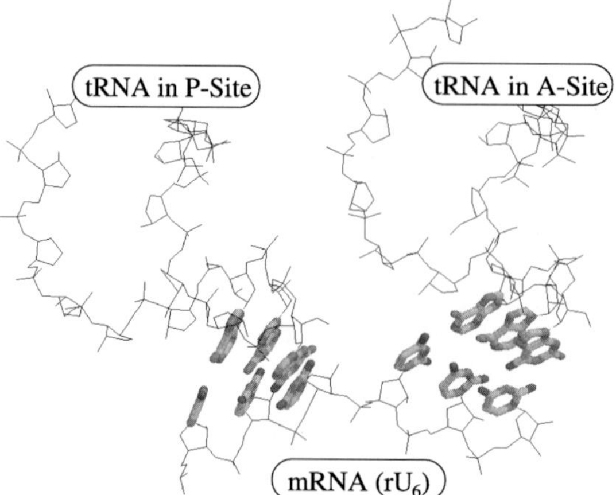

Fig. 5 Configuration of the codon/anticodon pairings in *Thermus thermophilus* ribosome. Taken from the X-ray crystallographic data of Yusupov et al. (2001).

oriented but somewhat twisted to accomplish perfect codon/anticodon pairing and close proximity between the peptide end and the amino acid ester. The twisting is also needed to avoid overlapping of the bodies of the two tRNAs. As expected from the figure, four-base codon/anticodon pairs in the A-site or in the P-site do not cause serious overlapping of the anticodon loops and the whole bodies of the two tRNAs.

The above idea is made more convincing by the successful translation of five-base codons into nonnatural amino acids (Hohsaka et al., 2001b). A streptavidin mRNA containing a CGGUA five-base codon at the Tyr54 position was successfully read by a $tRNA_{UACCG}$ that is chemically aminoacylated with a nonnatural amino acid in an *E coli* in-vitro translation system. High-performance liquid chromatography (HPLC) analysis of the tryptic fragment of the translation product revealed that the nonnatural amino acid was incorporated corresponding to the CGGUA codon, without affecting the reading frame adjacent to the CGGUA codon. Another fifteen five-base codons $CGGN_1N_2$, where N_1 and N_2 indicate one of four nucleotides, were also successfully decoded by the aminoacyl-tRNAs containing the complementary five-base anticodons.

7 Nonnatural Base Pairs for Codon Extension

An alternative approach towards the extension of codon/anticodon pairs is the addition of nonnatural nucleobases to the existing four types of nucleobase. The artificial nucleobases must be orthogonal to the existing nucleobases, and must form a complementary pair with each other at the highest fidelity. Hopefully, the artificial pair may be replicated and transcribed by DNA polymerases and RNA polymerases, respectively. Clearly, these requirements are not easy to satisfy, but several artificial base pairs have been reported to fulfil, at least partially, the conditions. Some of these artificial bases are listed in Figure 6. Benner and coworkers reported the first successful nonnatural base pair (Switzer et al., 1989; Piccirilli et al., 1990). These authors proposed that *iso*G and *iso*C may function as an orthogonal base pair,

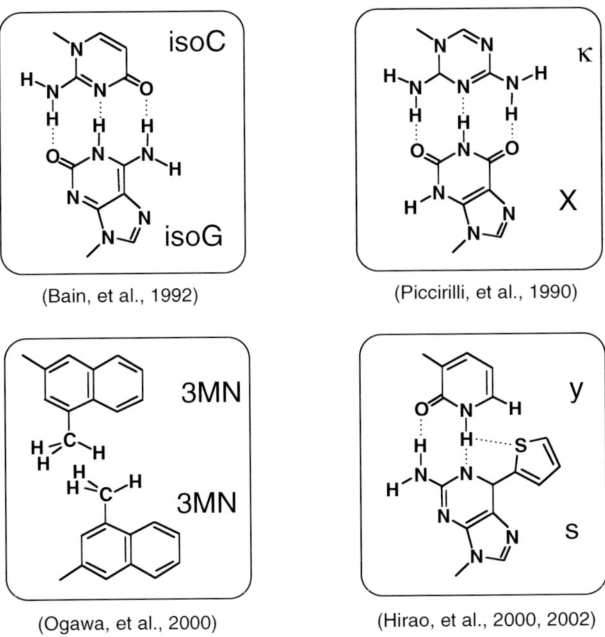

Fig. 6 Nonnatural base pairs that are orthogonal to the existing A–T and G–C pairs and which function, at least partly, as substrates of RNA polymerase.

although the fidelity in the replication and the transcription needs to be improved. The nonnatural codon/anticodon pair, *iso*C-A-G/C-U-*iso*G, has been utilized to incorporate iodotyrosine (Bain et al., 1992). Hirao et al. reported that 2-amino-6-(*N*,*N*-dimethylamino)purine(x)-pyridin-2-one (y) pair was practically orthogonal to the existing pairs and could be used in the RNA synthesis with T7 RNA polymerase (Hirao et al., 2000; Ohtsuki et al., 2001). Very recently, these authors also reported that another base pair, s-y, is effective enough to direct a nonnatural amino acid into proteins (Hirao et al., 2002).

Schultz's group proposed that substituted naphthyl groups incorporated into DNAs form a complementary pair only by hydrophobic interactions (Ogawa et al., 2000). Although at present the orthogonality and fidelity of the nonnatural base pairs seems still insufficient for practical applications, the approach should offer a promising alternative to the nonsense suppression or the four-base codon/anticodon strategy.

8
In-vitro Protein Synthesis in the Presence of tRNAs that are Amino-acylated with Nonnatural Amino Acids

Messenger RNAs that contain four-base codons are expressed in the extracts of protein-synthesizing components from *E. coli*, rabbit reticulocytes, or wheat germ. In addition to the ingredients for conventional in-vitro synthesis, a target mRNA with four-base codon(s), tRNA(s) with the complementary four-base anticodon(s) that are aminoacylated with nonnatural amino acid(s) ($tRNA_{NNNN}$-aa*) must be added. The tRNA must be free from aminoacylation by the existing ARSs in the in-vitro system. The advantages of the in-vitro synthesis are its simple, quick operation and its tolerance to harmful proteins, whilst its very low productivity represents a major and sometimes serious drawback. Under average laboratory conditions, the protein yields range from 0.1 to 10 μg, though recently yields have been markedly improved through optimization of the reaction conditions and the exclusion of waste compounds by the use of membranes (Roche Molecular Biochemicals, RTS500).

In the case of nonnatural mutants, however, the protein yield tends to saturate at an earlier stage than the conventional in-vitro synthesis. Typically, expression of mRNA of ^{83}CGGG-streptavidin in an *E. coli* in-vitro system in the presence of $tRNA_{CCCG}$-nitrophenylalanine became saturated after 5–10 min, whereas expression of the wild-type streptavidin continued up to ~30 min. One possible reason for the early saturation is hydrolysis of the $tRNA_{NNNN}$-aa*. Because the free tRNA with a four-base anticodon cannot be aminoacylated again by any ARSs, the irreversible hydrolysis causes termination of the protein synthesis. The half-lives of the $tRNA_{NNNN}$-aa*s depend on the types of amino acids, and range from 10 to 30 min (Hentzen et al., 1972).

The hydrolysis of aminoacylated tRNA may not be a major reason for the early saturation of the protein synthesis however, because even if fresh tRNA-aa* is added to the saturated reaction mixture, the protein synthesis did not recommence. Neither free $tRNA_{CCCG}$ nor free $tRNA_{CCCG}$(-CA) functions as an inhibitor. Another reason for the early saturation might be a three-base reading of the four-base codons to produce a truncated protein. Although this situation might lead to early saturation when inefficient four-base codons were used, it cannot occur when efficient four-base codons such as CGGG and GGGU were used, especially where there is a sufficiently high concentration of the tRNA with a four-base anticodon.

In a preliminary experiment, some byproducts were that were found to be associated with chemical aminoacylation of the $tRNA_{NNNN}$(-CA) with pdCpA-aa* might have been responsible for the inhibitory effect. Although the possible inhibitor has not yet been identified, one possible candidate is a cyclic tRNA(-CA) that may be produced by self-ligation of the 5′-phosphate end with the cytosine unit at the 3′ end (Bruce et al., 1978). The formation of cyclic tRNA might be facilitated by truncation of the CA dinucleotide unit at the 3′ end, due to the close proximity between the 5′-phosphate group and the 3′-OH group.

In any case, by removing the byproducts that functions as inhibitors, the yields of nonnatural mutants will reach the same level as the in-vitro synthesis of conventional proteins.

9 Purification and Identification of Nonnatural Mutants

Proteins synthesized in the in-vitro system are obtained as a mixture with various enzymes, tRNAs, mRNAs, ribosomes and other ingredients. Attachment of the histidine hexamer (His-tag) to the C-terminus of the mutant is often very effective for its selective binding to an affinity column that carries Ni or Co metal complexes. The His-tag strategy is particularly effective for the purification of nonnatural mutants obtained through the four-base codon strategy, as unsuccessful translation of four-base codons causes off-frame products that do not contain the His-tag. In particular, when the four-base codon is located close to the C-terminus, the truncated product from three-base translation and the full-length protein from four-base translation have similar molecular weights, but only the latter contains the His tag. The His-tag affinity method is especially effective for such cases.

The presence and position of a nonnatural amino acid in a mutant protein is sometimes very difficult to verify, although this is suggested indirectly from Western blotting of the products after His-tag affinity purification, which indicates the presence of a full-length protein with a His tag. If the nonnatural mutant can be totally purified, then mass spectroscopy will provide clear evidence for the incorporation of a particular amino acid. However, this technique does not provide any information on the position of mutation, unless a somewhat sophisticated method has been used. Unequivocal evidence of the incorporation of a nonnatural amino acid is provided by HPLC analysis of digestion products of the mutant. If the products of a trypsin digestion contain a peptide fragment that is identified as (Arg)–aa_1–aa_2– —· aa_n–Arg, where a nonnatural amino acid is positioned at one of the amino acids, then the presence and position of the nonnatural amino acid are verified. This approach has been used successfully by comparing the digestion product with a peptide fragment that was chemically synthesized (Hohsaka et al., 1999b).

10 Multiple Incorporation of Nonnatural amino acids by using Different Orthogonal Four-base Codons

Besides the high translation efficiency, the four-base codon strategy is advantageous for its potentiality of multiple incorporation of different nonnatural amino acids into single proteins. For example, an orthogonal pair of four-base codons, CGGG and AGGU, was used to direct Lys(NBD) and L-2-naphthylalanine to the 54th and 57th positions of streptavidin, respectively (Murakami et al., 2002b). As shown in Figure 7, full-length streptavidin was detected only when both the aminoacyl $tRNA_{CCCG}$ and aminoacyl $tRNA_{ACCU}$ were added to the in-vitro system. The multiple mutation was directly confirmed by HPLC analysis of the trypsin-digested fragment. It is conceivable that multiple mutation with different amino acids will open a way in which a path for electron transfer in proteins may be built, and indeed in a preliminary experiment in our group a sensitizer–quencher pair has been introduced into a single protein, and photoinduced electron transfer has been detected. It is likely that double mutations with electron donor–acceptor pairs or energy donor–acceptor pairs will open a new scope in protein engineering.

11 Mutants that Contain a Single Nonnatural Amino Acid at Random Positions

Nonnatural mutants that contain amino acids with a large functional side group often lose their original activity due to unfolding or misfolding, and this represents a serious potential problem when there is a need to introduce large specialty groups without changing or losing the protein's inherent activity. Because protein conformation and function depend on the positions of the nonnatural amino acids, the search for the best position is an important step in obtaining the optimum mutant. To this end, it has been necessary to synthesize a large number of mutants in which a nonnatural amino acid is incorporated at different positions. Clearly, this requires much elaboration, and a short-cut would be to build a

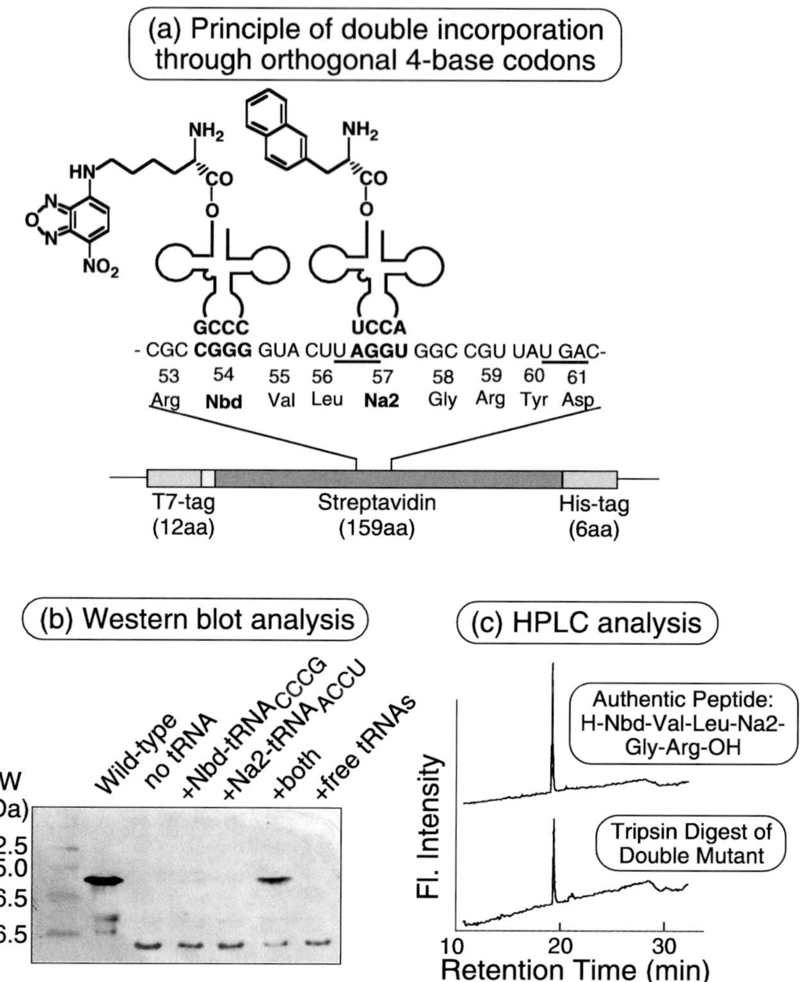

Fig. 7 Principle of double incorporation of nonnatural amino acids through two independent four-base codons, CGGG and AGGU. (a) The 54th four-base codon CGGG, is translated by a $tRNA_{CCCG}$ carrying an NBD-Lys and the 57th AGGU 4-base codon is translated by a $tRNA_{ACCU}$ carrying a naphthylalanine. When the 54th codon is read as a CGG three-base codon, protein synthesis will be stopped at the UAG codon; when the 57th codon is read as a AGG three-base codon, protein synthesis will stop at the UGA codon. Full-length streptavidin will be obtained only when both the CGGG and AGGU sequences were translated as the four-base codons. (b) Results of Western blotting. Full-length streptavidin was synthesized only when $tRNA_{CCCG}$ and $tRNA_{ACCU}$ aminoacylated with NBD-Lys and 2-napththylalanine, respectively, were added to the in-vitro system. (c) Results of HPLC analysis of the trypsin digests of the double mutant. The fluorescent component was identified as the fragment expected for cleavage at the arginine units.

"library" of proteins in which a nonnatural amino acid has been incorporated at different positions. Very recently, a new technique has been developed by which an arbitrary four-base codon is inserted at random positions on a gene in place of three consecutive bases (Sisido et al., 2000). The new technique, which is known as random

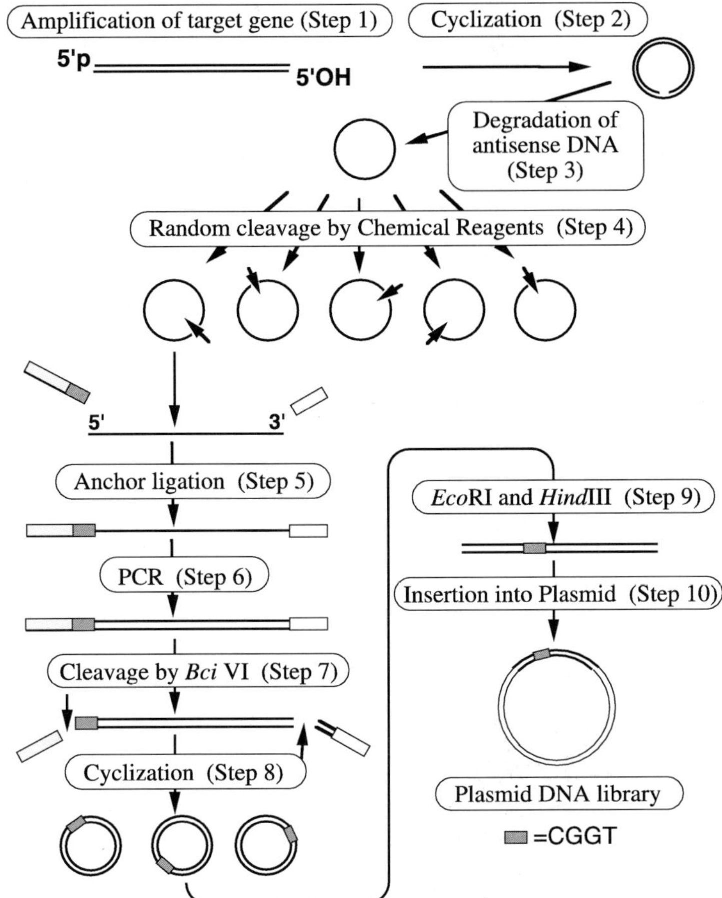

Fig. 8 A general scheme of the random insertion/deletion (RID) mutagenesis for the construction of a library of mutant genes.

insertion/deletion (RID) mutagenesis, enables the deletion of an arbitrary number of consecutive bases at random positions and, simultaneously, the insertion of a specific sequence or random sequences of an arbitrary number into the same position.

The key steps of RID mutagenesis are illustrated in Figure 8. The procedure consists of eight major steps.

Step 1: (1) The fragment obtained by digesting the original gene with *Eco*RI and *Hin*dIII is ligated to a linker. (2) The product is then digested with *Hin*dIII to make a linear double-stranded (ds) DNA with a nick in the antisense chain.

Step 2: The gene fragment is cyclized with T4 DNA ligase to make a circular dsDNA with a nick in the antisense chain.

Step 3: The circular dsDNA is treated with T4 DNA polymerase to produce a circular single strand (ss) DNA.

Step 4: The circular ssDNA is randomly cleaved at single positions by treating with Ce(IV)–EDTA complex.

Step 5: The linear ssDNAs, which have unknown sequences at both ends, are ligated to the 5′-anchor and the 3′-anchor, respectively.

Step 6: The DNAs that are linked to the two anchors at both ends are amplified by PCR.

Step 7: The PCR products are treated with *Bci*VI, leaving several bases from the 5′-anchor, at the 5′-end. The *Bci*VI treatment also deletes a specific number of bases at the 3′-end.

Step 8: The digested products are treated with Klenow fragment to create blunt ends, and cyclized again with T4 DNA ligase. The products are treated with *Eco*RI and *Hin*dIII, and the fragments are cloned into an *Eco*RI-*Hin*dIII site of modified pUC18 (pUM).

The applicability of RID mutagenesis was demonstrated by replacing randomly selected portions of three consecutive bases by the *Bgl*II recognition sequence (AGATCT) in the GFP_{UV} gene. Alternatively, the randomly selected three bases were replaced by a mixture of twenty codons for twenty naturally occurring amino acids. The DNA libraries were expressed in *E. coli*, and the mutants that showed different fluorescence properties from the wild-type GFP_{UV} were selected.

As described above, RID mutagenesis not only enables four-base codons to be introduced at random positions in DNA, but also provides a general method for codon-based mutagenesis among the twenty naturally occurring amino acids. To date, random point mutations have been introduced into genes, most commonly by an error-prone PCR method that is based on inaccurate copying of a DNA sequence by DNA polymerases. However, the error-prone PCR method has an inherent drawback of biasing the occurrence of amino acids because only single bases are replaced within the triplet codons. For example, mutation from AUG (Met) to UAG (Trp) is unlikely to occur by a single error-prone PCR mutation. The RID technique provides a general and flexible method for constructing a DNA library in which any specific three- or four-base codons are introduced in place of three-consecutive bases at random positions.

Very recently, Schultz's group built a library of four-base codons/anticodon pairs in the *E. coli* in-vivo system (Magliery et al., 2001). These authors have selected four-base codon/anticodon pairs that function in the living cell, and showed the AGGA/UCCU pair to be the best candidate. These studies showed that the four-base codon strategy operates even in the in-vivo system.

12
Combination of In-vitro Synthesis and Chemical Synthesis for Screening Nonnatural Mutants and for Large-scale Production

In-vitro protein synthesis is advantageous for its quick production of mutant proteins in a few tens of minutes, and for its easy purification procedures. However, under average laboratory conditions, the amount of protein produced is on the order of micrograms, or less. It is very difficult–if not possible–to obtain milligram or larger quantities of nonnatural mutants by conventional in-vitro synthesis, although significant progress has recently been made in this respect, as described above.

Chemical synthesis might be the best choice for preparing a few tens of milligrams or larger quantities of nonnatural mutants, if the number of amino acids is less than 100. However, because chemical synthesis requires higher cost, longer time and is more wasteful of chemicals, it is not practical to screen an optimum mutant by the repeated synthesis of various proteins. At present, the quickest way to obtain practical quantities of nonnatural mutants that are functionally optimized, is the in-vitro synthesis for screening of types and positions of nonnatural amino acids, followed by chemical synthesis of the optimized mutant.

Recent progress in solid-phase peptide synthesis has enabled the preparation of small proteins of less than 100 amino acids, although the purity of the crude product following cleavage from the solid support may vary widely. The use of the Fmoc N-protecting group and a highly efficient coupling reagent such as *o*-(7-azabenzotriazol-1-yl)-1,1,3,3-tetramethyluronium hexafluorophosphate (HATU) with purified solvents are strongly recommended. Selection of the solid support is important not only for determining the C-terminal group (-COOH or -$CONH_2$) but also for selecting conditions for safe cleavage. Purification by gel chromatography and preparative HPLC leads to 10 to 100-mg quantities of small proteins, the purity of which can be assessed using time-of-flight (TOF) mass spectroscopy. An example of the synthesis and purification of the λ-Cro repressor protein containing an electron-accepting amino acid is shown in Figure 9. The positions of a nonnatural amino acid have been screened for various mutants prepared through in-vitro synthesis to retain the DNA binding activity, and the optimum position of the nonnatural amino acid was determined to be the 64th position (Sisido et al., 2000). The mutant was then chemically synthesized in about 50-mg quantity (yield ~30%) with sufficiently high purity.

Longer proteins can be synthesized by repeated couplings of fragment peptides. For example, Kimura synthesized green fluorescent protein (GFP), a 238-amino acid protein, that fluoresces exactly as does the genetically engineered counterpart (Nishiuchi et al., 1998). Although fragment synthesis requires the sophisticated knowledge and skills of peptide chemistry, it should be used more widely in order to obtain large amounts of either natural or nonnatural mutants.

A combination of the screening of in-vitro products and large-scale chemical synthesis of the optimum mutant is currently the most practical way to obtain nonnatural mutants for commercial use.

13
In-vivo Synthesis of Nonnatural Mutants

In the living cell, if it is possible to control the entire process relating to the introduction of amino acids into proteins, then it should also be possible to introduce a nonnatural amino acid as the twenty-first member. As has been described above, several unique codon/anticodon pairs are available which function independently of existing codon/anticodon pairs. Yeast tRNAs carrying four-base anticodons also function independently of the tRNAs and ARSs in *E. coli*, with some exceptions. The aminoacylation of unique tRNAs with nonnatural amino acids is the final step to be taken towards the goal of a "synthetic organism".

As described earlier, Schultz's group has been using a $tRNA^{Tyr}$/TyrRS pair from *Methanococcus jannashii* as an orthogonal pair in the *E. coli* system. The *M. jannashii* TyrRS was then engineered to change its amino acid specificity, from charging tyrosine to charging either *O*-methyltyrosine (Wang et al., 2001) or 2-naphthylalanine (Wang et al., 2002). Thus, these authors succeeded in adding the latter amino acid as the twenty-first member. However, as the types of amino acid introduced into the living cell are dependent upon the mutation of ARSs, this technique seems to be limited to the analogs of naturally occurring amino acids.

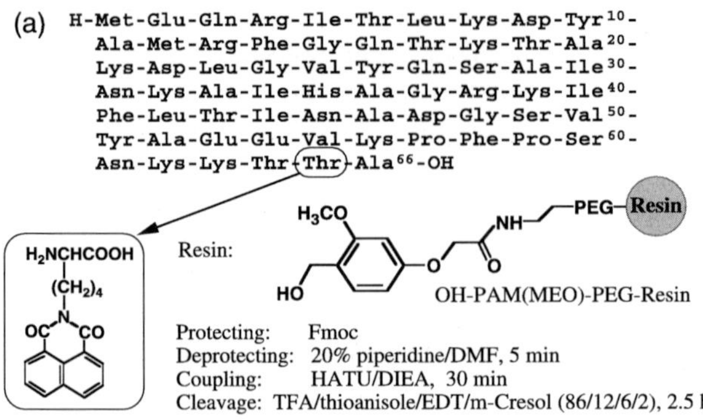

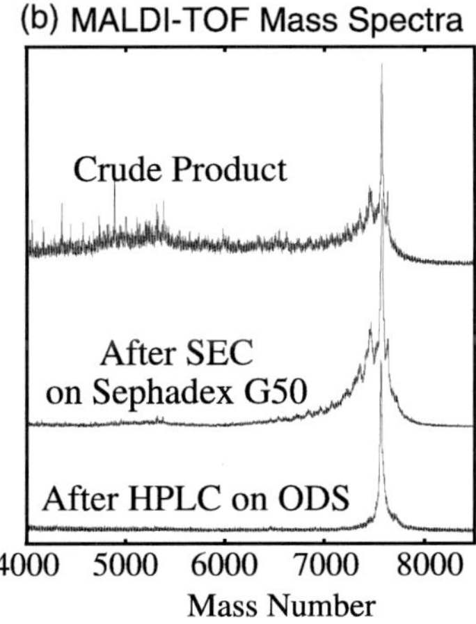

Fig. 9 Example of chemical synthesis of a small protein, λ-Cro repressor protein of 66 amino acids that contain a nonnatural amino acid. (a) Amino acid sequence and synthetic conditions. (b) Results of time-of-flight (TOF) mass analysis at each stage of purification.

The ribozyme-mediated aminoacylation of orthogonal tRNA represents another promising approach to the in-vivo synthesis. As stated above, Suga's group succeeded in aminoacylating phenylalanine derivatives to tRNAs, though in this case the types of amino acids were limited by the chemical diversity of RNAs (which could be narrower than that of proteins or peptides). In any case, either the mutation of existing ARSs or selection of ribozymes requires much further study in order to introduce all new nonnatural amino acids. Consequently, the development of a simple and widely applicable method to aminoacylate specific tRNAs with nonnatural amino acids carrying a wide variety of specialty functions has become a major target in this area.

14 Examples of Specialty Functions of Nonnatural Mutants

14.1 Fluorescence Labeling

A potential application of the nonnatural mutation is the fluorescence labeling of receptors, antibodies, and enzymes at specific positions. By using correctly labeled proteins it is possible to detect ligands, antigens, and inhibitors at very low concentrations. For this purpose, however, it is necessary to identify fluorescent nonnatural amino acids that can be incorporated into proteins in high yields. The fluorescent groups must be highly sensitive to small changes in the microenvironment that are caused by the binding of small molecules. Furthermore, the nonnatural amino acids must be introduced at a specific position where the binding activity of the protein is not suppressed.

To date, several types of nonnatural amino acids bearing highly fluorescent groups have been successfully incorporated through the *E. coli* in-vitro system. These include $^{\varepsilon}$*N*-dansyl-L-lysine (Steward et al., 1997), a fluorescein group linked by a keto-selective reaction (Cornish et al., 1996), 3-*N*-(7-nitro-2,1,3-benzooxadiazol-4-yl)-2,3-diamino-propionic acid (Turcatti et al., 1996), γ-(7-methoxycoumarin-4-yl)-L-homoalanine (Murakami et al., 2000) and β-anthraniloyl-L-α,β-diaminopropionic acid (atnDap) (Taki et al., 2001). Among these nonnatural amino acids, the latter is highly sensitive to the change in microenvironment and is sufficiently small to retain inherent binding activity of the protein. The fluorescence spectra of 120atnDap–streptavidin in the presence of biotin is shown in Figure 10.

The incorporation of a sensitive fluorescent amino acid into a specific position of various receptors, antibodies, and enzymes represents a promising approach to diagnostics. Hence it is likely that, in the future, fluorescence-modified proteins will form the basis of many applications in proteomics technology.

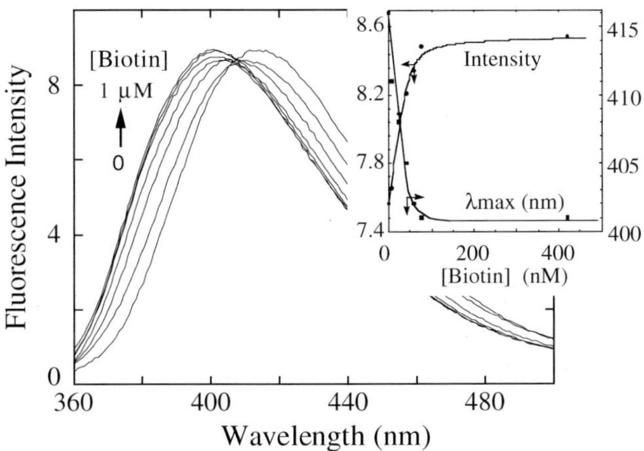

Fig. 10 Change in fluorescence intensity and wavelength of 120atnDap-streptavidin in the presence of different amounts of biotin.

14.2 Electron Transfers Inside Protein Frameworks

Electron transfer (ET) is a key process that governs photosynthetic and biological reduction-oxidation (reductoxidase) reactions. In nature, proteins which are relevant to ET processes contain specialized prosthetic groups such as metal porphyrins and quinones. However, if external redox groups in the form of nonnatural amino acids can be added at specific positions, then the scope of the reductoxidase reactions will be expanded. To this end, it is necessary to study ET rates within a protein framework and pathways through which ET occurs. The basic information can be obtained by placing electron donor–acceptor pairs at specific positions on a single protein, and this has been achieved using nonnatural mutants of streptavidin incorporated with single *p*-nitrophenylalanine as an electron-accepting amino acid located at various positions (Murakami et al., 1998). A single pyrenyl group was linked to biotin and introduced into the biotin-binding site. Thus, a pyrenyl group as a photosensitizer and nitrophenylalanine as an acceptor were introduced into single streptavidin with different interchromophore distances and donor–acceptor ET paths. Rates of photoinduced ET from the excited pyrenyl group to the nitrophenyl group were measured and plotted as the function of the interchromophore edge-to-edge distances (Figure 11). The plot in Figure 11 shows a linear relationship between the logarithm of the ET rate constants and the edge-to-edge distances, indicating that ET on a single protein is governed mainly by the interchromophore edge-to-edge distances. However, detailed consideration of the ET pathways between the donor and acceptor suggested that the protein ET is also reasonably interpreted in terms of the tunneling pathway model, including an electron jump through hydrogen bonds. The mechanism is supported by experiments on model peptides that take α-helical conformation (Sisido et al., 2001) and β-sheet structures. (Sasaki et al., 2001)

Protein ET can be studied in more detail by placing both a donor and an acceptor in the forms of nonnatural amino acids. This has been possible by using two different four-base codons, one for the electron donor and another for the electron acceptor. In a preliminary experiment, a fluorescent amino acid, α-anthraniloyl-L-β-diaminopropionic acid was introduced at the 84th position of streptavidin, and its fluorescence was effectively quenched by L-*p*-nitrophenylalanine introduced at the 54th position.

The introduction of a strong electron acceptor into proteins may create an artificial oxidase. A single anthraquinonylalanine (anqAla) was introduced into λ-Cro repressor protein, retaining the DNA-binding activity of the latter. A 64anqAla-Cro showed a strand-specific and position-specific DNA photoscission when the mutant was bound to dsDNA and the complex was photoirradiated. The photoscission can be seen as the result of a photoinduced ET from a G unit in the dsDNA to the excited anthraquinonyl group.

15 Outlook and Perspectives

Currently, nonnatural mutation represent a developing technique that requires further improvements in several crucial areas, most notably in the aminoacylation step. Because living organisms have been evolving for a very long time under the constraints of twenty-type amino acids, four-type nucleic bases and three-base codons, it is not surprising that the synthetic approaches will require much effort to overcome these

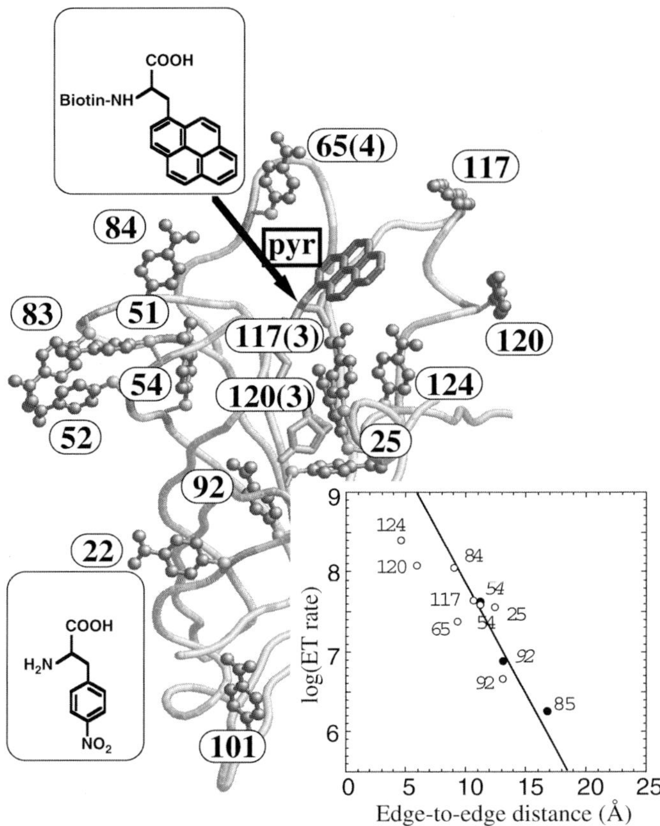

Fig. 11 Electron transfer (ET) rate on mutant streptavidins that contain single *p*-nitrophenylalanine at different positions and bind *N*-biotinyl-L-1-pyrenylalanine. The ET rates were evaluated from the fluorescence intensity of fluorescence decay curves of the pyrenyl group and plotted as a function of edge-to-edge distances between the pyrenyl and nitrophenyl groups.

constraints. Nonetheless, developments achieved during the past 10–15 years have been most noteworthy in that there is now a wide variety of nonnatural amino acids available, as well as several four-base codons that can assign different nonnatural amino acids in addition to the twenty naturally occurring ones, and that orthogonal tRNAs are available which can carry nonnatural amino acids into the ribosomes. Furthermore, a technique has been developed to create a DNA library in which any four-base codon is inserted in place of three consecutive bases.

Today, nonnatural mutagenesis is beginning to find a variety of applications, despite the current restrictions of poor incorporation efficiency and total yield. However, the range of uses will undoubtedly expand dramatically during the next few years as procedures become simpler and total yields of nonnatural mutants are increased, perhaps 100-fold or more.

Another goal of nonnatural mutagenesis is to create a "synthetic microorganism" in which nonnatural amino acids are automatically incorporated into specified positions in proteins. This novel microorganism will show specialty functions depending on the nonnatural amino acids incorporated, and will open a new stage in bioorganic chemistry.

Acknowledgments

Studies conducted by the author and colleagues were supported by a Grand-in-Aid for Specially Promoted Research from the Ministry of Education, Science, Sports and Culture, Japan (No.11102003).

16 References

Bain, J. D., Glabe, C. G., Dix, T. A., Chamberlin, A. R. (1989) Biosynthetic site-specific incorporation of a non-natural amino acid into a polypeptide, *J. Am. Chem. Soc.* **111**, 8013–8014.

Bain, J. D., Switzer, C., Chamberlin, A. R., Benner, S. A. (1992) Ribosome-mediated incorporation of a non-standard amino acid into a peptide through expansion of the genetic code, *Nature* **356**, 537–539.

Böck, A., Forchhammer, K., Heider, J., Baron, C. (1991) Seleno protein synthesis: an expansion of the genetic code, *Trends Biochem. Sci.* **16**, 463–466.

Bruce, A. G., Olke, C., Uhlenbeck, O. C. (1978) Reactions at the termini of tRNA with T4 RNA ligase, *Nucleic Acids Res.* **10**, 3665–3677.

Chapeville, F., Lipmann, F., von Ehrenstein, G., Weisblum, B., Ray, W. J., Benzer, S. (1962) The role of soluble ribonucleic acid in coding for amino acids, *Proc. Natl. Acad. Sci. USA* **48**, 1086–1092.

Cornish, V. W., Hahn, K. M., Schultz, P. G. (1996) Site-specific protein modification using a ketone handle, *J. Am. Chem. Soc.* **118**, 8150–8151.

Fahnestock, S., Rich, A. (1971) Ribosome-catalyzed polyester formation, *Science* **173**, 340–343.

Gallivan, J. P., Lester, H. A., Dougherty, D. A. (1997) Site-specific incorporation of biotinylated amino acids to identify surface-exposed residues in integral membrane proteins, *Chem. Biol.* **4**, 739–749.

Heckler, T. G., Zama, Y., Naka, T., Hecht, S. M. (1983) Dipeptide formation with misacylated tRNAPhes, *J. Biol. Chem.* **258**, 4492–4495.

Hentzen, D., Mandel, P., Garel, J.-P. (1972) Relation between aminoacyl-tRNA stability and the fixed amino acid, *Biochim. Biophys. Acta* **281**, 228–232.

Hirao, I., Ohtsuki, T., Mtsui, T., Yokoyama, S. (2000) Dual specificity of the pyrimidine analogue, 4-methylpyridin-2-one, in DNA, *J. Am. Chem. Soc.* **122**, 6118–6119.

Hirao, I., Ohtsuki, T., Fujiwara, T., Mitsui, T., Yokogawa, T., Okuni, T., Nakayama, H., Takio, K., Yabuki, T., Kigawa, T., Kodama, K., Yokogawa, T., Nishikawa, K., Yokoyama, S. (2002) An unnatural base pair for incorporating amino acid analogs into proteins, *Nature Biotechnol.* **20**, 177–182.

Hohsaka, T., Sato, K., Sisido, M., Takai, K., Yokoyama, S. (1993) Adaptability of nonnatural aromatic amino acids to the active center of *E. coli* ribosomal A site, *FEBS Lett.* **335**, 47–50.

Hohsaka, T., Ashizuka, Y., Murakami, H., Sisido, M. (1996) Incorporation of nonnatural amino acids into streptavidin through *in vitro* frame-shift suppression, *J. Am. Chem. Soc.* **118**, 9778–9779.

Hohsaka, T., Kajihara, D., Ashizuka, Y., Murakami, H., Sisido, M. (1999a) Efficient incorporation of nonnatural amino acids with large aromatic groups into streptavidin in *in vitro* protein synthesizing systems, *J. Am. Chem. Soc.* **121**, 34–40.

Hohsaka, T., Ashizuka, Y., Sasaki, H., Murakami, H., Sisido, M. (1999b) Incorporation of two different nonnatural amino acids independently into a single protein through extension of the genetic code, *J. Am. Chem. Soc.* **121**, 12194–12195.

Hohsaka, T., Ashizuka, Y., Murakami, H., Sisido, M. (2001a) Incorporation of nonnatural amino acids into proteins by using various four-base codons in the *E. coli in vitro* translation system, *Biochemistry* **40**, 11060–11064.

Hohsaka, T., Ashizuka, Y., Murakami, H., Sisido, M. (2001b) Five-base codons for incorporation of nonnatural amino acids into proteins, *Nucleic Acids Res.* **29**, 3646–3651.

Lee, N., Bessho, Y., Wei, K., Szostak, J. W., Suga, H. (2000) Ribozyme-catalyzed tRNA aminoacylation, *Nature Struct. Biol.* **7**, 28–33.

Liu, D. R., Schultz, P. G. (1999) Progress toward the evolution of an organism with an expanded genetic code, *Proc. Natl. Acad. Sci. USA* **96**, 4780–4785.

Liu, D. R., Magliery, T. J., Schultz, P. G. (1997) Engineering a tRNA and aminoacyl-tRNA synthetase for the site-specific incorporation of unnatural amino acids into proteins *in vivo*, *Proc. Natl. Acad. Sci. USA* **94**, 10092–10097.

Lodder, M., Golowine, S., Hecht, S. M. (1997) Chemical deprotection strategy for the elaboration of misacylated transfer RNAs, *J. Org. Chem.* **62**, 778–779.

Lohse, P. A., Szostak, J. W. (1996) Ribozyme-catalysed amino-acid transfer reactions, *Nature* **381**, 442–444.

Magliery, T. J., Anderson, J. C., Schultz, P. G. (2001) Expanding the genetic code: selection of efficient suppressors of four-base codons and identification of "shifty" four-base codons with a library approach in *Escherichia coli*, *J. Mol. Biol.* **307**, 755–769.

Mino, K., Yamanoue, T., Ohno, K., Imamura, K., Sakiyama, T., Eisaki, N., Matsuyama, A., Nakanishi, K. (2000) Production of β-(pyrazol-1-yl)-L-alanine from L-serine and pyrazole using recombinant *Escherichia coli* cells expressing serine acetyltransferase and O-acetylserine sulfhydrylase-A, *Biotechnol. Lett.* **23**, 2051–2055.

Murakami, H., Hohsaka, T., Ashizuka, Y., Sisido, M. (1998) Site-directed incorporation of *p*-nitrophenylalanine into streptavidin and site-to-site photoinduced electron transfer from pyrenyl group to nitrophenyl group on the protein framework, *J. Am. Chem. Soc.* **120**, 7520–7529.

Murakami, H., Hohsaka, T., Ashizuka, Y., Hashimoto, K., Sisido, M. (2000) Site-directed incorporation of fluorescent nonnatural amino acids into streptavidin for highly sensitive detection of biotin, *Biomacromolecules* **1**, 118–125.

Murakami, H., Bonzagni, N. J., Suga, H. (2002a) Aminoacyl-tRNA synthesis by a resin-immobilized ribozyme, *J. Am. Chem. Soc.* **124**, 6834–6835.

Murakami, H., Hohsaka, T., Sisido, M. (2002b) Random insertion and deletion of arbitrary number of bases for codon-based random mutation of DNAs, *Nature Biotechnol.* **20**, 76–81.

Nishiuchi, Y., Inui, T., Nishio, H., Boedi, J., Kimura, T., Tsuji, F. I., Sakakibara, S. (1998) Chemical synthesis of the precursor molecule of the Aequorea green fluorescent protein, subsequent folding, and development of fluorescence, *Proc. Natl. Acad. Sci. USA* **95**, 13549–13554.

Noren, C. J., Anthony-Cahill, J. A., Griffith, M. C., Schultz, P. G. (1989) A general method for site-specific incorporation of unnatural amino acids into proteins, *Science* **244**, 182–188.

Ogawa, A. K., Wu, Y., Berger, M., Schultz, P. G., Romesberg, F. E. (2000) Rational design of an unnatural base pair with increased kinetic selectivity, *J. Am. Chem. Soc.* **122**, 8803–8804.

Ohno, S., Yokogawa, T., Nishikawa, K. (2001) Changing the amino acid specificity of yeast tyrosyl tRNA synthetase by genetic engineering, *J. Biochem.* **130**, 417–434.

Ohtsuki, T., Kimoto, M., Ishikawa, M., Mitsui, T., Hirao, I., Yokoyama, S. (2001) Unnatural base pairs for specific transcription, *Proc. Natl. Acad. Sci. USA* **98**, 4922–4925.

Pastrnak, M., Magliery, T. L., Schultz, P. G. (2000) A new orthogonal suppressor tRNA/aminoacyl-tRNA synthetase pair for evolving an organism with an expanded genetic code, *Helv. Chim. Acta* **83**, 2277–2286.

Piccirilli, J. A., Krauch, T., Moroney, S. E., Benner, S. A. (1990) Enzymatic incorporation of a new base pair into DNA and RNA extends the genetic alphabet, *Nature* **343**, 33–37.

Roche Molecular Biochemicals RTS500; http://biochem.roche. com/rts/

Saito, H., Watanabe, K., Suga, H. (2001) Concurrent molecular recognition of the amino acid and tRNA by a ribozyme, *RNA* **7**, 1867–1878.

Sasaki, H., Makino, M., Sisido, M., Smith, T. A., Ghiggino, K. P. (2001) Photoinduced electron transfer on β-sheet cyclic peptides, *J. Phys. Chem. B* **105**, 10416–10423.

Sisido, M., Hohsaka, T. (2001) Introduction of specialty functions by the position-specific incorporation of nonnatural amino acids into proteins through four-base codon/anticodon pairs, *Appl. Microbiol. Biotechnol.* **57**, 274–281.

Sisido, M., Tokunaga, S., Hohsaka, T. (2000) Synthesis of nonnatural mutants of λ-Cro repressor protein that contain an electron-accepting amino acid, *Proc. Japan Acad.* **B76**, 92–96.

Sisido, M., Hoshino, S., Kusano, H., Kuragaki, M., Makino, M., Sasaki, H., Smith, T. A., Ghiggino, K. P. (2001) Distance dependence of photoinduced electron transfer along α-helical polypeptides, *J. Phys. Chem. B.* **105**, 10407–10415.

Steward, L. E., Collins, C. S., Gilmore, M. A., Carlson, J. E., Alexander Ross, J. B., Chamberlin, A. R. (1997) In *vitro* site-specific incorporation of fluorescent probes into β-galactosidase, *J. Am. Chem. Soc.* **119**, 6–11.

Suga, H., Lohse, P. A., Szostak, J. W. (1998) Structural and kinetic characterization of an acyl transferase ribozyme, *J. Am. Chem. Soc.* **120**, 1151–1156.

Switzer, C., Moromey, S. E., Benner, S. A. (1989) Enzymatic incorporation of a new base pair into DNA and RNA, *J. Am. Chem. Soc.* **111**, 8322–8323.

Taki, M., Hohsaka, T., Murakami, H., Taira, K., Sisido, M. (2001) A nonnatural amino acid for efficient incorporation into proteins as a sensitive fluorescent probe, *FEBS Lett.* **507**, 35–38.

Turcatti, G., Nemeth, K., Edgerton, M. D., Meseth, U., Talabot, F., Peitsch, M., Knowles, J., Vogel, H., Chollet, A. (1996) Fluorescent labeling of NK2 receptor at specific sites *in vivo* and fluorescence energy transfer analysis of NK2 ligand–receptor complexes, *J. Biol. Chem.* **271**, 19991–19998.

Wang, L., Brock, A., Herberich, B., Schultz, P. G. (2001) Expanding the genetic code of *Escherichia coli*, *Science* **292**, 498–500.

Wang, L., Brock, A., Schultz, P. G. (2002) Adding L-3-(2-naphthyl)alanine to the genetic code of *E. coli*, *J. Am. Chem. Soc.* **124**, 125.

Williams, R. M. (1989) Synthesis of optically active α-amino acids, in: *Organic Chemistry Series*, Volume 7. Pergamon Press.

Yusupov, M. M., Yusupova, G. Z., Baucom, A., Lieberman, K., Earnest, T. N., Cate, J. H. D., Noller, H. F. (2001) Crystal structure of ribosome at 5.5 Å resolution, *Science* **292**, 883–896.

3
Non-ribosomal Biosynthesis of Linear and Cyclic Oligopeptides

Dr. Hans von Döhren

Arbeitsgruppe Biochemie und Molekulare Biologie, Institut für Chemie, Technische Universität Berlin, Franklinstr. 29, 10587 Berlin, Germany; Tel.: +49-30-31422697; Fax +49-303142-4783; E-mail: doehren@chem.tu-berlin.de

2mGln	2-methyl-L-glutamine
2mIle	2-methyl-L-isoleucine
Aad	aminoadipic acid
Ac	acetyl
ACV	δ-(L-α-aminoadipyl)-L-cysteinyl-D-valine
Aib	L-aminoisobutyric acid
Ala	L-alanine
AMP	adenosine-5′-monophosphate
aIle	L-allo-isoleucine
Ala	L-alanine
Asm	N-methyl-L-asparagine
Asp	L-aspartic acid
Asn	L-asparagine
aThr	*allo*-threonine
ATP	adenosine-5′-triphosphate
Blm	bleomycine
Cys	L-cysteine
Gln	L-glutamine
Glu	L-glutamic acid
Gly	glycine
His	L-histidine
hAsm	*threo*-2-hydroxy-N-methyl-L-asparagine
hVal	hydroxy-L-valine
htLeu	3-hydroxy-t-leucine
Ile	L-isoleucine
Iva	isovaline
kb	kilobase
Leu	L-leucine
MATP	$MgATP^{2-}$ or $MnATP^{2-}$
mGln	methylglutamine
mIle	2-methyl-isoleucine
Mta	myxothiazole
Nme	N-methyl
MPPi	$MgPPi^{2-}$ or $MnPPi^{2-}$
NRPS	non-ribosomal peptide synthetase
Orn	L-ornithine
PCR	polymerase chain reaction
Phe	L-phenylalanine
Pheol	L-phenylalaninol
PKS	polyketide synthase
PPi	pyrophosphate (inorganic)
Ser	L-serine
Thr	L-threonine
tLeu	t-leucine
Trp	L-tryptophan

Tyr L-tyrosine
Val L-valine
Valol L-valinol

1 Introduction

Structures of peptides already offer various clues to their biosynthetic origin (Kleinkauf and von Döhren, 1997a). Differences between the ribosomal peptide-forming system and nucleic acid independent multi-step systems have been compiled in Table 1.

There seem to be restrictions to the sizes of polypeptides; these, however, remain unexplained. The largest protein known, titin, contains 26,926 amino aids, corresponding to a molecular mass of 2993 MDa (Labeit and Kolmerer, 1995). The largest known enzyme, a non-ribosomal peptide synthetase from the ascomycete *Trichoderma virens* presumably producing a family of peptaibols, TVBI, II and IV, of 18 amino acids in length **(1)**, contains 20,925 amino acids, as deduced from the structure of its 63.5-kb sized intron-devoid gene (Wiest et al., 2002). Assuming an error frequency of 10^{-4} for the ribosomal process, no two of these large heteropolymers would be alike. Genome-sequencing efforts indicate that large open reading frames and large mRNA structures are rare.

Enzymatic condensation systems, on the other hand, require information input of at least 3 kb DNA for each residue encoded by the respective protein template in the case of heteropolymers of defined sequence (Kleinkauf and von Döhren, 1996). The largest known enzyme system, alamethicin synthetase from the filamentous fungus *Trichoderma viride*, catalyzes the formation of this 20-membered peptide **(2)**, including the acetylation of the N-terminal residue and reduction of the C-terminal carboxyl to the amino alcohol. The largest peptide structure presently predicted to be enzymatically formed is a 48-residue peptide isolated from the sponge *Theonella swinhoei* **(3)**, which would require about 150 kb of information. It remains to be shown whether this information is contained in the sponge DNA or in the genome of an associated organism.

The increasing size of such modular protein structures appears to be limited by an economical constraint balancing genetic load and usefulness of the product. This does not hold for homopolymers with their repeated use of a condensation device.

The constituents found in ribosomally derived peptides exceed the 20 protein

TVBI Ac^{1}AibGlyAlaValAibGlnAibAlaAibSerLeuAibProLeuAibAibGln^{18}Valol

TVBII Ac^{1}AibGlyAlaLeuAibGlnAibAlaAibSerLeuAibProLeuAibAibGln^{18}Valol **(1)**

TVBIV Ac^{1}AibGlyAlaLeuAlaGlnIvaAlaAibSerLeuAibProLeuAibAibGln^{18}Valol

AcAibProAibAlaAibAlaGlnAibVal^{10}AibGlyLeuAibProValAibAibGluGln^{20}Pheo **(2)**

Gly → 2mIle → Gly → tLeu → tLeu → tLeu → Ala → tLeu → tLeu → Ala^{10}
→ Gly → Ala → tLeu → Ala → Asm → hVal → Gly → Ala → Gly → tLeu^{20}
→ Asm → 2mGln → hVal → Ala → Gly → Gly → Asm → Ile → hAsm → tLeu^{30} **(3)**
→ hVal → Gly → Asm → Ile → Asm → Val → hAsm → Ala → Asm → Val^{40}
→ Ser → Val → Asn → htLeu → Asn → Gln → Thr → aThr^{48}

Tab. 1 Comparison of ribosomal and non-ribosomal systems

Property	***Ribosomal path***	***Non-ribosomal path/multienzymes***
Size (amino acids)	No size limitation	Two to about 50, 4 to 12 dominating, polymers (?)
Amino acid constituents	22 protein amino acids and modified amino acids	Various types of amino acids including 2-, 3-, and 4-amino compounds (more than 300 known)
D-amino acids	Not more than one, epimerized post-translationally; in antibiotics, epimerization of several residues by dehydration and thioether formation	Often several, either incorporated directly, or epimerized during synthesis
Non-amino-acid constituents	Acyl residues, amines originating from decarboxylation	Various acyl-residues, including aromatic acids, hydroxy acids
Cyclic structures	Rare; frequent disulfide cycles, often with several disulfide linkages; thioether cycles; artificially generated, employing intein/extein structures	More frequent than linear structures; various peptide-bond cyclizations, but also lactones
Modifications	Formylation; hydroxylation; dehydration (Ser, Thr); side chain cyclization (Cys to thiazoles, Thr to oxazoles, Glu to pyroGlu); glycosylation; phosphorylation; acylation; thioether linkages	N-methylation, formylation; hydroxylation; reduction; side chain cyclization (Cys to thiazole); glycosylation; side chain cross-linking (aromatic rings)
Unusual constituents	Not known	Urea type of peptide bond; phospho-amino acids; amino-modified fatty acids-derived components (lipopeptides); mixed polyketide structures
Biosynthesis	Gene can be identified; consider splicing, processing, and post-translational modification; prepropeptides often detected; proteolytic processing common	Non-ribosomal enzyme systems present; peptide families are frequent in the same or related organisms; biosynthesis of rare precursors needed
Sources	All types of organisms	Mainly bacteria and lower fungi, occasionally plants and insects

amino acids, selenocysteine and pyrrolysine (Srinivasan et al., 2002; Hao et al., 2002), because of post-translational modifications of side chains (Table 2). Cyclization modes include disulfide and thioether bridges, as well as yet unidentified linkages to form cyclic or multi-cyclic structures. Enzymes catalyzing rare cyclopeptide formations remain to be identified.

Enzymatic systems can be classified into two major groups: single-step and multi-step biocatalysts (Table 3). Both groups can be clearly differentiated by comparing amino acid sequence data of their structural genes. The group of single-step enzymes contains one well-defined superfamily of peptide synthetases including murein and folyl-γ-glutamate ligases (Eveland et al., 1997), cyanobacterial cyanophycin synthetase (Ziegler et al., 1998; Berg et al., 2000; Aboulmagd et al., 2001; and see Chapter 4 by Oppermann-Sanio and Steinbüchel in this volume), and as poly-γ-D-glutamate synthetases in *Bacillus anthracis* and *Bacillus subtilis* (Eveland et al., 1997; Ashiuchi et al., 1999, 2001; and see Chapter 6 by Ashiuchi and Misono in this volume). On the other hand, beta-lysine chains attached to the amino sugar moiety of the nourseothricins produced by Streptomycete strains are poly-

Tab. 2 Unusual features of ribosomally derived peptides

Modified amino acids	***Modification***	***Process***
Glu	PyroGlu	Terminal modification
Cys/Ser	Lanthionine	Enzymatic dehydration and thioether linkage
Cys/Thr	Methyl-lanthionine	
Ser/Thr	Oxazole-derivatives	Enzymatic cyclization
Cys	Thiazole-derivatives	
Pro	Hydroxylation	Enzymatic
Trp	Bromination	Enzymatic
Gly	Amidation	Terminal enzymatic modification

Tab. 3 Peptide-forming enzyme systems

	Single-step systems	***Multi-step systems***
Types of peptides made	Linear peptides, branched peptides	Linear and cyclic peptides; eptidolactones and depsipeptides; polymers (?)
Length of peptides made	2 to 5, or polymers	2 to 48 (?)
Activation of carboxyl groups	Phosphate or adenylate	Adenylate
Intermediates	Free intermediates (not clear for polymers)	Intermediates remain enzyme-bound

merized by an enzyme machinery related to the superfamily of multi-step systems described below (Grammel et al., 2002), which has been termed non-ribosomal peptide synthetases (NRPS).

Less clear is the structural grouping of single-step enzymes involved in glutathione and coenzyme A biosynthesis (Abbott et al., 2001). Contrary to multi-step systems, single-step systems activate *peptide* carboxyl groups as well.

The multi-step systems of NRPS are clearly described as a superfamily related to adenylate-forming enzymes like acetyl-CoA-synthetases and the related insect luciferases (Marahiel et al., 1997; von Döhren et al., 1997). In addition to this domain type, peptidyl carrier proteins, which are members of the acyl carrier protein superfamily, are obligatory. A third type of domain comprises peptide bond-forming and -epimerizing domains. Additional domains are unique N-methyl-transferases (Burmester et al., 1995; Hacker et al., 2000), reductase domains (Silakowski et al., 2001b), and thioesterases also implicated in cyclization reactions (Kohli et al., 2002; Shaw-Reid et al., 1999; Trauger et al., 2001). Current systems under study have been compiled in Table 4. The following discussion focuses on multi-step systems, since various single-step systems have been treated in Chapters 4, 5 and 6 in this volume. Other prominent multienzyme systems containing acyl carrier proteins are polyketide synthases (PKS). These systems also have been discussed by Schwecke elsewhere in these series (2002). Mixed systems combining structural features of both non-ribosomal peptides (NRP) and polyketides (PK) produce a number of prominent natural products, including rapamycin, epothilone, mycosubtilin, bleomycin, and microcystin. The respective biosynthetic clusters contain structural genes of the NRPS and PKS type, which sometimes are fused into NRPS–PKS hybrid structures

Tab. 4 Non-ribosomal peptide synthetase systems

Peptide	Organism	Structural type[a]	Gene(s) cloned	Enzymology	Reference
Linear peptides					
Bacilysin	*Bacillus subtilis*	P-2-M	(+)	(+)	Yazgan et al., 2001
Thaxtomin	*Streptomyces acidiscabies*	P-2-M	+	–	Healey et al., 2000
Pyochelin	*Pseudomonas aeruginosa*	R-P-2	+	+	Quadri et al., 1999; Patel and Walsh, 2001
ACV	*Streptomyces clavuligerus*	P-3	+	+	Martin, 2000;
	Nocardia lactamdurans		+	+	Von Döhren et al., 2001;
	Lysobacter lactamgenus		+	+	Sohn et al., 2001
	Aspergillus nidulans		+	+	
	Penicillium chrysogenum		+	+	
	Acremonium chrysogenum		+	+	
Bialaphos	*Streptomyces hygroscopicus*	P-3	+	+	Seto and Kuzuyama, 1999
Anguibactin	*Vibrio anguillarum*	R-P-2-M	+	–	Welch et al., 2000
Yersiniabactin	*Yersinia pestis*	R-P-3-M	+	+	Gehring et al., 1998a,b Keating et al., 2000c,d; Suo et al., 2000; Miller and Walsh, 2001; Miller et al., 2002
	Yersinia enterocolitica		+	–	Pelludat et al., 1998
Vibriobactin	*Vibrio cholerae*	R-P-3-M	+	+	Keating et al., 2000a,b; Marshall et al., 2001, 2002
Phaseolotoxin	*Pseudomonas syringae* pv. ph.	R-P-3-M	(+)	–	Hernandez-Guzman and Alvarez-Morales, 2000
Mycobactin	*Mycobacterium tuberculosis*	R-P-4-E-M	+	+	Quadri et al., 1998a
Exochelin	*Mycobacterium smegmatis*	P-5-M	+	–	Zhu et al., 1998; Yu et al., 1998
Balhimycin	*Amycolatopsis mediterranei*	P-7-M	+	+	Recktenwald et al., 2002
Complestatin	*Streptomyces lavendulae*	P-7-M	+	+	Chiu et al., 2001
Teicoplanin	*Actinoplanes teichomyceticus*	P-7-M	+	–	Sosio et al. 2000
Chloroeremomycin	*Amycolatopsis orientalis*	P-7-M	+	+	Van Wageningen et al., 1998; Trauger and Walsh, 2000
Pyoverdins	*Pseudomonas aeruginosa*	R-P-8-M to A-R-P11-C-4	(+)	–	Meyer, 2000
Bleomycin	*Streptomcyces verticillus*	R-P-8-M	+	+	Shen et al., 2001, 2002
Ampullosporin	*Sepedonium ampullosporum*	R-P-15-M	(+)	(+)	Reiber et al., 2002
TVB	*Trichoderma virens*	R-P-18-M	+	–	Wilhite et al., 2001; Wiest et al., 2002
Alamethicin	*Trichoderma viride*	R-P-20-M	–	(+)	Mohr and Kleinkauf, 1978
Cyclopeptides					
Thaxtomin A	*Streptomyces acidiscabies*	P-C-2-M	+	–	Healy et al., 2000

Tab. 4 (cont.)

Peptide	***Organism***	***Structural type***[a]	***Gene(s) cloned***	***Enzymology***	***Reference***
Enterobactin	*Escherichia coli*	P-C-E-3	+	+	Shaw-Reid et al., 1999; Ehmann et al., 2000a
Bacillibactin	*Bacillus subtilis*	P-C-E-3	+	–	May et al., 2001
Vicibactin	*Rhizobium leguminosrum*	R-C-3-M	+	–	Carter et al., 2002
HC-toxin	*Cochliobolus carbonum*	C-4	+	+	Panaccione et al., 1992; Cheng and Walton, 2000
Tentoxin	*Alternaria alternata*	C-4	–	+	Johnson et al., 2001
Ferrichrom	*Ustilago maydis*	C-6-M	+	–	Yuan et al., 2001
	Aspergillus quadricinctus		–	+	Siegmund et al., 1991
Microcystin	*Microcystis aeruginosa*	C-7	+	(+)	Tillett et al., 2000; Nishizawa et al., 1999, 2000, 2001
Mycosubtilin	*Bacillus subtilis*	C-7	+	+	Duitman et al., 1999
Iturin	*Bacillus subtilis*	C-8	(+)	–	Tsuge et al., 2001
Gramicidin S	*Bacillus brevis*	C-(P-5)$_2$	+	+	Luo and Walsh, 2001
Tyrocidin	*Bacillus brevis*	C-10	+	+	Mootz and Marahiel, 1997b
Cyclosporin	*Tolypocladium niveum*	C-11-M	+	+	Weber et al., 1994; Billich and Zocher, 1987
Mycobacillin	*Bacillus subtilis*	C-13	(+)	(+)	Majumder et al., 1988
Lactones					
Ergotpeptides	*Claviceps purpurea*	R-P-3-M	+	+	Tudzynski et al., 1999; Walzel et al., 1997; Riederer et al., 1996
Actinomycin	*Streptomyces chrysomallus*	R-(L-5)$_2$-M	+	+	Schauwecker et al., 1998; Keller and Schauwecker, 2001
Anabaenapep-to-lide	*Anabaena* strain 90	R-P-7-L-6	+	–	Rouhiainen et al., 2000
Destruxin[b]	*Metarhizium anisopliae*	L-6	–	(+)	Rabie, 1995
Pristinamycin	*Streptomyces pristineaspiralis*	R-L-6	+	+	De Crécy-Lagard et al., 1997a
Streptogramin B	*Streptomyces virginiae*	R-L-6	+	+	De Crécy-Lagard et al., 1997b
Lichenysin	*Bacillus licheniformis*	L-7	+		Yakimov et al., 1998; Konz et al., 1999
Etamycin	*Streptomyces griseoviridis*	R-L-7	–	(+)	Schlumbohm and Keller, 1990
Surfactin	*Bacillus subtilis*	L-8	+	+	Cosmina et al., 1993
Triostin	*Streptomyces triostinicus*	(R-P-4)$_2$	–	+	Glund et al., 1990
R106[c]	*Aureobasidium pullulans*	L-9	–	(+)	Von Döhren et al., unpublished
Syringomycin Syringostatin	*Pseudomonas syringae*	R-L-9	+	(+)	Guenzi et al., 1998; Scholz-Schroeder et al., 2001
SDZ90-215	*Septoria* sp.	L-10	–	+	Lee and Lawen, 1993

Tab. 4 (cont.)

Peptide	*Organism*	*Structural type*[a]	*Gene(s) cloned*	*Enzymology*	*Reference*
SDZ214-103	*Cylindrotrichum oligosporum*	L-11	(+)	+	Lawen and Traber, 1993; Bernhard et al., 1996
Depsipeptides					
Enniatin	*Fusarium* sp.	C-$(P_2)_3$-M	+	+	Billich and Zocher, 1987; Pieper et al., 1995
Beauvericin	*Beauveria bassiana*	C-$(P_2)_3$-M	–	+	Peeters et al., 1983
PF1022	*Mycelia sterilia*	C-$(P_2)_4$-M	–	(+)	Weckwerth et al., 2000
Branched polypeptides					
Fengycin	*Bacillus subtilis*	P-10-C-7	+	+	Steller et al., 1999
Bacitracin	*Bacillus licheniformis*	P-12-C-7	+	+	Konz et al., 1997
Nosiheptide	*Streptomyces actuosus*	R-P-13-C-10-M	–	(+)	Houck et al., 1988
Micrococcin	*Staphylococcus equorum*	R-P-14-C-9-M	+	–	Carnio et al., 2001
Thiostrepton	*Streptomyces laurentii*	R-P-17-C-10-M	–	(+)	Mocek et al., 1993
Branched peptidolactones					
Lysobactin	*Lysobacter* sp.	P-11-L-9	(+)	+	Bernhard et al., 1996
CDA	*Streptomyces coelicolor*		+	–	Chong et al., 1998
A21798A	*Streptomyces roseosporus*	R-P-13-L-10	(+)	+	McHenney et al., 1998; Wessels et al., 1996
A54145	*Streptomyces fradiae*	R-P-13-L-10	–	+	Wessels et al., 1996
Tolaasin	*Pseudomonas tolaasii*	R-P-18-L-5	(+)	–	Rainey et al., 1993
Mixed polyketide-peptide structures					
Rapamycin	*Streptomyces hygroscopicus*	Polyketide +a	+	+	Schwecke et al., 1995
FK-506	*Streptomyces* sp.	Polyketide +a	+	+	Motamedi and Shafiee, 1998
TA	*Myxococcus xanthus*	Polyketide +a	+	–	Paitan et al., 1999
Myxalamid	*Stigmatella aurantaica*	Polyketide +a	+		Silakowski et al., 2001b
Albicidin	*Xanthomonas albilineans*	Polyketide +a	+	–	Huang et al., 2001

[a] The abbreviations used are: a amino acid, P peptide, C cyclopeptide, L lactone, E ester, R acyl, M modified. The structural types are defined by the number of amino-, imino- or hydroxy acids in the precursor chain. The ring sizes of cyclic structures are indicated in the number following C, L, or E, defining the type of ring closure. The abbreviations used for unusual amino acids and other compounds are listed in the abbreviations footnote in the abbreviations section.

[b] A gene cluster obtained from *Metarhizium anisopliae* does not encode destruxin synthetase, but a yet unknown peptide product.

[c] A gene cluster associated with R106 (aureobasidin) biosynthesis does not encode this NRPS, but presumably a siderophore from *Aureobasidium pullulans*.

(Molnar et al., 2000; Tang et al., 2000; Du and Shen 2001; Du et al., 2001; Shen et al., 2001, 2002; Silakowski et al., 2001a,b; Duitman et al., 1999; Nishizawa et al., 2000; Tillett et al., 2000; Huang et al., 2001).

2
Historical Introduction

Early work on peptides of non-ribosomal origin, tyrothricin, penicillin and gramicidin S, had been motivated by their antibacterial potential (von Döhren and Kleinkauf, 1988). Cyclic structures and the presence of D-amino acids clearly implicated a different biosynthetic mechanism than the translation machinery becoming uncovered largely in the 1960s. By 1971, the groups of Kurahashi, Laland, and Lipmann described the features of the thiotemplate mechanism for multienzymes activating their substrate amino acids as adenylates, forming thioester intermediates, and utilizing 4′-phosphopantetheine as a carrier in analogy to fatty acid biosynthesis (Lipmann et al., 1971). In the following years, several multienzyme systems forming cyclic peptides, cyclodepsipeptides, peptidolactones, and linear peptides have been enzymatically characterized (Kleinkauf and von Döhren, 1990). Sequencing of the respective genes in the last decade revealed the presence of a multi-gene family, highly similar in prokaryotic and eukaryotic organisms, that has been termed non-ribosomal peptide synthetase (NRPS) system (Kleinkauf and von Döhren, 1996; Marahiel et al., 1997; von Döhren et al., 1997). It seems obvious that this system has the capacity to evolve enzyme systems for the synthesis of peptides, and to integrate other biosynthetic enzymes, for the formation of natural products with a wide variety of properties. Thus, current efforts focus on a deeper understanding of the structure-function relationships of the proteins involved, with the aim of utilizing this capacity for an enhanced evolution of useful compounds.

3
Proteins and Protein Domains in Non-ribosomal Peptide Synthetase Systems

Non-ribosomal peptide-forming synthetases are complex multienzyme systems catalyzing (1) activation of their amino(imino) acid substrates, (2) their sequential condensation, and (3) various modifications including epimerizations, N-methylations, and side-chain modifications. The term module has been introduced for a functional unit of domains leading to the extension of a peptide by one condensation reaction, and special modules combine initiation and termination functions. A minimal module consists of an activating adenylation domain, a carrier domain for the transport of the intermediate amino acid or peptide, and a condensation domain. Features and functions of these and additional domains are introduced below, while their interactions are addressed in Section 4.

3.1
Adenylate Domains

Differing from polyketide systems, *activation* of amino acid carboxyl groups is an integrated function in peptide synthetases. This activation function catalyzed by adenylate-forming domains is a pre-selection step in the events controlling fidelity of the peptide chain. Adenylate-forming domains share extensive homologies with acyl-CoA synthetases and the homologous insect luciferases, which likewise form acyl adenylates followed by transfer to the thiol group of CoA. These can be readily identified by their respective signature sequences or core motives (Fig-

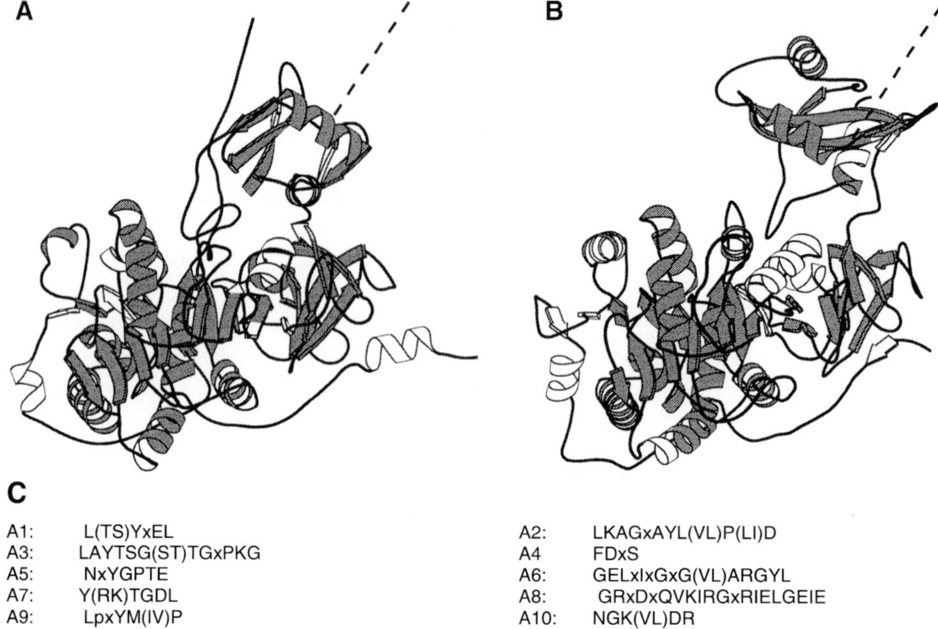

Fig. 1 (a) and (b): Crystal structures of the first adenylate domain of the tyrocidine biosynthetic system from *Bacillus brevis*, activating phenylalanine and the firefly luciferase. Note the large N-terminal and the small C-terminal subdomains (reproduced from Conti et al., 1997). (c): Core motifs of NRPS adenylate domains (Konz and Marahiel, 1999).

ure 1) and are easily accessible with the terms *AMP-binding domain* for adenylate domains (InterPro entry IPR000873, PROSITE motif PS00154, Pfam database entry PF00501).

A detailed analysis of the crystal structure of a gramicidin S synthetase domain has led Marahiel's group to the description of an amino-acid-binding pocket, lined by positionally defined side chains (Stachelhaus et al., 1999). This concept defines a non-ribosomal code and has been useful in the prediction of the substrate specificity of many adenylate domains by inspection of their sequence (Challis et al., 2000). It is occasionally possible to predict the structure of a non-ribosomally encoded peptide like coelichelin, a siderophore suspected in *Streptomyces coelicolor*, from the respective structure of the peptide synthetase gene (Challis and Ravel, 2000).

Amino acid selection at this adenylate step is controlled with varying selectivity (Kleinkauf and von Döhren, 1997b; von Döhren et al., 1999). Sometimes there is discrimination of leucine against valine or isoleucine, as in cyclosporine **(4)**, but occasionally all branched chain types are accepted.

DAla → NMeLeu → Val → NMeLeu → NMeGly → Abu
↑ ↓
Ala ← NMeLeu ← NMeLeu ← NmeVal ← NMeBmt

(4)

Likewise, phenylalanine might be discriminated against tyrosine and tryptophan, while some activation sites are known to accept all aromatic amino acids. Stereoselective discrimination of L- and D-residues is common, but both epimers may be direct precursors, as in the gramicidin S **(5)** and tyrocidine **(6)** systems.

DPhe → Pro → Val → Orn → Leu
↑ ↓
Leu ← Orn ← Val ← Pro ← Dphe

(5)

DPhe → Pro → Phe → DPhe → Asn
↑ ↓
Leu ← Orn ← Val ← Tyr ← Gln

(6)

In some cases, both epimers are substrates, but the subsequent condensation reaction is stereocontrolled (Belshaw and Walsh, 1999). Adenylates are thought to be stabilized by the interaction of the two subdomains of the adenylate domain. With low efficiency, they may be subject to aminolysis, leading to dipeptide formation without stereoselectivity (Byford et al., 1997; Dieckmann et al., 2001b). The reverse reaction of aminoacyl formation with labeled PPi is used as a sensitive reaction for the detection of NRPS systems. With ATP or polyphosphates, diadenosine tetraphosphate or adenyl polyphosphates are formed (Dieckmann et al., 2001a). Such products are known as signal molecules and might be involved in the intracellular regulation of these enzyme systems.

In NRPS systems acyl-, amino- or iminoacyl adenylates are transferred to 4′-phosphopantetheineyl-thiol groups of adjacent or sometimes nonadjacent carrier domains. Because peptide synthesis proceeds exclusively from these thiolesters, the term "thiotemplate mechanism" has been coined (Kleinkauf and von Döhren, 1999). These covalently attached intermediates are fairly stable and can be isolated (Luo and Walsh, 2001).

3.2 Carrier Domains for the Transport of Intermediates

Adjacent to each adenylate domain is the carrier domain, homologous to the well-known acyl carrier protein of fatty acid biosynthesis. Carrier domains are post-translationally modified with 4′-phosphopantetheine. This modification is catalyzed by specific CoA-4′-phosphopantetheine-protein transferases (Walsh et al., 1997; Quadri et al., 1998b; Reuter et al., 1999) (Figure 2).

The terminal cysteamine thiol of the cofactor is acceptor of the acyl-, aminoacyl- or iminoacyl moiety of the adenylate, with

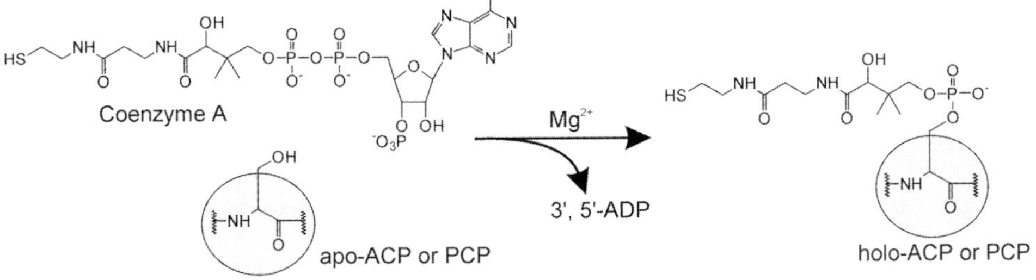

Fig. 2 Reaction catalyzed by the 4′-phosphopantetheine-protein transferase family (PPT). The 4′-phosphopantetheine moiety of CoA is transferred post-translationally onto a conserved serine residue contained in acyl carrier or peptidyl carrier domains or proteins in PKS and NRPS systems. Thus, inactive apo-forms are converted into active holo enzymes. The reaction is dependent on Mg^{2+} and yields 3′,5′-ADP as a second product, which may act as inhibitor of the reaction (reproduced from Mootz et al., 2001).

release of AMP. This thioester formation stabilizes the acyl intermediates, presumably by insertion in the conserved cleft formed by 3 helices (Crump et al., 1996; Dieckmann and von Döhren 1997; Weber et al., 2000). Some data indicate that this step permits proofreading, so that some adenylates are not processed (Schwecke et al., 1993). Thiol-bound intermediates can be isolated and characterized (Luo and Walsh, 2001). This system property has led to the name thiotemplate mechanism (Kleinkauf and von Döhren, 1999). Aminoacyl-thioesters have been used for mechanistic investigations to circumvent amino acid activation and acylation steps and to probe the specificity of condensation domains (Belshaw and Walsh, 1999; Ehmann et al., 2000a).

Aminoacyl/peptidyl carrier domains are structurally similar to acyl carrier proteins (ACPs) (Weber et al., 2000) (Figure 3). Structural data are easily accessible with the terms *pp-binding, phosphopantetheine,* or *ACP domain* (IPR000255, PS00455, PF00550). The solution structure of peptidyl carrier domain (PCP) from tyrocidine synthetase has been resolved by NMR analysis, defining domain boundaries and the 4-helix bundle structure with an extended loop for cofactor attachment (Weber et al., 2000).

When comparing amino acid sequences of PCPs, distinct subgroups can be identified (von Döhren et al., 1999). These include PCPs interacting with only condensation domains, with epimerization domains, and with N-methyl-transferase domains, while substrate specificity cannot be clearly defined. Therefore, this cluster formation upon alignment is presumably related to domain communication elements.

The charging or aminoacylation of carrier proteins often is performed by nonadjacent adenylation domains, implying the requirement of specific domain–domain recognition. Thus, a single cysteine-specific integrated adenylation domain serves three carrier proteins in yersiniabactin biosynthesis (Keating et al., 2000d). Likewise, non-integrated adenylate and carrier domains interact upon the addition of D-alaninyl residues to lipoteichoic acid (Volkman et al., 2001).

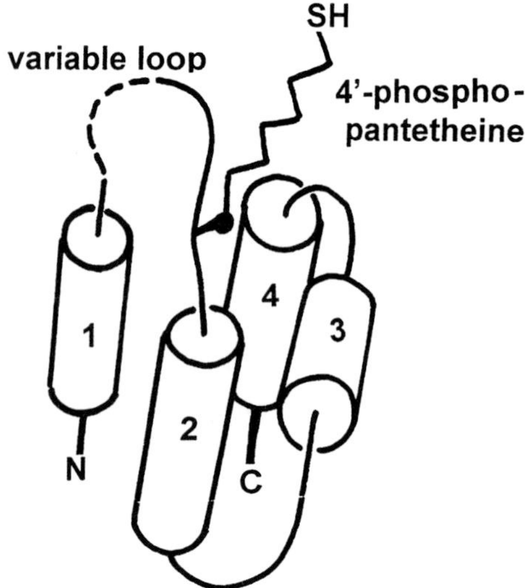

Fig. 3 Topology of acyl carrier proteins and peptidyl carrier proteins, as well as assumed for their respective integrated domains in polyketide synthase and non-ribosomal peptide synthetase systems (reproduced from Dieckmann and von Döhren, 1997).

3.3 Condensation Domains and Related Epimerization Domains

The directed condensation of either two aminoacyl intermediates (initiation) or of a peptidyl and an aminoacyl intermediate (elongation) is thought to take place at a condensation domain, interacting with the two carrier domains. This reaction is strictly stereospecific (Belshaw and Walsh, 1999; Ehmann et al., 2000b). If both epimers are present as intermediates because of the action of a reversible epimerase in the multienzyme, only the respective D-isomer is processed. In detailed studies recombining condensation domains in the gramicidin S and tyrocidine system, the specificity of condensation reactions has been investigated (Linne and Marahiel, 2000). Thus, a relaxed specificity is found with respect to the incoming "downstream" acyl compound transferred during peptide bond formation to the "upstream" carrier domain (Belshaw and Walsh, 1999; Linne and Marahiel, 2000; Ehmann et al., 2000b). An isolated condensation domain, vibH, has been characterized as amide synthase in the vibriobactin system (Keating et al., 2000a). The recently established crystal structure of vibH represents a monomeric pseudodimer. It is representative of both NRPS condensation and epimerization domains, as well as the condensation-variant cyclization domains, which are integrated domains. Surprisingly, despite favorable positioning in the active site, a universally conserved histidine that is important in the structurally related chloramphenicol acetyltransferase and dihydrolipoamide acyltransferase and in other C domains is not critical for general base catalysis in VibH (Keating et al., 2002).

A variant of the condensation domain catalyzes heterocyclization of cysteinyl side chains to thiazolines (Konz et al., 1997; Gehring et al., 1998b; Marshall et al., 2001; Keating and Walsh, 1999), and this is also expected for the respective oxazolyl derivatives originating from seryl and threonyl residues (Sinha Roy et al., 1999).

Epimerization domains are likewise highly similar to condensation domains, catalyzing the conversion of chiral thioester-attached aminoacyl- or peptidyl–carboxyl groups (Linne and Marahiel, 2000; Stachelhaus and Walsh, 2000). Epimerization domains have no significant similarity to amino acid and carboxylic acid racemases, but together with condensation domains they share the signature motif HHxxxDxxSW with a superfamily of acyl transferases (de Crécy-Lagard et al., 1995). In this so-called His-motif, the second His and the Asp residue are required for condensation activity (Stachelhaus et al., 1998), and at least the His is required for epimerization activity (Stachelhaus and Walsh, 2000). Both reactions require deprotonation of either amino groups for condensation or Cα-carbons for epimerizations, respectively.

These three types of domains are described in the protein databases as condensation and epimerization domains, *DUF4* (IPR001240), and can be identified by their specific signatures or core sequences (Figure 4), which also have been used for NRPS gene search by PCR (Chiu et al., 2001).

3.4
Thioesterase Domains

Termination of synthesis may proceed by hydrolysis in the case of linear peptides (thioesterase required) (de Ferra et al., 1997, Kallow et al., 2000) or various modes of cyclization, leading to cyclic or branched cyclic peptides, peptidolactones, or cyclodepsipeptides (Keating et al., 2001). A characteristic terminal domain, the integrated thioesterase, has been shown to specifically catalyze the condensation of peptide intermediates (Keating et al., 2001; Kohli et al., 2001). These thioesterase domains have been described in the IPR database with the accession number 001031. The structure of the excised domain of the surfactin system has been shown to contain a distinctive bowl-shaped hydrophobic cavity to promote the folding of the cyclic macrolactone from the linear peptidyl precursor (Bruner et al., 2002). The isolated thioesterase domain

C1	SxAQxR(LM)(WY)xL	E1:	LxPIQxWF
C2	RHExLRTxF	E2:	LxxxHD
C3	MHHxISDG(WV)S	E3:	HHxxVDxVSWxIL
C4	YxD(FY)AVW	E4:	VxxEGHGRE
C5	(IV)GxFVNT(QL)xR	E5:	TVGWFTxxxPxxL
C6	(HN)QD(YV)PFE	E6:	PxxGxGy
C7	RDxSRNPL	E7:	VxFNYLG

Cy1	FPL(TS)xxQxAYxxGR
Cy2	RHx(IM)L(PAL)x(ND)GxQ
Cy3	(LI)Pxx(PAL)x(LPF)P
Cy4	(TS)(PA)xxx(LAF)xxxxxx(IVT)LxxW
Cy5	(GA)(DNQ)FT
Cy6	P(IV)VF(TA)SxL
Cy7	QVx(LI)Dx(QH)x_{11}W(DYF)

Fig. 4 Core motifs of condensation domains (C1-5), condensation domains forming heterocyles (Cy1-2), and the related epimerization domains (E1-7) (Konz and Marahiel, 1999).

has promising applications in the cyclization of chemically synthesized peptidyl-thioesters (Trauger et al., 2001; Kohli et al., 2002).

The structural variability of these domains permits the processing of hydroxy acids as well, leading to depsipeptides or the repeated use of templates, leading to dimers, trimers, or tetramers. Known processes are the dimerization of two pentapeptides in gramicidin S formation, the trimerization of dihydroxybenzoylseryl- or hydroxyisovaleryl-N-methyl-valyl intermediates (Shaw-Reid et al., 1999; Haese et al., 1993), or their tetramerization in the cases of bassianolide and PF1022 (Weckwerth et al., 2000). As in polyketide biosynthesis, repositioning of a terminating thioesterase domain may promote the release of intermediates (de Ferra et al., 1997). Associated thioesterases may contribute to the rate of product formation (Schneider and Marahiel, 1998), presumably by releasing non-reactive intermediates. Three types of NRPS thioesterases have been described so far (Kallow et al., 2000) (see Table 5). Linking of the process with acylation reactions or polyketide-forming systems gives access to an enormous variation of peptide structures.

3.5 N-methyltransferase Domains

While C- and O-methyltransferases are key players in polyketide structural variations, NRPS systems frequently contain integrated N-methyltransferase domains. Methyl groups are transferred from S-adenosyl methionine onto amino groups of thioester-activated amino acids forming N-methylaminoacyl intermediates (Zocher and Keller, 1997). From these iminoacyl intermediates, N-methylated peptide bonds are formed. N-methylated peptide bonds are resistant to protease attack and provide conformational restrictions to these usually cyclic peptide and depsipeptide regions. Prominent examples of N-methylated cyclopeptides are enniatin, cyclosporin, actinomycin, pristinamycin, and microcystin. An example of a nonintegrated N-methyltransferase has been found in the chloroeremomycin biosynthetic cluster (van Wageningen et al., 1998; O'Brien et al., 2000). The function of this tailoring enzyme and the timing of this methylation step remain to be shown.

The N-methyl transferase domain sequences contain structural motifs related to

Tab. 5 Core motifs of thioesterases involved in NRPS systems[a]

Motif[b]	*Separate thioesterases*	*Integrated thioesterases following thiolation domains*	*Integrated thioesterases following epimerase domains (ACV synthetases)*
T1	PxAGG	lFxfxP(a/v)gg	LF(V/L)LPPGEGGAESY
T2	pgr	–	–
T3	pxxxfGHSmGa	GpyxxxG(W/Y)SxGg	QPxGPYxxxGWSFGG
T4	LfiSgxxxAP	–	–
T5	LPxLRAD	–	–
T6	Wr	W	–
T7	GgxHHΦl	gxgxh	–
ET1	–	–	VvFNN
ET2	–	–	LxxiDxΦF
ET3	–	–	LDPI
ET4	–	–	IvLFKA
ET5	–	–	QxxlΦEyy
ET6	–	–	NnLDxlLp

[a] Conserved residues are given in capital letters and not-completely conserved positions, in small letters; Φ represents aromatic amino acids (F, Y, W).
[b] Motifs designated as T may be found in several type of thioesterases, while motifs defined as ET may be restricted to thioesterases associated with epimerization domains.

substrate binding and catalysis (Figure 5) (Burmester et al., 1995, Hacker et al., 2000). These motifs have been used for degenerate primer design in the identification of NRPS genes (Burmester et al., 1995; Chiu et al., 2001). Biochemical studies have shown S-adenosyl-L-homocysteine and sinefungin to inhibit the N-methyltransferase reaction (Billich and Zocher, 1987). While sinefungin acts competitively with respect to S-adenosyl-L-methionine, S-adenosyl-L-homocysteine appears to bind to a second site as well. In the presence of inhibitors or in the absence of S-adenosyl-L-methionine, unmethylated enniatins are formed at a reduced rate. Site-specific affinity labeling can be achieved by UV radiation in the presence of S-adenosyl-L-methionine, involving a tyrosine residue in motif 2 (Hacker et al., 2000).

M1: vLEiGxG(TS)G(LM)(LIV)(LM)xx(LV)
M2: (IV)(IV)xNSV(AIV)QYFPxxxYL
M3: (IV)fxGD(IVM)R
M4: (ED)xEllxxPx(FY)F
M5: NE(LM)xxxRY

Fig. 5 N-methyl-transferases core motifs derived from 19 fungal and bacterial domains.

Engineering of N-methylated peptides has been studied with actinomycin synthetase II (Schauwecker et al., 2001). This synthetase catalyzes the aryl-primed formation of 4-toluyl-Thr-D-Val intermediates. Replacement of the Val adenylate domain with the Val-methyltransferase didomain from actinomycin synthetase III yielded the aryl-Thr-N-methyl-Val product, indicating that the condensation domain does catalyze reactions of both unmethylated and methylated amino acids. Aryl capping, however, was required for condensation. Insertion of the didomain, leaving the terminal epimerization domain of actinomycin synthetase II, did not lead to epimerized products. This indicates that both epimerization and N-methylation are not compatible within a single module, in agreement with the failure to detect N-methylated D-amino acids in natural peptides.

3.6 Oxidation and Reduction Domains

While a structural variant of the condensation domain has been shown to catalyze thiazoline formation from cysteine side chains, many peptides contain thiazol or thiazolidine moieties, thus requiring additional oxidation or reduction steps.

Respective integrated domains have been identified within the bleomycin synthetase system (Du et al., 2000; Shen et al., 2002), as well as biosynthetic clusters forming myxothiazol (Silakowski et al., 1999) and epothilone (Molnar et al., 2000; Tang et al., 2000; Julien et al., 2000). This domain has been localized either downstream of the PCP domain (in the case of BlmIII and MtaC) or between the A8 and A9 motifs of an adenylation domain (EposP and MtaD). The BlmIII domain has been shown to contain one molar equivalent FMN as a prosthetic group (Du et al., 2000). A similar protein is contained in a multienzyme complex modifying the ribosomal peptide antibiotic microcin B17 (Milne et al., 1999).

Alternative termination reactions involve the reduction of the terminal thioester-attached carboxyl group to aldehydes or amino alcohol functions. Integrated domains of 40–50 kDa with homology to NAD(P)-dependent reductases have been identified in aminoadipate semialdehyde formation in lysine biosynthesis (Ehmann et al., 1999), in the saframycin precursor tetrapeptide aldehyde (Pospiech et al., 1996), in alaninol formation in mycobacterial peptidolipid synthesis (Billman-Jacobe et al., 1999), and in amino alcohol formation in peptaibol biosynthesis termination (Wiest et al., 2002).

4 The Reaction Cycle

Non-ribosomal peptide formation is a multi-step process exclusively catalyzed by interacting protein domains. In analogy to ribosomal peptide formation initiation, elongation and termination can be assigned to individual reactions, but the overall process is often more complex. Not only activation of substrates but also modifications of substrates may be part of the catalytic cycle. In some cases, evidence for prominent side reactions leading to two or more products implies more complex cycles; in addition, the formation of product families indicates differences in rates of formation of distinct steps of the process. Finally, the overall rates for non-ribosomal peptide synthesis are fairly small, ranging between 10 and 100 moles of peptide per mole of enzyme and minute (von Döhren et al., 2001).

4.1 The Multi-step Assembly Line

NRPS systems catalyze activation of substrate amino acids, and possibly modifications; the directed irreversible condensation of residues, with possible modification of intermediates; and termination by peptide release (Marahiel et al., 1997; von Döhren et al., 1997; Mootz and Marahiel, 1997a; Stein et al., 1996; Keating and Walsh, 1999; Walsh et al., 2001; Weber and Marahiel, 2001). The total number of reactions in this assembly-line process is considerable. The evaluation of kinetic properties of NRPS systems is a fairly complex problem. The basic path was established in 1971, which defined activation, thiolation, and peptidyl transfers as basic reactions. The catalytic cycle is described in analogy to other polymerization processes with initiation, elongation, and termination steps. Structural data have established the multiple carrier model and will help to evaluate the dynamics of domain interactions. As an example for the catalytic cycle, a single multienzyme tripeptide synthetase that forms δ-(L-α-aminoadipyl)-L-cysteinyl-D-valine (ACV), the precursor of β-lactam antibiotics, will be discussed. The ACV synthetase operates with 4 different substrates at 6 binding sites releasing 3 moles of AMP and 3 moles of MgPPi for each ACV formed at optimal conditions (Kallow et al., 1998). A sequence of 10 reactions has been proposed in analogy to other NRPS systems (note that the enzyme-attached intermediates are bound as thioesters with their carboxyl group, so these are written from C- to N-terminal direction, while the product ACV corresponds to the N- to C-terminal convention):

The synthetase consists of the three modules E1, E2, and E3. Each module is composed of an activation site forming the acyl or aminoacyl adenylate, a carrier domain that is post-translationally modified with 4′-phosphopantetheine (S^p), and a condensation domain (C1, C2) or, alternatively, a structurally similar epimerization domain (Ep). Activation of aminoadipate (Aad) leads to an acylated enzyme intermediate, where Aad is attached to the terminal cysteamine of the cofactor (E1-S^{p1}-Aad) (Reactions 1 and 2). Likewise, activation of cysteine (Cys) leads to cysteinylated module 2 (Reactions 3 and 4). For the condensation reaction to occur between aminoadipate as donor and cysteine as acceptor, both intermediates are thought to react at the condensation site of module 1 (C1). Each condensation site is composed in analogy to the ribosomal peptide formation of an aminoacyl and a peptidyl site, and, in this case of initiation, the thioester of Aad enters the P site, while the thioester of Cys enters the A site. Condensation occurs and leaves the dipeptidyl intermediate Aad-Cys at the carrier protein of the second module (Reaction 5).

The third amino acid valine is activated on module 3, and Val is attached to the carrier protein 3 (Reactions 6 and 7). Formation of the tripeptide occurs at the second condensation site C2, with the dipeptidyl intermediate entering the P site and the valinyl intermediate entering the A site (Reaction 8).

Finally, epimerization of the tripeptide (or dipeptide) intermediate occurs at the epimerization site of module 3 (Ep3) (Reaction 9), and the stereospecific peptide release is controlled by the thioesterase (TE) (Reaction 10) (Kallow et al., 2000). This process is illustrated schematically in Figure 6. A detailed discussion is provided in a recent review (von Döhren et al., 2001).

4.2 Side Reactions

Two types of side reactions have been observed so far: (1) the aminolysis of amino-

$$\text{E1} + \text{Aad} + \text{ATP} \longrightarrow \text{E1(Aad-AMP)} + \text{PPi} \longrightarrow \text{E1-S}^{p1}\text{-Aad} + \text{AMP} \qquad (1, 2)$$

$$\text{E2} + \text{Cys} + \text{ATP} \longrightarrow \text{E2(Cys-AMP)} + \text{PPi} \longrightarrow \text{E2-S}^{p2}\text{-Cys} + \text{AMP} \qquad (3, 4)$$

$$\text{E1-S}^{p1}\text{-Aad} + \text{E2-S}^{p2}\text{-Cys} \xrightarrow{\text{C1}} \text{E2-S}^{p2}\text{-Cys-Aad} + \text{E1-S}^{p1}\text{H} \qquad (5)$$

$$\text{E3} + \text{Val} + \text{ATP} \longrightarrow \text{E3(Val-AMP)} + \text{PPi} \longrightarrow \text{E3-S}^{p3}\text{-Val} + \text{AMP} \qquad (6, 7)$$

$$\text{E2-S}^{p2}\text{-Cys-Aad} + \text{E3-S}^{p3}\text{-Val} \xrightarrow{\text{C2}} \text{E3-S}^{p3}\text{-Val-Cys-Aad} + \text{E2-S}^{p2}\text{H} \qquad (8)$$

$$\text{E3-S}^{p3}\text{-Val-Cys-Aad} \xrightarrow{\text{Ep3}} \text{E3-S}^{p3}\text{-D-Val-Cys-Aad} \xrightarrow{\text{TE}} \text{Aad-Cys-D-Val} + \text{E3-S}^{p3}\text{H} \qquad (9, 10)$$

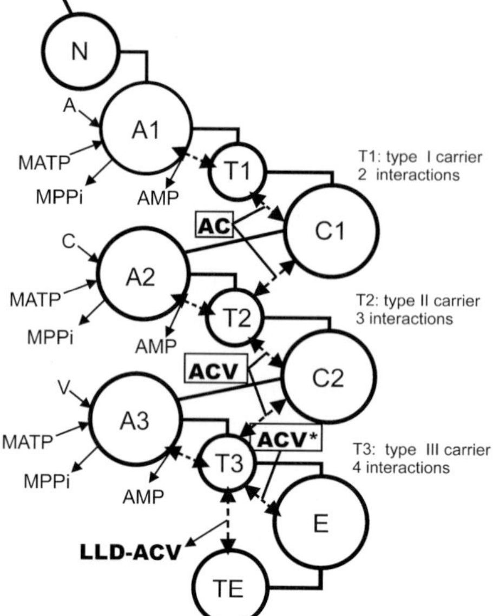

Fig. 6 Scheme of domain interactions in ACV synthetases. Domains are presented as circles, and linkers are indicated as thin lines. Descriptions are as in Figure 3. Substrate amino acids (A1-A3) and MATP enter activation domains A1 to A3, with MPPi leaving as product. Interactions of the A-domains with their adjacent thiolation domains (T1-T3) leads to (amino)acylation with release of AMP. Peptidyl intermediates (AC, LLL-ACV(ACV) and LLD-ACV(ACV*) are boxed and are produced upon interactions of the thiolation domains with the respective condensation domains. The third thiolation domain interacts with four domains, and the final interactions lead to epimerization and hydrolytic release.

acyladenylates as reactive mixed anhydrides leading to dipeptides (Byford et al., 1997; Dieckmann et al., 2001a) and (2) reactions of thioester intermediates. As in solid-phase peptide synthesis, certain steric conditions favor abortive cyclization reactions, especially of proline-containing or N-methylated dipeptides to the respective piperazinediones (Kleinkauf and von Döhren 1997b; Stachelhaus et al., 1998) and ornithyl side chains to oxopiperazines (Gadow et al., 1983). Less expected have been out-of-sequence dipeptide formations in ACV synthetase (Byford et al., 1997; Shiau et al., 1995), actinomycin synthetase II (Schauwecker et al., 1998), and cyclosporin synthetase (Glinski et al., 1996). Such side reactions are promoted when lack of substrates prevents the completion of the full catalytic cycle. However, given the significance of cyclodipeptides as signaling molecules (Holden et al., 1999, 2000), these apparent side reactions may be intrinsic properties of the peptide synthetases, providing an additional product with a defined function.

4.3 Tailoring Reactions

Most modifications of non-ribosomally formed peptides and polyketides are catalyzed by integrated enzyme functions on covalently attached intermediates. These intermediates are thus checked before further processing, although alternative products may be generated *in vitro* under specified conditions. Nonintegrated tailoring enzymes have been occasionally detected in NRPS clusters. Tailoring reactions include methylation, epimerization, hydroxylation, halogenation, heterocyclization, oxidative cross-linking, and glycosylation (Solenberg et al., 1997; Süßmuth et al., 1999; Walsh et al., 2001; Puk et al., 2002). Reactions of autonomous enzymes studied in some detail include epimerases in the biosynthesis of cyclosporin, HC-toxin, and microcystin, leading to the direct incorporation of D-amino acids (Cheng and Walton, 2000; Nishizawa et al., 2001) and the oxidative cross-linking of aromatic side chains of vancomycin-type heptapeptide precursors (Bischoff et al., 2001; Süßmuth et al., 1999).

5 Isolation and Cloning of Biosynthetic Gene Clusters

The genetic information to form polycondensation products like peptides or polyketides by complex enzyme systems is generally found to be clustered into polyenzymes composed of covalently linked domains. Such polygenes are found especially in prokaryotes clustered into polyenzyme clusters, associated with additional genes for modification or transport functions. The clustering of genes and their frequent occurrence in unrelated organisms implies the possibility of intergenic transfer of this information.

5.1 Occurrence and Detection of NRPS Clusters

Non-ribosomal systems can be identified by their respective core motifs, experimentally detectable by degenerated primers, followed by the comparative analysis of the primary structures of the respective coding regions.

A survey of the completed genomes shows their absence in higher eukaryotes and yeast, while PKS–NRPS genes have been found in only 8 of 70 bacterial genomes sequenced so far, belonging to the *Actinobacteria*, gram-positive, δ- and γ-Proteobacteria, Cyanobacteria, and the Cytophaga/Flavobacterium groups. Strains of well-known secondary metabolite producers have the capacity for the production of up to 25 different types of PK and NRP products, corresponding to about 5% of their genomic coding capacity. Most of these products are still unknown and, given the functionality of the respective genes and gene clusters, provide a significant reservoir of new natural products.

A survey of NRPS genes in the databases provides a more comprehensive view. The number of orphan clusters is steadily increasing.

5.2 Structure of NRPS Clusters in Prokaryotes

Bacterial biosynthetic enzymes are usually clustered, while bacterial genomes often do not contain such activities. If activities are present, there may be just a single siderophore cluster, as in *E. coli*, or more than 20 clusters, as in the industrial producer strain of *Streptomyces avermitilis*. Actinomycetes, Myxobacteria, and many Cyanobacteria are especially prominent producers of NRP and PK compounds. Given their morphological variety involving cell–cell interactions, it is speculated that NRPs not only serve ecological competition as antibacterial or antifungal agents but also function in the signaling and regulation of differentiation processes (Holden et al., 2000).

5.3 Eukaryotic NRPS Systems

In contrast to prokaryotic biosynthetic genes, the respective eukaryotic systems are only partially clustered. No system has been described yet that contains a 4′-phosphopantetheine-protein transferase gene linked to NRPS genes. While the β-lactam antibiotic biosynthetic genes for penicillin and cephalosporin in prokaryotes are clustered, tailoring genes for the transformation of penicillin into cephalosporins reside on a different chromosome in *Acremonium chrysogenum*. Because most fungi are known to produce several siderophores, they obviously contain several NRPS clusters. Preliminary surveys of currently studied fungal genomes indicate the presence of multiple NRP genes, including many orphan peptide synthetases. So far, such genes have been restricted to fungi and have not been detected in higher eukaryotes, although the octapeptide lophyrotomin known from insects and several N-methylated peptides from plants are presumably of non-ribosomal origin. It is not an easy task to differentiate the true source of such small peptides in complex organisms containing microbial symbionts and associated microbes. In several marine animals, metabolites have been traced to associated prokaryotes (Kobayashi and Ishibashi, 1993).

6 Manipulation of NRPS Systems

The manipulation of NRPS systems includes not only the NRPS structural genes but also genes involved in their regulation; the respective regulatory DNA regions; genes involved in the biosynthesis of the respective precursor molecules; the over-expression and post-translational modification of NRPS, often linked to the subcellular localization of the systems; the presence and level of tailoring enzymes; and various transport phenomena, including the import of precursors and the export of products.

6.1 Metabolic Engineering and Heterologous Expression

Extensive experiences in metabolic engineering approaches have been collected in industrial β-lactam antibiotic production processes. These can be grouped into manipulating the expression levels of NRPS and tailoring enzymes, the biosynthesis and degradation of the respective precursors, and the introduction of a new tailoring function. In addition, the involvement of transport processes in yield improvement in industrial strains has been demonstrated.

The expression of NRPS genes and clusters in new hosts has been accomplished (Smith et al., 1990; Eppelmann et al., 2001; Lu et al., 2002), and the heterologous expression of orphan clusters of even unknown origin is currently being investigated (Handelsman et al., 2002). The complexity of the design of high-producer strains is just being appreciated. Alterations in the expression of genes in highly efficient over-producers generated by random mutagenesis are fairly complex. We are still far from the rational design of these strains. However, significant improvements are often achieved by alteration of the expression levels of single genes.

6.2 Genetic Alterations of Programming

The modular arrangement of most polycondensation pathways implies reprogramming by modular exchange (Keating and Walsh, 1999; Mootz and Marahiel, 1999). Several attempts have been successful in the polyketide field using the erythromycin system (for reviews see Khosla, 1997; Staunton and Wilkinson, 1997). In the peptide field, functional hybrid synthetases have been constructed by didomain exchange into the surfactin synthetase system from the gramicidin S synthetase system and ACV synthetase (Stachelhaus et al., 1995; Schneider et al., 1998). Further studies on the tyrocidine system and actinomycin system have led to functional dipeptide and tripeptide synthetase systems (Mootz et al., 2000; Schauwecker et al., 2001). At present, the often low-catalytic efficiencies of the constructs present a problem difficult to approach without knowledge of the detailed spatial structures involved. Particularly challenging are the required interactions of nonadjacent domains during the multi-step process.

7 Outlook and Perspectives

Three principal processes to arrive at polypeptides are known: (1) the ribosomal path utilizing a nucleic acid template combined with protein-RNA catalytic structures; (2) a stepwise peptide carboxyl-activating process responsible for compounds like glutathione, coenzyme A, peptidoglycan, and cyanophycin polymers; and (3) an assembly process utilizing pure protein templates in the directed condensation of activated amino acids. This latter process, referred to as a non-ribosomal system, is related to various complex biosynthetic processes of directed condensations of carboxyl-activated compounds, including, besides peptides, various types of polyketides. Both enzymatic systems can be readily identified by highly conserved amino acid sequences. The multistep systems display a modular structure and integrate most of the reactions involved into large, multifunctional proteins. Similar to the ribosomal system, no intermediate peptides are released. The basic features of peptide-forming systems have been resolved almost completely. The modular systems now need detailed functional and structural studies to permit reprogramming and combinatorial strategies. A major future directive will be to understand selection strategies toward natural targets and to employ these for drug development.

8
References

Abbott, J. J., Pei, J., Ford, J. L., Qi, Y., Grishin, V. N., Pitcher, L. A., Phillips, M. A., Grishin, N. V.(2001) Structure prediction and active site analysis of the metal binding determinants in g-glutamylcysteine synthetase, *J. Biol. Chem.*.**276**, 42099–42107.

Aboulmagd, E., Oppermann-Sanio, F. B., Steinbüchel, A. (2001) Purification of *Synechocystis sp. strain* PCC6308 cyanophycin synthetase and its characterization with respect to substrate and primer specificity, *Appl. Environ. Microbiol.* **67**, 2176–2182.

Ashiuchi, M., Soda, K., Misono, H. (1999) A poly-gamma-glutamate synthetic system of *Bacillus subtilis* IFO 3336: gene cloning and biochemical analysis of poly-gamma-glutamate produced by *Escherichia coli* clone cells, *Biochem. Biophys. Res. Comm.* **263**, 6–12.

Ashiuchi, M., Nawa, C., Kamei, T., Song, J. J., Hong, S. P., Sung, M. H., Soda, K., Misono, H. (2001) Physiological and biochemical characteristics of poly gamma-glutamate synthetase complex of *Bacillus subtilis, Eur. J. Biochem.* **268**, 5321–5328.

Belshaw, P., Walsh, C. T. (1999) Aminoacyl-CoAs as probes of condensation domain selectivity in non-ribosomal peptide synthesis, *Science* **284**, 486–489.

Berg, H., Ziegler, K., Piotukh, K., Baier, K., Lockau, W., Volkmer-Engert, R. (2000) Biosynthesis of the cyanobacterial reserve polymer multi-L-arginyl-poly-aspartic acid (cyanophycin): mechanism of the yanophycin synthetase reaction studied with synthetic primers, *Eur. J. Biochem.* **267**, 5561–5570.

Bernhard, F., Demel, G., Soltani, K., von Döhren, H., Blinov, V. (1996) Identification of genes encoding for peptide synthetases in the gram-negative bacterium *Lysobacter sp.* ATCC 53042 and the fungus *Cylindrotrichum oligospermum, DNA Sequ.* **6**, 319–330.

Billich, A., Zocher, R. (1987) Enzymatic synthesis of cyclosporin A, *Biochemistry* **26**, 8417–8423.

Billman-Jacobe, H., McConville, M. J., Haltes, R. E., Kovacevic, S., Coppel, R. L. (1999) Identification of a peptide synthetase involved in the biosynthesis of glycopeptidolipids of *Mycobacterium smegmatis, Mol. Microbiol.* **33**, 1244–1253.

Bischoff, D., Pelzer, S., Holtzel, A., Nicholson, G.J., Stockert, S., Wohlleben, W., Jung, G., Süßmuth, R. D. (2001) The biosynthesis of vancomycin-type glycopeptide antibiotics: New insights into the cyclization steps, *Angew. Chem. Int. Edn.* **40**, 1693–1696.

Bruner, S. D., Weber, T., Kohli, R. M., Schwarzer, D., Marahiel, M. A., Walsh, C. T., Stubbs, M. T. (2002) Structural basis for the cyclization of the lipopeptide antibiotic surfactin by the thioesterase domain SrfTE, *Structure* **10**, 301–310.

Burmester, J., Haese, A., Zocher, R. (1995) Highly conserved N-methyltransferases as an integral part of peptide synthetases, *Biochem. Mol. Biol. Int.* **37**, 201–207.

Byford, M. F., Baldwin, J. E., Shiau, C.-Y., Schofield, C. J. (1997) The mechanism of ACV synthetase, *Chem. Rev.* **97**, 2631–2649.

Carnio, M. C., Stachelhaus, T., Francis, K. P., Scherer, S.(2001) Pyridinyl polythiazole class peptide antibiotic micrococcin P1, secreted by foodborne *Staphylococcus equorum* WS2733, is biosynthesized non-ribosomally, *Eur. J. Biochem.* **268**, 6390–6401.

Carter, R. A., Worsley, P. S., Sawers, G., Challis, G. L., Dilworth, M. J., Carson, K. C., Lawrence, J. A., Wexler, M., Johnston, A. W., Yeoman, K. H. (2002) The vbs genes that direct synthesis of the siderophore vicibactin in *Rhizobium leguminosum*: their expression in other genera requires

ECF sigma factor Rpol, *Mol. Microbiol.* **44**, 1153–1166.

Challis, G. L., Ravel, J. (2000) Coelichelin, a new peptide siderophore encoded by the *Streptomyces coelicolor* genome: structure prediction from the sequence of its non-ribosomal peptide synthetase, *FEMS Lett.* **187**, 111–114.

Challis, G. L., Ravel, J., Townsend, C. A. (2000) Predictive, structure-based model of amino acid recognition by non-ribosomal peptide synthetase adenylation domains, *Chem. Biol.* **7**, 211–224.

Cheng, Y. Q., Walton, J. D. (2000) A eukaryotic alanine racemase gene involved in cyclic peptide biosynthesis, *J. Biol. Chem.* **275**, 4906–4911.

Chiu, H. T., Hubbard, B. K., Shah, A. N., Eide, J., Fredenburg, R. A., Walsh, C. T., Khosla, C. (2001) Molecular cloning and sequence analysis of the complestatin biosynthetic gene cluster, *Proc. Natl. Acad. Sci. USA* **98**, 8548–8553.

Chong, P. P., Podmore, S. M., Kieser, H. M., Redenbach, M., Turgay, K., Marahiel, M., Hopwood, D. A., Smith, C. P.(1998) Physical identification of a chromosomal locus encoding biosynthetic genes for the lipopeptide calcium-dependent antibiotic (CDA) of *Streptomyces coelicolor* A3(2),.*Microbiology.* **144**, 193–199.

Conti, E., Stachelhaus, T., Marahiel, M. A., Brick, P. (1997) Structural basis for the activation of phenylalanine in the non-ribosomal biosynthesis of gramicidin S, *EMBO J.* **16**, 4174–4183.

Cosmina, P., Rodriguez, F., de Ferra, F., Gandi, G., Peego, M., Venema, G., van Sinderen, D. (1993) Sequence and analysis of the genetic locus responsible for surfactin synthesis in *Bacillus subtilis, Mol. Microbiol.* **8**, 821–831.

Crump, M. P., Crosby, J., Dempsey, C. E., Murray, M., Hopwood, D. A., Simpson, T. J. (1996) Conserved secondary structure in the actinorhodin polyketide synthase acyl carrier protein from *Streptomyces coelicolor* A3(2) and the fatty acid synthase acyl carrier protein from *Escherichia coli, FEBS Lett.* **391**, 302–306.

De Crécy-Lagard, V., Marlière, P., Saurin, W. (1995) Multienzymatic non-ribosomal peptide biosynthesis: identification of the functional domains catalysing peptide elongation and epimerisation, *C. R. Acad. Sci. Ser. II :Life Sci.* **318**, 927–936.

De Crécy-Lagard, V., Blanc, V., Gil, P., Naudin, L., Lorenzon, S., Famechon, A., Bamas-Jacques, N., Crouzet, J., Thibaut, D. (1997a) Pristinamycin I biosynthesis in *Streptomyces pristinaespiralis*: molecular characterization of the first two structural peptide synthetase genes, *J. Bacteriol.* **179**, 705–713.

De Crécy-Lagard, V., Saurin. W., Thibaut, D., Gil, P., Naudin, L., Crouzet, J., Blanc, V. (1997b) Streptogramin B biosynthesis in *Streptomyces pristinaespiralis* and *Streptomyces virginiae*: molecular characterization of the last structural peptide synthetase gene, *Antimcrob. Agents Chemother.* **41**, 1904–1909.

De Ferra, F., Rodriguez, F., Tortora, O., Tosi, C., Grandi, G. (1997) Engineering of peptide synthetases. Key role of the thioesterase-like domain for efficient production of recombinant peptides, *J. Biol. Chem.* **272**, 25304–25309.

Dieckmann, R., von Döhren, H. (1997), in : *Developments in Industrial Microbiology* – GMBIM 1996 (Baltz, R, Hegemann, G., Skatrud, P., Eds.) Fairfax, VA : Society of Industrial Microbiology, 79–85.

Dieckmann, R., Neuhof, T., Pavela-Vrancic, M, von Döhren, H. (2001a) Dipeptide synthesis by an isolated adenylate forming domain of non-ribosomal peptide synthetases (NRPS), *FEBS Lett.* **498**, 42–45

Dieckmann, R., Pavela-Vrancic, M., von Döhren H. (2001b) Synthesis of adenosine 5′-polyphosphates and diadenosine polyphosphates catalyzed by tyrocidine synthetase 1, *Bichim. Biophys. Acta* **1546**, 234–241.

Du, L., Shen, B. (2001) Biosynthesis of hybrid peptide-polyketide natural products, *Curr. Opin. Drug Discov. Devel.* **4**, 215–228.

Du, L., Chen, M., Sanchez, C., Shen, B. (2000) An oxidation domain in the BlmIII non-ribosomal peptide synthetase probably catalyzing thiazole formation in the biosynthesis of the anti-tumor drug bleomycin in *Streptomyces verticillus* ATCC15003, *FEMS Microbiol. Lett.* **189**, 171–175.

Du, L., Sanchez, C., Shen, B. (2001) Hybrid peptide-polyketide natural products: biosynthesis and prospects toward engineering novel molecules, *Metab. Eng.* **3**, 78–95.

Duitman, E. H., Hamoen, L. W., Rembold, M., Venema, G., Seitz, H., Saenger, W., Bernhard, F., Reinhardt, R., Schmidt, M., Ullrich, C., Stein, T., Leenders, F., Vater, J. (1999) The mycosubtilin synthetase of *Bacillus subtilis* ATCC6633 : a multifunctional hybrid between a peptide synthetase, an amino transferase, and a fatty acid synthase, *Proc. Natl. Acad. Sci. USA* **96**, 13294–13299.

Ehmann, D. E., Gehring, A. M., Walsh, C. T. (1999) Lysine biosynthesis in Saccharomyces cerevisiae: mechanism of alpha-aminoadipate reductase (Lys2) involves post-translational phosphopantetheinylation by Lys5, *Biochemistry* **38**, 6171–6177

Ehmann, D. E., Shaw-Reid, C. A., Losey, H. C., Walsh, C. T. (2000a) The EntF and EntE adenylation domains of *Escherichia coli* enterobactin synthetase: sequestration and selectivity in acyl-AMP transfers to thiolation domain cosubstrates, *Proc. Natl. Acad. Sci. USA* **97**, 2509–2514.

Ehmann, D. E., Trauger, J. W., Stachelhaus, T., Walsh, C. T. (2000b) Aminoacyl-SNACs as small-molecule substrates for the condensation domains of non-ribosomal peptde synthetases, *Chem. Biol.*7, 765–772.

Eppelmann, K., Doekel, S., Marahiel, M. A. (2001) Engineered biosynthesis of the peptide antibiotic bacitracin in the surrogate host *Bacillus subtilis*, *J. Biol. Chem.* **276**, 34824–34831.

Eveland, S. S., Pompliano, D. L., Anderson, M. S. (1997) Conditionally lethal *Escherichia coli* murein mutants contain point defects that map to regions conserved among murein and folyl poly-gamma-glutamate ligases: identification of a ligase superfamily, *Biochemistry* **36**, 6223–6229.

Gadow, A., Vater, J., Schlumbohm, W., Palacz, Z., Salnikow, J. Kleinkauf, H. (1983)? Gramicidin S synthetase. Stability of reactive thioester intermediates and formation of 3-amino-2-piperidone, *Eur. J. Biochem.* **132**, 229–234.

Gehring, A. M., DeMoll, E., Fetherston, J. D., Mori, I., Mayhew, G. F., Blattner, F. R., Walsh, C. T., Perry, R. D. (1998a) Iron acquisition in plague: modular logic in enzymatic biogenesis of yersiniabactin by *Yersinia pestis*, *Chem. Biol.* **5**, 573–586.

Gehring, A. M., Mori, I. I., Perry, R. D., Walsh, C. T. (1998b) The non-ribosomal peptide synthetase HMWP2 forms a thiazoline ring during biogenesis of yersiniabactin, an iron-chelating virulence factor of *Yersinia pestis*, *Biochemistry.* **37**, 17104.

Glinski, M., Walther, M., Zocher, R. (1996) Symp. Enzymol.Biosynthesis of Natural Products, Technical University Berlin, 65 p.

Glund, K., Schlumbohm, W., Bapat, M., Keller , U. (1990) Biosynthesis of quinoxaline antibiotics: purification and characterization of the quinoxaline-2-carboxylic acid activating enzyme from *Streptomyces triostinicus*, *Biochemistry* **29**, 3522–3527.

Grammel, N., Pankevych, K., Demydchuk, J., Lambrecht, K., Saluz, H. P., Krügel, H. (2002) A beta-lysine adenylating enzyme and a beta-lysine binding protein involved in poly beta-lysine chain assembly in nourseothricin synthesis in *Streptomyces noursei*, *Eur. J. Biochem.* **269**, 347–357.

Guenzi, E., Galli, G., Grgurina, I., Gross, D. C., Grandi, G. (1998) Characterization of the syringomycin synthetase gene cluster. A link between prokaryotic and eukaryotic peptide synthetases, *J. Biol. Chem.* **273**, 32857–32863.

Hacker, C., Glinski, M., Hornbogen, T., Doller, A., Zocher, R. (2000) Mutational analysis of the N-methyltransferase domain of the multifunctional enzyme enniatin synthetase, *J. Biol. Chem.* **275**, 30826–30832.

Haese, A., Schubert, M., Hermann, M., Zocher, R. (1993) Molecular characterization of the enniatin synthetase gene encoding a multifunctional enzyme catalysing N-methyldepsipeptide formation in *Fusarium scirpi*, *Mol. Microbiol.* **7**, 905–914.

Handelsman J., Wackett L. P. (2002) Ecology and industrial microbiology: Microbial diversity – sustaining the Earth and industry, *Curr. Opin. Microbiol.* **5**, 237–239.

Hao, B., Gong, W., Ferguson, T. K., James, C. M., Krzycki, J. A., Chan, M. K. (2002) A new UAG-encoded residue in the structure of a methanogen methyltransferase, *Science* **296**, 1462–1466.

Healy, F. G., Wach, M., Krasnoff, S. B., Gibson, D. M., Loria, R. (2000) The txtAB genes of the plant pathogen *Streptomyces acidiscabies* encode a peptide synthetase required for phytotoxin thaxtomin A production and pathogenicity, *Mol. Microbiol.* **38**, 794–804.

Hernandez-Guzman, G., Alvarez-Morales, A. (2000) Isolation and characterization of the gene coding for the amidinotransferase involved in the biosynthesis of phaseolotoxin in *Pseudomonas syringae pv.* Phaseolicola, *Mol. Plant Microbe Interact.* **14**, 545–554.

Holden, M. T. G., Chhabra, S. R., de Nys, R., Stead, P., Bainton, N. J., Hill, P. J., Manefield, M., Kumar, N., Labatte, M., England, D., Rice, S., Givskov, M., Salmond, G. P. C., Stewart, G. S. A. B., Bycroft, B. W., Kjelleberg, S., Williams, P. (1999) Quorum-sensing cross talk: isolation and chemical characterization of cyclic dipeptides from *Pseudomonas aeruginosa* and other Gram-negative bacteria, *Mol. Microbiol.* **33**, 1254–1266.

Holden, M., Swift, S., Williams, P. (2000) New signal molecules on the quorum-sensing block, *Trends Microbiol.* **8**, 101–103.

Houck, D. R., Chen, L. C., Keller, P. J., Beale, J. M., Floss, H. G. (1988) Biosynthesis of the modified peptide antibiotic nosiheptide in *Streptomyces actuosis*, *J. Am. Chem. Soc.* **110**, 5800–5806.

Huang, G., Zhang, L., Birch, R. G. (2001) A multifunctional polyketide-peptide synthetase essential for albicidin biosynthesis in *Xanthomonas albilineans, Microbiology* **147**, 631–642.

Johnson, L. J., Johnson, R. D., Akamatsu, H., Salamiah, A., Otani, H., Kohmoto, K., Kodama, M. (2001) Spontaneous loss of a conditionally dispensable chromosome from the *Alternaria alternata* apple pathotype leads to loss of toxin production and pathogenicity, *Curr. Genet.* **40**, 65–72.

Julien, B., Shah, S., Ziermann, R., Goldman, R., Katz, L., Khosla, C. (2000) Isolation and characterization of the epothilone biosynthetic gene cluster from *Sorangium cellulosum, Gene* **249**, 153–160.

Kallow, W., von Döhren, H., Kleinkauf, H. (1998) Penicillin biosynthesis: energy requirement for tripeptide precursor formation by delta-(L-alpha-aminoadipyl)-L-cysteinyl-D-valine synthetase from *Acremonium chrysogenum, Biochemistry* **37**, 5947–5952.

Kallow, W., Kennedy, J., Arezi, B., von Döhren, H., Turner, G. (2000) Thioesterase domain of δ-(L-α-aminoadipyl)-L-cysteinyl-D-valine synthetase: alteration of stereospecificity by site-directed mutagenesis, *J. Mol. Biol.* **297**, 395–408.

Keating, T. A., Walsh, C. T. (1999) Initiation, elongation, and termination strategies in polyketide and polypeptide antibiotic biosynthesis, *Curr. Opin. Chem. Biol.* **3**, 598–606.

Keating, T. A, Marshall, C. G, Walsh, C. T. (2000a) Vibriobactin biosynthesis in *Vibrio cholerae*: VibH is an amide synthase homologous to non-ribosomal peptide synthetase condensation domains, *Biochemistry* **39**, 15513–15521.

Keating, T. A., Marshall, C. G, Walsh, C. T. (2000b) Reconstitution and characterization of the *Vibrio cholerae* vibriobactin synthetase from VibB, VibE, VibF, and VibH, *Biochemistry* **39**, 15522–15530.

Keating, T. A., Miller, D. A., Walsh, C. T. (2000c) Expression, purification, and characterization of HMWP2, a 229 kDa, six domain protein subunit of yersiniabactin synthetase, *Biochemistry.* **39**, 4729–4739.

Keating, T. A., Suo, Z., Ehmann, D. E., Walsh, C. T. (2000d) Selectivity of the yersiniabactin synthetase adenylation domain in the two-step process of amino acid activation and transfer to a holo-carrier protein domain, *Biochemistry* **39**, 2297–2306.

Keating, T. A., Ehmann, D. E., Kohli, R. M., Marshall, C. G., Trauger, J. W., Walsh, C. T. (2001) Chain termination steps in non-ribosomal peptide synthetase assembly lines: directed acyl-s-enzyme breakdown in antibiotic and siderophore biosynthesis, *Chembiochem. Europ. J. Chem. Biol.* **2**, 99–107.

Keating, T. A., Marshall, C. G., Walsh, C. T., Keating, A. E. (2002) The structure of VibH represents non-ribosomal peptide synthetase condensation, cyclization and epimerization domains, *Nat. Struct. Biol.* **9**, 522–526.

Keller, U., Schauwecker, F. (2001) Non-ribosomal biosynthesis of microbial chromopeptides, *Prog. Nucleic Acid Res. Mol. Biol.* **70**, 33–89.

Khosla, C. (1997) Harnessing the biosynthetic potential of modular polyketide synthases, *Chem. Rev.* **97**, 2577–2589.

Kleinkauf, H., von Döhren, H. (1990) Non-ribosomal biosynthesis of peptide antibiotics, *Eur. J. Biochem.* **192**, 1–15.

Kleinkauf, H., von Döhren, H. (1996) A non-ribosomal system of peptide biosynthesis, *Eur. J. Biochem.* **236**, 335–351.

Kleinkauf, H., von Döhren, H. (1997a) Peptide antibiotics. In: *Biotechnology, Volume 7: Products of Secondary Metabolism* (Kleinkauf, H., von Döhren, H., Eds.) 2nd edition, Weinheim: Wiley-VCH, 277–322.

Kleinkauf, H., von Döhren, H. (1997b) Applications of peptide synthetases in the synthesis of peptide analogues, *Acta Biochim. Polonica* **44**, 839–848.

Kleinkauf, H., von Döhren, H. (1999) Thiotemplate mechanism, In: *The Encyclopedia of Molecular Biology* (Creighton, T. E., Ed.) New York: Wiley, 2539–2541.

Kobayashi, J., Ishibashi, M. (1993) Bioactive metabolites of symbiotic marine microorganisms, *Chem. Rev.* **93**, 1753–1769.

Kohli, R. M., Trauger, J. W., Schwarzer, D., Marahiel, M. A., Walsh, C. T. (2001) Generality of peptide cyclization catalyzed by isolated thioesterase domains of non-ribosomal peptide synthetases, *Biochemistry* **40**, 7099–7108.

Kohli, R. M., Takagi, J., Walsh, C. T. (2002) The thioesterase domain from a non-ribosomal peptide synthetase as a cyclization catalyst for integrin binding peptides, *Proc. Natl. Acad. Sci. USA* **99**, 1247–1252.

Konz, D., Marahiel, M. A. (1999) How do peptide synthetases generate structural diversity? *Chem. Biol.* **6**, R39–48.

Konz, D., Klens, A., Schörgendörfer, K., Marahiel, M. A. (1997) The bacitracin biosynthesis operon of *Bacillus licheniformis* ATCC 10716: molecular characterization of three multi-modular peptide synthetases, *Chem. Biol.* **4**, 927–937.

Konz, D., Doekel, S., Marahiel, M. A. (1999) Molecular and biochemical characterization of the protein template controlling the biosynthesis of the lipopeptide lichenysin, *J. Bacteriol.* **181**, 133–140.

Labeit, S., Kolmerer, B. (1995) Titins: giant proteins in charge of muscle ultrastructure and elasticity, *Science* **270**, 293.

Lawen, A., Traber, R. (1993) Substrate specificities of cyclosporin synthetase and peptolide SDZ 214–103 synthetase. Comparison of the substrate specificities of the related multifunctional polypeptides, *J. Biol. Chem.* **268**, 20452–20465.

Lee, C., Lawen, A. (1993) *In vitro* biosynthesis of peptolide SDZ 90–215 by a 1.2 MDa multienzyme polypeptide, *Biochem. Mol. Biol. Int.* **31**, 797–805.

Linne, U., Marahiel, M. A. (2000) Control of directionality in non-ribosomal peptide synthesis: the role of the condensation domain in preventing misinitiation and timing of epimerization, *Biochemistry* **39**, 10439–10442.

Lipmann, F., Gevers, W., Kleinkauf, H., Roskoski, R., Jr., (1971) Polypeptide synthesis on protein templates: the enzymatic synthesis of gramicidin S and tyrocidine, *Adv. Enzymol.* **35**, 1–33.

Lu, S. E., Scholz-Schroeder, B. K., Gross, D. C. (2002) Construction of pMEKm12, an expression vector for protein production in *Pseudomonas syringae, FEMS Microbiol. Lett.* **210**, 115–121.

Luo, L., Walsh, C. T. (2001) Kinetic analysis of three activated phenylalanyl intermediates generated by the initiation module PheATE of gramicidin S synthetase, *Biochemistry* **40**, 5329–53237.

Majumder. S., Mukhopadhyay, N. K., Ghosh, S. K., Bose, S. K. (1988) Genetic analysis fo the mycobacillin biosynthetic pathway in *Bacillus subtilis* B3, *J. Gen. Microbiol.* **134**, 1147–1153.

Marahiel, M. A., Stachelhaus, T., Mootz, H. D. (1997) Modular peptide synthetases involved in non-ribosomal peptide synthesis, *Chem. Rev.* **97**, 2651–2673.

Marshall, C. G., Burkart, M. D., Keating, T. A., Walsh, C. T. (2001) Heterocycle formation in vibriobactin biosynthesis: alternative substrate utilization and identification of a condensed intermediate, *Biochemistry* **40**, 10655–10663.

Marshall, C. G., Hillson, N. J., Walsh, C. T. (2002) Catalytic mapping of the vibriobactin biosynthetic enzyme VibF, *Biochemistry* **41**, 244–250.

Martin, J. F.(2000) Alpha-aminoadipyl-cysteinyl-valine synthetases in beta-lactam producing organisms. From Abraham's discoveries to novel concepts of non-ribosomal peptide synthesis, *J. Antibiot. (Tokyo)*. **53**, 1008–1021.

May, J. J., Wendrich, T. M., Marahiel, M. A. (2001) The dhb operon of *Bacillus subtilis* encodes the biosynthetic template for the catecholic siderophore 2,3-dihydroxybenzoate-glycine-threonine trimeric ester bacillibactin, *J. Biol. Chem.* **276**, 7209–7217.

McHenney, M. A., Hosted, T. J., Dehoff, B. S., Rosteck, P. R. Jr, Baltz, R. H.(1998) Molecular cloning and physical mapping of the daptomycin gene cluster from *Streptomyces roseosporus, J. Bacteriol.* **180**,143–151.

Meyer, J. M. (2000) Pyoverdines: pigments, siderophores and potential taxonomic markers of fluorescent *Pseudomonas* species, *Arch. Microbiol.* **174**, 135–142.

Miller, D. A., Walsh, C. T. (2001) Yersiniabactin synthetase: probing the recognition of carrier protein domains by the catalytic heterocyclization domains, Cy1 and Cy2, in the chain-initiating HWMP2 subunit, *Biochemistry.* **40**, 5313–5321.

Miller, D. A., Lu, L., Hillson, N., Keating, T. A., Walsh, C. T. (2002) Yersiniabactin synthetase. A four-protein assembly line producing the non-ribosomal peptide/polyketide hybrid siderophore of *Yersinia pestis, Chem. Biol.* **9**, 333–344.

Milne, J. C., Sinha Roy, R., Eliot, R. C., Kelleher, N. L., Wokhlu, A., Nickels, B., Walsh, C. T. (1999) Cofactor requirements and reconstitution of microcin B17 synthetase: a multienzyme complex that catalyzes the formation of oxazoles and thiazoles in the antibiotic microcin B17, *Biochemistry* **38**, 4768–4781.

Mocek, U., Zeng, Z., O'Hagan, D., Zhou, P., Fans, L. D. G., Beale, J. M., Floss, H. G. (1993) Biosynthesis of the modified antibiotic thiostrepton in *Streptomyces azureus* and *Streptomyces laurentii, J. Am. Chem. Soc.* **115**, 7992–8001.

Mohr, H., Kleinkauf, H. (1978) Alamethicin biosynthesis: acetylation of the amino terminus and attachment of phenylalaninol, *Biochim. Biophys. Acta* **526**, 375–386.

Molnar, I., Schupp, T., Ono, M., Zirkle, R., Milnamow, M., Nowak-Thompson, B., Engel, N., Toupet, C., Stratmann, A., Cyr, D. D., Gorlach, J., Mayo, J. M., Hu, A., Goff, S., Schmid, J., Ligon, J. M. (2000) The biosynthetic gene cluster for the microtubule-stabilizing agents epothilones A and B from *Sorangium cellulosum* So ce90, *Chem. Biol.* 7, 97–109.

Mootz, H. D., Marahiel, M. A.(1997a) Biosynthetic systems for non-ribosomal peptide antibiotic assembly, *Curr. Opin. Chem. Biol.* **1**, 543–551.

Mootz, H. D., Marahiel, M. A. (1997b) The tyrocidine biosynthesis operon of *Bacillus brevis*:

complete nucleotide sequence and biochemical characterization of functional internal adenylation domains, *J. Bacteriol.* **179**, 6843–6850.

Mootz, H. D., Marahiel, M. A. (1999) Design and application of multimodular peptide synthetases, *Curr. Opin. Biotechnol.* **10**, 341–348.

Mootz, H. D., Schwarzer, D., Marahiel, M. A. (2000) Construction of hybrid peptide synthetases by module and domain fusions, *Proc. Natl. Acad. Sci. USA* **97**, 5848–5853.

Mootz, H. D., Finking, R., Marahiel, M. A. (2001) 4'-phosphopantetheine transfer in primary and secondary metabolism of *Bacillus subtilis*, *J. Biol. Chem.* **276**, 37289–37298.

Motamedi, H., Shafiee, A.(1998) The biosynthetic gene cluster for the macrolactone ring of the immunosuppressant FK506, *Eur. J. Biochem.* **256**, 528–534.

Nishizawa, T., Asayama, M., Fujii, K., Harada, K., Shirai, M. (1999) Genetic analysis of the peptide synthetase genes for a cyclic heptapeptide microcystin in *Microcystis* spp., *J. Biochem. (Tokyo)* **126**, 520–529.

Nishizawa, T., Ueda, A., Asayama, M., Fujii, K., Harada, K., Ochi, K., Shirai, M. (2000) Polyketide synthase gene coupled to the peptide synthetase module involved in the biosynthesis of the cyclic heptapeptide microcystin, *J. Biochem. (Tokyo)* **127**, 779–789.

Nishizawa, T., Asayama, M., Shirai, M. (2001) Cyclic heptapeptide microcystin biosynthesis requires the glutamate racemase gene, *Microbiology* **147**, 1235–1241.

O'Brien, D. P., Kirkpatrick, P. N., O'Brien, S. W., Staroske, T., Richardson, T. I., Evans, D. A., Hopkinson, A., Spencer, J. B., Williams, D. H. (2000) Expression and assay of an N-methyltransferase involved in the biosynthesis of a vancomycin group antibiotic, *Chem. Comm.* **2000**, 103–104.

Paitan, Y., Alon, G., Orr, E., Ron, E. Z., Rosenberg, E. (1999) The first gene in the biosynthesis of the polyketide antibiotic TA of *Myxococcus xanthus* codes for a unique PKS module coupled to a peptide synthetase, *J. Mol. Biol.* **286**, 465–474.

Panaccione, D. G., Scott-Craig, J. S., Pocard, J. A., Walton, J. D. (1992) A cyclic peptide synthetase gene required for pathogenicity of the fungus *Cochliobolus carbonum* on maize, *Proc. Natl. Acad. Sci. USA* **89**, 6590–6594.

Patel, H. M., Walsh, C. T. (2001) In vitro reconstitution of the *Pseudomonas aeruginosa* nonribosomal peptide synthesis of pyochelin: characterization of backbone tailoring thiazoline reductase and N-methyltransferase activities, *Biochemistry* **40**, 9023–9031.

Peeters, H., Zocher, R., Madry, N., Oelrichs, P. B., Kleinkauf, H., Kraeoelin, G. (1983) Synthesis of beauvericin by a multifunctional enzyme, *J. Antibiot.* **36**, 1762–1766.

Pelludat, C., Rakin, A., Jacobi, C. A., Schubert, S., Heesemann, J. (1998) The yersiniabactin biosynthetic gene cluster of *Yersinia enterocolitica*: organization and siderophore-dependent regulation, *J. Bacteriol.* **180**, 538–546.

Pieper, R., Haese, A., Schröder, W., Zocher, R. (1995) Arrangement of catalytic sites in the multifunctional enzyme enniatin synthetase, *Eur. J. Biochem.* **230**, 119–126.

Pospiech, A., Bietenhader, J., Schupp, T. (1996) Two multifunctional peptide synthetases and an O-methyltransferase are involved in the biosynthesis of the DNA-binding antibiotic and antitumour agent saframycin Mx1 from *Myxococcus xanthus*, *Microbiology* **142**, 741–746.

Puk, O., Huber, P., Bischoff, D., Recktenwald, J., Jung, G., Süßmuth, R. D:, van Pée, K.-H., Wohlleben, W., Pelzer, S. (2002) Glycopeptide biosynthesis in *Amycolatopsis mediterranei* DSM 5908: function of a halogenase and a haloperoxidase/haloperhydrolase, *Chem. Biol.* **9**, 225–235.

Quadri, L. E., Sello, J., Keating, T. A., Weinreb, P. H., Walsh, C. T. (1998a) Identification of a *Mycobacterium tuberculosis* gene cluster encoding the biosynthetic enzymes for assembly of the virulence-conferring siderophore mycobactin, *Chem. Biol.* **5**, 631–645.

Quadri, L. E. N., Weinreb, P. H., Lei, M., Nakano, M. M., Zuber, P., Walsh, C. T. (1998b) Characterization of Sfp, a *Bacillus subtilis* phosphopantetheinyl transferase for peptidyl carrier protein domains in peptide synthetases, *Biochemistry* **37**, 1585–1595.

Quadri, L. E., Keating, T. A., Patel, H. M., Walsh, C. T. (1999) Assembly of the Pseudomonas aeruginosa non-ribosomal peptide siderophore pyochelin: *in vitro* reconstitution of aryl-4,2-bisthiazoline synthetase activity from PchD, PchE, and PchF, *Biochemistry* **38**, 14941–14954.

Rabie, M. (1995) Investigations on the occurence and activity of the entomotoxins of *Metarhizium anisopliae* with reference to the biological control of *Spodoptera littoralis*, Thesis, TU Berlin.

Rainey, P. B., Brodey, C. L., Johnstone, K. (1993) Identification of a gene cluster encoding three high-molecular-weight proteins, which is required for synthesis of tolaasin by the mushroom pathogen *Pseudomonas tolaasii*, *Mol. Microbiol.* **8**, 643–652.

Recktenwald, J., Shawky, R., Puk, O., Pfennig, F., Keller, U., Wohlleben, W., Pelzer, S. (2002) Non-ribosomal biosynthesis of vancomycin-type antibiotics: a heptapeptide backbone and eight peptide synthetase modules, *Microbiology* **148**, 1105–1118.

Reiber, K., Schwecke, T., Neuhof, T., Gräfe, U., von Döhren, H. (2002) Characterization of a peptaibol synthetase from *Sepedonium* involved in the production of Ampullosporin, Presentation Eur. Conf. Fungal Genetics Pisa 2002.

Reuter, K., Mofid, M. R., Marahiel, M. A., Ficner, R. (1999) Crystal structure of the surfactin synthetase-activating enzyme Sfp: a prototype of the 4′-phosphopantetheinyl transferase superfamily, *EMBO J.* **18**, 6823–6831.

Riederer, B., Han, M., Keller, U. (1996) D-Lysergyl peptide synthetase from the ergot fungus *Claviceps purpurea*, *J. Biol. Chem.* **271**, 27524–27530.

Rouhiainen, L., Paulin, L., Suomalainen, S., Hyytiainen, H., Buikema, W., Haselkorn, R., Sivonen, K. (2000) Genes encoding synthetases of cyclic depsipeptides, anabaenopeptilides, in *Anabaena strain* 90. *Mol. Microbiol.* **37**, 156–167.

Schauwecker, F., Pfennig, F., Schröder, W., Keller, U. (1998) Molecular cloning of the actinomycin synthetase gene cluster from *Streptomyces chrysomallus* and functional heterologous expression of the gene encoding actinomycin synthetase II, *J. Bacteriol.* **180**, 2468–2474.

Schauwecker, F., Pfennig, F., Grammel, N., Keller, U. (2001) Construction and *in vitro* analysis of a new bi-modular polypeptide synthetase for synthesis of N-methylated acyl peptides, *Chem. Biol.* **7**, 287–297.

Schlumbohm, W., Keller, U. (1990) Chromophore activating enzyme involved in the biosynthesis of the mikamycin B antibiotic etamycin from *Streptomyces griseoviridis*, *J. Biol. Chem.* **265**, 2156–2161.

Schneider, A., Marahiel, M. A. (1998) Genetic evidence for a role of thioesterase domains, integrated in or associated with peptide synthetases, in non-ribosomal peptide biosynthesis in *Bacillus subtilis*, *Arch. Microbiol.* **169**, 404–410.

Schneider, A., Stachelhaus, T., Marahiel, M. A. (1998) Targeted alteration of substrate specificity of peptide synthetases by rational module swapping, *Mol. Gen. Genet.* **257**, 308–318.

Scholz-Schroeder, B. K., Soule, J. D., Lu, S. E., Grgurina, I., Gross, D. C. (2001) A physical map of the syringomycin and syringopeptin gene clusters localized to an approximately 145-kb DNA region of *Pseudomonas syringae pv.* syringae strain B301D, *Mol. Plant Microbe Interact.* **14**, 1426–1435.

Schwecke, T., Tobin, M. B., Kovacevic, S., Miller, J. R., Skatrud, P. L., Jensen, S. E. (1993) The A module of ACV synthetase: expression, renaturation and functional characterization, in: *Industrial Microorganisms* (Baltz, R., Hegemann, G. D., Skatrud, P. L., Eds.), Washington DC: American Society of Microbiology, 291.

Schwecke, T., Aparicio, J. F., Molnar, I., König, A., Khaw, L. E., Haydock, S. F., Olyinyk, M., Caffrey, P., Cortes, J., Lester, J. B. et al. (1995) The biosynthetic gene cluster for the polyketide immunosuppressant rapamycin, *Proc. Natl. Acad. Sci. USA* **92**, 7839–7843.

Seto, H., Kuzuyama, T. (1999) Bioactive natural products with carbon-phosphorus bonds and their biosynthesis, *Nat. Prod. Rep.* **16**, 589–596.

Shaw-Reid, C. A., Kelleher, N. L., Losey, H. C., Gehring, A. M., Berg, C., Walsh, C. T. (1999) Assembly line enzymology by the multimodular non-ribosomal peptide synthetases: the thioesterase domain of *E. coli* EntF catalyzes both elongation and cyclolactonization, *Chem. Biol.* **6**, 385–400.

Shen, B., Du, L., Sanchez, C., Edwards, D. J., Chen, M., Murrell, J. M. (2001) The biosynthetic gene cluster for the anticancer drug bleomycin from *Streptomyces verticillus* ATCC15003 as a model for hybrid peptide-polyketide natural product biosynthesis, *J. Ind. Microbiol. Biotechnol.* **27**, 378–385.

Shen, B., Du, L., Sanchez, C., Edwards, D. J., Chen, M., Murrell, J. M. (2002) Cloning and characterization of the bleomycin biosynthetic gene cluster from *Streptomyces verticillus* ATCC15003(1), *J. Nat. Prod.* **65**, 422–431.

Shiau, C. Y., Baldwin, J. E., Byford, M. F., Schofield, C. J. (1995) Delta-L-(alpha-aminoadipoyl)-L-cysteinyl-D-valine synthetase: isolation of L-cysteinyl-D-valine, a "shunt" product, and implications for the order of peptide bond formation, *FEBS Lett.* **373**, 303–306.

Siegmund, K. D., Plattner, H. J., Diekmann, H. (1991) Purification of ferrichrome synthetase from *Aspergillus quadricinctus* and characterization as a phosphopantetheine containing multienzyme complex, *Biochim. Biophys. Acta* **1076**, 123–129.

Sinha Roy, R., Gehring, A. M., Milne, J. C., Belshaw, P. J., Walsh, C. T. (1999) Thiazole and oxazole

peptides: biosynthesis and molecular machinery, *Nat. Prod. Rep.* **16**, 249–263.
Silakowski, B., Schairer, H. U., Ehret, H., Kunze, B., Weinig, S., Nordsiek, G., Brandt, P., Blöcker, H., Höfle, G., Beyer, S., Müller, R. (1999) New lessons for combinatorial biosynthesis from myxobacteria. The myxothiazol biosynthetic gene cluster of *Stigmatella aurantiaca* DW4/3-1, *J. Biol. Chem.* **274**, 37391–37399.
Silakowski B, Kunze B, Müller R. (2001a) Multiple hybrid polyketide synthase/non-ribosomal peptide synthetase gene clusters in the myxobacterium *Stigmatella aurantiaca, Gene* **275**, 233–240.
Silakowski, B., Nordsiek, G., Kunze, B., Blöcker, H., Müller, R. (2001b) Novel features in a combined polyketide synthase/non-ribosomal peptide synthetase: the myxalamide biosynthetic gene cluster of the myxobacterium *Stigmatella aurantiaca* Sga15, *Chem. Biol.* **8**, 59–69.
Smith, D. J., Burnham, M. K. R., Edwards, J., Earl, A., Turner, G. (1990) Cloning and heterologous expression of the penicillin biosynthetic gene cluster from *Penicillium chrysogenum, Biotechnology* **8**, 39–41.
Sohn, Y. S., Nam, D. H., Ryu, D. D. (2001) Biosynthetic pathway of cephabacins in *Lysobacter lactamgenus*: molecular and biochemical characterization of the upstream region of the gene clusters for engineering of novel antibiotics, *Metab. Eng.* **3**, 380–392.
Solenberg, P. J., Matsushima, P., Stack, P. R., Wilkie, S. C., Thompson, R. C., Baltz, R. H. (1997) Production of hybrid glycopeptide antibiotics *in vitro* and in *Streptomyces toyocaensis, Chem. Biol.* **4**, 195–202.
Sosio, M., Bianchi, A., Bossi, E., Donadio, S. (2000) Teicoplanin biosynthesis genes in *Actinoplanes teichomyceticus, Antonie Van Leeuwenhoek.* **8**, 379–384.
Srinivasan, G., James, C. M., Krzycki, J. A. (2002) Pyrrolysine encoded by UAG in Archae: charging of a UAG-decoding specialized tRNA, *Science* **296**, 1459–1462.
Stachelhaus, T., Walsh, C. T. (2000) Mutational analysis of the epimerization domain in the initiation module PheATE of gramicidin S synthetase, *Biochemistry* **39**, 5775–5787.
Stachelhaus, T., Schneider, A., Marahiel, M. A. (1995) Rational design of peptide antibiotics by targeted replacement of bacterial and fungal domains, *Science* **269**, 69–72.
Stachelhaus, T., Mootz, H. D., Bergendahl, V., Marahiel, M. A. (1998) Peptide bond formation in nonribosomal peptide biosynthesis. Catalytic role of the condensation domain, *J. Biol. Chem.* **273**, 22773–22781.
Stachelhaus, T., Mootz, H. D., Marahiel, M. A. (1999) The specificity code of adenylation domains in non-ribosomal peptide synthetases, *Chem. Biol.* 6, 493–505.
Staunton, J., Wilkinson, B. (1997) Biosynthesis of erythromycin and rapamycin, *Chem. Rev.* **97**, 2611–2629.
Stein, T., Vater, J., Kruft, V., Otto, A., Wittmann-Liebold, B., Franke, P., Panico, M., McDowell, R., Morris, H. R. (1996) The multiple carrier model of non-ribosomal peptide biosynthesis at modular multienzymatic templates, *J. Biol. Chem.* **271**, 15428–15435.
Steller, S., Vollenbroich, D., Leenders, F., Stein, T., Conrad, B., Hofemeister, J., Jacques, P., Thonart, P., Vater, J. (1999) Structural and functional organization of the fengycin synthetase multienzyme system from *Bacillus subtilis* B213 and A1/3, *Chem. Biol.* **6**, 31–41.
Süßmuth, R. D., Pelzer, S., Nicholson, G., Walk, T., Wohlleben, W., Jung, G. (1999) New advances in the biosynthesis of glycopeptide antibiotics of the vancomycin type from *Amycolatopsis mediterranei, Ang. Chem Int. Ed.* **38**, 1976–1979.
Suo, Z., Chen, H., Walsh, C. T. (2000) Acyl-CoA hydrolysis by the high molecular weight protein 1 subunit of yersiniabactin synthetase: mutational evidence for a cascade of four acyl-enzyme intermediates during hydrolytic editing, *Proc. Natl. Acad. Sci. USA* **97**, 14188–14193.
Tang, L., Shah, S., Chung, L., Carney, J., Katz, L., Khosla, C., Julien, B. (2000) Cloning and heterologous expression of the epothilone gene cluster, *Science* **287**, 640–642.
Tillett, D., Dittmann, E., Erhard, M., von Döhren, H., Börner, T., and Neilan, B. (2000) Structural organization of microcystin biosynthesis in *Microcystis aeruginosa* PCC7806: an integrated peptide-polyketide synthetase system, *Chem. Biol.* **7**, 753–764.
Trauger, J. W., Walsh, C. T. (2000) Heterologous expression in *Escherichia coli* of the first module of the non-ribosomal peptide synthetase for chloroeremomycin, a vancomycin-type glycopeptide antibiotic, *Proc. Natl. Acad. Sci. USA* **97**, 3112–3117.
Trauger, J. W., Kohli, R. M., Walsh, C. T. (2001) Cyclization of backbone-substituted peptides catalyzed by the thioesterase domain from the tyrocidine non-ribosomal peptide synthetase, *Biochemistry* **40**, 7092–7098.

Tsuge, K., Akiyama, T., Shoda, M. (2001) Cloning, sequencing, and characterization of the iturin A operon, *J. Bacteriol.* **183**, 6265–6273.

Tudzynski, P., Holter, K., Correia, T., Arntz, C., Grammel, N., Keller, U. (1999) Evidence for an ergot alkaloid gene cluster in *Claviceps purpurea, Mol. Gen. Genet.* **261**, 133–141.

van Wageningen, A. M., Kirkpatrick, P. N., Williams, D. H., Harris, B. R., Kershaw, J. K., Lennard, N. J., Jones, M., Jones, S. J., and Solenberg, P. J.(1998) Sequencing and analysis of genes involved in the biosynthesis of a vancomycin group antibiotic, *Chem. Biol.* **5**, 155–162.

Volkman, B. F., Zhang, Q., Debabov, D. V., Rivera, E., Kresheck, G. C., Neuhaus, F. C. (2001) Biosynthesis of D-alanyl-lipoteichoic acid: the tertiary structure of apo-D-alanyl carrier protein, *Biochemistry* **40**, 7964–7972.

von Döhren, H., Kleinkauf, H. (1988) Research on non-ribosomal systems: Biosynthesis of peptide antibiotics, in: *The Roots of Modern Biochemistry* (Kleinkauf, H., von Döhren, H., Jaenicke, R., Eds.) Berlin: de Gruyter, 355–368.

von Döhren, H., Keller, U., Vater, J., Zocher, R. (1997) Multifunctional peptide synthetases. *Chem. Rev.* **97**, 2675–2705

von Döhren, H., Pavela-Vrancic, M., Dieckmann, R. (1999) The non-ribosomal code, *Chem. Biol.* **6**, R273–R279.

von Döhren, H., Kallow, W., Tavanlar, M. A., Schwecke, T. S., Dieckmann, R., Uhlmann, V. (2001) δ-(L-α-Aminoadipyl)-L-cysteinyl-D-valine Synthetase as a model tripeptide synthetase, in: *Enzyme Technologies for Pharmaceutical and Biotechnological Applications* (Kirst, H. A., Yeh, W.-K., Zmijewski, M. J., Eds.), New York: Marcel Dekker, 1–38.

Walsh, C. T., Gehring, A. M., Weinreb, P. H., Quadri, L. E. N., Flugel, R. S. (1997) Post-translational modification of polyketide and nonribosomal peptide synthases, *Curr. Opin. Chem. Biol.* **1**, 309–315.

Walsh, C. T., Chen, H., Keating, T. A., Hubbard, B. K., Losey, H. C., Luo, L., Marshall, C. G., Miller, D. A., Patel, H. M. (2001) Tailoring enzymes that modify non-ribosomal peptides during and after chain elongation on NRPS assembly lines, *Curr. Opin. Chem. Biol.* **5**, 525–534.

Walzel, B., Riederer, B., Keller, U. (1997) Mechanism of alkaloid cyclopeptide synthesis in the ergot fungus *Claviceps purpurea, Chem. Biol.* **4**, 223–230.

Weber, T., Marahiel, M. A. (2001) Exploring the domain structure of modular non-ribosomal peptide synthetases, *Structure* **10**, R3–R9.

Weber, G., Schörgendörfer, K., Schneider-Scherzer, E., Leitner, E. (1994) The peptide synthetase catalyzing cyclosporine production in *Tolypocladium niveum* is encoded by a giant 45.8-kilobase open reading frame, *Curr. Genet.* **26**, 120–125.

Weber, T., Baumgartner, R., Renner, C., Marahiel, M. A., Holak, T. A. (2000) Solution structure of PCP, a prototype for the peptidyl carrier domains of modular peptide synthetases, *Structure* **8**, 407–418.

Weckwerth, W., Miyamoto, K., Iinuma, K., Krause, M, Glinski, M., Storm, T., Bonse, G., Kleinkauf, H., Zocher, R. (2000) Biosynthesis of PF1022 and related cyclooctadepsipeptides, *J. Biol. Chem.* **275**, 17909–17915.

Welch, T. J., Chai, S., Crosa, J. H. (2000) The overlapping angB and angG genes are encoded within the trans-acting factor region of the virulence plasmid in *Vibrio anguillarum*: essential role in siderophore biosynthesis, *J. Bacteriol.* **182**, 6762–6773.

Wessels, P., von Döhren, H., Klerinkauf, H. (1996) Biosynthesis of acylpeptidolactones of the daptomycin type: a comparative analysis of peptide synthetases forming A21978C and A54145, *Eur. J. Biochem.* **242**, 665–673.

Wiest, A., Grzegroski, D., Xu, B. W., Goulard, C., Rebouffat, S., Ebbole, D. J., Bodo, B., Kenerley, C. M. (2002) Identification of peptaibols from *Trichoderma virens* and cloning of a peptaibol synthetase, *J. Biol. Chem.* **277**, 20862–20868.

Wilhite, S. E., Lumsden, R. D., Straney, D. C. (2001) Peptide synthetase gene in *Trichoderma virens, Appl. Environ. Microbiol.* **67**, 5055–5062.

Yakimov, M. M., Kroger, A., Slepak, T. N., Giuliano, L., Timmis, K. N., Golyshin, P. N. (1998) A putative lichenysin A synthetase operon in *Bacillus licheniformis*: initial characterization, *Biochim. Biophys. Acta.* **1399**, 141–153.

Yazgan, A., Ozcengiz, G., Marahiel, M. A. (2001) Tn10 insertional mutations of *Bacillus subtilis* that block the biosynthesis of bacilysin, *Biochim. Biophys. Acta* **1518**, 87–94.

Yu, S., Fiss, E., Jacobs, W. R. Jr. (1998) Analysis of the exochelin locus in *Mycobacterium smegmatis*: biosynthesis genes have homology with genes of the peptide synthetase family, *J. Bacteriol.* **180**, 4676–4685.

Yuan, W. M., Gentil, G. D., Budde, A. D., Leong S. A. (2001) Characterization of the *Ustilago maydis* sid2 gene, encoding a multidomain peptide synthetase in the ferrichrome biosynthetic gene cluster, *J. Bacteriol.* **183**, 4040–4051.

Zhu, W., Arceneaux, J. E., Beggs, M. L., Byers, B. R., Eisenach, K. D., Lundrigan, M. D. (1998) Exochelin genes in *Mycobacterium smegmatis*: identification of an ABC transporter and two non-ribosomal peptide synthetase genes, *Mol. Microbiol.* **29**, 629–639.

Ziegler, K., Diener, A., Herpin, C., Richter, R., Deutzmann, R., Lockau, W. (1998) Molecular characterization of cyanophycin synthetase, the enzyme catalyzing the biosynthesis of the cyanobacterial reserve material multi-L-arginyl-poly-L-aspartate (cyanophycin), *Eur. J. Biochem.* **254**, 154–159.

Zocher, R., Keller, U. (1997) Thiol template peptide synthesis systems in bacteria and fungi, *Adv. Microb. Physiol.* **38**, 85–131.

4
Cyanophycin

Dr. Fred Bernd Oppermann-Sanio[1], **Prof. Dr. Alexander Steinbüchel**[2]
[1] Institut für Mikrobiologie der Westfälischen Wilhelms-Universität, Corrensstraße 3, 48149 Münster, Germany; Tel.: ++49-251-8339823; Fax: ++49-251-8338388; E-mail: opperma@uni-muenster.de
[2] Institut für Mikrobiologie der Westfälischen Wilhelms-Universität, Corrensstraße 3, 48149 Münster, Germany; Tel.: ++49-251-8339821; Fax: ++49-251-8338388; E-mail: steinbu@uni-muenster.de

CD circular dichroism
CGP cyanophycin granule polypeptide
DSM Deutsche Sammlung von Mikroorganismen und Zellkulturen
ESI electrospray ionization mass spectrometry
ORF open reading frame
PCC Pasteur Culture Collection

1 Introduction

Cyanophycin, which is also referred to as cyanophycin granule polypeptide (CGP), was discovered microscopically as highly refractile cell inclusions of cyanobacteria (blue-green algae) more than 100 years ago (Borzi, 1887). Until now, it had been thought that CGP was unique to the oxygenic phototrophic bacteria, and so it was surprising when Krehenbrink et al. (2002) demonstrated the occurrence of the polyamide in chemotrophic bacteria such as *Acinetobacter* sp. The accepted main function of CGP, at least in cyanobacteria, is that of a (transient) nitrogen reserve.

Besides poly(glutamic acid) and poly(lysine), both of which are described in detail in other chapters of this series (Ashiuchi and Misono, 2002; Yoshida et al., 2002), CGP represents the third poly(amino acid) known to occur in nature. In contrast to the former compounds, the constituents of CGP are aspartic acid-arginine dipeptides arranged as a poly(aspartic acid) backbone with pendent arginine residues linked to the β-carboxyl groups of the aspartyl residues (Simon and Weathers, 1976). CGP shares with the other two poly(amino acids) the following three characteristics, which distinguish it from ribosomally synthesized proteins (Oppermann-Sanio and Steinbüchel, 2002):

1) CGP contains only one type of amino acid in the polymer backbone, whereas proteins are copolyamides composed of a mixture of up to 21 amino acids.
2) The biosynthesis of CGP is performed by the comparatively simple enzyme CGP synthetase in a template-independent manner, whereas protein synthesis is catalyzed in a template-dependent mode involving the highly complex ribosomal translation apparatus. Hence, in contrast to protein synthesis, inhibitors of translation such as chloramphenicol do not inhibit CGP formation.
3) As a consequence of point (2), the polypeptide strands of proteins exhibit exactly defined lengths and are monodisperse, whereas CGP shows a remarkable size distribution and is polydisperse.

As CGP is completely biodegradable (Obst et al., 2002), it may be considered as a basis for environmentally friendly polyamides. Purified CGP can be chemically converted to a derivative with reduced arginine content (Joentgen et al., 1998), and this might be applied in the future to a variety of technical processes as a biodegradable substitute for polyacrylate (Schwamborn, 1998).

2 Historical Outline

The first microscopic description of large granules of "cianoficina" appeared in a report from the Italian botanist Borzi which provided details of intracellular inclusions in filamentous cyanobacteria (Borzi, 1887) (Figure 1). Later, Hegler (1901) reported the gradual disappearance of CGP when cyanobacterial cultures were transferred to darkness. In subsequent studies conducted by Geitler (1932), the granules were considered as proteinaceous reserve material, while Fogg (1951) described the development of heterocysts in filaments of *Anabaena cylindrica*. By staining with the arginine-specific Sakaguchi reagent, Fogg was able to demonstrate the appearance of large CGP areas at the junctions between heterocysts and adjoining vegetative cells.

During the 1970s, Simon and coworkers – in a series of elegant studies–elucidated not only the structure of the polymer (Figure 2) but also the principles of the enzymatic reaction leading to its production (Simon, 1971, 1973a, b, 1976; Simon and Weathers, 1976). The CGP-synthesizing enzyme CGP synthetase was for the first time isolated from *A. variabilis* by Ziegler et al. (1998). Subsequently, CGP synthetase-encoded *cphA* genes were first cloned by Ziegler et al. (1998) and Oppermann-Sanio et al. (1999) from *A. variabilis* and *Synechocystis* sp. strain PCC 6803, respectively. Meanwhile, *cphA* sequences from other cyanobacteria have been published (Aboulmagd et al., 2000; Berg et al. 2000; Hai et al., 2002). Quite recently, by performing a systematic analysis of published genome sequences, Krehenbrink et al. (2002) found *cphA* homologues and CGP as well as CGP synthetase in noncyanobacterial bacteria such as the chemoorganoheterotrophic *Acinetobacter* sp. (Figure 1), thus annulling the long-standing adage that CGP occurred uniquely in the cyanobacteria (Lawry and Simon, 1982; Allen, 1984, 1988; Simon, 1987).

3 Chemical Structure

The structure of CGP, as shown in Figure 2, was proposed by Simon and Weathers (1976) and soon became generally accepted. The strands of the purified polymer consist of roughly equimolar amounts of arginine and aspartic acid arranged as a poly(aspartic acid) backbone, with arginine moieties linked to

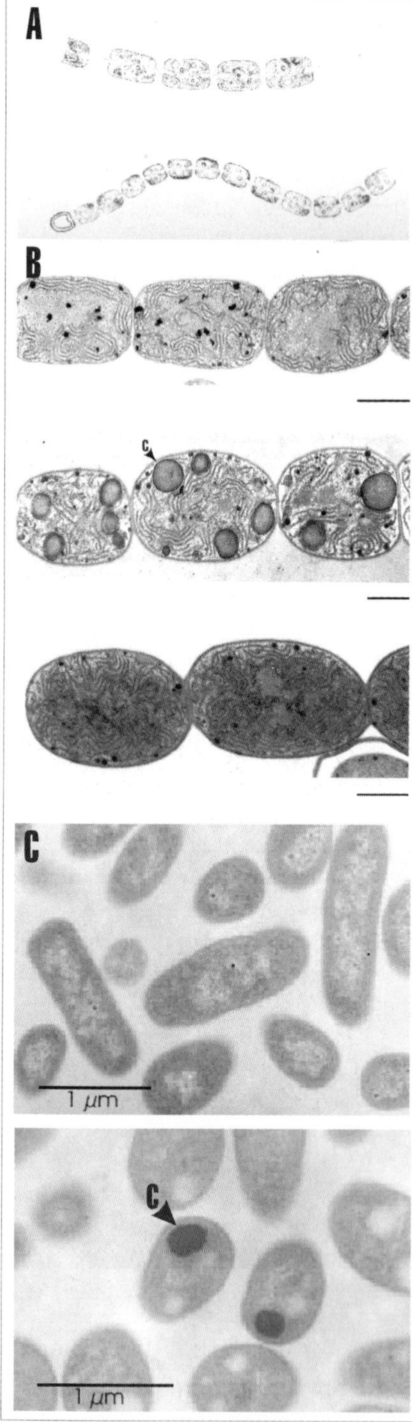

Fig. 1 Occurrence of CGP in cyanobacteria and *Acinetobacter* sp. (A) First published drawing of cyanophycin granules ("cianoficina") in filamentous cyanobacteria. The figure was taken from Borzi (1878). (B) Dependence of CGP accumulation in the filamentous cyanobacterium *Anabaena* sp. on culture conditions and on the CGP synthetase gene *cphA*. Wild-type cells grown in phosphate-rich and phosphate-reduced medium are depicted in the upper (0.4 mM) and central (10 μM) figures, respectively. The bottom figure shows cells of a *cphA* knock-out mutant grown in phosphate-reduced medium. (Figure B reprinted from *FEMS Microbiol. Lett.* **196**, Ziegler, K., Stephan, D. P., Pistorius, E. K., Ruppel, H. G., Lockau, W., A mutant of the cyanobacterium *Anabaena variabilis* ATCC 29413 lacking cyanophycin synthetase: growth properties and ultrastructural aspects, pp. 13–18. © 2001, with permission from Elsevier Science.) (C) Appearance of cyanophycin granules in cells of the chemoorganoheterotrophic *Acinetobacter* sp. under phosphate limitation (Krehenbrink et al., 2002). Cells were grown either in the presence of 1.3 mM phosphate (upper figure) or 64 μM (bottom figure). (Figure C was a gift from M. Krehenbrink; unpublished.) C, CGP.

the β-carboxyl group of almost every aspartic acid residue. All known analogues of arginine, which have been detected as minor constituents of purified CGP (Ziegler et al., 1998; Frey et al., 2002), or which have been found to replace part of arginine during the in-vitro synthesis of CGP (Berg et al., 2000; Aboulmagd et al., 2001a), are shown in Figure 3. A CGP-like material, which had been isolated from nitrogen-limited cultures of the unicellular cyanobacterium *Synechocystis* sp. strain PCC 6308, contained glutamic acid instead of arginine (Merritt et al., 1994). Glutamic acid was recently also detected in significant amounts (15 mol.%) in CGP from a marine halotolerant *Synechococcus* strain (Wingard et al., 2002).

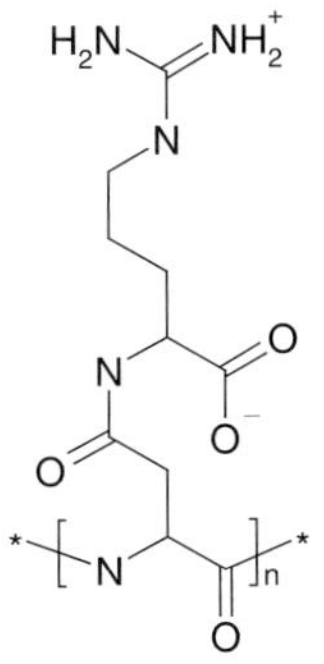

Fig. 2 Chemical structure of the aspartic acid-arginine building block of CGP.

surrounding membrane (Allen and Weathers, 1980).

4 Chemical Analysis and Detection

The cyanophycin granules represent only one type among a variety of cell inclusions that can be found in cyanobacterial cells. Among the other simple inclusions, which consist of either polyglucans, poly(3-hydroxybutyrate), lipids or polyphosphate and the more complex carboxysomes, phycobilisomes and gas vacuoles (Allen, 1984), CGP can be recognized by light microscopy because of its affinity for different reagents and dyes [i.e. Sakaguchi reagent, acetocarmine, neutral red, methylene blue, amido black (Fogg, 1951; Lang, 1968)]. Detailed electron microscopic analyses of native CGP revealed a radiating pattern of substructure known as "structured granules" (Lang et al., 1972) and the absence of a surrounding membrane (Allen and Weathers, 1980).

4.1 Solubility Properties of CGP and Isolation

Intact or native CGP was first isolated from *A. cylindrica* by differential centrifugation (Simon, 1971). The orange-colored upper part of the pellet from a 10-min centrifugation (20,000 g) of disrupted cells was resuspended in distilled water and re-centrifuged for 30 s at 1000 g; centrifugation of the supernatant at 1000 g for 15 min produced a pellet in which the lower white layer consisted almost entirely of CGP. Further purification was achieved by alternating centrifugal washing steps with concomitant dissolution of the granules in 100 mM HCl and reprecipitation at neutral pH (Simon and Weathers, 1976). This procedure takes advantage of the characteristic solubility properties of CGP (Fogg, 1951; Simon, 1971): The polypeptide is insoluble at neutral pH and physiological (low) ionic strength, but is soluble under acidic (pH < 2) and alkaline (pH > 9) conditions. Furthermore, the polymer can be solubilized in 1% (w/v) sodium dodecylsulfate and 4.0 M urea, but remains insoluble in organic solvents (e.g., methanol, ethylene glycol, dimethylsulfoxide and formamide), 2% (v/v) Triton X-100, and 1% (w/v) sodium deoxycholate (Lang et al., 1972).

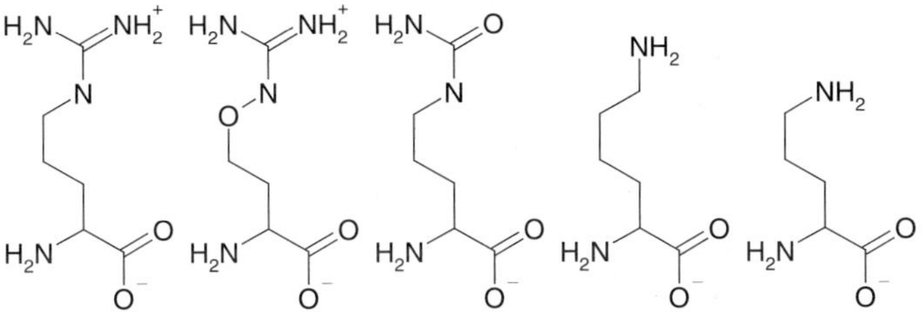

Fig. 3 Arginine and analogues thereof, which had been found in CGPs from recombinant *Escherichia coli* cells harboring *cphA* from different cyanobacteria (i.e., lysine; Ziegler et al., 1998; Frey et al., 2002), or shown to be incorporated into CGP *in vitro* by purified CGP synthetases (i.e., canavanine, citrulline, lysine, ornithine; Berg et al., 2000; Aboulmagd et al., 2001b).

4.2 Molecular Weight and Structure of CGP

The primary structure of purified CGP (see Figure 2) and the following characteristics were first revealed by the detailed investigations of Simon and coworkers (Simon, 1971, 1973a; Simon and Weathers, 1976). The polyamide as isolated from cyanobacteria is highly polydisperse, with a molecular weight ranging from 25 to 125 kDa as estimated with SDS–polyacrylamide gel electrophoresis, and corresponding to degree of polymerization of 90 to 400; CGP is unaffected by a variety of proteases such as pronase, pepsin, chymotrypsin, α- or β-carboxypeptidase, leucine aminopeptidase, and clostridiopeptidase B. Further examination of the polypeptide with circular dichroism (CD) and laser Raman spectroscopy (Simon et al., 1980) revealed an ordered structure which was environmentally sensitive: it exhibited a defined secondary structure in acidic solution, but not under alkaline conditions. CD spectra suggested the existence of β-sheets in the acid-soluble form and in the insoluble material at neutral pH. Data obtained from ^{1}H-, ^{13}C- and ^{15}N-nuclear magnetic resonance (NMR) spectroscopy were consistent with the proposed polypeptide structure (Suarez et al., 1999).

4.3 Quantification of CGP

Standard methods for quantifying levels of CGP in the cell mass are based on the chemical assay for arginine either with the Sakaguchi reagent (Fogg, 1951; Messineo, 1966, Simon, 1973a) or with high-performance liquid chromatography (HPLC) to assay both amino acid constituents (Aboulmagd et al., 2000) after the isolation of CGP using the method described above. The staining procedure that uses the Sakaguchi reagent is based on the formation of a red complex between 2,4-dichloro-1-naphthol and arginine in alkaline solution upon addition of an oxidizing agent (NaOCl). Recently, a more rapid and sensitive method for the quantification of the polymer from crude preparations on the basis of ^{1}H-NMR has been described (Erickson et al., 2001).

4.4 Variations in Composition and CGP-like Polymers

Variations in the amino acid composition of CGP have been reported to occur in *Synechocystis* sp. strain PCC 6308 as a function of growth conditions and physiological background, respectively (Merritt et al., 1994). As determined by gas chromatography-mass spectrometry of ^{15}N-labeled CGP-like material, which had been isolated from nitrogen-limited cultures, this polymer contained glutamic acid rather than arginine. After transfer to full medium, glutamic acid was replaced during growth by arginine to form the "normal" CGP as depicted in Figure 2. As the incorporation of glutamic acid into CGP is not catalyzed by CGP synthetase (Aboulmagd et al., 2000, 2001a; also see below), the synthesis and physiological relevance of this glutamate-containing CGP-like polymer remains obscure. Lysine was identified as an additional minor amino acid component in the range of 3 to 18 mol.% in CGP, which had been synthesized by recombinant cells of *Escherichia coli* harboring the CGP synthetic genes from *A. variabilis* (Ziegler et al., 1998) or *Synechocystis* sp. strain PCC 6803 (Frey et al., 2002; see Figure 3). In addition, it was found that CGP either synthesized by purified CGP synthetase *in vitro* or isolated from recombinant *E. coli* exhibited a remarkably lower average molecular weight and lower polydispersity than the authentic material isolated from the parental cyanobacterium (Ziegler et al., 1998; Aboulmagd et al., 2000).

5 Occurrence

Since the discovery of CGP until the present day it has generally been accepted, and stated in microbiology textbooks (e.g., Schlegel, 1993), that CGP is a form of storage material that is unique to cyanobacteria. This was confirmed by the absence of any ultrastructural evidence for the occurrence of similar inclusions in other organisms, and by the failure to identify material with the typical characteristics of CGP in other sources (Simon, 1987). Hence, the first demonstration by Krehenbrink et al. (2002) of the ability of noncyanobacteria to synthesize CGP came as something of a shock!

5.1 Occurrence in Cyanobacteria

Among the cyanobacteria, cyanophycin granules occur in all groups, including the unicellular and filamentous, nitrogen-fixing and non-nitrogen-fixing (Lawry and Simon, 1982; Allen, 1984, 1988; Simon, 1987; Golecki and Heinrich, 1991). Even strains of the genus *Synechococcus*, which had been reported to be incapable of accumulating CGP (Lawry and Simon, 1982; Simon, 1987), have been shown recently to synthesize this polymer under laboratory conditions (Hai et al., 1999; Wingard et al., 2002). Whereas CGP is normally barely detectable in the cells during exponential growth (e.g., 0.1%, wt/wt, in *A. cylindrica*; Simon, 1973a, b), CGP can accumulate during the transition to the stationary phase to levels of between 8 and 18% (wt/wt) of the cell dry mass (Figure 1; Simon, 1973a; Lawry and Simon, 1982; Mackerras et al., 1990). As the CGP content depends not only on the growth phase and cultivation regime (see below) but also on the purification protocol and the determination method used, a comparative survey of published CGP contents is both difficult to compile and of minor significance.

In nonheterocyst-forming cyanobacteria, the cyanophycin granules are distributed in the protoplasts (Lang, 1968; Tomaselli Feroci et al., 1976). The inclusions, which are roughly spherical, may vary in size depending both on cultivation period and condition (Dembinska and Allen, 1988); whilst the average diameter of granules from cells of *Synechocystis* sp. strain PCC 6308 during exponential growth was 181 nm, it increased to 446 nm at 48 h after the addition of chloramphenicol; purified granules from untreated cells in the late stationary phase had a diameter as much as 1390 nm. As the number of granules seemed to remain more or less constant during the entire incubation period, the increase in CGP content under those conditions was clearly the result of an enlargement of existing granules rather than of the formation of additional granules (Dembinska and Allen, 1988).

In the filamentous heterocyst-forming cyanobacteria, the granules are more often found in the heterocysts, these being specialized for the fixation of molecular nitrogen. In those cells, the granules are associated primarily with the polar nodule and the neck, which connects the heterocyst to the adjoining vegetative cell (Fogg, 1951; Rippka and Stanier, 1978; Sherman et al., 2000; Ziegler et al., 2001). In akinetes, a spore-like cell occurring in some filamentous cyanobacteria, the CGP content is drastically increased during cell differentiation (see below).

5.2 Occurrence in Chemotrophic Bacteria

Since CGP can be seen as an ideal nitrogen reserve material with a relatively simple structure (see below), the question arose as to why its presence should be restricted to the special group of oxygenic phototrophic bacteria. The availability of both, the sequences from cloned cyanobacterial *cphA* genes encoding the CGP-synthesizing enzyme CGP synthetase (see below) and the sequences of dozens of total and partial microbial genome sequences, enabled Krehenbrink et al. (2002) to screen for the occurrence of the genetic information for CGP synthesis outside of the cyanobacteria. Among 65 genome sequences available at the start of the study (in August 2001), *cphA* homologous genes were found in strains of *Acinetobacter* sp. (encoding a protein similar to CphA from *Synechocystis* sp. strain PCC 6803 with 40% amino acid identity), *Bordetella* sp. (39%), *Clostridium botulinum* (39%), *Desulfitobacterium hafniense* (38%) and *Nitrosomonas europaea* (37%) (Figure 4).

Among this collection of noncyanobacterial *cphA* homologous putative genes, the gene from *Acinetobacter* sp. strain DSM 587 was chosen for detailed investigations. Transformants of *E. coli* harboring a PCR-generated *cphA*-containing genomic fragment expressed CphA activity to up to 1.2 U per mg protein and accumulated CGP up to 7.5% (wt/wt) of the cellular dry matter, indicating that CphA of *Acinetobacter* sp. strain DSM 587 was functionally active. In addition, CGP was accumulated by cells of *Acinetobacter* sp. DSM 587 under suitable conditions (see Figure 1), which clearly showed the physiological relevance of CGP for this chemotrophic bacterium.

Whether the other strains mentioned above also contain active CGP synthetase or CGP remains to be determined. However, the number of microbial noncyanobacterial species harboring *cphA* will surely rise in future in line with the increasing number of genome sequences that will be published.

6 Functions

From the microorganism's point of view, CGP fulfills perfectly the criteria for an intracellular nitrogen reserve (Simon, 1987): five nitrogen atoms are held in each building block of the polymer, which is insoluble at cellular pH and ionic strength. Although polyarginine would contain even more nitrogen per mass, polyarginine is soluble under physiological conditions and would have a severe adverse effect on the cell's internal milieu either by binding to polyanionic DNA or by altering the osmolarity.

In cyanobacteria, CGP functions as a temporary nitrogen reserve as it accumulates during the transition from the exponential to the stationary growth phase, and disappears when balanced growth resumes (Mackerras et al., 1990). In the nitrogen-fixing heterocysts, CGP appears in the late stages of heterocyst maturation (Fogg, 1951), where it is thought to function as a dynamic buffer for newly fixed nitrogen (Carr, 1988; Sherman et al., 2000; Li et al., 2001). However, based on investigations with a mutant of *A. variabilis* that is unable to synthesize CGP, it was concluded that

Fig. 4 Multialignment of cyanophycin synthetase ▶ amino acid sequences from the representative cyanobacterium *Synechocystis* sp. strain PCC 6803 and the following chemotrophic bacteria (Aboulmagd et al., 2000; Krehenbrink et al., 2002): *Bordetella bronchiseptica*, *Nitrosomonas europaea*, *Desulfitobacterium hafniense*, *Clostridium botulinum*, *Acinetobacter* sp. strain ADP1. Amino acid residues conserved in all published cyanobacterial CphA sequences are compiled in the line *Cyanobact. Consensus*. Amino acid residues that are identical in all CphA sequences are shaded. The B-loop of the ATP-grasp fold (see text) is indicated; amino acid residues involved in ATP-binding are underlined. Consensus sequences of the Regions I, II, III, and IV of enzymes belonging to the superfamily of three substrate ligases (see text) are marked.

```
SynechocystisPCC6803       1MKILKTLTLRGPNYWSIRRKKLIVMRLDLEDLAERPSNSIP-GFYEGLIRVLPSLVEH
Cyanobact. Consensus        M ILKL TLRGPNYWSIRR KLI MRLDLE  A  PSN I  GF  GL   LPSL  H
Bordetella bronchisept.     MEVSRIRALRGPNLWSRN--TAIEAIAACSD-AECAINGMP-DFEARLRARFPQLGVL
Nitrosomonas europaea       MKVLQIRALRGPNVWSKL--AAIEATLVFEQ-NECLPDSIP-GFDTRLREYFPDIALL
Desulfitobacterium haf.     MEILKIQAIPGANVYSYR--PVIRAVVDLQEWTERTSDTFG-DFNTRLVQCLPSLYEH
Clostridium botulinum       MKIENIRVFEGRNIYSHK--KCIRMDVDLEGYSNISSKEIQ-EFNKTLLNYVPELREH
Acinetobacter sp.           MNIISTSVYVGPNVYASIP--LIRLVIDLNPHYITQLASMGSEVLENLEKVIPTLKTE

SynechocystisPCC6803     59FCSPGHRGGFLARVREGTYMGHIVEHVALELQELVGMTAGFGRTRETSTP-GIYNVVYEY
Cyanobact. Consensus       CS     GGFL R   EGT   GH VEH ALELQEL   M   GFGRTR T TP    Y VV EY
Bordetella bronchisept.  QPE---------GQRDAISMAHVLQATALGLQAHAGCPVTFGRTSPTIEP-GVFQVVVEY
Nitrosomonas europaea    QP----------VDWQETTLAHILAFITLKLQERAGCSVSFSRVIKMAEA-NTWRVVVEY
Desulfitobacterium haf.  FCSRGKPGGFVERLKEGTLVGHIIEHVTIELLTRAGQNIPYGKTLCLPEHPGHYEIIFNY
Clostridium botulinum    CCCVGKKGGFVERLYEGTYLSHICEHVIIALQNRIGIDVSYGKAREIEGE--KYYIIYQY
Acinetobacter sp.        QDAKLQHKLEELRQAPQQQIGELVAILALHLQRLAGQKGGAAFSAYCHED--ETEILYSY

SynechocystisPCC6803    120VDEQAGRYAGRAAVRLCRSLVDTGDYS-LTELEKDLEDLRDLGANSALGPSTETIVTEAD
Cyanobact. Consensus        E AGRYAGR AVRLC S    G Y        D  DL          LGPST      EA
Bordetella bronchisept.  SEEDVGRLAFELAEALIRSAQDDTPF----DLPQALQRLRDLDEDTRLGPSTGSIVNAAV
Nitrosomonas europaea    SEEAVGRLALEQSLALCRAVAEAAPF----DTSEAVNRLRELYEDIRLGPSTNSIVQAAV
Desulfitobacterium haf.  DSLEGGLEGFKQGYALVQELLAGQKP----NVTNRIERIREVIQRFELGASTRAIIEAAE
Clostridium botulinum    KYKNMAIECGKIAVDLINNIINGKRY----NIKTKTRELVCLLKTEELGPSTLSIIQEAK
Acinetobacter sp.        ESEEIGIEAGEVVCDMLVALAKAHEAGDQIDLNRDVKGFLRYADRFALGPSALALVQAAE

SynechocystisPCC6803    180ARKIPWMLLSARAMVQLGYGVHQQRIQATLSSHSGILGVELACDKEGTKTILQDAGIPVP
Cyanobact. Consensus        R IPW  L  R    QLGYGAR  R QAT        IL VELA DKE   K  L    G PVP
Bordetella bronchisept.  ARSIPYHRLTQGSMVQFGWGSKQRRIQAAETDMTSAISESIAQDKDLTKMLLDAAGVPVP
Nitrosomonas europaea    RRKIPYRRLTDGSLVQFGWGSRQRRILASESDLTSVVAESIVQDKDLTKMLLHTAGIPVP
Desulfitobacterium haf.  GRGIPVIRLNDSSLLQLGYGRNQKRVQAAMSDQTSCIGVDIACDKGLTKKLLYEGGIPVP
Clostridium botulinum    KRNIPVTKIGEDSMFQLGYGIKGKTIEATICNSTSAVSVDIACDKLLSSKNILMDQCIPVA
Acinetobacter sp.        ERNIPWYRLNDASLIQVGQGKYQKRIEAALTSGTSHIAVEIAGDKNVCNQLLQDLGLPVP

SynechocystisPCC6803    240RGTTIQYFDDLEEAINDVGGYPVVIKPLDGNHGRGITINVRHWEEAIAAYDLAAEESKSR
Cyanobact. Consensus        GT I       L  AI    GGYP  V  KPL  GNHGRGITI          A   A     A     S
Bordetella bronchisept.  MGSSVDSAAAAWQAAQALG-GSVVVKPRDGSQGRGVAVNIETRDRIESAFEAAAEIS--S
Nitrosomonas europaea    TGRPVISADDAWAAACEIG-APIVIKPQDGNQGKGVTANLTDRDQIKAAYHVAAERS--R
Desulfitobacterium haf.  DGVVTRNEDEAVEVFRQLD-RLVVVKPYNGNQGKGVTLKLGTEAEVRAAFRVAQTYE--E
Clostridium botulinum    EGYKVKNYIDLLFKAEKLG-YPVVLKPRFGNQGKGVVVNIKNQKELVNAYSIVNKKF--Q
Acinetobacter sp.        KQRVVYDIDDAVRAARRVG-FPVVLKPLDGNHGRGVSVNLTTDEAVEAAFDIAMSEG--S
                                                 ILLKPLDGMGGASI
                                                 ----B-loop----

SynechocystisPCC6803    300SIIVERYYEGSDHRVLVVNGKLVAVAERIPAHVTGDGTSTITELIDKTNQDPNRGDGHAN
Cyanobact. Consensus        IVER     G  D  R  LV            AV  ER  PAHV  G       T    ELI     N    P   RG  GH
Bordetella bronchisept.  EAIVERYIPGHDFRLLVVGDTLVAAARRDPPQVTGDGTHTIAELVAQVNADPLRGDGHAT
Nitrosomonas europaea    NVLVERYISGHDYRLLVVGNKLVAAARRDPPQVVGDGIHSIAQLVKQINSNPLRSEGHAN
Desulfitobacterium haf.  QVVVEEYIEGKNYRLLVVDGKMAAAAERIPAHVIGDGVSTVGELVQLANSDPQRGEDHEK
Clostridium botulinum    DIMIEKYINGKDYRACVVDGKVVAVAQRIPPYIIGNGKSTIYELIKELNRDERRGDGHEK
Acinetobacter sp.        AVIVESMLYGDDHRLLVVNGELVAAARRVPGHIVGDGKHNVEALIEIVNQDPRRGVGHEN

SynechocystisPCC6803    360ILTKIVVNKTAIDVMERQGYNLDSVLPKDEVVYLRATANLSTGGIAIDRTDDIHPENIWL
Cyanobact. Consensus        LT  I                  G                   YLRATANLSTGG  A  DRTD  IHP  N  W
Bordetella bronchisept.  SLTKIRFDDIAIATLRKQGYEADSVAAAGALVVLRNNANLSTGGAATDVTDEVHPELAAR
Nitrosomonas europaea    LLTRIHLDEISLAHLALQGLNAASVPDKGKLVTLRNNANLSTGGTATDVTDEVHPDIAEC
Desulfitobacterium haf.  ALTKIKIDPVVLMTLTQKKIALETVPADGEVVYLRDSANLSTGGISVDVTERVHPDNAAL
Clostridium botulinum    PLTKVKIDKDLKNNINKEGYTLGYILPEEYKLELRHNANLSTGGVAIDCTDLICTETREV
Acinetobacter sp.        MLTKIELDEQALKLLAEKGYDKDSIPAKDEVVYLRRTANISTGGTAIDVTDTIHPENKLM

SynechocystisPCC6803    420MERVAKVIGLDIAGIDVVTSDISKPLRETNGVIVEVNAAPGFRMHVAPSQGLPRNVAAPV
Cyanobact. Consensus               GLDI  G  D  V    DI     PL          GV   VEVNAAPGFRMH     PSQG     RNVA     V
Bordetella bronchisept.  AVTAARMIGLDICGVDVVAETVHLPLEDQHGGVVEVNAAPGLRMHLNPSFGKGRAVGEAI
Nitrosomonas europaea    AVMAARMTGIDICGIDVICSSLSRPLGEQGGAVIEVNAAPGLRMHLQPSYGKPRAVGEAI
Desulfitobacterium haf.  AEYAARIVGLDIAGVDMVLEDIERPHQEQRGAIIEVNAAPGLRMHQYPTVGRPLDVGKII
Clostridium botulinum    CERAAKAIGLNICGIDICCSDISKPLKENEG-IMEVNAAPGIRMHQYPYKGKSRNVAKAI
Acinetobacter sp.        AERAIRAVGLDIGAVDFLTTDITKSYRDIGGGICEVNAGPGLRMHISPSEGPSRDVGGKI
```

```
SynechocystisPCC6803 480LDMLFPSGTPSRIPILAVTGTNGKTTTTRLLAHIYRQTGKTVGYTSTDAIYINEYCVEKG
Cyanobact. Consensus      MLFP       PI    GTNGKTTTTRL AHI  QTGQ VGYTTTD  Y      VEG
Bordetella bronchisept. ISHMFAERDDGRIPVVAVAGTNGKTTTVRLTAHILDCAGHRVGMTNSDGVYVGKQRIDTG
Nitrosomonas europaea   IDHLFAPGENARIPVIAVTGTNGKTTTVRLIANMLENNRLRVGIACTDGVFVNGQCVDTG
Desulfitobacterium haf. VDHVMP-KGNGRIPVISVTGTNGKTTTTRMIGKMLTDRELAVGMTTTDGIYVGGKLLLKG
Clostridium botulinum   VDMMFK-EYDGNIPIISITGTNGKTTTTRLIAHILSFSGKKVGMTTTGGIYINNKCINKG
Acinetobacter sp.       MDMLFPQGSQSRVPIAAITGTNGKTTCSRMLAHILKMAGHVVGQTSTDAVYIDGNVTVKG
                                           GTNGKTT
                                           Region I

SynechocystisPCC6803 540DNTGPQSAAVILRDPTVEVAVLETARGGILRAGLAFDTCDVGVVLNVA-ADHLGLGDIDT
Cyanobact. Consensus    D TGP SA  IL DPTVE AVLE ARGGILR GL F    VG VLNV  DDH G GDI T
Bordetella bronchisept. DCSGPRSARRLLLHPDVDAAVLETARGGVLREGLAFDRCNVAIVTNIGMGDHFGLGYIST
Nitrosomonas europaea   DCSGPQSARNILFHPEVDAAVLETARGGILREGLGFDYCDVAVVTNIGRGDHLGLANINT
Desulfitobacterium haf. DTTGPESAQIVLRHPDVQVAVLETARGGILRAGLAYDYADVAVVTNVA-NDHLGQYGMES
Clostridium botulinum   DTTGYYSAKTVLTNKEAEVAVLELARGGLIRDGLPYDLADVGIITNVT-EDHLGLGGINT
Acinetobacter sp.       DMTGPVSAKMVLRDPSVDIAVLETARGGIVRSGLGYQFCDVGAVLNVS-SDHLGLGGVDT
                                            V E                      N    DH D
                                           -------------------Region II ------------

SynechocystisPCC6803 600IEQMAKVKSVIAEVVDPS-GYAVLNADDPLVAAMADKVKAK-VAYFSMNPDNPVIQNHIR
Cyanobact. Consensus      Q A  A V  E      GYA LNADD  VA M    K     YF M         H
Bordetella bronchisept. VEDLAVVKRVIVQYVQPD-GMAVLNAADPMVAEMASACPGS-ITYFAEDRNHPVMATHRA
Nitrosomonas europaea   AEELAAVKRTIVENVNPKTGVAVLNADDPLVLGMASHCPGN-VTFFSRNHRHPVILEQRV
Desulfitobacterium haf. LEDIAHVKSLIAEVVRPH-SYVVLNADDPLVASFARKTKGK-VIFFSTEKDNLTIRKHLA
Clostridium botulinum   LEDMAYVKALVGEAIKKD-GYVVINADDEASISIINRMKSK-IILFTKNKNNPIISQYLD
Acinetobacter sp.       LDGLAEVKRVIAEVTKDT---VVLNADNAYTLKMAGHSPAKHIMYVTRDAENKLVREHIR
                               K                  DD
                        --------             Region III

SynechocystisPCC6803 660RNGIAAVYES----GYVSILEGSWTLRVEEATLIPMTMGGMAPFMIANALAACLAAFVNG
Cyanobact. Consensus        AAVYE     G  SI  G  T R E A   P T    A FMIANALAA LAAF
Bordetella bronchisept. QGHRVVYRDG----DSLVAAQGG-AEVAFALADIPLTRGGAIAFQVENTMAALAAAWALG
Nitrosomonas europaea   QGKRVIYMED----HHIIVAEAG-TERRISLSQIRLTKNGMISFQIDNAMASIGAGLAIE
Desulfitobacterium haf. VGGIAVFVRR----GNILLCQGDQSHKICGVKDLPVTWNGKALHNLQNALAAIAVGWSLG
Clostridium botulinum   NKNLVLYLDE----DTIYLKKLNKNEEIINVNKIPITLGGKLIYNVENAMAAIAALIALG
Acinetobacter sp.       LGKRAVVLEKGLNGDQIVIYENGTQIPLIWTHLIPATLEGKAIHNVENAMFAAGMAYALG
                                                                 G  N  N  AA   A    G
                                                                 ---------Region IV-

SynechocystisPCC6803 719LDVEVIRQGVRTFTTSAEQTPGRMNLFNLGRYHALVDYAHNPAGYRAVGDFVKNWH-GQ
Cyanobact. Consensus       E IR     F     QTPGRMNLF       L DYAHN   Y A G FV  W   G
Bordetella bronchisept. IDWDTIRHALAIFVNDAQTAPGRFNVFDFRGATLIADYGHNPDAIQALVRAIDTMPA-RR
Nitrosomonas europaea   LDWTTICAGLADFVSDAQTVPGRFNLFNYREATLIADYGHNPDAMEALVCAIDHIPA-KK
Desulfitobacterium haf. LKAEGIRTSLSEFTSDPECNRGRLNPYTIGGVQVFIDYGHNAAGIKAIAQTLRKFKA-PA
Clostridium botulinum   IDVNTIRQGLESFSNE-EQNPGRFNMYNVHGTNVILDYGHNIEGYKVVLESIKKIKH-KR
Acinetobacter sp.       KNLDQIRIGLRTFDNTFFQSPGRMNVFDKHGFRVILDYGHNEAAVGAMTELVDRLNPRGR
                                              R
                        -----------------------

SynechocystisPCC6803 780RFGVVGGPGDRRDSDLIELGQIAAQVFDRIIVKEDDDKRGRSGGETADLIVKGILQENPG
Cyanobact. Consensus    R GV GGPGDRRD D   LG      FD    KEDDD RGR  G     I  G
Bordetella bronchisept. RSVVISGAGDRRDEDLRQQTEILGGAFDDVILYQDQCQRGRADGEVVALLRQGLQQAPRT
Nitrosomonas europaea   RTVVISAAGDRRNEDIRLQTRILGDVFDEVVLFQDKCQRGRADGEVLGLLREGLENAKRV
Desulfitobacterium haf. VVGCVTVPGDRPDETIREVARVAARGFHRLIIREDGDLRGRRPGEIAGMIMEEAIASGMD
Clostridium botulinum   IIGVVGVPGDRTNSSTLKVGNICGENFDYVYIKEDRDKRGRKHGEIADLLKKGILETGFK
Acinetobacter sp.       RLLGVTCPGDRRDEDVVAIAAKVAGHFDEYYCHRDDDLRGRAPDETPKIMRDALIQLGVP

SynechocystisPCC6803 840-AAYEVILDETVALNKALDQVEEKGLVVVFPESVSKAIELIKARKPIG------------
Cyanobact. Consensus       Y  I DE  A    L       LVV  P  V   I  I
Bordetella bronchisept. -RHIDEIQGEFVAIDAALERLAPGDLCLILVDQVEEALAHIASRVTGQP-----------
Nitrosomonas europaea   -RKVSEIRGEFKAIDTALTNLEAGELCLILIDQVEQALGYIHSRIAVA------------
Desulfitobacterium haf. PRRISVVLPEREAFCHGLDTCKPGEIFVMFYEHLEPIEEEIALRLESGPLAKEEEGFLEV
Clostridium botulinum   NSKLNIILDEEEALKKAIEFSNPGDLVIMFFEEFEPAENIVKDKIKKGKITKRETALA--
Acinetobacter sp.       ESRIHIVEQEEDSLAAVLTEAQVDDLVLFFCENITRSWKQIVHFTPEFNIENDHETLELK
```

CGP metabolism is neither directly coupled with, nor essential for, aerobic nitrogen fixation (Ziegler et al., 2001). In akinetes, CGP appears to serve as energy and nitrogen sources for akinete germination. In *A. cylindrica*, the polyamide content of the cell dry mass increased from 0.1% to 7% (wt/wt) in the akinetes during the differentiation and maturation processes, but then fell to below the limit of detection before akinete outgrowth (Miller and Lang, 1968; Simon, 1987).

Under natural conditions, in the typical surrounding aquatic medium, cyanobacteria are often faced with nitrogen limitation or depletion (Howarth and Cole, 1985), and competition with respect to nitrogen characterizes the natural community. When cyanobacterial cells were inoculated into fresh medium containing a limited amount of ammonia, CGP accumulated before the extracellular ammonia was completely depleted (Mackerras et al., 1990). Thus, a declining concentration of extracellular nitrogen seems to be a signal for CGP production, and the rate of synthesis is maximal when the ammonia concentration reaches a low level but is not completely exhausted (Liotenberg et al., 1996). This "forward planning" gives cyanobacteria a competitive advantage over other organisms (Mackerras et al., 1990). The effect can be explained by the mathematical model of Parnas and Cohen (1976), which states that the synthesis of a storage material should occur before the end of a growth period is reached.

The same is probably true for noncyanobacterial CGP-producing bacteria, and this is supported by the observation that CGP accumulation in *Acinetobacter* sp. strain DSM 587 began in the late exponential growth phase (Krehenbrink et al., 2002). In addition to the growth phase, accumulation of CGP in *Acinetobacter* sp. DSM 587 was shown to depend on the phosphate concentration in the medium (see Figure 1). Under phosphate-limited conditions, the CGP content rose about eight-fold, corresponding to a maximum CGP content of approx. 1% (wt/wt) of the cell's dry mass (Krehenbrink et al., 2002).

7 Physiology

In cyanobacteria, the biosynthesis of CGP begins with the ATP-dependent fixation of NH_4^+ into glutamic acid, forming glutamine, the reaction being catalyzed by glutamine synthetase. From glutamine, the amide group can be transferred by glutamate synthase (EC 1.4.1.13) to 2-oxoglutarate, producing two glutamic acid molecules per reaction. Whereas aspartic acid synthesis can be achieved simply by amino group transfer from glutamic acid to the central metabolite oxaloacetate (by the action of glutamate-oxaloacetate aminotransferase), the route from glutamic acid to arginine follows the acetyl-group recycling variant of the arginine biosynthetic pathway via *N*-acetylglutamic acid, *N*-acetylglutamyl phosphate, *N*-acetylglutamate semialdehyde, *N*-acetylornithine, ornithine, citrulline, and argininosuccinate (Cunin et al., 1986). A recent study indicated that cyanobacteria possess isoenzymes of the arginine biosynthetic pathway to allow regulatory differentiation between arginine supply for either protein or CGP synthesis (Leganés et al., 1998). In addition to the fixation of ammonium into arginine and aspartic acid, the carbon flux from CO_2 fixation towards aspartic acid and arginine synthesis is enhanced to meet the demands of CGP accumulation during the transition from exponential to stationary phase (Weathers and Allen, 1978).

Various laboratory conditions have been reported to promote the synthesis and accumulation of CGP in cyanobacteria (Lawry and Simon, 1982): the addition of inhibitors acting on transcription (e.g., rifamycin; Rodriguez-Lopez et al., 1971) or on translation (e.g., chloramphenicol; Simon, 1973b; Pandey and Talpasayi, 1982; Allen and Hawley, 1983), and the deprivation of sulfur (Allen et al., 1980; Ariño et al., 1995) or phosphorus (see Figure 1; Lawry and Simon, 1982; Stephan et al., 2000). Depending on the species, the polymer also accumulates following the addition of a variety of nitrogen-containing compounds to the medium, including ammonium, urea, glycine, aspartic acid, and arginine (Rippka and Stanier, 1978; Allen et al., 1980; Lawry and Simon, 1982; Sarma and Khattar, 1986; Liotenberg et al., 1996). In addition, it was found that cultivation under reduced light and temperature raised the CGP content in most cyanobacteria (van Eykelenburg, 1980). In *Scytonema* sp., both CGP synthesis and glycogen formation are induced in response to salt stress (Page-Sharp et al., 1998). A link between salt tolerance and CGP content is also evident from the study of salt-sensitive mutants of *Synechocystis* sp. PCC 6803, which showed increased CGP accumulation (Zuther et al., 1998).

8 Biochemistry

The biosynthesis of CGP is catalyzed by a single enzyme, referred to as CGP synthetase. Knockout of the CGP synthetase-encoding gene *cphA* in *A. variabilis* resulted in complete loss of the ability to form both CGP and the polar nodules of heterocysts (Ziegler et al., 2001). The enzyme was first characterized by Simon (1976), who enriched it 92-fold from the soluble cell fraction of *A. cylindrica*. *In vitro*, the enzyme-catalyzed polymerization is dependent on the presence of both monomers, ATP, Mg^{2+}, K^{+}, a sulfhydryl reagent and small amounts of CGP as primer. The same dependencies apply to CGP synthetases purified from *A. variabilis* (Ziegler et al., 1998), from the thermophilic *Synechococcus* sp. strain MA19 (Hai et al., 1999), and from *Synechocystis* sp. strain PCC 6308 (Aboulmagd et al., 2001a). CGP synthetases from these sources consist of identical subunits of 90 to 130 kDa, respectively, which most probably assemble two homodimers under physiological conditions (Ziegler et al., 1998; Hai et al., 1999; Aboulmagd et al., 2001a).

8.1 Primary Structures and Motifs Occurring in CGP Synthetases

Sequence analysis of the translational products of several CGP synthetase-encoding genes revealed high degrees of similarity in the primary structure among all CGP synthetases, and also identified similarities to different ATP-dependent ligases (see Figure 4). The following five regions were identified. The first region, which is located in the N-terminal half, matched the B-loop sequences of the ATP-grasp folds of several members of the superfamily of enzymes with ATP-dependent carboxylate-amine/thiol ligase activity (Thoden et al., 1999; i.e. D-alanine-D-alanine ligase, glutathione synthetase, biotin carboxylase, carbamoyl phosphate synthetase and the ribosomal protein S6 modification enzyme RimK). In these enzymes, the B-loop is directly involved in ATP/ADP-binding (Galperin and Koonin, 1997). The other four regions (I to IV in Figure 4) are located in the C-terminal portion and were identified due to similarities to the four corresponding regions of proteins from the superfamily of three substrate ligases (Eveland et al., 1997; Dementin et al., 2001; i.e. the murein ligases MurC, MurD, MurE and MurF and the folyl poly-γ-glutamate ligases). Region I has been shown to occur in MurD as an ADP-binding P-loop (Bertrand et al., 1999). In addition, remarkable sequence overlaps have been located with the *capB*-translational product of *Bacillus anthracis*, which is involved in synthesis of extracellular poly(glutamic acid) (Makino et al., 1989; Ziegler et al., 1998; Aboulmagd et al., 2000; Berg et al., 2000, Ashiuchi and Misono, 2002). Whereas two potential ATP-binding sites have been identified, the locations of the residues involved in amino acid binding have still to be elucidated. No significant similarities were obtained to conserved motifs of the non-ribosomal peptide synthetases, which are described in detail elsewhere in this series (von Döhren, 2002).

8.2 Assay of CGP Synthetase Activity and Substrate Specificity

By improving the technique of activity measurement (Aboulmagd et al., 2000, 2001a), it could be shown that CGP synthetase from *Synechocystis* sp. strain PCC 6308 can only elongate pre-existing CGP primers in the presence of both amino acid substrates L-aspartic acid and L-arginine. No prolonged incorporation of either amino acid occurred in the absence of the other. The K_m values for arginine (49 μM), aspartic acid (0.45 mM), and ATP (0.20 mM) indicated a high affinity of the enzyme towards these substrates. During CGP synthesis *in vitro*, 1.3 mol ATP was converted to ADP per mole of amino acid incorporated. The incorporation of glutamic acid into CGP, which was demonstrated *in vivo* in the same cyanobacterium by Merritt et al. (1994) under nitrogen-limited conditions, was not catalyzed by the CGP synthetase, which did not accept glutamic acid as a substitute for arginine *in vitro* (Aboulmagd et al., 2000). However, a detailed study with different aspartic acid and arginine-analogous compounds (Aboulmagd et al., 2001a) revealed some acceptance of the enzyme towards other substrate analogs (depicted in Figure 3), thus indicating a capacity to synthesize also other CGP-like polyamides.

8.3 Mechanism of Catalysis

Knowledge of the mechanism of catalysis has been gained recently from analyses of the reaction products with synthetic primers by MALDI-TOF mass spectrometry (Berg et al., 2000), and this led to the two following proposals:

1) The primer, which at the least must consist of three CGP building blocks [(β-Asp-Arg)$_3$], is elongated at its C-terminus by stepwise incorporation of the amino acid substrates.
2) The reaction cycle starts at one of two active sites by ATP-dependent phosphorylation at the α-carboxylic acid group of the poly(aspartic acid) backbone, which is followed by amide linkage formation with the α-amino group of the incoming aspartic acid and subsequent transfer to the second active site, where the ATP-dependent phosphorylation of the β-carboxylic acid group is catalyzed before the cycle is finished by amide linkage formation to the α-amino group of the incoming arginine. No evidence was provided for the existence of a thioester intermediate during this catalytic cycle.

The lower average molecular weight and the lower polydispersity of CGP synthesized by *E. coli* expressing the CGP synthetase gene (in comparison with the authentic material from cyanobacteria; see above) might be due to insufficiencies and differences in the physiological background of *E. coli*, including:

- the absence of a further catalytic factor;
- an insufficient supply of substrates; or
- an unfavorable enzyme/substrate ratio (Aboulmagd et al., 2000).

Furthermore, the synthesis of CGP in recombinant strains of *E. coli* provoked a "chicken and egg" conflict. On the one hand, synthesis of CGP requires CGP as primer, but on the other hand no genetic information has to be introduced in addition to the CGP synthetase-encoding gene. By using primer-acting studies with different fractions and compounds from bacterial cells (e.g., from the cell wall), evidence has been obtained that other cellular molecules may serve as unspecific primers during long-term incubation *in vitro* or during cultivation of recombinants *in vivo* (Hai et al., 2002).

9 Molecular Genetics

The structural genes encoding the CGP synthetase have been cloned or sequenced from seven different cyanobacterial species. In all cases the CGP synthetase-encoding gene *cphA* is located in colinear orientation close to and downstream from the gene *cphB*, which encodes the intracellular CGPase (see below), thus forming the *cph* gene cluster (Figure 5). In *Synechocystis* sp. strain PCC 6308, an additional open reading frame (ORF), designated slr2003, is located downstream of the *cph* genes, the function of which remains unclear.

Analysis of the genomic neighborhood of the corresponding *cphA* homologous genes in the chemotrophic bacteria, revealed two additional types of organization (Krehenbrink et al., 2002): Whereas in the Gram-positive anaerobic bacteria *D. hafniense* and *C. botulinum* the same organization was found as in cyanobacteria, the *cphA* homologous genes in the Gram-negative *N. europaea* and *Bordetella* sp. are preceded by a further putative *cphA* homologue, which is referred to as *cphA'*. A unique situation was found in the genome of *Acinetobacter* sp., where *cphA* is followed by the so-called *cphI*. The 80-kDa *cphI* translational product can be seen as a fusion of two cyanobacterial-type *cphB* gene products. Although no catalytic function has yet been assigned to CphI, it can be assumed from the sequence similarities that the gene product is somehow involved in the mobilization of CGP (Krehenbrink et al., 2002).

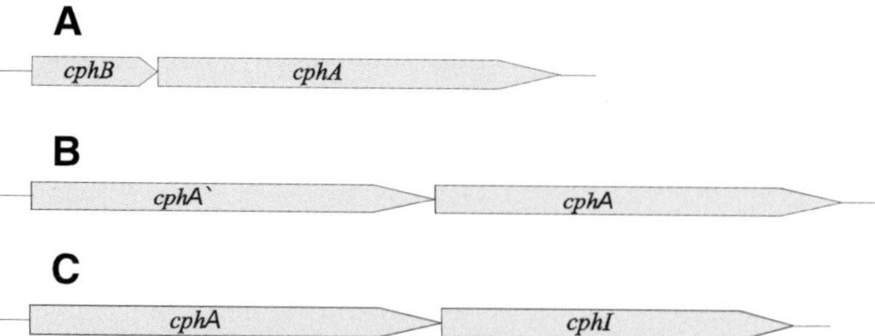

Fig. 5 Organization of *cph* genes found in cyanobacteria and chemotrophic bacteria, respectively (Krehenbrink et al., 2002). (A) Cyanobacteria, *Clostridium botulinum*, and *Desulfitobacterium hafniense*; (B) different *Bordetella* species, and *Nitrosomonas europaea*; (C) *Acinetobacter* sp.

10 Biodegradation

The ability of a CGP-accumulating bacterium to catalyze intracellular degradation of the polymer and to strictly control this mobilization is essential for the function of temporary nitrogen reservation. In addition to this intracellular breakdown, extracellular CGP-degrading enzymes must be present in nature, because CGP is an often-released biopolymer during biomass degradation, both under aerobic and anaerobic conditions. Thus, it is not surprising that a variety of CGP-degrading bacteria could have been isolated, all of which are able to use this polymer as sole extracellular carbon and nitrogen sources.

10.1 Intracellular Mobilization

The intracellular degradation of CGP by cyanobacteria was reported by Simon et al. (1980), Gupta and Carr (1981), and Allen et al. (1984) to occur in crude extracts from cells of cyanobacteria. Gel filtration of the reaction products showed the major product of the enzymatic degradation to be a β asp-arg dipeptide (Gupta and Carr, 1981). The CGPase catalyzing the initial cleavage of CGP into β asp-arg dipeptides was later purified from *Synechocystis* sp. strain PCC 6308 and characterized in detail (Richter et al., 1999). Based on the sequence similarity to the dipeptidase E from *Haemophilus influenzae*, the presence of the conserved active site motif Gly-Xaa-Ser-Xaa-Gly and the sensitivity towards phenylmethanesulfonyl fluoride and other similarly acting inhibitors, the CGPase was classified as a serine-type exopeptidase. The CGPase-encoding gene *cphB* was cloned, sequenced and localized upstream of and colinearly orientated to the CGP synthetase-encoding *cphA* (Figure 5; Richter et al., 1999). Evidence was recently obtained for an additional biosynthetic function of CphB in CGP metabolism from *cphB* knock-out mutants, which resembled Δ*cphA* mutants in their inability to form CGP (Li et al., 2001).

In order for complete mobilization of the amino acids from CGP to occur, the β asp-arg dipeptides must be cleaved. The enzyme which presumably catalyzes this step was recently identified in *Synechocystis* sp. PCC 6803 as "plant-type asparaginase" encoded by ORF sll0422 (Hejazi et al., 2002). This enzyme exhibited relatively high affinity for isoaspartyl dipeptides containing a basic amino acid such as β asp-arg as well as β asp-lys. Arginine and aspartic acid are presumably catabolized on a special arginine catabolic route, combining the arginase pathway and the urea cycle, thereby making the fixed nitrogen available to the cell during CGP mobilization (Quintero et al., 2000).

Based on the gene location and sequence similarities to cyanobacterial CphBs, it seems likely that CphI from the heterotrophic *Acinetobacter* sp. strain DSM 587 (see Figure 5) is also involved in the initial step of intracellular CGP degradation (Krehenbrink et al., 2002), although no catalytic function has yet been experimentally assigned to the *cphI* gene product.

10.2 Extracellular Breakdown

Recently, several bacterial strains were isolated that were able to use CGP as the sole carbon and energy sources for growth (Obst et al., 2002). To screen for these CGP-degrading bacteria, samples from typical habitats of cyanobacteria were plated on solid mineral medium containing CGP as sole carbon source. Due to the insolubility of CGP at neutral pH, the agar was turbid. Colonies of CGP-degrading microorganisms were recognized due to the formation of degradation halos, which appeared 12–18 h after inoculation and incubation at 30°C (Figure 6). Based on this feature, axenic cultures of nine bacterial strains were finally isolated from Baltic Sea water, different pond sediments and sewage sludge.

All isolates were Gram-negative rods, which grew under aerobic conditions. Three isolates were identified as different strains of *Pseudomonas alcaligenes*; strain BI was determined as *P. anguilliseptica*. Cells of strain BI, which was investigated in more detail, produced the extracellular CGPase only when CGP was present in the medium, thus indicating a specific induction of the enzyme by CGP or its degradation products. The extracellular CGPase of this strain was purified to electrophoretic homogeneity from supernatants of CGP-grown cultures by application of anion-exchange chromatography and by affinity chromatography on a L-arginine agarose matrix, the latter showing high selectivity towards the enzyme.

In contrast to the intracellular CGPase from *Synechocystis* sp. strain PCC 6803,

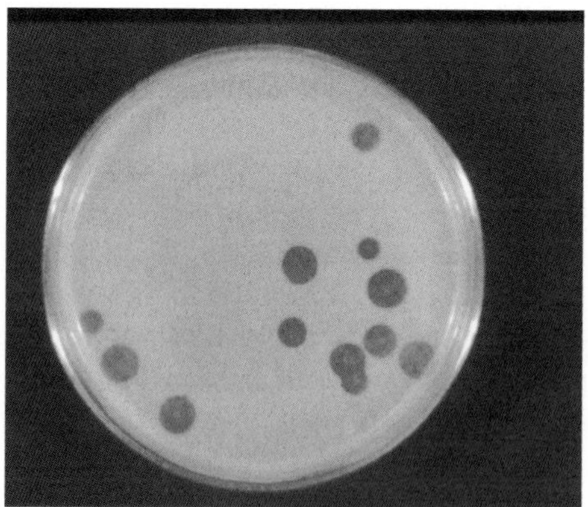

Fig. 6 Halo formation during isolation of CGP-degrading bacteria on CGP-containing mineral medium (Obst et al., 2002). Sample from a sediment of a pond was spread on solid mineral medium overlaid with 0.5% (wt/wt) agar containing 0.2% (wt/wt) of CGP.

which has an apparent molecular subunit mass of 27 kDa (Richter et al., 1999), the molecular mass of CphE from *P. anguilliseptica* BI was considerably higher (43 kDa). Studies on the substrate specificity of the CGPase revealed high specificity for CGP. HPLC and electrospray ionization (ESI) mass spectrometry analysis identified dipeptides as degradation products from cyanobacterial CGP (Figure 7). Based on results from kinetic studies on the degradation of carboxy-terminal L-[U-^{14}C]-arginyl-labeled CGP, it was concluded that the CGPase-catalyzed release of β asp-arg dipeptides proceeds via an exo-mechanism by successive cleavage of the α-amide bonds in the polymer backbone. Identification and cloning of the CGPase-encoding gene *cphE* was achieved by screening an *E. coli* cosmid gene library for heterologous expression of *cphE*. Molecular characterization of *cphE* revealed a DNA sequence that encodes a protein with a similarity of only 27–28% to intracellular CGPases (CphB) from cyanobacteria in the conserved region (see above). The amino acids Ser169, Glu185 and His222 were identified as candidates involved in the catalytic mechanism by matching the pattern of the catalytic triad of serine type proteases, which is in accordance with the sensitivity of CphE towards serine protease inhibitors. Detailed sequence analysis of the deduced amino acid sequence identified an N-terminal leader peptide, which presumably directs the enzyme outside the cell.

In conclusion, two different types of enzymes exist for CGP degradation:

1) Intracellular CGPases (CphB and CphI) of CGP-accumulating bacteria, which catalyze the intracellular mobilization of the N-reserve material.
2) Extracellular CGPases (CphE) of bacteria, which use CGP as extracellular carbon source for growth.

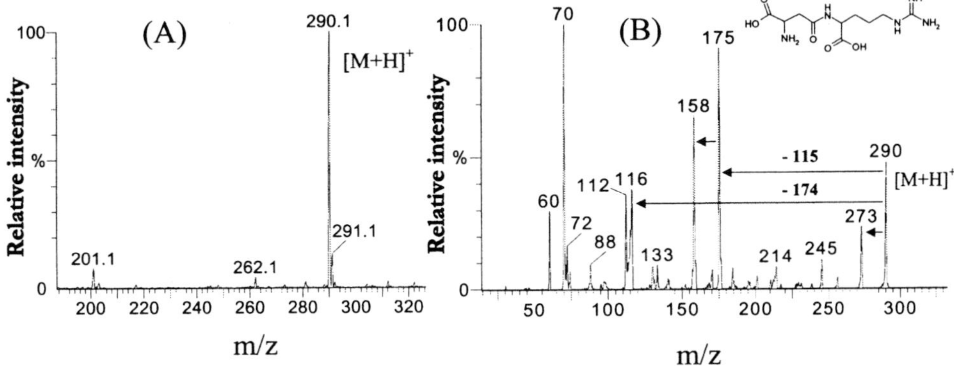

Fig. 7 Identification of the products of the degradation of CGP catalyzed by purified CphE from *Pseudomonas anguilliseptica* strain BI (Obst et al., 2002). (A) Positive ion ESI/MS analysis of the CGP degradation product; (B) ESI/MS/MS analysis of the m/z 290 peak from panel (A). Both the mass peak at m/z 290 in panel (A) and the fragmentation pattern of the latter given in panel (B) is consistent with the structure of the β asp-arg dipeptide depicted in panel (B).

11 Biotechnological Production of CGP

Because of the low cell yield, the low content of CGP, the long cultivation period and the highly sophisticated fermentation processing (Hai et al., 2000), cyanobacteria are unsuitable sources for CGP in terms of cost-effective production.

11.1 Fermentative Production after Heterologous Expression of *cphA* in *E. coli*

CGP synthesis at the laboratory scale has been established recently in several recombinant *E. coli* strains harboring *cphA* from different sources (Ziegler et al., 1998; Oppermann-Sanio et al., 1999; Aboulmagd et al., 2000). Although these result in a maximum content of 26% (wt/wt) of cell dry mass after growth on rich complex media, the reactions are very expensive to conduct (Aboulmagd et al., 2000). In order to identify the physical and material properties of CGP, a major effort has been made to develop an efficient biotechnological process for the production of CGP in kilogram quantities. With this in mind, the CGP synthetase gene (*cphA*) was inserted into a vector, which controls transcription of *cphA* by a thermosensitive lambda repressor (cI857), thus enabling induction of *cphA* by a simple temperature shift in the culture fluid (Frey et al., 2002). *E. coli* DH1 as host exhibited best stability of heterologous expression. With this system, it was possible to avoid the use of expensive agents such as isopropyl-β-D-thiogalactopyranoside to induce product synthesis. The time of temperature shift was shown to be crucial for product formation, with the best yields being obtained when induction was carried out during the early exponential growth phase. Cultivation experiments in different media showed that cyanophycin synthesis in the recombinant *E. coli* cells was dependent on complex nutrient components such as casamino acids. During growth on mineral media, both cell density and CGP content declined; this may be due to draining of the intracellular pool of amino acids (i.e., aspartic acid, arginine and its biosynthetic precursor glutamic acid) as a result of CGP biosynthesis, and lead to a derangement of the host's anabolism. This indicates that these strains must be provided with amino acids or small peptides that can serve as substrates or primers for CGP synthetase and CGP formation. In order to minimize process costs, cultivation in a mineral medium is desirable, and further investigations using metabolically engineered strains of *E. coli* are required to induce CGP production in such an environment.

11.2 Isolation of CGP

Besides cultivating the CGP-producing bacteria, the isolation of the polymer from the cell mass is the most crucial point in CGP production. The protocol for the purification of CGP on an analytical scale as described by Simon and Weathers (1976) starts with cell disruption at neutral pH, using sonification in the presence of Triton X-100. This is followed by collection of the CGP-containing debris by centrifugation, and five-fold washing to remove cellular components and the detergent. Purification of the crude CGP-containing pellet is then carried out in stepwise manner:

1) The pellet is dissolved at pH 1.0.
2) The solution is centrifuged at high speed to remove debris.
3) The CGP is precipitated from the supernatant at neutral pH.
4) The CGP is sedimented by centrifuging at high speed.

Each step is usually repeated once. After washing with water, the product is usually lyophilized. Although this pH-dependent solubilization/precipitation method showed promise, the laborious cell disruption and washing stages proved too time-consuming. Therefore, attempts were undertaken by Frey et al. (2002) to elaborate a simple modification of the method of Simon and Weathers (1976). Cell disruption by sonification could be replaced by stirring whole cells in diluted HCl at pH 1.0 for 6 h at room temperature (higher pH values or shorter incubation times reduced the final yield of cyanophycin). Although the cells appeared almost intact after this treatment (as evaluated by microscopic inspection), the number of intracellular cyanophycin granules was seen to be greatly reduced. This indicated that an exogenous pH of 1 resulted in: (i) dissolution of CGP inside the cells; and (ii) release of the dissolved polymer from the cells. Approximately 69% of the CGP was extracted from the cells by this first step, while a second acid extraction led to ~ 97% CGP extraction. Hence, two acid extractions were deemed sufficient for economic isolation of the polymer. Acid extraction of whole cells rendered mechanical cell disruption unnecessary, and eliminated the need for centrifugation washing of the crude CGP (Simon and Weathers, 1976). Attempts to replace other centrifugation methods by filtering through cotton wool, mull, and paper or glass-fiber filters resulted in insufficient separation.

The molecular size distribution and polydispersity of the CGP prepared using the acid extraction procedure were comparable with those of CGP obtained by the method of Simon and Weathers (1976). Both polymer preparations were also almost free from contaminating proteins. When compared with the established method, the acid extraction procedure yielded less CGP, but contamination by other cellular components (e.g., proteins and nucleic acids) was also less, most likely due to the avoidance of a cell-disruption step. Ultrasonic disruption of the cells releases not only the entire CGP content from the cells, but also a variety of undesired components. In addition, the molar ratio of the three amino acid constituents of both preparations differed with respect to the arginine content, which was slightly lower after acid extraction. This might be due to the long-term incubation required under acidic conditions, as this is known to cause some release of arginyl residues from the polymer (Simon and Weathers, 1976). Traces of long-chain fatty acids were detected by GC/MS analysis in the product after acid extraction (data not shown), whereas no fatty acids were found after the procedure of Simon and Weathers (1976); this was most likely due to the Triton X-100 treatment used in the latter protocol.

The acid extraction procedure abolished a bottleneck in the biotechnological production of CGP. In contrast to the labor-intensive, costly and time- consuming laboratory-scale method described by Simon and Weathers (1976), the acid extraction of whole cells is simple, and both time- and cost-saving. The latter technique offers the possibility of purifying CGP from a bulk cell mass on a large scale; the minor disadvantage of a lower yield was offset by there being fewer cellular impurities in the final product.

11.3 Fermentative Production after Heterologous Expression of *cphA* in other Bacteria

Recently, other widely used industrial bacterial hosts [e.g., *Corynebacterium glutamicum*, *Ralstonia eutropha*, *Pseudomonas* sp. (Byrom, 1992; Eggeling and Sahm, 1999; Füchtenbusch et al., 2000)] have been transformed by using *Synechocystis* sp. strain PCC 6308 *cphA*, to produce CGP at low cost (Aboulmagd et al., 2001b). The heterologous synthesis of CGP in these bacteria proved the functionality of the *cphA* translational product in these hosts. Due to the high CGP productivity of the recombinant strains of *C. glutamicum*, *R. eutropha*, and *P. putida* during cultivation on mineral salts media in comparison with *E. coli*, these recombinant strains might be suitable candidates for the biotechnological production of CGP.

11.4 In-Vitro Biosynthesis of CGP

Currently, attempts are under way to identify suitable conditions for the in-vitro synthesis of CGP on a "semi-technical scale". Indeed, it should be possible *in vitro* to use the ability of CGP synthetases to incorporate analogues of arginine or aspartate into the polyamide (see Figure 3). Materials produced by such in-vitro syntheses may then be further investigated for their suitability in technical applications. The translational product of *cphA* from the thermophilic cyanobacterium *Synechococcus* sp. strain MA19 was shown to maintain its parental thermophilic character and to remain active even after 2-h incubation periods at temperatures up to 50°C. Moreover, in the presence of ectoine, the purified enzyme retained 90% of its activity over the same time period (Hai et al., 2002). These features make this enzyme a promising candidate for the production of CGP and its derivatives, even under very harsh reaction conditions.

12 Outlook and Perspectives

Following the discovery of CGP in chemotrophic bacteria (Krehenbrink et al., 2002), and with the number of available CGP synthetase structural genes continuing to increase, it seems that the commercial life of this long-known polymer is set to begin. CGP has recently become available on a sufficiently large scale to permit investigations to be made into its material properties and applications. Future research will undoubtedly result in the detection of novel CGPs and CGP-like polyamides within a broad spectrum of noncyanobacterial organisms. As a consequence, it is likely that a veritable "treasure chest" will be opened of new polyamide-producing biosystems that will lead to a generation of CGP and CGP-like polymers.

13 Patents

Today, although CGP is beginning to attract attention from both the biotechnology and chemical industries alike, few patents are pending:

- Joentgen, W., Groth, T., Steinbüchel, A., Hai, T., Oppermann, F. B. (1998) Polyaspartic acid homopolymers and copolymers, biotechnical production and use thereof. International patent application WO 98/39090.
- Wohlleben, W., Lockau, W., Ziegler, K., Pistorius, E., Ruppel, H.-G., Stephan, D. P., Broer, I. (1999) Cyanophycinsynthetasegene zur Erzeugung von Cyanophycin oder Cyanophycinderivaten, und ihre Verwendung. German patent application DE 19813692.
- Ziegler, K., Lockau, W., Ebert, J., Piotukh, K., Berg, H., Volkmer-Engert, R. (2002) Method for improved production of cyanophycin and the secondary products thereof. International patent application WO 02/12459.
- Joentgen, W., Steinbüchel, A., Oppermann-Sanio, F. B., Aboulmagd, E. (2002) Method for improved production of cyanophycin and the secondary products thereof. International patent application WO 02/12508.

14 References

Aboulmagd, E., Oppermann-Sanio, F. B., Steinbüchel, A. (2000) Molecular characterization of the cyanophycin synthetase from *Synechocystis* sp. strain PCC 6308, *Arch. Microbiol.* **174**, 297–306.

Aboulmagd, E., Oppermann-Sanio, F. B., Steinbüchel, A. (2001a) Purification of *Synechocystis* sp. strain PCC 6308 cyanophycin synthetase and its characterization with respect to substrate and primer specificity, *Appl. Environ. Microbiol.* **67**, 2176–2182.

Aboulmagd, E., Voss, I., Oppermann-Sanio, F. B., Steinbüchel, A. (2001b) Heterologous expression of cyanophycin synthetase and cyanophycin synthesis in the industrial relevant bacteria *Corynebacterium glutamicum* and *Ralstonia eutropha* and in *Pseudomonas putida*, *Biomacromolecules* **2**, 1338–1342.

Allen, M. M. (1984) Cyanobacterial cell inclusions, *Annu. Rev. Microbiol.* **38**, 1–25.

Allen, M. M. (1988) Inclusions: cyanophycin, *Methods Enzymol.* **167**, 207–213.

Allen, M. M., Hawley, M. A. (1983) Protein degradation and synthesis of cyanophycin granule polypeptide in *Aphanocapsa* sp., *J. Bacteriol.* **154**, 1480–1484.

Allen, M. M., Weathers, P. J. (1980) Structure and composition of cyanophycin granules in the cyanobacterium *Aphanocapsa* 6308, *J. Bacteriol.* **141**, 959–962.

Allen, M. M., Hutchison, F., Weathers, P. J. (1980) Cyanophycin granule polypeptide formation and degradation in the cyanobacterium *Aphanocapsa* 6308, *J. Bacteriol.* **141**, 687–693.

Allen, M. M., Morris, R., Zimmerman, W. (1984) Cyanophycin granule polypeptide protease in a unicellular cyanobacterium, *Arch. Microbiol.* **138**, 119–123.

Ariño, X., Ortega-Calvo, J. J., Hernandez-Marine, M., Saiz-Jimenez, C. (1995) Effect of sulfur starvation on the morphology and ultrastructure of the cyanobacterium *Gloeothece* sp. PCC 6909, *Arch. Microbiol.* **163**, 447–453.

Ashiuchi, M., Misono, H. (2002) Poly-γ-glutamic acid, in: *Biopolymers* (Steinbüchel, A., Ed.), Weinheim: Wiley-VCH, pp. 123–173, Vol. 7.

Berg, H., Ziegler, K., Piotukh, K., Baier, K., Lockau, W., Volkmer-Engert, R. (2000) Biosynthesis of the cyanobacterial reserve polymer multi-L-arginyl-poly-L-aspartic acid (cyanophycin). Mechanism of the cyanophycin synthetase reaction studied with synthetic primers, *Eur. J. Biochem.* **267**, 5561–5570.

Bertrand, J. A., Auger, G., Martin, L., Fanchon, E., Blanot, D., Le Beller, D., van Heijenoort, J., Dideberg, O. (1999) Determination of the MurD mechanism through crystallographic analysis of enzyme complexes, *J. Mol. Biol.* **289**, 579–590.

Borzi, A. (1887) Le comunicazioni intracellulari delle Nostochinee, *Malpighia* **1**, 74–203.

Byrom, D. (1992) Production of poly-β-hydroxybutyrate–poly-β-hydroxyvalerate copolymers, *FEMS Microbiol. Rev.* **103**, 247–250.

Carr, N. G. (1988) Nitrogen reserves and dynamic reservoirs in cyanobacteria, in: *Biochemistry of the Algae and Cyanobacteria* (Rogers, L. J., Gallon, J. R., Eds.), Oxford: Clarendon Press, 13–21.

Cunin, R., Glansdorff, N., Piérard, A., Stalon, V. (1986) Biosynthesis and metabolism of arginine in bacteria, *Microbiol. Rev.* **50**, 314–352.

Dembinska, M. E., Allen, M. M. (1988) Cyanophycin granule size variation in *Aphanocapsa*, *J. Gen. Microbiol.* **143**, 295–298.

Dementin, S., Bouhss, A., Auger, G., Parquet, C., Mengin-Lecreulx, D., Dideberg, O., van Heijenoort, J., Blanot, D. (2001) Evidence of a functional requirement for a carbamoylated lysine residue in MurD, MurE and MurF synthetases as established by chemical rescue experiments, *Eur. J. Biochem.* **268**, 5800–5807.

Eggeling, L., Sahm, H. (1999) L-Glutamate and L-lysine: traditional products with impetuous developments, *Appl. Microbiol. Biotechnol.* **52**, 146–153.

Erickson, N. A., Kolodny, N. H., Allen, M. M. (2001) A rapid and sensitive method for the analysis of cyanophycin, *Biochim. Biophys. Acta* **1526**, 5–9.

Eveland, S. S., Pompliano, D. L., Anderson, M. S. (1997) Conditionally lethal *Escherichia coli* murein mutants contain point defects that map to regions conserved among murein and folyl poly-γ-glutamate ligases: identification of a ligase superfamily, *Biochemistry* **36**, 6223–6229.

Fogg, G. E. (1951) Growth and heterocyst production in *Anabaena cylindrica* Lemm. III. The cytology of heterocysts, *Ann. Bot.* **15**, 23–35.

Frey, K. M., Oppermann-Sanio, F. B., Steinbüchel, A. (2002) Technical scale production of cyanophycin with recombinant strains of *Escherichia coli, Appl. Environ. Microbiol.* **68**, 3377–3384.

Füchtenbusch, B., Wullbrandt, D., Steinbüchel, A. (2000) Production of polyhydroxyalkanoic acids by *Ralstonia eutropha* and *Pseudomonas oleovorans* from an oil remaining from biotechnological rhamnose production, *Appl. Microbiol. Biotechnol.* **53**, 167–172.

Galperin, M. Y., Koonin, E. V. (1997) A diverse superfamily of enzymes with ATP-dependent carboxylate-amine/thiol ligase activity, *Protein Sci.* **6**, 2639–2643.

Geitler, L. (1932) Cyanophyceae von Europa unter Berücksichtigung der anderen Kontinente, in: *Dr. L. Rabenhorst's Kryptogamen-Flora von Deutschland, Östereich und der Schweiz* (Kolkwitz, R., Ed.), Leipzig: Akademische Verlagsgesellschaft, Vol. 14.

Golecki, J. R., Heinrich, U. R. (1991) Ultrastructural and electron spectroscopic analyses of cyanobacteria and bacteria, *J. Microsc.* **162**, 147–154.

Gupta M., Carr, N. G. (1981) Enzyme activities related to cyanophycin metabolism in heterocysts and vegetative cells of *Anabaena* sp., *J. Gen. Microbiol.* **125**, 17–23.

Hai, T., Oppermann-Sanio, F. B., Steinbüchel, A. (1999) Purification and characterization of cyanophycin and cyanophycin synthetase from the thermophilic *Synechococcus* sp. MA19, *FEMS Microbiol. Lett.* **181**, 229–236.

Hai, T., Ahlers, H., Gorenflo, V., Steinbüchel, A. (2000) Axenic cultivation of anoxygenic phototrophic bacteria, cyanobacteria, and microalgae in a new closed tubular glass photobioreactor, *Appl. Microbiol. Biotechnol.* **53**, 383–389.

Hai, T., Oppermann-Sanio, F. B., Steinbüchel, A. (2002) Molecular characterization of a thermostable cyanophycin synthetase from the thermophilic cyanobacterium *Synechococcus* sp. strain MA19 and *in vitro* synthesis of cyanophycin and related polyamides, *Appl. Environ. Microbiol.* **68**, 93–101.

Hegler, R. (1901) Untersuchungen über die Organisation der Phycochromaceenzelle, *Jahrbücher für wissenschaftliche Botanik* **36**, 229–354.

Hejazi, M., Piothukh, K., Mattow, J., Deutzmann, R., Volkmer-Engerts, R., Lockau, W. (2002) Isoaspartyl dipeptidase activity of plant-type asparaginases, *Biochem. J.* **364**, 129–136.

Howarth, R. W., Cole, J. J. (1985) Molybdenum availability, nitrogen limitation, and phytoplankton growth in natural waters, *Science* **229**, 653–655.

Joentgen, W., Groth, T., Steinbüchel, A., Hai, T., Oppermann, F. B. (1998) Polyaspartic acid homopolymers and copolymers, biotechnical production and use thereof, International patent application WO 98/39090.

Krehenbrink, M., Oppermann-Sanio, F.-B., Steinbüchel, A. (2002) Evaluation of non-cyanobacterial genome sequences for occurrence of genes encoding proteins homologous to cyanophycin synthetase and cloning of an active cyanophycin synthetase from *Acinetobacter* sp. strain DSM 587, *Arch. Microbiol.* **177**, 371–380.

Lang, N. J. (1968) The fine structure of blue-green algae, *Annu. Rev. Microbiol.* **22**, 15–46.

Lang, N. J., Simon, R. D., Wolk, C. P. (1972) Correspondence of cyanophycin granules with structured granules in *Anabaena cylindrica, Arch. Mikrobiol.* **83**, 313–320.

Lawry, N. H., Simon, R. D. (1982) The normal and induced occurrence of cyanophycin inclusion bodies in several blue-green algae. *J. Phycol.* **18**, 391–399.

Leganés, F., Fernández-Piñas, F., Wolk, C. P. (1998) A transposition-induced mutant of *Nostoc ellipsosporum* implicates an arginine-biosynthetic gene in the formation of cyanophycin granules and of functional heterocysts and akinetes, *Microbiology* **144**, 1799–1805.

Li, H., Sherman, D. M., Bao, S., Sherman, L. A. (2001) Pattern of cyanophycin accumulation in nitrogen-fixing and non-nitrogen-fixing cyanobacteria, *Arch. Microbiol.* **176**, 9–18.

Liotenberg, S., Campbell, D., Rippka, R., Hourmard, J., Tandeau de Marsac, N. (1996) Effect of the nitrogen source on phycobiliprotein synthesis and cell reserves in a chromatically adapting filamentous cyanobacterium, *Microbiology* **142**, 611–622.

Mackerras, A. H., de Chazal, N. M., Smith, G. D. (1990) Transient accumulation of cyanophycin in *Anabaena cylindrica* and *Synechocystis* 6308, *J. Gen. Microbiol.* **136**, 2057–2065.

Makino, S. I., Uchida, I., Terakado, N., Sasakawa, C., Yoshikawa, M. (1989) Molecular characterization and protein analysis of the *cap* region, which is essential for encapsulation in *Bacillus anthracis*, *J. Bacteriol.* **171**, 722–730.

Merritt, M. V., Sid, S. S., Mesh, L., Allen, M. M. (1994) Variations in the amino acid composition of cyanophycin in the cyanobacterium *Synechocystis* sp. PCC 6308 as a function of growth conditions, *Arch. Microbiol.* **162**, 158–166.

Messineo, L. (1966) Modification of the Sakaguchi reaction: spectrophotometric determination of arginine in proteins without previous hydrolysis, *Arch. Biochem. Biophys.* **117**, 534–540.

Miller, M. M., Lang, N. J. (1968) The fine structure of akinete formation and germination in *Cylindrospermum*, *Arch. Mikrobiol.* **60**, 303–313.

Obst, M., Oppermann-Sanio, F. B., Luftmann, H., Steinbüchel, A. (2002) Isolation of cyanophycin-degrading bacteria–cloning and characterization of an extracellular cyanophycinase gene (*cphE*) from *Pseudomonas anguilliseptica* strain BI, *J. Biol. Chem.* **277**, 25096–25105.

Oppermann-Sanio, F. B., Hai, T., Aboulmagd, E., Hezayen, F. F., Jossek, S., Steinbüchel, A. (1999) Biochemistry of microbial polyamide metabolism, in: *Biochemical Principles and Mechanisms of Biosynthesis and Biodegradation of Polymers* (Steinbüchel, A., Ed.), Weinheim: Wiley-VCH, 185–193.

Oppermann-Sanio, F. B., Steinbüchel, A. (2002) Occurrence, functions and biosynthesis of polyamides in microorganisms and biotechnological production, *Naturwissenschaften* **89**, 11–22.

Page-Sharp, M., Behm, C. A., Smith, G. D. (1998) Cyanophycin and glycogen synthesis in a cyanobacterial *Scytonema* species in response to salt stress, *FEMS Microbiol. Lett.* **160**, 11–15.

Pandey, R. K., Talpasayi, E. R. S. (1982) Spore differentiation in relation to certain antibiotics in the blue-green alga *Nodularia spumigena* Mertens, *Z. Allg. Mikrobiol.* **22**, 191–196.

Parnas, H., Cohen, D. (1976) The optimal strategy for the metabolism of reserve materials in microorganisms, *J. Theor. Biol.* **56**, 19–55.

Quintero, M. J., Muro-Pastor, A. M., Herrero, A., Flores, E. (2000) Arginine catabolism in the cyanobacterium *Synechocystis* sp. strain PCC 6803 involves the urea cycle and arginase pathway, *J. Bacteriol.* **182**, 1008–1015.

Richter, R., Hejazi, M., Kraft, R., Ziegler, K., Lockau, W. (1999) Cyanophycinase, a peptidase degrading the cyanobacterial reserve material multi-L-arginyl-poly-L-aspartic acid (cyanophycin). Molecular cloning of the gene of *Synechocystis* sp. PCC 6803, expression in *Escherichia coli*, and biochemical characterization of the purified enzyme, *Eur. J. Biochem.* **263**, 163–169.

Rippka, R., Stanier, R. Y. (1978) The effects of anaerobiosis on nitrogenase synthesis and heterocyst development by nostocean cyanobacteria, *J. Gen. Microbiol.* **105**, 83–94.

Rodriguez-Lopez, M., Muñoz Calvo, M. L., Gomez-Acebo, J. (1971) The effect of rifamycins in the ultrastructure of *Anacystis montana*, *J. Ultrastruct. Res.* **36**, 595–602.

Sarma, T. A., Khattar, J. I. S. (1986) Accumulation of cyanophycin and glycogen during sporulation in the blue-green alga *Anabaena torulosa*, *Biochem. Physiol. Pflanz.* **181**, 155–164.

Schlegel, H. G. (1993) *General Microbiology*, Cambridge: Cambridge University Press.

Schwamborn, M. (1998) Chemical synthesis of polyaspartates: a biodegradable alternative to currently used polycarboxylate homo- and copolymers, *Polym. Degrad. Stab.* **59**, 39–45.

Sherman, D. M., Tucker, D., Sherman, L. A. (2000) Heterocyst development and localization of cyanophycin in N_2-fixing cultures of *Anabaena* sp. PCC 7120 (Cyanobacteria), *J. Phycol.* **36**, 932–941.

Simon, R. D. (1971) Cyanophycin granules from the blue-green alga *Anabaena cylindrica*: a reserve material consisting of copolymers of aspartic acid and arginine, *Proc. Natl. Acad. Sci. USA* **68**, 265–267.

Simon, R. D. (1973a) Measurement of the cyanophycin granule polypeptide contained in the blue-green alga *Anabaena cylindrica*, *J. Bacteriol.* **114**, 1213–1216.

Simon, R. D. (1973b) The effect of chloramphenicol on the production of cyanophycin granule polypeptide in the blue-green alga *Anabaena cylindrica*, *Arch Mikrobiol.* **92**, 115–122.

Simon, R. D. (1976) The biosynthesis of multi-L-arginyl-poly(L-aspartic acid) in the filamentous cyanobacterium *Anabaena cylindrica*, *Biochim. Biophys. Acta* **422**, 407–418.

Simon, R. D. (1987) Inclusion bodies in the cyanobacteria: cyanophycin, polyphosphate, polyhedral bodies, in: *The Cyanobacteria* (Fay, P., van Baalen, C., Eds.), Amsterdam: Elsevier, 199–225.

Simon, R. D., Weathers, P. J. (1976) Determination of the structure of the novel polypeptide containing aspartic acid and arginine which is found in cyanobacteria, *Biochim. Biophys. Acta* **420**, 165–176.

Simon, R. D., Lawry, N. H., McLendon, G. L. (1980) Structural characterization of the cyanophycin granule polypeptide of *Anabaena cylindrica* by circular dichroism and Raman spectroscopy, *Biochim. Biophys. Acta* **626**, 277–281.

Stephan, D. P., Ruppel, H. G., Pistorius, E. K. (2000) Interrelation between cyanophycin synthesis, L-arginine catabolism and photosynthesis in the cyanobacterium *Synechocystis* sp. strain PCC 6803, *Z. Naturforsch.* **55c**, 927–942.

Suarez, C., Kohler, S. J., Allen, M. M., Kolodny, N. H. (1999) NMR study of the metabolic ^{15}N isotopic enrichment of cyanophycin synthesized by the cyanobacterium *Synechocystis* sp. strain PCC 6308. *Biochim. Biophys. Acta* **1426**, 429–438.

Thoden, J. B., Kappock, T. J., Stubbe, J., Holden, H. M. (1999) Three-dimensional structure of N^5-carboxyaminoimidazole ribonucleotide synthetase: a member of the ATP grasp protein superfamily, *Biochemistry* **38**, 15480–15492.

Tomaselli Feroci, L., Margheri, M. C., Pelosi, E. (1976) Ultrastructure of *Spirulina* in comparison with *Oscillatoria*, *Zentralbl. Bakteriol. Abt. II* **131**, 592–601.

Van Eykelenburg, C. (1980) Ecophysiological studies on *Spirulina platensis*. Effect of temperature, light intensity and nitrate concentration on growth and ultrastructure, *Antonie Van Leeuwenhoek* **46**, 113–127.

Von Döhren, H. (2002) Non-ribosomal biosynthesis of linear and cyclic oligopeptides, in: *Biopolymers* (Steinbüchel, A., Ed.), Weinheim: Wiley-VCH, pp. 51–81, Vol. 7.

Weathers, P. J., Allen, M. M. (1978) Variations in short term products of inorganic carbon fixation in exponential and stationary phase cultures of *Aphanocapsa* 6308, *Arch. Microbiol.* **116**, 231–234.

Wingard, L. L., Miller, S. R., Sellker, J. M. L., Stenn, E., Allen, M. M., Wood, A. M. (2002) Cyanophycin production in a phycoerythrin-containing marine *Synechococcus* strain of unusual phylogenetic affinity, *Appl. Environ. Microbiol.* **68**, 1772–1777.

Yoshida, T., Hiraki, J., Nagasawa, T. (2002) ε-Poly-L-lysine, in: *Biopolymers* (Steinbüchel, A., Ed.), Weinheim: Wiley-VCH, pp. 107–121, Vol. 7.

Ziegler, K., Diener, A., Herpin, C., Richter, R., Deutzmann, R., Lockau, W. (1998) Molecular characterization of cyanophycin synthetase, the enzyme catalyzing the biosynthesis of the cyanobacterial reserve material multi-L-arginyl-poly-L-aspartate (cyanophycin), *Eur. J. Biochem.* **254**, 154–159.

Ziegler, K., Stephan, D. P., Pistorius, E. K., Ruppel, H. G., Lockau, W. (2001) A mutant of the cyanobacterium *Anabaena variabilis* ATCC 29413 lacking cyanophycin synthetase: growth properties and ultrastructural aspects, *FEMS Microbiol. Lett.* **196**, 13–18.

Zuther, E., Schubert, H., Hagemann, M. (1998) Mutation of a gene encoding a putative glycoprotease leads to reduced salt tolerance, altered pigmentation, and cyanophycin accumulation in the cyanobacterium *Synechocystis* sp. strain PCC 6803, *J. Bacteriol.* **180**, 1715–1722.

5
ε-Poly-L-Lysine

Dr. Toyokazu Yoshida[1], Dr. Jun Hiraki[2], Dr. Toru Nagasawa[3]

[1] Department of Biomolecular Science, Faculty of Engineering, Gifu University, Yanagido 1-1, Gifu 501-1193, Japan; Tel.: +81-58-293-2648; Fax: +81-58-230-1893; E-mail: toyosida@biomol.gifu-u.ac.jp

[2] Yokohama Research Center, Chisso Corporation, Okawa 5-1, Kanazawa-ku, Yokohama 236-8605, Japan; Tel.: +81-45-786-5504; Fax: +81-45-784-6247; E-mail: junhiraki@chisso.co.jp

[3] Department of Biomolecular Science, Faculty of Engineering, Gifu University, Yanagido 1-1, Gifu 501-1193, Japan; Tel.: +81-58-293-2647; Fax: +81-58-293-2647; E-mail: tonagasa@biomol.gifu-u.ac.jp

AEC	S-(2-aminoethyl)-L-cysteine
DP	degree of polymerization
ε-PL	ε-poly-L-lysine
MIC	minimum inhibitory concentration
NRPSase	nonribosomal peptide synthetase
ORF	open reading frame

1
Introduction

ε-Poly-L-lysine (ε-PL) is a homo-poly amino acid characterized by the peptide linkage between the carboxyl and ε-amino groups of L-lysine (Shima and Sakai, 1977). Some *Streptomyces* strains are able to produce ε-PL, and the representative strain *S. albulus* has been mainly used for a variety of studies (Shima and Sakai, 1981a, b). ε-PL shows a wide antimicrobial spectrum, and the minimum inhibitory concentration (MIC) for the growth of many bacteria is indicated as below 100 μg mL^{-1} (Shima et al., 1984; Hiraki, 2000). The mechanism of the inhibitory effect of ε-PL on microbial growth is the electrostatic adsorption of ε-PL to the cell surface of microorganisms on the basis of its polycationic properties (Shima et al., 1984). Although the physiological function of ε-PL has not been determined, ε-PL is applied practically as a food additive due to its strong antimicrobial activity. Indeed, when used together with other food additives, ε-PL has a significant synergistic effect on the antimicrobial activity of such compounds (Hiraki, 2000).

Most reports on the application of ε-PL have been published in Japanese, and several important papers dealing with the biochemical and microbiological studies on ε-PL are not available in the English language. This chapter, therefore, includes the results described in these reports, along with a wide spectrum of the latest data on ε-PL presented at Japanese meetings.

2
Historical Outline

During the 1970s, while screening for Dragendorff-positive substances of microbial origin, Shima and Sakai (1977) found ε-PL to be one such material in the culture filtrate of an actinomycete isolated from soil. The actinomycete was identified as *Streptomyces albulus* (Shima and Sakai, 1981a), and the Dragendorff-positive substance was verified as a homo-poly amino acid consisting only of L-lysine (Shima and Sakai, 1981b). These authors also showed that ε-PL exhibited both antimicrobial (Shima et al., 1984) and antiphage activities (Shima et al., 1982), and consequently the culture conditions for ε-PL production by *S. albulus* were investigated (Shima et al., 1983). The notable biological activity of ε-PL against various microorganisms attracted a great deal of attention for its possible use as a food preservative. The safety of ε-PL as a food additive has been demonstrated experimentally using rats (Neda et al., 1999), and in Japan ε-PL – when produced industrially by fermentation using a certain strain of *S. albulus* (Hiraki, 2000) – has already entered the commercial market. Several practical difficulties have been encountered when ε-PL is used as a food additive, though the biosynthetic mechanisms of the material have not yet been clarified.

In order to improve the productivity of ε-PL, several mutants have been derived from the wild strain of *S. albulus* by using a classical mutation technique (Hiraki et al., 1998). As the strain used for the industrial production of ε-PL also shows enzymatic ε-PL-degrading activity, the enzyme responsible has also been extensively studied. Recently, an enzyme catalyzing ε-PL degradation has been isolated and characterized from *S. albulus* (Kito et al., 2002b). In addition, some proteases from bacteria and fungi have been found to hydrolyze ε-PL efficiently (Kito et al., 2000a). Until now, the enzymes and genes involved in the biosynthesis of ε-PL have not been identified, and several research groups are currently continuing their investigations in this area.

3 Chemical Structure and Stability

ε-PL is a homo-poly amino acid characterized by the peptide bond between the carboxyl and ε-amino groups of L-lysine (Figure 1). The ε-PL produced by *S. albulus* contains between 25 and 35 L-lysine residues (Shima and Sakai, 1983), and the excretion of various lengths of ε-PL ($n = 8-36$) has recently been reported in several *Streptomyces* strains (Saimura et al., 2002). The ε-PL of various lengths are characterized only by their primary structure, and no definitive secondary and tertiary structures have been proposed. The water-solubility of ε-PL in water is high, and its isoelectric point with 25–35 residues is about pH 9. When an ε-PL solution was boiled at 100°C for 30 min or autoclaved at 120°C for 20 min, no degradation was observed and the polymer length was maintained. Neither was any structural change caused by heat treatment at pH 3 (Hiraki, 2000).

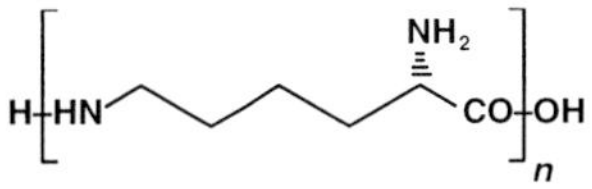

Fig. 1 Structure of ε-PL.

In this chapter, unless otherwise specified, ε-PL indicates a mixture of polymers with 25–35 L-lysine residues, as the studies on ε-PL have been mainly carried out using *S. albulus*. By contrast, the term "poly-L-lysine" has frequently appeared in various reports; this term – without the "ε-"–generally indicates a homo-poly L-lysine linked via the α-amino group, and this shows entirely different characteristics from ε-PL.

4 Chemical Analysis and Detection

The best-known simple and rapid method to determine ε-PL concentration is a colorimetric procedure that uses the anionic dye, methyl orange (Itzhaki, 1972). In this reaction, ε-PL interacts with methyl orange on the basis of its cationic property and forms a water-insoluble complex. ε-PL concentration can be estimated from the absorbance at 465 nm of the methyl orange remaining in solution. Free L-lysine and short lengths of ε-PL do not associate with methyl orange, and so the colorimetric method provides a rough estimate of the total amount of long lengths of ε-PL.

The degree of polymerization (DP) of ε-PL can be assessed using high-performance liquid chromatography (HPLC) (Kito et al., 2002a). When using a reverse-phase column for HPLC analysis, various lengths of ε-PL are individually detected at 215 nm by gradient elution of acetonitrile in phosphate buffer (pH 2.6) containing $NaClO_4$ and sodium octane sulfonate. Although the HPLC analysis is time-consuming, the estimation of ε-PL size is essential in connection with various studies on its physiological and biological properties. The HPLC analysis can also be applied to monitor the concentration of ε-PL, by use of a previously created standard curve. The determination of ε-PL of definitive molecular size is difficult, because authentic polymers are not available commercially as standards. Hence, in reports on ε-PL the concentration is always expressed as mg mL^{-1} or ppm, but not as mmoles mL^{-1}.

5 Occurrence

ε-PL is produced by several strains of *Streptomyces*, typically *S. albulus* (Shima and Sakai, 1981a). When the strains are cultivated in a nutrient medium, ε-PL is produced and excreted in the culture medium. Since the first identification of ε-PL by Shima and Sakai (1977) in the culture filtrate of *S. albulus* No. 236, no other naturally occurring source of ε-PL has been reported.

6 Functions

The physiological function of ε-PL in the ε-PL-producing strains remains unknown. However, ε-PL shows a wide antimicrobial spectrum (Shima et al., 1984) and is applied in practical circumstances as a food additive on the basis of its strong antimicrobial activity (Hiraki, 2000).

The inhibitory effects of ε-PL on microbial growth have been evaluated (Table 1). ε-PL inhibits the growth of both Gram-positive and -negative bacteria, with MICs for bacteria generally below 100 μg mL^{-1} (Hiraki, 2000). Although MICs for fungi are high, the inhibitory effect of ε-PL is superior to that of other antimicrobial compounds used as food preservatives. Shima et al. (1984) investigated the relationship between the molecular size of ε-PL and the antimicrobial activity for *E. coli* K-12. ε-PLs with more than nine L-lysine residues were shown to severely inhibit microbial growth, but the MIC of the ε-octamer of L-lysine was >100 μg mL^{-1}. When the amino groups of e-PL were partially amidated using aromatic carboxylic acids such as 4-chlorobenzoic acid, the antimicrobial activity of e-PL was remarkably diminished (Shima et al., 1984). The proposed mechanism of the inhibitory effect of ε-PL on microbial growth is electrostatic adsorption of ε-PL to the cell surface of microorganisms. This is based on the molecule's cationic properties, and leads to stripping of the outer membrane and abnormal distribution of cytoplasm as observed by electron microscopy (Shima et al., 1984).

Inhibition of spore germination by ε-PL was investigated in *Bacillus* strains; the inhibitory concentration against spores of *B. stearothermophilus* was 2.5 μg mL^{-1}, compared with 12.5 μg mL^{-1} for both *B. coagulans* and *B. subtilis* (Hiraki, 1995).

Likewise, a concentration of 500 μg mL^{-1} ε-PL led to a significant inactivation of morphological type bacteriophage that were long tailed, noncontractile and contained double-stranded DNA, with survival rates of

Tab. 1 Minimum inhibitory concentration (MIC) for growth of microorganisms

Microbial strain		*MIC [µg mL⁻¹]*	*Medium no.*
Fungi	*Aspergillus niger* IFO4416	250	1
	Trichophyton mentagrophytes IFO7522	60	1
Yeasts	*Candida acutus* IFO1912	6	2
	Phaffia rhodozyma IFO10129	12	2
	Pichia anomala IFO0146	150	2
	Pichia membranaefaciens IFO0577	<3	2
	Rhodotorula lactase IFO1423	25	1
	Sporobolomyces roseus IFO1037	<3	2
	Saccharomyces cerevisiae	50	2
	Zygosaccharomyces rouxii IFO1130	150	1
Gram-positive bacteria	*Bacillus stearothermophilus* IFO12550	5	6
	Bacillus coagulans IFO12583	10	6
	Bacillus subtilis IAM1069	<3	7
	Bacillus cereus IFO3514	30	7
	Clostridium acetobutylicum IFO13948	32	8
	Leuconostoc mesenteroides IFO3832	50	3
	Lactobacillus brevis IFO3960	10	3
	Lactobacillus plantarum IFO12519	5	3
	Micrococcus luteus IFO12708	16	6
	Staphylococcus aureus IFO13276	12	4
	Streptococcus lactis IFO12546	100	3
Gram-negative bacteria	*Aerobacter aerogenes* IFO3317	8	4
	Campylobacter jejuni	100	9
	Escherichia coli IFO13500	50	4,5
	Pseudomonas aeruginosa IFO3923	3	4
	Salmonella typhimurium	16	4

Medium number is as follows: 1, potato dextrose broth (pH 5.6); malt broth (pH 5.0); 3, glucose malt broth (pH 6.0); nutrient broth (pH 7.0); deoxycholate broth (pH 7.4); 6, brain heart infusion broth (pH 7.0); trypto-soya broth (pH 7.3); 8, thioglycollate medium without indicator (pH 7.1); 9, brucella broth.

between 0 and 27% (Shima et al., 1982); phage of other types of morphology were damaged to a lesser extent, however. The same antiphage spectrum was observed with α-poly-L-lysine, though e-PL was less active than α-poly-L-lysine for all types of bacteriophage.

The above antimicrobial and antiphage activities were observed at a low ε-PL concentration, and at a pH below 9. ε-PL is considered as a type of polycationic polymer, with an isoelectric point at 9.0; therefore, under alkaline conditions its activity would be significantly reduced. For example, the MIC for *E. coli* over a pH range of 5–8 was 25–50 µg mL⁻¹, while at pH 8.0 the MIC was >200 µg mL⁻¹ (Hiraki, 2000). Similarly, the coexistence of anionic polymers such as metaphosphoric acid reduces the antimicrobial activity of ε-PL due to the loss of its anionic charge.

7 Production

Currently, ε-PL is produced industrially using an aerobic fermentation system with a mutant derived from *S. albulus* No. 346, which was first isolated from soil by Shima and Sakai (1977). During the first fermentation study with a wild strain of this bacterium, and under optimum culture conditions, the accumulated concentration of ε-PL in the culture medium was 0.5 g L^{-1} (Shima and Sakai, 1981a). The optimum culture medium comprised 50 g glycerol, 10 g $(NH_4)_2SO_4$, 5 g yeast extract, 0.5 g $MgSO_4 \cdot 7H_2O$, 0.03 g $FeSO_4 \cdot 7H_2O$, 0.04 g $ZnSO_4 \cdot 7H_2O$, 1.36 g KH_2PO_4 and 3.58 g $Na_2HPO_4 \cdot 12H_2O$ per 1 L. The maximum accumulation of ε-PL was observed after fermentation initiated at pH 6.0. When cell growth had reached a maximum, an accumulation of ε-PL in the culture broth was observed, and lowering of the medium pH value (to pH 3–5) was essential for ε-PL production. During this fermentation process, the addition of L-lysine to the culture medium had no enhancing effect on ε-PL productivity.

Subsequently, the production of ε-PL by resting cells was examined. After cultivation in medium containing 20 g glycerol, 5 g yeast extract, 0.5 g $MgSO_4 \cdot 7H_2O$, 1.4 g KH_2PO_4 and 3.6 g $Na_2HPO_4 \cdot 12H_2O$ per 1 L, the resting cells were incubated using glucose and $(NH_4)_2SO_4$ as substrates (Shima et al., 1983). An accumulation of ε-PL was observed under acidic pH conditions, as well as during the fermentation process, and 4–5 g L^{-1} of ε-PL was produced. When incubated with the cultivated cells, ε-PL was rapidly degraded over a pH range of 5–6, indicating that *S. albulus* also possesses a ε-PL-degrading enzyme.

In order to improve ε-PL productivity, *S*-(2-aminoethyl)-L-cysteine (AEC) plus glycine-resistant mutants were derived from *S. albulus* No. 346 by Hiraki et al. (1998). Whilst bacterial growth was not inhibited by AEC at a concentration of 2 mg mL^{-1}, further addition of 1 mg mL^{-1} glycine inhibited cell growth, though the strain was able to tolerate glycine at this level. Some 99% of the mutants were found to be high producers of ε-PL; indeed, one of them (No. 11011A) showed a maximum productivity of 2.11 mg mL^{-1} in a test-tube culture, which was 10-fold higher than that of the wild strain. The mutants showed higher specific activity of aspartokinase, which is assumed to be a key enzyme for L-lysine biosynthesis of the strain. Using a 3-L mini jar fermentor, *S. albulus* No. 11011A produced 20 mg mL^{-1} of ε-PL with an 8.9% yield against the consumed glucose in 120 h cultivation with glucose and ammonium sulfate feeding and under continuous pH control (Figure 2).

Kahar et al. (2001) demonstrated the importance of strict pH control and glucose concentration in ε-PL production using *S. albulus* No. 410 (S410), which is one of the ε-PL producers that exhibits high productivity. During ε-PL production by fermentation, the pH of the culture broth generally decreased from its initial value of 6.8 to 4.2 by 36 h, and then gradually decreased further to 3.2 at 96 h. ε-PL began to accumulate in the broth when the pH was <4.2, and so its productivity under strict pH control was evaluated. The optimized culture conditions were divided into two control phases. In the first phase, cell growth was accelerated by maintaining the pH value at >5.0. In the second phase, ε-PL-producing activity was highest at pH 4.0, and the glucose concentration was maintained at ~10 g L^{-1}. Glucose depletion caused an increase in the pH of the culture broth, resulting in degradation of the ε-PL product. In this batch culture control system, ε-PL productivity was enhanced from 5.7 to 48.3 g L^{-1}.

All ε-PL-producing microorganisms belong to the genus *Streptomyces*, and reports on such activity have been made in the following strains: *S. albulus* No. 346 and its

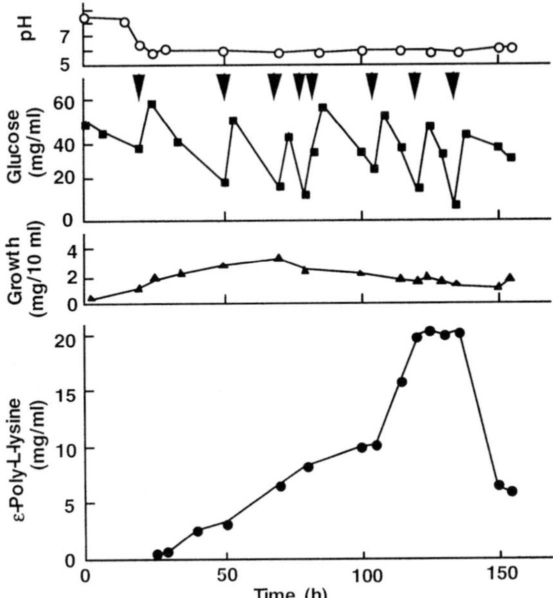

Fig. 2 Time course of ε-PL production in a 3-L jar fermentor. Glucose and ammonium sulfate were fed at the times indicated by the arrowheads

mutants, *S. noursei* IFO15452, *S. virginiae* IFO12827 and several unidentified actinomycetes (Shima and Sakai, 1977; Hiraki et al., 1998; Takagi et al., 2000; Kahar et al., 2001; Kito et al., 2002b; Saimura et al., 2002), which are probably classified in the genus *Streptomyces*. Hiraki et al. have continuously derived mutants which exhibit high ε-PL productivity, and ε-PL has now been produced industrially by the Chisso Company, Japan using a mutant derived from *S. albulus* No. 11011A.

8 Applications

The application of ε-PL has mainly focused on its use as a food additive (Hiraki, 2000). Although the addition of a large quantity of ε-PL results in a bitter taste, its high antimicrobial activity means that only low concentrations are required for food preservation is low, and so the unpleasant taste is not problematic. The safety of ε-PL as a food additive has been confirmed by experiments conducted in rats (Neda et al., 1999), whereby ε-PL has been shown to have no adverse reproductive toxicological effects, nor to affect neurological and immunological function, embryonic and fetal development and growth of offspring, and the development of embryos or fetuses for two generations. By combining ε-PL with other food additives, its preservative activity against microorganisms is greatly enhanced (Hiraki, 2000), and combined use of ε-PL with glycine, vinegar, ethanol and thiamine laurylsulfonate is effective in preserving various types of food. When the effect of ε-PL and glycine on the preservation of condensed milk was examined, a multiplicative inhibition of microbial growth was observed (Figure 3), and this led to smaller amounts of preser-

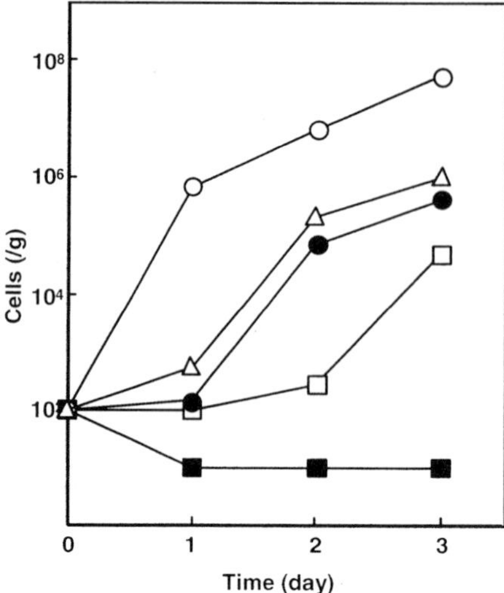

Fig. 3 Effect of ε-PL and glycine on the preservation of condensed milk. ○, no additions; r, 2 mg mL^{-1} glycine; ●, 0.2 mg mL^{-1} ε-PL; ❑, 2 mg mL^{-1} glycine plus 0.1 mg mL^{-1} ε-PL; ■, 2 mg mL^{-1} glycine plus 2 mg mL^{-1} ε-PL.

Fig. 4 Rice balls containing e-PL as a food preservative. Rice balls are often wrapped with dried seaweed.

vatives being added to the food. In Japan, it is normal to see a variety of food products containing ε-PL, especially the boiled rice in lunch boxes purchased from supermarkets or drugstores (Figure 4).

A different type of application of ε-PL has been developed by Kuraray Co., Ltd. (2001). Here, the amino groups of ε-PL are used in the synthesis of macromolecular gels with water-binding capacity; the hydrogels are synthesized by chemical cross-linking of polysaccharide, using low molecular weight materials and macromolecular cross-linking reagents. When propylene glycol alginate ester and proteins such as gelatin are mixed, amidation occurs between the ester site and ε-amino groups originating in the lysine residues of proteins, and this results in the formation of hydrogels. The water-binding capacity of gels containing proteins is low, however. When ε-PL is used to cross-link the polysaccharide, its α-amino groups become involved in the amidation, and the gel that forms exhibits high water-binding capacity, gel strength and swelling rate. The biological safety of ε-PL in humans is also important, as macromolecular gels with high water-binding capacity are used extensively in fields of agriculture, food processing and medicine.

9
Biodegradation

ε-PL shows a wide range of antimicrobial and antiphage activities, and its use as a food additive is steadily increasing in Japan. In this regard, several reports have been made on its biological degradation in a natural environment or *in vivo*. On the subject of its microbial degradation, several enzymes that catalyze ε-PL degradation have been characterized (Kito et al., 2002a, b; Takimoto et al., 2002). Whilst *S. albulus* and its mutants produce and excrete ε-PL in culture medium at about pH 4, ε-PL is immediately degraded by this strain over a pH range of 5–8 (Shima et al., 1983; Hiraki et al., 1998; Kahar et al., 2001).

The presence of an ε-PL-degrading enzyme is disadvantageous in the industrial production of ε-PL using by *S. albulus*, though this enzyme has recently been purified and characterized (Kito et al., 2002b). Likewise, ε-PL-tolerant bacteria have only recently been isolated from soils, and their ε-PL-degrading enzymes characterized (Kito et al., 2002a; Takimoto et al., 2002).

The ε-PL-degrading enzyme of *S. albulus* is tightly bound to the cell membrane and solubilized by 1.2 M NaSCN in the presence of Zn^{2+} (Kito et al., 2002b). The enzyme has been purified to homogeneity, the subunit molecular mass being 54 kDa. The mode of ε-PL degradation is of the exo-type, with the enzyme releasing N-terminal L-lysine residues one by one. L-Lysyl-*p*-nitroanilide, L-arginyl-*p*-nitroanilide and L-leucyl-*p*-nitroanilide are efficiently hydrolyzed, and the enzyme is classified as a type of aminopeptidase. The enzyme activity is inhibited by *o*-phenanthroline, and can be restored in the presence of Mg^{2+}, Ca^{2+}, Fe^{3+} or Zn^{2+}. The highest activity is observed at pH 7.0, and this pH profile is in agreement with the finding that a pH of about 6.0 leads to degradation of ε-PL produced in the culture medium (Shima et al., 1983; Hiraki et al., 1998; Kahar et al., 2001) (see Section 7). In studies on the distribution of membrane-bound ε-PL-degrading aminopeptidases of various *Streptomyces* strains, an interesting correlative distribution between ε-PL-degrading activity and its productivity was recently found (Kito et al., 2002b). *S. virginiae* IFO12827 and *S. noursei* IFO15452 each exhibited significant ε-PL-degrading aminopeptidase activity, and were also able to produce ε-PL in the culture medium. However, in other strains which showed considerably lower ε-PL-degrading activity, ε-PL-producing activity was not detected. These data suggest that ε-PL-producing strains commonly possess the membrane-bound ε-PL-degrading enzyme.

ε-PL-degrading enzymes were found in ε-PL-tolerant microorganisms isolated via an enrichment culture medium containing 0.1–1 mg mL^{-1} ε-PL as the sole carbon source (Kito et al., 2002a; Takimoto et al., 2002). Two bacteria exhibiting high ε-PL-degrading activity were identified as *Sphingobacterium multivorum* OJ10 and *Chryseobacterium* sp. OJ7. Both bacteria were able to grow well in the presence of 10 mg mL^{-1} ε-PL without any prolonged lag phase. Shima et al. (1984) reported that the MICs for bacteria were generally 1–8 μg mL^{-1}, whilst those for fungi and yeast were ~250 μg mL^{-1}. This suggests the presence of a relationship between high ε-PL-degrading activity and tolerance to ε-PL. Activity of the ε-PL-degrading enzyme of *Sph. multivorum* OJ10 was detected in the cell-free extract, in contrast to the specific membrane localization of the ε-PL-degrading enzyme of *S. albulus* (Kito et al., 2002a). The purified enzyme catalyzed exo-type degradation of ε-PL, released L-lysine, and was activated by Co^{2+} and Ca^{2+}. The ε-PL-degrading enzyme of *Sph. multivorum* degraded various peptides, and the hydrolysis of ε-PL was not a specialized function of this enzyme. The ε-PL-degrading enzyme of *Chryseobacterium sp.* OJ7, which is uniquely excreted in the culture medium, has also been purified to homogeneity (Takimoto et al., 2002). The purified enzyme catalyzes the endo-type degradation of ε-PL, in contrast to those of *S. albulus* and *Sph. multivorum* OJ10.

ε-PL-degrading activities of commercially available proteases have also been examined (Kito et al., 2002a). Protease A, Protease P and Peptidase R (all supplied by Amano Enzyme Co.) are each derived from fungi, and catalyze the endo-type degradation of ε-PL. The ε-PL-degrading activity of these proteases might contribute to the generally

Tab. 2 Properties of ε-PL-degrading enzymes

	Streptomyces albulus	*Chryseobacterium* sp. OJ7	*Sphingobacterium multivorum* OJ10
Localization	Membrane	Extracellular	Intercellular
ε-PL degradation	Exo type	Endo type	Exo type
Molecular mass (kDa)		38.4	80–110
Subunit molecular mass (kDa)	54	19.5	80–110
Metal requirement	Zn^{2+}, Ca^{2+}, Mg^{2+}, Fe^{3+}		Co^{2+}, Ca^{2+}
Optimum temperature (°C)	50	55	30
Thermal stability (°C)	<50	<40	<30
Optimum pH	7	7–7.5	7
pH stability	6–10	>7	5–9

high MIC of ε-PL for fungi. The characteristics of enzymes catalyzing ε-PL degradation are summarized in Table 2.

10 Biosynthesis of ε-PL

Various antimicrobial compounds which are composed of amino acids and have peptide linkages in their structures are produced by actinomycetes, and the diverse primary structures of many of these compounds has now been revealed. The peptide linkages in these antimicrobial compounds are formed independently of the ribosome. In contrast to the concept of protein biosynthesis based on the genetic information of mRNA, low molecular-weight peptides showing antimicrobial activity are generally synthesized by nonribosomal peptide synthetases (NRPSase) (Lipmann, 1980; Zocher and Keller 1997). Amino acids are first activated by ATP to produce adenylate which, being unstable, is subsequently converted to a thioester. The thioesterified amino acid is then incorporated into the peptide as another substrate. The precursor of ε-PL biosynthesis has been identified as L-lysine by an incorporation experiment using [^{14}C]-L-lysine in *S. albulus* No. 346 (Shima and Sakai, 1983). During the course of deriving AEC-resistant mutants from *S. albulus* No. 11211A, the enhancement of aspartokinase activity was accompanied by high productivity of ε-PL. Aspartokinase is a key enzyme in L-lysine biosysnthesis by fermentation carried out by strains such as *Brevibacterium flavum*, *B. lactofermentum* and *Corynebacterium glutamicum* (Hiraki et al., 1998). Therefore, with regard to ε-PL, a NRPSase system might be similarly involved in the biosynthesis using L-lysine as a substrate. ε-PL produced by several *Streptomyces* strains varies in length from 10 to 35 residues, depending on the producing strain. For example, *S. albulus* No. 346 produces and excretes L-lysine polymers with 25–35 residues. The existence of regulatory mechanisms for the elongation reaction or excretion system has been suggested in the biosynthesis of ε-PL by NRPSase.

11 Possible Physiological Role of the ε-PL-degrading Enzyme

The high activity of the ε-PL-degrading enzyme, which specifically distributes in the cell membrane, has been detected in all ε-PL-producing *Streptomyces* strains (Kito et al., 2002b), whereas non-ε-PL-producing strains of *Streptomyces* do not possess the ε-PL-degrading enzyme. This close correlation suggests a physiological role for the ε-PL-degrading enzyme. Excreted ε-PL is adsorbed to the cell surface by virtue of its ionic affinity (Shima et al., 1984), and consequently those microorganisms that are sensitive to ε-PL cannot grow due to this adsorbent action. The ε-PL-degrading enzyme of ε-PL producers might play a role in self-protection against ε-PL, and this is suggested by the fact that ε-PL-degrading enzymes are found in ε-PL-tolerant microorganisms. The ε-PL-tolerant bacteria, *Sph. multivorum* OJ10 and *Chryseobacterium* sp. OJ7, grow well in the presence of 10 mg mL^{-1} ε-PL without any prolonged lag phase, and growth is not completely inhibited even at 100 mg mL^{-1} (Kito et al., 2002a). As for *S. albulus*, good growth is observed even at 100 mg mL^{-1}. ε-PL-tolerant bacteria have no ε-PL-producing ability, and so their adventitious possession of proteases that catalyze ε-PL degradation enables them to grow at a high concentration of ε-PL. Generally, the MICs for bacteria and for fungi and yeasts are 1–8 μg mL^{-1} and ~250 μg mL^{-1}, respectively (Shima et al., 1983; Hiraki, 2000). This difference in inhibitory concentration might derive from the substrate specificity of proteases or cell surface conditions.

12 Molecular Genetics

All enzymes involved in ε-PL biosynthesis in *S. albulus* remain to be verified, and consequently few reports exist on the analysis of genes from this bacterium. The genes necessary for ε-PL production have not yet been directly isolated and analyzed, though a high molecular-weight plasmid (pNO33) of 37 kbp, which was related to the ε-PL-producing activity of *S. albulus*, has been isolated and its nucleotide sequence partially revealed (Takagi et al., 2000). The plasmid was not detected in the non-ε-PL-producing *S. albulus* and *S. noursei* by Southern blotting using pNO33 DNA as a probe. The sequence analysis of a 4.6-kbp fragment in pNO33 revealed four putative open reading frames (ORFs). The detailed function of pNO33 is unknown, since the sequence analysis of other regions has not been reported; however, the fact that pNO33 has been found only in the ε-PL-producing *S. albulus* suggests its possible involvement with ε-PL biosynthesis, or its tolerance against ε-PL.

Recently, the gene encoding the ε-PL-degrading enzyme of *S. albulus* has been cloned and analyzed (Kito et al., 2002c). As described in Section 9, the distribution of ε-PL-degrading activity in *Streptomyces* is closely related to ε-PL-producing activity, and tolerance against ε-PL is most likely required for ε-PL producers. As for pNO33, the plasmid has been presumed to encode the gene(s) involved in ε-PL production or resistance (Takagi et al., 2000). However, the gene for the ε-PL-degrading enzyme was cloned from genomic DNA of *S. albulus*. The deduced amino acid sequence from nucleotide sequence exhibited a significant homology (70%) with a putative metallopeptidase of *S. coelicolor*, and further analysis revealed that the ε-PL-degrading enzyme belongs to a certain group of zinc-containing aminopeptidases. The flanking regions of the gene for the ε-PL-degrading enzyme have not yet been analyzed. It is not known whether the gene(s) for the biosynthesis of ε-PL are localized in the region adjacent to the gene for its biodegradation.

13
Outlook and Perspectives

Since the discovery of ε-PL in the 1970s (Shima and Sakai, 1977), many investigators have noted its wide spectrum of antimicrobial activity, and ε-PL is now widely used as a food additive (Hiraki, 2000). Today, PL is produced industrially, and studies on the development of further applications are continuing apace. The unique physicochemical properties of ε-PL as a polycationic compound, as well as its high antimicrobial activity, will in time inevitably be applied to fields other than the food industry. Moreover, with regard to its biodegradation, proteases that catalyze its hydrolysis have now been found in microorganisms (Kito et al., 2002a). One potential problem is that in order to meet future demands for ε-PL, the current production system might be inadequate, and consequently new production process should be investigated.

Despite the widespread use of ε-PL, the mechanism of its biosynthesis and the nature of the related enzymes have not yet been elucidated. However, during the course of studies on the development of both ε-PL productivity and degradability, a number of clues have been obtained and details of these reactions are currently the subject of investigation of several laboratories. Indeed, following the revelation of enzymes (and their genes) involved in ε-PL biosynthesis, it is highly likely that an efficient process for ε-PL production will be developed in the near future.

14
Patents

The main claims of most patents on ε-PL are based on its antimicrobial activity, and may be classified into two main categories: (1) ε-PL-producing strains and the conditions for ε-PL production; and (2) applications of ε-PL as a superior food additive or antimicrobial compound. The relevant patents on ε-PL are listed in Table 3.

Tab. 3 Patents on ε-PL production and its application

*Patent No.**	*Holder*	*Inventors*	*Title*	*Date*
1245361	Sakai, H	Sakai, H. Shima, S.	Production of polylysine	28.06.1978
1678766	Chisso Co.	Hiraki, J. Morita, Y.	Microbial strain producing ε-poly-L-lysine	01.03.1988
1678767	Chisso Co.	Morita, Y. Hiraki, J.	Production of ε-poly-L-lysine	01.03.1988
1713630	Chisso Co.	Hiraki, J. Morita, Y.	Production of ε-poly-L-lysine	18.06.1991
1887522	Chisso Co.	Fujii, M. Morita, Y.	Production of ε-polylysine and ε-polylysine producing microorganisms	26.07.1989
2005392	Chisso Co.	Hiraki, J. Sugano, E. Fujii, M.	Preservative for food	23.03.1993
2030601	Chisso Co.	Hiraki, J. Morita, Y. Suzuki, E.	Microbial strain producing ε-poly-L-lysine	23.03.1993

Tab. 3 (cont.)

*Patent No.**	*Holder*	*Inventors*	*Title*	*Date*
2649650	Chisso Co.	Noaki, N.	Readily degradable molding and its production	09.10.1995
2710006	Yamamoto, T.	Yamamoto, T.	Method for enhancing elasticity of fishery paste	27.01.1995
2748102	Sakai, M.	Takahashi, K. Hattori, M.	Method for modifying starch	09.04.1996
2762335	Sakai, M.	Katusta, K.	Gelatinous composition of whey protein and its production	24.03.1993
2849802	Sakai, M.	Yamamoto, T. Fujii, M.	Production of frozen ground fish meat	04.02.1997
2858231	Sakai, M.	Tananito, T. Fujii, M.	Quality improvement and production of fishery paste product	07.01.1997
2906113	Sakai, M.	Otsuka, N.	Method for improving taste of seasoning	19.03.1996
2920491	Sakai, M.	Shima, S. Fujii, M.	Card composition	12.08.1997
2982044	Fujii, M.	Fujii, M.	Production of camembert cheese	17.09.1997
3037605	T Paul Co.	Hamamichi, Y. Ri, B.	Germicidal cleaner composition	08.07.1997
3120607	Chisso Co.	Shindo, T. Hiraki, J.	Stabilization of enzyme reaction	03.06.1993
3154826	Taiyo Kagaku Co. Ltd.	Murata, M. Hori, T. Takahashi, K. Kato, T.	Production of packed and thermally sterilized Chinese noodles	15.02.1994
3158610	Chisso Co.	Hiraki, J. Watanabe, S.	Preservative for food with high salt content	31.08.1993
3185423	Chisso Co.	Hiraki, J. Shindo, T.	Stabilizer of enzymes	03.06.1994
3257181	Chisso Co.	Iwazawa, Y. Hiraki, J. Watanabe, S. Sato, S.	Food preservative	14.03.1995
3257738	Mercian Co.	Sato, M. Kuroda, S. Tone, H.	Antimicrobial pharamaceutical preparation containing polylysine	11.07.1995
3282271	Chisso Co.	Hiraki, J. Fukshi, H., Suzuki, E.	Production of ε-poly-L-lysine	29.03.1993

*Japan patent numbers are indicated.

15
References

Hiraki, J. (1995) Basic and applied studies on ε-polylysine, *J. Antibact. Antifung. Agents* **23**, 349–354.

Hiraki, J. (2000) ε-Polylysine; its development and utilization, *Fine Chemicals* **29**, 18–25.

Hiraki, J., Hatakeyama, M., Morita, H., Izumi, Y. (1998) Improved ε-poly-L-lysine production of an *S*-(2-aminoethyl)-L-cysteine resistant mutant of *Streptomyces albulus, Seibutsu-kogaku Kaichi* **76**, 487–493.

Itzhaki, F. R. (1972) Colorimetric method for estimating polylysine and polyarginine, *Anal. Biochem.* **50**, 569–574.

Kahar, P., Iwata, T., Hiraki, J., Park, Y. E., Okabe, M. (2001) Enhancement of ε-polylysine production by *Streptomyces albulus* strain 410 using pH control, *J. Biosci. Bioeng.* **91**, 190–194.

Kito, M., Onji, Y., Yoshida, T., Nagasawa, T. (2002a) Occurrence of ε-poly-L-lysine-degrading enzyme in ε-poly-L-lysine tolerant *Sphingobacterium multivorum* OJ10: purification and characterization, *FEMS Microbiol. Lett.* **207**, 147–151.

Kito, M., Takimoto, R., Yoshida, T., Nagasawa, T. (2002b) Purification and characterization of ε-poly-L-lysine-degrading enzyme from an ε-poly-L-lysine-producing strain *Streptomyces albulus, Arch. Microbiol.* (in press).

Kito, M., Yoshida, T., Nagasawa, T. (2002c) Cloning and analysis of the gene for ε-poly-L-lysine-degrading enzyme of an ε-poly-L-lysine producing *S. albulus*, Annual Meeting of the Society for Biosciences, Biotechnology and Biochemistry, Japan, 3–6Bp22, p. 208.

Kuraray Co., Ltd. (2001) Water swelling polymer gel and its production method, Japanese patent 2001–278984.

Lipmann, F. (1980) Bacterial production of antibiotic polypeptides by thiol-linked synthesis on protein templates, *Adv. Microbiol. Physiol.* **21**, 227–266.

Neda, K., Sakurai, T., Takahashi, M., Aiuchi, M., Ohgushi, M. (1999) Two-generation reproduction study with teratology test of ε-poly-L-lysine by dietary administration in rats, *Jpn. Pharmacol. Ther.* **27**, 1139–1159.

Saimura, M., Ikezaki, A., Takahara, M., Hirohara, H. (2002) Structures of ε-polylysine produced by several actinomycetes and their classification on the basis of productivity of ε-polylysine, Annual Meeting of the Society for Biosciences, Biotechnology and Biochemistry, Japan, 2–5Da07, p. 54.

Shima, S., Sakai, H (1977) Polylysine produced by *Streptomyces, Agric. Biol. Chem.* **41**, 1807–1809.

Shima, S., Sakai, H. (1981a) Poly-L-lysine produced by *Streptomyces*. Part II. Taxonomy and fermentation studies, *Agric. Biol. Chem.* **45**, 2497–2502.

Shima, S., Sakai, H. (1981b) Poly-L-lysine produced by *Streptomyces*. Part III. Chemical studies, *Agric. Biol. Chem.* **45**, 2503–2508.

Shima, S., Fukuhara, Y., Sakai, H. (1982) Inactivation of bacteriophages by ε-poly-L-lysine produced by *Streptomyces, Agric. Biol. Chem.* **46**, 1917–1919.

Shima, S., Oshima, S., Sakai, H. (1983) Biosynthesis of ε-poly-L-lysine by washed mycelium of *Streptomyces albulus* No. 346, *Nippon Nogeikagaku Kaishi* **57**, 221–226.

Shima, S, Matsuoka, H., Iwamoto, T., Sakai, H. (1984) Antimicrobial action of ε-poly-L-lysine, *J. Antibiotics* **37**, 1449–1455.

Takagi, H., Hoshino, Y., Nakamori, S., Inouye S. (2000) Isolation and sequence analysis of plasmid pNO33 in the ε-poly-L-lysine-producing actinomycete *Streptomyces albulus* IFO14147, *J. Biosci. Bioeng.* **89**, 94–96.

Takimoto, R., Kito, M., Yoshida, T., Onji, Y., Nagasawa, T. (2002) Characterization of microbial enzymes catalyzing the degradation of ε-poly-L-lysine, Annual Meeting of the Society for Biosciences, Biotechnology and Biochemistry, Japan, 4–5Fa06, p. 297.

Zocher, R., Keller, U. (1997) Thiol template peptide synthesis systems in bacteria and fungi, *Adv. Microbiol. Physiol.* **38**, 85–131.

6
Poly-γ-glutamic Acid

Dr. Makoto Ashiuchi[1], Prof. Dr. Haruo Misono[2]
[1] Department of Bioresources Science, Kochi University, Nankoku, Kochi 783-8502, Japan; Tel.: +81-888645215; Fax: +81-888645200;
E-mail: ashiuchi@cc.kochi-u.ac.jp
[2] Department of Bioresources Science, Kochi University, Nankoku, Kochi 783-8502, Japan; Tel.: +81-888645187; Fax: +81-888645200;
E-mail: hmisono@cc.kochi-u.ac.jp

α-PGA	poly-α-glutamic acid
α,γ-PGA	poly-α,γ-glutamic acid
ATCC	American Type Culture Collection
cap	encapsulation genes
CD	circular dichroism
DAT	D-amino acid aminotransferase
dep	poly-γ-glutamic acid depolymerase-encoding gene
DMSO	dimethylsulfoxide
ε-PL	poly-ε-lysine
ggt	γ-glutamyltranspeptidase-encoding gene
GGT	γ-glutamyltranspeptidase
γ-PGA	poly-γ-glutamic acid
γ-D-PGA	γ-PGA composed only of D-glutamic acid
γ-DL-PGA	γ-PGA composed of D- and L-glutamic acids
γ-L-PGA	γ-PGA composed only of L-glutamic acid
GPC	gel permeation chromatography
HPLC	high-performance liquid chromatography
IFO	Institute of Fermentation Osaka (Japan)
IR	infrared absorption
NMR	nuclear magnetic resonance
ORD	optical rotatory dispersion
pgd	poly-γ-glutamic acid degradation gene
PGE_1	prostaglandin E_1
pgs	poly-γ-glutamic acid synthesis genes
PgluU	poly-α,β-1,4-D-glucuronic acid
SDS–PAGE	sodium dodecyl sulfate-polyacrylamide gel electrophoresis
TCA cycle	tricarboxylic acid cycle
TUA	teichuronic acid
TUP	teichuronopeptide

1 Introduction

Today, most plastics and synthetic polymers are produced from petrochemicals. These long-lasting polymers are used even for short-lived applications, and this leads to a profound influence on the environment. Plastic materials that are incorrectly disposed of are serious sources of environmental pollution. The elimination of waste plastics is, therefore, of value in the following disciplines: surgery, health, catering, packing, agriculture, fishing, environmental protection, and other technical endeavors. Increased knowledge of the preservation of environmental systems has provoked a complete change in the production of conventional and nondegradable polymers. The interest is now in producing environment-friendly biodegradable polymers, the production and use of which will contribute to savings in energy and resources, thereby curbing the greenhouse effect, developing ecocompatible processes and products, and diversifying agriculture for food production.

Since poly-γ-glutamic acid (γ-PGA) is biodegradable and has a high degree of water absorbency, it serves as a potent substitute for nondegradable thermoplastics and hydrogels. γ-PGA is an unusual anionic polypeptide (a polyamino acid) in which glutamic acid is polymerized via γ-amide linkages. This biopolymer, therefore, can protect cells from the action of proteases. In addition, because of its many carboxyl groups, γ-PGA may be used as a cryoprotective material, like the insect cryoprotectant sericin (Tsujimoto et al., 2001). Current studies on γ-PGA biosynthesis strongly support its physiological functions as an adaptation agent in various environments (see Section 6). Among these functions, the polymer produced by extremely halophilic archaea (Niemetz et al., 1997; Hezayen et al., 2001) is of some interest as it is able to absorb large amounts of water even under high-saline conditions. By contrast, the water-absorption capability of γ-PGA hydrogels so far studied declines sharply under the same conditions (Kunioka, 1997). Structural analyses of the archaeal polymers may provide insight into the functional modification of γ-PGA, and should also contribute to the development of new applications of this promising biopolymer. In order that γ-PGA be used on a practical basis, it is essential that a mass production process be established; however, to achieve this the biosynthetic mechanism of γ-PGA must first be elucidated. Until quite recently, there was little agreement about this biosynthetic mechanism, mainly due to the structural variety of γ-PGA (Table 1), the diversity of γ-PGA production (see Section 5), the complexity of the regulation of γ-PGA metabolism (see Sections 7 and 8), and the extreme instability of γ-PGA synthetase (see Section 8). However, recent studies on γ-PGA biosynthesis in *Bacillus subtilis* (Ashiuchi et al., 1999a, 2001b) have helped to find a solution to this problem.

This chapter first briefly reviews the historical aspects of the structural analysis and chemical synthesis of γ-PGA, and subsequently discusses the current state of γ-PGA research including molecular genetics, biosynthesis, biodegradation, and potential applications. Patents that contain important techniques toward the practical use of γ-PGA are listed. The potential industrial and scientific benefits that might accrue from extensive investigations of structurally and functionally unique γ-PGAs and their synthetases are also described.

Tab. 1 Molecular sizes and stereochemical compositions of γ-PGAs produced by various organisms (Ashiuchi and Misono, 2002)

Producers	*Molecular mass [kDa]*	*Content [%]*	
		D-Enantiomer	*L-Enantiomer*
Bacillus subtilis (natto)	10–1000	50–80	20–50
Bacillus subtilis (chungkookjang)	>1000	60–70	30–40
Bacillus licheniformis	10–1000	10–100	0–90
Bacillus anthracis	ND	100	0
Bacillus megaterium	>200	50	50
Bacillus halodurans	10–15	0	100
Natrialba aegyptiaca	>1000	0	100
Hydra	3–25	0	100

ND, not determined.

2 Historical Outline

Since the discovery of γ-PGA (Ivánovics and Erdös, 1937), its structure and function have been the subjects of many investigations over a long period of time. Some simple methods for the detection, purification, and quantification of γ-PGA have been established (Ogawa et al., 1997; Do et al., 2001b), most of which are useful, even in genetic and biochemical studies on γ-PGA.

2.1 Chemical Analysis

The structural analyses of γ-PGA and crucial data for the determination of its primary structure are summarized in this section. Techniques that are useful in γ-PGA research are also described.

2.1.1 Linkage formula

Ivánovics and Bruckner (1937), in their earlier studies on the capsular polypeptide of *Bacillus anthracis,* had proposed four types of stable peptide linkage of the polymer: α-peptide linkages (α-PGA; Figure 1a); γ-peptide linkages (γ-PGA; Figure 1b); α-peptide linkages with internal pyrrolidone rings (α-*pyro*-PGA; Figure 1c); and α,γ-crossed (or branched) linkages (α,γ-PGA; Figure 1d). All data from ninhydrin and biuret reactions (Bovarnick, 1942), Van Slyke cleavage (Hanby and Rydon, 1946), p*K* measurements (Waley, 1955; Appel and Yang, 1965), stability at alkaline pH (Levene et al., 1931), resistance to proteases such as trypsin (Haurowitz and Bursa, 1949), solubility in acidic solution (Hanby et al., 1950), and infrared absorption (IR) spectra (Hanby et al., 1950) indicated that the polypeptide was γ-PGA. Chemical degradation may be useful for determining the linkage formula of a wide range of naturally occurring polypeptides (Bruckner et al., 1953; Kovaca et al., 1953; Nitecki and Goodman 1971).

The strategy is illustrated in Figure 2; α- and γ-PGAs should yield α-amino-δ-hydroxyvaleric acid (Figure 2, upper) and γ-amino-δ-hydroxyvaleric acid (Figure 2, lower), respectively. All the polypeptides of *B. anthracis* (Bruckner et al., 1955), *B. subtilis* (Chibnall et al., 1958), and *Bacillus licheniformis* ATCC 9945A (formerly *B. subtilis* ATCC 9945A) (Troy, 1973b) were shown to

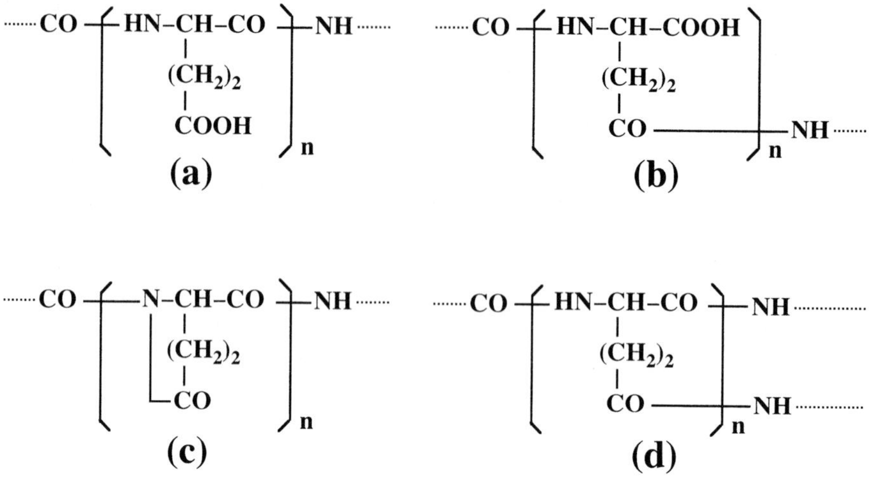

Fig. 1 Presumed linkage formula of naturally occurring glutamyl polypeptides (Nitecki and Goodman, 1971). Linkages: (a) α-linkage; (b) γ-linkage; (c) α-linkage with internal pyrrolidone rings; (d) α,γ-crossed (or branched) linkages.

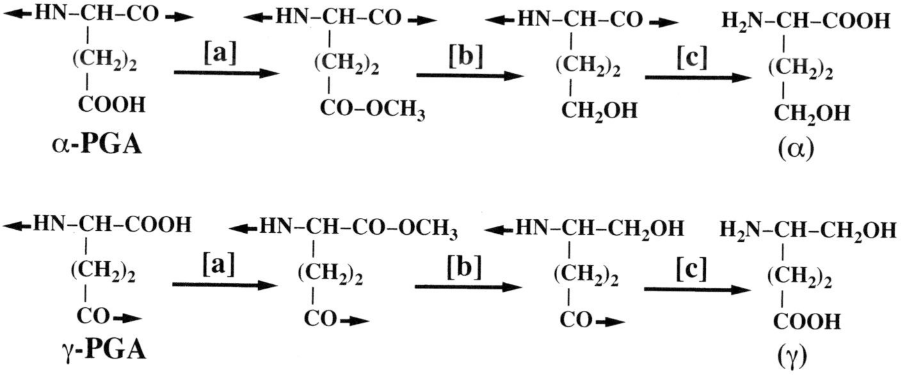

Fig. 2 Characterization of amide linkages in glutamyl polypeptides by chemical digestion (Troy, 1973b). Reactions: (a) methylation of free carboxyl groups of a glutamyl polypeptide with diazomethane; (b) reduction of the modified groups with lithium borohydride; (c) hydrolysis of the polymer with a high concentration of HCl. Reaction products: Upper (α): α-amino-δ-hydroxyvaleric acid; lower (γ): γ-amino-δ-hydroxyvaleric acid.

be γ-PGA. The *Bacillus megaterium* polypeptide was also characterized as γ-PGA by hydrazinolysis (Torii, 1959). ^{1}H- and ^{13}C-nuclear magnetic resonance (NMR) spectroscopic studies were employed in the structural analysis of γ-PGA (Borbély et al., 1994; Pérez-Camero et al., 1999).

Both α-L-PGA and an α-linked copolymer consisting of L-glutamic acid and L-glutamine were found in the cell walls of pathogenic mycobacteria (Wietzerbin et al., 1975; Harth et al., 2000), but naturally occurring α-D-PGA, α-*pyro*- and α,γ-PGAs have not yet been identified.

2.1.2
Stereochemistry

The stereochemical properties of γ-PGA have been investigated using chemical, enzymatic, and immunochemical methods. Among these techniques, chemical analysis is the most readily available (Kubota et al., 1993a; Cromwick and Gross, 1995a; Ito et al., 1996). Besides optical configuration measurements (Bovarnick, 1942), the development of a post-labeling high-performance liquid chromatography (HPLC) (Kubota et al., 1993c; Cromwick et al., 1996) and the application of chiral carriers (Kunioka 1995; Ashiuchi et al., 2001b) allowed the stereochemical analysis of γ-PGA to be made. As shown in Table 1, *B. subtilis, B. licheniformis,* and *B. megaterium* produce stereochemically differing γ-PGAs (D-isomer contents $>50\%$), unlike stereochemically homogeneous γ-PGA producers such as *B. anthracis* (only D) and *B. halodurans* (only L). There was a problem as to whether the γ-PGAs of the former organisms are a mixture of γ-D- and γ-L-PGAs (γ-D-PGA/total γ-PGA, $>50\%$) or a copolymer consisting of both D-and L-glutamic acid. Tanaka et al. (1997) showed that the γ-PGA of *B. subtilis* was a DL-copolymer by using a γ-PGA *endo*-depolymerase that cleaves only the L-chain regions in γ-PGA (Figure 3). Ashiuchi et al. (2001b) characterized a *B. subtilis* mutant that was defective in the γ-PGA synthetase complex. It is now accepted that *B. subtilis* possesses the sole machinery for γ-DL-PGA synthesis, but not two distinct systems for the syntheses of γ-D- and γ-L-PGAs.

Antibodies should be useful in monitoring the stereochemistry of γ-PGA due to their high structural and configuration specificities. A carrier (e.g., methylated bovine serum albumin) was required to trigger the immune response and produce a highly specific antibody against γ-PGA due to its nonimmunogenic characteristics. Ivánovics (1958) examined the immunochemical properties of the γ-PGAs of *B. anthracis* and *B. licheniformis,* and of chemically synthesized glutamyl polymers. The specificity of an antibody (anti-A) against the γ-D-PGA of *B. anthracis* for glutamyl polymers is summarized in Table 2. Utsumi et al. (1958, 1959) suggested that the anti-A recognizes the D-configuration in γ-PGA. An antibody (anti-

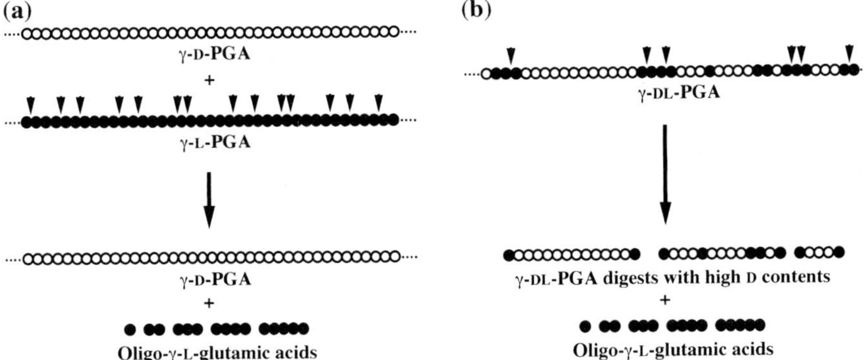

Fig. 3 Characterization of amide linkages in glutamyl polypeptides by enzymatic digestion (Tanaka et al., 1997). Schematic diagram of the digestion of (a) a mixture of γ-D- and γ-L-PGAs and (b) a γ-DL-PGA copolymer by the γ-PGA *endo*-L-depolymerase. D- and L-glutamyl residues of γ-PGA are shown by open and filled circles, respectively.

Tab. 2 Serological activities of naturally occurring and chemically synthesized γ-PGAs. These properties were investigated with both the antibodies against capsule components of *B. anthracis* (anti-A) and those of *B. megaterium* (anti-M)

Glutamyl polypeptides	*Serological activities*	
	Anti-A	*Anti-M*
B. anthracis γ-PGA (D form)	+	±
B. licheniformis γ-PGA (DL form)	+	+
B. megaterium γ-PGA (DL form)	+	+
Synthetic γ-D-PGA	+	ND
Synthetic γ-DL-PGA	+	ND
Synthetic γ-L-PGA	–	ND
Synthetic α-D-PGA	–	ND
Synthetic α-L-PGA	–	ND
Synthetic α,γ-D-PGA	–	ND
Synthetic α,γ-L-PGA	–	ND

+, positive; ±, slightly positive; –, negative. ND, not determined.

M) against γ-PGA of *B. megaterium* was also prepared, and this recognized the L-configuration in γ-PGA. In the immunochemical study, the γ-PGA of *B. megaterium* was found to be a DL-copolymer.

2.1.3 Purification

γ-PGA may be purified to homogeneity via five steps: (1) removal of cells by centrifugation or filtration; (2) acidification of the supernatant; (3) precipitation with alcohol; (4) protease treatment; and (5) dialysis of the resulting solution (Ito et al., 1996; Ashiuchi et al., 1998). Anion-exchange column chromatography can also be applied (Kubota et al., 1993a). Steps (2), (4), and (5) are crucial for the purification due to a need for the digestion of polysaccharides and α-polypeptides (mainly proteins) and the removal of any impurities of low molecular mass (< 10 kDa). *B. subtilis* synthesizes only γ-DL-PGA, whereas *B. licheniformis* produces a mixture of γ-D-, γ-L-, and γ-DL-PGAs in various ratios (Thorne and Leonard, 1958). Do et al. (2001b) devised an efficient strategy for the separation and recovery of γ-PGA from the culture broth of *B. licheniformis*.

2.1.4 Quantification

In order to quantify γ-PGA, it is normally lyophilized and the dry matter weighed. For highly elongated γ-PGAs (> 1000 kDa), concentrations can be estimated densitometrically from basic dye-stained bands corresponding to the γ-PGAs after sodium dodecyl sulfate–polyacrylamide gel electrophoresis (SDS–PAGE) (Ashiuchi et al., 2001a) or by use of gel permeation chromatography (GPC) with a refractive index detector (Kunioka and Goto, 1994). Kanno and Takamatsu (1995) established a simple spectrophotometric method to quantify γ-PGA of *B. subtilis* (*natto*) using cetyltrimethylammonium bromide.

2.1.5
Molecular size

For macromolecules such as γ-PGA, the molecular size may seriously affect the properties as a polymer. The molecular size of γ-PGA is usually determined by GPC and SDS-PAGE. With GPC, the size is characterized by the number-averaged molecular mass (M_n), weight-averaged molecular mass (M_w), and polydispersity (M_w/M_n) (Ko and Gross, 1998; Yoon et al., 2000). γ-PGA can be visualized with basic dyes such as methylene blue or alcian blue (Ogawa et al., 1997; Ashiuchi et al., 2001a), or by autoradiography of the ^{14}C-labeled γ-PGA (Ogawa et al., 1997). The molecular size is estimated from its mobility on a gel after being subjected to SDS–PAGE.

2.2
Chemical Synthesis

During the early stages of γ-PGA research, the chemical synthesis of glutamyl polymers, α-D-, α-L-, α-DL-, γ-D-, γ-L-, γ-DL-, α,γ-D-, α,γ-L-, and α,γ-DL-PGAs, was established and summarized by Nitecki and Goodman (1971). Although both a linkage formula and stereochemical compositions of the polymers can be designed without restrictions, the molecular sizes of the synthetic polymers (<9 kDa) are clearly low compared with those of naturally occurring γ-PGAs. Moreover, the chemical synthesis of γ-PGA is complex compared with that of α-PGA. To obtain γ-PGA, a γ-linked dipeptide with both α-carboxyl groups suitably protected, the γ-glutamyl-glutamic acid-α,α′-dimethyl ester, must be first synthesized via the following three steps: (1) the successful condensation between N-Z-glutamyl-γ-azide and glutamic acid-γ-benzyl ester; (2) esterification of the carboxyl groups with diazomethane; and (3) removal of the N-Z and γ-benzyl groups of the intermediate by hydrogenolysis. These polymerization methods have been established (Sheehan and Hess, 1955; Nitecki and Goodman, 1971). The physico-chemical properties of the PGAs, as demonstrated using optical rotatory dispersion (ORD) spectra and IR spectra, were seen to differ between the synthetic α- and γ-PGAs. These synthetic polymers have not yet been prepared commercially because their preparation requires state-of-the-art, costly technology when compared with the straightforward isolation of naturally occurring γ-PGA. However, they should be very useful for understanding the correlation between the structure and function of γ-PGA, the analysis of reaction mechanisms of γ-PGA synthetases, and the development of combinatorial engineering techniques towards practical uses for γ-PGA.

3
Molecular Structure

The conformation of naturally occurring γ-PGAs has been investigated using viscosity and ORD measurements, IR spectroscopy, and attenuated total reflectance Fourier-transform IR spectroscopy.

- γ-D-PGA from *B. anthracis*: an important determinant of the conformation is the ratio of protonation/deprotonation of functional carboxyl groups. The protonated γ-D-PGA (free acids) exhibits an α-helical conformation, but the deprotonated γ-D-PGA (sodium salt) transits in a randomly coil state. Rydon (1964) suggested that the protonated γ-D-PGA takes either a 3_{17} or a 3_{19} α-helix. Intramolecular hydrogen bonds are crucial to the stability of these helical structures. Ohki et al. (1981) examined the circular dichroism (CD) spectra of both chemically synthesized α-D- and α-L-PGAs and concluded that α-D-PGA forms a left-handed α-helix and the L-isomer takes a right-handed α-helix. This indicates that naturally occurring γ-D-PGA also exhibits a left-handed α-helical conformation. In fact, a combination of molecular dynamics and quantum mechanical calculation demonstrated that the left-handed 3_{19} α-helix is most stable in protonated γ-D-PGA and suggested that intramolecular hydrogen bonds set between the CO of the amide group *i* and the NH of the amide group $i + 3$ play an important role in the construction of this structure (Zanny et al., 1998). Thus, hydrogen bonds between side carboxylic oxygens and the NH of backbone amide groups support an unexpectedly high stability seen in the left-handed α-helix.

- γ-PGA from *B. licheniformis*: He et al. (2000) found that pH, ionic strength, and polymer concentration in solution were crucial determinants for the conformation of the γ-D-PGA of *B. licheniformis*. The γ-D-PGA had an α-helical form at low pH, low ionic strength or low polymer concentration, but a β-sheet form at neutral-to-high pH, high ionic strength or high polymer concentration. The conformation of the γ-D-PGA probably changes from α-helix into β-sheet when the influence of intermolecular interactions on the structure surpasses that of intramolecular interactions. Borbély et al. (1994) showed the effects of solvents on the conformation of γ-PGA (probably the γ-DL-PGA) of *B. licheniformis*: the γ-PGA formed a random coil in water, an α-helix in dimethylsulfoxide (DMSO), and a β-sheet in formic acid.
- γ-DL-PGA from *B. subtilis* (*natto*): Unlike γ-D-PGA (also γ-L-PGA), γ-DL-PGA took a parallel β-sheet form in an acidic solution, a random coil form at near neutral pH, and a more contracted random form in an alkaline solution (Sawa et al., 1973; Saito et al., 1974). γ-DL-PGA undergoes substantial conformational changes as the pH of the solution changed.

3.1 Molecular Spring

The effects of the conformational changes of γ-PGA on its physical properties, including metal-binding affinity, must be significant. The high molecular transformability and plasticity of γ-PGA is of great interest due to its possible application, for example, to nano-machinery in nano-technology, a most promising new technology.

α-Helical structures are considered to be some of the simplest and best mechanical elements by which to understand the structure–function relationship of macromolecules such as proteins and to design nano-mechanical devices. In order to characterize the helical behavior of polyamino acids in aqueous solution, various model peptides have been synthesized and investigated with respect to their tendencies toward α-helix formation (Muñoz and Serrano, 1995; Idiris et al., 2000). Among them, glutamyl polymers have a high tendency to form an α-helix. Idiris et al. (2000) examined the mechanical behavior of chemically synthesized α-L-PGA throughout a pH range (pH 3.0–8.0) during its α-helix-random coil transition and analyzed the spring-like behavior of the α-helix of α-L-PGA using atomic force microscopy. Based on the results that α-L-PGA, in its helical conformation, could be stretched almost fully with a continuous increase in the stretching force, these authors suggested that α-L-PGA might be used as a reliable coil-spring in the future design of spring-loaded molecular machinery, and indeed these data suggest that γ-PGA has great potential in such technology. α-PGA consists of a C_2 unit-backbone (hereafter, C_3-side chains), whereas the main chain of γ-PGA is a C_4-unit

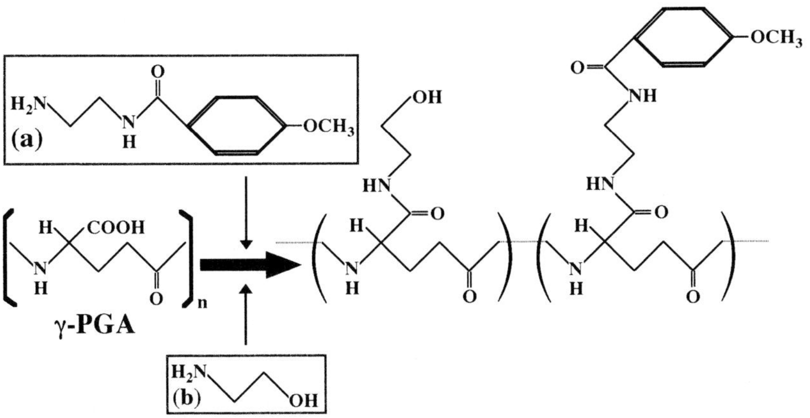

Fig. 4 Schematic diagram of the covalent modification of γ-PGA (King et al., 1998). Reagents: (a) *N*-(2-aminoethyl)-4-methoxy-benzamide, i.e., amine (1); (b) ethanolamine.

backbone (hereafter, C_1-side chains). Therefore, the γ-PGA spring may possess more advantageous and unique molecular characteristics than the α-PGA spring.

4 Chemical Modification

The development of viable methods to chemically modify γ-PGA has extended the range of possible applications of this polymer, including environmental uses. This section details in particular the techniques of esterification and crosslinking.

4.1 Esterification

This technique was developed for γ-PGA to be used as a thermoplastic (Kubota et al., 1993b). Borbély et al. (1994) established a simple, effective esterification method of γ-PGA: the α-esters of γ-PGA were synthesized in DMSO in the presence of Na_2HCO_3 and a different alkyl bromide. King et al. (1998) developed the method to synthesize a water-soluble modified γ-PGA with amine in the presence of water-soluble *N*-ethyl-*N*-(dimethylaminopropyl)carbodiimide hydrochloride and hydroxybenzotriazole as a catalyst (Figure 4). These authors also suggested the value of water-soluble γ-PGAs modified with a variety of functional groups.

4.2 Crosslinking

This technique was developed for γ-PGA to be used as a hydrogel. Some universal methods for creating hydrogels from macromolecules have been established (Hoffman, 1987), and both irradiation and chemical crosslinking have been shown to be effective in producing hydrogels from γ-PGA.

4.2.1 Irradiation

Kunioka (1997) established a simple method for preparing highly water-absorbent γ-PGA hydrogels by γ-irradiation of γ-PGA solution (5%, w/w) with ^{60}Co (1.6 kGy h^{-1}), and showed the specific water content (weight of water/weight of polymer) to reach 3500–5000. Kunioka proposed a crosslinking mechanism by intermolecular combination

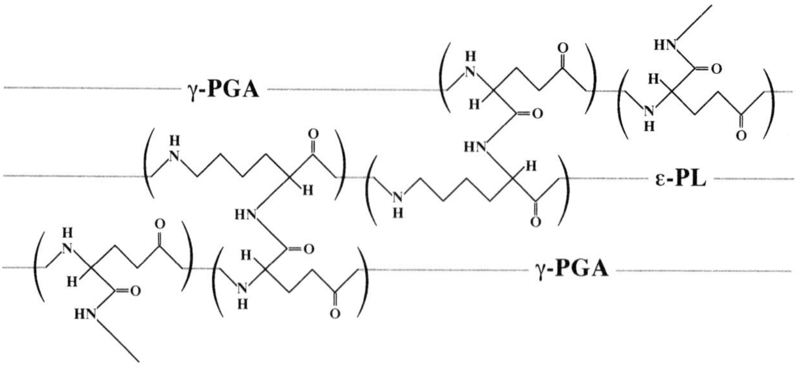

Fig. 5 A promising biodegradable ε-PL-crosslinked γ-PGA hydrogel.

between free radicals being generated on methylene carbons of γ-PGA via the γ-ray-induced cleavage of C–H bonds. However, this extremely high water-absorbing ability disappeared in the presence of NaCl or $CaCl_2$. γ-PGA hydrogels swelled at high pH (presumably due to the ionic repulsion of deprotonated carboxyl groups) and collapsed at low pH (presumably due to loss of the ionic repulsion with an increase in the protonated carboxyl groups), implying that negatively charged carboxyl groups were contained in the polymer network. One hydrogel which was synthesized from a mixture of γ-PGA and poly-ε-lysine (ε-PL) (Shima and Sakai, 1981) by γ-irradiation showed improved pH sensitivity compared with the γ-PGA hydrogels (Choi et al., 1995). γ-PGA hydrogels also showed chemical degradability at 100°C, though their biodegradability remained obscure. Even if these hydrogels are biodegradable, it remains to be shown that the use of radioactivity is acceptable in the manufacture of environment-friendly materials. Clearly, it is essential to develop a safe and biodegradable crosslinking agent before γ-PGA hydrogels can be put to practical use.

4.2.2 Chemical crosslinking

γ-PGA hydrogels can be prepared by the addition of a crosslinking agent such as hexamethylene diisocyanate and dihalogenoalkanes in DMSO or dimethylsulfoxide (Kakinoki et al., 1993; Gonzales et al., 1996). They may also be synthesized from γ-PGA and an alkanediamine (as a crosslinking agent) in the presence of a water-soluble carbodiimine (as a catalyst) in aqueous media (Kunioka, 1997). The specific water contents of the product ranged from 300 to 2000. This finding indicates that a promising biohydrogel (Figure 5), which should be biodegradable and even edible, could be synthesized from a mixture of γ-PGA and ε-PL (a basic polyamino acid, which is used as a biopreservative in foods) in aqueous solution, without the need for γ-irradiation.

5 Producers

As shown in Table 1, several γ-PGA-producing organisms have been identified to date. *B. anthracis* was the first to be discovered (Kramar, 1921), and its capsule was shown to consist mainly of γ-D-PGA (Hanby and

Rydon, 1946). Previously, Sawamura (1913) had isolated *B. subtilis* (*natto*), and studies on the extracellular viscous material of this organism showed the presence of γ-DL-PGA (Bovarnick, 1942; Kubota et al., 1993a; Ashiuchi et al., 1998). Since γ-PGA depolymerase accumulates during γ-PGA production (Abe et al., 1997), γ-PGAs with a variety of molecular masses (10–1000 kDa) may be obtained from the culture. *B. licheniformis* produces stereochemically differing γ-PGAs (D-isomer content 10–100%) (Pérez-Camero et al., 1999). *B. megaterium* also synthesizes γ-DL-PGA (D-isomer content 50%) (Torii, 1956). γ-L-PGA with a low molecular mass (~10 kDa) is found as an essential component of the cell walls of the alkalophilic *Bacillus halodurans* (Aono, 1987). A highly elongated γ-L-PGA (> 1000 kDa) was isolated from the culture of an extremely halophilic archaeon, *Natrialba aegyptiaca* (Hezayen et al., 2001; Ashiuchi and Misono, 2002). *Natronococcus occultus* is also a γ-L-PGA-producing archaeon (Niemetz et al., 1997). Among the eukarya, *Hydra* produces γ-L-PGA (Weber 1990).

Several *Bacillus* strains which produce γ-PGA in abundance have potential industrial application, and have formed the focus of many studies. On examining the nutrient requirements of these *Bacillus* strains, they were classified as being either glutamic acid-dependent or -independent. In addition to the effect of glutamic acid, other conditions such as ionic strength, aeration, temperature, and culture time, were also shown to significantly affect both the productivity and quality of γ-PGA. Optimal γ-PGA production has been determined for some *Bacillus* strains (Table 3).

5.1 Glutamic Acid-dependent Producers

γ-PGA-producing *Bacillus* strains that require glutamic acid addition to the media to stimulate γ-PGA production (and also sometimes cell growth) include *B. anthracis*, *B. subtilis* (*natto*) IFO 3335, *B. licheniformis* ATCC 9945A, *B. subtilis* (*natto*) MR-141, *B. subtilis* (*chungkookjang*), and *B. subtilis* F-2-01.

5.1.1 *B. subtilis* (*natto*) IFO 3335

This strain has been the most used to study γ-PGA production. Kunioka and Goto (1994) showed that *B. subtilis* (*natto*) IFO 3335 produced abundant γ-PGA without any byproducts (e.g., fructan) in a medium containing L-glutamic acid, citric acid (as a carbon source), and $(NH_4)_2SO_4$ (as a nitrogen source). These authors also found that, during γ-PGA production, the glutamic acid concentration in the medium was apparently unchanged and suggested this material was in fact produced in the medium during γ-PGA production. Later, Kunioka (1997) concluded that γ-PGA is mainly produced from α-ketoglutaric acid, which is formed from citric acid via the tricarboxylic acid (TCA) cycle, and NH_3. In this strain, exogenous glutamic acid functions as an activator for the γ-PGA precursor and/or the γ-PGA synthesis but not as a direct precursor of γ-PGA synthesis. In fact, γ-PGA was produced abundantly after the addition to the medium of a small amount of L-glutamine, rather than a large amount of L-glutamic acid (Kunioka, 1995). Hence, this strain is a de-novo γ-PGA producer and belongs to the glutamic acid-independent group. This also implies a need to re-classify *B. subtilis* (*natto*) strains that have previously been included in this category. A schematic pathway for the synthesis of γ-PGA from L-

Tab. 3 γ-PGA overproducers and optimal culture conditions

Strain	Important components in media	Culture conditions	Productivity [g L^{-1}]
Glutamic acid-dependent γ-PGA producers			
B. subtilis IFO 3335	Glutamic acid (30 g L^{-1}); $(NH_4)_2SO_4$(30 g L^{-1}); citric acid (20 g L^{-1})	37°C, 2 days	10–20[a]
B. licheniformis ATCC 9945A	Glutamic acid (20 g L^{-1}); NH_4Cl (7 g L^{-1}); citric acid (12 g L^{-1}); $CaCl_2$ (0.2 g L^{-1}); $MnSO_4 \cdot 7H_2O$ (0.3 g L^{-1})	37°C, 2–3 days (Fed-batch culture with pure oxygen supply)	35[b]
B. subtilis (*natto*) MR-141	Glutamic acid (30 g L^{-1}); maltose (60 g L^{-1}); soy sauce (70 g L^{-1})	40°C, 3–4 days	35[c]
B. subtilis (*chungkookjang*)	Glutamic acid (20 g L^{-1}); sucrose (50 g L^{-1}); NaCl (0.5–5.0 g L^{-1})	30°C, 5 days	13.5–16.5[d]
B. subtilis F-2-01	Glutamic acid (70 g L^{-1}); glucose (1 g L^{-1}); veal infusion broth (20 g L^{-1})	30°C, 2–3 days	50[e]
Glutamic acid-independent γ-PGA producers			
B. subtilis TAM-4	NH_4Cl (18 g L^{-1}); fructose (75 g L^{-1})	30°C, 4 days	20[f]
B. licheniformis A35	NH_4Cl (18 g L^{-1}); glucose (75 g L^{-1}); $MnSO_4 \cdot 87H_2O$ (0.04 g L^{-1}); HNO_3 (20 g L^{-1})	30°C, 3–5 days	8–12[g]
B. licheniformis S173	NH_4Cl (4 g L^{-1}); citric acid (20 g L^{-1}); Mn^{2+}, Fe^{2+}, Ca^{2+}, Zn^{2+} (1 mM each)	37°C, 30 h	1.27[h]

References: [a]Kunioka and Goto, 1994; [b]Yoon et al., 2000; [c]Ogawa et al., 1996; [d]Ashiuchi et al., 2001a; [e]Kubota et al., 1993a; [f]Ito et al., 1996; [g]Cheng et al., 1989; [h]Kambourova et al., 2001.

glutamic acid (or L-glutamine), citric acid, and NH_4^+, which is entirely based on a plausible mechanism of γ-PGA synthesis in *B. licheniformis* ATCC 9945A, was proposed (Kunioka, 1997). In this pathway, endogenous L-glutamic acid is formed from α-ketoglutaric acid in two different ways (Figure 6). In the absence of glutamine, L-glutamic acid is synthesized by the L-glutamic acid dehydrogenase pathway (Belitsky and Sonenshein, 1998), whereas in the presence of glutamine, an alternative pathway involving L-glutamic acid synthase and L-glutamine synthetase is activated (Belitsky et al., 2000). The main component of γ-PGA, however, is D-glutamic acid (83%) (Goto and Kunioka, 1992). Kunioka and co-workers believe that the alanine racemase-coupled D-amino acid aminotransferase (DAT) pathway (Figure 7) is crucial for D-glutamic acid supply in γ-PGA production by *B. subtilis* (*natto*) IFO 3335.

5.1.2
B. licheniformis ATCC 9945A

γ-PGA production by this organism has been studied systematically. Initially, Thorne et al. (1954) showed that various factors, such as glutamic acid, inorganic salts, carbon sources, cell density, and aeration, affected the yield and quality of γ-PGA. Subsequently, Leonard et al. (1958) prepared the E medium which is most frequently used for γ-PGA production. Yoon et al. (2000) established a

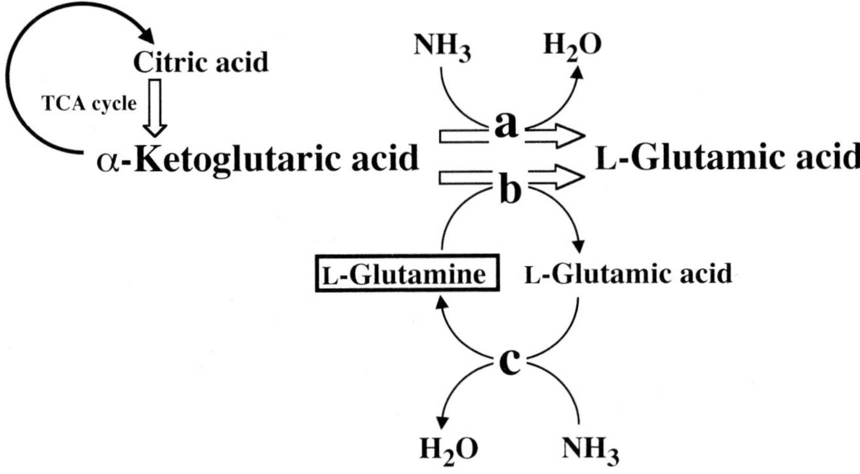

Fig. 6 L-Glutamic acid synthetic pathways in *Bacillus*. Enzymes: (a) L-glutamic acid dehydrogenase; (b) L-glutamic acid synthase; (c) L-glutamine synthetase.

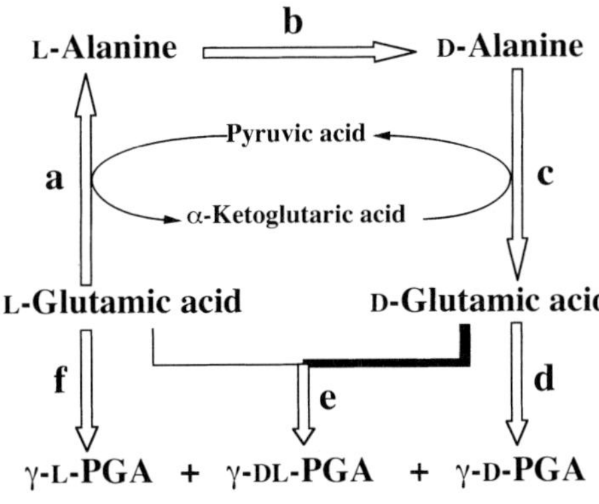

Fig. 7 A proposed pathway of γ-PGA synthesis in *B. licheniformis* ATCC 9945A. Enzymes: (a) L-amino acid aminotransferase; (b) alanine racemase; (c) D-amino acid aminotransferase; (d) γ-D-PGA synthetase; (e) γ-DL-PGA synthetase; (f) γ-L-PGA synthetase.

high production of γ-PGA by the fed-batch culture using pulsed-feeding of both citric acid and L-glutamic acid.

There were two diametrically opposite opinions as to whether the stereochemical composition of γ-PGA produced by this strain was modulated by Mn^{2+} (Thorne and Leonard et al., 1958; Gardner and Troy 1979). Cromwick and Gross (1995a) confirmed that both the yield and the D-glutamic acid content of γ-PGA increased significantly with an increase in Mn^{2+} concentration of the medium. Pérez-Camero et al. (1999) also showed that both the D-isomer content and the molecular size of γ-PGA were dramatically changed by the Mn^{2+} concentration. Cromwick and Gross (1995b) examined the γ-PGA synthetic pathway in this strain using L-[^{13}C] glutamic acid and [^{13}C]citric acid, and showed that exogenous L-glutamic acid was directly incorporated into γ-PGA (L-isomer content 50%) at low Mn^{2+} concentration, while endogenous precursors synthesized from α-ketoglutaric acid were incorporated into γ-PGA produced (L-isomer content $< 10\%$) at high Mn^{2+} concentration. It seems likely that *B. licheniformis* possesses two or more distinct γ-PGA synthetic systems, and if these were to operate irregularly, then stereochemically differing γ-PGAs would be produced. The γ-PGA syntheses in *B. licheniformis* must be largely different from those in *B. subtilis*, because a dramatic response to Mn^{2+} (as seen in γ-

PGA production by *B. licheniformis*) was not seen with *B. subtilis* (Kunioka, 1995; Ito et al., 1996; Nagai et al., 1997; Ashiuchi et al., 1999a), despite *B. licheniformis* and *B. subtilis* being phenotypically and genotypically very similar.

Further studies on carbon source utilization in the E medium indicated a relatively high overall usage of citric acid and glycerol, though glutamic acid remained unchanged. In addition, removal of citric acid (but not L-glutamic acid) from the medium resulted in a dramatic decrease in γ-PGA production. Cromwick et al. (1996) examined the effects of pH on γ-PGA production by *B. licheniformis* ATCC 9945A and suggested that a de-novo system via the TCA cycle was more active for γ-PGA synthesis. Surprisingly, Kambourova et al. (2001) showed that γ-PGA production by several *B. licheniformis* strains was strongly inhibited by an increased concentration of exogenous L-glutamic acid. These data clearly contradicted previously published results (Cromwick and Gross, 1995a).

5.1.3
B. subtilis (*natto*) MR-141

The culture media of γ-PGA producers become highly viscous due to polymer accumulation, with maximum viscosity being observed at neutral pH (Cromwick et al., 1996). A significant increase in viscosity results in limited volumetric oxygen mass transfer, and this leads to insufficient cell growth and a reduction in γ-PGA yield. Although a high concentration of NaCl (>5%) serves as an antifoaming agent and also suppresses the increase in viscosity, *B. subtilis* (*natto*) generally does not grow well under high-saline conditions. Initially, it was thought that to establish the mass production of γ-PGA by using this strain would be difficult, but Ogawa et al. (1997) subsequently isolated *B. subtilis* (*natto*) MR-141 which could be used in the large-scale production of γ-PGA; moreover, these authors also showed that endogenous L-glutamic acid serves directly as the precursor of γ-PGA.

5.1.4
B. subtilis (*chungkookjang*)

This strain was isolated from the traditional Korean fermented soybean paste, chung-kook-jang (Ashiuchi et al., 2001a), which is a highly salty seasoning containing abundant γ-PGA. The strain can grow, forming a highly mucous colony, on a high-saline medium containing L-glutamic acid. As described above, the apparent viscosity of the culture medium decreased dramatically as the salt concentration was increased. γ-PGAs with relatively high molecular masses (>1000 kDa) were synthesized at low NaCl concentrations (<0.5%), while those with comparatively low molecular masses (10–200 kDa) were produced preferentially at high NaCl concentrations (>10%). Thus, it is possible to control the molecular size of γ-PGA, and also the viscosity of the medium, by using this characteristic of *B. subtilis* (*chungkookjang*).

This strain produced abundant γ-PGA in the medium containing amino acids of the glutamic acid family, such as L-glutamic acid, L-glutamine, and L-arginine (Ashiuchi et al., 2001b). Without these amino acids, elongated γ-PGAs (>1000 kDa) are not produced. This characteristic is of interest because γ-PGA production can be easily controlled by altering the concentration of L-glutamic acid in the medium. L-Alanine, which is the best substrate for the alanine racemase-coupled DAT pathway, did not induce γ-PGA production, implying that this pathway is not important in γ-PGA production. As this strain did not synthesize any byproducts, its γ-PGA production could be analyzed kinetically. The Lineweaver-Burk plot produced a straight line: the apparent K_m (for exogenous L-glutamic acid) and V_{max} (as γ-PGA production rate) were 0.7% (41 mM) and 8×10^{-13} g cell^{-1} day^{-1}, respectively. The turnover number of the glutamic acid ligation by one cell was estimated as 240 min^{-1}.

Most strains of plasmid-bearing *B. subtilis* (*natto*) are generally unsuitable for genetic manipulation because of their little or no genetic competence (Ashikaga et al., 2000), but *B. subtilis* (*chungkookjang*) harbored no plasmid and was shown to be the first *B. subtilis* strain with naturally high γ-PGA productivity and high genetic competence (Ashiuchi et al., 2001a). Such capability should, in time, allow the design of genetically and metabolically engineered γ-PGA producers.

Among *Bacillus* γ-PGA producers, there is a close relationship between γ-PGA synthesis and sporulation (Guex-Holzer and Tomcsik, 1956; McCuen and Thorne, 1971; Ito et al., 1996). This strain, however, was shown to have an absolute requirement for Mn^{2+} for sporulation, but produced abundant γ-PGA without Mn^{2+}. γ-PGA production by *B. subtilis* (*natto*) is not reproducible, and often effectively disappears after several rounds of subcultivation. However, the γ-PGA productivity of *B. subtilis* (*chungkookjang*) remains unchanged even after long-term storage and repeated subcultivation (Ashiuchi et al., 2001a). It is likely that *B. subtilis* (*chungkookjang*) will serve as a potent tool not only in the development of a mass-producer of γ-PGA but also in the elucidation of the γ-PGA synthetic mechanism in *B. subtilis* in its entirety.

5.1.5 *B. subtilis* F-2-01

This strain was isolated from soil (Kubota et al., 1993a), and shows the highest γ-PGA productivity (50 g L^{-1}) among γ-PGA producers identified to date. The highest productivity resulted from the addition of L-glutamic acid to the medium. The D-glutamic acid content in γ-PGA increased significantly with culture time.

5.2 Glutamic Acid-independent Producers

Among γ-PGA-producing *Bacillus*, the strains that require little glutamic acid for γ-PGA production include *B. subtilis* (*natto*) 5E, *B. subtilis* TAM-4, *B. licheniformis* A35, and *B. licheniformis* S173. Among these organisms, *B. subtilis* 5E is indeed a glutamic acid-independent producer but not a de-novo γ-PGA producer due to a requirement for L-proline (a member of the glutamic acid family) (Shih and Van, 2001). The de-novo γ-PGA producers among glutamic acid-independent producers have been selected for industrial use, despite little being known about them compared with the glutamic acid-dependent producers.

5.2.1 *B. subtilis* TAM-4

This strain was isolated from soil (Ito et al., 1996) and produced a large amount of γ-PGA from organic/inorganic nitrogen sources and various sugars under aerobic conditions, indicating that it is a de-novo γ-PGA producer. Because *B. subtilis* TAM-4 can grow without addition of biotin to the medium, this strain is probably more suited to industrial uses than many *B. subtilis* (*natto*) strains (Fujii, 1963). *B. subtilis* TAM-4 also possesses a long-lasting γ-PGA production ability, similar to *B. subtilis* (*chungkookjang*). High γ-PGA productivity could be obtained by using an M medium, but no byproducts were accumulated in sugar media. Most *B. subtilis* (*natto*) strains produce considerable amounts of polysaccharides in the medium, and this may be an advantage for the isolation of γ-PGA. The D-glutamic acid content of the γ-PGA produced remained constant at 73%.

5.2.2 *B. licheniformis* A35

This is the only γ-PGA producer that can synthesize the polymer under nitrate-respiration conditions (Cheng et al., 1989). Oxygen availability is quite important for the cell growth and γ-PGA productivity of many *Bacillus* strains (Cromwick et al., 1996). A strictly controlled aeration system that provides a continuous supply of oxygen-enriched air to the culture medium, or an ionic substitute for gaseous oxygen (e.g., NO_3^-), is required for effective γ-PGA production. As the latter is more practicable and economical than the former, this strain is a candidate as an industrial γ-PGA producer. For γ-PGA production, this organism could utilize various sugars, but not organic acids. High γ-PGA productivity could be obtained by using a culture medium containing NH_4Cl, glucose, and NO_3^- under nitrate-respiration conditions– in other words, a de-novo γ-PGA synthesis. Mn^{2+} was shown to affect the D-glutamic acid content of γ-PGA, this characteristic being similar to that seen in *B. licheniformis* ATCC 9945A.

5.2.3 *B. licheniformis* S173

Although the γ-PGA productivity of this strain is not very high, its characteristics provide a clue to the crucial difference between the γ-PGA production systems of *B. licheniformis* and *B. subtilis*. Kambourova et al. (2001) examined the effects of nitrogen and carbon sources on γ-PGA production by this strain. γ-PGA yields increased in the order of $NH_4Cl > NH_4NO_3 > (NH_4)_2SO_4$ > L-glutamine > L-arginine > L-aspartic acid > $NaNO_3$. γ-PGA production, however, was clearly inhibited by L-glutamic acid. The inhibition of γ-PGA productivity by the amino acid was observed in all *B. licheniformis* strains used. Among the carbon sources investigated, citric acid was the best for γ-PGA production, and was readily metabolized into α-ketoglutaric acid via the TCA cycle. Since inorganic ammonium salts are effective for γ-PGA production, L-glutamic acid as the precursor of γ-PGA is preferably synthesized from α-ketoglutaric acid and NH_3 (see Figure 7). Both L-glutamine and L-arginine are metabolized into L-glutamic acid via the L-glutamine synthetase-coupled L-glutamic acid synthase pathway and the L-arginine catabolic pathway (Belitsky and Sonenshein, 1998), respectively. L-Aspartic acid aminotransferase converts L-aspartic acid and α-ketoglutaric acid into oxaloacetic acid and L-glutamic acid, respectively. It thus seems likely that the maintenance of correct levels of endogenous glutamic acid is important for γ-PGA production in *B. licheniformis*, and that an incorrect increase in glutamic acid concentration *in vivo* triggers a change in the metabolic flow from γ-PGA production into another type of production.

6 Physiology

Identifying the physiological reason for γ-PGA production is important. Because γ-PGA is generally produced in the stationary phase with the development of spores, it was originally thought that the material might participate in spore formation or was involved in the physiology of spores. However, as described in Section 5.1, *B. subtilis* (*chungkookjang*) produces a large amount of γ-PGA without sporulation having been induced. The γ-PGA-synthesis genes of *B. subtilis* (*natto*), designated as *pgsB*, -*C*, and -*A*, are almost identical to the unidentified genes of non-γ-PGA-producing *B. subtilis* (Marburg) 168, *ywsC*, *ywtA*, and *ywtB*, respectively (Ashiuchi et al., 1999a). γ-PGA is probably not involved in primary functions for cell viability. Birrer et al. (1994) reported that the use of cryogenically frozen vegetative cells of *B. licheniformis* ATCC 9945A overcame the problem whereby this strain is readily transformed into non-γ-PGA-producing variants by repeated subcultivation (Leonard et al., 1958). A sharp low-temperature shift led the probable γ-PGA-transporter moiety (PgsA) of the membranous γ-PGA synthetase complex (PgsBCA) of *B. subtilis* (see Section 8.2) into an active form (M. Ashiuchi et al., unpublished results). It was believed that *B. subtilis* (*natto*) originated from a cold, dry region of China (Hara, 1990); hence, it is possible that some *B. subtilis* and *B. licheniformis* strains may have used the high cryoprotectivity of γ-PGA (Mitsuiki et al., 1998) in order to adapt to environments at low temperature, though further studies are required to verify such a claim. Recent studies on bacterial, archaeal, and animal γ-PGA producers other than *B. subtilis* and *B. licheniformis* have suggested that this polymer functions as an adaptation agent in various environments. These novel physiological functions of γ-PGA are reviewed in the following section.

6.1 Nullification of Immunity in Infectious *B. anthracis*

B. anthracis is the causative agent of anthrax, and two major virulence factors of *B. anthracis* – a capsule (mainly γ-D-PGA) and a tripartite toxin–are known (Thorne et al., 1960). Although γ-D-PGA itself is avirulent in mammals, it strongly inhibits the phagocytosis of hosts and has properties that enhance the virulence of the *B. anthracis* toxin *in vivo* (Keppie et al., 1963). Makino et al. (1989) first identified *B. anthracis* encapsulation genes (*cap*; see Section 7.1) and showed that the gene-disruptant was highly sensitive to phagocytosis and almost lost the virulence to hosts. In addition, γ-D-PGA depolymerase (Dep) was essential for exhibiting the virulence of this pathogenic agent (Uchida et al., 1993). γ-PGA plays an important role in evading mammalian immune defense mechanisms.

6.2 Neutralization of Near-cell Surface in Alkalophiles

The cell walls of Gram-positive bacteria usually consist of a peptidoglycan layer and some specific polymers such as proteins and polysaccharides. In alkalophilic strains of *Bacillus*, the cell surfaces are surrounded with two acidic polymers: one is a teichuronic acid (TUA), and the other is a teichuronopeptide (TUP) consisting of poly-α,β-1,4-D-glucuronic acid (PGlcU) and γ-L-PGA (Aono, 1987). Aono and co-workers (Aono and Ohtani, 1990; Ito et al., 1994) constructed three mutants of a facultative alkalophilic *B. halodurans* C-125 (formerly *B. lentus* C-125 or *Bacillus* sp. C-125). These organisms are defective in the production of TUA and/or TUP. The non-TUP-producing *B. halodurans* C-125 mutant lost its alkalophily. The cell walls of the TUP mutant contained TUA and PGlcU and thus lacked only γ-L-PGA, suggesting that γ-L-PGA contributes to maintaining the pH homeostasis of *B. halodurans* at high alkaline pH. It has also been assumed that a high density of fixed anion charges, which are derived from the deprotonated carboxyl groups of γ-L-PGA, function as obstacles repulsing surplus hydroxyl ions in the environment and as reservoirs capturing hydrogen ions. The *tupA* gene of *B. halodurans* has been cloned by genetic complementation of alkalophily of the TUP mutant. The gene product is involved in the formation of a TUP complex from PGlcU and γ-L-PGA (Aono et al., 1999). The γ-L-PGA-synthesis genes and enzymologic properties of these gene products are discussed in Section 8.2.

6.3 Prevention of Drastic Dehydration under High-saline Conditions in Halophiles

Some extremely halophilic archaea are able to synthesize γ-PGA (Niemetz et al., 1997). Hezayen et al. (2000) identified a new γ-PGA-producing archaeon, *N. aegyptiaca* which can grow in a medium containing >10% NaCl, though γ-PGA production was observed only under extremely high-saline conditions (>20% NaCl). The water-binding capacity of this polymer contributes to the protection of the microorganism from drastic dehydration under extremely high-saline conditions. In fact, the moisture of the surrounding area of this sticky colony has been maintained as expected after long-term cultivation (e.g., over 1 month at 40°C) in a medium containing 25% NaCl (M. Ashiuchi et al., unpublished results). In contrast, the water-binding capacities of γ-PGAs of *Bacillus* strains usually decrease dramatically in the presence of salts (Kunioka, 1997). The *N. aegyptiaca* polymer may possess a structural feature that can contain water in remarkable abundance even in the presence of a compound showing high ionic strength – a property which might prove advantageous for future environmental uses of γ-PGA. This polymer consisted mainly of L-glutamic acid and contained little D-glutamic acid, unlike most γ-PGAs produced by *Bacillus* (Hezayen et al., 2001). This γ-L-PGA had a fairly high molecular size (Ashiuchi and Misono, 2002). This is the sole example of naturally occurring γ-PGA with both stereochemical homogeneity and large molecular size. Polymer production was strongly induced by casamino acid but not by L-glutamic acid; this suggests that the γ-PGA synthetic system of *N. aegyptiaca* is entirely different from those of *Bacillus* strains. The identification of this system from *N. aegyptiaca* may be urgently required for structural analyses and applications of its unique γ-PGA.

6.4 Regulation of Osmotic Pressure in Cnidarians

The polymer is the major constituent of the sticky substances in the nematocysts (the capsular secretory products of stringing cells) of *Hydra*, and also plays a crucial role in the function of cytes (Weber, 1989, 1990). This sticky substance, which is not related to the acute toxic effects of nematocysts (Weber et al., 1988), serves to adhere to the surfaces of aggressor or prey animals to prevent them from escaping when *Hydra* injects biologically active substances, such as highly toxic venoms or lytic enzymes, into them using its inverted tubule (Hessinger and Lenhoff, 1988). The material consists mainly of γ-L-PGAs with various molecular masses, and is, in cooperation with major bioactive cations such as Ca^{2+}, Mg^{2+}, and K^{+}, responsible for the generation and regulation of an internal osmotic pressure. γ-PGA is widely distributed over the cnidarians. It is important physiologically and evolutionarily to conduct a thorough investigation and comparison of the structure and function of bacterial, archaeal, and cnidarian γ-PGA synthetic systems.

7 Molecular Genetics

For many years, groups conducting research into γ-PGA have shown a strong interest in the genes concerned with γ-PGA synthesis. The first report made was on *B. licheniformis* ATCC 9945A when the loci of two *pep* genes, the transduction of which resulted in a deficiency of γ-PGA production, were mapped on the chromosome. These *pep* mutants became highly transformable, suggesting that *pep* genes are also involved in the development of genetic competence for transformation in the bacterium (McCuen and Thorne, 1971). Many γ-PGA-producing *B. subtilis* (*natto*) strains possess small cryptic plasmid(s) (Meijer et al., 1998; Thorsted et al., 1999), whereas *B. subtilis* 168 and its derivatives, which harbor substantially no plasmid, do not produce γ-PGA (Hara et al., 1982). Hara et al. (1982) also analyzed the function of the *B. subtilis* (*natto*) plasmid pUH1 in γ-PGA production by plasmid curing and transformation, and concluded that the gene(s) involved in γ-PGA production lay on the plasmid DNA. Later, these authors reported that the gene encodes a novel γ-glutamyltranspeptidase (GGT) (Hara et al., 1992), which has no sequence similarity to GGTs identified so far, even to the enzyme from *B. subtilis* (Xu and Strauch, 1996). In response to this contradiction, Hara et al. (1995) claimed again that this gene encoded a positive regulator of the *ggt* gene but not the enzyme itself, and designated it the *psf* gene. However, both the *psf* gene and its surrounding region are highly homologous to the *mob*/*pre* genes and its recognition site RS_A for the mobilization and site-specific recombination of mobilizable plasmids of a variety of Gram-positive bacteria (Bates and Gilbert 1989; Parini et al., 1991; LeBlanc et al., 1993). Nagai et al. (1997) showed that plasmid curing has no effect on γ-PGA production and concluded that, in *B. subtilis* (*natto*), plasmids do not encode any gene that is important for γ-PGA synthesis. *B. subtilis* 168 cannot be transformed into a γ-PGA producer by the introduction of *B. subtilis* (*natto*) plasmids. Moreover, plasmid-free PGA producers of *B. subtilis* have been isolated (Onodera et al., 1994; Ashiuchi et al., 2001a). This dispute has been finally resolved by a recent study of the γ-PGA synthetase complex-encoding genes on the *B. subtilis* chromosome (Ashiuchi et al., 1999a, 2001b) whereby crucial results were presented on γ-PGA synthesis from the point of view of molecular genetics.

7.1 Encapsulation (*cap*) Genes

The exhibition of full virulence by *B. anthracis* requires the production of two major factors: the secreted anthrax toxins, and a surface γ-D-PGA capsule. This serious pathogen harbors two comparatively large plasmids, pXO1 (180 kb) and pXO2 (95 kb). For most animals, strains cured of either plasmid are either greatly attenuated (carrying only pXO1) or avirulent (carrying only pXO2) (Mikesell et al., 1983; Green et al., 1985; Ivins et al., 1986). Uchida et al. (1985) showed that almost all genes responsible for the encapsulation lie on the plasmid pXO2. By using genetic complementation techniques, Makino et al. (1988, 1989) identified three cistrons, *capB*, *capC*, and *capA* (*cap*: en*cap*sulation; DDBJ/EMBL/GenBank accession no. M24150). The organization of *capBCA* genes is illustrated in Figure 8a. All three genes were essential for the encapsulation of the pXO2-cured *B. anthracis* mutant and the *Escherichia coli* clone. This is the first example of cloning of the genes responsible for γ-PGA synthesis. The minicell analysis using [^{35}S]methionine, i.e., detection of proteins that have been newly synthesized *in vivo* during a short-term incubation with the radioactive amino acid by a combination of SDS–PAGE and autoradiography, suggested that all of these gene products, CapB

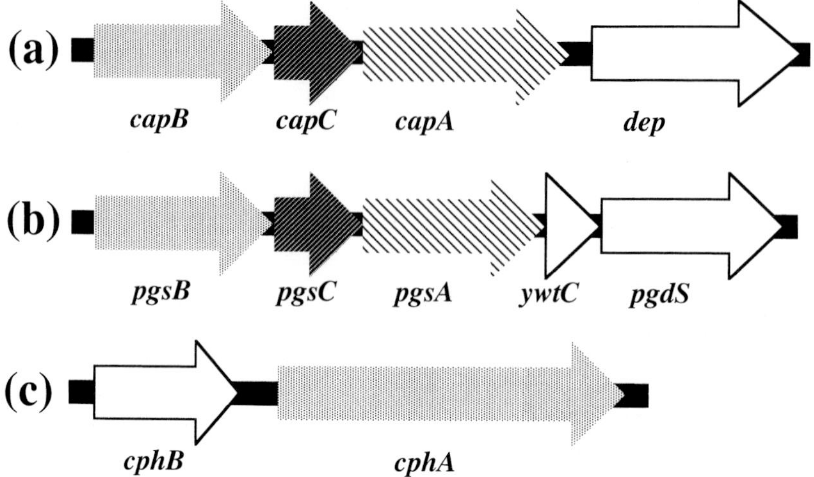

Fig. 8 Organization of open reading frames (ORFs) concerned with the metabolism of polyamino acids (Ashiuchi and Misono, 2002). (a) ORFs for the capsular metabolism on the *B. anthracis* plasmid pXO2: *capB*, gene encoding the CapB component of the γ-D-PGA synthetase complex; *capC*, gene encoding the CapC component; *capA*, gene encoding the CapA component; *dep*, gene encoding the γ-D-PGA depolymerase. (b) ORFs for the γ-DL-PGA metabolism on the *B. subtilis* chromosome: *pgsB*, gene encoding the PgsB component of the γ-DL-PGA synthetase complex; *pgsC*, gene encoding the PgsC component; *pgsA*, gene encoding the PgsA component; *pgdS*, gene encoding the γ-PGA depolymerase; *ywtC*, a functionally unidentified gene (may be encoding a small leader protein for the *pgdS*-gene attenuation system). (c) ORFs for the cyanophycin metabolism on the *Synechocystis* PCC6308 chromosome (Aboulmagd et al., 2000): *cphA*, cyanophycin synthetase gene; *cphB*, cyanophycinase (cyanophycin depolymerase) gene.

(44 kDa), CapC (16 kDa), and CapA (46 kDa), are associated with membranes. These authors further concluded that *cap* genes are unique in *B. anthracis*, and therefore no orthologue is inherited in other γ-PGA-producing *Bacillus* such as *B. subtilis*, *B. megaterium*, and *B. licheniformis*.

7.2
Poly-γ-glutamic Acid Synthesis (*pgs*) Genes

The purification of the genuine γ-PGA synthetic system (γ-PGA synthetase complex) of *Bacillus*, which can synthesize highly elongated γ-PGA *in vitro* in the same way as *in vivo*, has not been successful due to the extreme instability of the system. Ashiuchi et al. (1999a) therefore introduced a genetic approach into γ-PGA research. A gene cluster encoding the γ-PGA synthetase complex of *B. subtilis* (*natto*) IFO 3336 was identified by screening a γ-PGA-producing *E. coli* clone which harbored three genes, designated as *pgsB*, *pgsC*, and *pgsA* (*pgs*: poly-γ-glutamic acid synthesis; accession no. AB016245), which were highly homologous to *B. anthracis cap* genes (overall identities of PgsB, -C, and -A to CapB, -C, and -A are 66, 77, and 50%, respectively; these similarities are 80, 89, and 70%, respectively) contrary to what had been proposed by Makino et al. (1989). As shown in Figure 8b, the *pgs* genes form a gene cluster (probably an operon structure) on the *B. subtilis* chromosome. All *pgsB*, -*C*, and -*A* genes were indispensable for γ-PGA production by the *E. coli* clone (Table 4). Mn^{2+} is a good activator for γ-PGA production by the *E. coli* clone cells but

Tab. 4 γ-PGA production by *E. coli* clones and the stereochemistry of the γ-PGAs (Ashiuchi and Misono, 2002)

E. coli clones	Metal ions[a]	γ-PGA productivity [mg L^{-1}]	Content [%]	
			D-Enantiomer	L-Enantiomer
JM109/pPGS1	Mn^{2+}	12 ± 5	13 ± 7	87 ± 7
	Mg^{2+}	2 ± 1	10 ± 5	90 ± 5
JM109/pPGS1+pBSGR3	Mn^{2+}	24 ± 2	64 ± 4	36 ± 4
	Mg^{2+}	10 ± 3	59 ± 6	41 ± 6

[a] 1 mM. The γ-PGA productivities of both *E. coli* JM109/pPGS1 (harboring the *pgsBCA* genes) and JM109/pPGS1+pBSGR3 (harboring both the *pgsBCA* genes and the *glr* gene) clones are shown. Other *E. coli* clones, JM109/pTrc99A (harboring vector only), JM109/pBSGR3 (harboring the *glr* gene only), JM109/pPGB1 (harboring the *pgsB* gene only), JM109/pPGC1 (harboring the *pgsC* gene only), JM109/pPGA1 (harboring the *pgsA* gene only), JM109/pPGBC1 (harboring the *pgsBC* genes), JM109/pPGBA1 (harboring the *pgsBA* genes), and JM109/pPGCA1 (harboring the *pgsCA* genes), could not produce γ-PGA.

is not a determinant of the D-glutamic acid content.

In order to clarify whether a γ-PGA synthetic system other than the PgsBCA system operates in *B. subtilis*, the *pgsBCA* genes-disruptant of *B. subtilis* (*chungkookjang*) was constructed (Ashiuchi et al., 2001b). The *pgs* null mutant could not produce γ-PGA under any conditions. Urushibata et al. (2002) also reported the significance of all of the genes, *ywsC* (corresponding to *pgsB*), *ywtA* (corresponding to *pgsC*), and *ywtB* (corresponding to *pgsA*), in γ-PGA production by *B. subtilis* (*natto*). The PgsBCA system is thus the sole machinery of the γ-PGA synthesis.

7.3 Regulatory Genes

The reproducibility of γ-PGA productivity of *B. subtilis* (*natto*) is not high, and it unexpectedly and readily changes (or is even lost) while under strictly controlled culture conditions. The identification and systematic understanding of the regulatory systems of γ-PGA production are important not only in basic molecular genetics for the study of γ-PGA producers but also in its practical application for mass production of the polymer. Nagai et al. (1997, 2000) demonstrated two new genetic factors that were responsible for such a dramatic decrease in γ-PGA productivity: a generalized transducing phage ϕBN100 and a transposable insertion sequence IS*4Bsu1* (accession no. AB031551). In particular, the latter is important as the first finding of the IS element occurring naturally in *B. subtilis*. These authors found that, in *B. subtilis* (*natto*) which was defective in γ-PGA production, the element was very frequently inserted into the region encoding the sensor domain of the ComP protein. There are structural and functional variations in ComPXQA quorum-sensing systems between *B. subtilis* 168 and *B. subtilis* (*natto*) (accession no. AB039951), and *B. subtilis* cells sharing the same pheromone system establish a specific intercellular communication (Tran et al., 2000). In *B. subtilis*, the ComPXQA system functions as the most primary global regulator, which plays crucial roles in cell density-dependent phenotypes including the development of genetic competence. Ito (1999) proposed a regulation mechanism for the genes responsible for γ-PGA production. First, the ComX pheromone is processed from the native ComX protein by the ComQ protein and released extracellularly as a quorum-sensing signal. The interaction with the externally accumulated pheromone and the sensor domain of ComP activates the ComP-ComA two-component signal transduction system. The resulting phosphorylated ComA protein mediates up- and down-regulation of the expression of second transcriptional regulators. The *pgs* genes are activated via subsequent cascade reactions that are as yet unidentified. Therefore, identification and characterization of the specific regulator(s) of the *pgs* gene expression must be indispensable for elucidation of the γ-PGA-production system in its entirety, and the clue may be found in a characteristic of *B. subtilis* Marburg 168. This strain's transformant harboring an expression vector of the *pgs* genes under the control of the *xylA* promoter could not increase γ-PGA productivity even in the presence of an adequate concentration of xylose (M. Ashiuchi et al., unpublished results). The same phenomenon has been found in the expression test of the *cap* genes (Makino et al., 1988). In addition to transcriptional regulations, post-translational regulations are likely to be important in the exhibition of γ-PGA productivity. *B. subtilis* 168 may produce in abundance a factor which disrupts an active form of the *B. subtilis* Pgs system for γ-PGA synthesis such as specific proteases for membrane clearance, or γ-PGA-producing *B. subtilis* may synthesize a factor leading newly synthesized Pgs polypeptides into an active form such as molecular chaperones.

Capsule synthesis in *B. anthracis* is dependent upon the presence of CO_2. Makino et al. (1988) isolated a DNA fragment containing both *cap* genes and a gene responsible for its CO_2 regulation from the pXO2 plasmid. γ-PGA production by both the *B. anthracis* and *E. coli* clones was significantly induced by addition of CO_2 to the media. Later, Vietri et al. (1995) cloned the *acpA* gene of pXO2, which encodes a positive *trans*-acting protein for the bicarbonate-mediated regulation of γ-PGA synthesis (accession no. U02535) and showed that the transcription of *acpA* is bicarbonate-mediated. This is probably the first identification of the specific regulatory gene for the expression of γ-PGA-synthesis genes. The AcpA protein shows a structural feature as the bacterial transcriptional regulator; the amino acid sequence is highly similar to those of two virulent regulators, i.e., the Mry protein, a *trans*-acting regulator of M protein synthesis of *Streptococcus pyogenes* (identity 21.6%; similarity 46%) (Pérez-Casal et al., 1991) and the AtxA protein, a *trans*-activator of the anthrax toxins on the pXO1 plasmid (identity 27.8%; similarity 51.1%).

8
Biosynthesis

In view of the structural features of γ-PGA, including the linkage formula, the stereochemistry, and the variation in molecular size, it has been suggested that the polymer is synthesized in a ribosome-independent manner. Multienzyme systems mostly operate on the basis of the thiotemplate mechanism (the best-studied ribosome-independent mechanism), which generally offers high stereoselectivity for amino acid precursors and productivity of comparatively small peptides in which amino acid arrangements are strictly regulated (Kleinkauf and Von Döhren, 1996). A synthetase of highly elongated γ-PGA should involve an extremely unique reaction mechanism compared with those of peptide synthetic machineries.

This section deals with some of the enzymes involved in γ-PGA synthesis.

8.1 Poly-γ-glutamic Acid Precursor Biosynthesis

The main component of naturally occurring γ-PGAs is usually D-glutamic acid, and two different types of enzymes synthesizing D-glutamic acid have been identified to date.

8.1.1 D-Amino Acid Aminotransferase (DAT)

In *B. anthracis*, L-alanine is formed by L-amino acid aminotransferase that catalyzes the transamination between pyruvic acid and various L-amino acids such as L-glutamic acid (Housewright and Thorne, 1950), whilst D-glutamic acid (the γ-PGA precursor) and pyruvic acid are produced by DAT catalyzing the transamination between α-ketoglutaric acid and D-alanine (Thorne and Molnar, 1955). D-Glutamic acid was synthesized with the cell extract of *B. licheniformis* ATCC 9945A (Thorne et al., 1955). The alanine racemase-coupled DAT pathway involved in the synthesis of a mixture of γ-D-, γ-L-, and γ-DL-PGAs of *B. licheniformis* ATCC 9945A is shown in Figure 7.

8.1.2 Glutamic acid racemase

Although, in *Bacillus*, DAT has been assumed to produce D-glutamic acid, its activity was not detected in the cell extract of *B. subtilis* (*natto*) IFO 3336. Alternatively, a high activity of glutamic acid racemase was found in this strain (Ashiuchi et al., 1998). DAT and glutamic acid racemase coexist in *B. subtilis* (*chungkookjang*) cells in the exponential phase. An enzymologic analysis of the aminotransferase before and during γ-PGA production (in the stationary phase) showed a dramatic decrease in its activity in the γ-PGA-production stage (Ashiuchi et al., 2001a). This is consistent with the result that L-alanine has no effect on γ-PGA production (see Section 5.1). DAT most likely does not participate in the D-glutamic acid supply for γ-PGA production. In contrast, the activity of glutamic acid racemase increased somewhat during γ-PGA production. The attenuation of glutamic racemase gene expression (Balikó and Venetianer, 1993) and the regulation of glutamic acid racemase production (Ho et al., 1995) are physiologically significant, and probably result in extremely low activity of glutamic acid racemase in almost all bacteria (Ashiuchi et al., 1999a). In fact, the enforced production of MurI, the glutamic acid racemase of *E. coli* responsible for peptidoglycan synthesis, and other bacterial MurI-type racemases results in the suppression of cell proliferation of *E. coli* clones through characteristic changes such as aberration in nucleoid separation (Balikó and Venetianer, 1993; Ashiuchi et al., 1999b). In this respect, the glutamic acid racemase of the γ-PGA-producing *B. subtilis* is peculiar. Its enzymologic characteristics are summarized in Table 5, and its gene and the gene product (the enzyme) were designated *glr* and Glr, respectively. In addition to Glr, the glutamic acid racemase-isozyme, YrpC (encoded by the *yrpC* gene of *B. subtilis*), was identified (Ashiuchi et al., 1999b). The enzymologic features of YrpC, such as the primary structure, substrate specificity, and cofactor independence, were very similar to those of Glr, but their kinetic parameters were quite different (see Table 5). The expression of the *yrpC* gene resulted in particular growth inhibition of the *E. coli* clone, whereas that of the *glr* gene did not affect the cell proliferation of the *E. coli* clone. YrpC more closely resembles the MurI-type glutamic acid racemases, which function in peptidoglycan synthesis, than Glr. The low catalytic efficiency and high affinity of YrpC for L-glutamic acid is

Tab. 5 Characterization of glutamic acid racemase isozymes of *Bacillus subtilis*, Glr and YrpC (MurI) (Ashiuchi and Misono, 2002)

Property *Molecular mass [kDa]*	*Glr*	*YrpC (MurI)*
SDS–PAGE	30	30
Gel filtration	30 (monomer)	30 (monomer)
Optimal pH	8.0	7.5–8.0
Optimal temperature [°C]	37	30–37
Thermostability [°C]	60	30
Cofactor requirement	No	No
Effect of UDP-MurNAc-L-Ala[a]	No	No
Kinetic parameters		
[D-Glutamate]		
K_m [mM]	2.5	1.2
V_{max} [μmol min^{-1} mg^{-1}]	56	0.33
V_{max}/K_m	22.4	0.275
K_m [mM]	50	0.18
V_{max} [μmol min^{-1} mg^{-1}]	1150	0.05
V_{max}/K_m	23.0	0.278
[K_{eq} (D/L)]	0.974	0.989
Effects of enforced expression on cell growth	No	Toxic

[a]UDP-MurNAc-L-Ala, UDP-*N*-acetylmuramyl-L-alanine, which functions as the activator of the *E. coli* glutamate racemase (Ho et al., 1995).

probably advantageous to provide a limited amount of D-glutamic acid required for cell growth at a low concentration of L-glutamic acid. In contrast, the high catalytic efficiency and low affinity of Glr for L-glutamic acid are important to maintain a high concentration of D-glutamic acid in cells where L-glutamic acid is accumulated in abundance (e.g., in γ-PGA-producing cells). The major enzyme supplying D-glutamic acid for γ-PGA synthesis in *B. subtilis* (at least the glutamic acid-dependent producers) is the Glr-type glutamic acid racemase.

In fact, an increase in both the productivity and the D-isomer content of γ-PGA has been found in the *E. coli* clone having both the *pgsBCA* genes and the *glr* gene of *B. subtilis* (see Table 4). This is due to an increase in the D-glutamic acid productivity of *E. coli* clone cells by co-production of the Glr-type glutamic acid racemase.

8.2
Poly-γ-glutamic Acid Biosynthesis

Until recently, only limited knowledge was available of the enzymes involved in γ-PGA synthesis. The membranous γ-D-PGA synthetase complex (Troy, 1973a) and the cytosolic γ-L-PGA synthetase (Leonard and Housewright, 1963) from *B. licheniformis* ATCC9945A and the extracellular GGTs of *B. licheniformis* and *B. subtilis* (*natto*) (Williams et al., 1955; Noda et al., 1980) were reported. These reaction mechanisms were also proposed, but all had little reproducibility of γ-PGA synthesis. The current state of these mysterious enzymes is described in the next

$$\text{(a) L-Glu} + \text{ATP} \rightleftharpoons [\gamma\text{-L-Glu-AMP}] + \text{PPi}$$

$$\text{(b) } [\gamma\text{-L-Glu-AMP}] + \text{HS-Enz}_1 \rightleftharpoons [\gamma\text{-D/L-Glu-S-Enz}_1] + \text{AMP}$$

$$\text{(c) } [\gamma\text{-D/L-Glu-S-Enz}_1] + [(\gamma\text{-D-Glu})_n\text{-S-Enz}_2] \rightleftharpoons [(\gamma\text{-D-Glu})_{n+1}\text{-S-Enz}_2] + \text{HS-Enz}_1$$

$$\text{(d) } [(\gamma\text{-D-Glu})_{n+1}\text{-S-Enz}_2] + \text{Acceptor} \rightleftharpoons (\gamma\text{-D-Glu})_{n+1}\text{-acceptor} + \text{HS-Enz}_2$$

Fig. 9 A proposed reaction mechanism of the membranous γ-D-PGA synthetase complex of *B. licheniformis* (Gardner and Troy, 1979). (a) L-Glutamic acid is first activated by ATP to form L-glutamyl-γ-adenylate. (b) This activated L-glutamic acid is bound to a catalytically essential sulfhydryl group in the enzyme (or in an acceptor) and isomerized into the D-isomer. (c) The γ-D-glutamyl moiety was transferred to a growing poly-γ-D-glutamyl chain bound to another sulfhydryl group of the synthetase (or an acceptor). (d) The eventually elongated γ-D-PGA chain was transferred to the NH_2 terminus of an endogenous γ-D-PGA acceptor.

section, after which recent studies of *B. subtilis* γ-DL-PGA synthetase and of *B. halodurans* γ-L-PGA synthetase, in which novel synthetic mechanisms of γ-PGA have been proposed, are reviewed.

8.2.1
γ-D-PGA Synthetase of *B. licheniformis*

Troy (1973a) showed that [^{14}C]γ-PGA was synthesized only from L-[^{14}C]glutamic acid but not D-[^{14}C]glutamic acid by an unidentified membranous enzyme from *B. licheniformis* ATCC 9945A and found that a sulfhydryl group-modifying reagent readily inhibited this reaction. Based on these data, Gardner and Troy (1979) suggested the presence of an extremely unstable γ-PGA synthetase complex. As shown in Figure 9, their proposed reaction mechanism was similar to those of multienzyme systems (Vater et al., 1985; Stein et al., 1995). This mechanism is unique in the following respects: (1) the formation of highly elongated peptide chains (>10^4 of average linkage number) compared with those by non-ribosomal peptide synthetases identified so far (Kleinkauf and Von Döhren, 1996); and (2) the requirement for an endogenous protein other than components of the enzyme complex as the thiotemplate acceptor. Since it is anticipated that the structural and functional analyses of the enzyme complex and a comparison of those of ribosomal and non-ribosomal peptide synthetic systems will provide insights into the evolution and molecular design of polyamino acids and polypeptides, many attempts have been made to isolate the enzyme complex. However, others were unable to reconfirm this interesting activity in any γ-PGA producers, including *B. licheniformis* ATCC 9945A.

8.2.2
γ-L-PGA Synthetase of *B. licheniformis*

This enzyme differs from the γ-D-PGA synthetase complex in its enzymologic properties (Leonard and Housewright, 1963); for example, it has high metal ion (Mn^{2+}) selectivity and broad nucleotide-cofactor specificity. Unfortunately, the γ-L-PGA synthetase and its unique activity have not yet been reconfirmed.

8.2.3
γ-Glutamyltranspeptidase (GGT)

Williams and Thorne (1954) were the first to report the significance of GGT in γ-DL-PGA synthesis by *B. licheniformis* ATCC 9945A. The transpeptidase was also found in the culture filtrate of *B. subtilis* (*natto*) (Noda et al., 1980), and subsequently purified to homogeneity and characterized by Ogawa et al. (1991). However, notwithstanding the many inconsistencies in the reports of GGT by Hara et al. (1992, 1995) (see Section 7), it remained unclear whether the enzyme was involved in γ-PGA synthesis. Although the enzyme is extracellular, cofactor-independent, and very stable, the reaction product that elongates more than a tripeptide could not be found. γ-PGA consists only of glutamic acid, but the substrate specificity of the enzyme is very broad. In contradicting the previously published results of Noda et al. (1980), Abe et al. (1997) reported a rapid digestion of γ-PGA by the purified GGT of *B. subtilis* (*natto*). This enzyme is thus an extracellular γ-PGA depolymerase of *B. subtilis* (*natto*). An inconsistency between GGT activities and γ-PGA productivities in various γ-PGA producers was also reported (Kambourova et al., 2001). Urushibata et al. (1997) reported a *B. subtilis* (*natto*) mutant that was defective in the enzyme gene yet produced abundant γ-PGA, though in previous biochemical studies the significance of GGT in γ-PGA synthesis had been denied (see Section 7.2).

8.2.4
γ-DL-PGA Synthetase of *B. subtilis*

Ashiuchi and co-workers (Ashiuchi et al., 2001a; Kamei et al., 2001) focused on *B. subtilis* (*chungkookjang*), which maintains a constant production of highly elongated γ-PGA, and for the first time succeeded in achieving an enzymatic synthesis of highly elongated γ-PGA (> 1000 kDa) with cell membranes of the bacterium in the presence of ATP and D-glutamic acid. No PGA synthetic activity was found in the cytosolic fraction. The γ-PGA synthetic activity was readily lost in the presence of a high concentration of detergent, suggesting that the enzyme is probably maintained in an active form through interaction with the cell membranes. The best substrate was DL-glutamic acid (D-isomer content 60–80%), followed by D-glutamic acid (55% of maximum activity) (Ashiuchi et al., 2002a). γ-PGA productivity was not found with cell membranes of the *pgs* null mutant, implying that *pgs* gene products, i.e., PgsBCA, are responsible for this membranous γ-PGA synthetic activity. It was also reported that elongated γ-D-PGA (> 100 kDa) was synthesized from D-glutamic acid (but not the L-enantiomer) in the presence of ATP by cell membranes of the *pgsBCA*-bearing *E. coli* clone, and that the membranous PgsBCA system contained neither a glutamic acid racemase nor a glutamic acid-isomerizing component (Ashiuchi et al., 2001b).

The γ-PGA synthesis is a ligase reaction for glutamic acid; coincident ATPase activity can be detected in the presence of glutamic acid. The D-glutamic acid-dependent ATPase activity was higher than the L-glutamic acid-dependent ATPase activity, and ADP (but not AMP) was formed by the ATPase reaction. In addition to these observations, an experiment using hydroxylamine indicated that the reaction proceeded in an amide ligase-like fashion (Figure 10) and not by the thiotemplate-based mechanism (Gardner and Troy, 1979). Conclusions have been drawn regarding the function of components of the γ-PGA synthetase complex, PgsB, -C, and -A, from the kinetic data of

$$\mathrm{PGA_n\overset{\gamma}{-}\overset{O}{\overset{\|}{C}}\text{-}OH \xrightarrow{ATP} \left[PGA_n\overset{\gamma}{-}\overset{O}{\overset{\|}{C}}\text{-}OPO_3^{2-}\right] + ADP \xrightarrow{H_2\ddot{N}\text{-}Glu} PGA_{n+1} + Pi}$$

Fig. 10 A proposed reaction mechanism of the membranous γ-DL-PGA synthetase complex of *B. subtilis* (Ashiuchi and Misono, 2002). According to the general principle of amide ligation, the phosphoryl group of ATP is first transferred to a terminal carboxyl group of elongated γ-PGA (as an acceptor) through substrate-dependent ATP hydrolysis; then, an amide linkage is formed by nucleophilic attack of an amino group of glutamic acid (as a donor) into the phosphorylated carboxyl group. A main chain of γ-PGA would be formed and highly elongated when a series of the reaction occurred iteratively and successively at an active site of the γ-DL-PGA synthetase complex (PgsBCA) due to the cooperative operation of the PgsBC components as a glutamic acid ligase and the PgsA component as a PGA transporter.

newly synthesized proteins by an in-vitro transcription/translation system. The results described in Table 6 indicate that there is little difference in affinity for glutamic acid between PgsBC and PgsBCA, and that the turnover number is apparently increased in the presence of PgsA. After storage for 24 h at 4°C, the activity of PgsBCA was reduced to the level of that of PgsBC, suggesting that PgsB and PgsC probably associate tightly with each other, whereas the interaction between PgsBC and PgsA is comparatively loose (Ashiuchi et al., 2001b). Instability of the enzyme complex solubilized with a high concentration of detergent is most likely due to the weak association between PgsBC and PgsA. Elongated γ-PGA could not be synthesized by the cell membrane of the *E. coli* clone producing only the PgsBC proteins (and therefore lacking the PgsA protein). Based on these results and the structural features of PgsA (Ashiuchi et al., 2001b), it has been concluded that the PgsA protein effectively removes the reaction products that are highly negatively charged (eventually γ-PGA) from an active site of the enzyme complex, and that the complex is stably anchored into cell membranes by the action of PgsA. PgsA may function as a γ-PGA transporter, and this is important for the enlargement (elongation) of γ-PGA. A structural feature seen in amide ligases (Eveland et al., 1997) has been found in PgsB (Ashiuchi et al., 2001b). As all the ligases identified so far have been cytosolic enzymes (Huang et al., 1993; Bertrand et al., 1999), this is the first example of a membranous amide ligase. Orthologues of PgsC, the most hydrophobic component (Ashiuchi et al., 2001b), have not been found in organisms other than γ-PGA producers. Although the active sites of amide ligases are generally constructed with only a single polypeptide, an active site of the enzyme complex is probably constituted mainly of PgsB and PgsC. Amide ligases show strict stereospecificity for amino acid substrates (McGuire and Coward 1984; Eveland et al., 1997), but PgsBC and PgsBCA acted on both enantiomers of glutamic acid as the substrate. The γ-PGA synthetase complex is thus conformationally and functionally unique and atypical among the enzymes of the amide ligase superfamily.

By contrast, Urushibata et al. (2002) reported that the γ-PGA synthetase (EC 6.3.2.-) of *B. subtilis* (*natto*) IFO 16449 consists of both the native YwsC protein (44 kDa; corresponding to the PgsB protein) and its in-phase overlapped 33-kDa protein. The 33-kDa protein completely lacks the catalytically essential region in the native YwsC (PgsB) protein at the N-terminus, e.g., the ATP-binding sequences (Ashiuchi et al., 2001b). These authors reported that this extremely stable enzyme produced γ-PGA only from L-

Tab. 6 Glutamic acid-dependent ATPase activity of each component of the PgsBCA system. The ATPase activity of each of the in-vitro translated proteins (50 ng) was measured in the presence of 50 mM glutamic acid (Ashiuchi et al., 2000b)

Proteins &Bild&	V_{max} **[U mg^{-1}]**		K_m **[mM]**	
	D-Glu	***L-Glu***	***D-Glu***	***L-Glu***
None	0	0	–	–
Control (thioredoxin)	0	0	–	–
PgsB	0	0	–	–
PgsC	0	0	–	–
PgsA	0	0	–	–
PgsBC	148 ± 28	54 ± 19	3.6	17
PgsBA	0	0	–	–
PgsCA	0	0	–	–
PgsBCA	495 ± 95	113 ± 48	4.2	15

glutamic acid, and not from the D-enantiomer. If the enzyme catalyzes the polymer synthesis in *B. subtilis* (*natto*) IFO 16449, then highly elongated γ-DL-PGA should be formed in the reaction. However, data on the enzymatically produced γ-PGA, including the stereochemistry, molecular size, and yield were not reported.

8.2.5
γ-L-PGA Synthetase of *B. halodurans*

B. halodurans is genotypically close to *B. subtilis*. The entire genomic sequence of *B. halodurans* C-125 was recently published (Takami et al., 2000), but no orthologue with the *pgs* genes of *B. subtilis* could be found. This suggests the existence of another γ-PGA synthetic system that is entirely different from the PgsBCA system. Ashiuchi and co-workers (Komatsu et al., 2001; Ashiuchi et al., 2002b) found some candidates of the γ-L-PGA synthetase of *B. halodurans* C-125 based on sequence homology with the synthetase of other polyamino acids, i.e., cyanophycin (water-insoluble multi-L-arginyl-poly-α-L-apspartic acid) (see Figure 8c). This enzyme consists of two major domains (Aboulmagd et al., 2000): the C-terminal domain shows high sequence similarities to various amide ligases, including the PgsB component of the γ-PGA synthetase complex of *B. subtilis* (Ashiuchi et al., 2001b), and mediates the arginylation of β-side chains of this polymer, whereas the N-terminal domain resembles ATP-grasp peptide synthetases, including γ-glutamylcysteine synthetase and D-alanyl-D-alanyl ligase (Galperin and Koonin, 1997), in primary structure. Although the function of the N-terminal domain has not yet been identified, it is presumably involved in the synthesis of the poly-α-L-aspartic main chain. Three orthologues of the N-terminal domain, designated tentatively as Plg1 (identity 27%; similarity 47%; accession no. AB071408), Plg2 (identity 31%; similarity 47%; accession no. AB071407), and Plg3 (identity 30%; similarity 48%; accession no. AB071409), were identified from numerous functionally unknown proteins from *B. halodurans*. These protein-encoding genes are arranged in tandem and most likely form an operon structure (Figure 11). These Plg proteins were highly purified and characterized. In the presence of oligo-γ-L-glutamic acid, the Plg3 enzyme produced γ-L-PGA with a comparatively high molecular mass (50–100 kDa). Additionally, an L-glutamic

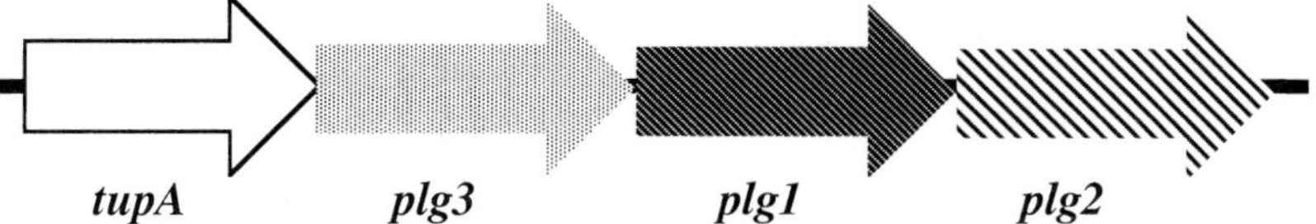

Fig. 11 Organization of the γ-L-PGA synthetase-encoding genes on the *B. halodurans* chromosome. The three genes were termed *plg1*, *plg2*, and *plg3* on the basis of predicted molecular sizes of the gene products. *tupA* has been identified as the TUP-synthesis gene in this bacterium by Aono et al. (1999).

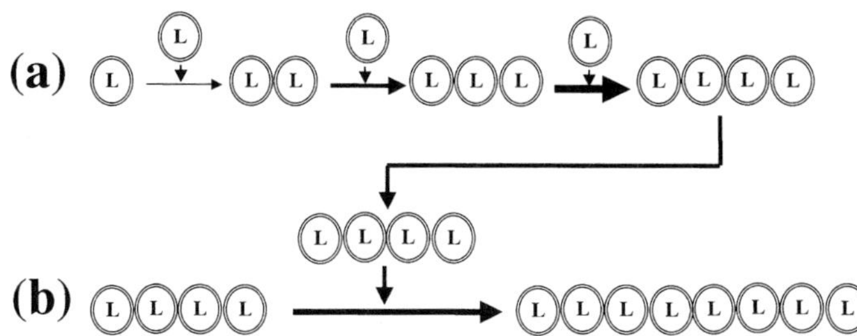

Fig. 12 A proposed mechanism of γ-L-PGA synthesis in *B. halodurans*. Schematic diagram of possible reactions catalyzed by the Plg2 (a) and the Plg3 (b) enzymes. First, the Plg2 enzyme synthesizes oligo-γ-L-glutamic acid from L-glutamic acid in the presence of ATP. The oligo-γ-L-glutamic acid thus formed serves as the substrate of an ATP-dependent amide ligation catalyzed by the Plg3 enzyme, and eventually γ-L-PGA (>10 kDa) is produced.

acid-dependent ATPase activity (i.e., by-reaction of the amino acid ligation) was found in the Plg2 enzyme, whereas an oligo-γ-L-glutamic acid-dependent ATPase activity was found in the Plg3 enzyme. D-Glutamic acid and oligo-γ-D-glutamic acid were inert. Based on these results, a novel mechanism of glutamic acid polymerization has been proposed (Figure 12). The Plg1 protein has no ATPase activity, presumably due to its lacking essential amino acid residue(s) in catalysis, and thus will serve as a useful tool for understanding this unique reaction mechanism. The data also imply that the system synthesizing γ-PGA from only L-glutamic acid is substantially different in structure and catalytic properties from that for γ-PGA containing D-glutamic acid.

9 Biodegradation

This is the most significant phenomenon seen in biopolymers and bioplastics, but not in most petrochemically synthesized polymers and thermoplastics. In order to prove the biodegradability of γ-PGA, many investigations have been made into the enzyme responsible for such biodegradation, namely γ-PGA depolymerase. Here, the information currently available on this enzyme is reviewed.

9.1 Occurrence

Thorne et al. (1954), in their earlier studies with *B. licheniformis* ATCC 9945A, demonstrated a dramatic decrease in the viscosity of the culture medium and in γ-PGA yields after long-term cultivation, and on this basis predicted the existence of a depolymerase which was responsible for the breakdown of γ-PGA. Later, Birrer et al. (1994) reported that the enzyme was either cytosolic or membranous in nature, while Nomura et al. (1973) demonstrated γ-PGA-depolymerizing activity (*exo*-type) in a lysate of *B. subtilis* (*natto*) after infection with a bacteriophage NP-38. Later, the depolymerization (fragmentation) of newly synthesized γ-PGA was observed during the polymer production by *B. subtilis* (*natto*) without the phage infection, indicating that *B. subtilis* (*natto*) itself produces γ-PGA depolymerase(s). Kambourova et al. (2001) showed that extracellular γ-PGA produced during a 30-h cultivation of *B. licheniformis* S173 disappeared completely after another 20-h cultivation period. This implies a rapid production of γ-PGA depolymerase(s) and/or a high catalytic activity of the enzyme (probably *exo*-type). γ-PGA producers thus substantially possess the *exo*-type of γ-PGA depolymerase(s). In addition, depolymerase activities have also been found in some apparently non-γ-PGA-producing microorganisms. *exo*-Depolymerase activities were found in the culture medium of *Flavobacterium polyglutamicum* (Volcani and Margalith, 1957), *Micromonospora melanosporea* IFO 12515 (Muro et al., 1990), and *Aspergillus oryzae* N-2 (Ueda et al., 1988), while *endo*-depolymerase activity was found in the culture medium of *Myrothecium* sp. TM-4222 (ATCC 201200) (Tanaka et al., 1993b).

9.2
Enzymology

A γ-PGA *exo*-depolymerase of *B. subtilis* (*natto*) and two γ-PGA *endo*-depolymerases from *Myrothecium* sp. TM-4222 and *B. licheniformis* ATCC 9945A were purified to homogeneity from each culture medium. These depolymerases were therefore identified as extracellular enzymes, contradicting the suggestion made previously by Birrer et al. (1994). Abe et al. (1997) investigated the enzymologic properties of GGT-isozymes A [a complex of 23 kDa (P-I), 39 kDa (P-II), and 40 kDa (P-III)] and B (a complex of the P-I and the P-II) from *B. subtilis* (*natto*). Besides GGT activities, both enzymes hydrolyzed effectively the γ-PGA of *B. subtilis* (*natto*) in turn from either the N- or C-terminus. GGTs of *B. subtilis* (*natto*) most likely function physiologically as γ-PGA *exo*-depolymerases. These enzymes hydrolyzed more effectively the fragmentized γ-PGAs than the highly elongated γ-PGAs, suggesting that *B. subtilis* (*natto*) produces concurrently a γ-PGA *endo*-depolymerase that catalyzes the hydrolysis of γ-PGAs with very high molecular masses. Tanaka et al. (1993a, 1997) identified the γ-PGA *endo*-depolymerase (molecular mass, 68 kDa) from *Myrothecium* sp. TM-4222. The enzyme acted exclusively on γ-DL-PGA of *B. subtilis* (*natto*) (and probably also on γ-L-PGA), but α-D- and α-L-PGAs, γ-L-glutamyl-L-glutamic acid, reduced glutathione, and γ-L-glutamyl-*p*-nitroanilide were inert. The inhibition experiments revealed that the enzyme does not belong to serine, cysteine, or metal proteases, requires no cofactor in the catalysis, and probably has functional sulfhydryl group(s). The digestion of highly elongated γ-DL-PGA (1000 kDa; D-isomer content 50%) by the enzyme yielded a mixture of 62% of γ-polypeptides consisting mainly of D-glutamic acid (average 100 kDa) and 38% of small γ-L-peptides (average 0.5 kDa). This enzyme may be tentatively designated γ-PGA *endo*-L-depolymerase. King et al. (2000) characterized the γ-PGA depolymerase of *B. licheniformis* ATCC 9945A. The enzyme was tightly (though not covalently) associated with γ-PGA, and was activated by both Zn^{2+} and Ca^{2+} ions. It could be isolated from the γ-PGA-enzyme complex with excess SDS and

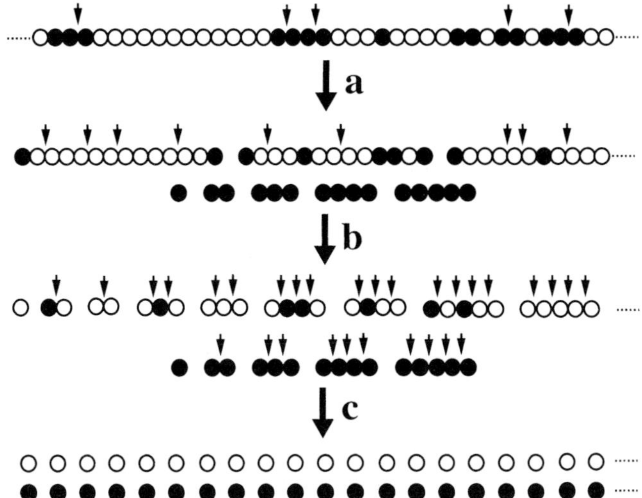

Fig. 13 A proposed pathway of γ-PGA digestion in *Bacillus*. Highly elongated γ-PGA is first digested by γ-PGA *endo*-L- (a) and -D-depolymerases (b) to yield γ-PGA with comparatively low molecular masses, which are further hydrolyzed to either oligo-γ-glutamic acids or the amino acid monomer by γ-PGA *exo*-depolymerase (c). D- and L-Glutamyl residues of γ-PGA are shown by open and filled circles, respectively.

Zn^{2+}, and its activity was restored by dilution in the presence of the γ-D-PGA of *B. licheniformis*. The enzyme was highly purified by affinity chromatography with a γ-PGA-immobilized Sepharose, and both 44- and 50-kDa proteins were isolated. The analysis of reaction products from [^{14}C]γ-D-PGA indicated that the enzyme is an *endo*-type. This enzyme may be tentatively designated γ-PGA *endo*-D-depolymerase. These suggest the biodegradation mechanism of γ-PGA (Figure 13).

9.3 Molecular Genetics

In *B. subtilis*, *ggt* was first identified as the extracellular GGT-encoding gene (Xu and Strauch, 1996). The N-terminal sequence of the γ-PGA *exo*-depolymerase overlaps on the predicted amino acid sequence of the *ggt* gene product (Abe et al., 1997). The γ-PGA *exo*-depolymerase of *B. subtilis* (*natto*) is probably a *ggt* gene product. In *B. anthracis*, the *dep* gene, which lies immediately downstream of the encapsulation (*capBCA*) genes at the same direction (see Figure 8a), is responsible for the degradation of the capsular polymer, γ-D-PGA (Uchida et al., 1993). The coexistence of the *cap* genes with the *dep* gene resulted in increased production of fragmentized γ-D-PGAs in the *E. coli* clone, whereas the *dep* gene-disruptant of *B. anthracis* produced highly elongated γ-D-PGA, as did the *E. coli* clone harboring the *cap* genes only. The gene encodes a GGT-like enzyme (molecular mass 51 kDa). The newly synthesized Dep enzyme, however, showed no GGT activity, indicating that it acts exclusively on γ-D-PGA. Its biochemical characteristics, such as the cleavage mechanism (*endo* or *exo*) and cofactor requirements, remain almost obscure. As shown in Figure 8b, the structural gene of the PGA depolymerase, *pgdS* (*pgd*: poly-γ-glutamic acid degradation; accession no. AB085821), maps immediately downstream of the γ-PGA synthesis (*pgsBCA*) genes on the *B.*

subtilis chromosome (Ashiuchi and Misono, 2002). It seems likely that the *pgdS* gene product hydrolyzes γ-PGA in an *endo*-cleavage fashion. The physiological significance and biochemical characteristics of the *pgdS* gene product of γ-PGA-producing *B. subtilis* are currently under investigation.

One interesting feature is apparent among polyamino acid metabolism genes (see Figure 8). Although the γ-PGA metabolism genes of *B. subtilis* and *B. anthracis* are clearly different in gene loci from each other, these synthetase and depolymerase genes organize a cluster. Moreover, even in the case of cyanophycin (Oppermann-Sanio and Steinbüchel, 2002) a polyamino acid that differs from γ-PGA, the genes encoding cyanophycin synthetase and cyanophycinase (cyanophycin depolymerase) form an operon structure (see Figure 8c; accession no. AF220099). The loci of genes involved in the synthesis and degradation of the third polyamino acid, ε-PL, on the *Streptomyces albulus* genome remain unidentified, but these facts imply that such a gene arrangement is probably preferable for polyamino acid metabolism and its regulation.

9.4
Applications

Control of the molecular sizes of γ-PGA is particularly significant because polymers of different molecular size are required for different purposes. For example, in the investigation of drug-delivery systems or polymeric drugs, a polymer of variable molecular size is needed to control delivery in the tissues. Chemical methods (e.g., alkaline hydrolysis; Kubota et al., 1996), physical methods (e.g., ultrasonic degradation; Pérez-Camero et al., 1999), and biological and biochemical methods (e.g., microbial degradation; Tanaka et al., 1993b) and molecular size-controlled production (Ashiuchi et al., 2001a) have each been attempted, with enzymatic digestion by γ-PGA *endo*-depolymerases appearing to be the most promising. Kageyama (1998) devised a unique use of γ-PGA depolymerase as a cleansing agent.

10
Potential Applications

It is considered that γ-PGA, regardless of the D/L ratio of the glutamic acid units, is nontoxic to humans and the environment, and is even edible. Therefore, the potential applications of γ-PGA and its derivatives have been the focus of study by several industries. In addition to the environmental uses of γ-PGA as a biodegradable substitute and for bioremediation, a wide range of unique applications, including use as cryoprotectants, bitterness-relieving agents, thickeners, animal feed additives, osteoporosis-preventing agents, humectants, drug delivery agents, gene vectors, curable biological adhesives, dispersants, and enzyme-immobilizing materials, have been developed (Shih and Van, 2001). These applications are summarized in Table 7.

10.1
Biodegradable Plastics and Hydrogels

Chemically synthesized thermoplastics and hydrogels have been widely used; indeed, today they are somewhat of a necessity in several industries in terms of convenience and economy. The abuse of these conveniences, however, often gives rise to environmental and health problems, as some of these thermoplastics and hydrogels are barely degraded in nature, whilst some of the monomers produced are harmful to humans. An example is acrylamide and its monomeric derivatives which are formed by

Tab. 7 Potential applications of γ-PGA and its derivatives

Category	*Applications*	*Details*
Biodegradable materials	Thermoplastics, fibers, films	Substitution for chemically synthesized, nonbiodegradable plastics as industrial and daily necessity
	Hydrogels	Substitution for non-biodegradable water-absorbents such as polyacrylate in diaper; potential application for desert greening
Bioremediation	Flocculants	Substitution for nonbiodegradable flocculants such as polyacrylamide
	Metal absorbents	Removal of heavy metals and radionuclides
Others	Cryoprotectants	Preservation of cryolabile nutrients
	Bitterness-relieving agents	Relief of bitter taste by amino acids, peptides, quinine, caffeine, minerals, etc.
	Thickeners	Viscosity enhancement for drinks; prevention of aging of foods such as bakery products and noodles; improvement of textures
	Mineral absorbents	Promotion of absorption of bioavailable minerals such as Ca^{2+}: live stocks, increase in egg-shell strength, decrease in body fat, etc.; human, prevention of osteoporosis
	Humectants	Use for skin-care in cosmetics
	Drug delivers	Improvement of anticancer drugs
	Gene vectors	Use for gene therapy
	Curable biological adhesives	Substitution of fibrin
	Membranes	Separation of heavy metals; enantioselection of amino acids
	Dispersants	Uses in detergents, cosmetics, paper making, etc.
	Biomacromolecule-immobilizing materials	

the degradation of polyacrylamide derivatives; these are both neurotoxic and carcinogenic in humans (Dearfield and Abermathy, 1988). Worse still, the incineration of waste materials that contain present-day plastics often generates compounds with endocrine-disrupting properties, such as dioxins. Hence, the development of safe, biodegradable plastics and hydrogels that will minimize environmental and health risks is urgently required.

Ester derivatives of γ-PGA have potential as biodegradable substitutes for currently used nonbiodegradable conveniences: thermoplastics, fibers, films, and membranes with acceptable strength, transparency, and elasticity (Yahata et al., 1992). These γ-PGA derivatives can be further improved, as the γ-DL-PGA that is currently available shows irregular stereochemistry. The thermoplasticity of degradable polyesters, including polylactic acid, is strictly dependent on the homogeneity of the stereochemical composition (Engelberg and Kohn, 1991), and so γ-D-PGA is a strong candidate. γ-L-PGAs produced by *B. halodurans* and *N. aegyptiaca* are also of great interest in this respect.

New hydrogels with high water absorption or stimulus–response capacity have been developed from biopolymers in recent years. Choi and Kunioka (1995) were the first to identify the most important characteristic of a γ-PGA derivative when they identified a new crosslinker that was prepared by γ-

Fig. 14 A highly water-absorbent gel prepared from natto mucilage (Hara, 2000b). (a) Natto, the Japanese fermented soybean food. Its mucilage which contains γ-PGA in abundance is indicated by the white arrow. (b) The dry powder of the γ-PGA crosslinker (10 mg) is encircled by a white line. The material can absorb thousands of times its own weight of water, thereafter transforms into a transparent gel; the hydrogel formed is placed in the 50-mL beaker.

irradiation of an aqueous solution of γ-PGA of *B. subtilis* (*natto*) and displayed very high water-absorption capabilities (Figure 14). Such hydrogels will have great potential in agricultural, environmental, and biomedical applications, for example, as a material for biodegradable diapers, as water reservoirs in agriculture, and as hydrogel implants for the slow release of drugs and fertilizers. Hara (2000b) recently introduced a new recycling plan using the γ-PGA hydrogel, and also referred to a hopeful prospect for "desert greening".

10.2 Bioremediation

Since the Industrial Revolution, man has been releasing into the environment a wide range of pollutants, including heavy metals, radionuclides, and excess chemicals (agricultural chemicals, artificial manure, etc.), all of which threaten public health and increase the likelihood of a universal shortage of provisions due to profound contamination leading to reduced agricultural output, contaminated (and hence unusable) water, and effects such as acid rain. Today, the remediation of contaminated soils, sediments, and waters is a major challenge facing mankind, and an understanding of how these pollutants interact with biomacromolecules should provide a basis for the development of new remediation technologies. Macaskie and Basnakova (1998) reported a bioremediation process which reduced, by virtue of enhanced chemisorption, environmental risks due to heavy metals and long-lived isotopes of neptunium and plutonium. Likewise, Pötter et al. (2001) established a new biological technique to solve the serious environmental problems caused by excessive use of liquid manure in intensified agriculture, namely by reducing excess NH_3 in the soil and converting nitrogen into γ-PGA. γ-PGA functions not only as a transit depot for waste nitrogen but also as an earth-friendly fertilizer by which naturally occurring bioavailable cations (e.g., Fe^{2+}, Fe^{3+}, Ca^{2+}, Zn^{2+}, Mg^{2+}, Mn^{2+}) can be temporarily condensed and more efficiently transferred to the rhizospheres of plants (Kinnersly et al., 1994).

10.2.1
Flocculants

Various flocculants have been developed and widely used in wastewater treatment, dredging, and industrial downstream processes. Shih and Van (2001) categorized these flocculants into three groups: inorganic agents; chemically synthesized polymers; and naturally occurring biopolymers (e.g., proteins, polysaccharides, glycoproteins, γ-PGA). Among these materials, chemically synthesized polymers are most frequently used as they are economical and highly effective. However, as described earlier, the abuse of these convenient and nonbiodegradable polymers might eventually lead to the destruction of living systems. In aiming to minimize these environmental and health risks, studies on the isolation and utilization of biodegradable flocculants produced by microorganisms have made rapid progress during recent years. Yokoi et al. (1995, 1996) have characterized γ-PGA flocculants from both *Bacillus* sp. PY-90 and *B. subtilis* (*natto*) IFO 3335. The former is activated and inhibited by divalent and trivalent cations, respectively, whereas flocculating activity of the latter increases in the presence of trivalent cations. γ-PGA of *B. licheniformis* CCRC 12826 was identical in its flocculating properties to that of *B. subtilis* (*natto*) IFO 3335 (Shih et al., 2001). γ-PGA flocculants are useful for various inorganic (e.g., active carbon, acid clay solid soil) and organic materials (e.g., cellulose, yeast). In time, γ-PGA may be used not only for wastewater treatments but also for water purification and processing (e.g., deep seawater processing) and downstream processing in food and fermentation industries.

10.2.2
Heavy Metals and Radionuclide-binding Agents

During their initial studies of remediation using biopolymers, Cassity and Kolodziej (1984) found that the capsule of *B. megaterium* ATCC 19213, which contains γ-PGA in abundance, was capable of binding both bioavailable cations and toxic metals (e.g., Cu^{2+}, Hg^{2+}, Ag^{+}). Likewise, γ-PGA of *B. licheniformis* ATCC 9945A was shown to form a complex with various metal ions such as Ni^{2+}, Cu^{2+}, Mn^{2+}, Fe^{2+}, Fe^{3+}, Al^{3+}, and Cr^{3+}. This polymer can bind even U^{4+}, an ionic radionuclide, in a binuclear, bidentate fashion (He et al., 2000). Bhattacharyya et al. (1998) established an innovative approach to make highly efficient metal-binding microfiltration membranes that are functionalized by the covalent attachment of chemically synthesized α-PGA. The metal-binding capacities were shown to be 2.5, 0.8, and 0.8 mol of metal per mol repeat units for Pb, Cd, and Ni, respectively–all much higher than for commercially available carboxylic ion-exchange resins. As the functions of naturally occurring γ-PGA (e.g., metal-binding affinity, water-absorption capacity) are superior to those of synthetic α-PGA (Kunioka, 1997; Tanimoto et al., 2001), a more advantageous performance may be realized by the application of γ-PGA as the functionalizer.

10.3
Other Applications

10.3.1
Cryoprotectants

Frequent freezing and thawing causes undesirable deterioration in living cells and biologically active substances such as enzymes, and this leads to unstable nutrients in food. The extreme antifreeze activity of highly negatively charged peptides was found accidentally when Mitsuiki et al. (1998) examined the antifreeze activities of various α- and γ-glutamyl oligomers and γ-PGA of *B. licheniformis* ATCC 9945A. Using differential scanning calorimetry, these authors showed the antifreeze activities of the glutamyl polymers to be greater than that of glucose, which is recognized as a potent antifreeze material. The modified γ-PGA production, in which its reproducibility was markedly improved by using pre-frozen cells of *B. licheniformis* (Birrer et al., 1994), is the first biological application of the cryoprotective effect of γ-PGA.

10.3.2
Food Applications

Several applications of γ-PGA in the food industries have been established (Shih and Van, 2001); these include use to relieve bitter tastes (amino acids, peptides, quinine, caffeine, minerals), to improve the quality of foods and drinks, as an ice cream stabilizer, as a thickener, and as a functional supplement in animal feeds.

Bone mass is known to decrease more and less with increasing age, and osteoporosis – which affects mainly elderly women–is the result of a dramatic deterioration in bone density (Price, 1985). Natto has been shown to have beneficial effects on osteoporosis (Tsukamoto et al., 2000; Yamaguchi et al., 2000). Tanimoto et al. (2001) recently reported that γ-PGA increased Ca^{2+} solubility *in vitro* and *in vivo*, as well as intestinal Ca^{2+} absorption. Natto is also excellent as a source of vitamin K, which plays an important role in bone formation. Although vitamin K is usually fat-soluble and water-insoluble, *B. subtilis* (*natto*) strains can produce water-soluble vitamin K which is easily and rapidly available in the human body. Ikeda and Doi (1990) identified a vitamin K-binding factor responsible for the formation of water-soluble vitamin K from the culture medium of *B. subtilis* (*natto*) and showed this to consist mainly of glutamic acid. These findings imply that naturally occurring γ-PGA and functional foods supplemented by correct quantities of γ-PGA might serve as a therapeutic agent to prevent osteoporosis.

10.3.3
Pharmaceuticals

Although at present studies on the applications of glutamyl polymers in medicine are limited to α-PGA, practical uses of γ-PGAs should become increasingly significant as the various molecular sizes of this material become available commercially. Some drugs show great potential as chemotherapeutic agents against human malignancies but are difficult to use clinically due to their insolubility in water (Rowinsky and Donehower, 1995). With taxol, this defect might be overcome by its conjugation to α-PGA (Shih and Van, 2001), which may improve tumor-targeting and also prolong the drug's plasma half-life (Li et al., 1999). This method has also been used to improve the efficacy and reduce the toxicity (by increasing solubility and controlling release) of other anti-cancer drugs such as 5-fluorouracil derivatives (Kishida et al., 1998; Kim et al., 1999). α-PGA was also shown to be useful in the functional improvement of prostaglandin E_1 (PGE_1). Hashida et al. (1999) conjugated PGE_1 with galactocylated α-L-PGA hydrazine (α-L-PGA-HZ-Gal-PGE_1); administration of the conjugate resulted in a gradual release of PGE_1 and superior therapeutic efficacy against hepatic damage. Dekie et al. (2000) also established the use of glutamyl polymers as a vector for gene therapy.

Many synthetic and semi-synthetic surgical adhesives and fibrin glues that are currently used in surgery are prepared from human blood (Spotniz, 1996; Shih and Van, 2001), and the quest for new and safe biological adhesives has been extensive in recent years. A new biological adhesive produced by crosslinking of gelatin and α-PGA showed much higher bonding strength to soft tissue and better hemostatic capacity than fibrin glues; moreover, it was also slowly degraded in the body without inducing any inflammatory response (Otani et al., 1998). Sekine et al. (2000) succeeded in producing a potent adhesive from porcine collagen and α-PGA, which was superior to fibrin in sealing air leakage from the lungs.

10.3.4 Biochemical Applications

In the future, γ-PGA may be applied to the processing of functional biomacromolecules such as enzymes, DNA, and polysaccharides (Tachaboonyakiat et al., 2000). As discussed in Section 6, γ-PGA has the characteristics of an adaptation agent in various environments, and some γ-PGA-immobilized enzymes may be useful under extreme conditions where enzymes are normally inactivated, for example, high salt concentration, a dry milieu, very low temperature, and high pH. In addition, γ-PGA will also most likely contribute to practical uses of extremophilic enzymes. For example, as the increase in the amount of free-water present is dependent upon the cryoprotective function of γ-PGA, psychrophilic enzymes may exhibit their genuine abilities below the freezing point. Recently, freeze-labile enzymes were stabilized below the freezing point in the presence of sericin (poly-α-L-serine) having a high content of hydroxyl groups (Tsujimoto et al., 2001). γ-PGA with a large content of anionic groups might also allow the stabilization and reuse of labile biomacromolecules. As γ-PGA is a γ-linked and essentially nonimmunogenic polypeptide, a γ-PGA-coating technique might lead to a dramatic improvement in the in-vivo stability of enzyme-pharmaceuticals such as a nattokinase (Fujita et al., 1993). By contrast, specific inhibitors of γ-PGA synthetase might well function as pharmaceuticals to treat infectious *B. anthracis* which itself produces γ-PGA to avoid attack by macrophages (see Section 6). The *B. subtilis* PgsBCA system (Ashiuchi et al., 1999a, 2001b), which closely resembles the *B. anthracis* CapBCA system that is responsible for γ-PGA synthesis (Makino et al., 1989), might therefore represent a potent tool in the development of drugs.

10.4 Manufacturers

At present, γ-PGA is not available commercially on account of its high cost. Rather, γ-PGA is normally prepared on a comparatively small scale in the laboratory, solely for fundamental research purposes. However, Meiji Seika Kabushiki Kaisha (Odawara, Japan), which has experience in γ-PGA fermentation, will produce γ-PGA of *B. subtilis* (*natto*) F-2-01 to special order. More recently, BioLeaders Co. (Daejeon, Korea) have developed a manufacturing system for highly elongated γ-PGA by using a jar fermentor with *B. subtilis* (*chungkookjang*) (Sung et al., 2001b), and intend to provide a commercial source of γ-PGA in the near future.

11 Outlook and Perspectives

Although γ-PGA possesses vast potential as a new macromolecular material, several problems remain to be solved before its practical use. The first problem is that of cost, which is estimated under existing conditions to be several tens to several hundred-fold higher than that of conventional materials in current use. The second problem relates to the great difficulties encountered when synthesizing artificially hyper-elongated peptides such as γ-PGA by means of modern industrial (organic) chemistry. At present, the most important step is to construct a mass-production system for γ-PGA by applying molecular biology techniques, and hence an extensive knowledge of both the producers and the synthetases of γ-PGA is indispensable. It is likely that public opinion will also push for the development of such environment-friendly technology. Moreover, resolution of the ternary structures of γ-PGA synthetases and, subsequently, detailed analyses of their catalytic actions at the molecular level will provide major insights into these highly elongated, structurally and functionally modified peptides. In turn, this should allow the design of an unprecedented protein-synthesis system that is independent of living bodies, of cells, and even of DNA, and presumably even the creation of atypical but functional proteins, e.g., an enzyme consisting only of D-amino acids.

We believe and hope that further developments in both fundamental and practical aspects of these naturally occurring poly-amino acids will open new fields in biotechnology and bioengineering.

12 Patents

Patents in this area of research can be classified into two categories: (1) improvements in γ-PGA production; and (2) extension of γ-PGA uses.

The first category deals mainly with the isolation of valuable γ-PGA producers, improved conditions for γ-PGA production, and effective purification of γ-PGA from exceedingly viscous cultures (Thorne et al., 1959; Siraishi et al., 1995; Kinouchi et al., 1997; Kim and Ko, 1999; Do et al., 2001a). γ-PGA can be used for different purposes, depending upon the difference in the molecular size of the polymer. Alex and Louis (2000) and Sung et al. (2001b) hold individual patents on the production of high molecular-weight γ-PGA. In contrast, Hiruta et al. (1995) claim the patent of low molecular-weight γ-PGA. The patent for the knowledge to control the molecular size of γ-PGA is held by Kubota et al. (1992), while that covering the genetically modified γ-PGA synthesis has been issued by Ashiuchi et al. (2001) and is crucial for the establishment of an improved γ-PGA mass production.

The second category deals mainly with the processing of γ-PGA to thermoplastics and highly water-absorbent hydrogels (Hachiman et al., 1992; Taketa et al., 1995; Hara, 2000a), the design of foods with γ-PGA (Hoshino and Masatomi, 1998; Kai et al., 1998), and uses of γ-PGA for drug delivery and in medicine (Inagaki et al., 1992; Giannous et al., 2000). Key patents in environmental and medical uses of γ-PGA are held by Nagatomo et al. (1996) and by Swadesh and Sevoian (1999). The patent issued by Ogawa et al. (1994) is advantageous in the development of foods and pharmaceuticals to prevent osteoporosis.

More recently, a very interesting patent about the potential application of the membranous γ-PGA synthetase (PgsBCA) complex itself, but not of the enzymatically synthesized γ-PGA, was published (Sung et al., 2001a). This novel technology, which is concerned with cell-surface display and edible antibody, was developed by making good use of the characteristic of the PgsA component, namely, that it moves readily to the cell surfaces (or outer membranes) of various bacteria. These authors succeeded in displaying certain important antibodies on cell surfaces of fermented food starters, e.g., edible *Bacillus* or *Lactobacillus*, and claimed that the probiotic antibody should become a crucial near-future technology.

The patent list below includes many of the key patents in this field in addition to those discussed in the previous sections.

13
Patent List

Alex, D., Louis, S. (2000) High molecular weight γ-poly(glutamic acid), ER Patent no. 1,032,699.
Ashiuchi, M., Misono, H., Soda, K. (2001) Production of poly-γ-glutamic acid, JP Patent no. 2,001,017,182.
Dainippon Pharmaceutical Co, Ltd. (1972) Ice cream stabilizer, JP Patent no. 19,735/72.
Do, J. H., Kwon, S. H., Chag, H. N., Lee, S. Y. (2001a) Process for preparing γ-glutamic acid from high-viscous culture broth, US Patent no. 2,001,016,341.
Fujita, Y., Nakamura, Y., Takeuchi, M., Takeda, H. (1998) Poly(γ-glutamic acid) salt complex and its production, JP Patent no. 10,077,342.
Giannous, S. A., Diah, S. M., Berner, B. (2000) Temporally controlled drug delivery systems, US Patent no. 6,068,853.
Gross, R. A., McCarthy, S. P., Shah, D. T. (1995) γ-Poly(glutamic acid) esters, US Patent no.5,378,807.
Hachiman, K., et al. (1992) Poly-γ-glutamic acid ester fiber and production thereof, JP Patent no. 4,370,217.
Hara, T. (2000a) Production of crosslinked poly-γ-glutamic acid and apparatus therefore, JP Patent no. 2,000,327,795.
Hara, T., Kaneko, N., Iwamoto N. (1996) Quick monitoring system of poly-γ-glutamic acid, JP Patent no. 8,163,993.
Hasebe, K., Inagaki, M. (1999) Preparation composition for external use containing γ-polyglutamic acid and vegetable extract in combination, JP Patent no. 11,240,827.
Hiruta, O., et al. (1995) Production of low-molecular γ-poly(glutamic acid), JP Patent no.7,316,286.
Hoshino, M., Masatomi, F. (1998) New *Bacillus natto* and its acquisition and poly-γ-glutamic acid produced with new *Bacillus natto* and seasoning utilizing the same, JP Patent no. 10,262,655.
Inagaki, M., et al. (1992) Production of poly-γ-glutamic acid grafted product, JP Patent no. 4,298,533.
John R. T., Louis, D. V. (1964) Poly-γ-esters of optically active glutamic acid, US Patent no. 3,119,794.
Kai, T., Ikisu, T., Kaneko, N., Harada, M., Yamamoto, S., Shiraishi, A. (1998) Method for decreasing protease activity coexisting in poly-γ-glutamic acid noodles and cakes containing this agent, JP Patent no. 10,155,412.
Kakigi, N., et al. (1994) High polymer as carrier of medicine and sustained release carcinostatic agent, JP Patent no. 6,092,870.
Karasawa, M., Tanimoto, H., Toride, Y. (1998) The use of poly-γ-glutamic acid for preparing an agent for increasing the phosphorus assimilation, EP Patent no. 0,838,160.
Kim, C. J., Ko, Y. H. (1999) Medium for γ-poly glutamic acid preparation, KR Patent no. 217,512.
Kinouchi, N., Kino, K., Kusatsu, M., Tanekawa, T., Watanabe, K., Ogawa, T. (1997) Isolation of poly-γ-glutamic acid, JP Patent no. 9,252,794.
Konno, A., Taguchi, T., Yamaguchi, T. (1988a) Bakery products and noodles containing polyglutamic acid, US Patent no. 4,888,193.
Konno, A., Taguchi, T., Yamaguchi, T. (1988b) New use of polyglutamic acid for food, EP Patent no. 284,386.
Kubota, H., et al. (1992a) Method for controlling molecular weight of poly-γ-glutamic acid, JP Patent no. 4,300,860.
Kubota, H., Nanbu, Y., Endo, T. (1992b) Poly-γ-glutamic acid ester and shaped body thereof, US Patent no. 5,118,784.
Matsukawa, M., Hirasawa, H., Mikami, F., Yamasoe, K. (1996) Hydrophobic surface-treating aqueous solution and hydrophilic surface-treat-

ing method using polyglutamic acid, US Patent no. 5,527,854.

Nagatomo, A., Tamatani, H., Ajioka, M., Yamaguchi, A. (1996) Superabsorbent polymer and process for producing same, US Patent no. 5,525,682.

Ogaki, A., et al. (1993) Poly-γ-glutamic acid ester and its production, JP Patent no. 5,117,388.

Ogawa, Y., et al. (1994) Production of calcium ion solubilization agent, JP Patent no. 6,032,742.

Okuma, Y., Toida, T., Tsunoda, I. (1976) Electro-optical cell, US Patent no. 3,977,767.

Sato, Y., Uda, S., Kamata, S. (2000) Production of dried poly-γ-glutamic acid (salt), JP Patent no. 2,000,226,450.

Sikes, C. S. (1994) Polyamino acid dispersants, US Patent no. 5,328,690.

Siraishi, A., et al. (1995) Production of poly-γ-glutamic acid, JP Patent no. 7,135,991.

Sonoda, C., Sakai, K., Murase, K. (2000) Bitterness relieving agent, WO Patent no. 0,021,390.

Sumita, T., Tomioka, K., Ariki, Y., Kitano, M., Takeda, H. (1999) Material for medical care made of poly(γ-glutamic acid) salt complex, JP Patent no. 11,276,572.

Sung, M. H., et al. (2001a) Construction of a novel vector by the use of poly-γ-glutamic acid synthetase complex-encoding genes of *Bacillus subtilis* and its application to the bacterial cell-surface display of proteins, KR Patent no. 2,001/004,837.

Sung, M. H., et al. (2001b) *Bacillus subtilis* var. *chungkookjang* producing high molecular weight poly-γ-glutamic acid, KR Patent no. 01/01,372.

Swadesh, J. K., Sevoian, M. (1999) Antigen-processing cell-targeted conjugates, US Patent no. 5,898,033.

Taketa, H., et al. (1995) Molded poly-γ-glutamic acid and method for molding poly-γ-glutamic acid, JP Patent no. 7,138,364.

Takeda, H., Furumoto, H. (1996) Polymer composition comprising poly-γ-glutamic acid and production of polymer molding, JP Patent no. 8,319,421.

Tanimoto, H., Kido, K., Kuraishi, C., Sato, H., Seguro K. (1994) Compositions and goods containing minerals and poly-γ-glutamic acid, EP Patent no. 0,605,757.

Tanimoto, H., Sato, H., Karasawa, M., Iwasaki, K., Oshima, A., Adachi, S. (2001) Method for reducing body fat accumulation in livestock or poultry, US Patent no. 6,251,422.

Tanimoto, H., Sato, H., Karasawa, M., Iwasaki, K., Oshima, A., Adachi, S. (2000) Feed composition containing poly-γ-glutamic acid, WO Patent no. 9,635,339.

Tanimoto, H., Sato, H., Kuraishi, C., Kido, K., Seguto, K. (1995) High absorption mineral-containing composition and foods, US Patent no. 5,447,732.

Thorne, C. B., Housewright, R. D., Leonard, C. G. (1959) Production of glutamyl polypeptide by *Bacillus subtilis*, US Patent no. 2,895,882.

Yamaguchi, N., et al. (1995) Quantitative determination of γ-polyglutamic acid, JP Patent no. 7,120,431.

Yamanaka, S., et al. (1991) New γ-polyglutamic acid, production therefore and drinking agent containing the same, JP Patent no. 3,047,087.

14
References

Abe, K., Ito, Y., Ohmachi, T., Asada, Y. (1997) Purification and properties of two isozymes of γ-glutamyltranspeptidase from *Bacillus subtilis* TAM-4, *Biosci. Biotechnol. Biochem.* **61**, 1621–1625.

Aboulmagd, E., Oppermann-Sanio, F. B., Steinbüchel, A. (2000) Molecular characterization of the cyanophycin synthetase from *Synechocystis* sp. strain PCC6308, *Arch. Microbiol.* **174**, 297–306.

Aono, R. (1987) Characterization of structural component of cell walls of alkalophilic strain of *Bacillus* sp. C-125, *Biochem. J.* **245**, 467–472.

Aono, R., Ohtani, M. (1990) Loss of alkalophily in cell-wall-component-defective mutants derived from alkalophilic *Bacillus* C-125. Isolation and partial characterization of the mutants, *Biochem. J.* **266**, 933–936.

Aono, R., Ito, M., Machida, T. (1999) Contribution of the cell wall component teichuronopeptide to pH homeostasis and alkalophily in the alkalophile *Bacillus lentes* C-125, *J. Bacteriol.* **181**, 6600–6606.

Appel, P., Yang, J. T. (1965) Helix-coil transition of poly-L-glutamic acid and poly-L-lysine in D_2O, *Biochemistry* **4**, 1244–1249.

Ashikaga, S., Nanamiya, H., Ohashi, Y., Kawamura, F. (2000) Natural genetic competence in *Bacillus subtilis* OK2, *J. Bacteriol.* **182**, 2411–2415.

Ashiuchi, M., Misono, H. (2002) Biochemistry and molecular genetics of poly-γ-glutamate synthesis, *Appl. Microbiol. Biotechnol.* **59**, 9–14.

Ashiuchi, M., Tani, K., Soda, K., Misono, H. (1998) Properties of glutamate racemase from *Bacillus subtilis* IFO 3336 producing poly-γ-glutamate, *J. Biochem.* **123**, 1156–1163.

Ashiuchi, M., Soda, K., Misono, H. (1999a) A poly-γ-glutamate synthetic system of *Bacillus subtilis* IFO3336: gene cloning and biochemical analysis of poly-γ-glutamate produced by *Escherichia coli* clone cells, *Biochem. Biophys. Res. Commun.* **263**, 6–12.

Ashiuchi, M., Soda, K., Misono, H. (1999b) Characterization of *yrpC* gene product of *Bacillus subtilis* IFO 3336 as glutamate racemase isozyme, *Biosci. Biotechnol. Biochem.* **63**, 792–798.

Ashiuchi, M., Kamei, T., Baek, D. H., Shin, S. Y., Sung, M. H., Soda, K., Yagi, T., Misono, H. (2001a) Isolation of *Bacillus subtilis* (*chungkookjang*), a poly-γ-glutamate producer with high genetic competence, *Appl. Microbiol. Biotechnol.* **57**, 764–769.

Ashiuchi, M., Nawa, C., Kamei, T., Song, J. J., Hong, S. P., Sung, M. H., Soda, K., Misono, H. (2001b) Physiological and biochemical characteristics of poly γ-glutamate synthetase complex of *Bacillus subtilis*, *Eur. J. Biochem.* **268**, 5321–5328.

Ashiuchi, M., Kamei, T., Sung, M. H., Soda, K., Misono, H. (2002a) Biochemical analysis of low-molecular-mass poly-γ-glutamyl peptide and poly-γ-glutamate synthetase in cell membrane of *Bacillus subtilis* (chungkookjang), (in preparation).

Ashiuchi, M., Komatsu, K., Yamamoto, T., Nakamura, H., Misono, H. (2002b) Identification of poly-γ-L-glutamate synthetic system of alkalophilic *Bacillus halodurans*, (in preparation).

Balikó, G., Venetianer, P. (1993) An *Escherichia coli* gene in search of a function: phenotypic effects of the gene recently identified as *murI*, *J. Bacteriol.* **175**, 6571–6577.

Bates, E. E. M., Gilbert, H. J. (1989) Characterization of a cryptic plasmid from *Lactobacillus plantarum*, *Gene* **85**, 253–258.

Belitsky, B. R., Sonenshein, A. L. (1998) Role and regulation of *Bacillus subtilis* glutamate dehydrogenase genes, *J. Bacteriol.* **180**, 6298–6305.

Belitsky, B. R., Wray, L. V., Jr., Fisher, S. H., Bohannon, D. E., Sonenshein, A. L. (2000) Role of TnrA in nitrogen source-dependent repression of *Bacillus subtilis* glutamate synthase gene expression, *J. Bacteriol.* **182**, 5939–5947.

Bertrand, J. A., Auger, G., Martin, L., Fanchon, E., Blanot, D., Le Beller, D., Van Heijenoort, J., Dideberg, O. (1999) Determination of the MurD mechanism through crystallographic analysis of enzyme complexes, *J. Mol. Biol.* **289**, 579–590.

Bhattacharyya, D., Hestekin, J. A., Brushaber, P., Cullen, L., Bachas, L. G., Sikder, S. K. (1998) Novel poly-glutamic acid functionalized microfiltration membranes for sorption of heavy metals at high capacity, *J. Membrane Sci.* **141**, 121–135.

Birrer, G. A., Cromwick, A. M., Gross, R. A. (1994) γ-Poly(glutamic acid) formation by *Bacillus licheniformis* 9945A: physiological and biochemical studies, *Int. J. Biol. Macromol.* **16**, 265–275.

Borbély, M., Nagasaki, Y., Borbély, J., Fan, K., Bhogle, A., Sevoian, M. (1994) Biosynthesis and chemical modification of poly(γ-glutamic acid), *Polymer Bull.* **32**, 127–132.

Bovarnick, M. (1942) The formation of extracellular D-glutamic acid polypeptide by *Bacillus subtilis*, *J. Biol. Chem.* **145**, 415–424.

Bruckner, V., Kovaca, J., Kovaca, J. (1953) The structure of native poly-D-glutamic acid. IV. The synthesis of poly-L-glutamine and the Hofmann degradation thereof, *J. Chem. Soc.* (*London*) **1953**, 1512–1514.

Bruckner, V., Kovaca, J., Kandel, I., Denes, G. (1955) Über die Struktur der natürlichen D-Polyglutaminsäure. V. Mitteilung, *Acta Chim. Acad. Sci. Hung.* **7**, 223–232.

Cassity, T. R., Kolodziej, B. J. (1984) Role of the capsule produced by *Bacillus megaterium* ATCC 19213 in the accumulation of metallic cations, *Microbios* **41**, 117–125.

Cheng, C., Asada, Y., Aida, T. (1989) Production of γ-polyglutamic acid by *Bacillus subtilis* A35 under denitrifying conditions, *Agric. Biol. Chem.* **53**, 2369–2375.

Chibnall, A. C., Rees, W. M., Richard, F. M. (1958) Structure of the polyglutamic acid from *Bacillus subtilis*, *Biochem. J.* **68**, 129–135.

Choi, H. J., Kunioka, M. (1995) Preparation conditions and swelling equilibria of hydrogel prepared by γ-irradiation from microbial poly γ-glutamic acid, *Radiat. Phys. Chem.* **46**, 175–179.

Choi, H. J., Yang, R., Kunioka, M. (1995) Synthesis and characterization of pH-sensitive and biodegradable hydrogels prepared by γ-irradiation using microbial poly(γ-glutamic acid) and poly(ε-lysine), *J. Appl. Polym. Sci.* **58**, 807–814.

Cromwick, A. M., Gross, R. A. (1995a) Effect of manganese (II) on *Bacillus licheniformis* ATC-C9945A: physiology and γ-poly(glutamic acid) formation, *Int. J. Biol. Macromol.* **17**, 259–267.

Cromwick, A. M., Gross, R. A. (1995b) Investigation by NMR of metabolic routes to bacterial γ-poly(glutamic acid) using ^{13}C-labeled citrate and glutamate as media carbon sources, *Can. J. Microbiol.* **41**, 902–909.

Cromwick, A. M., Birrer, G. A. Gross, R. A. (1996) Effects of pH and aeration on γ-poly(glutamic acid) formation by *Bacillus licheniformis* in controlled batch fermentor cultures, *Biotechnol. Bioeng.* **50**, 222–227.

Dearfield, K. L., Abermathy, C. O. (1988) Acrylamide: its metabolism, development and reproductive effects, genotoxicity, and carcinogenicity, *Mutat. Res.* **195**, 45–77.

Dekie, L., Toncheve, V., Dubruel, P., Schacht, E. H., Baarrett, L., Seymour, L. W. (2000) Poly-L-glutamic acid derivatives as vectors for gene therapy, *J. Control. Release* **65**, 187–202.

Do, J. H., Chang, H. N., Lee, S. Y. (2001b) Efficient recovery of γ-poly (glutamic acid) from highly viscous culture broth, *Biotechnol. Bioeng.* **76**, 219–223.

Engelberg, I., Kohn, J. (1991) Physico-mechanical properties of degradable polymers used in medical applications: a comparative study, *Biomaterials* **12**, 292–304.

Eveland, S. S., Pompliano, D. L., Anderson, M. S. (1997) Conditionally lethal *Escherichia coli* murein mutants contain point defects that map to regions conserved among murein and folyl poly-γ-glutamate ligases: identification of a ligase superfamily, *Biochemistry* **36**, 6223–6229

Fujii, H. (1963) On the formation of mucilage by *Bacillus natto*. III. Chemical constitutions of mucilage in natto (1), *Nippon Nougeikagaku Kaishi* **37**, 407–411.

Fujita, M., Nomura, K., Hong, K., Ito, Y., Asada, A., Nishimuro, S. (1993) Purification and characterization of a strong fibrinolytic enzyme (nattokinase) in the vegetable cheese *natto*, a popular soybean fermented food in Japan., *Biochem. Biophys. Res. Commun.* **197**, 1340–1347.

Galperin, M. Y., Koonin, E. V. (1997) A diverse superfamily of enzymes with ATP-dependent carboxylate-amine/thiol ligase activity, *Protein Sci.* **6**, 2639–2643.

Gardner, J. M., Troy, F. A. (1979) Chemistry and biosynthesis of the poly(γ-D-glutamyl) capsule in *Bacillus licheniformis*, *J. Biol. Chem.* **254**, 6262–6269.

Goto, A., Kunioka, M. (1992) Biosynthesis and hydrolysis of poly(γ-glutamic acid) from *Bacillus subtilis* IFO3335, *Biosci. Biotechnol. Biochem.* **56**, 1031–1035.

Gonzales, D., Fan, K., Sovoian, M. (1996) Synthesis and swelling characterizations of a poly(γ-glutamic acid) hydrogel, *J. Polym. Sci. Part A: Polym. Chem.* **34**, 2019–2027.

Green, B. D., Battisti, L., Koehler, T. M., Thorne, C. B., Ivins, B. E. (1985) Demonstration of a capsule plasmid in *Bacillus anthracis, Infect. Immun.* **49**, 291–297.

Guex-Holzer, S., Tomcsik, J. (1956) The isolation and chemical nature of capsular and cell-wall haptens in a *Bacillus* species, *J. Gen. Microbiol.* **14**, 14–25.

Hanby, W. E., Rydon, H. N. (1946) The capsule substance of *Bacillus anthracis. Biochem. J.* **40**, 297–309.

Hanby, W. E., Waley, S. G., Watson, J. (1950) Synthetic polypeptides. II. Polyglutamic acid, *J. Chem. Soc.* (*London*) **1950**, 3239–3249.

Hara, T. (1990) The history of Natto, *Kagaku To Seibutsu* **28**, 676–681.

Hara, T. (2000b) Desert greening. Greening by utilization of microbial macromolecules, *Kobunshi* **49**, 367–370.

Hara, T., Aumayr, A., Fujio, Y., Ueda, S. (1982) Elimination of plasmid-linked polyglutamate production by *Bacillus subtilis* (*natto*) with acridine orange, *Appl. Environ. Microbiol.* **44**, 1456–1458.

Hara, T., Nagatomo, S., Ogata, S., Ueda, S. (1992) The DNA sequence of γ-glutamyltranspeptidase gene of *Bacillus subtilis* (*natto*) plasmid pUH1. *Appl. Microbiol. Biotechnol.* **37**, 211–215.

Hara, T., Saito, H., Iwatomo, N., Kaneko, S. (1995) Plasmid analysis in polyglutamate-producing *Bacillus* strain isolated from non-salty fermented soybean food, "Kinema", in Nepal, *J. Gen. Appl. Microbiol.* **41**, 3–9.

Harth, G., Zamecnik, P. C., Tang, J. Y., Tabatadze, D., Horwitz, M. A. (2000) Treatment of *Mycobacterium tuberculosis* with antisense oligonucleotides to glutamine synthetase mRNA inhibits glutamine synthetase activity, formation of the poly-L-glutamate/glutamine cell wall structure, and bacterial replication, *Proc. Natl. Acad. Sci. USA* **97**, 418–423.

Hashida, M., Akamatsu, K., Nishikawa, M., Yamashita, F., Takakura, Y. (1999) Design of polymeric prodrugs of prostaglandin E_1 having galactose residues for hepatocyte targeting, *J. Control. Release* **62**, 253–262.

Haurowitz, F., Bursa, F. (1949) The linkage of glutamic acid in protein molecules, *Biochem. J.* **44**, 509–512.

He, L. M., Neu, M. P., Vanderberg, L. A. (2000) *Bacillus licheniformis* γ- glutamyl exopolymer: physicochemical characterization and U(VI) interaction, *Environ. Sci. Technol.* **34**, 1694–1701.

Hessinger, D. A., Lenhoff, H. M. (1988) *The Biology of Nematocysts.* San Diego, CA: Academic Press.

Hezayen, F. F., Rehm, B. H. A., Eberhardt, R., Steinbüchel, A. (2000) Polymer production by two newly isolated extremely halophilic archaea: application of a novel corrosion-resistant bioreactor, *Appl. Microbiol. Biotechnol.* **54**, 319–325.

Hezayen, F. F., Rehm, B. H. A., Tindall, B. J., Steinbüchel, A. (2001) Transfer of *Natrialba asiatica* B1T to *Natrialba taiwanensis* sp. nov., a novel extremely halophilic, aerobic, non-pigmented member of the *Archaea* from Egypt that produces extracellular poly(glutamic acid), *Int. J. Syst. Evol. Microbiol.* **51**, 1133–1142.

Ho, H. T., Falk, P. J., Ervin, K. M., Krishnan, B. S., Discotto, L. F., Dougherty, T. J., Pucci, N. J. (1995) UDP-*N*-acetylmuramyl-L-alanine functions as an activator in the regulation of the *Escherichia coli* glutamate racemase activity, *Biochemistry* **34**, 2464–2470.

Hoffman, A. S. (1987) Applications of thermally reversible polymers and hydrogels in therapeutics and diagnostics, *J. Control. Release* **6**, 297–305.

Housewright, R. D., Thorne, C. B. (1950) Synthesis of glutamic acid and glutamyl polypeptide by *Bacillus anthracis, J. Bacteriol.* **60**, 89–100.

Huang, C. S., Chang, L. S., Anderson, M. E., Meister, A. (1993) Catalytic and regulatory properties of the heavy subunit of rat kidney γ-glutamylcysteine synthetase, *J. Biol. Chem.* **268**, 19675–19680.

Idiris, A., Alam, M. T., Ikai, A. (2000) Spring mechanics of α-helical polypeptide, *Protein Eng.* **13**, 763–770.

Ikeda, H., Doi, Y. (1990) A vitamin-K_2-binding factor secreted from *Bacillus subtilis, Eur. J. Biochem.* **192**, 219–224.Ito, M., Tabata, K., Aono, R. (1994) Construction of a new teichronopeptide-defective derivative from alkaliphilic *Bacillus* sp. C-125 by cell fusion, *Biosci. Biotechnol. Biochem.* **58**, 2275–2277.

Ito, M., Tabata, K., Aono, R. (1994) Contraction of a new teichronopeptide-defective derivative from alkaliphilic *Bacillus* sp. C-125 by cell fusion, *Biosci. Biotechnol. Biochem.* **58**, 2275–2277.

Ito, Y. (1999) Molecular regulation and genetic instability of γ-glutamic acid production in

Bacillus subtilis (natto), *Bioind. Soc. Jpn.: Bioscience and Industry* **57**, 247–250.

Ito, Y., Tanaka, T., Ohmachi, T., Asada, Y. (1996) Glutamic acid independent production of poly(γ-glutamic acid) by *Bacillus subtilis* TAM-4, *Biosci. Biotechnol. Biochem.* **60**, 1239–1242.

Ivánovics, G. (1958) Vergleichende serologische untersuchungen synthetischer polyfkutaminsäuren verschiedener konstitution, *Tetrahedron* **2**, 236–240.

Ivánovics, G., Bruckner, V. (1937) Chemische und immunologische Studien über den Mechnimus der Milzbrandinfektion und Immunitat; die chemische Struktur der Kapdelsubstanz des Milzbrandbazillus und der serologisch identischen spezifischen Substanz des *Bacillus mesentericus*, *Z. Immunitatsforsch.* **90**, 304–318.

Ivánovics, G., Erdös, L. (1937) Ein Beitrag zum Wesen der Kapselsubstanz der Milzbrandbazillus, *Z. Immunitatsforsch.* **90**, 5–19.

Ivins, B. E., Ezzell, J. W., Jr., Jemski, J., Hedlund, K. W., Ristroph, J. D., Leppla, S. H. (1986) Immunization studies with attenuated strains of *Bacillus anthracis*, *Infect. Immun.* **52**, 454–458.

Kageyama, M. (1998) Cleansing agent, Japanese Patent no. 10,183,187.

Kakinoki, K., Kishida, A., Akashi, M., Endo, T. (1993) Synthesis of poly(γ-glutamic acid) hydrogel, *Chem. Soc. Jpn. Prep.* **65II**, 258.

Kambourova, M., Tangney, M., Priest, F. G. (2001) Regulation of polyglutamic acid synthesis by glutamate in *Bacillus licheniformis* and *Bacillus subtilis*, *Appl. Environ. Microbiol.* **67**, 1004–1007.

Kamei, T., Ashiuchi, M., Sung, M. H., Soda, K., Misono, H. (2001) Characteristics of a poly-γ-glutamate producer newly isolated from Chungkookjang, in: *Nippon Seibutsukogakukai Taikai 2001*, Yamanashi, Japan.

Kanno, A., Takamatsu, H. (1995) Determination of γ-polyglutamic acid in "Natto" using cetyltrimethylammonium bromide, *Nippon Shokuhin Kagaku Kaishi* **42**, 878–886.

Keppie, J., Harris-Smith, P. W., Smith, H. (1963) The chemical basis of the virulence of *Bacillus anthracis*. IX. Its aggressins and their mode of action, *Br. J. Exp. Pathol.* **44**, 446–453.

Kim, K. S., Kim, T. K., Graham, N. B. (1999) Controlled release behavior of prodrugs based on the biodegradable poly(L-glutamic acid) microspheres, *Polym. J.* **31**, 813–816.

King, E. C., Watkins, W. J., Blacker, A. J., Bugg, T. D. M. (1998) Covalent modification in aqueous solution of poly-γ-D-glutamic acid from *Bacillus licheniformis*, *J. Polym. Sci. Part A: Polym. Chem.* **36**, 1995–1999.

King, E. C., Blacker, A. J., Bugg, T. D. M. (2000) Enzymatic breakdown of poly-γ-D-glutamic acid in *Bacillus licheniformis*: identification of a polyglutamyl-γ-hydrolase enzyme, *Biomacromolecules* **1**, 75–83.

Kinnersley, A, Strom, D., Meah, R. Y., Koskan, C. P. (1994) Composition and method for enhanced fertilizer uptake by plants, WO Patent no. 94/09,628.

Kishida, A., Murakami, K., Goto, H., Akashi, M. (1998) Polymer drugs and polymeric drugs X. Slow release of 5-fluorouracil from biodegradable poly(γ-glutamic acid) and its benzyl ester matrices. *J. Bioact. Comp. Polym.* **13**, 270–278.

Kleinkauf, H., Von Döhren, H. (1996) A nonribosomal system of peptide biosynthesis, *Eur. J. Biochem.* **236**, 335–351.

Ko, Y. H. Gross, R. A. (1998) Effects of glucose and glycerol on γ-poly(glutamic acid) formation by *Bacillus licheniformis* ATCC9945A, *Biotechnol. Bioeng.* **57**, 430–437.

Komatsu, K., Yamamoto, T., Ashiuchi, M. (2001) Expression of putative poly-γ-L-glutamate synthetases of *Bacillus halodurans* C-125 in *Escherichia coli* and characterization of their recombinant enzymes, in: *Nippon Seibutsukogakukai Taikai 2001*, Yamanashi, Japan.

Kovaca, J., Bruckner, V., Denes, G. (1953) The structure of native poly-D-glutamic acid. II. Synthesis of α-poly-L-glutamic acid hydrazide and the Curtius degradation, *J. Chem. Soc. (London)* **1953**, 145–147.

Kramar, E. (1921) in: *Zentr. Bakteriol. Parastitenk., Abt. 1*, vol. 87, pp. 401.

Kubota, H., Matsunobu, T., Uotani, K., Takebe, H., Satoh, A., Tanaka, T., Taniguchi, M. (1993a) Production of poly(γ-glutamic acid) by *Bacillus subtilis* F-2-01, *Biosci. Biotechnol. Biochem.* **57**, 1212–1213.

Kubota, H., Nanbu, Y., Endo, T. (1993b) Convenient and quantitative esterification of poly(γ-glutamic acid) produced by microorganism, *J. Polym. Sci. Part A: Polym. Chem.* **31**, 2877–2878.

Kubota, H., Nanbu, Y., Endo, T. (1993c) Some properties of poly(γ-glutamic acid) produced by microorganism, *Chem. Soc. Jpn.* **8**, 973–977.

Kubota, H., Nanbu, Y., Endo, T. (1996) Alkaline hydrolysis of poly γ-glutamic acid produced by microorganism, *J. Polym. Sci. Chem.* **34**, 1347–1351.

Kunioka, M. (1995) Biosynthesis of poly(γ-glutamic acid) from L-glutamine, citric acid, and ammonium sulfate in *Bacillus subtilis* IFO3335, *Appl. Microbiol. Biotechnol.* **44**, 501–506.

Kunioka, M. (1997) Biosynthesis and chemical reactions of poly(amino acid)s from microorganisms, *Appl. Microbiol. Biotechnol.* **47**, 469–475.

Kunioka, M., Goto, A. (1994) Biosynthesis of poly(γ-glutamic acid) from L-glutamic acid, citric acid, and ammonium sulfate in *Bacillus subtilis* IFO3335, *Appl. Microbiol. Biotechnol.* **40**, 867–872.

LeBlanc, D. J., Chen, Y. Y. M., Lee, L. N. (1993) Identification and characterization of a mobilizable gene in the streptococcal plasmid, pVA380-1, *Plasmid* **30**, 296–302.

Leonard, C. G., Housewright, R. D. (1963) Polyglutamic acid synthesis by cell-free extracts of *Bacillus licheniformis, Biochim. Biophys. Acta* **73**, 530–532.

Leonard, C. G., Housewright, R. D., Thorne, C. B. (1958) Effects of some metallic ions on glutamyl polypeptide synthesis by *Bacillus subtilis, J. Bacteriol.* **76**, 499–503.

Levene, P. A., Steiger, R. E., Marker, R. E. (1931) Studies on racemization. X. Action of alkali on ketopiperazines and peptides, *J. Biol. Chem.* **93**, 605–621.

Li, C., Price, J. E., Milas, L., Hunter, N. R., Ke, S., Tansey, W., Charnsagavej, C., Wallace, S. (1999) Antitumor activity of poly(L-glutamic acid)-paclitaxel on syngeneic and xenografted tumors, *Clin. Cancer Res.* **5**, 891–897.

Macaskie, L. E., Basnakova, G. (1998) Microbially-enhanced chemisorption of heavy metals: a method for the bioremediation of solutions containing long-lived isotopes of neptunium and plutonium, *Environ. Sci. Technol.* **32**, 184–187.

McCuen, R. W., Thorne, C. B. (1971) Genetic mapping of genes concerned with glutamyl polypeptide production by *Bacillus licheniformis* and a study of their relationship to the development of competence for transformation, *J. Bacteriol.* **76**, 499–503.

McGuire, J. J., Coward, J. K. (1984) Pteroylpolyglutamates: biosynthesis, degradation, and function, in: *Folates and Proteins. Chemistry and Biochemistry of Folates* (Blankey, R.L., Benkovic, S.J., Eds.), New York: Wiley, 135–190.

Makino, S., Sasakawa, C., Uchida, I., Terakado, N., Yoshikawa, M. (1988) Cloning and CO_2-dependent expression of the genetic region for encapsulation from *Bacillus anthracis, Mol. Microbiol.* **2**, 371–376.

Makino, S., Uchida, I., Terakado, N., Sasakawa, C., Yoshikawa, M. (1989) Molecular characterization and protein analysis of the *cap* region, which is essential for encapsulation in *Bacillus anthracis, J. Bacteriol.* **171**, 722–730.

Meijer, W. J. J., Wisman, G. B. A., Terpstra, P., Thorsted, P. B., Thomas, C. M., Holsappel, S., Venema, G., Bron, S. (1998) Rolling-circle plasmids from *Bacillus subtilis*: complete nucleotide sequences and analyses of genes of pTA1015, pTA1040, pTA1050 and pTA1060, and comparisons with related plasmids from Gram-positive bacteria, *FEMS Microbiol. Rev.* **21**, 337–368.

Mikesell, P., Ivins, B. E., Ristroph, J. D., Dreier, T. M. (1983) Evidence for plasmid-mediated toxin production in *Bacillus anthracis, Infect. Immun.* **39**, 371–376.

Mitsuiki, M., Mizuo, A., Tanimoto, H., Motoki, M. (1998) Relationship between the antifreeze activities and the chemical structures of oligo- and poly(glutamic acid)s, *J. Agric. Food Chem.* **46**, 891–895.

Muñoz, V., Serrano, L. (1995) Elucidating the folding problem of helical peptides using empirical parameters. III. Temperature and pH dependence, *J. Mol. Biol.* **245**, 297–308

Muro, T., Nagamori, Y., Okada, S., Tominaga, Y. (1990) Some properties and action of poly(glutamic acid) hydrolase II from *Micromonospora melanosporea* IFO 12515, *Agric. Biol. Chem.* **54**, 1065–1067.

Nagai, T., Koguchi, K., Ito, Y. (1997) Chemical analysis of poly-γ-glutamic acid produced by plasmid-free *Bacillus subtilis* (*natto*): evidence that plasmids are not involved in poly-γ-glutamic acid production, *J. Gen. Appl. Microbiol.* **43**, 139–143.

Nagai, T., Phan Tran, L. S., Inatsu, Y., Itoh, Y. (2000) A new IS4 family insertion sequence, IS4*Bsu1*, responsible for genetic instability of poly-γ-glutamic acid production in *Bacillus subtilis, J. Bacteriol.* **182**, 2387–2392.

Nitecki, D. E., Goodman, J. W. (1971) Polyglutamic acid, in: *Chemistry and Biochemistry of Amino Acids, Peptides and Proteins* (Weinstein, B., Ed.), New York: Marcel Dekker, I87–I126.

Niemetz, R., Kärcher, U., Kandlera, O., Tindall, B. J., König, H. (1997) The cell wall polymer of the extremely halophilic archaeon, *Natronococcus occultus, Eur. J. Biochem.* **249**, 905–911.

Noda, K., Igata, K., Horikawa, Y., Fujii, H. (1980) Synthesis of γ-glutamyl peptides catalyzed by transamidase from *Bacillus natto, Agric. Biol. Chem.* **44**, 2419–2423.

Nomura, S., Hongo, M., Yoshimoto, A. (1973) γ-Polyglutamic acid depolymerase induced by infections of *natto* and *subtilis* phages and its further properties, *Agric. Biol. Chem.* **37**, 83–90.

Ogawa, Y., Hosokawa, H., Hamano, M., Motai, H. (1991) Purification and properties of γ-glutamyltranspeptidase from *Bacillus subtilis* (*natto*), *Agric. Biol. Chem.* **55**, 2971–2977.

Ogawa, Y., Yamaguchi, F., Yuasa, K., Tahara, Y. (1997) Efficient production of γ-polyglutamic acid by *Bacillus subtilis* (*natto*) in jar fermenters, *Biosci. Biotechnol. Biochem.* **61**, 1684–1687.

Ohki, K., Kondo, Y., Fujii, T., Iizuka, E. (1981) Circular dichroism of liquid crystalline solution of polyglutamic acid, *Kobunshi Ronbunshu* **38**, 493–496.

Onodera, T., Ohmachi, T., Asada, Y. (1994) Plasmid-independent poly(γ-glutamic acid) production in bacteria producing this acid *de novo, Nippon Nogeikagaku Kaishi* **68**, 1475–1478.

Oppermann-Sanio, F. B., Steinbüchel, A. (2002) Occurrence, functions and biosynthesis of polyamides in microorganisms and biotechnological production, *Naturwissenschaften* **89**, 11–22.

Otani, Y., Tabata, Y., Ikeda, Y. (1998) Hemostatic capability of rapidly curable from gelatin, poly(L-glutamic acid), and carbodiimide, *Biomaterials* **19**, 2091–2098.

Parini, C., Fortina, M. G., Manachini, P. L., De Rossi, E., Riccardi, G. (1991) Detection and characterization of naturally occurring plasmids in *Bacillus licheniformis, FEMS Microbiol. Lett.* **81**, 329–334.

Pérez-Camero, G., Congregado, F., Bou, J. J., Muñoz-Guerra, S. (1999) Biosynthesis and ultrasonic degradation of bacterial poly(γ-glutamic acid), *Biotechnol. Bioeng.* **63**, 110–115.

Pérez-Casal, J., Caparon, M. G., Scott, J. R. (1991) Mry, a *trans*-acting positive regulator of the M protein gene of *Streptococcus pyogenes* with similarity to the receptor proteins of two-component regulatory systems, *J. Bacteriol.* **173**, 2617–2624.

Pötter, M., Oppermann-Sanio, F. B., Steinbüchel. A. (2001) Cultivation of bacteria producing polyamino acids with liquid manure as carbon and nitrogen source, *Appl. Environ. Microbiol.* **67**, 617–622.

Price, P. A. (1985) Vitamin K-dependent formation of bone gla protein (osteocalcin) and its function, *Vitam. Horm.* **42**, 65–108.

Rowinsky, K. E., Donehower, R. C. (1995) Paclitaxel (Taxol), *N. Engl. J. Med.* **332**, 1004–1014.

Rydon, H. N. (1964) Polypeptides. X. The optical rotatory dispersion of poly γ-D-glutamic acid, *J. Chem. Soc.*, 1328–1333.

Saito, T., Iso, N., Mizuno, H., Kaneda, H., Suyama, Y., Kawamura, S., Osawa, S. (1974) Conformational change of a natto mucin in solution, *Agric. Biol. Chem.* **38**, 1941–1946.

Sawa, S., Murakawa, T., Watanabe, T., Murao, S., Omata, S. (1973) Isolation and purification of polyglutamic acid produced by *Bacillus subtilis* no. 5E, and studies on its chemical properties, *Nippon Nogeikagaku Kaishi* **47**, 159–165.

Sawamura, S. (1913) On *Bacillus natto, J. Coll. Agric.* (*Tokyo*) **5**, 189–191.

Sekine, T., Nakamura, T., Shimizu, Y., Ueda, H., Matsumoto, K., Takimoto, Y., Kiyotani, T. (2000) A new type of surgical adhesive made from porcine collagen and polyglutamic acid, *J. Biomed. Mater. Res.* **35**, 305–310.

Sheehan, J. C., Hess, G. P. (1955) A new method of forming peptide bonds, *J. Am. Chem. Soc.* **77**, 1067–1068.

Shih, I. L., Van, Y. T. (2001) The production of poly-(γ-glutamic acid) from microorganisms and its various applications, *Bioresour. Technol.* **79**, 207–225.

Shih, I. L., Van, Y. T., Yeh, L. C., Lin, H. G., Chang, Y. N. (2001) Production of a biopolymer flocculant from *Bacillus licheniformis* and its flocculation properties, *Bioresour. Technol.* **78**, 267–272.

Shima, S., Sakai, H. (1981) Poly-L-lysine produced by *Streptomyces.* III. Chemical studies, *Agric. Biol. Chem.* **45**, 2503–2508.

Spotniz, W. D. (1996) History of tissue adhesives, in: *Surgical Adhesives and Sealants, Current Technology and Application* (Sierra, D., Saits, R., Eds.), USA: Technomic, 3–11.

Stein, T., Kluge, B., Vater, J. (1995) Gramicidin S synthetase 1 (phenylalanine racemase), a prototype of amino acid racemases containing the cofactor 4′-phosphopantetheine, *Biochemistry* **34**, 4633–4642.

Tachaboonyakiat, W., Serizawa, T., Endo, T., Akashi, M. (2000) The influence of molecular weight over the ultrathin films of biodegradable polyion complexes between chitosan and poly(γ- glutamic acid), *Polym. J.* **32**, 481–485.

Takami, H., Nakasone, K., Takaki, Y., Maeno, G., Sasaki, R., Masui, N., Fuji, F., Hirama, C., Nakamura, Y., Ogasawara, N., Kuhara, S., Horikoshi, K. (2000) Complete genome sequence of the alkaliphilic bacterium *Bacillus halodurans* and genomic sequence comparison with *Bacillus subtilis, Nucleic Acids Res.* **28**, 4317–4331.

Tanaka, T., Hiruta, O., Futamura, T., Uotani, K., Satoh, A., Taniguchi, M., Oi, S. (1993a) Purification and characterization of poly(γ-glutamic acid) hydrolase from a filamentous fungus, *Myrothecium* sp. TM-4222, *Biosci. Biotechnol. Biochem.* **57**, 2148–2153.

Tanaka, T., Yamaguchi, T., Hiruta, O., Futamura, T., Uotani, K., Satoh, A., Taniguchi, M., Oi, S. (1993b) Screening for microorganism having poly(γ-glutamic acid) endohydrolase activity and the enzyme production by *Myrothecium* sp. TM-4222, *Biosci. Biotechnol. Biochem.* **57**, 1809–1810.

Tanaka, T., Fujita, K., Takenishi, S., Taniguchi, M. (1997) Existence of an optically heterogeneous peptide unit in poly(γ-glutamic acid) by produced by *Bacillus subtilis, J. Ferment. Bioeng.* **84**, 361–364.

Tanimoto, H., Mori, M., Motoki, M., Torii, K., Kadowaki, M., Noguchi, T. (2001) Natto mucilage containing poly-γ-glutamic acid increases soluble calcium in the rat small intestine, *Biosci. Biotechnol. Biochem.* **65**, 516–521.

Thorne, C. B., Leonard, C. G. (1958) Isolation of D- and L-glutamyl polypeptides from culture filtrate of *Bacillus subtilis, J. Biol. Chem.* **233**, 1109–1112.

Thorne, C. B., Molnar, D. M. (1955) D-Amino acid transamination in *Bacillus anthracis, J. Bacteriol.* **70**, 420–426.

Thorne, C. B., Gómez, C. G., Noyes, H. E., Housewright, R. D. (1954) Production of glutamyl polypeptide by *Bacillus subtilis, J. Bacteriol.* **68**, 307–315.

Thorne, C. B., Gómez, C. G., Housewright, R. D. (1955) Transamination of D-amino acid by *Bacillus subtilis, J. Bacteriol.* **69**, 357–362.

Thorne, C. B., Molnar, D. M., Strange, R. E. (1960) Production of toxin *in vitro* by *Bacillus anthracis* and its separation into two components, *J. Bacteriol.* **79**, 450–455.

Thorsted, P. B., Thormas, C. M., Poluektova, E. U., Prozorov, A. A. (1999) Complete sequence *of Bacillus subtilis* plasmid p1414 and comparison with seven other plasmid types found in Russian soil isolates of *Bacillus subtilis, Plasmid* **41**, 274–281.

Torii, M. (1956) Optical isomers of glutamic acid comprising bacterial glutamyl polypeptides, *Med. J. Osaka Univ.* **6**, 1043–1046.

Torii, M. (1959) Studies on the chemical structure of bacterial glutamyl polypeptides by hydrazinolysis, *J. Biochem.* **46**, 189–200.

Tran, L. S., Nagai, T., Itoh, Y. (2000) Divergent structure of the ComQXPA quorum-sensing components: molecular basis of strain-specific communication mechanism in *Bacillus subtilis, Mol. Microbiol.* **37**, 1159–1171.

Troy, F. A. (1973a) Chemistry and biosynthesis of the poly(γ-D-glutamyl) capsule in *Bacillus licheniformis.* I. Properties of the membrane-mediated biosynthesis reaction, *J. Biol. Chem.* **248**, 305–316.

Troy, F. A. (1973b) Chemistry and biosynthesis of the poly(γ-D-glutamyl) capsule in *Bacillus licheniformis.* II. Characterization and structural properties of the enzymatically synthesized polymer, *J. Biol. Chem.* **248**, 316–324.

Tsujimoto, K., Takagi, H., Takahashi, M., Yamada, H., Nakamori, S. (2001) Cryoprotective effect of the serine-rich repetitive sequence in silk protein sericin, *J. Biochem.* **129**, 979–986.

Tsukamoto, Y., Ichise, H., Kakuda, H., Yamaguchi, M. (2000) Intake of fermented soybean (*natto*) increases circulating vitamin K_2 (menaquinone-7) and γ-carboxylated osteocalcin concentration in normal individuals, *J. Bone Miner. Metab.* **18**, 216–222.

Uchida, I., Sekizaki, T., Hashimoto, K., Terakado, H. (1985) Association of the encapsulation of *Bacillus anthracis* with a 60 megadalton plasmid, *J. Gen. Microbiol.* **131**, 363–367.

Uchida, I., Makino, S., Sawamura, C., Yoshikawa, M., Sugimato, C., Terakado, M. (1993) Identification of a novel gene, *dep*, associated with depolymerization of the capsular polymer in *Bacillus anthracis, Mol. Microbiol.* **9**, 487–496.

Ueda, S., Ohba, R., Ijiri, S. (1988) Screening of Koji mold with hydrolytic activity on γ-poly glutamic acid, *Nippon Shokuhin Kogyo Gakkaishi* **35**, 302–308.

Urushibata, Y., Tokuyama, S., Tahara, Y. (1997) γ-Polyglutamic acid productivity of *Bacillus subtilis* NR-1 mutant defective of γ-glutamyltranspeptidase gene, in: *Nippon Seibutsukogakukai Taikai 1997,* Tokyo, Japan.

Urushibata, Y., Tokuyama, S., Tahara, Y. (2002) Characterization of the *Bacillus subtilis ywsC* gene, involved in γ-polyglutamic acid production, *J. Bacteriol.* **184**, 337–343.

Utsumi, S., Torii, M., Kurimura, O., Yamamuro, H., Amano, T. (1958) Heterogeneity of antibodies evoked by a glutamyl polypeptide of *Bacillus megaterium, Biken's J.* **1**, 201–202.

Utsumi, S., Torii, M., Kurimura, O., Yamamuro, H., Amano, T. (1959) Immunochemical studies of bacterial glutamyl polypeptides, *Biken's J.* **2**, 165–176.

Vater, J., Mallow, N., Gerhardt, S., Gadow, A., Kleinkauf, H. (1985) Gramicidin S synthetase. Temperature dependence and thermodynamic parameters of substrate amino acid activation reactions, *Biochemistry* **24**, 2022–2027.

Vietri, N. J., Marrero, R., Hoover, T. A., Welkos, S. L. (1995) Identification and characterization of a trans-activator involved in the regulation of encapsulation by *Bacillus anthracis*, *Gene* **159**, 1–9.

Volcani, B. E., Margalith, P. (1957) A new species (*Flavobacterium polyglutamicum*) which hydrolyzes the γ-glutamyl bond in polypeptides, *J. Bacteriol.* **74**, 646–655.

Waley, S. G. (1955) The structure of bacterial polyglutamic acid, *J. Chem. Soc.* (*London*) **1955**, 517–522.

Weber, J. (1989) Nematocysts (stinging capsules of *Cnidaria*) as Donnan-potential-dominated osmotic systems, *Eur. J. Biochem.* **184**, 465–476.

Weber, J. (1990) Poly(γ-glutamic acid)s are the major constituents of Nematocysts in *Hydra* (*Hydrozoa, Cnidaria*), *J. Biol. Chem.* **265**, 9664–9669.

Weber, J., Klug, M., Tardent, P. (1988) In: *The Biology of Nematocysts* (Hessinger, D. A., Lenhoff, H.M., Eds.), San Diego, CA: Academic Press, 427–444.

Wietzerbin, J., Lederer, F., Petit, J. F. (1975) Structural study of the poly-L-glutamic acid of the cell wall of *Mycobacterium tuberculosis* var. *hominis*, strain Brevannes, *Biochem. Biophys. Res. Commun.* **62**, 246–252.

Williams, W. J., Litwin, J., Thorne, C. B. (1955) Further studies on the biosynthesis of γ-glutamyl peptides by transfer reaction, *J. Biol. Chem.* **212**, 427–438.

Williams, W. J., Thorne, C. B. (1954) Biosynthesis of glutamyl peptides from glutamine by a transfer reaction, *J. Biol. Chem.* **210**, 203–217.

Xu, K., Strauch, M. A. (1996) Identification, sequence, and expression of the gene encoding γ-glutamyltranspeptidase in *Bacillus subtilis*, *J. Bacteriol.* **178**, 4319–4322.

Yahata, K., Sadanobu, J., Endo, T. (1992) Preparation of poly-α-benzyl-γ- polyglutamate fiber, *Polym. Prep. Jpn.* **41**, 1077.

Yamaguchi, M., Kakuda, H., Gao, Y. H., Tsukamoto, Y. (2000) Prolonged intake of fermented soybean (*natto*) diets containing vitamin K_2 (menaquinone-7) prevents bone loss in ovariectomized rats, *J. Bone Miner. Metab.* **18**, 71–76.

Yokoi, H., Arima, T., Hirose, J., Hayashi, S., Takasaki, Y. (1996) Flocculation properties of poly(γ-glutamic acid) produced by *Bacillus subtilis*, *J. Ferment. Bioeng.* **82**, 84–87.

Yokoi, H., Natsuda, O., Hirose, J., Hayashi, S., Takasaki, Y. (1995) Characteristics of a biopolymer flocculant produced by *Bacillus* sp. PY-90, *J. Ferment. Bioeng.* **79**, 378–380.

Yoon, S. H., Do, J. H., Lee, S. Y., Chang, H. N. (2000) Production of poly-γ-glutamic acid by fed-batch culture of *Bacillus licheniformis*, *Biotechnol. Lett.* **22**, 585–588.

Zanny, D., Alemán, C., Muñoz-Guerra, S. (1998) On the helical conformation of un-ionized poly(γ-D-glutamic acid), *Int. J. Biol. Macromol.* **23**, 175–184.

7
Polyaspartic Acids

Dr. Winfried Joentgen[1], Dr. Nikolaus Müller[2], Dr. Alfred Mitschker[3], Dr. Holger Schmidt[4]

[1] Bayer AG, 42096 Wuppertal, Germany; Tel.: +49 202/36-2661; Fax.: +49 202/36-2566; E-mail: winfried.joentgen.wj@bayer-ag.de

[2] Bayer AG, 51368 Leverkusen, Germany; Tel.: +49 214/30-65962; Fax.: +49 214/30-72098; E-mail: nikolaus.mueller.nm@bayer-ag.de

[3] Bayer AG, 51368 Leverkusen, Germany; Tel.: +49 214/30-54043; Fax.: +49 214/30-81079; E-mail: alfred.mitschker.am@bayer-ag.de

[4] Bayer AG, 51368 Leverkusen, Germany; Tel.: +49 214/30-61318; Fax.: +49 214/30-81079; E-mail: holger.schmidt.hs2@bayer-ag.de

bPP	bovine phosphophoryn
DOC	dissolved organic carbon
GPC	gel-permeation chromatography
HEMA	hydroxyethylmethacrylate
NMR	nuclear magnetic resonance
OECD	Organisation for Economic Cooperation and Development
PASP	polyaspartic acid
SSNC	Swedish Society for Nature Conservation

1 Introduction

The presence of deposits of inorganic crystals in living organisms is a ubiquitous phenomenon, and the processes of formation of the crystalline phases are grouped together under the term "biomineralization", where "bio" suggests the active involvement of the cells of the host organism in the process.

Control of this inorganic crystallization is essential to living systems of virtually all types. For example, inorganic crystals that have been modified by association with organic components are used to produce skeletons, shells, and teeth, as well as a fascinating array of other structures and cellular inclusions. The study of such composites is inherent in the more general field of biomineralization, which has been the subject of numerous volumes and some excellent recent reviews (Addadi and Weiner, 1992; Mann, 1993; Allemand and Cuif, 1994).

A multitude of organic molecules from biological systems potentially interact with and modify inorganic crystallization. Principal among these is a class of polyanionic proteins, the main anionic residues of which are aspartic acid and serine, with up to 90% of the serines being phosphorylated. Aspartic acid and phosphoserine together account for approximately 80 mol.% of the total amino acid composition (Wheeler et al., 1988; Rusenko et al., 1991). One example of such a highly anionic protein is bovine phosphophoryn (bPP) which is found in mineralizing tissues such as dentine tissues. bPP is a large molecule (M_w 150,000 Da) which contains about 1130 amino acid residues of which aspartic acid accounts for some 450 residues and serine (90% phosphorylated) for another 550. Thus, about 950 of the total 1130 residues (84%) are potentially anionic (Stetler-Stevenson and Veis, 1983).

When dissolved in solution, these proteins can inhibit crystal nucleation and growth as well as alter the morphology of the crystals that are formed. On the other hand, when the proteins are immobilized, perhaps by cross-linking to a structural framework such as collagen or chitin (as are found in tissues), they can promote nucleation and the growth of crystals, again in specific orientations. These compounds influence processes not only in vivo but also in vitro, with the latter effects including the dispersion, nucleation inhibition or modification of the crystal growth of minerals or weakly soluble salts.

Whilst the formation of insoluble mineral precipitates or incrustations is common in nature, it is also a widespread problem in many large-scale industrial processes. Nowadays, in aqueous systems in particular, this can lead to severe environmental problems because the organic polymers used in such processes as either scale inhibitors or dispersing agents (namely polyacrylic acid and copolymers with maleic acid, respectively) suffer from a lack of biodegradability (Schwamborn, 1996). The latter problem becomes increasingly important if these substances are released directly as effluents, for example as one component of the so-called "blow-down" of cooling water circuits.

In order to improve the environmental acceptability of these water-soluble scale-controlling agents, sequestrants and dispersants that ultimately reach the environment (generally via the surface water), several attempts have been made to develop new biodegradable alternatives. It was clear that although the above-mentioned proteinogenic structures might represent an interesting approach for the control of such processes, these proteins themselves are not accessible for industrial use.

Hence, an attempt was made to mimic the structural and functional properties of these surface active proteins (Sikes et al., 1991), and this led to the development of polyaspartic acids (PASPs) – a new class of synthetic polyamides which are structural and functional analogues of subdomains of biomineralization-controlling proteins.

As yet, PASPs as homo polypeptides have not been found in nature; nor are they accessible via biotechnological procedures, unlike polyglutamic acid (Goto and Kunioka, 1992) or polylysine (Kurshwaha, 1980).

Today, PASPs can be synthesized via two different chemical pathways. Whilst peptide chemical procedures mostly allow the synthesis of uniformly directed α- or β-polyaspartic acid of various molecular masses (Rao et al., 1993), thermal polymerization techniques are preferred for production on an industrial scale. In this way, the sodium salts of PASP can be synthesized via thermal polymerization of aspartic acid (Figure 1). Aspartic acid itself can be manufactured chemically starting from maleic acid anhydride, or by fermentation starting from fumaric acid or glucose (Eyal and Cami, 1997).

2
Historical Outline

The first experiments with PASP were carried out during the second half of the nineteenth century. Both French and German scientists heated ammonium salts of maleic acids (Dessaignes, 1850; Wolff, 1850) or asparagine (Schaal, 1871), and separated an organic material which was insoluble in water. Hydrolysis of this unknown substance with hydrochloric acid or with an aqueous solution of ammonia yielded optically inactive aspartic acids. At this time, Schaal supposed this material to be a condensation product of several monomers.

At the turn of that century, Schiff heated aspartic acid at temperatures of 200°C for 20 h, and separated two different types of condensation product from the mixture; these he called octa- and tetra-aspartide (Schiff, 1897). The treatment of these compounds with dilute caustic or ammonia yielded the corresponding PASPs.

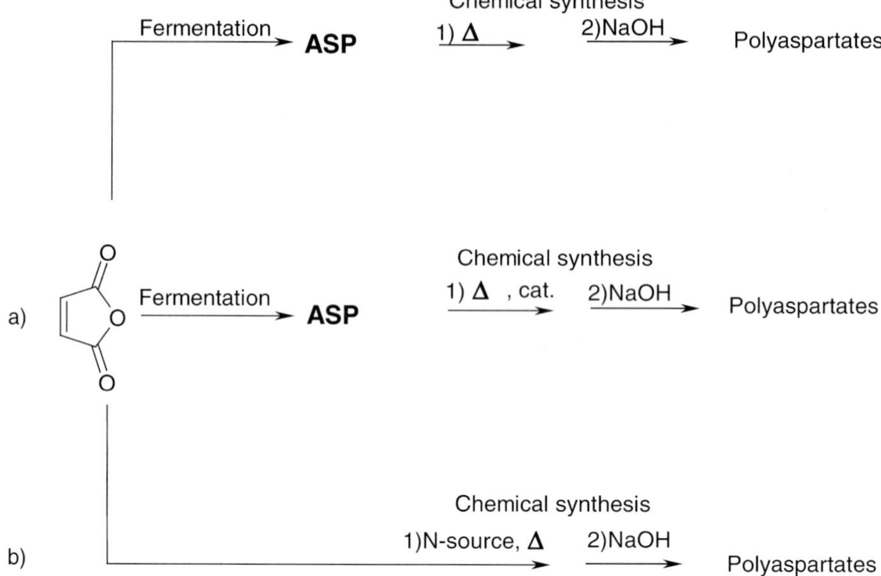

Fig. 1 Main synthetic pathways to thermal polyaspartic acids

About 50 years later, Kovacs et al. proposed that heating aspartic acid results in the formation of a reactive internal anhydride which condenses to a polyimide (Kovacs et al., 1953). This polyimide, when exposed to an alkaline solution, readily undergoes a ring-opening reaction to PASPs. A transformation to the corresponding amides, followed by a Hofmann degradation, leads to acetaldehyde and 1,2-diamino-propanic acid. Based on this result, Kovacs and Könyves (1954) postulated a 1:1.3 ratio of the α,β-amide bonds in these PASPs.

Several investigations have dealt with the thermal polycondensation of amino acids or their precursors, such as maleamic acid to yield proteinoids. Only homopolymers of aspartic acid or glycine can be prepared using this method (Meggy, 1956; Fox et al., 1959; Harada, 1959). Heating of the other natural amino acids alone results ordinarily in products such as ketopiperazines, tars, and other pyrolytic material. However, in the presence of large proportions of aspartic acid, glutamic acid or lysine, it is possible to synthesize copolymeric peptides by pyrocondensation of amino acid mixtures (Fox and Harada, 1960b). The formation of such proteinoids can also be catalyzed by phosphorus compounds such as orthophosphoric acid (Fox and Harada, 1960a) or polyphosphoric acid (Harada and Fox, 1964). During the late 1960s, Fox and coworkers discussed these pathways to proteinoids in detail with regard to questions about the origin of life (Fox et al., 1970).

During the 1970s and 1980s, Pivcova and colleagues were investigating the structure of the PASPs, and found both α- and β-linkages of the aspartic acid units in all thermally condensed derivatives (Pivcova et al., 1981; Pivcova and Saudek, 1985).

Some of the first publications concerning the applications of PASPs involved the inhibition of calcium sulfate scale formation (Sarig and Shifrin, 1977). Syntheses of

PASPs starting from aspartic and phosphonic acids, as well as their application in health care, have been described by P. Neri and coworkers (Neri et al., 1973). From an economical viewpoint, one of the most important properties of PASP is its ability to inhibit calcium carbonate scale formation, and this was investigated fully by Sikes and coworkers (Sikes and Wheeler, 1985, 1988). As a result, a large number of patents were filed during the 1990s which reflected the commercial interest of the chemical industry in PASPs or their derivatives starting at that time (see Section 9).

3
Chemical Structure

Naturally occurring PASP consists of α-linked aspartic acid units up to a chain length of 50 subunits (Rusenko et al., 1991), these occurring as part of polyanionic proteins rich in aspartic acid and phosphoserine (Sikes and Wierzbicki, 1996). In contrast to the linkage of aspartic acid in native materials, PASPs obtained by a thermal polymerization process contain α- and β-linked moieties in a constant molar ratio of 30:70 (Low et al., 1996) in a random distribution over the polymer chain according to determinations by nuclear magnetic resonance (NMR) spectroscopy (Pivcova et al., 1981, 1982) (Figure 2).

4
Chemical Analysis

The molecular structure of PASP has mainly been investigated using spectroscopic methods (Matsubara et al., 1997). In particular, NMR was the most often used technique to obtain information about the structural composition of PASP. Alternative methods such as infrared spectroscopy also proved valuable in the analysis of PASP and/or its copolymers.

4.1
^{1}H-NMR Spectroscopy

PASP can be identified from its NMR spectrum (Figure 3); in alkaline solution the following signals are measured (pH 13; 400 MHz), as shown in the spectrum in Figure 3. Due to the polymer structure of PASP there is an absence of sharp peaks, and the signals appear in the form of broader multiplets. In a mixture, of which the resonance signals of PASP and of other compounds do not overlap, it is possible to analyze quantitatively PASP by using integrals. It is also possible to identify different copolymer units in the molecule. Matsuyama et al. (1980) were able to determine the ratio of α- and β-amide units in the main chain by integrating the separated methine signals in the ^{1}H-NMR spectrum.

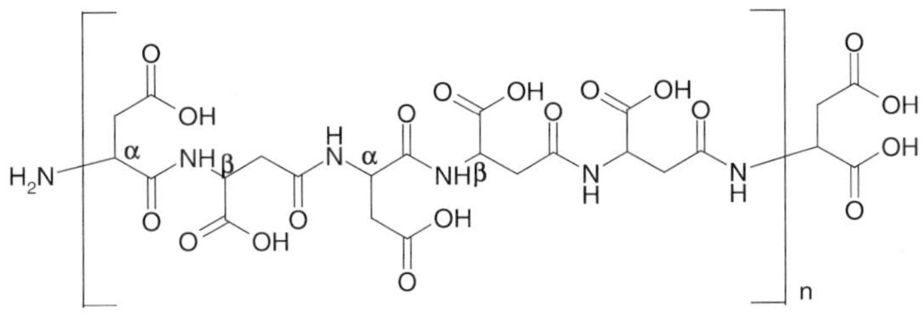

Fig. 2 Structures of poly-α-aspartic acid and poly-β-aspartic acid.

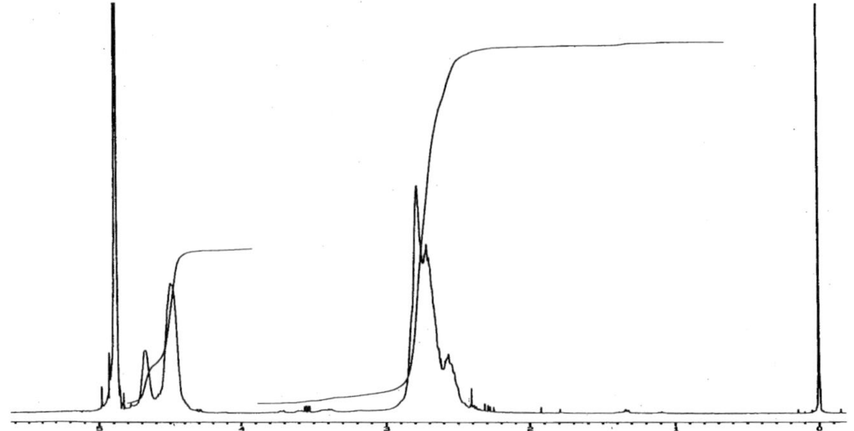

Fig. 3 ^{1}H-NMR spectrum of thermal PASP from aspartic acid in D_2O at pH 13 (Source: Bayer AG.)

4.2
^{13}C-NMR Spectroscopy

^{13}C-NMR spectroscopy has also been used to provide information about the molecular structure of PASP. Pivcova et al. (1981) evaluated the ratio of α- and β-amide units using the methylene signals in the ^{13}C-NMR spectrum, and then analyzed the amide bond sequence using the amide carbonyl signals. Their conclusion was that the distribution of the α- and β-bonds was random (Pivcova et al., 1982).

4.3
^{15}N-NMR Spectroscopy

A further spectroscopic technique used to obtain greater detail of the PASP structure, and especially the three-dimensional structure of PASPs of different origin, is that of ^{15}N-NMR spectroscopy, both solid state and in solution.

According to previous publications (Freeman et al., 1994; Morall et al., 1997), the limited biodegradation of PASP is dependent upon its process of manufacture.

It has been suggested that the different biodegradation behaviors of various PASPs might be assigned to the different tertiary structures, and especially to branching sites within the polymer (Wolk et al., 1994).

By using ^{15}N-NMR spectroscopy of ^{15}N-enriched model compounds, it was possible to visualize the N- terminal groups and N-branching sites (Figure 4). It could also be shown in this way that maleic imide terminal groups (1) and amide branching sites (2) have an unfavorable influence on biodegradation. A comparison of ^{15}N-NMR spectra of PASP prepared either by catalytic polycondensation or by thermal condensation is shown in Figure 5 (see also Figure 1).

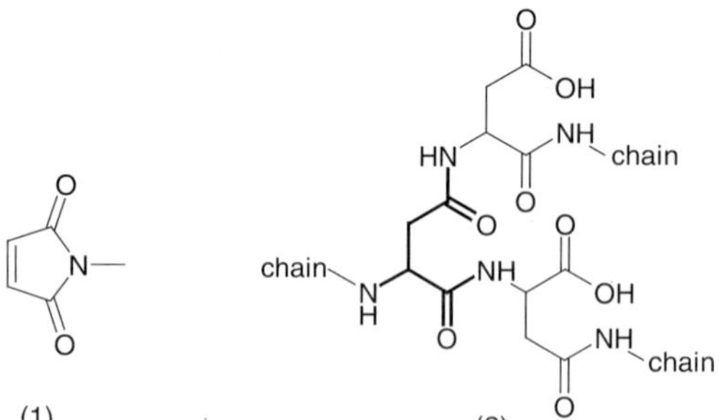

(1) (2)

Fig. 4 N-terminal group (1) and N-branching site (2) of polyaspartic acid produced by thermal condensation.

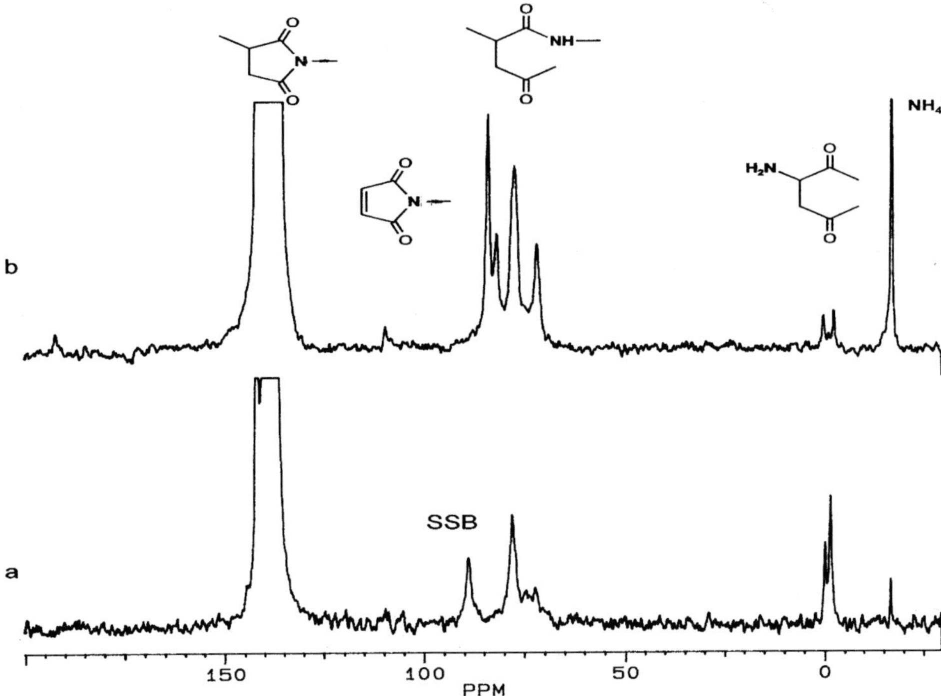

Fig. 5 Comparison of catalytic polyaspartic acid and thermal polyaspartic acid. (a) Polyaspartic acid from catalytic polycondensation. (b) Polyaspartic acid from thermal polycondensation. SSB, spinning side band. (Source: Bayer AG.)

4.4 Gel-permeation Chromatography

The molecular weight of PASP was determined in aqueous solution using gel-permeation chromatography (GPC). With this method it is possible to use polyacrylates or polystyrene sulfonic acids as a standard polymer, though other PASPs of defined and unique chain length are commercially available and can be used as standard materials. During these measurements it is important to maintain a defined pH value of the solution as the molecule extension is heavily dependent on charge density, which in turn is influenced by the degree of dissociation. For the GPC procedure a crosslinked polyhydroxyethylmethacrylate (poly-HEMA) can be used as the column material.

4.5 Isotachophoresis

Isotachophoresis is a method which is both environmentally friendly and inexpensive to operate, mainly because it uses only very small amounts of material and only dilute aqueous buffer solutions are required. The method is well suited for complex mixtures, and also under conditions of extreme concentration. The process of isotachophoresis is based on the fact that charged particles may be characterized by their migration behavior in an electric field, with different dissolved ions being sharply separated into defined fractions that can be detected in succession. Basic devices have been described (Everaerts et al., 1979; Kaniansky and Havasi 1983) that allow quantitative analysis of PASP to be carried out, even in the presence of a variety of other ions.

In isotachophoresis, the sample is injected between two electrolytes, from which one has the highest, and the other the lowest mobility of all ions present in the system (these are the “leading” and “terminating” electrolytes). By applying an electric field, anions and cations are separated according to their different mobilities, if these lie between the mobilities of the two electrolytes. When a stationary state has been reached, all distinguished zones move through the solution at the same speed (hence the name “iso-tacho”), thereby remaining in direct contact with each other and maintaining the electrical circuit. Either ultraviolet radiation or conductivity can be used as a detection method.

5 Detection in Technical Applications

The detection of PASP is mainly important in technical applications where the user requires information about actual product concentration within the process. This is especially the situation in water conditioning, where the process control must include a permanent feedback of how much PASP must be added to a system in order to compensate for any decrease in product.

5.1 Fluorescence Spectroscopy

Thermally synthesized PASP can incorporate traces of chromophore structures into the polymer chain, thus allowing it to be analyzed photometrically. These chromophores are generated during the manufacturing process and are similar to Maillard reaction products (Angrick and Rewicki, 1980). Clearly, they consist of heterocyclic unsaturated groups without any defined structure. PASP has an excitation maximum at 336 nm, and emits fluorescent light with a maximum at 411 nm. The intensity of the emitted radiation is transformed directly into a value of concentration, as this is proportional to the intensity of the fluorescence. This method of analysis is suitable for any application where rapid and reliable online control of the PASP concentration is required. The measured signal determines the amount of product to be added, and in this way a minimum concentration can be maintained without uneconomic overdosing. Nowadays, portable instruments which can be used as mobile measurement stations are available commercially, and these can be specially adapted where frequent changes of the analysis are needed, at both indoor and outdoor locations. An example of this is in drainage systems, where PASP concentrations must be screened at different distances from the dosing equipment.

5.2 Precipitation Titration

One method which can be carried out using basic laboratory equipment is precipitation titration with iron trichloride (Heise and Wendt, 2000). The sample solution is mixed with hydrochloric acid to adjust the pH to 8; subsequently a 0.05 M solution of $FeCl_3$ is added and the resultant turbidity is detected photometrically. The collected data are evaluated using calibration curves constructed with defined concentrations of polymers and with a similar distribution of molecular weights.

5.3 Polyelectrolyte Titration

The basis of polyanion determination is the formation of a polysalt by reacting the polyanion with a counterpolycation, against which it is titrated (Wassmer et al., 1991). The polysalt reaction meets the conditions for a titrimetric determination; namely that it proceeds essentially quantitatively and is very rapid. In many cases it is a reaction in the sense of a 1:1 reaction (Horn, 1980; Domard and Rinaudo, 1981). The end-point of the 1:1 reaction between a polyanion and a polycation is indicated by the formation of an associate between the exceeding polycation and anionic metachromic dye, e.g., eriochrome black T.

6 Biodegradation

Different approaches have been taken to enhance the biodegradability of water-soluble sequestrants and dispersants discharged into effluent waters, and hence new biodegradable alternatives to the currently used polycarboxylates were needed. The introduction of predetermined breaking points in the carbon backbone of the polymer should favor biodegradation and mineralization by specific or unspecific chemical or microbial attacks (Schwamborn, 1998). Polypeptides contain bonds that can be cleaved, and in-vitro degradation studies with different endoproteases have shown that protein-like polymers are degraded by a random mechanism, similar to that seen with proteins or polypeptides (Miller, 1964a; Gonsalves and Mungara, 1996).

Many methods have been established to test the biodegradability of industrial chemicals (OECD, 1981; Karsa and Porter, 1995). These techniques have been developed in order to highlight the behavior of a substance under different environmental conditions, and are adjusted to the requirements of specific fields of application. If data relating to biodegradability are required, then the following methods are most often used, although the method used might depend upon the degree of biodegradability encountered.

- Readily biodegradable: This is an arbitrary classification of chemicals which have passed certain specified screening tests for ultimate biodegradability. The conditions in these tests are so stringent–relatively low density of nonacclimatized bacteria, short duration, absence of other organic compounds–that such chemicals will rapidly and completely biodegrade in aquatic environments under aerobic conditions.
- Inherently biodegradable: This is a classification of chemicals for which there is unequivocal evidence of biodegradation (primary or ultimate) in any test for biodegradability. No limits are placed on the conditions under which the tests are carried out.

In the OECD 301 E-test–which was designed specially for detergent industry–a chemical or material may be classified as readily biodegradable when 70% of it has been removed (biodegraded) after 28 days. For the test to be conducted, the chemical is added as the sole source of carbon at 10–40 mg carbon per liter ($C\ L^{-1}$) to a mineral salt medium containing Na, K, Mg, Ca, Fe, NH_4, Cl, SO_4, and PO_4. The medium is buffered (using a phosphate buffer) at pH 7.4. The medium is then inoculated to provide an increased cell density of 10^2–10^3 per mL. Duplicate vessels are set up containing the chemical plus inoculum, and another inoculated pair containing no added test chemical is used as a control. Another inoculated flask is set up containing a reference chemical (aniline, benzoate, acetate) at 20 mg C L^{-1} to monitor the efficacy of the procedure. Abiotic controls are set up (if required) which contain a sterilized (e.g., with $HgCl_2$) uninoculated solution of the chemical. An inoculated vessel containing both the test and reference chemicals is prepared if the chemical's inhibitory properties are to be determined. The flasks are incubated in the dark at 22 ± 2°C, with continuous shaking. Samples are taken at regular intervals during the 28-day incubation period to enable an adequate biodegradation curve to be constructed. To remove the bacterial cells the samples are either filtered using membranes or centrifuged, after which the concentration of dissolved organic carbon (DOC) is determined in duplicate on the filtrate or supernatant. Filtered samples may be stored at 2–4°C for up to 2 days, or below –18°C for a longer period. In all cases the DOC values are corrected for those of the blank controls.

In the modified Sturm test (OECD 301 B), 3-L batches of the mineral medium (as described above) are enclosed in vessels and inoculated with 30 mg L^{-1} active sludge at 1% (v/v), thereby giving a cell concentration of approximately 10^5–10^6 cells per mL. The inoculated mixtures are aerated with CO_2-free air overnight to purge the system of CO_2. The test and reference chemicals are added to provide separately 10–20 mg DOC L^{-1}. Abiotic controls may be set up if required. CO_2-free air is bubbled through the liquid at rates of 30 to 100 mL min^{-1} as the vessels are held at 22 ± 2°C in the dark. CO_2 generated in the process of degradation is trapped in 100 mL of 0.125 M barium hydroxide solution and then quantitated by titration of the unreacted barium hydroxide. On the 28th day, 1 mL of concentrated HCl is added to the test medium, which is then aerated overnight to drive off any residual CO_2 into the trap. Typical results from this test are shown in Figure 6, where PASP is compared with a common polycarboxylate, with glucose as the readily biodegradable material. The degradation behavior of PASP was seen to be closely related to that of glucose.

In order to conduct the OECD 302 B test (Zahn–Wellens test), a 2-L batch of the above-mentioned mineral medium is prepared containing the test chemical at 50–400 mg DOC per liter and 200–1000 mg of

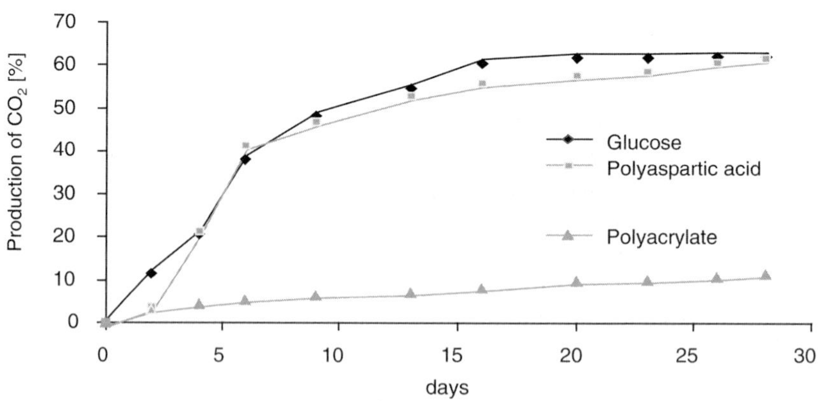

Fig. 6 Degradation curve of thermally synthesized polyaspartic acid (from maleic anhydride and ammonia) according to the modified Sturm test (according to OECD 301 B) with polyaspartic acid, polyacrylate and glucose. (Source: Bayer AG.)

dry solids per liter of previously washed active sludge. The ratio of inoculum to test chemical is maintained between 2.5:1 and 4:1. Additional batches contain either no added chemical or a reference chemical as a control agent. The suspensions are kept in glass cylinders, each of which is equipped with a stirrer. The contents are aerated with purified, humidified air at 20–25°C in the dark or diffuse light for 28 days. The mixture is stirred to keep the sludge in suspension and to maintain the concentration of dissolved oxygen above 1 mg L^{-1}. The pH value is frequently monitored and adjusted to 6.5–8. The samples from the medium are filtered or centrifuged and the concentration of DOC is determined in the filtrate or supernatant.

The mechanism of PASP hydrolysis during biodegradation remains unclear. Based on results from a series of enzymatic studies conducted with poly(α,L-glutamic acid) and poly(α,L-lysine), Miller concluded that these homopolymers were subject to both exo- and endopeptidase activity (Miller, 1961, 1964b). Since the structure of thermally produced PASP is atypical for a protein, the rate and extent of biological degradation may differ considerably from polypeptides based on "natural" sources. Being constituted of one amino acid would limit the field of proteases that might attack the structure (Alford et al., 1994). Microbes with the ability to degrade PASP were identified as *Sphingomonas* sp. KT-1 (JCM 10459) and *Pedobacter* sp. KP-2 (JCM 10638) (Doi et al., 2000). Wolk et al. and Freeman et al. also investigated the structure–biodegradability relationship for PASP, and determined that branching sites actually inhibit biodegradation (Wolk et al., 1994; Freeman et al., 1996). It is possible that the structures of the end groups, as well as their concentration, might also influence biodegradability of the material.

7 Production of PASP

Several different routes of synthesis have been investigated for the manufacture of PASP (Figure 7). As the starting point of synthesis, maleic anhydride is the most economic material, as it can be transformed into the ammonium salts of maleic acid or maleic amide acid which can be thermally polymerized to polysuccinimide. Alternatively, maleic acid can be converted with ammonia to yield aspartic acid, which in turn can be thermally polymerized to polysuccinimide. In both cases, polysuccinimide is ultimately hydrolyzed to the sodium salt of PASP.

A number of synthetic processes have been published in patent applications by several chemical manufacturers, and three basic developments are described below.

One synthesis starting with aspartic acid has been described in a patent of Donlar Corporation (USA) (Koskan et al., 1995) in which a DVT-130 drier-mixer (Littleford Brothers, Inc., Florence, Kentucky, USA) is used. The jacketed drier utilizes oil as thermal fluid and a plough-blade impeller. The drier-mixer has a stack which is open to the atmosphere, and a heat transfer area of 0.9 m^2. The reactor's oil reservoir is preheated to 290 °C to provide an oil inlet at a temperature of ~260 °C. The reactor is charged with 50 kg of powdered L-aspartic acid. Hot oil begins to flow through the jacket and the impeller speed is set at 155 r.p.m. The temperature of both product and oil steadily rises when, at a product temperature of 200 °C, there is a sudden endothermic reaction which causes the product temperature to drop. A loss of water is evidenced by the evolution of steam. A sample taken reveals that the color of the powder is changed from white to pink. Thereafter, the product temperature begins

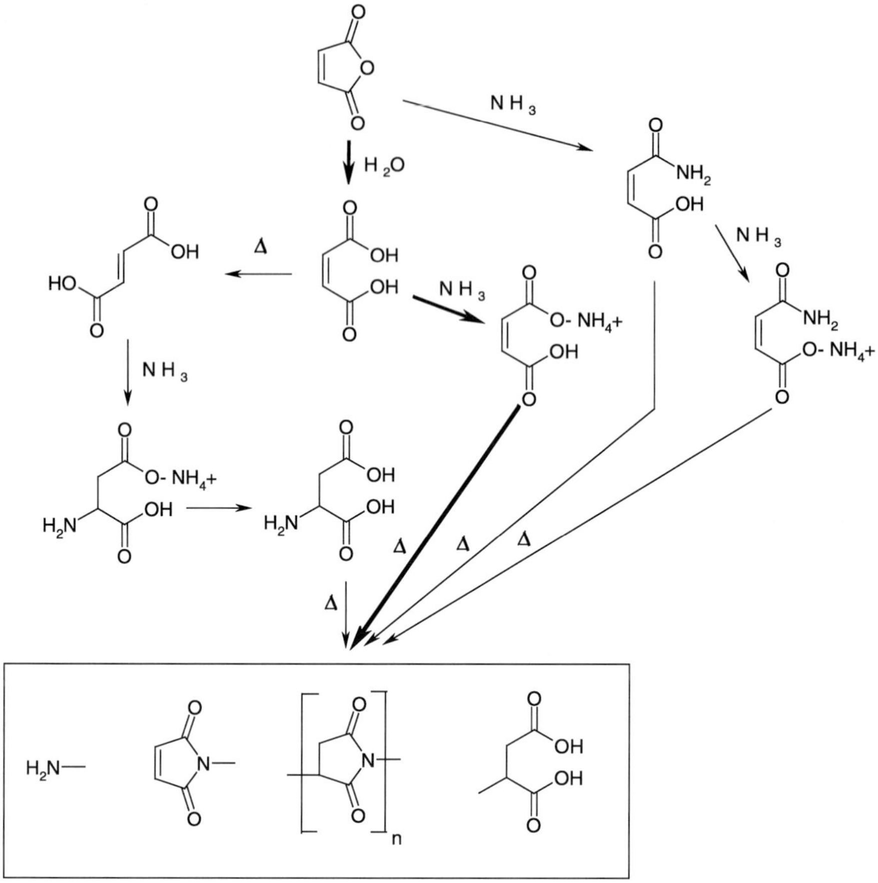

Fig. 7 Different routes for the thermal synthesis of polysuccinimide, the precursor of polyaspartic acid. C- or N-termini of the polymer chain can consist of the structural elements shown. The bold arrows marks the synthetic route carried out at Bayer AG, Leverkusen.

to rise steadily until it reaches a plateau at 220°C, and this continues for 1 h. During that reaction, steam is evolved and the conversion increases in linear mode. After 1 h, the temperature has risen to 230°C, at which the reaction undergoes a second endothermic step, immediately after which the steam ceases to evolve. Shortly after this point the reaction is at least 88% complete. Following the second endothermic reaction the product slowly changes from a pink to a yellow color. The final conversion has been measured as 97% complete.

Thermal condensation can also be performed in a Krauss Maffei VTA 12/8 plate drier fitted with eight stainless steel plates, each of 1200 mm diameter. The drier is operated at ambient atmosphere, but the first two plates are maintained at a relatively lower temperature than the next five. The last plate is not utilized due to temperature control problems. Each plate has 0.35 m^2 of heat transfer surface and four arms of plows. Except for the first plate, the fourth arm of each plate has the plows moving in reverse in order to increase the retention time and turnover rate. L-aspartic acid powder is delumped and fed to the first plate of the drier. The plows then distribute the deposited L-aspartic acid evenly until thin ridges cover the plate. The deposited powder follows a spiral across the plate until it is pushed off the edge onto the next plate. There, the process is repeated until the powder is swept from the center of the plate. When the conditions of the process have stabilized, samples are taken from each plate to measure extent of reaction. The conversion of L-aspartic acid to polysuccinimide can be followed by noting the color change of the powder.

Polysuccinimide produced in the foregoing manner can be readily converted to PASP as a result of base hydrolysis.

According to a description in a patent of Rohm and Haas (Adler et al., 1995), the synthesis can also be carried out by thermal condensation of aspartic acid under addition of orthophosphoric acid: 950 g of L-aspartic acid is mixed with 50 g of a 85 wt.% orthophosphoric acid in a stainless steel pan. The bed depth of the reaction mixture is ~4 cm. The reaction mixture is placed in a muffle furnace at atmospheric pressure, and this is preheated to 240°C. After being held for 1 h at 240°C, the slightly wet white powder turns pink, with some yellow clumps. This reaction mixture is then moved from the muffle furnace and ground using a mortar and pestle to simulate a mechanical means of maintaining an intimate admixture. After grinding, the reaction mixture is returned to the muffle furnace. After a total of 6 h at 240°C the mostly pink powder becomes tan in color. After a total of 10 h at 240°C, the reaction mixture is removed from the muffle furnace, whereby the process yield is 728 g of tan powder that is identified as polysuccinimide with a M_w of 5560 g mol^{-1} and a M_n of 4050 g mol^{-1}.

Hydrolysis of the polysuccinimide is achieved as follows: 136 g of deionized water is placed into a 1-L four-neck flask equipped with a mechanical stirrer, thermocouple, condenser, pH probe, and an inlet for addition of base. Poly(anhydroaspartic acid) (25 g) is added to the flask and the resulting mixture is heated to 90°C while a 50 wt.% solution of sodium hydroxide is added dropwise. The feed is controlled using a Chem-Cadet® (Cole Palmer Instrument Inc., Chicago, USA) such that the feed is stopped when the mixture reaches a pH of 10.8. A total of 16 g of 50 wt.% sodium hydroxide solution is required to attain a stable pH of 10.8. After complete addition of sodium hydroxide, the mixture is held at 90°C for 30 min, after which the mixture is cooled to room temperature. The mixture is then lyophilized to produce a dried sample of PASP.

In the Bayer process (Wagner et al., 1998) an aqueous solution of an ammonium salt of maleic acid with low molecular weight adducts is first prepared from maleic anhydride and ammonia and continuously polymerized. The prereactor (Figure 8) is intended to produce the above-mentioned salt solution. In the next step of the polymerization it essential to ensure an efficient and intensive exchange of heat, and any type of suitable device for heat exchange can be used for this (e.g., a tube-bundle heat exchanger, a falling film evaporator, a plate heat exchanger, a temperature-controlled static mixer, a mixing vessel with specific stirrer geometry for viscous products).

In the later stages of polymerization the supply of heat and mechanical energy must be sufficient to keep the reaction mixture in motion and to renew the surface of the mass. The reactor must also ensure the duration of reaction and bring the highly viscous material to dryness. Particular preference is given to the break-up of the resultant solid in this process to give a large number of small particles. This break-up considerably im-

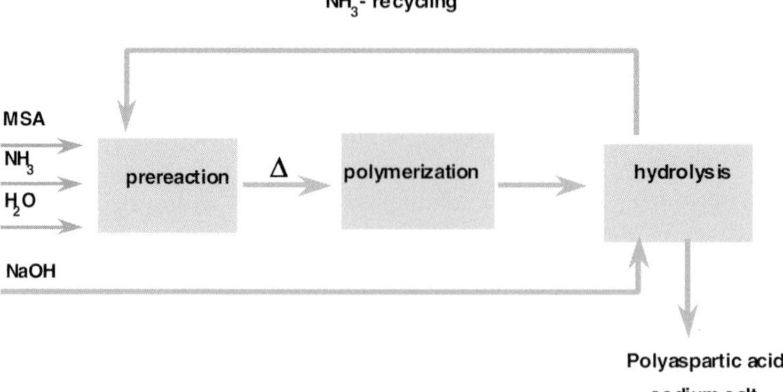

Fig. 8 Thermal synthesis of polyaspartic acid sodium salt from maleic anhydride, ammonia and sodium hydroxide.

proves the evaporation, removal of the substance eliminated during the condensation, and also considerably reduces the diffusion path lengths for the substances eliminated. Alongside the more effective removal of the substances eliminated, the large surface of the solid particles markedly improves the transfer of heat, leading to a fully reacted product. The resulting material has a considerably smaller amount of residual monomers and better characteristics in chemical analysis and in use.

During polymerization, the viscosity and temperature rise further, and in some cases the material becomes a solid which is no longer flowable. As with other technical polycondensations (e.g., the manufacture of polyesters), efficient removal of the released water is necessary in order to achieve the desired build-up of molecular weight (Elias, 1997). Any commercially available kneading reactor may be used for this novel process, as long as it is capable of achieving the above-mentioned objectives. Polymers obtained in this way are subjected to hydrolysis with sodium hydroxide solution, so that the resultant polymer has recurring aspartic acid sodium salt units.

8 World Market and Applications

From the chemical point of view, PASPs are to be integrated into the class of polycarboxylates, which are used in large amounts as water-soluble anionic dispersing agents. In 1994, the annual world market of these compounds was estimated at about 265,000 tons, and it is expected that this market will grow in line with the gross domestic product.

The main uses of water-soluble polycarboxylates are also shown in Figure 9, the main use being as scale inhibitors or dispersing agents.

Scale inhibitors prevent poorly soluble salts from crystallizing, even if used in substoichiometric concentrations. This effect is based on their adsorption at the surface of microscopically small salt crystals, e.g., calcium carbonate. This property is used in industrial applications such as cooling water circuits where water hardness reduces the efficiency of the processes applied.

Dispersions or suspensions are thermodynamically unstable systems. They exhibit the tendency eventually to divide into two

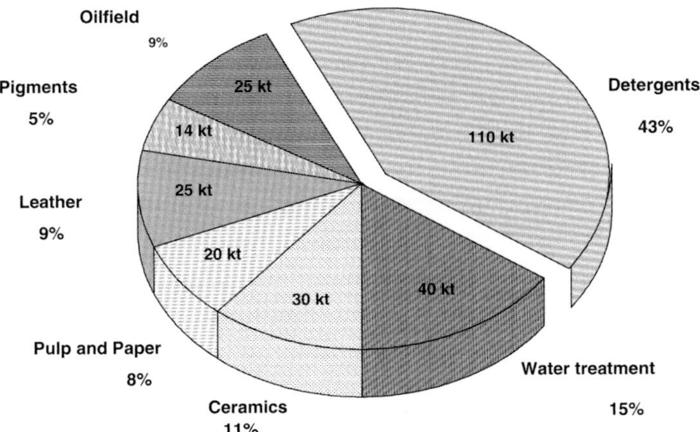

Fig. 9 Market volume of water-soluble polycarboxylate homo- and copolymers (from Schwamborn, 1997).

separate continuous phases–a coherent solid phase and a liquid phase. Due to their tendency to adsorb onto the surface of fine particles, polyanionic polymers are able to stabilize dispersions because the repulsion forces of the particles are increased.

Polyacrylate homo- and copolymers (e.g., with maleic acid) are well established in this market. Although these polyacrylates are partially eliminated in sewage plants by adsorption, they cannot be classified as biodegradable (Larson et al., 1996).

In order to improve their environmental acceptance, one alternative might be to create polyaspartates that are biodegradable (see Figure 6). It is known that biomineralization processes, e.g., the building of skeletons or shells of crustaceae, are preferably controlled by polyanionic proteins rich in aspartic acid and phosphoserine (Sikes and Wierzbicki, 1996), and which contain substructures of up to 50 aspartic acid subunits (Rusenko et al., 1991).

These proteins influence the aggregation and growth of crystals during the process of crystallization. Modern analytical methods, e.g., atomic force microscopy (Wierzbicki et al., 1994) and molecular modeling methods (Sikes and Wierzbicki, 1994; Fizwater, 1995) have been applied in an attempt to understand these interaction phenomena. The protein chain interacts not only in vivo but also *in vitro* with certain surfaces of a microcrystal, and thereby hinders or delays crystal growth. Thus, the presence of polyaspartate leads to a different type of crystal morphology (Sikes and Wierzbicki, 1996) (Figure 10).

Furthermore, new seed crystals can be formed on the polymer chains so that ultimately many small crystals are formed which can move relative to each other and which are separated by polyaspartate molecules. On a macroscopic scale, this collection of crystallites has quite different mechanical properties from those of large crystals.

Since 1997, technical quantities of polyaspartic acid have been available commercially and have been used successfully in several areas, including scale inhibitors in cooling water circuits, in environmentally benign detergents, as corrosion inhibitors, in the liquefaction of mineral slurries, and as a calcium binder in textile processes.

Fig. 10 Calcite crystal growth without (left) and with (right) polyaspartic salts. (reproduced from Biomimetic Material Chemistry, p. 256 Figure 9.6(b) (Mann, S., Ed.), ISBN: 1–56801-669-4 (1996). © Wiley & Sons Inc.

9 Patents

Patents relating to PASP deal mainly with commercial applications where the adsorptive and dispersing properties of the product are used; hence, the majority of products relate to surface effects such as cleaning and protection (Table 1).

- Production of PASP: In the production of PASP the patents of Donlar, BASF, Rohm & Haas and Bayer are of most relevance. Donlar and Rohm & Haas deal mainly with the manufacture of PASP from monomer aspartic acid; these processes enable a production based on renewable resources such as carbohydrates, which can be biologically transformed into aspartic acid. The Bayer process is based on maleic anhydride and ammonia, and consists of a thermal polymerization with subsequent hydrolysis by sodium hydroxide. Among a range of special patent applications, several have described discrete polymerization techniques and compositions of the resultant products (see Table 1).
- Laundry: Laundry applications refer mainly to the dispersing and scale-inhibiting effects of PASP in textile cleaning processes. These patents focus on the use of PASP as a detergent builder in different combinations with varying tensides, bleaching agents, and inorganic compounds. PASP is described as water softener, inhibitor of the redeposition of removed soil, and crystal growth inhibitor. The graying of white textiles and damage to the launderette heating devices are prevented by PASP. The biodegradability of PASP is especially in laundry applications – a fact highlighted by the diverse range of patents.
- Hard surface cleaning and dishwashing: In dishwashing, PASP is described as having comparable effects to laundry formulations. Additional surface effects are achieved by the ability of PASP to adsorb onto mineral surfaces, thereby inhibiting corrosive attack which generally appears as turbidity on glass. A slight calcium-binding effect supports the efficiency of tensides and leads to better cleaning results, notably the removal of stains such as tea, milk, egg, oat flakes, or spinach.
- Corrosion: Due to the affinity of PASP towards the surfaces of crystalline solids, the polymer is able to reduce corrosive

influences on metals in aqueous media. Additives to protect steel, copper and brass have been investigated by specialty chemical manufacturers in combination with other inhibiting ingredients such as benzotriazoles.

- Water treatment: Prevention of scale build-up is economically important in cooling water circuits, boilers, drainage systems and other plants where mineral deposits can only be removed at high cost. Permanent administration of a scale-inhibiting agent is generally the most efficient way to maintain such a plant's reliable function. In all applications where treated water in released into the environment, biodegradability is a necessary precondition for the use of any additive. PASP and its precursor polysuccinimide have special potential in tunnel drainage treatments as outlets flow directly into surface water in most cases. The toxicological profile of PASP permits safe use for the surrounding biosphere.
- Agriculture: PASP has two main benefits in agriculture and plant protection: (i) It can be used to provide effective dispersion of mineral ingredients in fertilizers; and (ii) it can be used in the slow release of nitrogen to plants, thus causing more rapid growth.
- Oil fields: In most cases, drilling for oil also involves an extended escape of water from underground reservoirs. Mineral materials which emerge with the water can cause scaling and heavy sedimentation. Addition of PASP to the water stream maintains homogeneity of the mineral sludge and also helps to prevent corrosion of the drilling device.
- Leather: Leather can be made biodegradable by preparation without the use of Cr, Ti, Fe, Al or Zr, including post-tanning with PASP or its salts. PASP-treated leather degraded completely after 3 months of composting, compared with only 5% degradation for chrome-tanned leather. PASP can also be used as an auxiliary component for leather-dyeing, and as additive in formulations to remove fat from raw leather.
- Personal care: As PAPS consists of a naturally occurring amino acid, it has a close chemical relationship to the proteins of skin and hair. As with any polymeric material, skin penetration is low and this leads to high tolerance. A possible derivatization of PASP with oleochemical compounds can impart a mild tenside effect to cosmetic formulations. The calcium-binding properties of PASP lead to better foam behavior and low tenside consumption due to precipitation. Biodegradability allows extended use, even in regions where sewage disposal is limited.

Tab. 1 Patents relating to polyaspartic acid (PASP) production and applications

Patent no.	*Publication date*	*Patent holder*	*Inventors*	*Patent title*
Production of PASP				
WO 95 / 00479	05.01.95	Donlar Corp.	Koskan et al.	Polyaspartic acid manufacture
US 5 457 176	10.10.95	Rohm and Haas Company	Adler et al.	Acid catalyzed process for preparing amino acid polymers
US 6 187 898	27.08.98	Bayer AG	Wagner et al.	Method for carrying out polycondensation reactions

Tab. 1 (cont.)

Patent no.	*Publication date*	*Patent holder*	*Inventors*	*Patent title*
WO 98 48032	29.10.98	Degussa AG	Tirrell et al.	Poly-(α-L-aspartic acid), poly-(α-L-glutamic acid) and copolymers of L-asp and L-glu, Method for their production and their use
EP 0 256 366	24.02.88	Bayer AG	Boehmke et al.	Verfahren zur Herstellung von PAS und ihren Salzen
EP 0 578 448	12.01.94	Rohm and Haas Company	Paik, Swift	Method for preparing polysuccinimides using a rotary tray dryer
EP 0 578 449	12.01.94	Rohm and Haas Company	Swift, Simon	Process for preparing polysuccinimides from aspartic acid
EP 0 578 450	12.01.94	Rohm and Haas Company	Paik, Swift	Process for preparing polysuccinimides from maleamic acid
EP 0 604 813	06.07.94	Bayer AG	Joentgen, Groth	Verfahren zur Herstellung und Verwendung von Polysuccinimid, Polyasparaginsäure und ihre Salze
EP 0 612 784	31.08.94	Bayer AG	Joentgen, Groth	Verfahren zur Herstellung von Polysuccinimid und Polyasparaginsäure
EP 0 613 920	07.09.94	Bayer AG	Joentgen, Groth	Verfahren zur Herstellung von Polyasparaginsäure
EP 0 660 852 WO 94/06843	17.09.93	Donlar Corp.	Koskan et al.	Production of polysuccinimide and polyaspartic acid from maleic anhydride and ammonia
EP 0 660 853	05.07.95	Donlar Corp.	Koskan et al.	Production of polysuccinimide and polyaspartic acid from maleic anhydride and ammonia
EP 0 689 555 WO 94/21695	21.05.97	BASF AG	Kroner et al.	Polycondensate auf Basis von Asparaginsäure, Verfahren zu ihrer Herstellung und ihre Verwendung
EP 786 487	30.07.97	Bayer AG	Joentgen, Groth	Preparation of polymers containing repeating succinyl units
US 5 373 086	13.12.94	Donlar Corp.	Koskan et al.	Polyaspartic acid having more than 50 % β form and less than 50 % α form
US 5 449 748	12.09.95	Monsanto Company	Ramsey	Preparation of anhydropolyamino acids at temperatures of 350°C or above
US 5 457 176	10.10.95	Rohm and Haas Company	Adler et al.	Acid catalyzed process for preparing amino acid polymers
US 5 459 234	17.10.95	Rohm and Haas Company	Bortnick et al.	Continuous thermal polycondensation for preparing polymers
US 5 463 017	31.10.95	Rohm and Haas Company	Paik, Swift	Process for preparing polysuccinimides from maleamic acid
US 5 466 779 WO 95/31494	14.11.95	Donlar Corp.	Ross	Production of polysuccinimide
US 5 468 838	21.11.95	Bayer AG	Boehmke, Schmitz	Process for the preparation of polysuccinimide polyaspartic acid and their salts
US 5 491 212 EP 578 448	13.02.96	Rohm and Haas Company	Paik, Swift	Method for preparing polysuccinimides using a rotary tray dryer

Tab. 1 (cont.)

Patent no.	Publication date	Patent holder	Inventors	Patent title
WO 96/19524 EP 0 747 417	11.12.96	Mitsubishi Chemical Corp.	Nakato et al.	High molecular wt. Polyaspartic acid with good calcium ion chelating capacity-useful as chelating agent flocculant scale inhibitor, detergent builder etc. Produced by hydrolysis of polysuccinimide in high yield
WO 98/34976	13.08.98	Solutia Inc.	Martin	Production of solid polyaspartate salt
Laundry applications				
EP 0 324 595	19.07.89	Procter & Gamble Company	Heinzman et al.	Amino-functional compounds as builders/dispersants in detergent compositions
EP 0 454 126	22.11.91	Rohm and Haas Company	Du Vosel et al.	Polyamino acids as builders for detergent formulations
EP 0 497 611	05.08.92	Rohm and Haas Company	Swift, Weinstein	Biodegradable polymers, process for preparation of such polymers and compositions containing such polymers
EP 0 511 037	28.10.92	Rhone-Poulenc S.A.	Ponce, Tournlhac	Composition detergente contenant un biopolymere polymide hydrolysable en milieu lessiviel
EP 0 859 825 DE 19 540 086	30.04.97	Henkel KG aA	Kottwitz et al.	Verwendung von polymeren Aminodicarbonsäuren in Waschmitteln
US 5 538 671	23.07.96	Procter & Gamble Company	Morrall	Detergent compositions with builder system comprising aluminosilicates and polyaspartate
WO 96/22352	25.07.96	Procter & Gamble Company	Hall et al.	Detergent compositions comprising stabilized poylamino acid compounds
Hard surface cleaning and dishwashing				
EP 0 561 452	22.09.93	Unilever N.V.	Van Dijk et al.	Machine dishwashing composition containing polyaminoacids as builder
EP 0 561 464	22.09.93	Unilever N.V.	Van Dijk et al.	Polyaminoacids as builder for rinse aid compositions
EP 0 651 052	03.05.95	Procter & Gamble Company	MacBeath,	Machine dishwashing detergent compositions
US 5 770 553	23.06.98	BASF AG	Kroner et al.	Use of Polyaspartic Acid in Detergents and Cleaners
WO 93/21243	28.10.93	Unilever N.V.	Hales et al.	Polymers and detergent compositions containing them
Corrosion				
EP 0 514 376 WO 91/12354	27.09.00	Monsanto/ Solutia S.A.	Kalota, Silverman	Compositions and process for corrosion inhibition of ferrous metals
EP 0 700 987	11.09.95	Rohm and Haas Company	Freeman et al.	Method of inhibiting corrosion in aqueous systems
EP 0 710 733 DE 04439193	23.10.95	Bayer AG	Rother, Kunisch	Mischung zur Korrosionshemmung von Metallen

Tab. 1 (cont.)

Patent no.	*Publication date*	*Patent holder*	*Inventors*	*Patent title*
US 5 531 934	02.07.96	Rohm and Haas Company	Freeman et al.	Method of inhibiting corrosion in aqueous systems using poly(amino acids)
Water treatment				
EP 0 305 283 US 4 868 287	01.03.89	University of South Alabama	Sikes, Wheeler	Inhibition of mineral deposition by polyanionic/hydrophobic peptides
EP 0 391 629 US 5 051 936	10.10.90	University of South Alabama	Sikes	Inhibition of mineral deposition by phosphorylated and related polyanionic peptides
EP 0 530 358 WO 92/16462	10.03.93	Donlar Corp.	Koskan et al.	Polyaspartic acid as a calcium carbonate and a calcium phosphate inhibitor
EP 0 530 359 WO 92/16463	10.03.91	Donlar Corp.	Koskan et al.	Polyaspartic acid as a calcium sulfate and barium sulfate inhibitor
EP 0 672 625	20.09.95	Bayer AG	Kleinstück et al.	Mittel zur Wasserbehandlung mit Polyasparaginsäure oder einem Derivat davon und Phosphonsäure
EP 0 819 653	28.01.98	Nalco Chemical Company	Tang, Davis	Use of biodegradable polymers in preventing scale build up
EP 0 821 054	28.01.98	Bayer AG	Groth et al.	Verwendung von Polyasparaginsäure und ihrer Salze
EP 0 950 641 CH 689 452	03.04.99	Wegmueller	Wegmüller	Process for preventing deposits caused by cement interactions in a construction dewatering system
WO 94/19288	01.09.94	Deutsche Nalco-Chemie GmbH	Wegmüller	Verfahren zur Verhinderung von Ablagerungen in einem Bauwerksentwässerungssystem
Agriculture				
US 5 593 947	14.01.97	Donlar Corp.	Sanders	Method for more efficient uptake of plant growth nutrients
US 5 646 133	08.07.97	Donlar Corp.	Sanders	Polyaspartic acid and its analogues in combination with insecticides
US 5 783 523	21.07.98	Donlar Corp.	Sanders	Method and composition for enhanced hydroponic plant productivity with polyamino acids
WO 97/34481	25.09.97	Donlar Corp.	Sanders	Polyorganic acids and their analogues to enhance herbicide effectiveness
Oil fields				
US 6 02 2401	08.02.00	Nalco Chemical Company	Tang et al.	Modified polyaspartic acid copolymers for biodegradable corrosion inhibitors and scale control in aqueous petroleum media
WO 96/29502	26.09.96	BP Explo-ration Opera-ting Company Ltd.	Duncum et al.	Hydrate Inhibitors for drilling fluids and oil and natural gas pipelines
WO 97/28230	07.08.97	Baker Hughes Ltd.	Ehle et al.	Scale inhibition in petroleum and natural gas production

Tab. 1 (cont.)

Patent no.	*Publication date*	*Patent holder*	*Inventors*	*Patent title*
Leather				
EP 0 711 842	27.10.95	Bayer AG	Joentgen et al.	Ledergerbstoffe und Stellmittel für Farbstoffe
WO 94/01587	20.01.94	Henkel KGaA	Ritter et al.	Neue Lederfettungsmittel und ihre Verwendung
WO 98/38340	03.09.98	Bayer AG	Träubel et al.	Biologically degradable leather
Personal care				
EP 0 767 191 US 5686066	30.09.96	Mitsu Chem. Inc., Ajinomoto Co. Inc.	Harada et al.	Polymer, preparation process thereof, hairtreating compositions and cosmetic compositions
EP 0 774 247 JP 296432	21.05.97	Mitsui Chem. Ajinomoto	Matsuzawa et al.	Hair care products
EP 0 821 932 US 5961965	04.02.98	BASF AG	Nguyen et al.	Wasserlösliche und wasserdispergierbare Polyasparaginsäurederivate, ihre Herstellung und ihre Verwendung
EP 0 884 344	16.12.98	Goldschmidt AG	Grüning et al.	Milde tensidhaltige Zubereitungen mit copolymeren Polyasparaginsäurederivaten für kosmetische oder Reinigungszwecke
FR 2 403 076	14.09.77	L'Oreal S.A.	Jacquet et al.	Nouvelles compositions cosmetiques a base d'amides de l'acide polyaspartique
US 5 686 066 EP 0 767 191	11.11.97	Mitsui Chem. Inc., Ajinomoto Co. Inc.	Harada et al.	Polymer, preparation process thereof, hairtreating compositions and cosmetic compositions
WO 2000012054	09.03.00	Wella	Birkel et al.	Use of polyaspartic acid in hair treatment preparations

10 Outlook and Perspectives

When used as polyanionic dispersants, PASPs – in contrast to the well-established polyacrylates–are biodegradable and therefore suitable for improving the environmental compatibility of water-soluble dispersing agents and scale inhibitors that might enter the surface water.

On condition that the price–performance relationship is competitive, the more environmentally benign performance chemicals such as PASP will gain market shares that equate with acceptance of concepts of sustainable care. Although the emphasis placed by different nations varies widely at present, it is assumed that such acceptance will increase in the future. Consequently, regulations are already in place in several countries relating to the use of environmentally benign ingrediants in detergents and cleaners, and most recently for the European Union (European Eco-label, 1992). The first laundry detergent based on PASP received the environmental label of the "EU Flower" (this is awarded to environmentally friendly consumer products in member states of the EU) in February 2001.

The Scandinavian countries have strong, longstanding interests in environmental protection, and have specified in their regulations for detergents whether the materials should be produced from petrochemical resources or from renewable carbon sources. Only detergents with ingredients basically produced from natural raw materials will obtain the highest classification (Swedish Society for Nature Conservation, 2001). To what extent this development will be accepted for example by the EU or North America–which represent the major markets – is difficult to predict in the near future. Successful penetration of these markets will depend upon a competitive price and good price–performance relationship; moreover, the safe supply of the raw materials used will be essential.

11 References

Addadi, L., Weiner, S. (1992) Kontroll- und Designprinzipien bei der Biomineralisation, *Angew. Chem. Int. Ed. Engl.* **31**, 153–169.

Adler, D. E., Freeman, M. B., Lipovsky, J. M., Paik, Y. H. (1995) Acid catalyzed process for preparing amino acid copolymers, Rohm and Haas Company, US Patent 5 457 176.

Alford, D. D., Wheeler, A. P., Pettigrew, C. A. (1994) Biodegradation of thermally synthesized polyaspartate, *J. Environ. Polymer Degrad.* **2**, 225–236.

Allemand, D., Cuif, J.-P. (1994) Biomineralisation 93, 7th International Symposium of Biomineralization. Bulletin de l'institut oceánographique Monaco, 5–10.

Angrick, M., Rewicki, D. (1980) *Chemie in unserer Zeit* **14**, 149–157.

Dessaignes, M. (1850) Nouvelles recherches sur la production de l'acide succinique au moyen de la fermentation, *C. R. Seances* **31**, 432–433.

Doi, Y., Tabata, K., Abe, H. (2000) Microbial degradation of poly(aspartic acid) by two isolated strains of *Pedobacter* sp. and *Sphingomonas* sp., *Biomacromolecules* **1**, 157–161.

Domard, A., Rinaudo, M. (1981) Polyelectrolyte complexes: interaction between poly(L-lysine) and polyanions with various charge densities and degrees of polymerization, *Macromolecules* **14**, 620–625.

Elias, G. E. (1997) *An Introduction to Polymer Science*, Weinheim, New York: Verlag Chemie, 120–127.

European Eco-label, (1992) http://europa.eu.int./ecolabel.

Everaerts, F. M., Verheggen, T. P., Mikkers, F. E. (1979) Determination of substances at low concentrations in complex mixtures by isotachophoresis with column coupling, *J. Chromatogr.* **169**, 21–38.

Eyal, A., Cami, P. (1997) Process for the preparation of aspartic acid, Amylum, WO/ 97 27312.

Fizwater, S. (1995) Computer simulation of polyelectrolyte adsorption on mineral surface, *ACS Symp. Ser.* **589**, 316–325.

Fox, S. W., Harada, K. (1960a) Thermal copolymerization of amino acids in the presence of phosphoric acid, *Arch. Biochem. Biophys.* **86**, 281–285.

Fox, S. W., Harada, K. (1960b) The thermal copolymerization of amino acids to protein, *J. Amer. Chem. Soc.* **82**, 3745–3751.

Fox, S. W., Harada, K., Vegotsky, A. (1959) Thermal polymerization of amino acids and a theory of biochemical origins, *Experientia* **15**, 81–84.

Fox, S. W., Harada, K., Krampitz, G., Mueller, G. (1970) Chemical origins of cells, *Chem. Eng. News* **48**, 80–94.

Freeman, M. B., Paik, Y. H., Swift, G., Wilczynski, R., Wolk, S. K., Yocom, K. M. (1994) Biodegradability of polycarboxylates: structure–activity studies, *Polym. Prep. (Am. Chem. Soc.)* **35**, 423–424

Freeman, M. B., Swift, G., Paik, Y. H., Wilczynski, R., Wolk, S. K., Yokom, K. (1996) Biodegradability of polycarboxylates: structure–activity studies, in: *Hydrogels and Biodegradable Polymers for Bioapplications* (Ottenbrite, R. M., Huang, S. J., Park, K., Eds.), Washington, DC: American Chemical Society, Chapter 10, 118–136.

Gonsalves, K. E., Mungara, P. M. (1996) Synthesis and properties of degradable polyamides and related polymers, *Trends Polym. Sci.* **4**, 25–31.

Goto, A., Kunioka, M. (1992) Effective fermentation of poly(gamma-glutamic acid) from a medium containing citric acid, *Polymer Reprints* **41**, 217–219.

Harada, K. (1959) Polycondensation of thermal precursors of aspartic acid, *J. Org. Chem.* **24**, 1662–1666.

Harada, K., Fox, S. W. (1964) Thermal polycondensation of free amino acids with polyphosphoric acid, *Principles Med. Biol.* **1**, 289.

Heise, K.P., Wendt, H. (2000) Method for the quantitative determination of poly aspartic acid, Bayer AG, WO 00/28305.

Horn, D. (1980) *Polymeric Amines and Ammonium Salts.* (Goethals, E. J., Ed.), Oxford, New York: Pergamon Press, 333.

Kaniansky, D., Havasi, P. (1983) Instrumentation for capillary isotachophoresis, *Trends Anal. Chem.* **2**, 197–202.

Karsa, D. R., Porter, M. R. (Eds.) (1995) *Biodegradability of Surfactants*, Glasgow: Blackie, 95–97.

Koskan, L. P., Low, K. C., Meah, A., Atencio, A. M., (1995) Polyaspartic acid manufacture, Donlar Corporation, WO 95/00479.

Kovacs, J., Könyves, I. (1954) Über DL-α,β-Polyasparaginsäure, *Naturwissenschaften* **41**, 333.

Kovacs, J., Könyves, I., Pusztai, A. (1953) Über DL-α, β-Polyasparaginsäure, *Experientia* **9**, 459–460.

Kurshwaha, D. R. S. (1980) Poly-ε-lysine: synthesis and conformation, *Biopolymers* **19**, 219–229.

Larson, R. J., Bookland, E. A., Williams, R. T., Yocom, K. M., Saucy, D. A., Freeman, M. B., Swift, G. (1997) Biodegradation of acrylic acid polymers and oligomers by mixed microbial communities in activated sludge, *J. Environ. Polym. Degrad.* **5**, 41– 48.

Low, K. C., Wheeler, A. P., Koskan, L. P. (1996) Commercial poly(aspartic acid) and its uses, *Adv. Chem. Ser.* **248**, 99–111.

Mann, S. J. (1993) Biomineralization: the hard part of bioinorganic chemistry, *J. Chem. Soc., Dalton Trans.* **1993**, 1–9.

Matsubara, K, Nakato, T., Tomida, M. (1997) ^{1}H and ^{13}C NMR characterization of polysuccinimide prepared by thermal condensation of L-aspartic acid, *Macromolecules* **30**, 2305–2312.

Matsuyama, M., Kokufuta, E., Kusumi, T., Harada, K. (1980) On the Poly(β-DL- aspartic acid), *Macromolecules* **13**, 196–198.

Meggy, A. B. (1956) Glycine peptides. II. Heat and entropy of formation of the peptide bond in polyglycine, *J. Chem. Soc.* **1956**, 1444–1445.

Miller, W. G. (1961) Degradation of Poly-α,L-glutamic Acid. I. Degradation of high molecular weight PGA by papain, *J. Am. Chem. Soc.* **83**, 259–265.

Miller, W. G. (1964a) Degradation of synthetic polypeptides. II. Degradation of poly-α, L-glutamic acid by proteolytic enzymes in 0.20 M sodium chloride, *J. Am. Chem. Soc.* **86**, 3913–3922.

Miller, W. G. (1964b) Dynamics of the helix-coil transition in poly-α,L-glutamic acid, *Biopolymers* **2**, 489–500.

Morall, S. W., Bookland, E. A., Larson, R. J., Rothgelb, T. M., Yocum, K. M., Swift, G. (1997) Biodegradation and chemical structure of synthetic polyaspartic acids, Gordon Research Conference, Barga, Italy, 5–9 May 1997, 1–14.

Neri, P., Antoni, G., Benvenuti, F., Cocola, F., Gazzei, G. (1973) Synthesis of α,β-Poly[(2-hydroxyethyl)-DL-aspartamide], a new plasma expander, *J. Med. Chem.* **16**, 893–897.

Organization for Economic Cooperation and Development (1981) *Guidelines for Testing of Chemicals.*

Pivcova, H., Saudek, V. (1985) Carbon-^{13}C-NMR relaxation study of poly(aspartic acid), *Polymer* **26**, 667–672.

Pivcova, H., Saudek, V., Drobnik, J. (1981) NMR study of poly(aspartic acid). I. α- and β- peptide bonds in poly(aspartic acid) prepared by thermal polycondensation, *Biopolymers* **20**, 1605–1614.

Pivcova, H., Saudek, V., Drobnik, J. (1982) 13C n.m.r. study of the structure of poly(aspartic acid), *Polymer* **23**, 1237–1241.

Rao, V. S., Lapointe, P., McGregor, D. N. (1993) Synthesis of uniform polyaspartic acid, *Makromol. Chem.* **194**, 1095–1104.

Rusenko, K. W., Donachy, J. E., Wheeler, A. P. (1991) Purification and characterization of a shell matrix phosphoprotein from the American oyster, in: *Surface Reactive Peptides and Polymers: Discovery and commercialisation* (Sikes, C. S., Wheeler, A. P., Eds.), Washington DC: ACS Books, 107–121.

Sarig, S., Shifrin, F. (1977) The use of polymers for retardation of scale formation, *National Counc. Res. Dev. [Rep] NCRD (Isr.)*,150–157.

Schaal, E. (1871) Über einige aus Asparaginsäure entstehende Produkte, *Leibig's Annalen der Chemie* **157**, 24–34.

Schiff, H. (1897) Über Polyaspartsäuren, *Chem. Ber.* **30**, 2449–2459.

Schwamborn, M. (1996) Polyasparaginsäuren, *Nachr. Chem. Tech. Lab.* **44**, 1167–1170.

Schwamborn, M. (1997) *Neue biologisch abbaubare Funktionspolymere*, GDCH Umwelttagung Heidelberg.

Schwamborn, M. (1998) Chemical synthesis of polyaspartates: a biodegradable alternative to currently used polycarboxylate homo- and copolymers, *Polym. Degrad. Stab.* **58**, 39–45.

Sikes, C. S., Wheeler, A. P. (1985) Inhibition of inorganic or biological $CaCO_3$ deposits by poly amino acid derivatives, University of South Alabama, US Patent 4 534 881.

Sikes, C. S., Wheeler, A. P. (Eds.) (1988) Chemical aspects of regulation of mineralization. Proceedings of a Symposium sponsored by the Division of Industrial and Engineering Chemistry of the American Chemical Society, New Orleans, September 3–4, 1987.

Sikes, C. S., Wierzbicki, A. (1994) Mechanistic studies of polyamino acids as antiscalants, *Corrosion*, **94**, Paper No. 193, 1–19.

Sikes, C. S., Wierzbicki, A. (1996) Biomimetic material chemistry, in: *Polyamino Acids as Antiscalants, Dispersants, Antifreezes, and Absorbent Gelling Material* (Mann, S., Ed.), New York: Verlag Chemie, Chapter 9, 249–278.

Sikes, C. S., Yeung, M. L., Wheeler, A. P. (1991) Inhibition of calcium carbonate and phosphate crystallization by peptides enriched in aspartic acid and phosphoserine, in: *Surface Reactive Peptides and Polymers: Discovery and commercialisation* (Sikes, C. S., Wheeler, A. P., Eds.), ACS Symposium Series, Washington DC: ACS Books, 50–71.

Stetler-Stevenson, W. G., Veis, A. (1983) Bovine dentin phosphophoryn: composition and molecular weight, *Biochemistry* **22**, 4326–4335.

Swedish Society for Nature Conservation (SSNC) (2001). http://www.SNF.se.

Wagner, P., Döbert, F., Menzel, T., Groth, T., Joentgen, W., Liesenfelder, U., Weinschenck, J., Heise, K. P. (1998) Method for carrying out polycondensation reactions, Bayer AG, US Patent 6 187 898.

Wassmer, K. H., Schroeder, U., Horn, D. (1991) Characterization and detection of polyanions by direct polyelectrolyte titration, *Makromol. Chem.* **192**, 553–565.

Wheeler, A. P., Rusenko, K. W., Swift, D. M., Sikes, C. S. (1988) Regulation of *in vitro* and in vivo calcium carbonate crystallization by fractions of oyster shell organic matrix, *Mar. Biol.* **98**, 71–80.

Wierzbicki, A., Sikes, C. S., Madura, J. D., Drake, B. (1994) Atomic force microscopy and molecular modeling of protein and peptide binding to calcite, *Calcif. Tissue Int.*, **54**, 133–141.

Wolff, J. (1850) Über Asparaginsäure aus Äpfelsäure, *Ann.* **75**, 293–297.

Wolk, S. K., Swift, G., Paik, Y. H., Yokom, K. M. Smith, R. L., Simon, E. S. (1994) One- and two-dimensional nuclear magnetic resonance characterization of poly(aspartic acid) prepared by thermal polymerization of L-aspartic acid, *Macromolecules* **27**, 7613–7620.

8
Cell Membranes: Protein Components and Functions

Dr. William C. Wimley
Tulane University Health Sciences Center, Department of Biochemistry SL43, 1430 Tulane Avenue, New Orleans, LA 70112, USA; Tel.: +1-504-988-7076; Fax: +1-504-584-2739; E-mail: wwimley@tulane.edu

ATP adenosine triphosphate
EGF epidermal growth factor
GPI glycerophosphatidylinositol
NCAM neural cell adhesion molecule
OmpX outer membrane Protein X

1 Introduction

The existence of known life is dependent on the compartmentalization provided by lipid membranes. Even in the simplest cells, membranes separate cellular contents from the environment, retain the molecules that are essential for life, and exclude the molecules that are harmful to it. Membrane-dependent compartmentalization must have been an essential early step in the development of life on Earth, and could have been utilized by ancient self-replicating systems even before the rise of protein-based life (Segre et al., 2001). In more complex cells, lipid membranes also delineate the numerous subcellular organelles in which specialized processes occur and in which important molecules are stored. Microscopic images of eukaryotic cells reveal complex and very membrane-rich environments.

While lipid membranes are required to delineate and compartmentalize cells and organelles, living organisms require much more than a simple passive barrier. Membrane composition must be actively regulated, membranes must allow for the dynamic exchange of information, nutrients and metabolic waste products, and membranes must allow for specific interactions with other cells, surfaces and the environment. Membrane proteins (polypeptides that reside in lipid membranes) carry out these functions, and more. This chapter describes the chemical, physical, and biological properties of membrane proteins. It will begin with a discussion of the unique physical properties of the lipid bilayer membrane and how the physico-chemical and topological constraints of the lipid bilayer influence the structure and function of membrane proteins. The various biological functions of membrane proteins will then be detailed, after which the future of membrane proteins in biotechnology and the fields of genomics and proteomics will be discussed.

2 Historical Outline

In 1774, Benjamin Franklin poured a small volume of olive oil onto a pond at Clapham Common, UK, and noted (Franklin et al., 1774) that the oil "... though not more than a teaspoonful, produced an instant calm over a space several yards square, which spread amazingly and extended itself gradually till it reached the lee side, making all that quarter of the pond, perhaps half an acre, as smooth as a looking glass."

Although neither Franklin nor any of his contemporaries made the connection at the time, a simple calculation of the height of this surface film from the known volume and area would have been the first measurement of the thickness of a lipid film. In fact, it would have been the first measurement of the dimensions of any molecule. In metric units, a teaspoonful of oil (~2 cm^3) spread over a half-acre (~2000 m^2) would have a thickness of approximately 10^{-7} cm, or about 10 Å. It would take more than a century before the measurement of lipid film thickness was made carefully, and the calculation done properly, but this estimate would have been remarkably close to the true dimensions. The first accurate descriptions and measurements of lipid films on water were made Lord Rayleigh (Lord Rayleigh, 1890), Agnes Pockels (Pockels, 1891) and most importantly, Irving Langmuir (Langmuir, 1917). Based in part on surface film experiments, in 1917 Langmuir made a convincing argument that lipid molecules such as olive oil (trioleoyl glycerol) form a film on aqueous surfaces that is one molecule deep and has a thickness between 13 and 16 Å. With hindsight, we can now see that these experiments formed the foundation for our modern understanding of biological membranes. However, it would be another half-century after Langmuir's classical work on lipid films before the scientific community unanimously recognized the lipid film as the physical basis for the biological membrane.

2.1 The Fluid Mosaic Model of Biological Membranes

In 1901, two researchers, Overton and Meyer, simultaneously published independent accounts (Meyer, 1901; Overton, 1901) of the relationship between the permeability of small organic anesthetic molecules into cells and their solubility in olive oil. The resulting Meyer–Overton hypothesis noted that the permeability barrier that prevents polar solutes from diffusing into cells has chemical properties which are very similar to the properties of olive oil. Based on the Meyer–Overton hypothesis and the work of Rayleigh, Pockels and Langmuir, various lipid-based models for the biological membrane were put forth between 1900 and 1960 among which was the Gorter–Grendel model (Gorter and Grendel, 1925). Although it was largely ignored, the Gorter–Grendel model was the first based correctly on the lipid *bilayer*, a membrane film that is two lipid molecules thick. The Danielli–Davson model (Danielli and Davson, 1935), was also a bilayer model and was one of the first to incorporate proteins, although in that model the proteins formed a surface that coated the lipid bilayer. Based on early electron microscopy, a similar model of a protein-coated lipid bilayer was proposed by Robertson (Robertson, 1957). During this same time period, other proposed membrane models were based on the notion that proteins formed the structural foundation for the biological membrane, and that the lipids coated the protein membrane, or filled in the spaces like the mortar of a brick wall.

These various models all had supporters into the early 1960s when an explosion of interest in biological membranes occurred and new techniques such as electron microscopy, X-ray diffraction and circular dichroism spectroscopy were applied to their study. Ultimately, these modern techniques proved that the biological membrane was, in fact, formed from a lipid bilayer and that the membrane proteins were mostly α-helical globular proteins that were inserted through the bilayer and spanned the two leaflets. The modern image of the biological membrane was produced in a synthesis of this membrane research in a 1972 *Science* article (Singer and Nicolson, 1972). These authors called the biological membrane a "fluid mosaic" of proteins dissolved in a fluid, two-dimensional, phospholipid bilayer membrane. The original image from that paper is shown in Figure 1. Since the first recognition of the fluid mosaic model of biological membranes, a vast amount of detail has been added, although the image of the basic structural principles of the biological membrane has not changed. Most importantly for this discussion, the first crystal structure of a membrane protein was published in 1985 (Deisenhofer et al., 1985). Since that time the three-dimensional structures of several dozen membrane proteins have been determined, and the library of structures is growing every year.

3 Chemical Structures

Biological membranes are complex systems composed of hundreds to thousands of species of proteins, carbohydrates, polar lipids, sterols (e.g., cholesterol), and other amphipathic molecules. Some of the important chemical and physical properties of these molecules and how they function together to form biological membranes, are discussed below.

3.1 The Lipid Bilayer Milieu

Studies with model membrane systems have shown that only the polar lipids, when dispersed in aqueous media, can spontaneously self-assemble into membranes that reproduce the structure and permeability barrier properties of biological membranes (Bangham et al., 1967; Pagano and Thompson, 1968; Huang, 1969). No other compo-

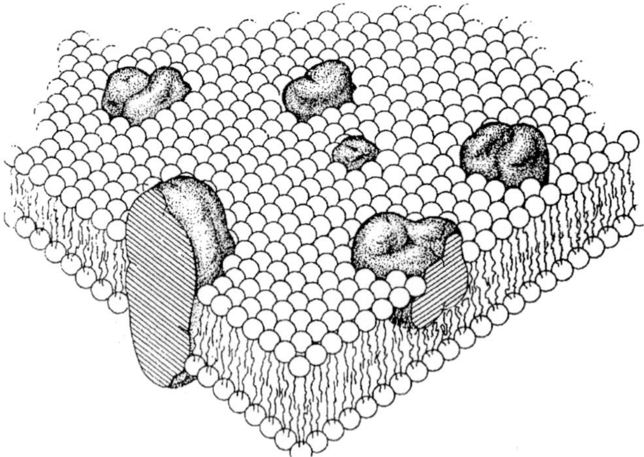

Fig. 1 The fluid mosaic model of cell membranes. In 1972, the fluid mosaic model was invoked to describe the biological membrane as a two-dimensional fluid lipid bilayer phase in which membrane proteins are dissolved. Over the intervening decades, many details have been added to our understanding of biological membranes, but this image and the report from which it is taken correctly defined the true structural and dynamic properties, and biological function of membranes and membrane proteins. (From Singer and Nicolson, 1972, with permission.)

nents of a biological membrane will form membranes when present in pure form in water.

Despite there being a vast molecular diversity among known membrane-forming lipids, several properties are conserved across all types. The chemical structure of some representative membrane-forming lipids are shown in Figure 2 to demonstrate these features. All membrane-forming lipids have two long hydrophobic chains ranging from 14 to 24 carbons in length; these nonpolar chains form the core of the biological membrane. Typical bacterial and eukaryotic lipids have linear, aliphatic chains, with a mixture of saturated chains and chains with nonconjugated double bonds. In some lipid species the chains are fatty acids and are connected to a backbone glycerol moiety though ester linkages. In other lipids, such as the sphingolipids (Hakomori and Igarashi, 1993), the two long aliphatic chains are attached to a serine molecule. In either case, polar lipids always have a polar group attached to the backbone moiety. The polar groups on bilayer-forming lipids can range from a single hydroxyl group, to small polar or charged moieties, to complex branched carbohydrate moieties, as in the glycosphingolipids. Many bacterial and eukaryotic membrane-forming lipids are glycerophospholipids in which polar moieties such as choline, ethanolamine, glycerol, or serine are linked to the carbon 3 of the backbone glycerol through a phosphate group. A diacyl glycerophosphocholine is shown on the left in Figure 2.

Archaeal lipids are also two-chain lipids, and may have branched polyisoprenyl or isopranyl chains such as the 20-carbon molecules phytane (Koga et al., 1993; Albers et al., 2000), with the two chains attached to a glycerol moiety through ether linkages. Interestingly, archaeal hyperthermophiles (Tolner et al., 1997; Albers et al., 2000) also utilize lipid species in their bilayers with chains that are approximately 40 carbons long and are linked to glycerol at both ends. These single molecules span both monolayers of the lipid bilayer and help to form thermostable membranes with properties very similar to those of bilayer membranes formed near room temperature from glycerophospholipids.

3.2
Physical Properties of Lipid Bilayer Membranes

Nonpolar or amphipathic molecules self-assemble into multimeric aggregates in water because of the unfavorable interaction of the nonpolar groups with water (Tanford, 1980). There are three main types of phases that can occur, and the phase that forms when a lipid is dispersed in water depends on the polarity and molecular shape of the lipid (Wieslander et al., 1980; Chernomordik et al., 1985). Bulk phases (e.g., droplets or crystals) form when the lipids are nonpolar. For example, fats (i.e., triacyl glycerols) form droplets when mixed with water, while pure cholesterol in water forms crystals. Oblate, ellipsoidal or even tubular micelles form from amphipathic lipids (those with both polar and nonpolar parts) when the cross-sectional shape of the lipid molecule is conical. Finally, planar bilayers form from amphipathic (i.e., polar) lipids when the cross-section of the lipid molecule is cylindrical.

The lipid bilayer is a planar liquid crystal (Chapman, 1966; Fergason and Brown, 1968) in which the component polar lipid molecules have their nonpolar chains segregated into a nonpolar core and have their polar headgroups exposed to the aqueous phase. As a liquid crystal, the components of a bilayer membrane are free to diffuse laterally (Huang, 1973; Edidin et al., 1976;

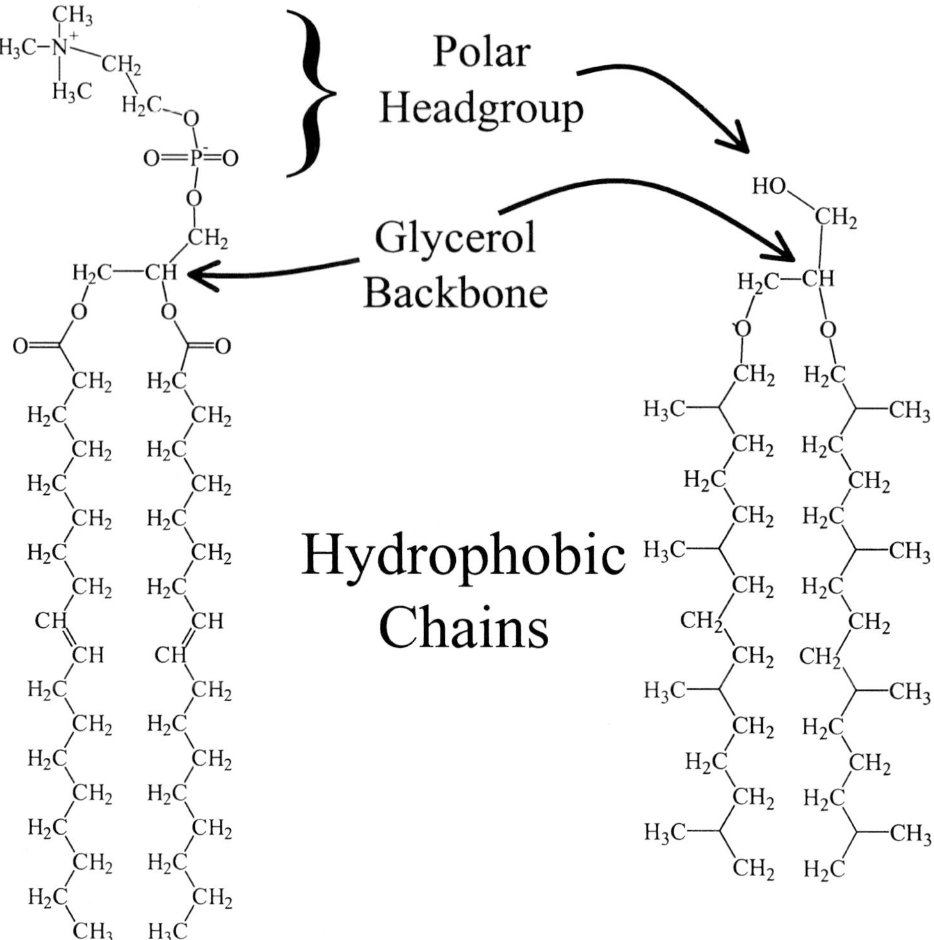

Fig. 2 Chemical structure of some membrane-forming lipids. Of the many types of molecules that comprise a cell membrane, only the two-chained, polar lipids will form bilayer membranes when dispersed into water in pure form. Thus, the two-chain lipids form the structural basis for all biological membranes. The illustration shows two typical membrane-forming lipids. (Left) A diacyl glycerophospholipid (phosphatidylcholine); this has ester-linked fatty acid chains connected to the 2 and 3 carbons of a glycerol moiety, and a phosphocholine headgroup linked to the C1 glycerol carbon. (Right) An archaeal two-chain lipid. Archaeal lipids have two isopranyl- or isoprenyl- chains that are ether-linked to a glycerol moiety. Archaeal lipids often do not have extra polar headgroups attached to the glycerol moiety; the hydroxyl group serves as the lipid headgroup. The structure illustrated is a diphytane-glycerol lipid, which has two chains, each containing 20 carbons.

Edidin and Weaver, 1985; Tocanne et al., 1994), but they are constrained by hydrophobic interactions along the axis perpendicular to the plane of the membrane. Recent studies have generated a detailed image of the structural and dynamic properties of the fluid lipid bilayer (Wiener et al., 1991; Wiener and White, 1992a, b). The transbilayer profiles of the lipid molecules were found to be best represented by sums of Gaussian distributions. These distributions are several fold broader than the molecular size of the “quasi molecular” groups they represent. This apparent broadening is due to thermal fluctuations of the lipids in the bilayer. The experimentally determined transbilayer profile of a diacylglycerophospholipid bilayer is shown along with the Gaussian distributions of the constituent quasimolecular groups in Figure 3.

To understand the structure and function of proteins in membranes, the most important feature of the bilayer is the nonpolar core. The dimensions of the core are shown by the profile of partial charge density superimposed on the transbilayer distributions; it occupies about half the thickness of the membrane (White and Wimley, 1998, 1999). This nonpolar core is very similar to a bulk alkane in its charge density (i.e., polarity) and mass density. It is the region of the membrane that imposes the most stringent constraints on possible protein structures, and is also the region in which membrane proteins experience an environment most different from the typical aqueous environment of soluble proteins. The bilayer profile also shows a heterogeneous

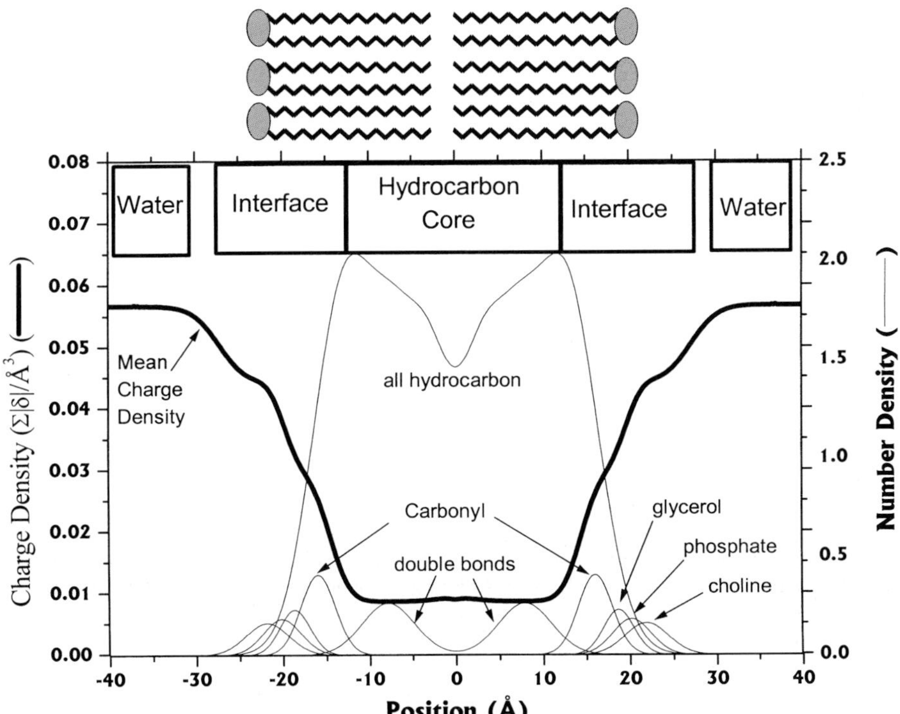

Fig. 3 Transbilayer properties of a fluid bilayer membrane. Number density is the distribution of quasimolecular groups in a fluid diacyl glycerophospholipid, 1,2-dioleoyl-glycerol-3-phosphatidylcholine bilayer membrane obtained with neutron and X-ray diffraction (Wiener and White, 1992b). Each Gaussian peak represents the number density of one group. The areas under the peaks equal the number of each group in the lipid. The half-widths of the distributions are a measurement of the thermal disorder in a fluid lipid bilayer. Charge density across the membrane is calculated from the partial charges on each atom of each quasimolecular group divided by the volume occupied by that group. Note that between the high charge density of bulk water (left and right) and the low charge density of the hydrocarbon core (center), there is a broad heterogeneous “interfacial” region, which occupies about half the thickness of the membrane.

and anisotropic "interface" which comprises about half the thickness of the membrane. The interface contains both polar and nonpolar groups whose transbilayer distributions overlap broadly, and has been called a region of "tumultuous chemical heterogeneity" (Wiener and White, 1992b). The interfacial region of a bilayer offers a more permissive environment for unfolded or amphipathic polypeptides, and is the region in which many peptide hormones and toxins form their native structure (Kaiser and Kezdy, 1983; White and Wimley, 1994).

3.3 Polypeptides in Membranes

Proteins are linear polypeptides composed of 20 of the naturally occurring amino acids. Membrane proteins are those proteins that are intimately associated with the bilayer membranes of a cell, and are diverse in their architecture, ranging in length from a few dozen to a few thousand residues. They can have as few as 20 and as many as several hundred residues in their membrane-spanning segments, comprising as little as 1% or as much as 90% of the total protein. An example is the visual pigment protein rhodopsin (Palczewski et al., 2000), the structure of which is shown in Figure 4 as both a schematic diagram and a spacefilling image. Rhodopsin has seven membrane-spanning segments comprising about 90% of the total protein.

3.3.1 Physical Forces Acting on Membrane Proteins

A protein's biological function is dependent on its folding into specific, compact, three-dimensional structure (Dill, 1990). Protein folding is driven by noncovalent interactions

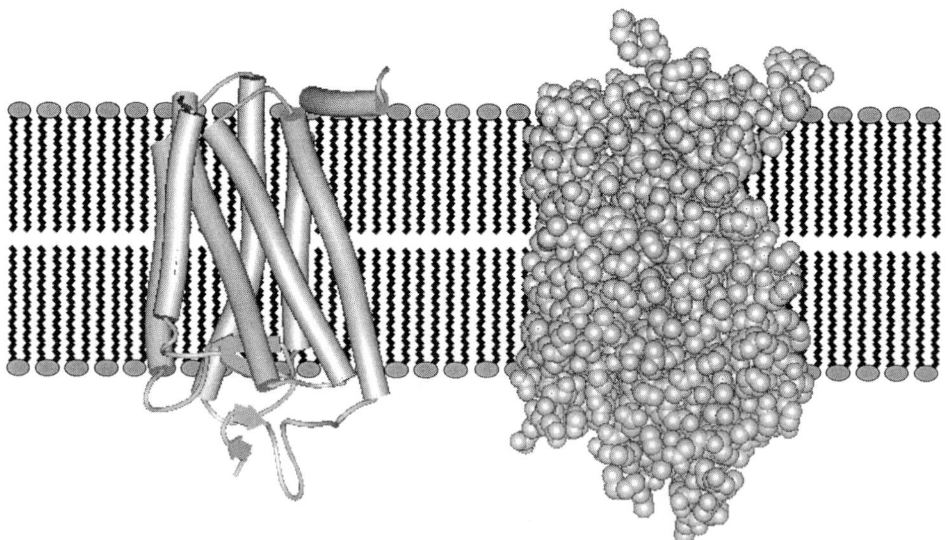

Fig. 4 Structure of a typical membrane protein. An integral membrane protein has elements of secondary structure that span the membrane, connecting the extracellular and intracellular domains. This protein is bovine rhodopsin, a visual pigment protein (Palczewski et al., 2000). In this protein, most of the mass is in the membrane-spanning segments. The roughly parallel transmembrane helices, shown as rods on the left, span the 30 Å thickness of the bilayer. In the space-filling model shown at the right, the close packing of the protein in the membrane is visible.

including hydrophobic, van der Waals, hydrogen bonding and electrostatic interactions, and is opposed mainly by the large configurational entropy change of folding (Dill, 1990). Our current understanding of the interactions that drive protein folding in membranes is derived mainly from model system studies which have shown that the contributions to protein folding in membranes are different from those that drive folding of water-soluble proteins (Engelman et al., 1986; Liu and Deber, 1998; White and Wimley, 1999). In water, hydrophobic interactions drive water-soluble proteins to collapse into compact globular structures (Sturtevant, 1977; Privalov and Gill, 1988). In membrane proteins the hydrophobic effect drives a polypeptide to reside in the membrane where the assembly of those membrane-bound segments is then driven by hydrogen bonding and van der Waals interactions. Whether a protein folds into a globular, water-soluble protein or a membrane protein depends on its amino acid composition and on the detailed arrangement of its hydrophobic residues.

Understanding the interactions that drive membrane protein folding, or predicting the structure of membrane proteins must begin with a measure of the propensity of a given protein segment to reside in the membrane. To achieve this, the hydrophobicities of the amino acids have been quantitated with various experimental or theoretical hydrophobicity scales (Kyte and Doolittle, 1982; Engelman et al., 1986; Wimley and White, 1996). Two experimentally determined hydrophobicity scales that have been shown to be useful in understanding the energetics of protein structure and folding in membranes are shown in Table 1 (Wimley and White, 1996; Wimley et al., 1996). These scales are "whole-residue" scales that include the contribution of the backbone polar groups. One of the important features revealed in the measurement of these whole-residue hydrophobicity scales is that the energetic cost of partitioning an open, nonhydrogen-bonded –CO–NH– peptide backbone group into the membrane interface is at least 1–2 kcal per mol per residue (White and Wimley, 1998, 1999). In the nonpolar core, the cost is even higher (Engelman et al., 1986; White and Wimley, 1999). These costs are similar in magnitude to the hydrophobicities of the most hydrophobic amino acids (see Table 1), and so unstructured polypeptides with open H-bond groups will, at best, partition weakly into a membrane's interfacial region and be excluded from the nonpolar core. Because the energetic cost of hydrogen-bonded peptide groups in the membrane is expected to be much smaller than for nonhydrogen-bonded groups, polypeptides that are hydrophobic enough to stably partition from water into membranes will also strongly favor the formation of hydrogen-bonded secondary structure in membranes. In this sense, membrane binding and structure formation in the membrane are thermodynamically coupled (Kaiser and Kezdy, 1983; White and Wimley, 1999), and any protein sequence that is bound to a membrane is likely to have secondary structure. Many experiments confirm this partitioning-folding coupling (e.g., Ladokhin and White, 1999; Wieprecht et al., 1999).

A major consequence of the high energetic cost of nonhydrogen-bonded peptide bonds in the membrane is that membrane-spanning proteins are constrained to include only those structural motifs that can stably reside in the lipid bilayer with all potential backbone hydrogen bond groups involved in hydrogen bonds. For the same reasons, the side chains exposed to the lipid bilayer will be predominantly hydrophobic; lipid exposure of polar side chains will be very costly. The diversity of structural motifs found in membrane proteins is thus limited by the

Tab. 1 Hydrophobicities of the amino acids for partitioning into the membrane interface (Wimley and White, 1996) and for partitioning into octanol (Wimley et al., 1996).

Amino Acid	Sidechain Structure	ΔG Octanol (kcal/mol)	ΔG Bilayer (kcal/mol)
Tryptophan TRP (W)	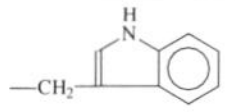 $-CH_2-$ (indole, N–H)	-2.09	-1.85
Phenylalanine PHE (F)	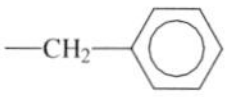 $-CH_2-$ (phenyl)	-1.71	-1.13
Leucine LEU (L)	$-CH_2-CH(CH_3)CH_3$	-1.25	-0.56
Isoleucine ILE (I)	$-CH(CH_3)-CH_2-CH_3$	-1.12	-0.31
Tyrosine TYR (Y)	$-CH_2-$ (phenyl) $-OH$	-0.71	-0.94
Methionine MET (M)	$-CH_2-CH_2-S-CH_3$	-0.67	-0.23
Valine VAL (V)	$-CH(CH_3)CH_3$	-0.46	0.07
Cysteine CYS (C)	$-CH_2-SH$	-0.02	-0.24
Histidine His (H)	$-H_2C-$ (imidazole, NH, N)	0.11	0.17
Proline PRO (P)	COO^-, $C\alpha$, ^+H_2N, CH_2, H_2C-CH_2	0.14	0.45
Threonine THR (T)	$-CH(OH)CH_3$	0.25	0.14
Serine SER (S)	$-CH_2-OH$	0.46	0.13
Alanine ALA (A)	$-CH_3$	0.50	0.17
Glutamine GLN (Q)	$-CH_2-CH_2-C(=O)NH_2$	0.77	0.58
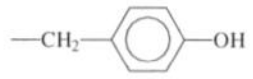 Asparagine ASN (N)	$-CH_2-C(=O)NH_2$	0.85	0.42
Glycine GLY (G)	$-H$	1.15	0.01
Arginine ARG (R)	$-CH_2-CH_2-CH_2-NH-C(HN_2)NH_2^+$	1.81	0.81
Lysine LYS (K)	$-CH_2-CH_2-CH_2-CH_2-NH_3^+$	2.80	0.99
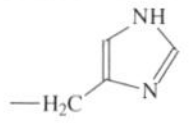 Glutamate GLU (E)	$-CH_2-CH_2-C(=O)O^-$	3.63	2.02
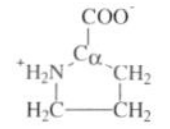 Aspartate ASP (D)	$-CH_2-C(=O)O^-$	3.64	1.23

In the first column we give the name of the amino acid and its three-letter and one-letter code. The second column is the chemical structure of the sidechain attached to the amino acid backbone –NH2–CH–COO– at the alpha-carbon. Note that the structure of proline includes the backbone atoms because the backbone and sidechain are covalently connected at the a carbon and at the nitrogen. The third and fourth columns are the octanol and interfacial hydrophobicities for each amino acid. The amino acids are ranked according to their octanol hydrophobicity. These are absolute, whole-residue scales that include the contribution of the backbone groups and are not normalized. Thus, a negative number for a summation of hydrophobicity values over a protein sequence signifies a net preference of that sequence for the bilayer. The octanol hydrophobicity has been shown to be highly accurate in the prediction of membrane-spanning helices (Jayasinghe et al., 2001). Octanol is thus apparently a good mimic of the average properties of the bilayer.

energetic constraints imposed on membrane protein structure by the membrane. Consequently, while many classes of structure motifs are known for proteins overall (Berman et al., 2000a, b), only two are known for membrane-spanning proteins. These motifs are discussed in the next section.

3.3.2
α-Helical Membrane Proteins

The most abundant structural motif for membrane proteins is the α-helical bundle (see Figure 4). Genomic analysis algorithms have indicated that as many as 25% of all proteins are α-helical bundle membrane proteins (Wallin and von Heijne, 1998). This corresponds to 8000–10 000 membrane proteins in the human genome. The basic structural unit, the α-helix, is a right-handed helix with i to $i+4$ hydrogen internal hydrogen-bonding pattern, as shown in Figure 5. Minimally, 19 amino acid residues are required to span the ~ 30 Å hydrophobic thickness of a lipid membrane if the

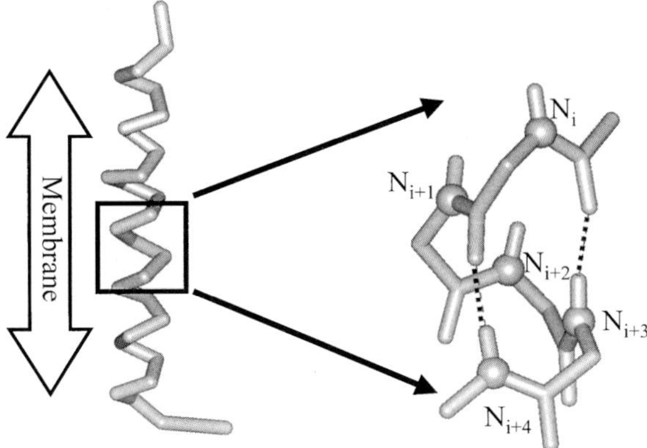

Fig. 5 Hydrogen-bonding pattern of an α-helix. The majority of known membrane-spanning protein segments fold into α-helices, which are right-handed helices that have an internal hydrogen bonding pattern in which the carbonyl (C=O) of amino acid i is hydrogen-bonded to the amide (NH) of residue i + 4. (Left) The backbone atoms of a 30 Å-long α-helix. A membrane-spanning helix is about 20 residues long and contains five to six turns of the helix. (Right) Details of the hydrogen-bonding pattern of an α-helix. The backbone nitrogens are highlighted and two hydrogen bonds are shown by the dotted lines.

sequence folds into an α-helix. As predicted by the hydrophobicity scales in Table 1, the minimum hydrophobicity for a sequence to be stable as a membrane-spanning segment is roughly equal to that of valine. In agreement with these predictions, polyalanine peptides do not form membrane-spanning helices (Moll and Thompson, 1994; Lewis et al., 2001), while polyleucine or mixed leucine-alanine peptides readily do so (Zhang et al., 1995a, b).

Because the backbone hydrogen bonds are internally satisfied within each helix, helical membrane proteins can have a single membrane-spanning α-helix. Figure 6 shows part of the sequence of a single helix membrane protein. This example is the human epidermal growth factor (EGF) receptor (Carpenter, 2000), a large protein in which an extracellular domain of 620 residues is connected to an intracellular domain of 540 residues by a single, membrane-spanning α-helical segment of 20 amino acids. At the other extreme, there are also membrane proteins with 20 or more membrane-spanning α-helices or with most of the protein involved in the membrane-spanning helices. Rhodopsin, an example of a seven-helix membrane protein, is shown in Figure 4. Rhodopsin is a member of a particularly widespread and abundant type of membrane protein, the G-protein-coupled receptor, which always have seven trans-membrane helices (Okada and Palczewski, 2001). As in rhodopsin (see Figure 4) these receptors have most of their amino acids in the membrane-spanning segments.

The architecture of multi-helical membrane proteins is further organized by lateral interactions between the membrane-spanning helices. The helices in membrane proteins are generally close-packed and parallel. Lateral interactions in the membrane can be driven by hydrogen bonds or electrostatic interactions, but they can also be driven by van der Waals interactions between the hydrophobic surfaces of the helices. In any case, the interactions can be strong and specific, and these drive the lateral organization of membrane-spanning α-helices in biological membranes. For example, the seven-helix protein bacterio-

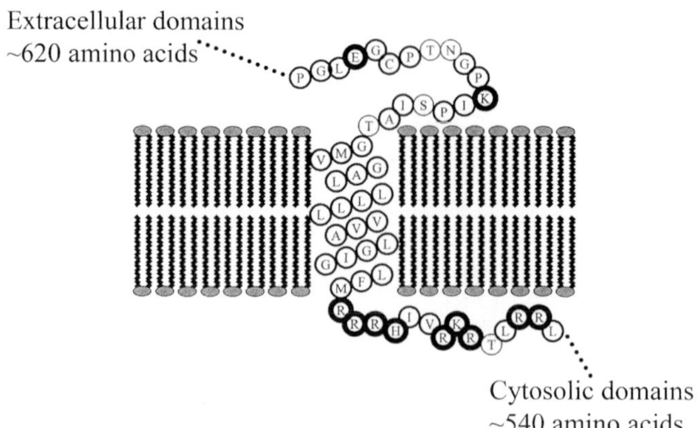

Fig. 6 Amino acid sequence of the membrane-spanning helix of the human epidermal growth factor (EGF) receptor. The EGF receptor has a single membrane-spanning α-helix of about 20 residues connecting large extracellular and intracellular domains. The sequence of the transmembrane helix and nearby amino acids are shown. Hydrophobic amino acids are shown as plain circles with thin lines, charged residues are shown by circles with thick lines and polar residues are shown by shaded circles. Notice that the membrane-spanning sequence is hydrophobic while the flanking sequences, like most protein segments contain both charged and polar amino acids.

rhodopsin remains in its native, functional conformation in the membrane even after the proteolytic cleavage of several of the extracellular loops between the membrane-spanning helices (Hunt et al., 1997; Marti, 1998). The specific lateral interactions between helices hold the structure together in the membrane.

The best-studied transmembrane helix–helix interaction is the homodimerization of the single transmembrane helix of the erythrocyte protein glycophorin A. The single membrane-spanning helix of glycophorin A is both necessary and sufficient for dimerization (Schulte and Marchesi, 1978; Lemmon et al., 1992a, b; Treutlein et al., 1992). The sequence of this helix is shown in Figure 7. The boxed residues have been shown by mutation studies to be important in dimerization, and follow the "heptad repeat" pattern expected for residues that lie on a contiguous surface of an α-helix. The location of these boxed residues on the surface of an ideal α-helix is shown in the

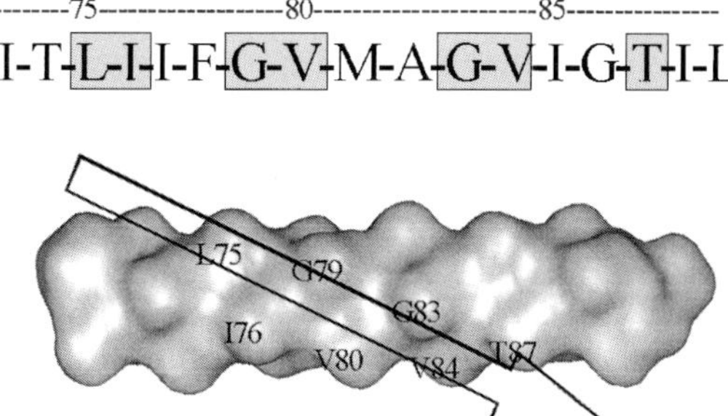

Fig. 7 Amino acid sequence of the membrane-spanning helix of the erythrocyte protein glycophorin A. Interactions between the membrane-spanning α-helices are responsible for the strong homodimerization of this protein in the membrane. Mutation studies have highlighted the residues that are important for dimerization (MacKenzie et al., 1997; Fleming et al., 1997), and these are shown in shaded boxes at the top. Because this structure is a α-helix with 3.6 residues per turn, these boxed residues fall on a contiguous surface of the helix. The helix dimer thus has a right-handed crossing angle, as shown by the arrow at the bottom, and is driven by highly specific surface-surface packing interactions.

lower image, and initially suggested a right-handed crossing between helices (Treutlein et al., 1992); this was later proven to be correct when the structure of the dimer was determined (MacKenzie et al., 1997). Note how the interacting surfaces are composed primarily of hydrophobic residues. The specific interaction between glycophorin A helices, and many other membrane protein helices, is due to the tight packing of these hydrophobic surfaces against each other in the membrane.

3.3.3
β-Barrel Membrane Proteins

The cells of Gram-negative bacteria are bounded by two lipid membranes; an inner membrane in which the integral protein constituents are mainly helical proteins, and an outer membrane in which the integral proteins utilize the "β-barrel" folding motif (Schulz, 2000). The β-barrel is the only other motif known that than can span a lipid membrane while satisfying the requirement that all backbone polar groups be involved in hydrogen bonds. The basic structural unit of a β-barrel is the β-strand, an extended, membrane-spanning polypeptide chain that is hydrogen-bonded to neighboring chains. All known β-barrels have an anti-parallel, β-meander topology in which neighboring strands in the structure are also nearest neighbors in the sequence. Most strands are part of a β-hairpin, formed when two adjacent β-strands are connected by a short loop of five or fewer residues (Schulz, 2000). Hairpins are laterally hydrogen bonded to each other and to neighboring hairpins. The backbone trace and hydrogen bonding pattern of the "outer membrane protein X" (OmpX) from *Escherichia coli* is shown in Figure 8.

As their name implies, β-barrels are circular structures in which all membrane-spanning strands are laterally hydrogen bonded. Known β-barrel structures contain between eight

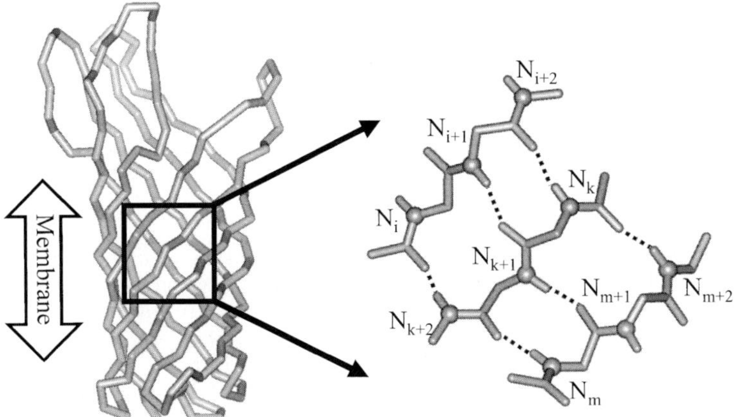

Fig. 8 Structure and hydrogen bonding pattern of a β-barrel membrane protein, the outer membrane protein X (OmpX) of *Escherichia coli* (Pautsch et al., 1999). In addition to α-helical membrane proteins, the only other class of membrane-spanning protein is the β-barrel. These structures span the membrane with extended polypeptide chains that are hydrogen bonded to neighboring chains. (Left) Backbone trace of a β-barrel membrane protein. This protein has eight membrane-spanning β-strands that are circularly hydrogen bonded. Because β-strands are extended chains, only about ten residues are required to span the thickness of the membrane. (Right) Hydrogen-bonding pattern of a β-barrel. Three strands of the β-barrel are shown here with the nitrogens highlighted and the NH to C=O hydrogen bonds shown as dotted lines. Note that the strands are antiparallel and that all backbone polar groups are involved in hydrogen bonds.

and 22 β-strands. Because of steric constraints in the barrel core, eight strands is the minimum number that can form a β-barrel (Sansom and Kerr, 1995), whereas the largest β-barrels of known structure–the TonB-dependent importers–contain 22 β-strands each (Buchanan et al., 1999). However, there is evidence that some bacterial toxins assemble on membranes into β-barrels containing dozens of β-strands or more (Rossjohn et al., 1998; Shatursky et al., 1999).

Because β-strands are in an extended conformation, their rise per residue is about twice that for an α-helix. Thus, only about 10 residues are required for a β-sheet to span a lipid bilayer (Seshadri et al., 1998; Schulz, 2000; Wimley, 2002). An important feature of the extended β-strands is that they span membranes with an alternating, or dyad repeat (Schulz, 2000; Wimley, 2002), motif in which every other residue is either on the external surface of the barrel interacting with the membrane lipids or on the interior of the barrel.

Other β-barrel membrane proteins are also found in the outer membranes of mitochondria, chloroplasts and mycobacteria. Many peptide antibiotics and bacterial toxins also utilize β-sheet motifs to assemble in and permeabilize target membranes. These include the defensins (White et al., 1995), the anthrax protective antigen (Petosa et al., 1997), the α-toxin of *Staphylococcus aureus* (Song et al., 1996), and the cholesterol-dependent cytolysins (Rossjohn et al., 1998; Gilbert et al., 1999).

3.3.4 Monotopic Membrane Proteins

A third structural motif has been identified for membrane proteins; the "monotopic" membrane proteins which insert deeply into one leaflet, or monolayer, of the bilayer membrane. Although not a membrane-spanning motif, monotopic membrane proteins are classified as an "integral" membrane protein by virtue of the fact that they cannot be removed from a membrane by any treatment except those that solubilize the membrane itself. Known examples have a domain composed of α-helices that are oriented parallel to the membrane surface. These helices have very hydrophobic surfaces that interact with membranes. Pharmaceutically important monotopic membrane proteins include the cyclooxygenase enzymes that are involved in inflammation processes. Cyclooxygenases are the pharmacological targets of the nonsteroidal anti-inflammatory drugs such as aspirin, acetaminophen, and ibuprofen. Interestingly, in 1972 Singer and Nicholson imaginatively drew monotopic membrane proteins in their image of a fluid mosaic membrane (see Figure 1) without having much data supporting the possibility that such structures existed. Indeed, that particular motif remained in the realm of conjecture for 20 years until it was observed in the first crystal structures of monotopic membrane proteins in 1994 (Garavito et al., 1994; Picot et al., 1994).

3.3.5 Lipid-Linked Membrane Proteins

There is a fourth class of proteins that are, in some ways, also membrane proteins. These proteins are associated with membranes by virtue of a covalent linkage to a lipid molecule, and three varieties have been identified. First, there are fatty acylated proteins, that have either myristate (a 14-carbon linear, saturated fatty acid) attached to an N-terminal glycine residue, or that have a palmitate (a 16-carbon linear, saturated fatty acid) attached to the side chain of a cysteine. Second, there are isoprenyl-linked proteins that have either a 15-carbon isoprenyl chain or a 20-carbon chain attached at a C-terminal cysteine residue. Most acylated or isoprenylated proteins are associated with the inner surface of a eukaryotic cell's outer membrane, and many of them are involved in receptor-dependent signal transduction. Finally, there are proteins linked through their terminal carboxyl group to a two-chain glycerophospholipid with an inositol-containing headgroup. These are the glycerophosphatidylinositol (GPI)-linked proteins, and they are found exclusively attached to the outer surface of cells.

4 Composition and Organization of Biological Membranes

The amount of protein in biological membranes ranges from 20% to 80% of the membrane's dry weight (Guidotti, 1972), with the majority of membranes containing ~50% protein by weight. The image in Figure 1 shows (schematically) the packing density of a membrane with ~50% protein by weight, although it is now known that many membrane proteins have much more mass outside the membrane than those shown in Figure 1. A weight fraction of 50% is equivalent to 2 mol.% protein or 50 lipids per protein. Studies with spin-labeled lipids have shown that the lipid molecules that immediately surround proteins–the so-called "boundary lipids" – often interact specifically with the proteins and have special motional properties (Lavialle et al., 1980; Van Gorkom et al., 1990; Okada and Palczewski, 2001). Lipids beyond the boundary lipids have properties very similar to lipids in protein-free bilayers.

There are also higher-order interactions between proteins in some biological membranes; these may be either transient interactions within the two-dimensional liquid crystal matrix of the bilayers, as occur between the electron transport proteins, or they can be stable, long-lived interactions. For example, the protein bacteriorhodopsin, found in *Halobacterium halobium* is a seven-helix bundle that is present in the membrane in the form of a noncovalently assembled protein trimer (Luecke et al., 1998). Under certain conditions, the bacteriorhodopsin trimers can further assemble into large two-dimensional arrays (Watts, 1995).

5 Genomics and Proteomics

The number of proteins in the sequence databases has been increasing exponentially for several years (Benson et al., 2002), and one of the more important tasks for bioscientists is to assign structure, function and cellular localization to these mostly unknown proteins. For helical membrane proteins this task is accomplished, in part, using sliding window hydrophobicity measurements as shown in Figure 9. The basic principle of the algorithm is that a membrane-spanning helix will have a stretch of about 20 hydrophobic amino acids. The identification of helical membrane proteins is one of the most accurate genomic structure predictions, with accuracy in identifying membrane-spanning helices exceeding 99% in recent applications that use the hydrophobicity scales shown in Table 1 (Jayasinghe et al., 2001). From these types of analyses it is known that at least 20–30% of all proteins are helical membrane proteins (Wallin and von Heijne, 1998), and that there is a broad distribution in the number of transmembrane helices per protein (Arkin et al., 1997).

The identification of β-barrel membrane proteins from physical principles is more difficult because the external hydrophobic surfaces are cryptic, hidden within the alternating inside-outside dyad repeat motif

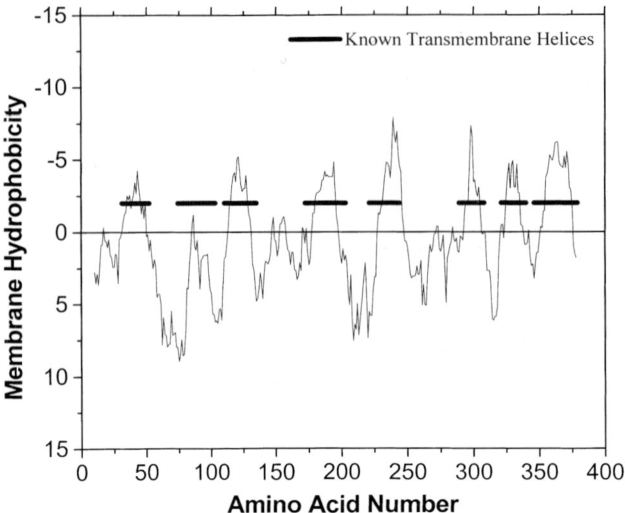

Fig. 9 Sliding window hydropathy analysis of the L-subunit of the photosynthetic reaction center, a membrane protein of known three-dimensional structure (Chang et al., 1991). The octanol hydrophobicity values (Table 1) were summed over a sliding 19-residue window to generate this plot. The line is the score for each sliding window hydrophobicity, plotted against the center residue of the window. The *known* transmembrane α-helices of this protein are delineated by the horizontal bars. Note that the success of this hydrophobicity scale is high; the seven high peaks in the analysis correspond exactly to seven of the eight transmembrane helices in the protein.

(Seshadri et al., 1998; Schulz, 2000; Wimley, 2002). Recent progress has been made in developing algorithms for identifying β-barrels (Wimley, 2002) and for predicting their structure and topology (Schirmer and Cowan, 1993; Gromiha et al., 1997; Jacoboni et al., 2001). In Gram-negative bacteria, the β-barrels comprise at least several hundred proteins (Molloy et al., 2000) or at least 5% of the total proteins in the genome.

6
Biological Functions

Membrane proteins perform many diverse and important biological functions, and their importance is exemplified by the fact that they are the target of 50% of all drugs (Drews, 2000), and of many toxins, venoms, and antibiotics. For example, the botulinum toxin–one of the most potent natural toxins known – functions by targeting specific membrane proteins in the neuromuscular junction (Hambleton, 1992). The various classes of membrane protein function are discussed in the next sections.

6.1
Information Transfer

Any protein that spans the lipid membrane and transfers information across it is termed a "receptor". In fact, receptors are one of the most abundant types of membrane protein and have essential roles in all organisms because they form the link between a cell and its environment; accordingly, almost 50% of drugs are targeted to receptor proteins (Drews, 2000). Receptors form an extremely diverse group of proteins, and comprise dozens of broad classes (Bray,

1998). In all cases, receptors interact with the environment outside of a cell or membrane-encapsulated compartment and transmit a signal through the membrane to the interior compartment. Transmembrane signal transduction is carried out by conformational changes in the receptor protein, by changes in its oligomerization state in the membrane, or by changes in the conduction of ions through the protein. These changes affect the interactions or function of the intracellular domains or other proteins within the cell which then lead to further "downstream" effects via complex signaling pathways.

One well-studied receptor class is that of the bacterial chemotaxis receptors (Bray, 1998; Bren and Eisenbach, 2000), such as the aspartate receptor (Koshland, 1996). These receptor proteins have extracellular domains that bind to a ligand or are affected by environmental signals (e.g., pH). The aspartate receptor has an extracellular domain that specifically binds aspartate, an essential nutrient. As a consequence, conformational changes in the extracellular domains that are induced by aspartate binding are coupled to conformational changes in an intracellular domain via the membrane-spanning segments undergoing small changes in orientation and position of the membrane-spanning helices (Koshland, 1996). Other proteins inside the cell respond to changes in the conformation of the aspartate receptor's intracellular domain, and ultimately modulate the behavior of the flagellum. The physiological outcome is that the bacterium moves towards a source of aspartate.

Other receptors function by changing their oligomerization state when ligand is bound. For example, the mammalian EGF receptors (see Figure 6) are induced to form dimers in the presence of specific hormone growth factors (Carpenter, 2000). The dimeric form is an activated state in which the cytosolic tyrosine kinase domains are active. Finally, some receptor proteins are ion channels; these include many neurotransmitter receptors (e.g., acetylcholine receptor) in which ion conduction through the channel receptor is affected by ligand binding.

6.2 Material Transfer

In any living cell, a large amount of material must constantly pass across cellular membranes. Nutrients, minerals, cofactors and other molecules must be imported into a cell across its outer membrane, while waste products, secreted proteins and chemical signals must be exported from the cell. Similarly, protons and other ions must be actively pumped across membranes to maintain compartment-specific pH values and ion concentration gradients. Because the lipid bilayer is highly impermeable to polar solutes, the movement of these molecules is dependent on membrane proteins.

6.2.1 Transport Down Concentration Gradients

There are several classes of membrane transporters, the simplest being the protein pores, or porins. These are "filter" proteins that have nonspecific pores containing a central, water-filled channel which is large enough to pass molecules below a certain molecular weight. Protein pores are found in membranes where tight permeability control is not necessary, for example the outer membrane of Gram-negative bacteria, the outer membrane of mitochondria, or the nuclear membrane of eukaryotic cells. No metabolic energy is used to move molecules through a porin; they simply flow through the pore in response to the concentration gradient across the membrane.

Similarly, carriers and channels are proteins through which molecules flow down a concentration gradient, and are sometimes called uniporters or facilitated transporters. Carriers and channels are highly specific, and typically allow the passage of only one molecular species. A good example is the Glut1 transporter in erythrocytes, which specifically allows the entry of glucose from plasma into red blood cells.

Many channel proteins (uniporters that specifically transport ions) are tightly modulated and can exist in either "closed" or open conformational states. There are many receptor proteins that are also ion channels; these proteins, such as the acetylcholine receptor, open or close their conductance pore in response to ligand binding.

6.3 Transport Against Concentration Gradients

Many membrane proteins that are involved in material transport across membranes use metabolic energy to transport molecules against concentration gradients. Symporters and antiporters are one way in which the movement of one molecule against a concentration gradient is coupled to the movement of another down a concentration gradient. The movements can be in the same (sym-) or in the opposite (anti-) direction. For example, there is a large family of solute:sodium symporters (Jung, 2001) in which the transport of metabolites into cells is coupled to the cotransport of sodium down its concentration gradient into the cell.

Pumps or active transporters are transport proteins that use adenosine triphosphate (ATP) hydrolysis as a source of metabolic energy to pump molecules across a membrane, the movement being against a concentration gradient if necessary. All cells have sodium pumps that are used to maintain low internal sodium concentrations even in high sodium environments. In mammals, the sodium concentration of extracellular fluid is ~150 mM, while the cell cytosol contains about 10 mM sodium. Potassium concentrations are also actively maintained with pumps and have the opposite gradient, being high inside the cell and low outside.

Calcium is even more stringently pumped out of most cells; external and internal calcium concentrations are in the millimolar and micromolar concentration ranges, respectively. Many cells take advantage of the calcium gradient by using tightly regulated calcium influx through calcium channels to initiate events within the cell, followed by rapid removal of the calcium through the action of calcium pumps. For example, muscle contraction is triggered by calcium release into the muscle cell cytosol and is stopped by the removal of the calcium from the cytosol through the action of calcium pumps. Cardiac muscle cells have cyclic fluctuations of calcium levels that correspond to the beat frequency of the heart muscle.

The molecular mechanism of pump proteins has been extensively studied, and the three-dimensional structures of one has recently been solved (Toyoshima et al., 2000). These proteins undergo a set of conformational changes, coupled to ATP binding and hydrolysis, in which solutes bind to the protein on one side of the membrane, followed by a large-scale ATP-dependent conformational change that exposes the solute to the opposite side of the membrane for release.

6.3.1 Electron Transport Chains

Another large group of membrane proteins are the electron transport proteins, which transfer high-energy electrons derived from glycolysis or photosynthesis down a chain of proteins with decreasing electrochemical potentials. In doing so, some of the electron energy is used to pump protons across membranes. The "chemiosmotic" (Mitchell and Moyle, 1965) energy of the subsequent transmembrane proton gradient is the primary metabolic energy source for life on earth because it is the driving force for the production of ATP by the F1-F0 ATP synthase (Fillingame et al., 2000). Electron transport proteins contain cofactors such as flavins, heme or iron-sulfur clusters which serve as the electron acceptor/donor moieties. In biological membranes, electron transport chains do not exist as fixed complexes, but rather rely on their ability to diffuse within the fluid mosaic of the membrane to interact transiently with their obligate partners in the electron transfer chains.

6.4 Membrane Protein Enzymes

Not surprisingly, enzymes that catalyze reactions involving lipid substrates or other membrane-associated substrates are themselves often membrane proteins. Biological membranes are complex and dynamic structures with many lipid species. Membrane biogenesis and homeostasis is performed, in part, by membrane protein enzymes. The cyclooxygenases are examples of membrane protein enzymes, and specifically are involved in the conversion of the long-chain fatty acid arachidonic acid to prostaglandins, which are important signaling molecules for the inflammation process. Other membrane protein proteases are involved in post-translational processing of membrane proteins; one medically important example is presenilin (Dewji and Singer, 1997), mutations of which in humans can cause incorrect protein processing that leads to Alzheimer's disease (Selkoe, 2001).

6.5 Membrane Anchors

One important function of membrane proteins is to modulate the interactions of a cell with its environment. Some membrane proteins, such as glycophorin A in red blood cells, act as anchoring points for cell-surface carbohydrates or cytoskeletal proteins which stabilize cells. Other membrane proteins, for example the neural cell adhesion molecule (NCAM) (Kiss and Muller, 2001), are involved in cell–cell interactions during development. Pathogenic or symbiotic organisms also use membrane receptors as adhesion molecules for site-specific attachment to host cells.

7 Outlook and Perspectives

In this so-called "post-genomic" era, several important tasks lie ahead, one of the most important being the assignment of structure, function, and cellular localization to the millions of protein sequences that have been generated since genomic DNA sequencing was started. Another task is to design new drugs that are targeted to membrane proteins and receptor-initiated signaling pathways in cells. Although at least 25% of all proteins are regarded as membrane proteins (Wallin and von Heijne, 1998), current knowledge of these important biopolymers is limited, and genome annotation for these molecules is especially sparse. In part, the lack of information on membrane proteins is due to experimental difficulties in obtaining structural information; indeed, although up to one-fourth of all proteins are membrane proteins, only 0.1% of the resolved protein structures in the Protein Data Bank (Berman et al., 2000a) are those of membrane proteins. In the face of these difficulties, the biomedical research community is turning, in part, to proteomics and genomics to address the structure and function of membrane proteins on a genome-wide scale, and some progress has been made in bacteria (Gromiha et al., 1997; Alm et al., 2000; Wimley, 2002). In this respect, the wide-scale application of genomic and proteomic research to the study of membrane proteins will undoubtedly provide many insights to their structure and function in the near future.

8
Patents

Given the broad biomedical importance of membrane proteins, it should not be surprising that there are many hundreds of US patents focused on this subject. These patents fall into several broad categories such as novel membrane protein sequences, membrane protein-targeted drugs, biotechnological applications of membrane proteins, and novel methods of purification or expression.

9
References

Albers, S. V., van de Vossenberg, J. L., Driessen, A. J., Konings, W. N. (2000) Adaptations of the archaeal cell membrane to heat stress, *Front. Biosci.* **5**, D813–D820.

Alm, R. A., Bina, J., Andrews, B. M., Doig, P., Hancock, R. E., Trust, T. J. (2000) Comparative genomics of *Helicobacter pylori*: analysis of the outer membrane protein families, *Infect. Immun.* **68**, 4155–4168.

Arkin, I. T., Brünger, A. T., Engelman, D. M. (1997) Are there dominant membrane protein families with a given number of helices? *Proteins* **28**, 465–466.

Bangham, A. D., de Gier, J., Greville, G. D. (1967) Osmotic properties and water permeability of phospholipid liquid crystals, *Chem. Phys. Lipids* **1**, 225–246.

Benson, D. A., Karsch-Mizrachi, I., Lipman, D. J., Ostell, J., Rapp, B. A., Wheeler, D. L. (2002) GenBank, *Nucleic Acids Res.* **30**, 17–20.

Berman, H. M., Bhat, T. N., Bourne, P. E., Feng, Z., Gilliland, G., Weissig, H., Westbrook, J. (2000a) The Protein Data Bank and the challenge of structural genomics, *Nature Struct. Biol.* **7** (Suppl.), 957–959.

Berman, H. M., Westbrook, J., Feng, Z., Gilliland, G., Bhat, T. N., Weissig, H., Shindyalov, I. N., Bourne, P. E. (2000b) The Protein Data Bank, *Nucleic Acids Res.* **28**, 235–242.

Bray, D. (1998) Signaling complexes: biophysical constraints on intracellular communication, *Annu. Rev. Biophys. Biomol. Struct.* **27**, 59–75.

Bren, A., Eisenbach, M. (2000) How signals are heard during bacterial chemotaxis: protein-protein interactions in sensory signal propagation, *J. Bacteriol.* **182**, 6865–6873.

Buchanan, S. K., Smith, B. S., Venkatramani, L., Xia, D., Esser, L., Palnitkar, M., Chakraborty, R., van der Helm, D., Deisenhofer, J. (1999) Crystal structure of the outer membrane active transporter FepA from *Escherichia coli*, *Nature Struct. Biol.* **6**, 56–63.

Carpenter, G. (2000) The EGF receptor: a nexus for trafficking and signaling, *BioEssays* **22**, 697–707.

Chang, C. H., Elkabbani, O., Tiede, D., Norris, J., Schiffer, M. (1991) Structure of the membrane-bound protein photosynthetic reaction center from *Rhodobacter sphaeroides*, *Biochemistry* **30**, 5352–5360.

Chapman, D. (1966) Liquid crystals and cell membranes, *Ann. N. Y. Acad. Sci.* **137**, 745–754.

Chernomordik, L. V., Kozlov, M. M., Melikyan, G. B., Abidor, I. G., Markin, V. S., Chizmadzhev, Y. A. (1985) The shape of lipid molecules and monolayer membrane fusion, *Biochim. Biophys. Acta* **812**, 643–655.

Danielli, J. F., Davson, H. (1935) A contribution to the theory of permeability of thin films, *J. Cell. Comp. Physiol.* **5**, 495–508.

Deisenhofer, J., Epp, O., Miki, K., Huber, R., Michel, H. (1985) Structure of the protein subunits in the photosynthetic reaction centre of *Rhodospeudomonas viridis* at 3Å resolution, *Nature* **318**, 618–624.

Dewji, N. N., Singer, S. J. (1997) The seven-transmembrane spanning topography of the Alzheimer disease-related presenilin proteins in the plasma membranes of cultured cells, *Proc. Natl. Acad. Sci. USA* **94**, 14025–14030.

Dill, K. A. (1990) Dominant forces in protein folding, *Biochemistry* **29**, 7133–7155.

Drews, J. (2000) Drug discovery: a historical perspective, *Science* **287**, 1960–1964.

Edidin, M., Weaver, F. (1985) Lateral diffusion of proteins in the membranes of epithelial cells, *Stud. Biophys.* **110**, 77–82.

Edidin, M., Zagyansky, Y., Lardner, T. J. (1976) Measurement of membrane protein lateral diffusion in single cells, *Science* **191**, 466–468.

Engelman, D. M., Steitz, T. A., Goldman, A. (1986) Identifying nonpolar transbilayer helices in amino acid sequences of membrane proteins, *Annu. Rev. Biophys. Biophys. Chem.* **15**, 321–353.

Fergason, J. L., Brown, G. H. (1968) Liquid crystals and living systems, *J. Am. Oil Chem. Soc.* **45**, 120–127.

Fillingame, R. H., Jiang, W., Dmitriev, O. Y., Jones, P. C. (2000) Structural interpretations of F(0) rotary function in the *Escherichia coli* F(1)F(0) ATP synthase, *Biochim. Biophys. Acta* **1458**, 387–403.

Fleming, K. G., Ackerman, A. L., Engelman, D. M. (1997) The effect of point mutations on the free energy of transmembrane α-helix dimerization, *J. Mol. Biol.* **272**, 266–275.

Franklin, B., Brownrigg, W., Farish, M. (1774) Of the Stilling of Waves by means of Oil. Extracted from Sundry Letters between Benjamin Franklin, LLD, FRS. William Brownrigg, MD, FRS and the Reverend Mr. Farish. *Philos. Trans.* **64**, 445–460.

Garavito, R. M., Picot, D., Loll, P. J. (1994) Prostaglandin H synthase, *Curr. Opin. Struct. Biol.* **4**, 529–535.

Gilbert, R. J., Jimenez, J. L., Chen, S., Tickle, I. J., Rossjohn, J., Parker, M., Andrew, P. W., Saibil, H. R. (1999) Two structural transitions in membrane pore formation by pneumolysin, the pore-forming toxin of *Streptococcus pneumoniae*, *Cell* **97**, 647–655.

Gorter, E., Grendel, F. (1925) On bimolecular layers of lipoids on the chromocytes of the blood, *J. Exp. Medicine* **41**, 439–443.

Gromiha, M. M., Majumdar, R., Ponnuswamy, P. K. (1997) Identification of membrane spanning beta strands in bacterial porins, *Protein Eng.* **10**, 497–500.

Guidotti, G. (1972) The composition of biological membranes, *Arch. Intern. Med.* **129**, 194–201.

Hakomori, S., Igarashi, Y. (1993) Gangliosides and glycosphingolipids as modulators of cell growth, adhesion, and transmembrane signaling, *Adv. Lipid Res.* **25**, 147–162.

Hambleton, P. (1992) *Clostridium botulinum* toxins: a general review of involvement in disease, structure, mode of action and preparation for clinical use, *J. Neurol.* **239**, 16–20.

Huang, C.-H. (1969) Studies on phosphatidylcholine vesicles. Formation and physical characteristics, *Biochemistry* **8**, 344–352.

Huang, H. W. (1973) Mobility and diffusion in the plane of cell membrane, *J. Theor. Biol.* **40**, 11–17.

Hunt, J. F., Earnest, T. N., Bousché, O., Kalghatgi, K., Reilly, K., Horváth, C., Rothschild, K. J., Engelman, D. M. (1997) A biophysical study of integral membrane protein folding, *Biochemistry* **36**, 15156–15176.

Jacoboni, I., Martelli, P. L., Fariselli, P., De Pinto, V., Casadio, R. (2001) Prediction of the transmembrane regions of beta-barrel membrane proteins with a neural network-based predictor, *Protein Sci.* **10**, 779–787.

Jayasinghe, S., Hristova, K., White, S. H. (2001) Energetics, stability, and prediction of transmembrane helices, *J. Mol. Biol.* **312**, 927–934.

Jung, H. (2001) Towards the molecular mechanism of Na(+)/solute symport in prokaryotes, *Biochim. Biophys. Acta* **1505**, 131–143.

Kaiser, E. T., Kezdy, F. J. (1983) Secondary structures of proteins and peptides in amphiphilic environments (a review), *Proc. Natl. Acad. Sci. USA* **80**, 1137–1143.

Kiss, J. Z., Muller, D. (2001) Contribution of the neural cell adhesion molecule to neuronal and synaptic plasticity, *Rev. Neurosci.* **12**, 297–310.

Koga, Y., Nishihara, M., Morii, H., Akagawa-Matsushita, M. (1993) Ether polar lipids of methanogenic bacteria: structures, comparative aspects, and biosyntheses, *Microbiol. Rev.* **57**, 164–182.

Koshland, D. E., Jr. (1996) The structural basis of negative cooperativity: receptors and enzymes, *Curr. Opin. Struct. Biol.* **6**, 757–761.

Kyte, J., Doolittle, R. F. (1982) A simple method for displaying the hydropathic character of a protein, *J. Mol. Biol.* **157**, 105–132.

Ladokhin, A. S., White, S. H. (1999) Folding of amphipathic α-helices on membranes: energetics of helix formation by melittin, *J. Mol. Biol.* **285**, 1363–1369.

Langmuir, I. (1917) The shapes of group molecules forming the surfaces of liquids, *Proc. Natl. Acad. Sci. USA* **3**, 251–257.

Lavialle, F., Levin, I. W., Mollay, C. (1980) Interaction of melittin with dimyristoylphosphatidylcholine liposomes. Evidence for boundary lipid by Raman spectroscopy, *Biochim. Biophys. Acta* **600**, 62–71.

Lemmon, M. A., Flanagan, J. M., Hunt, J. F., Adair, B. D., Bormann, B. J., Dempsey, C. E., Engelman, D. M. (1992a) Glycophorin-A dimerization is driven by specific interactions between transmembrane alpha-helices, *J. Biol. Chem.* **267**, 7683–7689.

Lemmon, M. A., Flanagan, J. M., Treutlein, H. R., Zhang, J., Engelman, D. M. (1992b) Sequence specificity in the dimerization of transmembrane alpha-helices, *Biochemistry* **31**, 12719–12725.

Lewis, R. N., Zhang, Y. P., Hodges, R. S., Subczynski, W. K., Kusumi, A., Flach, C.R, Mendelsohn, R., McElhaney, R. N. (2001) A polyalanine-based peptide cannot form a stable transmembrane alpha-helix in fully hydrated phospholipid bilayers, *Biochemistry* **40**, 12103–12111.

Liu, L. P., Deber, C. M. (1998) Guidelines for membrane protein engineering derived from de novo designed model peptides, *Biopolymers* **47**, 41–62.

Lord Rayleigh (1890) Measurements of the Amount of Oil Necessary in Order to Check the Motions of Camphor upon Water, *Proc. R. Soc. Lond.* **47**, 364–367.

Luecke, H., Richter, H. T., Lanyi, J. K. (1998) Proton transfer pathways in bacteriorhodopsin at 2.3 angstrom resolution, *Science* **280**, 1934–1937.

MacKenzie, K. R., Prestegard, J. H., Engelman, D. M. (1997) A transmembrane helix dimer: structure and implications, *Science* **276**, 131–133.

Marti, T. (1998) Refolding of bacteriorhodopsin from expressed polypeptide fragments, *J. Biol. Chem.* **273**, 9312–9322.

Meyer, H. (1901) Zur theorie der alkolnarkose: Der einfluss wechselnder temperatur sur wirkungsstarke und theilungscoefficient dar narcotica, *Naunyn Schmiedebergs Arch. Pharmacol.* **46**, 388–396.

Mitchell, P., Moyle, J. (1965) Evidence discriminating between the chemical and the chemiosmotic mechanisms of electron transport phosphorylation, *Nature* **208**, 1205–1206.

Moll, T. S., Thompson, T. E. (1994) Semisynthetic proteins: model systems for the study of the insertion of hydrophobic peptides into preformed lipid bilayers, *Biochemistry* **33**, 15469–15482.

Molloy, M. P., Herbert, B. R., Slade, M. B., Rabilloud, T., Nouwens, A. S., Williams, K. L., Gooley, A. A. (2000) Proteomic analysis of the *Escherichia coli* outer membrane, *Eur. J. Biochem.* **267**, 2871–2881.

Okada, T., Palczewski, K. (2001) Crystal structure of rhodopsin: implications for vision and beyond, *Curr. Opin. Struct. Biol.* **11**, 420–426.

Overton, E. (1901) *Studien uber die Narkose*, Jena: Gustav Fischer.

Pagano, R., Thompson, T. E. (1968) Spherical lipid bilayer membranes: electrical and isotopic studies of ion permeability, *J. Mol. Biol.* **38**, 41–57.

Palczewski, K., Kumasaka, T., Hori, T., Behnke, C. A., Motoshima, H., Fox, B. A., Le Trong, I., Teller, D. C., Okada, T., Stenkamp, R. E., Yamamoto, M., Miyano, M. (2000) Crystal structure of rhodopsin: a G protein-coupled receptor, *Science* **289**, 739–745.

Pautsch, A., Vogt, J., Model, K., Siebold, C., Schulz, G. E. (1999) Strategy for membrane protein crystallization exemplified with OmpA and OmpX, *Proteins* **34**. 167–172.

Petosa, C., Collier, R. J., Klimpel, K. R., Leppla, S. H., Liddington, R. C. (1997) Crystal structure of the anthrax toxin protective antigen, *Nature* **385**, 833–838.

Picot, D., Loll, P. J., Garavito, R. M. (1994) The x-ray crystal structure of the membrane protein prostaglandin H_2 synthase-1, *Nature* **367**, 243–249.

Pockels, A. (1891) Surface tension, *Nature* **43**, 437–441.

Privalov, P. L., Gill, S. J. (1988) Stability of protein structure and hydrophobic interaction, *Adv. Protein Chem.* **39**, 191–234.

Robertson, J. D. (1957) New observations on the ultrastructure of the membranes of frog peripheral nerve fibers, *J. Biophys. Biochem. Cytol.* **3**, 1043–1047.

Rossjohn, J., Gilbert, R. J., Crane, D., Morgan, P. J., Mitchell, T. J., Rowe, A. J., Andrew, P. W., Paton, J. C., Tweten, R. K., Parker, M. W. (1998) The molecular mechanism of pneumolysin, a virulence factor from *Streptococcus pneumoniae*, *J. Mol. Biol.* **284**, 449–461.

Sansom, M. S. P., Kerr, I. D. (1995) Transbilayer pores formed by β-barrels: molecular modeling of pore structures and properties, *Biophys. J.* **69**, 1334–1343.

Schirmer, T., Cowan, S. W. (1993) Prediction of membrane-spanning β-strands and its application to maltoporin, *Protein Sci.* **2**, 1361–1363.

Schulte, T. H., Marchesi, V. T. (1978) Self-association of human erythrocyte glycophorin A: appearance of low mobility bands on sodium dodecyl sulfate gels, *Biochim. Biophys. Acta* **508**, 425–430.

Schulz, G. E. (2000) β-Barrel membrane proteins, *Curr. Opin. Struct. Biol.* **10**, 443–447.

Segre, D., Ben Eli, D., Deamer, D. W., Lancet, D. (2001) The lipid world, *Orig. Life Evol. Biosph.* **31**, 119–145.

Selkoe, D. J. (2001) Presenilin, Notch, and the genesis and treatment of Alzheimer's disease, *Proc. Natl. Acad. Sci. USA* **98**, 11039–11041.

Seshadri, K., Garemyr, R., Wallin, E., von Heijne, G., Elofsson, A. (1998) Architecture of beta-barrel membrane proteins: analysis of trimeric porins, *Protein Sci.* **7**, 2026–2032.

Shatursky, O., Heuck, A. P., Shepard, L. A., Rossjohn, J., Parker, M. W., Johnson, A. E., Tweten, R. K. (1999) The mechanism of membrane insertion for a cholesterol-dependent cytolysin: a novel paradigm for pore-forming toxins, *Cell* **99**, 293–299.

Singer, S. J., Nicolson, G. L. (1972) The fluid mosaic model of the structure of cell membranes, *Science* **175**, 720–731.

Song, L., Hobaugh, M. R., Shustak, C., Cheley, S., Bayley, H., Gouaux, J. E. (1996) Structure of staphylococcal α-hemolysin, a heptameric transmembrane pore, *Science* **274**, 1859–1866.

Sturtevant, J. M. (1977) Heat capacity and entropy changes in processes involving proteins, *Proc. Natl. Acad. Sci USA* **74**, 2236–2240.

Tanford, C. (1980) The Hydrophobic Effect: Formation of Micelles and Biological Membranes. New York: John Wiley & Sons.

Tocanne, J. F., Dupou-Cezanne, L., Lopez, A. (1994) Lateral diffusion of lipids in model and natural membranes, *Prog. Lipid Res.* **33**, 203–237.

Tolner, B., Poolman, B., Konings, W. N. (1997) Adaptation of microorganisms and their transport systems to high temperatures, *Comp. Biochem. Physiol. A Physiol.* **118**, 423–428.

Toyoshima, C., Nakasako, M., Nomura, H., Ogawa, H. (2000) Crystal structure of the calcium pump of sarcoplasmic reticulum at 2.6 Å resolution, *Nature* **405**, 647–655.

Treutlein, H. R., Lemmon. M. A., Engelman, D. M., Brünger, A. T. (1992) The glycophorin A transmembrane domain dimer: sequence-specific propensity for a right-handed supercoil of helices, *Biochemistry* **31**, 12726–12733.

Van Gorkom, L. C. M., Horvath, L. I., Hemminga, M. A., Sternberg, B., Watts, A. (1990) Identification of trapped and boundary lipid binding sites in M13 coat protein lipid complexes by deuterium NMR spectroscopy, *Biochemistry* **29**, 3828–3834.

Wallin, E., von Heijne, G. (1998) Genome-wide analysis of integral membrane proteins from eubacterial, archaean, and eukaryotic organisms, *Protein Sci.* **7**, 1029–1038.

Watts, A. (1995) Bacteriorhodopsin: the mechanism of 2D-array formation and the structure of retinal in the protein, *Biophys. Chem.* **55**, 137–151.

White, S. H., Wimley, W. C. (1994) Peptides in lipid bilayers: structural and thermodynamic basis for partitioning and folding, *Curr. Opin. Struct. Biol.* **4**, 79–86.

White, S. H., Wimley, W. C. (1998) Hydrophobic interactions of peptides with membrane interfaces. *Biochim. Biophys. Acta* **1376**, 339–352.

White, S. H., Wimley, W. C. (1999) Membrane protein folding and stability: physical principles, *Annu. Rev. Biophys. Biomol. Struct.* **28**, 319–365.

White, S. H., Wimley, W. C., Selsted, M. E. (1995) Structure, function, and membrane integration of defensins, *Curr. Opin. Struct. Biol.* **5**, 521–527.

Wiener, M. C., King, G. I., White, S. H. (1991) Structure of a fluid dioleoylphosphatidylcholine bilayer determined by joint refinement of x-ray and neutron diffraction data. I. Scaling of neutron data and the distribution of double-bonds and water, *Biophys J.* **60**, 568–576.

Wiener, M. C., White, S. H. (1992a) Structure of a fluid dioleoylphosphatidylcholine bilayer determined by joint refinement of x-ray and neutron diffraction data. II. Distribution and packing of terminal methyl groups, *Biophys. J.* **61**, 428–433.

Wiener, M. C., White, S. H. (1992b) Structure of a fluid dioleoylphosphatidylcholine bilayer determined by joint refinement of x-ray and neutron diffraction data. III. Complete structure, *Biophys. J.* **61**, 434–447.

Wieprecht, T., Apostolov, O., Beyermann, M., Seelig, J. (1999) Thermodynamics of the alpha-helix-coil transition of amphipathic peptides in a membrane environment: implications for the peptide-membrane binding equilibrium, *J. Mol. Biol.* **294**, 785–794.

Wieslander, A., Christiansson, A., Rilfors, L., Lindblom, G. (1980) Lipid bilayer stability in membranes. Regulation of lipid composition in *Acholeplasma laidlawii* as governed by molecular shape, *Biochemistry* **19**, 3650–3655.

Wimley, W. C. (2002) Toward genomic identification of beta-barrel membrane proteins: composition and architecture of known structures, *Protein Sci.* **11**, 301–312.

Wimley, W. C., Creamer, T.P., White. S. H. (1996) Solvation energies of amino acid sidechains and backbone in a family of host-guest pentapeptides, *Biochemistry* **35**, 5109–5124.

Wimley, W. C., White, S. H. (1996) Experimentally determined hydrophobicity scale for proteins at membrane interfaces, *Nature Struct. Biol.* **3**, 842–848.

Zhang, Y.-P., Lewis, R. N. A. H., Henry, G. D., Sykes, B. D., Hodges, R. S., McElhaney, R. N. (1995a) Peptide models of helical hydrophobic transmembrane segments of membrane proteins. 1. Studies of the conformation, intrabilayer orientation, and amide hydrogen exchangeability of Ac-K_2-$(LA)_{12}$-K_2-amide, *Biochemistry* **34**, 2348–2361.

Zhang, Y.-P., Lewis, R. N. A. H., Hodges, R. S., McElhaney, R. N. (1995b) Peptide models of helical hydrophobic transmembrane segments of membrane proteins. 2. Differential scanning calorimetric and FTIR spectroscopic studies of the interaction of Ac-K_2-$(LA)_{12}$-K_2-amide with phosphatidylcholine bilayers, *Biochemistry* **34**, 2362–2371.

9 Bacterial Protein Secretion and Targeting

Prof. Dr. Arnold J.M. Driessen[1], **Dr. Chris van der Does**[2], **Dr. Nico Nouwen**[3]

[1] Department of Microbiology, Groningen Biomolecular Sciences and Biotechnology Institute, University of Groningen, Kerklaan 30, 9751 NN HAREN, The Netherlands; Tel.: +31-50-3632164; Fax: +31-50-3632154;
E-mail: a.j.m.driessen@biol.rug.nl

[2] Department of Microbiology, Groningen Biomolecular Sciences and Biotechnology Institute, University of Groningen, Kerklaan 30, 9751 NN HAREN, The Netherlands; Tel.: +31-50-3632309; Fax: +31-50-3632154;
E-mail: c.van.der.does@biol.rug.nl

[3] Department of Microbiology, Groningen Biomolecular Sciences and Biotechnology Institute, University of Groningen, Kerklaan 30, 9751 NN HAREN, The Netherlands; Tel.: +31-50-3632403; Fax: +31-50-3632154;
E-mail: n.nouwen@biol.rug.nl

CL	cardiolipin
Cs	cold-sensitive
ER	endoplasmic reticulum
PE	phosphatidylethanolamine
PG	phosphatidylglycerol
PMF	proton motive force
SRP	signal recognition particle
TMS	transmembrane segment
Ts	temperature-sensitive
$\Delta\Psi$	transmembrane electrical potential
ΔpH	transmembrane pH gradient

1 Introduction

In bacteria, the major route of protein translocation across the cytoplasmic membrane is the so-called "Sec-pathway" (Driessen et al., 2001). Secretory proteins that use this route are synthesized as precursors (preproteins) with a short amino-terminal extension, termed the signal peptide. This directs the preprotein to the translocation machinery at the membrane and is needed for the initiation of translocation. It also slows down the folding of the mature domain of the preprotein, and thereby increases the time window for molecular chaperones such as SecB to interact. SecB is a molecular chaperone that stabilizes preproteins in a translocation-competent (non-aggregated, loosely folded) state and targets them to the membrane (Figure 1) (for a review, see Fekkes and Driessen, 1999). Preproteins with a very hydrophobic signal peptide and most membrane proteins are targeted to the membrane via the signal recognition particle (SRP) pathway (see Figure 1). SRP binds to the signal peptide when it emerges from the ribosome, and subsequently targets the ribosome-bound nascent chain complex via the SRP receptor FtsY at the membrane to the translocation machinery.

Preprotein translocation is mediated by a multicomponent membrane complex that consists of the membrane-associated ATPase SecA, and a large integral membrane domain consisting of SecY, SecE, SecG, SecD, SecF, and YajC as separate subunits. The essential (also termed "core") subunits of this complex are the SecY, SecE, and SecA proteins. Energy in the form of ATP and the proton motive force (PMF) is used to drive the translocation of preproteins across the cytoplasmic membrane. During or shortly

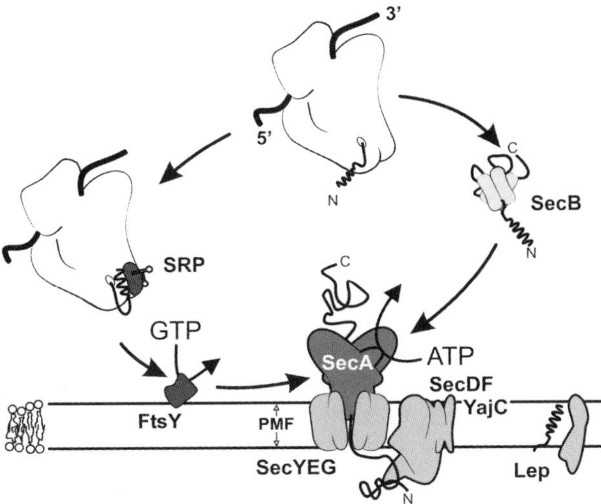

Fig. 1 Schematic overview of the components of bacterial translocase. See text for details.

after the translocation process, the signal peptide of the preprotein is removed by signal peptidase, and the mature domain is released into either the periplasm (Gram-negative bacteria) or medium (Gram-positive bacteria). In the case of Gram-negative bacteria, some proteins are subsequently targeted to the outer membrane and integrated into or translocated across this membrane. The components of the general secretion pathway are listed in Table 1.

In addition to the general secretion pathway, a number of specialized or minor translocation systems exist in the bacterial cytoplasmic membrane. The Tat (Twin arginine translocation) pathway mediates the translocation of proteins that are first folded in the cytoplasm. Usually, only a few substrates are translocated via this pathway, among which are generally several extracellular redox factor-containing proteins. These proteins are synthesized as precursors with an amino-terminal signal sequence that contains two consecutive arginine amino acyl residues. Type I–IV secretion systems are involved in the translocation of specific proteins to the external surface of the cell. These other systems and the Tat pathway will not be discussed here (for some excellent reviews on these translocation systems, see Berks et al., 2000; Robinson and Bolhuis, 2001 (Tat); Koronakis et al., 2001 (type I); Sandkvist, 2001 (type II); Plano et al., 2001 (type III); Christie, 2001 (type IV)).

2
Historical Outline

In 1971, Gunter Blobel published a hypothesis in which he suggested that secretory proteins share an amino-terminal signal peptide that directs these proteins to the translocation sites at the cytoplasmic membrane. Original evidence for this hypothesis was obtained in studies on protein translocation into the eukaryotic endoplasmic reticulum (ER). However, later studies demonstrated that analogous amino acid sequences are used in bacteria, and that these signal peptides fulfill a similar function. Importantly, the signal peptides that

Tab. 1 Components of the bacterial preprotein translocase

Component	*Localization*	*Essential for viability*	*Function*
SecA	Cytosol, peripheral membrane- associated	+	ATPase, motor protein, preprotein receptor
SecB	Cytosol	–	Molecular chaperone and targeting factor
SecD	Membrane protein	–	Subunit of SecDF complex; increases efficiency of translocation
SecE	Membrane protein	+	Subunit of translocation channel
SecF	Membrane protein	–	Subunit of SecDF complex; increases efficiency of translocation
SecG	Membrane protein	–	Increases efficiency of translocation
SecM	Secretory protein	–	Regulation of SecA expression
SecY	Membrane protein	+	Subunit of translocation channel
YajC	Membrane protein	–	Subunit of SecDF complex, no known function
YidC	Membrane protein	+	Subunit of SecDF complex; facilitates insertion of hydrophobic domains into the membrane
Ffh/4.5S RNA	Cytosol	+	Signal recognition particle, targeting factor, GTPase
FtsY	Peripheral membrane-associated	+	SRP receptor, GTPase

+, essential; –, not essential.

direct proteins to the ER of eukaryotes, the thylakoid of plant chloroplasts and the cytoplasmic membrane of bacteria are functionally exchangeable (von Heijne, 1998), suggesting similar mechanisms of protein translocation in these organelles.

During the early 1980s, sophisticated genetic screens were developed to select mutants of *Escherichia coli* in the genes for both exported proteins and the protein components that mediate export. Since the genes encoding the components of the translocase are essential for viability, conditional lethal mutant strains were selected that were incapable of growing at either low (cold-sensitive) or high (temperature-sensitive) temperature. In one of the many genetic selection protocols, mutant strains were obtained that carried a mutation in the Lamb signal sequence. Due to this mutation, the LamB protein is not transported to the outer membrane (Emr et al., 1981). As a consequence, these cells are unable to transport maltodextrins across the outer membrane and therefore cannot use this compound as a carbon source. Strains containing suppressor mutations were isolated that restored growth on maltodextrin; these suppressor mutants mapped at various loci, including the genes encoding major components of the translocase, *secA* and *secY*. Further advanced genetic studies led to the identification of the other components of the translocation machinery.

While the genetic studies on protein translocation in *E. coli* flourished, the biochemical analysis of protein translocation also began to develop, and an *in vitro* protein translocation assay with inverted bacterial membrane vesicles became available during the early 1980s. In this assay, precursors of secretory proteins were synthesized in a membrane-free extract from *E. coli* using plasmid DNA (Muller and Blobel, 1984). In the presence of the inverted plasma membrane vesicles, a major fraction of the newly synthesized precursor molecules were processed to yield mature forms that could resist externally added proteinase K unless the membrane vesicles were disrupted by mechanical means or by detergents. Using this *in vitro* system, it was shown that in *E. coli*, protein translocation is uncoupled from translation, and confirmed that protein export is a post-translational event. Further use of this in-vitro system led to the discovery of various soluble (SecA, SecB) and membrane proteins (SecY, SecE, and SecG) that are needed for translocation. These are the major components of the protein translocase, of which most are conserved through the bacterial kingdom and are essential for life.

Finally, in the early 1990s, the bacterial SecYEG complex was purified and functionally reconstituted into artificial lipid vesicles. Since the cellular level of SecYEG protein in the native membrane is very low (it comprises $<0.1\%$ of the total membrane protein), a highly sensitive biochemical assay had to be used during the purification. This assay monitored the precursor protein-stimulated ATPase activity of SecA (translocation ATPase) (Lill et al., 1989), this activity being dependent on a functional association of SecA with the SecYEG complex. Membranes of *E. coli* were solubilized with a detergent and fractionated using column chromatography; the fractions were then reconstituted into liposomes (Brundage et al., 1990). These liposomes were supplemented with SecA and assayed for translocation ATPase activity. This led to the purification of an integral membrane protein complex that consisted of SecY and SecE (the genes of which were previously identified in a genetic screen for secretion factors) together with a third component, SecG. The major outcome of this study was the finding that the SecYEG complex, together with the SecA ATPase, represents the minimal constituents that allow the reconstitution of an authentic preprotein translocation reaction *in vitro*.

3 Protein Targeting to the Translocase

When a secretory protein with an amino-terminal signal sequence is translated by the ribosome, it can either be targeted co- or post-translationally to the translocase at the cytoplasmic membrane. During co-translational targeting, the preprotein is directed to the translocase as a ribosome-bound nascent chain complex, whilst in post-translational targeting the preprotein is fully synthesized in the cytosol before being targeted to the translocase (see Figure 1).

3.1 Signal Peptides

Secretory proteins are synthesized in the cytosol as precursors with a cleavable amino-terminal signal sequence (von Heijne, 1998). Bacterial signal sequences have a length that ranges from 18 to 30 amino acids. Signal peptides show no conservation in amino acid sequence, but they are equipped with the same physical properties and have a tripartite structure:

1) N-domain: the amino-terminal domain, with a size of one to five amino acids, contains a net positive charge. Preproteins that do not have this positive charge are still recognized by the translocase, but are translocated slowly. The N-domain

interacts with the translocation ATPase SecA and with negatively charged phospholipids. Due to the positive charge, translocation of the N-domain is prevented by the transmembrane electrical potential ($\Delta\psi$), which in *E. coli* is negative inside and positive outside. This results in the orientation of the signal sequence with the N-terminus in the cytosol.

2) H-domain: the hydrophobic core of the signal sequence consists of a stretch of 7–15 hydrophobic residues that may fold into an α-helical conformation. This domain is thought to insert into the lipid bilayer. Frequently, glycine and proline residues are found in the middle of this domain. In the cytosol, these residues may act as helix breakers, thereby enabling the formation of a hairpin-like structure that facilitates insertion of the signal sequence into the membrane or translocation channel. Inside the lipid bilayer, the glycine and proline residues adopt an α-helical conformation, causing the "unlooping" of the hairpin and further insertion of the signal sequence (van Dalen et al., 1999). The N- and H-domains have overlapping functions in protein export, and both are needed for recognition by SecA (Akita et al., 1990). With isolated signal peptides, the α-helical conformation of the H-domain is promoted by interaction of positively charged residues of the N-domain with anionic phospholipids.

3) C-domain: the polar C-domain consists of three to seven residues, and contains the signal peptidase cleavage site. Signal peptidase recognizes residues at the –1 and –3 positions relative to the cleavage site. These positions contain amino acids with small neutral side chains. *E. coli* contains a signal peptidase for regular signal sequences, and another enzyme that cleaves the signal sequence of lipid-modified precursors proteins (Dalbey et al., 1997). In some bacteria, multiple signal peptidases are found with overlapping substrate specificities.

The signal peptide targets the protein to the translocase and is removed during or after translocation of the protein. This event is required for the release of the mature domain from the membrane. Released signal peptides are degraded by various peptidases and removed from the membrane.

3.2 Co-translational Protein Targeting

The co-translational protein targeting pathway directs ribosome nascent chain complexes to the translocase, and this results in translocation of the growing polypeptide chain across the cytoplasmic membrane. The two key components in co-translational targeting route are the SRP and the SRP receptor. SRP is a ribonucleoprotein which, in *E. coli*, consists only of a 48 kDa GTPase called Ffh (for fifty-four-homologue) and a 4.5S RNA (for a review, see Herskovits et al., 2000); in eukaryotes and archaea, SRP has a much more complex structure. These proteins were first identified due to their homology with the well-studied eukaryotic counterparts, but in a recent genetic screen, pleiotropic secretion defects also mapped to the *ffh* and *ftsY* genes (Tian and Beckwith, 2002). Both SRP and its receptor are essential for cell viability. In *E. coli*, the SRP pathway is only needed for the transport of a subset of precursor proteins and seems mostly involved in the co-translational targeting of α-helical membrane proteins. Ffh and 4.5S RNA form a complex that interacts specifically with the signal sequence or transmembrane segment of nascent chain complexes (Luirink et al., 1992). FtsY is the bacterial homologue of the mammalian SRα, one of the subunits of the SRP receptor (Gill and Salmond, 1990). The other subunit of the mammalian SRP receptor, SRβ, appears to be absent in bacteria. In mammals, it is necessary for SRα to be membrane-associated for release of the nascent chain from SRP. FtsY appears to fulfill the functions of both the SRα and SRβ subunits as it can bind to membranes composed of only phospholipids (de Leeuw et al., 2000). Nevertheless, other evidence points to the presence of an unknown receptor protein in the cytoplasmic membrane (Millman et al., 2001). In some bacteria, FtsY is anchored to the membrane by means of an α-helical transmembrane domain. *E. coli* SRP binds to FtsY in a GTP-dependent manner; in this reaction, binding of the 4.5S RNA to Ffh controls the association and dissociation of SRP and FtsY (Peluso et al., 2000).

Ffh consists of three regions: an amino-terminal N domain; a middle G domain; and a carboxy-terminal M domain. The G domain is closely related to the $p21^{Ras}$ GTPase family and contains the GTP binding site. The N-domain senses or controls the nucleotide occupancy of the G-domain through interfacial contacts. The GTPase cycle of Ffh, however, differs from those of other GTPases. The active site side chains of the nucleotide-free form of the G-domain are effectively sequestered and provide a relatively stable non-nucleotide bound state (Freymann et al., 1999). Ffh can be chemically crosslinked to hydrophobic signal sequences. The signal sequence-binding groove in the M-domain of Ffh is lined up with a large number of clustered methionine residues whose hydrophobic side chains (Freymann et al., 1999), together with the conserved domain IV of the RNA, form a surface that by a combination of hydrophobic and electrostatic interactions recognizes the signal sequence (Batey et al., 2000). The signal sequence recognition surface thus consists of both protein and RNA.

The SRP RNA from Gram-negative bacteria is much shorter than the eukaryotic homologue and does not contain the 5′ and 3′ RNA Alu domain. In eukaryotes, this region associates with the SRP9/14 heterodimer that retards ribosomal elongation of preproteins before their engagement with the translocon. Translation arrest may not be essential in bacteria due to the shorter traffic distances and the faster translocation rates. However, in archaea and Gram-positive bacteria such as *Bacillus subtilis*, the SRP RNA has a more complex structure and includes the 5′ and 3′ RNA Alu domain. These organisms lack homologues of SRP9/14, but in *B. subtilis*, a histone-like protein HBsu exists that interacts with the Alu domain of the SRP-RNA. Translocation arrest may thus be a feature in Gram-positive bacteria (Herskovits et al., 2000). A scheme for the SRP-mediated targeting of preproteins is shown in Figure 2.

3.3
Post-translational Protein Targeting

In *E. coli*, most preproteins are translocated post-translationally (Remaut et al., 1981), which means that they are first synthesized to full length before being targeted to the translocase. These proteins are stabilized in the cytosol as full-length unfolded precursors with the help of molecular chaperones. SecB is a chaperone with a function dedicated to protein translocation (for reviews, see Fekkes and Driessen, 1999; Driessen, 2001). SecB was genetically identified in a screen for mutants with pleiotropic defects in protein secretion (Kumamoto and Beckwith, 1983). The secB locus maps at 80.5 min on the *E. coli* genome, and encodes an acidic

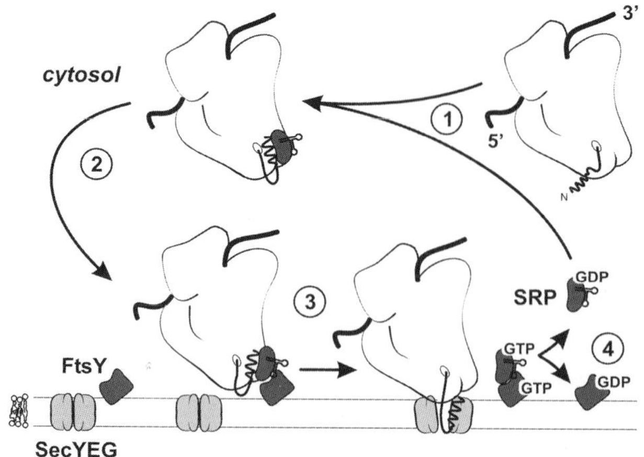

Fig. 2 Schematic overview of signal recognition particle (SRP)-mediated co-translational targeting of nascent preproteins. (1) Synthesis of preprotein begins with an unattached ribosome in the cytosol. The SRP complex recognizes the ribosome-emerged signal sequence when about 70 amino acids have been synthesized. (2) The ribosome nascent chain complex is subsequently targeted by SRP to the membrane-bound FtsY. (3) This binding event increases the GTP-binding affinity of Ffh and FtsY, and in an as yet unspecified order, binding of GTP to the membrane-bound FtsY and/or SRP dissociates the ribosome nascent chain complex from SRP and releases it to the SecYEG complex. The large ribosomal subunit also binds directly to the SecYEG complex via the 23S rRNA. (4) Subsequent hydrolysis of GTP dissociates SRP from FtsY, allowing it to recycle into the cytosol while FtsY remains bound to the membrane. At this stage the ribosome and SecA collaborate for the translocation of the preprotein across the membrane. The ribosome continues to elongate the polypeptide chain until translation is completed, while SecA drives the translocation of the polypeptide segments across the membrane (this will be discussed later). For simplicity, SecA is not shown.

17-kDa cytosolic protein (Kumamoto et al., 1989). SecB is present in most Gram-negative bacteria; it forms a homotetrameric protein, and is needed for the efficient translocation of preproteins (Kumamoto and Beckwith, 1983). Its function is, however, not required for the viability of *E. coli*. *In vivo*, SecB is very selective and found to interact with only a subset of preproteins, mainly outer membrane proteins that are rich in β-sheet structure and prone to aggregation (Kumamoto and Francetic, 1993). These proteins are kept in a translocation-competent state, i.e., they are loosely folded and nonaggregated. *In vitro*, SecB is rather unselective and interacts with any protein that is in a non-native conformation (Randall and Hardy, 1986).

The mechanism by which SecB differentiates between secretory and nonsecretory proteins is unknown. SecB does not associate with the signal sequence, but since the signal sequence retards folding of the mature domain, it assists in the binding of the preprotein to SecB. SecB preferentially binds the unfolded conformation of the mature part of preproteins. A typical SecB-binding motif is about nine amino acid residues long and enriched in aromatic and basic residues; acidic residues are strongly disfavored (Knoblauch et al., 1999). The slow folding, as dictated by the signal sequence, assists SecB in recognizing such peptide sequences that typically occur within internal regions of folded proteins (Liu et al., 1989). SecB associates with ribosome-bound nascent chains after they have reached a length of about 150 residues. These long binding regions simultaneously occupy multiple binding sites on the tetrameric SecB to allow a high affinity of interaction (K_d 5–50 nM). The three-dimensional structure of SecB has been determined without a peptide substrate in its binding site, and reveals a long surface-exposed channel on each side of the SecB tetramer that has all the characteristics of a peptide-binding site (Xu et al., 2000). In order to occupy the peptide-binding grooves on both sites, long unstructured polypeptide segments presumably wrap around the chaperone.

GroEL and DnaK can substitute for SecB in stabilizing preproteins in a translocation-competent state, but they are unable to target the proteins specifically to the translocase. The ability of secB to bind to SecA discriminates SecB from the general chaperones (Hartl et al., 1990). In the cytosol, SecA interacts only with poor affinity with SecB (den Blaauwen et al., 1997; Woodbury et al., 2000), but at the membrane, SecB binds with high affinity to SecA which itself is bound to the preprotein-conducting channel. A negatively charged solvent-exposed surface on each of the sides of the SecB tetramer (Xu et al., 2000) associates electrostatically with the positively charged carboxy-termini of the SecA dimer (Fekkes et al., 1997).

The preprotein transfer activity and the events that lead to the release of SecB from the membrane are not contained in the SecB structure. For these activities, SecB relies on the catalytic activity of SecA. The SecB–preprotein complex docks at the SecYEG-bound SecA, and subsequent binding of the exposed signal sequence region of the preprotein to SecA results in tightening of the SecB–SecA interaction (Fekkes et al., 1997). This interaction also results in a slight increase in the ATPase activity of SecA (Miller et al., 2002), and subsequently SecB releases the preprotein to SecA. SecB is released from this ternary complex when SecA binds ATP to initiate translocation. The released SecB recycles into the cytosol where it can bind a newly synthesized preprotein. The catalytic cycle of SecB-mediated preprotein targeting is shown schematically in Figure 3.

3.4
Converging Targeting Pathways

Preproteins decide between the SRP- and SecB-dependent targeting pathways at a very early stage. When the signal sequence protrudes from the ribosome, SRP binds specifically to long hydrophobic signal sequences, while less hydrophobic signal sequences associate with the ribosome-associated chaperone trigger factor, a cytosolic peptidyl-prolyl-*cis/trans*-isomerase which is capable of catalyzing protein folding *in vitro* (Beck et al., 2000). Preproteins that escape early SRP recognition in this way are either bound by SecB that interacts only with long nascent chains, or are targeted directly to the translocase by the signal sequence. Subtle differences in the hydrophobicity of signal peptides thus define the pathway that will be followed for targeting. The finding that the SRP dependency of preprotein translocation increases with the hydrophobicity of the signal sequence, while the requirement for SecB is reduced, is consistent with this hypothesis. Both targeting pathways converge at the translocase (Valent et al., 1998) (see Figure 1).

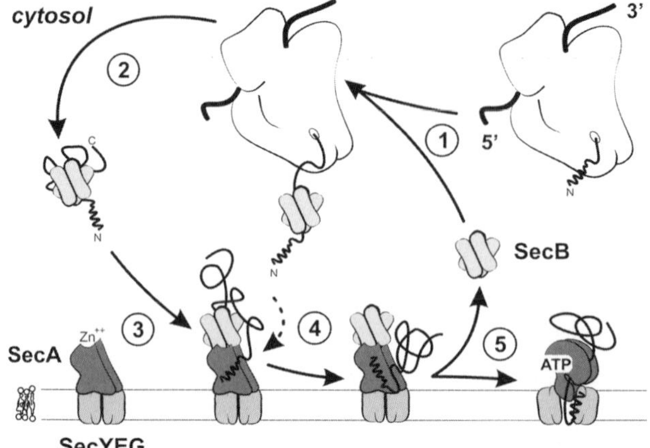

Fig. 3 Schematic overview of SecB-dependent targeting of preproteins to the translocase. (1) Cytosolic SecB binds to the mature domain of a nascent preprotein, and (2) stabilizes its unfolded state. (3) The binary SecB–preprotein complex is targeted to the SecYEG-bound SecA where SecB binds with high affinity to a carboxyl-terminal zinc-binding domain in SecA. Alternatively, it may first associate with low affinity with cytosolic SecA and remain in the cytosol until translocation sites at the membrane become available. (4) Binding of the signal sequence to SecA tightens the SecB–SecA interaction, stimulates SecA for ATPase activity and elicits the release of the preprotein from SecB and concomitant transfer to SecA. (5) The release of SecB from the membrane is coupled to the binding of ATP to SecA. Under these conditions, the carboxyl-termini of SecA are no longer available for SecB binding.

4 Translocase: A Multisubunit Integral Membrane Protein Complex

Bacterial translocase consists of the peripherally bound ATPase SecA and a complex of integral membrane proteins (see Figure 1). The core of the integral membrane domain is formed by the SecY and SecE polypeptides. SecYE forms a heterotrimeric complex with SecG, and this complex is the most abundant form in which SecYE exists. It can, however, also associate with other subunits, SecD, SecF and YajC that are possibly recruited by the SecYE complex to increase the fidelity of the preprotein translocation reaction.

4.1 SecA

SecA performs a central role in preprotein translocation. It is the ATP-driven motor protein that couples the hydrolysis of ATP to the stepwise translocation of preproteins across the membrane. SecA binds to the translocation channel, but also functions as a membrane receptor for preproteins and for the molecular chaperone SecB. The *secA* gene was discovered in a screen for proteins that cause a conditionally pleotropic defect in preprotein secretion (Oliver and Beckwith, 1981). It is a homodimeric 102 kDa polypeptide (Oliver and Beckwith, 1982) that distributes dynamically between the cytosol and the membrane. SecA binds with low affinity to phospholipids, while in its translocation-active state it associates with high affinity with the SecYEG complex. In the latter case, it can also bind the binary SecB–preprotein complex with high affinity–an association which involves a direct interaction of the exposed signal sequence of the preprotein with SecA. SecA is the only ATPase involved in preprotein translocation. It has a low endogenous ATPase activity that is increased by its interaction with the SecYEG complex and preproteins. This ATPase activity is termed "SecA Translocation ATPase" as it is coupled to preprotein translocation (Lill et al., 1989).

Some bacteria contain two copies of homologous SecA proteins, such as various *Mycobacterium*, *Bacillus* and *Staphylococcus* species (Owens et al., 2002). SecA functions as a homodimer (Akita et al., 1991; Driessen, 1993) in which each monomer of the protomer contains two thermodynamically independent folding domains (N- and C-domains) (den Blaauwen et al., 1996). In addition, the monomer contains two essential nucleotide-binding sites (NBS) (Mitchell and Oliver, 1993). NBS-1, which is located in the N-domain, is responsible for high-affinity ATP binding ($K_d = 0.13$ μM) and also contains the typical Walker A and Walker B motifs. NBS-2 was proposed to function with low-affinity ATP binding and is located in the C-domain, although this region does not appear to function as an independent nucleotide binding site, but rather acts as a regulatory domain which controls the hydrolysis of ATP at NBS-1 (Nakatogawa et al., 2000a; Sianidis et al., 2001). Both regions are essential for the preprotein-stimulated SecA ATPase activity and preprotein translocation (Klose et al., 1993; Mitchell and Oliver, 1993; van der Wolk et al., 1993; Sato et al., 1996). ADP-binding to SecA promotes interaction between the N- and C-domains of SecA (den Blaauwen et al., 1996; Sianidis et al., 2001), yielding a compact SecA conformation (den Blaauwen et al., 1996, 1999). Various biochemical methods indicate that the soluble SecA undergoes conformational changes upon nucleotide binding (Shinkai et al., 1991; den Blaauwen et al., 1996; Sianidis et al., 2001), but these changes do not appear to change the protein's overall shape (Shilton et al., 1998).

The signal sequence of the preprotein has been crosslinked to a region of SecA just adjacent to NBS-1 (Akita et al., 1990; Baud et al., 2002). Mutagenesis of Tyr326 in this region results in strong protein translocation defect and a loss of SecA translocation ATPase activity, and this has been attributed to a defect in preprotein release (Kourtz and Oliver, 2000). Isolated signal peptides also modulate the SecA ATPases (Cunningham and Wickner, 1989; Triplett et al., 2001) and cause a conformational change of the SecA protein (Ding et al., 2001; Baud et al., 2002). Mutations in SecA that suppress the export defect caused by signal sequence mutations, i.e., the so-called *prlD* suppressors (*pr*otein *l*ocalization) are located throughout the entire primary sequence of SecA. Remarkably, many *prlD* and *azi* mutations coincide (Huie and Silhavy, 1995). *Azi* mutations render SecA resistant to azide, an inhibitor of the translocation ATPase (Oliver et al., 1990). Azi- and PrlD-SecA proteins have a reduced affinity for ADP, an elevated membrane ATPase activity, and an altered conformation (Schmidt et al., 2000). These findings suggest that alterations in the turnover of SecA may lead to signal sequence suppression.

The extreme C-terminus of SecA has been implicated in various catalytic activities, such as SecB binding (Breukink et al., 1995; Fekkes et al., 1997, 1999) and lipid interaction (Breukink et al., 1995). In conjunction with N-terminal and central regions of SecA, the C-terminus of the SecYEG-bound SecA protein is accessible to membrane-impermeable reagents and proteases added from the periplasmic face of the membrane (van der Does et al., 1996; Ramamurthy and Oliver, 1997; Eichler and Wickner, 1998). This indicates a complex membrane topology of the SecYEG-bound SecA, and suggests that domains of SecA penetrate the membrane or are accessible from the outer surface of the membrane via the protein-conducting pore formed by the SecYEG complex.

4.2 SecY

The localization of the gene encoding SecY was discovered in a screen for mutations which could suppress signal sequence mutations. This genetic locus, termed *prlA* (Emr et al., 1981), is localized downstream of the ribosomal *spc* operon (Cerretti et al., 1983). The genetic organization of this locus is extremely well conserved in bacteria and even in archaea. Hydropathy analysis predicts that SecY harbors 10 hydrophobic transmembrane segments (TMS) (Akiyama and Ito, 1987). SecY is present in all eubacteria and archaea, in the ER (Sec61α) of eukaryotes, and in plant thylakoids. When the sequences of different SecY proteins are aligned, only the N- and C-terminal cytoplasmic domains, TMS 3, 4 and 6, and the periplasmic loops except P1 (periplasmic loop 1) are found not to be conserved. Dominant loss of function mutations have been identified in cytosolic loop 5, and this region–together with C6–appears important for a functional SecY–SecA interaction (Nakatogawa et al., 2000b; Mori and Ito, 2001). Conditionally lethal mutations are all located in the conserved domains. In addition, SecY contains a large number of strong *prl* mutations (Flower et al., 1995) which are mainly found in the conserved regions. Related to *prl* mutations are a number of mutations in TMS 7 that specifically block staphylokinase translocation (Sako and Iino, 1988; Iino and Sako, 1988).

The protein translocation activity associated with SecY was first purified and reconstituted from octylglucoside-solubilized *E. coli* inner membranes (Driessen and Wickner, 1990), and two other proteins– SecE and SecG – were found to co-purify with SecY (Brundage et al., 1990; Akimaru et al., 1991; Driessen et al., 1991). When purified separately, the translocation activity of reconstituted SecY increases proportionally with the amount of co-reconstituted SecE (Tokuda et al., 1991). SecY and SecE are interacting proteins; indeed, in the absence of SecE, SecY is unstable and becomes degraded by FtsH (Kihara et al., 1995). Overproduction of SecY in FtsH mutant cells has a deleterious effect on cell growth and protein export, which suggests that the elimination of uncomplexed SecY is important for cell viability (Kihara et al., 1995). Pulse-chase experiments demonstrated that there is no exchange of SecE molecules between SecYE complexes (Taura et al., 1993; Joly et al., 1994), indicating that SecY and SecE are tightly interacting proteins.

4.3 SecE

SecE has been found in a screen for mutations that increase the expression of the *secA* gene. The basis of this screening method is that mutations which cause protein secretion defects in general lead to elevated SecA levels in the cell. Further screens for extragenic suppressors of a LamB signal sequence mutation also showed that dominant suppressor mutations (called *prlG*) were linked to the *secE* gene (Stader et al., 1989). The gene is located at approximately 90 min on the *E. coli* chromosome and forms an operon with the *nusG* gene (Riggs et al., 1988). The *secE-nusG* genes are co-transcribed (Downing et al., 1990), and the genetic region surrounding *secE* seems well conserved in prokaryotes. Since *secE* encodes a rather small integral membrane protein, it is often overlooked in the annotation of bacterial genomes. In contrast to what has been suggested (Yang et al., 1997), *secE* is also present in the smallest bacterial genome of *Mycoplasma genitalium* where it is localized downstream from *nusG*.

Although the *E. coli* SecE contains three membrane-spanning domains (Schatz et al., 1989), most bacterial SecE proteins are much smaller and contain the conserved C2 region and the third TMS of *E. coli* SecE. In *E. coli*, this region also suffices to support growth (Schatz et al., 1991) and protein translocation (Nishiyama et al., 1992). SecE proteins from various bacteria with only one TMS domain complement the lethal *secE* deletion in *E. coli*. Replacement of the third TMS domain of the *E. coli* SecE for an unrelated TMS from the MalF protein yields a SecE protein that can complement the *ΔsecE* strain at 37°C but not at 30 or 42°C (Murphy and Beckwith, 1994). This suggests that this TMS mainly fulfills a passive function in maintaining the correct topological arrangement of the conserved cytoplasmic domain (C2) that precedes TMS3. However, mutations in TMS3 have been found that completely abolish the SecE function (Kontinen et al., 1996b; Pohlschroder et al., 1996). In addition, TMS3–like the P2 loop–contains strong suppressors of defective signal sequences (Flower et al., 1994), suggesting that TMS3 is involved in a catalytic function of SecE.

The eukaryotic ER homologue of SecE is Sec61γ, which interacts with Sec61α (Bairoch et al., 1997). The C2 domain is the most conserved sequence between SecE and Sec61γ. Most single amino acid substitutions in the C2 domain cause mild secretion defects and have little effect on growth at 37°C (Murphy and Beckwith, 1994). However, double substitutions or deletion of the conserved proline residues cause severe growth and translocation defects (Kontinen et al., 1996b; Pohlschroder et al., 1996). Several regions contribute to the formation of a stable SecE–SecY complex: the C-terminal part of SecE is essential for interaction with SecY (Nishiyama et al., 1992), and mutants in this region destabilize the SecE–SecY complex (Pohlschroder et al., 1996). Sites of direct contact between SecY and SecE have also been suggested by the synthetic lethality of certain combinations of *prlA* and *prlG* mutations (Flower et al., 1995). More recently, cysteine-scanning mutagenesis directly demonstrated interactions between the P1 loop of SecY and P2 loop of SecE (Harris and Silhavy, 1999), between TMS 2 of SecY and TMS 3 of SecE (Kaufmann et al., 1999), and between TMS 7 and 10 of SecY and TMS 3 of SecE (Veenendaal et al., 2001). Cytoplasmic loop C4 of SecY also plays an important role in SecY–SecE interaction; this was discovered using a dominant negative *secY* allele, *secY^{d}*, which contains a major deletion at the interface of C5 and TMS 9 of SecY (Shimoike et al., 1992). The deletion inactivates SecY but preserves its ability to interact with SecE, and thereby sequesters the SecE into an inactive complex. This lethal effect can be overcome by overexpressing SecE (Baba et al., 1994) or by second-site mutations that disrupt SecE binding. All of the mutations identified were clustered in the C4 loop of SecY. Cysteine-scanning mutagenesis of TMS 3 of SecE revealed that SecE of one SecYEG subunit is in close proximity to a SecE molecule in another SecYEG complex (Kaufmann et al., 1999; Veenendaal et al., 2001). These contacts are found all along the SecE TMS 3, suggesting that SecYEG complexes assemble into a higher oligomeric state.

4.4 SecG

SecG was not identified through a genetic screen, but by co-purification with the SecYE complex (Brundage et al., 1990; Nishiyama et al., 1993). SecG is not essential for protein translocation, as a basal translocation reaction can be reconstituted with purified SecA, SecY, SecE and acidic phospholipids (Akimaru et al., 1991). When SecG is co-reconstituted with SecYE, translocation is strongly enhanced (Nishiyama et al., 1993). The *secG* gene is located at 69 min on the *E. coli* chromosome, and encodes a hydrophobic protein of 11.4 kDa. Deletion of *secG* renders cells cold-sensitive (Cs) for growth and protein export, but this phenomenon depends on the genetic background of the *E. coli* strains (Bost and Belin, 1995) and was found to relate to a defect in phospholipid biosynthesis (Flower, 2001). SecG therefore seems to be important for the efficiency of protein translocation, particularly at lower temperatures (Nishiyama et al., 1994). In-vitro studies have also indicated that SecG is important in the absence of the PMF (Hanada et al., 1996). SecG does not affect the binding of SecA to SecYE (Hanada et al., 1994).

SecG contains two TMSs (Nishiyama et al., 1996), and proteolysis studies suggest that the topology of these completely reverses upon the membrane insertion of SecA (Nishiyama et al., 1996). It was speculated that this remarkable event is needed to facilitate membrane insertion of SecA, in particular at a lower temperature when the microviscosity is high. The latter suggestion is based on a genetic screen for suppressors of the Cs growth defect of a *ΔsecG* strain that yielded genes involved in phospholipid biosynthesis (Kontinen and Tokuda, 1995; Kontinen et al., 1996a; Shimizu et al., 1997). By increasing the total amount of anionic phospholipids, protein translocation becomes elevated and this would render cells less dependent on SecG (Suzuki et al., 1999). Mutations in the translocase membrane subunits often result in a Cs growth phenotype. It has been suggested that this reflects the cold sensitivity of protein export itself (Pogliano and Beckwith, 1993), which would be exacerbated by any mutations that interfere with the activity of the translocase. The mechanism underlying thermal sensitivity of this process is not known, but potential cold-sensitive steps could be the insertion of the signal sequence and/or SecA into the membrane, the topology inversion of SecG, and/or oligomerization of the SecYEG complex into a protein-conducting channel.

Bacterial SecG homologues have been identified by sequence analysis, but functional evidence has been obtained only in *B. subtilis* (van Wely et al., 1999). The *B. subtilis ΔsecG* strain shows a mild Cs phenotype for growth, which is exacerbated by high-level production of secretory proteins. Sequence comparison reveals that conserved amino acid residues are present throughout the entire SecG sequence, while the C-terminal periplasmic domain of Gram-negative SecG proteins is lacking in Gram-positive bacteria. The genetic localization of SecG is not conserved, and so far no homologues have been found in eukaryotes and archaea. These organisms contain another small membrane protein, Sec61β, but this is not homologous to SecG. Although no suppressors of signal sequence mutations were mapped to *secG* by the original screens, a direct selection of SecG mutants yielded *prl* alleles of *secG* (*prlH*). These mutations are distributed along the SecG sequence but do not map to the three residues that yield a defective SecG protein. Some of the *prlH* alleles are as strong as the strongest known *prlG* alleles of *secE*, and it has been suggested that SecG also contributes to signal sequence recognition (Bost and Belin, 1997).

Little is known about the interaction between SecG and the SecYE complex. SecG increases the stability of SecY (Nishiyama et al., 1995), of the SecYE complex (Homma et al., 1997), and of an unstable truncate of SecE (Nishiyama et al., 1995). Other studies have suggested that SecG interacts mainly with SecY (Bost and Belin, 1995; Homma et al., 1997) and SecA (Bost and Belin, 1995), and not with SecE. Although binding of SecA to the SecYE complex is not affected by SecG (Hanada et al., 1994), the topology inversion of SecG is hampered when a partially functional SecA truncate is used that lacks eight N-terminal amino acid residues (Mori et al., 1998). SecA is thought to interact with both SecY and SecG since a SecA mutant (SecA36) suppressed both the Cs-sensitive phenotype of the SecY205 and the Δ*secG* mutant (Matsumoto et al., 1998). Various lines of evidence suggest that the SecG and SecA conformational changes are coupled (Matsumoto et al., 1997), but such an interaction has not been demonstrated biochemically.

4.5 SecD, SecF, and YajC

SecD, SecF, and YajC are three integral membrane proteins that form a heterotrimeric complex that can associate with SecYEG to facilitate efficient translocation. *SecD* was identified in a screen for mutants altered in export of PhoA-LacZ and LamB-LacZ fusion proteins (Gardel et al., 1987). These mutants have a Cs phenotype and accumulate precursors in the cell. Analysis of the genomic region around the *secD* gene indicated the presence of another gene, *secF*, that when mutated exhibited the same phenotype as *secD* mutants. The first gene in the *secDF* operon, *yajC* or *orf12*, encodes a small membrane protein with one TMS and a large cytosolic domain. YajC is not essential for growth, nor does it affect secretion (Gardel et al., 1990). However, co-immunoprecipitation studies showed that YajC forms a complex with both SecYEG and SecDF (Duong and Wickner, 1997a). SecD, SecF and double-depletion strains exhibit a severe export defect at 37°C and are barely viable. SecD and SecF are integral membrane proteins with six TMSs and a large periplasmic domain (Gardel et al., 1990; Pogliano and Beckwith, 1994). In *B. subtilis* and some other bacteria, SecD and SecF form a single polypeptide with 12 putative TMS domains (Bolhuis et al., 1998). YajC is also conserved among bacteria, and its gene is often located in an operon together with *secD* and *secF*.

The exact role of SecD, SecF, and YajC in protein translocation is not clear. In *E. coli*, SecD has been suggested to be involved in the release of proteins from the periplasmic side of the cytoplasmic membrane (Matsuyama et al., 1993). This hypothesis is based on *in vivo* studies which showed that an antibody against SecD resulted in preprotein accumulation and prevented the release of mature MBP and OmpA from spheroplasts into the medium. The large periplasmic loops of SecD and SecF have also been suggested to act as valve that prevents the passage of protons/ions through the protein-conducting channel formed by the SecYEG complex (Arkowitz and Wickner, 1994). However, the experimental evidence supporting this last hypothesis appeared to be caused by glucose repression of the succinate dehydrogenase used to energize the membranes (Nouwen et al., 2001). SecDFYajC also has been suggested to control protein translocation by regulating the ATP-driven cycle of SecA membrane insertion and de-insertion (Duong and Wickner, 1997a). Overproduction of SecD and SecF stabilizes SecA in the membrane-inserted state (Kim et al., 1994; Economou et al., 1995). By stabilizing the SecA membrane-inserted form, the SecDFYajC complex slows the movement of preprotein in transit against both reverse and forward translocations (Duong and Wickner, 1997b). However, it is questionable that SecD and SecF directly affect membrane cycling of SecA as the latter is absent in archaea whereas these organisms contain SecD and SecF proteins (Tseng et al., 1999). Finally, SecD and SecF show some structural similarity to certain multidrug resistance RND (resistance, nodulation and division) -type pumps with 12 TMSs (i.e. AcrF and ActII-3) (Saier et al., 1998; Tseng et al., 1999). Based on this homology, it is possible that the SecDF complex fulfills a transport function, such as removal of the signal peptide or phospholipids from the aqueous protein-conducting pore formed by the translocase.

4.6 Conserved Protein Translocases in Bacteria, Eukaryotes, and Archaea

In recent years, homologues of the bacterial SecY and SecE have been identified in eukaryotes and archaea (Hartmann et al., 1994; Jungnickel et al., 1994; Schatz and Dobberstein, 1996; Pohlschroder et al., 1997). In eukaryotes, these homologues form a trimeric complex, called Sec61p, which is involved in protein translocation into the ER. The subunits of Sec61p are called Sec61α, Sec61β, and Sec61γ. Sec61α and Sec61γ are homologous to the bacterial SecY and SecE, respectively. Homologues of Sec61β have only been identified in eukaryotes (Esnault et al., 1993); however, although it shows no sequence similarity, Sec61β is probably functionally homologous to SecG. The ER of *Saccharomyces cerevisiae* contains two trimeric complexes, Sec61p and Ssh1p, which are involved in preprotein translocation. In this case, Sec61p consists of Sec61α, Sec61β and the Sss1p protein. In the Ssh1p complex, the α- and β-subunits, Ssh1p and Sbh2p are close homologues of Sec61α and Sec61β, while the γ-subunit, Sss1p is common for both trimeric complexes (Finke et al., 1996). In mammals, protein translocation occurs mainly co-translationally and is driven by chain elongation in the ribosome (pushing force). During this process, the Sec61α subunit of the Sec61p complex associates directly with the large subunit of the ribosome. Such interaction has also been demonstrated for the eubacterial SecY protein (Prinz et al., 2000), but it is not clear if chain elongation at the ribosome suffices to drive preproteins across the membrane (Neumann-Haefelin et al., 2000). In contrast, protein translocation in *S. cerevisiae* can occur post-translationally and is then driven by a lumenal chaperone called Kar2p/Bip ("pulling") that associates with the Sec62/63p complex. In contrast, preprotein translocation in eubacteria is driven by the hydrolysis of ATP by SecA which "pushes" the protein in a stepwise manner across the membrane, as will be discussed later (see Figure 4). (For an excellent review of protein translocation in eukaryotes, see Johnson and van Waes, 1999.)

Archaea contain homologues of Sec61α, Sec61β, and Sec61γ. Translocation in archaea may be co-translational and driven by chain elongation, since no homologues of SecA or BiP/Kar2p have been identified. SecE and SecY homologues have also been identified in the chloroplasts of higher plants, cyanobacteria, plastids and algae (Scaramuzzi et al., 1992b; Muller et al., 1994; Vogel et al., 1996), while higher plants such as *Arabidopsis thaliana* contain both a Sec61α and a SecY homologue. Sec61α is present in the ER, whereas the SecY protein is present in the thylakoid membrane of the chloroplast (Laidler et al., 1995). In thylakoids, translocation can be both co- and post-translational. Both homologues of SecA

(Scaramuzzi et al., 1992a; Nakai et al., 1994; Nohara et al., 1995) and components of SRP (Franklin and Hoffman, 1993; Packer and Howe, 1996; Kogata et al., 1999) have been identified in these organelles, which suggests that translocation occurs mainly according to the mechanisms defined in bacteria. It therefore appears that the basic structure of the translocation channel is remarkably conserved among all three kingdoms of life, whereas the energetic mechanism by which protein translocation takes place may differ.

5 Role of Lipids in Protein Translocation

The *E. coli* inner membrane contains about 70% phosphatidylethanolamine (PE) and 30% anionic phospholipids, mainly phosphatidylglycerol (PG) and cardiolipin (CL). These phospholipids play an important role in protein translocation. With the aid of an *E. coli* mutant in which the synthesis of the major anionic membrane phospholipids could be controlled, it was shown that the rate of preprotein translocation is directly proportional to PG content (Kusters et al., 1991). Only the negative charge is essential, as any anionic phospholipids can restore protein translocation in PG-depleted inner membrane vesicles (Kusters et al., 1991). Anionic phospholipids are also needed for functional reconstitution of the SecYE(G) complex into liposomes (Tokuda et al., 1990; Akimaru et al., 1991; Hendrick and Wickner, 1991; van der Does et al., 2000). Elevated levels of SecA can counteract the reduced translocation activity caused by low levels of anionic phospholipids (Kusters et al., 1992). On the other hand, anionic phospholipids stimulate the binding and penetration of SecA into the membrane (Breukink et al., 1992; Ulbrandt et al., 1992). Unlike the binding of SecA to the SecYEG complex, binding of SecA to the lipid surface occurs with low affinity. The catalytic function of the lipid-bound SecA in protein translocation remains to be elucidated since the SecYEG-bound SecA is shielded from lipids and so cannot be labeled with photoreactive lipid probes (Eichler et al., 1997; van Voorst et al., 1998).

Anionic phospholipids are needed for a functional interaction of the signal sequence with the membrane (Keller et al., 1992, 1996; Phoenix et al., 1993). As signal sequences contain at least one or two positive charges at their N-terminus, it has been hypothesized that the negatively charged phospholipid headgroups interact electrostatically with the positive charges in the signal sequence. Indeed, removal of the positive charges in the signal sequence reduces the translocation efficiency of preproteins, and this correlates with the ability of synthetic signal peptides to bind to membranes composed of anionic phospholipids. The loss of activity due to a reduction of positive charges in the N-domain can be compensated by an increase in the hydrophobicity of the H-region (Hikita and Mizushima, 1992). Strikingly, this phenomenon is paralleled by a reduction in the requirement for anionic phospholipids, suggesting that the signal sequence–membrane interaction is governed both by electrostatic and hydrophobic interactions.

Nonbilayer lipids, such as PE, are also needed for efficient protein transport (Rietveld et al., 1995). Depletion of PE in the *E. coli* cell is accompanied by an increase in the amount of PG and CL, though in the presence of high concentrations of Mg^{2+} or other divalent cations, this phenomenon is not lethal. Mg^{2+} forces the type II lipid CL into a nonbilayer conformation (Rand and Sengupta, 1972), thus compensating for the loss of PE. *In vitro*, preprotein translocation into PE-depleted inner membrane vesicles can be re-activated by the addition of Mg^{2+} or by the re-introduction of PE (Rietveld et al., 1995). In contrast, protein translocation can be reconstituted into liposomes that do not contain nonbilayer lipids (Hendrick and Wickner, 1991), but the inclusion of nonbilayer lipids such as PE or diacylglycerol strongly enhances the protein translocation efficiency (van der Does et al., 2000). Thus, unlike the absolute anionic phospholipid requirement, nonbilayer lipids only stimulate translocation. Although speculative, the nonbilayer lipids may facilitate insertion of the signal sequence into the membrane or affect the conformation or oligomeric state of the SecYEG complex.

6 Mechanism of Protein Translocation

Preprotein translocation is driven by ATP hydrolysis (Chen and Tai, 1986) and the PMF (Daniels et al., 1981; Enequist et al., 1981; Bayan et al., 1993). *In vivo*, both the PMF and ATP hydrolysis are essential for protein translocation (Bakker and Randall, 1984; Geller et al., 1986), and the components of the PMF, $\Delta\psi$, and the transmembrane gradient of protons, ΔpH, participate in the translocation process (Bakker and Randall, 1984; Yamane et al., 1987; Driessen and Wickner, 1991; Driessen, 1992). *In vitro*, however, whilst ATP hydrolysis is able to drive protein translocation alone (Yamane et al., 1987), the PMF is only stimulatory (Geller and Green, 1989; Yamada et al., 1989; Schiebel et al., 1991).

ATP and the PMF function at different stages of protein translocation (Geller and Green, 1989; Tani et al., 1989). ATP is essential for the initiation of preprotein translocation, while the PMF can drive translocation at the later stages and provide directionality to the process (Tani et al., 1990; Driessen and Wickner, 1991; Schiebel et al., 1991).

6.1 ATP-driven Translocation

The mechanism by which ATP hydrolysis drives the movement of a preprotein across the cytoplasmic membrane has been partially resolved. ATP-driven translocation is a stepwise process (Figure 4). First, the preprotein is targeted to SecA bound with high affinity to SecYEG. Preprotein binding to SecA promotes the exchange of ADP for ATP (Lill et al., 1989; Matsuyama et al., 1990b). Upon ATP-binding, the conformation of SecA changes from a compact ADP-bound state into a more extended ATP-bound state (den Blaauwen et al., 1996, 1999), and this may lead to the insertion of SecA domains into the membrane (Economou and Wickner, 1994; Economou et al., 1995). Concomitant with ATP binding, the chaperone SecB is released from its association with SecA (Fekkes et al., 1997), and a topological inversion of SecG takes place (Nishiyama et al., 1996). This initial event results in translocation of a loop of the signal sequence and early mature domain to the extent that the signal sequence can be processed by a signal peptidase (Schiebel et al., 1991). Subsequent hydrolysis of ATP releases the bound preprotein (Schiebel et al., 1991) and the SecA conformation reverses into its compact ADP-bound state, leading to membrane de-insertion. Rebinding of SecA to the partially translocated preprotein results in the translocation of another segment of about 2.0–2.5 kDa. Exchange of ADP for ATP leads to a new cycle of conformational changes, in turn allowing the translocation of another of 2.0–2.5 kDa of polypeptide mass (Schiebel et al., 1991; Uchida et al., 1995; van der Wolk et al., 1997). A complete catalytic cycle of the preprotein translocase thus permits the stepwise translocation of 5 kDa polypeptide segments by two consecutive events, i.e., ~2.5 kDa upon binding of the polypeptide by SecA, and another 2.5 kDa upon binding of ATP to SecA (Economou and Wickner, 1994; van der Wolk et al., 1997). The translocation inhibitor azide blocks the de-insertion step (van der Wolk et al., 1997). Although it is generally accepted that conformational changes in SecA drive protein translocation, one hypothesis proposes that these changes lead to a membrane insertion–deinsertion cycle of SecA (Economou and Wickner, 1994; Economou et al., 1995). Insertion of the SecA domain into the membrane has been inferred from the formation of a membrane-protected 30-kDa and 65-kDa N- and C-terminal protease-resistant fragments (Eichler and Wickner, 1997). These fragments cover almost the entire mass of SecA, and it has been argued that SecA penetrates the membrane to expose these domains to the periplasmic face of the membrane. Provided that the SecA–SecY interaction is maintained, the 30-kDa fragment can also be found in detergent solution, suggesting that such fragments are not membrane-protected (van der Does et al., 1998). There does not seem to be any strict correlation between translocation and the formation of these fragments. For instance, the overproduction of SecDFYajC, stabilizes the 30-kDa fragment (Kim et al., 1994; Economou et al., 1995) and slows down release of SecA from the membrane (Duong and Wickner, 1997b; Eichler and Wickner, 1997) without affecting the catalytic activity of the translocase. In contrast, PrlA mutants of SecY bind SecA more tightly and enhance the translocation activity (van der Wolk et al., 1998; de Keyzer et al., 2002), but a reduced amount of the 30-kDa fragment is formed (Nishiyama et al., 1999). Alternatively, the conformational changes in the SecA protein may drive membrane insertion of the preprotein domain only, and thus result in the stepwise movement of the polypeptide chain across the membrane.

6.2
Proton Motive Force-driven Translocation

The mechanism by which the PMF acts on protein translocation is largely unknown. Initially, it was believed that the $\Delta\psi$ functions via an electrophoretic mechanism, and drives the translocation of negatively charged residues in the preprotein across the membrane. The charge distribution of the mature domain, especially around the signal sequence, strongly affects the PMF requirement for protein translocation (Cao et al., 1995), but the translocation of a completely uncharged preprotein is also stimulated by the PMF (Yamada et al., 1989; Kato et al., 1992). It is therefore unlikely that the $\Delta\psi$ and the PMF act solely via an electrophoretic mechanism.

Several lines of evidence suggest that the PMF acts on the membrane components of the translocation machinery. First, the PMF stimulates SecA membrane retraction and release (Nishiyama et al., 1999). The latter corroborates with the observation that PMF-driven translocation occurs when SecA is no longer bound to the translocating preprotein (Schiebel et al., 1991), and is consistent with the observation that the PMF reduces the amount of ATP needed to translocate preproteins (Driessen, 1992). The PMF lowers the apparent K_m value of the translocation reaction for ATP (Shiozuka et al., 1990), possibly by stimulating the release of ADP from SecA. This step may be rate-limiting in the SecA catalytic cycle. Second, in PrlA strains, the translocation of preproteins becomes less dependent on the PMF

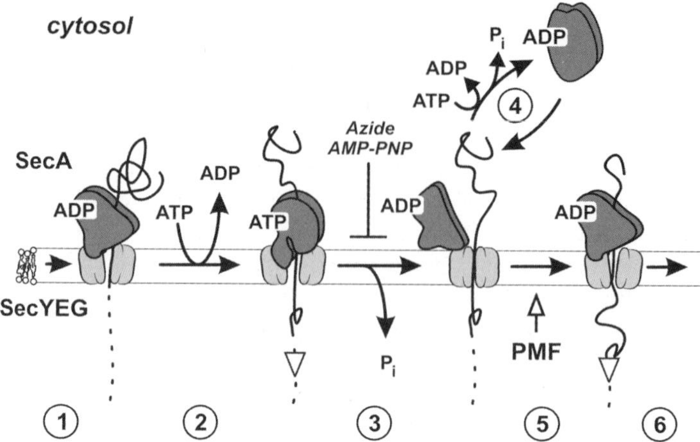

Fig. 4 Schematic model for SecA-mediated protein translocation. Protein translocation is a stepwise process. (1) A translocation intermediate is associated with ADP-bound SecA. (2) Next, the ADP is exchanged for ATP, resulting in a conformational change of SecA with the concomitant translocation of the bound preprotein segment of ~2.5 kDa (~20 amino acids). This stage of the translocation reaction is short-lived and only results in the formation of translocation intermediates when trapped with the nonhydrolyzable ATP analogue AMP-PNP or sodium azide. (3) Upon the hydrolysis of ATP, SecA reverses its conformational change and releases the preprotein. (4) The SecYEG-bound SecA can now be exchanged with cytosolic SecA in a process that requires the hydrolysis of another ATP molecule. (5) SecA can re-bind the translocation intermediate, and this suffices to translocate another 2.5 kDa of the polypeptide. The total step-size resulting from the SecA reaction cycle is ~5 kDa. Multiple repeats of the ATP-dependent catalytic cycle of SecA permit the step-wise translocation of the preprotein. The PMF can drive translocation when SecA is not associated with the preprotein, but the step-size is not known.

(Nouwen et al., 1996). It has been postulated that the PMF modulates the opening or formation of the translocation channel. It is not clear if this is the prime mechanism by which PrlA mutations function, since the *prlA4* mutation of SecY also causes an increased affinity for SecA (van der Wolk et al., 1998; de Keyzer et al., 2002) and a destabilization of the SecY–SecE interaction (Duong and Wickner, 1999). Tighter binding of SecA to the mutant SecYEG protein therefore will favor ATP-dependent translocation and in this way reduce the requirement for the PMF. However, whereas the PMF reduces the amount of membrane-inserted SecA (as monitored by a protease-protected fragment) in the wild-type protein, the amount of membrane-inserted SecA is low under all conditions (Nishiyama et al., 1999). This suggests that the mutant SecYEG complex has a conformation that is similar to the conformation of the wild-type complex in the presence of a PMF, and this results in an altered SecA binding affinity and possibly also the SecA conformational state.

6.3
Dynamics of the Protein-conducting Channel

Both the idle surface-bound state and the active "membrane-inserted" state of SecYEG-bound SecA are largely shielded from the phospholipid acyl chains (Eichler et al., 1997; van Voorst et al., 1998). Since dimeric SecA is a very large molecule (15 nm long and 8 nm wide; Shilton et al., 1998), this remarkable finding can only be understood if the SecA is part of an oligomeric assembly of the SecYEG protomer. Indeed, high-resolution electron microscopy of *B. subtilis* SecYE (Meyer et al., 1999) suggests an oligomeric assembly of translocase subunits. Two-dimensional crystals of the *E. coli* SecYEG show a dimeric arrangement (Collinson et al., 2001), and this state is also stable in blue native gel analysis where it is found to be stably associated with a preprotein translocation intermediate (Bessonneau et al., 2002). Using *E. coli* translocase, it has been shown that SecA recruits multiple SecYEG complexes upon binding of a nonhydrolyzable ATP analogue to form a large oligomeric structure (Manting et al., 2000). This large oligomeric structure has a diameter of 10.5 nm and exhibits an ~5 nm central depression, which is possibly a pore. Mass measurements and quantitative immunoblotting indicate that the oligomeric structure consists of a SecA dimer and a SecYEG tetramer, and that a similar structure is formed when the translocase is actively involved in preprotein translocation (Manting et al., 2000). Such an oligomeric organization of the SecYEG complex explains why SecY and SecE do not dissociate *in vivo* (Joly et al., 1994), whereas they appear to interact dynamically in a genetic screen based on suppressor-directed inactivation (Bieker-Brady and Silhavy, 1992). Within a single SecYEG protomer, SecY and SecE will remain physically associated, but within the oligomeric assembly they may appear as dissociable when the oligomer disassembles. The conditions that elicit oligomer formation also induce a pairwise contact between SecE subunits of the translocase (Kaufmann et al., 1999). The presence of a large pore in the translocase may be deleterious to the cells as, for its size, it might be predicted that this allows passage of small solutes and protons. Since the oligomerization of the SecYEG complex requires an active SecA, it might be speculated that the large channel structure accommodates part of the SecA protein. In addition, the large periplasmic domains that are present in SecD and SecF might also act as a valve to prevent leakage of small solutes and protons during translocation.

The translocase is a dynamic complex. When translocation is slow – for example at low temperature or at low ATP concentration – distinct preprotein translocation intermediates accumulate (Schiebel et al., 1991; Uchida et al., 1995; van der Wolk et al., 1997). However, the deletion or relocation of hydrophobic segments in the mature domain of the preprotein can significantly alter the pattern of translocation intermediates (Sato et al., 1997b). In this respect, intermediate stages of translocation are reversible (Schiebel et al., 1991), and in the absence of an energy source a reverse movement may take place (Driessen, 1992). This reverse translocation may also uncouple the SecA ATPase from the translocation progress, which is particularly notable when a non-translocatable polypeptide moiety is bound to the preprotein. The presence of hydrophobic segments in preproteins may result in a translocation arrest (Duong and Wickner, 1998; Saaf et al., 1998), as might the presence of positive charges around a hydrophobic domain (Yamane et al., 1990). Such polypeptide regions may elicit the retraction of SecA from the bound preprotein, and thereby allow the hydrophobic domain to partition into the lipid bilayer (Sato et al., 1997a). Small and moderate hydrophobic regions may escape membrane insertion as a result of rapid translocation (Duong and Wickner, 1998), while more hydrophobic regions may laterally leave the translocation pore formed by the SecYEG complex (von Heijne, 1997). It is not clear if this is a passive or active process that requires additional proteins. The recently discovered integral membrane protein YidC crosslinked to hydrophobic transmembrane segments (Houben et al., 2000; Samuelson et al., 2000; Scotti et al., 2000; van der Laan et al., 2001) would be a good candidate for the removal of hydrophobic segments from the SecYEG channel. Strikingly, YidC has been found to associate stably with the SecDFYajC complex (Nouwen and Driessen, 2002), suggesting a role for this complex in membrane protein insertion and partitioning. Alternatively, hydrophobic segments may simply be released into the lipid bilayer by dissociation of the oligomeric SecYEG pore complex once SecA has retracted.

6.4 Regulation of Protein Translocation

In *E. coli*, none of the integral membrane subunits of the translocase seems to be regulated in relation to the secretion demands of the cell (Schatz et al., 1991; Pogliano and Beckwith, 1994; Rensing and Maier, 1994; Bost and Belin, 1995), but the relative amounts seem well balanced. SecY not associated with SecE is readily proteolyzed by FtsH, and in this manner a balance of the components is maintained (Matsuyama et al., 1990a; Taura et al., 1993; Sagara et al., 1994). Estimates for the number of translocase subunits in the cell vary tremendously, the following values having been reported: SecY, 100–400 (Matsuyama et al., 1990a, 1992; Seoh and Tai, 1997); SecE, 100–600 (Matsuyama et al., 1990a, 1992; Schatz et al., 1991; Seoh and Tai, 1997); SecG, 100–1000 (Matsuyama et al., 1990a; Nishiyama et al., 1995); SecD, less than 30 (Pogliano and Beckwith, 1994) and up to 450–900 (Matsuyama et al., 1992); and SecF, 30–60 (Matsuyama et al., 1992; Pogliano and Beckwith, 1994). SecD and SecF thus appear to be present in substoichiometric amounts compared with SecY, SecE, and SecG.

SecA expression is tightly regulated in relation to the protein secretion demand and secretion stress. Normally, SecA is present at about 500 to 5000 copies per cell (Oliver and Beckwith, 1982; Matsuyama et al., 1992; Seoh and Tai, 1997), but mutations that interfere with the activity of the translocase, the high level expression of secretory proteins or expression of poorly translocated preproteins, may cause a 10- to 20-fold increase in cellular SecA (Riggs et al., 1988; Rollo and Oliver, 1988). SecA regulates its own translation (Rollo and Oliver, 1988; Dolan and Oliver, 1991) by binding to a large secretion-responsive element that spans a 96 nucleotide-long region (Salavati and Oliver, 1997) located near the 5′ end of the *secM-secA* mRNA (Schmidt and Oliver, 1989; Schmidt et al., 1991). As the ribosome and SecA binding site on the *secM-secA* mRNA are overlapping, a simple competition mechanism might regulate the translation initiation step (Salavati and Oliver, 1995). However, SecA expression seems to be more complex. *SecM* (previously termed *geneX*), the first gene in the *secM-secA* operon, encodes a nonessential secretory protein (Rajapandi et al., 1991; Oliver et al., 1998) that is rapidly degraded by the periplasmic tail-specific protease (Nakatogawa and Ito, 2001). Due to a C-terminal motif that delays the transit of SecM through the ribosomal protein exit tunnel, nascent SecM undergoes self-translation arrest in the cytosol (Nakatogawa and Ito, 2001; Sarker and Oliver, 2002). The translational arrest prevents formation of an RNA helix that normally blocks secA translational initiation. As the translational pause is controlled by the rate of translocation of nascent SecM, the latter thus functions as a type of secretion sensor. Although this may well be the main mechanism by which SecA expression is controlled, the Walker B motif of NBS-1 of SecA also shows some similarity to the DEAD-box found in RNA-binding proteins such as helicases (Koonin and Gorbalenya, 1992). SecA has been shown to possess RNA helicase activity (Park et al., 1997), and this may also help in unwinding the *secM-secA* mRNA structure, thereby autoregulating its own translation (Salavati and Oliver, 1997).

7 Outlook and Perspectives

During the past few decades, a detailed insight into the mechanism of bacterial preprotein translocation has been obtained through a combination of genetic and biochemical studies. In recent years, the structures have been solved of both SecB (Xu et al., 2000) and SecA (Weinkauf et al., 2001), whereas only a low-resolution structure has been obtained for the SecYEG dimer (Collinson et al., 2001). In future research, it will be possible to relate function to structure, thereby permitting a detailed insight into how preproteins cross the cytoplasmic membrane. It has also become apparent that although the general secretion system is responsible for the translocation of most proteins across the cytoplasmic membrane, other systems are involved in the translocation of a specific subset of proteins, including the folded redox proteins (Berks et al., 2000; Robinson and Bolhuis, 2001). It will be of particular interest to understand how these large folded proteins are able to cross the cytoplasmic membrane, without disrupting its barrier function.

Acknowledgments

These studies were supported by the Council for Chemical Sciences of the Netherlands Organization for Scientific Research (CW-NWO), and by the Division of Earth and Life Sciences (ALW).

8
References

Akimaru, J., Matsuyama, S., Tokuda, H., Mizushima, S. (1991) Reconstitution of a protein translocation system containing purified SecY, SecE, and SecA from *Escherichia coli*, *Proc. Natl. Acad. Sci. USA* **88**, 6545–6549.

Akita, M., Sasaki, S., Matsuyama, S., Mizushima, S. (1990) SecA interacts with secretory proteins by recognizing the positive charge at the amino terminus of the signal peptide in *Escherichia coli*, *J. Biol. Chem.* **265**, 8164–8169.

Akita, M., Shinaki, A., Matsuyama, S., Mizushima, S. (1991) SecA, an essential component of the secretory machinery of *Escherichia coli*, exists as homodimer, *Biochem. Biophys. Res. Commun.* **174**, 211–216.

Akiyama, Y., Ito, K. (1987) Topology analysis of the SecY protein, an integral membrane protein involved in protein export in *Escherichia coli*, *EMBO J.* **6**, 3465–3470.

Arkowitz, R. A., Wickner, W. (1994) SecD and SecF are required for the proton electrochemical gradient stimulation of preprotein translocation, *EMBO J.* **13**, 954–963.

Baba, T., Taura, T., Shimoike, T., Akiyama, Y., Yoshihisa, T., Ito, K. (1994) A cytoplasmic domain is important for the formation of a SecY-SecE translocator complex, *Proc. Natl. Acad. Sci. USA* **91**, 4539–4543.

Bairoch, A., Bucher, P., Hofmann, K. (1997) The PROSITE database, its status in 1997, *Nucleic Acids Res.* **25**, 217–221.

Bakker, E. P., Randall, L. L. (1984) The requirement for energy during export of beta-lactamase in *Escherichia coli* is fulfilled by the total protonmotive force, *EMBO J.* **3**, 895–900.

Batey, R. T., Rambo, R. P., Lucast, L., Rha, B., Doudna, J. A. (2000). Crystal structure of the ribonucleoprotein core of the signal recognition particle, *Science* **287**, 1232–1239.

Baud, C., Karamanou, S., Sianidis, G., Vrontou, E., Politou, A. S., Economou, A. (2002) Allosteric communication between signal peptides and the SecA protein DEAD motor ATPase domain, *J. Biol. Chem.* **277**, 13724–13731.

Bayan, N., Schrempp, S., Joliff, G., Leblon, G., Shechter, E. (1993). Role of the protonmotive force and of the state of the lipids in the *in vivo* protein secretion in *Corynebacterium glutamicum*, a gram-positive bacterium, *Biochim. Biophys. Acta* **1146**, 97–105.

Beck, K., Wu, L. F., Brunner, J., Muller, M. (2000) Discrimination between SRP- and SecA/SecB-dependent substrates involves selective recognition of nascent chains by SRP and trigger factor, *EMBO J.* **19**, 134–143.

Berks, B. C., Sargent, F., Palmer, T. (2000) The Tat protein export pathway, *Mol. Microbiol.* **35**, 260–274.

Bessonneau, P., Besson, V., Collinson, I., Duong, F. (2002) The SecYEG preprotein translocation channel is a conformationally dynamic and dimeric structure, *EMBO J.* **21**, 995–1003.

Bieker-Brady, K., Silhavy, T. J. (1992) Suppressor analysis suggests a multistep, cyclic mechanism for protein secretion in *Escherichia coli*, *EMBO J.* **11**, 3165–3174.

Bolhuis, A., Broekhuizen, C. P., Sorokin, A., van Roosmalen, M. L., Venema, G., Bron, S., Quax, W. J., van Dijl, J. M. (1998) SecDF of *Bacillus subtilis*, a molecular Siamese twin required for the efficient secretion of proteins, *J. Biol. Chem.* **273**, 21217–21224.

Bost, S., Belin, D. (1995) A new genetic selection identifies essential residues in SecG, a component of the *Escherichia coli* protein export machinery, *EMBO J.* **14**, 4412–4421.

Bost, S., Belin, D. (1997) *prl* mutations in the *Escherichia coli secG* gene, *J. Biol. Chem.* **272**, 4087–4093.

Breukink, E., Demel, R. A., de Korte-Kool, G., de Kruijff, B. (1992) SecA insertion into phospholipids is stimulated by negatively charged lipids and inhibited by ATP: a monolayer study, *Biochemistry* **31**, 1119–1124.

Breukink, E., Nouwen, N., van Raalte, A., Mizushima, S., Tommassen, J., de Kruijff, B. (1995) The C terminus of SecA is involved in both lipid binding and SecB binding, *J. Biol. Chem.* **270**, 7902–7907.

Brundage, L., Hendrick, J. P., Schiebel, E., Driessen, A. J. M., Wickner, W. (1990) The purified *E. coli* integral membrane protein SecY/E is sufficient for reconstitution of SecA-dependent precursor protein translocation, *Cell* **62**, 649–657.

Cao, G., Kuhn, A., Dalbey, R. E. (1995) The translocation of negatively charged residues across the membrane is driven by the electrochemical potential: evidence for an electrophoresis-like membrane transfer mechanism, *EMBO J.* **14**, 866–875.

Cerretti, D. P., Dean, D., Davis, G. R., Bedwell, D. M., Nomura, M. (1983) The *spc* ribosomal protein operon of *Escherichia coli*: sequence and cotranscription of the ribosomal protein genes and a protein export gene, *Nucleic Acids Res.* **11**, 2599–2616.

Chen, L., Tai, P. C. (1986) Effects of nucleotides on ATP-dependent protein translocation into *Escherichia coli* membrane vesicles, *J. Bacteriol.* **168**, 828–832.

Christie, P. J. (2001) Type IV secretion: intercellular transfer of macromolecules by systems ancestrally related to conjugation machines, *Mol. Microbiol.* **40**, 294–305.

Collinson, I., Breyton, C., Duong, F., Tziatzios, C., Schubert, D., Or, E., Rapoport, T., Kuhlbrandt, W. (2001) Projection structure and oligomeric properties of a bacterial core protein translocase, *EMBO J.* **20**, 2462–2471.

Cunningham, K., Wickner, W. (1989) Specific recognition of the leader region of precursor proteins is required for the activation of translocation ATPase of *Escherichia coli*, *Proc. Natl. Acad. Sci. USA* **86**, 8630–8634.

Dalbey, R. E., Lively, M. O., Bron, S., van Dijl, J. M. (1997) The chemistry and enzymology of the type I signal peptidases, *Protein Sci.* **6**, 1129–1138.

Daniels, C. J., Bole, D. G., Quay, S. C., Oxender, D. L. (1981) Role for membrane potential in the secretion of protein into the periplasm of *Escherichia coli*, *Proc. Natl. Acad. Sci. USA* **78**, 5396–5400.

de Keyzer, J., van der Does, C., Swaving, J., Driessen, A. J. M. (2002) The F286Y mutation of PrlA4 tempers the signal sequence suppressor phenotype by reducing the SecA binding affinity, *FEBS Lett.* **510**, 17–21.

de Leeuw, E., te Kaat, K., Moser, C., Menestrina, G., Demel, R., de Kruijff, B., Oudega, B., Luirink, J., Sinning, I. (2000) Anionic phospholipids are involved in membrane association of FtsY and stimulate its GTPase activity, *EMBO J.* **19**, 531–541

den Blaauwen, T., Fekkes, P., de Wit, J. G., Kuiper, W., Driessen, A. J. M. (1996) Domain interactions of the peripheral preprotein translocase subunit SecA, *Biochemistry* **35**, 11994–12004.

den Blaauwen, T., Terpetschnig, E., Lakowicz, J. R., Driessen, A. J. M. (1997) Interaction of SecB with soluble SecA, *FEBS Lett.* **416**, 35–38.

den Blaauwen, T., van der Wolk, J. P., van der Does, C., van Wely, K. H., Driessen, A. J. M. (1999) Thermodynamics of nucleotide binding to NBS-I of the *Bacillus subtilis* preprotein translocase subunit SecA, *FEBS Lett.* **458**, 145–150.

Ding, H., Mukerji, I., Oliver, D. (2001) Lipid and signal peptide-induced conformational changes within the C-domain of *Escherichia coli* SecA protein, *Biochemistry* **40**, 1835–1843.

Dolan, K. M., Oliver, D. B. (1991) Characterization of *Escherichia coli* SecA protein binding to a site on its mRNA involved in autoregulation, *J. Biol. Chem.* **266**, 23329–23333.

Downing, W. L., Sullivan, S. L., Gottesman, M. E., Dennis, P. P. (1990) Sequence and transcriptional pattern of the essential *Escherichia coli secE-nusG* operon, *J. Bacteriol.* **172**, 1621–1627.

Driessen, A. J. M. (1992) Precursor protein translocation by the *Escherichia coli* translocase is directed by the protonmotive force, *EMBO J.* **11**, 847–853.

Driessen, A. J. M. (1993) SecA, the peripheral subunit of the *Escherichia coli* precursor protein translocase, is functional as a dimer, *Biochemistry* **32**, 13190–13197.

Driessen, A. J. M. (2001) SecB, a molecular chaperone with two faces, *Trends Microbiol.* **9**, 193–196.

Driessen, A. J. M., Wickner, W. (1990) Solubilization and functional reconstitution of the protein-translocation enzymes of *Escherichia coli*, *Proc. Natl. Acad. Sci. USA* **87**, 3107–3111.

Driessen, A. J. M., Wickner, W. (1991) Proton transfer is rate-limiting for translocation of precursor proteins by the *Escherichia coli* translocase, *Proc. Natl. Acad. Sci. USA* **88**, 2471–2475.

Driessen, A. J. M., Brundage, L., Hendrick, J. P., Schiebel, E., Wickner, W. (1991) Preprotein translocase of *Escherichia coli*: solubilization, purification, and reconstitution of the integral membrane subunits SecY/E, *Methods Cell Biol.* **34**, 147–165.

Driessen, A. J. M., Manting, E. H., van der Does, C. (2001) The structural basis of protein targeting and translocation in bacteria, *Nature Struct. Biol.* **8**, 492–498.

Duong, F., Wickner, W. (1997a) Distinct catalytic roles of the SecYE, SecG and SecDFyajC subunits of preprotein translocase holoenzyme, *EMBO J.* **16**, 2756–2768.

Duong, F., Wickner, W. (1997b) The SecDFyajC domain of preprotein translocase controls preprotein movement by regulating SecA membrane cycling, *EMBO J.* **16**, 4871–4879.

Duong, F., Wickner, W. (1998) Sec-dependent membrane protein biogenesis: SecYEG, preprotein hydrophobicity and translocation kinetics control the stop-transfer function, *EMBO J.* **17**, 696–705.

Duong, F., Wickner, W. (1999) The PrlA and PrlG phenotypes are caused by a loosened association among the translocase SecYEG subunits, *EMBO J.* **18**, 3263–3270.

Economou, A., Wickner, W. (1994) SecA promotes preprotein translocation by undergoing ATP-driven cycles of membrane insertion and deinsertion, *Cell* **78**, 835–843.

Economou, A., Pogliano, J. A., Beckwith, J., Oliver, D. B., Wickner, W. (1995) SecA membrane cycling at SecYEG is driven by distinct ATP binding and hydrolysis events and is regulated by SecD and SecF, *Cell* **83**, 1171–1181.

Eichler, J., Wickner, W. (1997) Both an N-terminal 65-kDa domain and a C-terminal 30-kDa domain of SecA cycle into the membrane at SecYEG during translocation, *Proc. Natl. Acad. Sci. USA* **94**, 5574–5581.

Eichler, J., Wickner, W. (1998) The SecA subunit of *Escherichia coli* preprotein translocase is exposed to the periplasm, *J. Bacteriol.* **180**, 5776–5779.

Eichler, J., Brunner, J., Wickner, W. (1997) The protease-protected 30 kDa domain of SecA is largely inaccessible to the membrane lipid phase, *EMBO J.* **16**, 2188–2196.

Emr, S. D., Hanley-Way, S., Silhavy, T. J. (1981) Suppressor mutations that restore export of a protein with a defective signal sequence, *Cell* **23**, 79–88.

Enequist, H. G., Hirst, T. R., Harayama, S., Hardy, S. J., Randall, L. L. (1981) Energy is required for maturation of exported proteins in *Escherichia coli*, *Eur. J. Biochem.* **116**, 227–233.

Esnault, Y., Blondel, M. O., Deshaies, R. J., Scheckman, R., Kepes, F. (1993) The yeast *SSS1* gene is essential for secretory protein translocation and encodes a conserved protein of the endoplasmic reticulum, *EMBO J.* **12**, 4083–4093.

Fekkes, P., Driessen, A. J. M. (1999) Protein targeting to the bacterial cytoplasmic membrane, *Microbiol. Mol. Biol. Rev.* **63**, 161–173.

Fekkes, P., van der Does, C., Driessen, A. J. M. (1997) The molecular chaperone SecB is released from the carboxy- terminus of SecA during initiation of precursor protein translocation, *EMBO J.* **16**, 6105–6113.

Fekkes, P., de Wit, J. G., Boorsma, A., Friesen, R. H., Driessen, A. J. M. (1999) Zinc stabilizes the SecB binding site of SecA, *Biochemistry* **38**, 5111–5116.

Finke, K., Plath, K., Panzner, S., Prehn, S., Rapoport, T. A., Hartmann, E., Sommer, T. (1996) A second trimeric complex containing homologs of the Sec61p complex functions in protein transport across the ER membrane of *S. cerevisiae*, *EMBO J.* **15**, 1482–1494.

Flower, A. M. (2001) SecG function and phospholipid metabolism in *Escherichia coli*, *J. Bacteriol.* **183**, 2006–2012.

Flower, A. M., Doebele, R. C., Silhavy, T. J. (1994) PrlA and PrlG suppressors reduce the requirement for signal sequence recognition, *J. Bacteriol.* **176**, 5607–5614.

Flower, A. M., Osborne, R. S., Silhavy, T. J. (1995) The allele-specific synthetic lethality of *prlA-prlG* double mutants predicts interactive domains of SecY and SecE, *EMBO J.* **14**, 884–893.

Franklin, A. E., Hoffman, N. E. (1993) Characterization of a chloroplast homologue of the 54-kDa subunit of the signal recognition particle, *J. Biol. Chem.* **268**, 22175–22180.

Freymann, D. M., Keenan, R. J., Stroud, R. M., Walter, P. (1999) Functional changes in the structure of the SRP GTPase on binding GDP and Mg^{2+}GDP, *Nature Struct. Biol.* **6**, 793–801.

Gardel, C., Benson, S., Hunt, J., Michaelis, S., Beckwith, J. (1987) *secD*, a new gene involved in protein export in *Escherichia coli*, *J. Bacteriol.* **169**, 1286–1290.

Gardel, C., Johnson, K., Jacq, A., Beckwith, J. (1990) The *secD* locus of *E. coli* codes for two membrane proteins required for protein export, *EMBO J.* **9**, 3209–3216.

Geller, B. L., Green, H. M. (1989) Translocation of pro-OmpA across inner membrane vesicles of *Escherichia coli* occurs in two consecutive energetically distinct steps, *J. Biol. Chem.* **264**, 16465–16469.

Geller, B. L., Movva, N. R., Wickner, W. (1986) Both ATP and the electrochemical potential are required for optimal assembly of pro-OmpA into *Escherichia coli* inner membrane vesicles, *Proc. Natl. Acad. Sci. USA* **83**, 4219–4222.

Gill, D. R., Salmond, G. P. (1990) The identification of the *Escherichia coli ftsY* gene product: an unusual protein, *Mol. Microbiol.* **4**, 575–583.

Hanada, M., Nishiyama, K. I., Mizushima, S., Tokuda, H. (1994) Reconstitution of an efficient protein translocation machinery comprising SecA and the three membrane proteins, SecY, SecE, and SecG (p12), *J. Biol. Chem.* **269**, 23625–23631.

Hanada, M., Nishiyama, K., Tokuda, H. (1996) SecG plays a critical role in protein translocation in the absence of the proton motive force as well as at low temperature, *FEBS Lett.* **381**, 25–28.

Harris, C. R., Silhavy, T. J. (1999) Mapping an interface of SecY (PrlA) and SecE (PrlG) by using synthetic phenotypes and *in vivo* cross-linking, *J. Bacteriol.* **181**, 3438–3444.

Hartl, F. U., Lecker, S., Schiebel, E., Hendrick, J. P., Wickner, W. (1990) The binding cascade of SecB to SecA to SecY/E mediates preprotein targeting to the *E. coli* plasma membrane, *Cell* **63**, 269–279.

Hartmann, E., Sommer, T., Prehn, S., Gorlich, D., Jentsch, S., Rapoport, T. A. (1994) Evolutionary conservation of components of the protein translocation complex, *Nature* **367**, 654–657.

Hendrick, J. P., Wickner, W. (1991) SecA protein needs both acidic phospholipids and SecY/E protein for functional high-affinity binding to the *Escherichia coli* plasma membrane, *J. Biol. Chem.* **266**, 24596–24600.

Herskovits, A. A., Bochkareva, E. S., Bibi, E. (2000) New prospects in studying the bacterial signal recognition particle pathway, *Mol. Microbiol.* **38**, 927–939.

Hikita, C., Mizushima, S. (1992) The requirement of a positive charge at the amino terminus can be compensated for by a longer central hydrophobic stretch in the functioning of signal peptides, *J. Biol. Chem.* **267**, 12375–12379.

Homma, T., Yoshihisa, T., Ito, K. (1997) Subunit interactions in the *Escherichia coli* protein translocase: SecE and SecG associate independently with SecY, *FEBS Lett.* **408**, 11–15.

Houben, E. N., Scotti, P. A., Valent, Q. A., Brunner, J., de Gier, J. L., Oudega, B., Luirink, J. (2000) Nascent Lep inserts into the *Escherichia coli* inner membrane in the vicinity of YidC, SecY and SecA, *FEBS Lett.* **476**, 229–233.

Huie, J. L., Silhavy, T. J. (1995) Suppression of signal sequence defects and azide resistance in *Escherichia coli* commonly result from the same mutations in SecA, *J. Bacteriol.* **177**, 3518–3526.

Iino, T., Sako, T. (1988). Inhibition and resumption of processing of the staphylokinase in some *Escherichia coli prlA* suppressor mutants, *J. Biol. Chem.* **263**, 19077–19082.

Johnson, A. E., van Waes, M. A. (1999) The translocon: a dynamic gateway at the ER membrane, *Annu. Rev. Cell Dev. Biol.* **15**, 799–842.

Joly, J. C., Leonard, M. R., Wickner, W. T. (1994) Subunit dynamics in *Escherichia coli* preprotein translocase, *Proc. Natl. Acad. Sci. USA* **91**, 4703–4707.

Jungnickel, B., Rapoport, T. A., Hartmann, E. (1994) Protein translocation: common themes from bacteria to man, *FEBS Lett.* **346**, 73–77.

Kato, M., Tokuda, H., Mizushima, S. (1992) *In vitro* translocation of secretory proteins possessing no charges at the mature domain takes place efficiently in a protonmotive force-dependent manner, *J. Biol. Chem.* **267**, 413–418.

Kaufmann, A., Manting, E. H., Veenendaal, A. K., Driessen, A. J. M., van der Does, C. (1999) Cysteine-directed cross-linking demonstrates that helix 3 of SecE is close to helix 2 of SecY and helix 3 of a neighboring SecE, *Biochemistry* **38**, 9115–9125.

Keller, R. C., Killian, J. A., de Kruijff, B. (1992) Anionic phospholipids are essential for alpha-helix formation of the signal peptide of prePhoE upon interaction with phospholipid vesicles, *Biochemistry* **31**, 1672–1677.

Keller, R. C., ten Berge, D., Nouwen, N., Snel, M. M., Tommassen, J., Marsh, D., de Kruijff, B. (1996) Mode of insertion of the signal sequence of a bacterial precursor protein into phospholipid bilayers as revealed by cysteine-based site-directed spectroscopy, *Biochemistry* **35**, 3063–3071.

Kihara, A., Akiyama, Y., Ito, K. (1995) FtsH is required for proteolytic elimination of uncomplexed forms of SecY, an essential protein translocase subunit, *Proc. Natl. Acad. Sci. USA* **92**, 4532–4536.

Kim, Y. J., Rajapandi, T., Oliver, D. (1994) SecA protein is exposed to the periplasmic surface of the *E. coli* inner membrane in its active state, *Cell* **78**, 845–853.

Klose, M., Schimz, K. L., van der Wolk, J., Driessen, A. J. M., Freudl, R. (1993) Lysine 106 of the putative catalytic ATP-binding site of the *Bacillus subtilis* SecA protein is required for functional complementation of *Escherichia coli secA* mutants *in vivo*, *J. Biol. Chem.* **268**, 4504–4510.

Knoblauch, N. T., Rudiger, S., Schonfeld, H. J., Driessen, A. J. M., Schneider-Mergener, J.,

Bukau, B. (1999) Substrate specificity of the SecB chaperone, *J. Biol. Chem.* **274**, 34219–34225.

Kogata, N., Nishio, K., Hirohashi, T., Kikuchi, S., Nakai, M. (1999) Involvement of a chloroplast homologue of the signal recognition particle receptor protein, FtsY, in protein targeting to thylakoids, *FEBS Lett.* **447**, 329–333.

Kontinen, V. P., Tokuda, H. (1995) Overexpression of phosphatidylglycerophosphate synthase restores protein translocation in a *secG* deletion mutant of *Escherichia coli* at low temperature, *FEBS Lett.* **364**, 157–160.

Kontinen, V. P., Helander, I. M., Tokuda, H. (1996a) The *secG* deletion mutation of *Escherichia coli* is suppressed by expression of a novel regulatory gene of *Bacillus subtilis*, *FEBS Lett.* **389**, 281–284.

Kontinen, V. P., Yamanaka, M., Nishiyama, K., Tokuda, H. (1996b) Roles of the conserved cytoplasmic region and non-conserved carboxy-terminal region of SecE in *Escherichia coli* protein translocase, *J. Biochem. (Tokyo)* **119**, 1124–1130.

Koonin, E. V., Gorbalenya, A. E. (1992) Autogenous translation regulation by *Escherichia coli* ATPase SecA may be mediated by an intrinsic RNA helicase activity of this protein, *FEBS Lett.* **298**, 6–8.

Koronakis, V., Andersen, C., Hughes, C. (2001) Channel-tunnels, *Curr. Opin. Struct. Biol.* **11**, 403–407.

Kourtz, L., Oliver, D. (2000) Tyr-326 plays a critical role in controlling SecA-preprotein interaction, *Mol. Microbiol.* **37**, 1342–1356.

Kumamoto, C. A., Beckwith, J. (1983). Mutations in a new gene, *secB*, cause defective protein localization in *Escherichia coli*, *J. Bacteriol. 154*, 253–260.

Kumamoto, C. A., Francetic, O. (1993) Highly selective binding of nascent polypeptides by an *Escherichia coli* chaperone protein *in vivo*, *J. Bacteriol.* **175**, 2184–2188.

Kumamoto, C. A., Chen, L., Fandl, J., Tai, P. C. (1989) Purification of the *Escherichia coli secB* gene product and demonstration of its activity in an *in vitro* protein translocation system, *J. Biol. Chem.* **264**, 2242–2249.

Kusters, R., Dowhan, W., de Kruijff, B. (1991) Negatively charged phospholipids restore prePhoE translocation across phosphatidylglycerol-depleted *Escherichia coli* inner membranes, *J. Biol. Chem.* **266**, 8659–8662.

Kusters, R., Huijbregts, R., de Kruijff, B. (1992) Elevated cytosolic concentrations of SecA compensate for a protein translocation defect in *Escherichia coli* cells with reduced levels of negatively charged phospholipids, *FEBS Lett.* **308**, 97–100.

Laidler, V., Chaddock, A. M., Knott, T. G., Walker, D., Robinson, C. (1995) A SecY homolog in *Arabidopsis thaliana*. Sequence of a full-length cDNA clone and import of the precursor protein into chloroplasts, *J. Biol. Chem.* **270**, 17664–17667.

Lill, R., Cunningham, K., Brundage, L. A., Ito, K., Oliver, D., Wickner, W. (1989) SecA protein hydrolyzes ATP and is an essential component of the protein translocation ATPase of *Escherichia coli*, *EMBO J.* **8**, 961–966.

Liu, G., Topping, T. B., Randall, L. L. (1989) Physiological role during export for the retardation of folding by the leader peptide of maltose-binding protein, *Proc. Natl. Acad. Sci. USA* **86**, 9213–9217.

Luirink, J., High, S., Wood, H., Giner, A., Tollervey, D., Dobberstein, B. (1992) Signal-sequence recognition by an *Escherichia coli* ribonucleoprotein complex, *Nature* **359**, 741–743.

Manting, E. H., van der Does, C., Remigy, H., Engel, A., Driessen, A. J. M. (2000) SecYEG assembles into a tetramer to form the active protein translocation channel, *EMBO J.* **19**, 852–861.

Matsumoto, G., Yoshihisa, T., Ito, K. (1997) SecY and SecA interact to allow SecA insertion and protein translocation across the *Escherichia coli* plasma membrane, *EMBO J.* **16**, 6384–6393.

Matsumoto, G., Mori, H., Ito, K. (1998) Roles of SecG in ATP- and SecA-dependent protein translocation, *Proc. Natl. Acad. Sci. USA* **95**, 13567–13572.

Matsuyama, S., Akimaru, J., Mizushima, S. (1990a) SecE-dependent overproduction of SecY in *Escherichia coli*. Evidence for interaction between two components of the secretory machinery, *FEBS Lett.* **269**, 96–100.

Matsuyama, S., Kimura, E., Mizushima, S. (1990b) Complementation of two overlapping fragments of SecA, a protein translocation ATPase of *Escherichia coli*, allows ATP binding to its amino-terminal region, *J. Biol. Chem.* **265**, 8760–8765.

Matsuyama, S., Fujita, Y., Sagara, K., Mizushima, S. (1992) Overproduction, purification and characterization of SecD and SecF, integral membrane components of the protein translocation machinery of *Escherichia coli*, *Biochim. Biophys. Acta* **1122**, 77–84.

Matsuyama, S., Fujita, Y., Mizushima, S. (1993) SecD is involved in the release of translocated secretory proteins from the cytoplasmic membrane of *Escherichia coli*, *EMBO J.* **12**, 265–270.

Meyer, T. H., Ménétret, J. F., Breitling, R., Miller, K. R., Akey, C. W., Rapoport, T. A. (1999) The bacterial SecY/E translocation complex forms channel-like structures similar to those of the eukaryotic Sec61p complex, *J. Mol. Biol.* **285**, 1789–1800.

Miller, A., Wang, L., Kendall, D. A. (2002) SecB modulates the nucleotide-bound state of SecA and stimulates ATPase activity, *Biochemistry* **41**, 5325–5332.

Millman, J. S., Qi, H. Y., Vulcu, F., Bernstein, H. D., Andrews, D. W. (2001) FtsY binds to the *Escherichia coli* inner membrane via interactions with phosphatidylethanolamine and membrane proteins, *J. Biol. Chem.* **276**, 25982–25989.

Mitchell, C., Oliver, D. (1993) Two distinct ATP-binding domains are needed to promote protein export by *Escherichia coli* SecA ATPase, *Mol. Microbiol.* **10**, 483–497.

Mori, H., Ito, K. (2001) An essential amino acid residue in the protein translocation channel revealed by targeted random mutagenesis of SecY, *Proc. Natl. Acad. Sci. USA* **98**, 5128–5133.

Mori, H., Sugiyama, H., Yamanaka, M., Sato, K., Tagaya, M., Mizushima, S. (1998) Amino-terminal region of SecA is involved in the function of SecG for protein translocation into *Escherichia coli* membrane vesicles, *J. Biochem. (Tokyo)* **124**, 122–129.

Muller, M., Blobel, G. (1984) *In vitro* translocation of bacterial proteins across the plasma membrane of *Escherichia coli, Proc. Natl. Acad. Sci. USA* **81**, 7421–7425.

Muller, S. B., Rensing, S. A., Martin, W. F., Maier, U. G. (1994) cDNA cloning of a Sec61 homologue from the cryptomonad alga *Pyrenomonas salina, Curr. Genet.* **26**, 410–414.

Murphy, C. K., Beckwith, J. (1994) Residues essential for the function of SecE, a membrane component of the *Escherichia coli* secretion apparatus, are located in a conserved cytoplasmic region, *Proc. Natl. Acad. Sci. USA* **91**, 2557–2561.

Nakai, M., Goto, A., Nohara, T., Sugita, D., Endo, T. (1994) Identification of the SecA protein homolog in pea chloroplasts and its possible involvement in thylakoidal protein transport, *J. Biol. Chem.* **269**, 31338–31341.

Nakatogawa, H., Ito, K. (2001) Secretion monitor, SecM, undergoes self-translation arrest in the cytosol, *Mol. Cell* **7**, 185–192.

Nakatogawa, H., Mori, H., Ito, K. (2000a) Two independent mechanisms down-regulate the intrinsic SecA ATPase activity, *J. Biol. Chem.* **275**, 33209–33212.

Nakatogawa, H., Mori, H., Matsumoto, G., Ito, K. (2000b) Characterization of a mutant form of SecA that alleviates a SecY defect at low temperature and shows a synthetic defect with SecY alteration at high temperature, *J. Biochem.* **127**, 1071–1079.

Neumann-Haefelin, C., Schafer, U., Muller, M., Koch, H. G. (2000) SRP-dependent co-translational targeting and SecA-dependent translocation analyzed as individual steps in the export of a bacterial protein, *EMBO J.* **19**, 6419–6426.

Nishiyama, K., Mizushima, S., Tokuda, H. (1992) The carboxyl-terminal region of SecE interacts with SecY and is functional in the reconstitution of protein translocation activity in *Escherichia coli, J. Biol. Chem.* **267**, 7170–7176.

Nishiyama, K., Mizushima, S., Tokuda, H. (1993) A novel membrane protein involved in protein translocation across the cytoplasmic membrane of *Escherichia coli, EMBO J.* **12**, 3409–3415.

Nishiyama, K., Hanada, M., Tokuda, H. (1994) Disruption of the gene encoding p12 (SecG) reveals the direct involvement and important function of SecG in the protein translocation of *Escherichia coli* at low temperature, *EMBO J.* **13**, 3272–3277.

Nishiyama, K., Mizushima, S., Tokuda, H. (1995) Preferential interaction of SecG with SecE stabilizes an unstable SecE derivative in the *Escherichia coli* cytoplasmic membrane, *Biochem. Biophys. Res. Commun.* **217**, 217–223.

Nishiyama, K., Suzuki, T., Tokuda, H. (1996) Inversion of the membrane topology of SecG coupled with SecA-dependent preprotein translocation, *Cell* **85**, 71–81.

Nishiyama, K. I., Fukuda, A., Morita, K., Tokuda, H. (1999) Membrane deinsertion of SecA underlying proton motive force-dependent stimulation of protein translocation, *EMBO J.* **18**, 1049–1058.

Nohara, T., Nakai, M., Goto, A., Endo, T. (1995) Isolation and characterization of the cDNA for pea chloroplast SecA. Evolutionary conservation of the bacterial-type SecA-dependent protein transport within chloroplasts, *FEBS Lett.* **364**, 305–308.

Nouwen, N., Driessen, A. J. M. (2002) SecDFyajC forms a heterotetrameric complex with YidC, *Mol. Microbiol.* **44**, 1397–1405.

Nouwen, N., de Kruijff, B., Tommassen, J. (1996) *prlA* suppressors in *Escherichia coli* relieve the proton electrochemical gradient dependency of translocation of wild-type precursors, *Proc. Natl. Acad. Sci. USA* **93**, 5953–5957.

Nouwen, N., van der Laan, M., Driessen, A. J. M. (2001) SecDFyajC is not required for the maintenance of the proton motive force, *FEBS Lett.* **508**, 103–106.

Oliver, D. B., Beckwith, J. (1981) *E. coli* mutant pleiotropically defective in the export of secreted proteins, *Cell* **25**, 765–772.

Oliver, D. B., Beckwith, J. (1982) Regulation of a membrane component required for protein secretion in *Escherichia coli*, *Cell* **30**, 311–319.

Oliver, D. B., Cabelli, R. J., Dolan, K. M., Jarosik, G. P. (1990) Azide-resistant mutants of *Escherichia coli* alter the SecA protein, an azide-sensitive component of the protein export machinery, *Proc. Natl. Acad. Sci. USA* **87**, 8227–8231.

Oliver, D., Norman, J., Sarker, S. (1998) Regulation of *Escherichia coli secA* by cellular protein secretion proficiency requires an intact gene X signal sequence and an active translocon, *J. Bacteriol.* **180**, 5240–5242.

Owens, M. U., Swords, W. E., Schmidt, M. G., King, C. H., Quinn, F. D. (2002) Cloning, expression, and functional characterization of the *Mycobacterium tuberculosis secA* gene, *FEMS Microbiol. Lett.* **211**, 133–141.

Packer, J. C., Howe, C. J. (1996) The cyanobacterial genome contains a single copy of the *ffh* gene encoding a homologue of the 54 kDa subunit of signal recognition particle, *Plant Mol. Biol.* **31**, 659–665.

Park, S. K., Kim, D. W., Choe, J., Kim, H. (1997) RNA helicase activity of *Escherichia coli* SecA protein, *Biochem. Biophys. Res. Commun.* **235**, 593–597.

Peluso, P., Herschlag, D., Nock, S., Freymann, D. M., Johnson, A. E., Walter, P. (2000) Role of 4.5S RNA in assembly of the bacterial signal recognition particle with its receptor, *Science* **288**, 1640–1643.

Phoenix, D. A., Kusters, R., Hikita, C., Mizushima, S., de Kruijff, B. (1993) OmpF-Lpp signal sequence mutants with varying charge hydrophobicity ratios provide evidence for a phosphatidylglycerol-signal sequence interaction during protein translocation across the *Escherichia coli* inner membrane, *J. Biol. Chem.* **268**, 17069–17073.

Plano, G. V., Day, J. B., Ferracci, F. (2001) Type III export: new uses for an old pathway, *Mol. Microbiol.* **40**, 284–293.

Pogliano, K. J., Beckwith, J. (1993) The Cs *sec* mutants of *Escherichia coli* reflect the cold sensitivity of protein export itself, *Genetics* **133**, 763–773.

Pogliano, K. J., Beckwith, J. (1994) Genetic and molecular characterization of the *Escherichia coli secD* operon and its products, *J. Bacteriol.* **176**, 804–814.

Pohlschroder, M., Murphy, C., Beckwith, J. (1996) *In vivo* analyses of interactions between SecE and SecY, core components of the *Escherichia coli* protein translocation machinery, *J. Biol. Chem.* **271**, 19908–19914.

Pohlschroder, M., Prinz, W. A., Hartmann, E., Beckwith, J. (1997) Protein translocation in the three domains of life: variations on a theme, *Cell* **91**, 563–566.

Prinz, A., Behrens, C., Rapoport, T. A., Hartmann, E., Kalies, K. U. (2000) Evolutionarily conserved binding of ribosomes to the translocation channel via the large ribosomal RNA, *EMBO J.* **19**, 1900–1906.

Rajapandi, T., Dolan, K. M., Oliver, D. B. (1991) The first gene in the *Escherichia coli secA* operon, gene X, encodes a nonessential secretory protein, *J. Bacteriol.* **173**, 7092–7097.

Ramamurthy, V., Oliver, D. (1997) Topology of the integral membrane form of *Escherichia coli* SecA protein reveals multiple periplasmically exposed regions and modulation by ATP binding, *J. Biol. Chem.* **272**, 23239–23246.

Rand, R. P., Sengupta, S. (1972) Cardiolipin forms hexagonal structures with divalent cations, *Biochim. Biophys. Acta* **255**, 484–492.

Randall, L. L., Hardy, S. J. (1986) Correlation of competence for export with lack of tertiary structure of the mature species: a study *in vivo* of maltose-binding protein in *E. coli*, *Cell* **46**, 921–928.

Remaut, E., Stanssens, P., Fiers, W. (1981) Plasmid vectors for high-efficiency expression controlled by the PL promoter of coliphage lambda, *Gene* **15**, 81–93.

Rensing, S. A., Maier, U. G. (1994) The SecY protein family: comparative analysis and phylogenetic relationships, *Mol. Phylogenet. Evol.* **3**, 187–191.

Rietveld, A. G., Koorengevel, M. C., de Kruijff, B. (1995) Non-bilayer lipids are required for efficient protein transport across the plasma membrane of *Escherichia coli*, *EMBO J.* **14**, 5506–5513.

Riggs, P. D., Derman, A. I., Beckwith, J. (1988) A mutation affecting the regulation of a *secA-lacZ* fusion defines a new sec gene, *Genetics* **118**, 571–579.

Robinson, C., Bolhuis, A. (2001) Protein targeting by the twin-arginine translocation pathway, *Nat. Rev. Mol. Cell Biol.* **2**, 350–356.

Rollo, E. E., Oliver, D. B. (1988) Regulation of the *Escherichia coli secA* gene by protein secretion defects: analysis of *secA*, *secB*, *secD*, and *secY* mutants, *J. Bacteriol.* **170**, 3281–3282.

Saaf, A., Wallin, E., von Heijne, G. (1998) Stop-transfer function of pseudo-random amino acid segments during translocation across prokaryotic and eukaryotic membranes, *Eur. J. Biochem.* **251**, 821–829.

Sagara, K., Matsuyama, S., Mizushima, S. (1994) SecF stabilizes SecD and SecY, components of the protein translocation machinery of the *Escherichia coli* cytoplasmic membrane, *J. Bacteriol.* **176**, 4111–4116.

Saier, M. H. J., Paulsen, I. T., Sliwinski, M. K., Pao, S. S., Skurray, R. A., Nikaido, H. (1998) Evolutionary origins of multidrug and drug-specific efflux pumps in bacteria, *FASEB J.* **12**, 265–274.

Sako, T., Iino, T. (1988) Distinct mutation sites in *prlA* suppressor mutant strains of *Escherichia coli* respond either to suppression of signal peptide mutations or to blockage of staphylokinase processing, *J. Bacteriol.* **170**, 5389–5391.

Salavati, R., Oliver, D. (1995) Competition between ribosome and SecA binding promotes *Escherichia coli secA* translational regulation, *RNA* **1**, 745–753.

Salavati, R., Oliver, D. (1997) Identification of elements on *geneX-secA* RNA of *Escherichia coli* required for SecA binding and *secA* auto-regulation, *J. Mol. Biol.* **265**, 142–152.

Samuelson, J. C., Chen, M., Jiang, F., Moller, I., Wiedmann, M., Kuhn, A., Phillips, G. J., Dalbey, R. E. (2000) YidC mediates membrane protein insertion in bacteria, *Nature* **406**, 637–641.

Sandkvist, M. (2001) Biology of type II secretion, *Mol. Microbiol.* **40**, 271–283.

Sarker, S., Oliver, D. (2002) Critical regions of *secM* that control its translation and secretion and promote secretion-specific *secA* regulation, *J. Bacteriol.* **184**, 2360–2369.

Sato, K., Mori, H., Yoshida, M., Mizushima, S. (1996) Characterization of a potential catalytic residue, Asp-133, in the high affinity ATP-binding site of *Escherichia coli* SecA, translocation ATPase, *J. Biol. Chem.* **271**, 17439–17444.

Sato, K., Mori, H., Yoshida, M., Tagaya, M., Mizushima, S. (1997a) *In vitro* analysis of the stop-transfer process during translocation across the cytoplasmic membrane of *Escherichia coli*, *J. Biol. Chem.* **272**, 20082–20087.

Sato, K., Mori, H., Yoshida, M., Tagaya, M., Mizushima, S. (1997b) Short hydrophobic segments in the mature domain of ProOmpA determine its stepwise movement during translocation across the cytoplasmic membrane of *Escherichia coli*, *J. Biol. Chem.* **272**, 5880–5886.

Scaramuzzi, C. D., Hiller, R. G., Stokes, H. W. (1992a) Identification of a chloroplast-encoded *secA* gene homologue in a chromophytic alga: possible role in chloroplast protein translocation, *Curr. Genet.* **22**, 421–427.

Scaramuzzi, C. D., Stokes, H. W., Hiller, R. G. (1992b) Characterisation of a chloroplast-encoded *secY* homologue and *atpH* from a chromophytic alga. Evidence for a novel chloroplast genome organisation, *FEBS Lett.* **304**, 119–123.

Schatz, G., Dobberstein, B. (1996) Common principles of protein translocation across membranes, *Science* **271**, 1519–1526.

Schatz, P. J., Riggs, P. D., Jacq, A., Fath, M. J., Beckwith, J. (1989) The *secE* gene encodes an integral membrane protein required for protein export in *Escherichia coli*, *Genes Dev.* **3**, 1035–1044.

Schatz, P. J., Bieker, K. L., Ottemann, K. M., Silhavy, T. J., Beckwith, J. (1991) One of three transmembrane stretches is sufficient for the functioning of the SecE protein, a membrane component of the *E. coli* secretion machinery, *EMBO J.* **10**, 1749–1757.

Schiebel, E., Driessen, A. J. M., Hartl, F. U., Wickner, W. (1991) ΔH+ and ATP function at different steps of the catalytic cycle of preprotein translocase, *Cell* **64**, 927–939.

Schmidt, M. G., Oliver, D. B. (1989) SecA protein autogenously represses its own translation during normal protein secretion in *Escherichia coli*, *J. Bacteriol.* **171**, 643–649.

Schmidt, M. G., Dolan, K. M., Oliver, D. B. (1991) Regulation of *Escherichia coli secA* mRNA translation by a secretion-responsive element, *J. Bacteriol.* **173**, 6605–6611.

Schmidt, M., Ding, H., Ramamurthy, V., Mukerji, I., Oliver, D. (2000) Nucleotide binding activity of SecA homodimer is conformationally regulated by temperature and altered by *prlD* and *azi* mutations, *J. Biol. Chem.* **275**, 15440–15448.

Scotti, P. A., Urbanus, M. L., Brunner, J., de Gier, J. W., von Heijne, G., van der Does, C., Driessen, A. J. M., Oudega, B., Luirink, J. (2000) YidC, the *Escherichia coli* homologue of mitochondrial Oxa1p, is a component of the Sec translocase, *EMBO J.* **19**, 542–549.

Seoh, H. K., Tai, P. C. (1997) Carbon source-dependent synthesis of SecB, a cytosolic chaperone involved in protein translocation across *Escherichia coli* membranes, *J. Bacteriol.* **179**, 1077–1081.

Shilton, B., Svergun, D. I., Volkov, V. V., Koch, M. H., Cusack, S., Economou, A. (1998) *Escherichia coli* SecA shape and dimensions, *FEBS Lett.* **436**, 277–282.

Shimizu, H., Nishiyama, K., Tokuda, H. (1997) Expression of *gpsA* encoding biosynthetic sn-glycerol 3-phosphate dehydrogenase suppresses both the LB-phenotype of a *secB* null mutant and the cold-sensitive phenotype of a *secG* null mutant, *Mol. Microbiol.* **26**, 1013–1021.

Shimoike, T., Akiyama, Y., Baba, T., Taura, T., Ito, K. (1992) SecY variants that interfere with *Escherichia coli* protein export in the presence of normal SecY, *Mol. Microbiol.* **6**, 1205–1210.

Shinkai, A., Mei, L. H., Tokuda, H., Mizushima, S. (1991) The conformation of SecA, as revealed by its protease sensitivity, is altered upon interaction with ATP, presecretory proteins, everted membrane vesicles, and phospholipids, *J. Biol. Chem.* **266**, 5827–5833.

Shiozuka, K., Tani, K., Mizushima, S., Tokuda, H. (1990) The proton motive force lowers the level of ATP required for the *in vitro* translocation of a secretory protein in *Escherichia coli*, *J. Biol. Chem.* **265**, 18843–18847.

Sianidis, G., Karamanou, S., Vrontou, E., Boulias, K., Repanas, K., Kyrpides, N., Politou, A. S., Economou, A. (2001) Cross-talk between catalytic and regulatory elements in a DEAD motor domain is essential for SecA function, *EMBO J.* **20**, 961–970.

Stader, J., Gansheroff, L. J., Silhavy, T. J. (1989) New suppressors of signal-sequence mutations, *prlG*, are linked tightly to the *secE* gene of *Escherichia coli*, *Genes Dev.* **3**, 1045–1052.

Suzuki, H., Nishiyama, K., Tokuda, H. (1999) Increases in acidic phospholipid contents specifically restore protein translocation in a cold-sensitive *secA* or *secG* null mutant, *J. Biol. Chem.* **274**, 31020–31024.

Tani, K., Shiozuka, K., Tokuda, H., Mizushima, S. (1989) *In vitro* analysis of the process of translocation of OmpA across the *Escherichia coli* cytoplasmic membrane. A translocation intermediate accumulates transiently in the absence of the proton motive force, *J. Biol. Chem.* **264**, 18582–18588.

Tani, K., Tokuda, H., Mizushima, S. (1990) Translocation of ProOmpA possessing an intramolecular disulfide bridge into membrane vesicles of *Escherichia coli*. Effect of membrane energization, *J. Biol. Chem.* **265**, 17341–17347.

Taura, T., Baba, T., Akiyama, Y., Ito, K. (1993) Determinants of the quantity of the stable SecY complex in the *Escherichia coli* cell, *J. Bacteriol.* **175**, 7771–7775.

Tian, H., Beckwith, J. (2002) Genetic screen yields mutations in genes encoding all known components of the *Escherichia coli* signal recognition particle pathway, *J. Bacteriol.* **184**, 111–118.

Tokuda, H., Shiozuka, K., Mizushima, S. (1990) Reconstitution of translocation activity for secretory proteins from solubilized components of *Escherichia coli*, *Eur. J. Biochem.* **192**, 583–589.

Tokuda, H., Akimaru, J., Matsuyama, S., Nishiyama, K., Mizushima, S. (1991) Purification of SecE and reconstitution of SecE-dependent protein translocation activity, *FEBS Lett.* **279**, 233–236.

Triplett, T. L., Sgrignoli, A. R., Gao, F. B., Yang, Y. B., Tai, P. C., Gierasch, L. M. (2001) Functional signal peptides bind a soluble N-terminal fragment of SecA and inhibit its ATPase activity, *J. Biol. Chem.* **276**, 19648–19655.

Tseng, T. T., Gratwick, K. S., Kollman, J., Park, D., Nies, D. H., Goffeau, A., Saier, M. H. J. (1999) The RND permease superfamily: an ancient, ubiquitous and diverse family that includes human disease and development proteins, *J. Mol. Microbiol. Biotechnol.* **1**, 107–125.

Uchida, K., Mori, H., Mizushima, S. (1995) Stepwise movement of preproteins in the process of translocation across the cytoplasmic membrane of *Escherichia coli*, *J. Biol. Chem.* **270**, 30862–30868.

Ulbrandt, N. D., London, E., Oliver, D. B. (1992) Deep penetration of a portion of *Escherichia coli* SecA protein into model membranes is promoted by anionic phospholipids and by partial unfolding, *J. Biol. Chem.* **267**, 15184–15192.

Valent, Q. A., Scotti, P. A., High, S., de Gier, J. W., von Heijne, G., Lentzen, G., Wintermeyer, W., Oudega, B., Luirink, J. (1998) The *Escherichia coli* SRP and SecB targeting pathways converge at the translocon, *EMBO J.* **17**, 2504–2512.

van Dalen, A., Killian, A., de Kruijff, B. (1999). $\Delta\psi$ stimulates membrane translocation of the C-terminal part of a signal sequence, *J. Biol. Chem.* **274**, 19913–19918.

van der Does, C., den Blaauwen, T., de Wit, J. G., Manting, E. H., Groot, N. A., Fekkes, P., Driessen, A. J. M. (1996) SecA is an intrinsic subunit of the *Escherichia coli* preprotein translocase and exposes its carboxyl terminus to the periplasm, *Mol. Microbiol.* **22**, 619–629.

van der Does, C., Manting, E. H., Kaufmann, A., Lutz, M., Driessen, A. J. M. (1998) Interaction between SecA and SecYEG in micellar solution and formation of the membrane-inserted state, *Biochemistry* **37**, 201–210.

van der Does, C., Swaving, J., van Klompenburg, W., Driessen, A. J. M. (2000) Non-bilayer lipids stimulate the activity of the reconstituted bacterial protein translocase, *J. Biol. Chem.* **275**, 2472–2478.

van der Wolk, J. P., Klose, M., Breukink, E., Demel, R. A., de Kruijff, B., Freudl, R., Driessen, A. J. M. (1993) Characterization of a *Bacillus subtilis* SecA mutant protein deficient in translocation ATPase and release from the membrane, *Mol. Microbiol.* **8**, 31–42.

van der Wolk, J. P., de Wit, J. G., Driessen, A. J. M. (1997) The catalytic cycle of the *Escherichia coli* SecA ATPase comprises two distinct preprotein translocation events, *EMBO J.* **16**, 7297–7304.

van der Wolk, J. P., Fekkes, P., Boorsma, A., Huie, J. L., Silhavy, T. J., Driessen, A. J. M. (1998) PrlA4 prevents the rejection of signal sequence defective preproteins by stabilizing the SecA-SecY interaction during the initiation of translocation, *EMBO J.* **17**, 3631–3639.

van der Laan, M., Houben, E. N., Nouwen, N., Luirink, J., Driessen, A. J. M. (2001) Reconstitution of Sec-dependent membrane protein insertion: nascent FtsQ interacts with YidC in a SecYEG-dependent manner, *EMBO Rep.* **2**, 519–523.

van Voorst, F., van der Does, C., Brunner, J., Driessen, A. J. M., de Kruijff, B. (1998) Translocase-bound SecA is largely shielded from the phospholipid acyl chains, *Biochemistry* **37**, 12261–12268.

van Wely, K. H. M., Swaving, J., Broekhuizen, C. P., Rose, M., Quax, W. J., Driessen, A. J. M. (1999) Functional identification of the product of the *Bacillus subtilis yvaL* gene as a SecG homologue, *J. Bacteriol.* **181**, 1786–1792.

Veenendaal, A. K., van der Does, C., Driessen, A. J. M. (2001) Mapping the sites of interaction between SecY and SecE by cysteine scanning mutagenesis, *J. Biol. Chem.* **276**, 32559–32566.

Vogel, H., Fischer, S., Valentin, K. (1996) A model for the evolution of the plastid *sec* apparatus inferred from *secY* gene phylogeny, *Plant Mol. Biol.* **32**, 685–692.

von Heijne, G. (1997) Getting greasy: how transmembrane polypeptide segments integrate into the lipid bilayer, *Mol. Microbiol.* **24**, 249–253.

von Heijne, G. (1998) Life and death of a signal peptide, *Nature* **396**, 111–113.

Weinkauf, S., Hunt, J. F., Scheuring, J., Henry, L., Fak, J., Oliver, D. B., Deisenhofer, J. (2001) Conformational stabilization and crystallization of the SecA translocation ATPase from *Bacillus subtilis*, *Acta Crystallogr. D. Biol. Crystallogr.* **57**, 559–565.

Woodbury, R. L., Topping, T. B., Diamond, D. L., Suciu, D., Kumamoto, C. A., Hardy, S. J., Randall, L. L. (2000) Complexes between protein export chaperone SecB and SecA: evidence for separate sites on SecA providing binding energy and regulatory interactions, *J. Biol. Chem.* **275**, 24191–24198.

Xu, Z., Knafels, J. D., Yoshino, K. (2000) Crystal structure of the bacterial protein export chaperone SecB, *Nature Struct. Biol.* **7**, 1172–1177.

Yamada, H., Tokuda, H., Mizushima, S. (1989) Proton motive force-dependent and -independent protein translocation revealed by an efficient *in vitro* assay system of *Escherichia coli*, *J. Biol. Chem.* **264**, 1723–1728.

Yamane, K., Ichihara, S., Mizushima, S. (1987) *In vitro* translocation of protein across *Escherichia coli* membrane vesicles requires both the proton motive force and ATP, *J. Biol. Chem.* **262**, 2358–2362.

Yamane, K., Akiyama, Y., Ito, K., Mizushima, S. (1990) A positively charged region is a determinant of the orientation of cytoplasmic membrane proteins in *Escherichia coli*, *J. Biol. Chem.* **265**, 21166–21171.

Yang, Y. B., Yu, N., Tai, P. C. (1997) SecE-depleted membranes of *Escherichia coli* are active. SecE is not obligatorily required for the *in vitro* translocation of certain protein precursors, *J. Biol. Chem.* **272**, 13660–13665.

10
Self-assembling Symmetric Protein Materials

Jennifer E. Padilla[1], **Tianwei Yu**[2]

[1] Department of Chemistry and Biochemistry, 212 Boyer Hall, 611 Charles E. Young Dr. South, University of California, Los Angeles, Los Angeles, CA 90095-1569, USA; Tel. +1-310-825-8901; Fax +1-310-206-3914; E-mail: padilla@mbi.ucla.edu

[2] Department of Chemistry and Biochemistry, 212 Boyer Hall, 611 Charles E. Young Dr. South, University of California, Los Angeles, Los Angeles, CA 90095-1569, USA; Tel. +1-310-825-8901; Fax +1-310-206-3914; E-mail: tyu@mbi.ucla.edu

ADP	adenosine diphosphate
C_n	cyclic symmetry of degree n
D_n	dihedral symmetry of degree n
EM	electron microscopy
n-fold	denotes rotational axis of rotation 360/n degrees
n_m	denotes screw axis symmetry, a rotation by 360m/n degrees along the axis, followed by a translation along the axis
S-layer	bacterial surface layer
TEM	transmission electron microscopy

1 Introduction to Self-assembly

Many protein complexes assemble spontaneously from their protein subunits. The phenomenon of self-assembly results in the efficient creation of nanostructures within living cells. Without further expenditure of energy, or the synthesis of extra molecules, subunits arrange themselves in regular patterns to produce tubes, filaments, capsids, and other structures. Self-assembly is also a goal of nanotechnology–a way to build nanostructures from the ground up.

Symmetry plays an important role in the assembly of a great many molecular structures such as virus capsids, microtubules and clathrin cages. It is also a powerful tool for the design of self-assembling protein complexes in the laboratory. In this chapter, we will focus on the role of symmetry in the self-assembly of protein complexes, both natural and man-made.

1.1 Protein Properties that Promote Self-assembly

Proteins possess characteristics that make them naturally amenable to self-assembly. Unlike most other polymers, proteins adopt a unique, relatively fixed three-dimensional (3-D) conformation encoded by their amino acid sequence (Creighton, 1993). These defined structures allow them to associate with one another in a highly specific manner. Interfaces between proteins are highly evolved, and this often results in very tight binding. The combination of hydrophobic forces, hydrogen bonds, and salt bridges at a protein–protein binding interface provides both the energetic force that results in a low dissociation constant, and a specific interface that cannot be energetically satisfied by other protein faces, resulting in high specificity.

The specific binding between proteins drives self-assembly by ensuring that each subunit joins a growing protein complex in precisely the right orientation. No outside intervention is required to create the final assembly because each building block can only fit into the growing structure in the correct way. Furthermore, a protein is capable of having multiple binding sites on its surface. Because of its unique 3-D structure, these binding sites can be held in defined orientations with respect to one another. Thus, proteins need not rely on general properties of hydrophobicity or electrostatics in order to assemble into regular structures. Rather, they have specific interfaces designed to interlock like the pieces of a jigsaw puzzle.

2 Historical Outline

Symmetric protein complexes were identified with the advance of X-ray crystallography and electron microscopy. As early as in 1935, Crowfoot in Oxford, obtained X-ray diffraction data suggesting that insulin formed a complex with three-fold symmetry in the presence of zinc (Crowfoot, 1935). This was later confirmed when the atomic structure of insulin was determined (Adams et al., 1969).

In 1955, Franklin, while working at Birkbeck College in London, first found helical symmetry in tobacco mosaic virus through X-ray diffraction (Franklin, 1955), while in 1956 Crick and Watson proposed that spherical viruses should be built on cubic symmetry (Crick and Watson, 1956). Based on high-resolution micrographs of negatively stained spherical viruses, Caspar (at Harvard) and Klug (at Cambridge) subsequently established the model of icosahedral viral capsids in 1962 (Caspar and Klug, 1962).

The surface structure of clathrin-coated vesicles was revealed in 1969, at a time when little was known about clathrin itself. Kanaseki and Kadota (1969), in studies based on negatively stained electron micrographs, found the vesicles to be coated with pentagons and hexagons.

Hanson and Huxley, at Cambridge, established that actin and myosin were the major components of the protein assemblies that comprise thin and thick filaments in the muscle, respectively (Hanson and Huxley, 1953). In 1963, Hanson and Lowy established the helical structure of actin filament based on electron micrography and comparison with previous X-ray diffraction data, but multiple forms of myosin were observed before the first proposal of a general model of myosin organization (Squire, 1971).

The term "microtubule" was first suggested in 1963 by Slautterback, after a few observations of the tubular nature of some cytoplasmic fibers, and the subunit arrangement of tubulins in microtubules was revealed in detail in 1974 by Amos and Klug.

The occurrence of natural proteins in ordered two-dimensional (2-D) arrays was first observed with bacterial surface layers, termed S-layers, in 1952 (Houwink, 1953), and hundreds of S-layers were reported thereafter. In 1969, Brinton et al., while working at the University of Pittsburgh, were the first to find that an S-layer was made from a single protein (Sleytr, 1978).

The assembly of proteins by way of the mechanism which is now known as "domain swapping" was first observed in 1962 by Crestfield et al. using chemical modification methods (Liu and Eisenberg, 2002).

Anderson et al. demonstrated the first structural evidence in 1981 and, after a number of observations had been made available, Bennett et al., working at the University of California, first proposed that 3-D domain swapping was a general mechanism for switching between protein conformers (Bennett et al., 1994).

Rationally designed self-assembling peptide materials first appeared about 10 years ago when, in 1993, Ghadiri et al. first reported an eight-residue cyclic peptide that contained alternating D- and L-amino acids and was able to self-assemble into nanotubes when protonated (Ghadiri et al., 1993). Designs based on folded protein structures and specific protein–protein recognition started only recently however. For example, a protein filament formed from a heterodimeric coiled-coil of two polypeptides was reported by Pandya et al. (2000), while one year later Padilla et al. (2001) first designed a protein cage and proposed general principles of designing self-assembling protein complexes using symmetry.

3 Chemical Structures

The protein complexes described in this chapter are all assemblies of polypeptides synthesized biologically from the 20 standard amino acids, where each amino acid is connected to the next by a peptide bond, and each linear chain of amino acids is referred to as a subunit. The subunits in the described complexes are chemically identical to one another; that is, they each are made from the same amino acid sequence. Subunits can be linked covalently via disulfide bonds between cysteine residues.

4 Symmetry

Symmetry often underlies the beauty of a natural specimen or a work of art. The petals of a flower, the five arms of a starfish, or the repetition of purple spines on a sea urchin all catch the eye. Museums are filled with art objects featuring repeated motifs or mirror symmetry. However, symmetry does not occur only for aesthetic purposes–it is also functional, as it allows a larger object to be composed of identical subunits. This mode of construction, as well as being efficient, lends itself naturally to self-assembly, and the role of symmetry in self-assembly will be developed in the next few sections.

4.1 Symmetry Elements

Symmetry describes the way in which identical objects, or portions of an object, are related. The symmetry in the objects we see is described by "symmetry operations" and "symmetry elements". Some objects, such as lava rocks, possess no symmetry and any rotation or other shift in the lava rock's position is evident. On the other hand, a snowflake appears the same when rotated through 60° (Figure 1). The rotation of the snowflake is a symmetry operation, the axis about which it is rotated is a symmetry element, and is referred to as a "6-fold" axis of rotation. In general, an n-fold rotational axis of symmetry describes an object that

Fig. 1 A snowflake possessing 6-fold rotational symmetry. (From Hornung, 1959.)

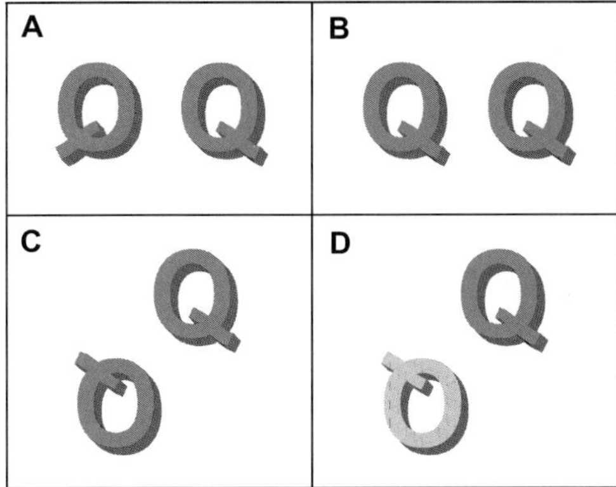

Fig. 2 The basic symmetry operations. (A) Reflection. The Qs are related by a mirror plane passing between them, perpendicular to the page. (B) Translation. The Q on the right is related to the Q on the left by a translation towards the right side of the page. (C) Rotation by 180°. The axis of rotation is perpendicular to the page, centered between the Qs. (D) Inversion. Inversion negates all the coordinates of an object. This can be accomplished by rotation in the plane of the page, followed by reflection in a plane below the Q and perpendicular to the page. This operation reveals that the back of this letter Q is a lighter shade of gray with a zigzag pattern.

looks the same when rotated by 360/n degrees. This symmetry is denoted C_n.

The basic symmetry operations are rotation, translation, reflection, and inversion (Figure 2). The elements that describe these symmetries are rotation axes, translation vectors, reflection planes and inversion points. Combining some of these basic elements yields screw axes and glide planes. A screw axis describes a combination of rotation and translation along the rotation axis, while a glide plane describes a reflection followed by a translation. Any symmetric object, no matter how complicated, is described by a combination of these basic elements.

4.2
Symmetric Contacts in Proteins

The majority of proteins form oligomers (Goodsell and Olson, 2000); these are complexes of a small number of proteins bound to one another, usually noncovalently, but sometimes with additional disulfide bonds. An oligomer with identical subunits–a "homo-oligomer"–usually has rotational symmetry. The rotational symmetry elements describing the oligomer can be imagined to be associated with a single subunit (Figure 3). The structure of one subunit together with all associated symmetry elements provides enough information to recreate the structure of the complete oligomer. Wherever proteins contact one another about a symmetry element, that contact can play a role in self-assembly, as will be discussed in the next section.

Proteins are chiral, so the only symmetries that can apply to them are rotation, translation, and screw symmetry. When considering the symmetry of protein complexes,

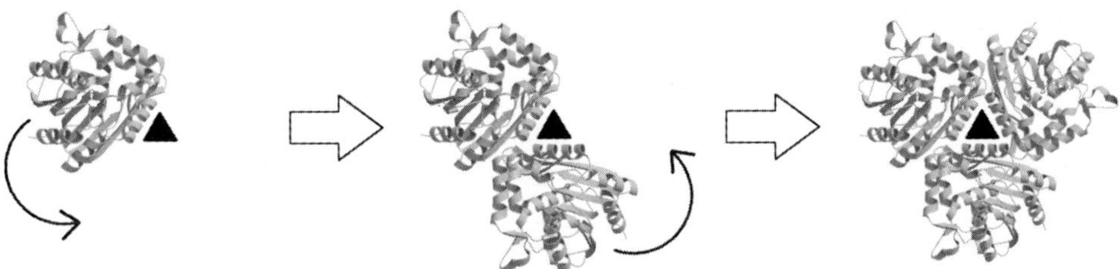

Fig. 3 Symmetry element associated with a protein. Repeated operation of the 3-fold rotation axis (black triangle), associated with this protein yields the structure of the complete trimer. The structure shown here is a trimer of bromoperoxidase (PDB ID code 1bro; Hecht et al., 1994). This figure was prepared using Molscript (Kraulis, 1991) and Raster3D (Merritt and Bacon, 1997).

one need not consider reflections or inversions. The chirality of proteins is derived from the L-amino acids, the mirror-related enantiomers of which–the D-amino acids–are distinct chemicals that are not used in protein synthesis.

4.3
Multiple Symmetry Elements

The fundamental difference between simple oligomers and large symmetric protein complexes is that the latter possess multiple symmetry elements. If a protein subunit has more than one symmetry element associated with it, it can assemble into a complex obeying one of many possible symmetries.

Protein complexes fall into four broad categories of structure: shells, filaments, layers, and crystals. While the shells are closed, individual particles, the rest of the architectures feature indefinite growth in one, two, or three dimensions, limited only by the finite nature of the sample, mechanical forces, or available space. All possible symmetric protein complexes fall into one of these architectures. These distinct architectures are built upon the point, helical, layer, and space group symmetries, respectively.

The point symmetries containing multiple symmetry elements are the dihedral and cubic symmetries. Dihedral symmetry, denoted D_n, consists of one n-fold rotational axis perpendicular to n 2-fold rotational axes (Figure 4A). Complexes of this symmetry

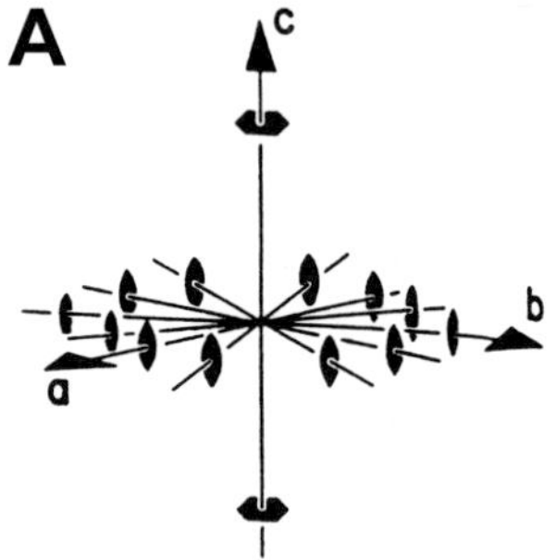

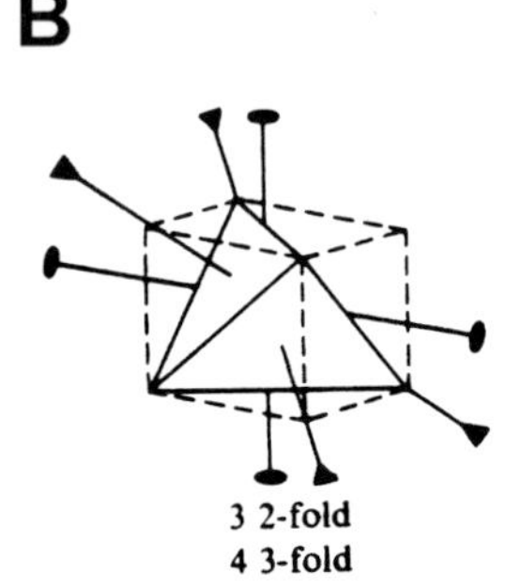

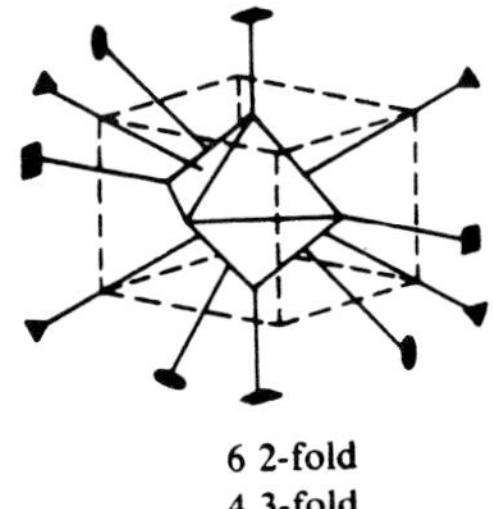

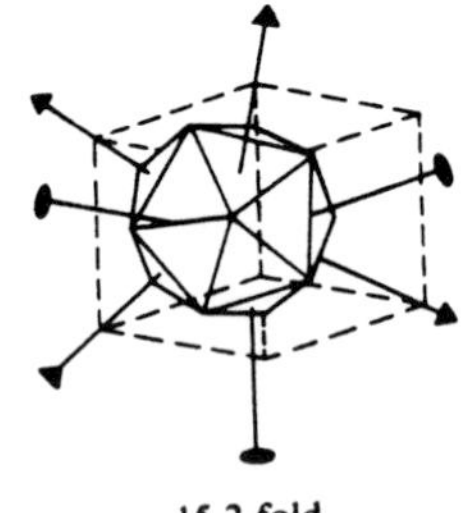

Fig. 4 Diagram of point group symmetries. (A) ▶ Dihedral symmetry. This image shows D_6 symmetry, with one 6-fold rotation axis perpendicular to six 2-fold rotation axes. (From Boisen and Gibbs, 1985; reprinted with permission from the Mineralogical Society of America.) (B) The cubic symmetries tetrahedral, octahedral, and icosahedral, are shown along with lists of their rotational axes. (From Wilson, 1966.)

appear as two stacked rings of protein subunits. The cubic symmetries are tetrahedral (T), octahedral (O), and icosahedral (I), consisting of 12, 24, and 60 subunits, respectively (Figure 4B). The cubic symmetries feature rotation axes that are arranged equidistantly about the central point.

Filaments break down into two subcategories: polar and bipolar. Polar filaments feature head-to-tail arrangement of their subunits (Figure 5A and B) in which it is possible to distinguish the two ends of the filament. Bipolar filaments have 2-fold symmetry perpendicular to the filament axis, making the two ends indistinguishable (Figure 5C). Helical symmetry is described by a screw axis, n_m; this notation indicates that each subunit is rotated 360m/n degrees about the helix axis with respect to the previous subunit, and translated along the helix axis.

Layers and crystals are made of repeated units, called unit cells, which are stacked next to one another like tiles or bricks. Despite their infinite extent, there are only a finite number of layer and space groups. In total, 230 space groups exist, 65 of which are "biological" and contain no mirror planes or inversions. Similarly, 17 of the 80 layer groups are available to proteins. Crystal

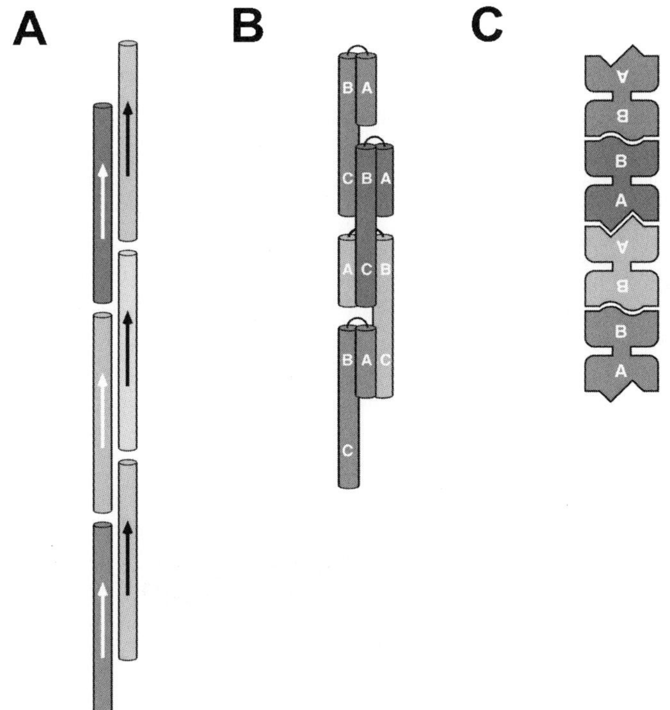

Fig. 5 Schematic representations of designed polar and bipolar filaments. (A) Heterodimeric coiled-coils (shown in different shades of gray), stack head-to-tail in a polar arrangement. (B) Domain-swapped three-helix bundles form a polar filament. (C) A bipolar filament is formed when there is 2-fold symmetry perpendicular to the filament axis. This schematic shows how two distinct dimeric interfaces, A and B, each with 2-fold rotational symmetry, may serve to drive the assembly of a bipolar filament. (Adapted with permission from Yeates and Padilla, 2002.)

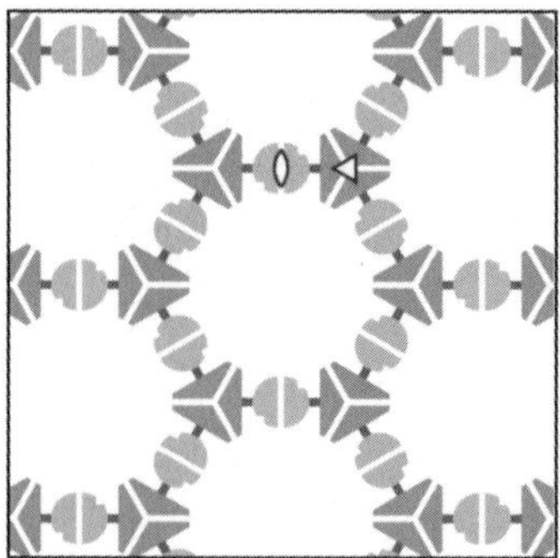

Fig. 6 A generating set for a layer with p6 symmetry. The generators shown are a 3-fold rotation (triangle), and 2-fold rotation (lens shape). The action of the 2-fold axis places another 3-fold axis in the plane. Similarly, the 3-fold axis places more 2-fold axes on the plane. Eventually, a 6-fold axis in the center of one hexagon is revealed. It becomes evident that all the symmetry elements in the p6 layer will appear as products of the two generators. (From Padilla et al., 2001.)

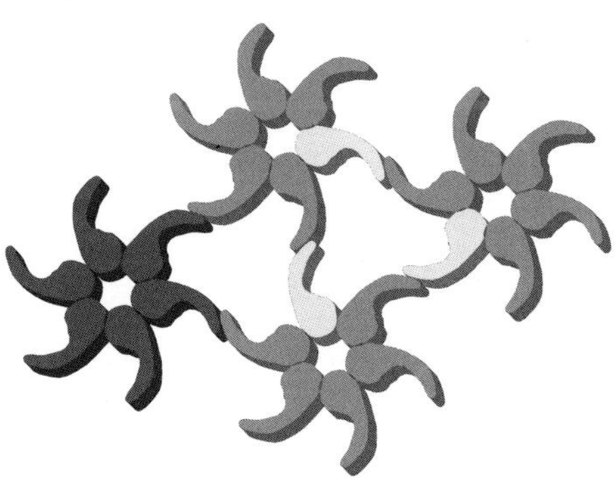

Fig. 7 Symmetric contacts. In this schematic portion of a p6 layer, the darkest subunits, centered about a 6-fold rotation axis, are in contact. That makes the 6-fold rotation a generator in this manifestation of a p6 symmetric layer. On the other hand, the lightly shaded subunits related by a 3-fold rotation axis are not in contact, therefore the 3-fold axis is not a generator here.

packing restricts the rotations within layer and space groups to 2-, 3-, 4-, and 6-fold rotations.

Sets of symmetry elements that describe a symmetric object form a mathematical entity called a "group". Although the symmetries that describe helices, layers and crystals are infinite, group theory tells us that all of these symmetries can be obtained from a finite subset of a few symmetry elements within that group (Figure 6). This is the basis for the self-assembly of symmetric objects. The symmetry elements in this subset are called group generators. Within a protein complex, any symmetry element about which the symmetry-related proteins are in contact is a generator. One can find the generators of a complex by observing all the points of contact between protein subunits (Figure 7). Many identical contacts occur repeatedly within the complex, but only one representative symmetry element from each is needed as a generator.

5 Naturally Occurring Symmetric Complexes

Nature makes extensive use of symmetric contacts to generate self-assembling protein complexes in the shell, filament and layer architectural classes. Using symmetry is highly efficient, because large structures can be built using identical building blocks. While protein crystals are rare *in vivo*, peroxisomes and hormone storage granules can hold crystalline arrays (Goodsell and Olson, 2000). Symmetry-breaking is also important in the assembly of protein complexes. With a little flexibility, Nature can create more structures than the rigid rules of group theory would ordinarily allow. The following are some examples of symmetric protein complexes from each architectural class.

5.1 Protein Oligomers

A well-known example of dihedral symmetry is the bacterial chaperonin, GroEL, which

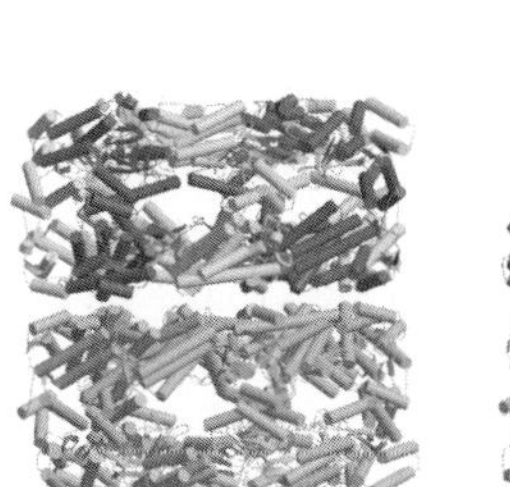
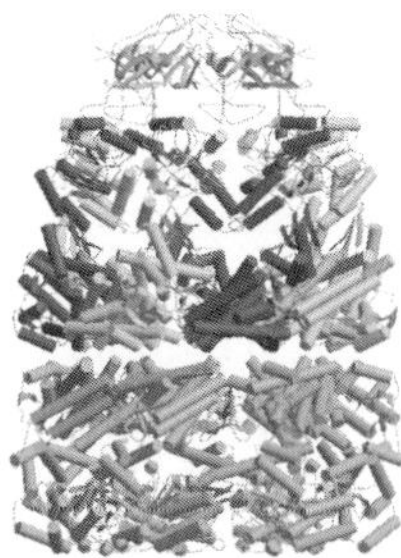

Fig. 8 The GroEL/GroES complex. D_7 symmetry is present in the absence of GroES (left, PDB ID code 1der; Boisvert et al., 1996). GroES binding reduces the symmetry to C_7 (right, PDB ID code 1aon; Xu et al., 1997). This figure was prepared using Molscript (Kraulis, 1991) and Raster3D (Merritt and Bacon, 1997).

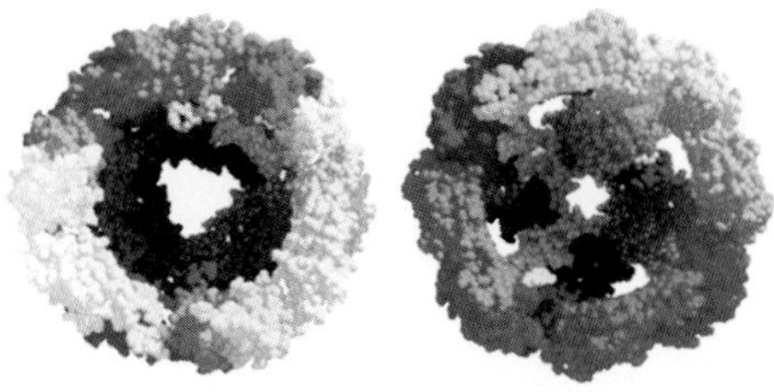

Fig. 9 An octameric small heat shock protein complex. (PDB ID code 1shs.) Reprinted with permission from *Nature* (Kim et al., 1998); © 1998 Macmillan Publishers Ltd.)

forms a 14-molecule cup with D_7 symmetry to house immature proteins and facilitate their folding (Figure 8). The co-chaperonin of GroEL, GroES, forms a conical seven-molecule cap with C_7 symmetry. The GroES cap binds to and dissociates from either end of the GroEL cup using a reciprocating mechanism. When bound, conformational changes occur between the two GroEL rings to break the dihedral symmetry, leaving the GroEL and GroES molecules to share a 7-fold rotational axis in the ADP state (Xu et al., 1997).

Some protein complexes adopt cubic symmetry for various purposes. Some enzyme complexes are tetrahedral, though in the case of *Pyrococcus furiosus* ornithine carbamoyltransferase, the tetrahedral symmetry of the enzyme underlies both its allostery and its thermostability (Villeret et al., 1998). In another example, 24 subunits of the small heat-shock protein from *Methanococcus jannaschii* form a hollow spherical complex with octahedral symmetry (Figure 9). It is suggested that this spherical structure is either responsible for the protection of critical proteins or RNAs, or is simply a dormant form of the chaperone (Kim et al., 1998).

5.2 Icosahedral Viral Capsids

Viral genomes are enclosed in a protein capsid, which plays a role in both genome protection and infection. The small size of the viral genome demands that the capsid be built from a very limited number of distinct protein sequences. Most viruses adopt either helical symmetry or cubic symmetry to construct their capsids, though some viruses also have an envelope made from lipid and protein molecules outside the capsids, and obtain the envelope when budding from the host cell. The symmetrical property of the capsids is retained when the envelope is added.

Icosahedral symmetry dominates spherical viral capsids (Figure 10). The icosahedron is particularly suitable for viral genome packaging due to the need to maximize storage volume. In an icosahedron, there is a 5-fold rotational axis at each of the 12 vertices, a 3-fold rotational axis at the center of each of the 20 faces, and a 2-fold rotational axis at the center of each of the 30 edges. To form an icosahedral capsid, the smallest possible number of subunits is 60 (five at each of the vertices). Using one type of protein subunit, the icosahedron can hold more genetic material than the tetrahedron with 12 subunits, or the octahedron with 24.

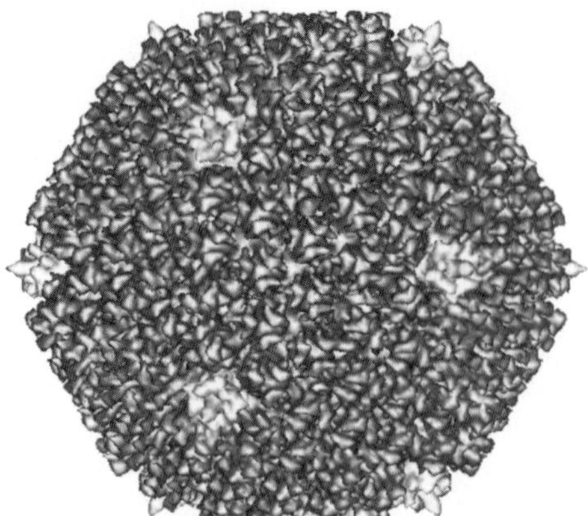

Fig. 10 A cryo-EM reconstruction of adenovirus structure. (From Stewart et al., 2000; reprinted with permission from Wiley-Liss, Inc.)

By breaking some of the symmetry, even larger capsids can be created. While some viruses do have 60 subunits, most have more subunits to form larger shells. The number of subunits is 60T; permissible values of T are given by $T = H^2 + HK + K^2$, where H and K are any integers, starting with $H = 1, K = 0$. The deviations from perfect symmetry occur in the local environment of each protein. Some viruses use nonidentical subunits to perform identical roles, which is termed "pseudosymmetry"; some viruses use identical subunits at different positions, which is termed "quasisymmetry" (Goodsell and Olson, 2000; Strauss and Strauss, 2002). The principle of quasi-equivalence was first developed by Caspar and Klug (1962).

5.3 Clathrin

Clathrin plays a key role in the formation of endocytic vesicles. Clathrin-coated vesicles are assembled from clathrin triskelions, each of which contains three heavy and three light polypeptide chains, joined at the C-terminus of the heavy chains, with a C_3 symmetry if not considering the flexibility of the arms. By using these arm–arm interactions, the triskelions are able to assemble into hexagonal arrays on the membrane. The introduction of local 5-fold vertices allows the 2-D array to fold up and make a coated vesicle (Figure 11). Each vertex of the vesicle is occupied by the center of a triskelion. Because of the flexibility of the arms, symmetry can be broken to create a wide range of vesicle geometries. Smaller structures are rich in 5-fold rings, and 6-fold rings are added to make larger structures (Goodsell and Olson, 2000). In theory, clathrin can form a dodecahedron having only 5-fold rings. The introduction of hexagonal faces increases the volume, but although these larger complexes roughly retain the local symmetry (3-fold on the vertices, 2-fold on the edges, 5- and 6-fold in the rings), the global symmetry is broken. The structure of a hexagonal barrel was studied in detail in which 36 triskelions are organized into a polyhedron with eight hexagon and 12 pentagon faces. The overall structure contained a 6-fold axis and three 2-fold axes (Smith et al., 1998).

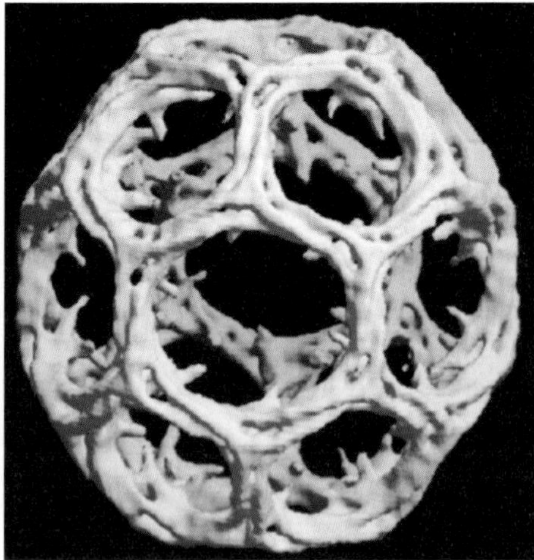

Fig. 11 Three-dimensional reconstruction of clathrin structure. (From Smith et al., 1998; reprinted with permission from Oxford University Press.)

5.4 Filaments

Actin is one of the major components of the cytoskeleton, and is also the major part of the thin filament in muscle. The actin filament (F-actin) is constructed through head-to-tail polymerization of globular actin monomers (G-actin). The morphology is two intertwined, right-handed helices. The overall symmetry of the filament is roughly 13_6, with the screw axis running through the middle of the filament (Figure 12). The head-to-tail arrangement in the filaments make them polar (Geeves and Holmes, 1999; Orlova et al., 2001).

Myosin assembles into the thick filaments in muscles. The coiled-coil domains of myosin molecules pack into the backbone of the myosin filament, while the myosin heads protrude from the filament. If not considering conformational changes, the myosin heads are arranged in a helical fashion. For example, vertebrate skeletal muscles normally have a morphology of three intertwined 9_1 helices, with the middle of the filament making a 3-fold rotation axis (Eakins et al., 2002).

Microtubules play an important role in organizing the distribution of organelles in the cell. Microtubules are hollow cylindrical structures made of heterodimers of α- and β-tubulin, which are highly homologous, both in sequence and 3-D structure. The tubulin heterodimers join linearly in a head-to-tail manner to form protofilaments, which are aligned in parallel to form microtubules (Figure 13). The number of protofilaments in a microtubule is usually 13, but a wide range is possible. If not considering the difference between α- and β-tubulin, the most common microtubule has 13_{12} helical symmetry, but if the difference is considered then only microtubules with an even number of protofilaments can possibly have helical symmetry. The protofilaments in some microtubules are skewed to absorb the mismatch caused by the number of protofilaments (Wade et al., 1998).

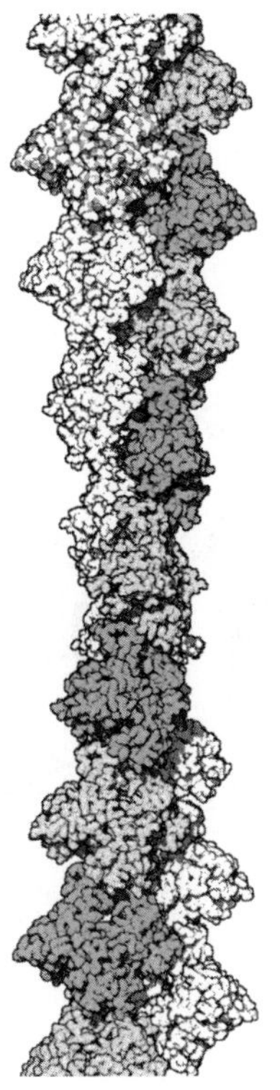

Fig. 12 A representation of actin filament structure. (Reprinted with permission from Dr. David S. Goodsell.)

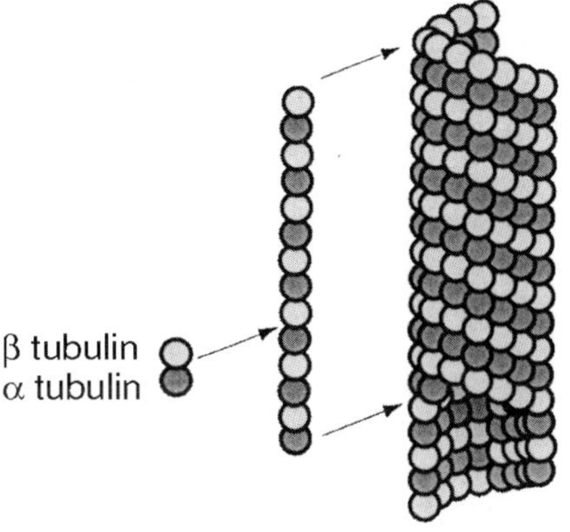

Fig. 13 The arrangement of subunits in a microtubule (From Downing, 2000).

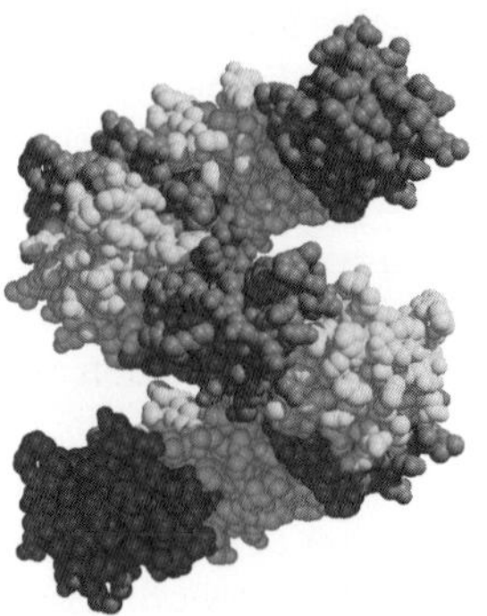

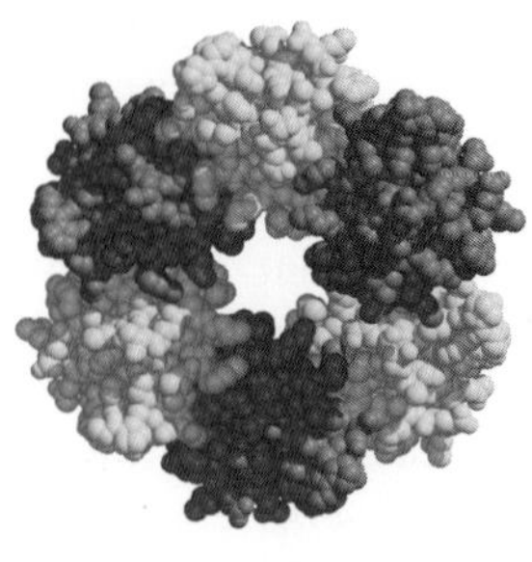

Fig. 14 The 6_5 helical symmetry of the polyhomeotic SAM polymer. (PDB ID code 1kw4; reprinted with permission from Kim et al., 2002; © 2002 Macmillan Publishers Ltd.)

Apart from actin-containing microfilaments and microtubules, the third type of cytoskeleton is made of intermediate filaments which are tight and stable structures. Although no high-resolution structural information is available, current evidence suggests that intermediate filaments are also constructed through hierarchical symmetric packing. A model was proposed in which protofilaments are composed of an antiparallel arrangement of two molecular strands, each of which consists of parallel molecules linked in a head-to-tail fashion. The surface of some intermediate filaments shows near-crystalline packing, as evidenced by X-ray diffraction. This indicates that protofilaments pack in a symmetrical manner to form the intermediate filaments (Parry and Steinert, 1999).

The SAM (sterile alpha motif) domains of two groups of transcriptional repressors were found to form fibers *in vitro* by joining head-to-tail with a 6_5 helical symmetry (Figure 14). It was suggested that the helical organization of the intact repressor proteins mediated by SAM domains could suppress transcription by inducing wrapping of the chromatin around the protein complex (Kim et al., 2001, 2002).

Many viruses are filament-shaped with viral capsids constructed using helical symmetry. An example of this is tobacco mosaic virus (TMV), the capsid of which is made of a single protein subunit. Its morphology is a right-handed helix with 16.3 subunits per turn (Figure 15), and each subunit interacts with six surrounding subunits and three nucleotides of the viral RNA. The protein subunits alone assemble into a double layered "disk" with C17 symmetry, which plays an important role in the nucleation of viral assembly when RNA is present (Klug, 1999).

Fig. 15 A schematic representation of tobacco mosaic virus (TMV) structure. (From Klug and Caspar, 1960.)

5.5 S-layers

Most prokaryotes have surface layers (S-layers) which lie outside the plasma membrane and serve as protective coats, molecular sieves, and enzyme scaffolding, among other functions. S-layers are 2-D arrays of a single protein or glycoprotein, with the arrangement of monomers being almost crystalline (Figure 16). Bending around the cell breaks the pure crystalline symmetry. Symmetries used by S-layers include p1, p2, p3, p4 and p6, depending on whether the subunits possess either no symmetry or 2-, 3-, 4-, or 6-fold symmetry, respectively. These symmetries are all oriented; the top of the layer differs from the bottom–an essential feature both for function and shape. One side is designed to attach to the underlying cell envelope layer, while the other side is exposed to the environment. The curvature of the layer is also favored by this symmetry. The regularity of the packing makes the S-layer lattice exhibit pores of the same size and same morphology. Isolated S-layer subunits can recrystallize into crystalline arrays, forming single- or double-layered sheets, open-ended cylinders or closed vesicles in solution. Monolayers can also form at an air/liquid surface, a liquid/solid surface, a lipid/water surface or on the envelope of the same or other prokaryotic cells. This behavior demonstrates that the information required for S-layer packing is contained within the structure of the subunits (Sleytr and Beveridge, 1999; Sleytr et al., 2001). Bacterial S-layers are discussed in Chapter 11 in Volume 7 of this series, and interested readers should consult Sleytr (2002).

Fig. 16 Atomic force microscopy image of a hexagonal bacterial S-layer. (From Scheuring et al., 2002; reprinted with permission from Blackwell Science Ltd.)

5.6 Protein Crystals

While protein crystals are not usually naturally occurring, their assembly provides further insight into the contribution of symmetry to self-assembly. The contacts between proteins within a crystal are much more tenuous than evolved biological contacts. Precipitants in solution alter the strength of van der Waals and electrostatic forces between protein surfaces to favor certain contacts. The contacts will occur at symmetric sites within the crystal, such as rotation axes and screw axes, or the proteins may contact head-to-tail to provide a purely translational component. As with natural symmetric assemblies, the necessary generators of the symmetry of the whole complex must be provided by the contacts between molecules. The number of generators needed may influence the frequency with which each space group is obtained (Wukovitz and Yeates, 1995).

6
Designed Self-assembling Protein Complexes

In order to create proteins that will self-assemble into a desired complex, it is necessary to engineer proteins that contain binding surfaces which are oriented to drive the assembly of the complex. We will examine some designed protein complexes from the point of view of how they were designed to promote proper assembly. The following section is categorized according to the methods used.

The driving force for assembly is the satisfaction of all the oligomeric interfaces of each protein subunit. It is energetically more favorable to satisfy all binding sites, rather than to leave some of them unpaired, and symmetric assemblies have the advantage of being able to do this.

6.1
Coiled-coil Extensions

"Sticky ends" can be used to drive the assembly of filaments. This strategy generates polar filaments, with subunits contacting one another head-to-tail, and two protein filaments have been created by designing overlapping coiled-coil helices.

In one design, a coiled-coil heterodimer was modified to have "sticky ends". These filaments assemble head-to-tail and also laterally to form fibers (Pandya et al., 2000) (see Figure 5A). Another design made use of a staggered assembly of five helices (Potekhin et al., 2001); this design provided greater control over lateral assembly, resulting in single filaments, and the peptide assembly was also able to undergo reversible transformations into spherical aggregates.

6.2
Domain Swapping

Domain swapping has been used to drive alpha-helical bundle proteins to assemble into filaments. This strategy is similar to the technique of using coiled-coil extensions (Ogihara et al., 2001) (see Figure 5B). A three-helix bundle was redesigned so that the third helix extended out and contacted the next subunit, rather than itself. The resulting symmetry is polar, as with the coiled-coil designs.

6.3
Symmetric Construction

The power of symmetry can be harnessed in the construction of protein complexes based on symmetric design rules. Known protein structures can be combined two at a time to create protein chimeras that possess two symmetry elements. The connection between protein oligomeric domains must be rigid enough to exclude other modes of symmetric assembly, yet flexible enough to conform to the exact angles required in a symmetric complex.

The symmetries available to a protein chimera possessing two symmetry elements are exactly those symmetries that can be generated by a set of two generators. All of the point groups can be generated by two elements (Table 1). The same is true for the helical symmetries, though not all biological plane and layer groups can be generated by two elements; some require three, four, or even five elements (Wukovitz and Yeates, 1995). Ten of the 17 layer groups, and 31 of the 65 biological space groups can be generated by two symmetry elements (Wukovitz and Yeates, 1995). Of these, there are further considerations in choosing generating sets.

Generating sets may be composed of rotations, translations, and screw symmetries. Practically speaking, if a protein binds naturally to another of its kind by translation or screw symmetry, then it will naturally form a filament, because it already contains a translational component. Therefore, it is easiest to generate layers and crystals from generating sets containing only rotations.

The design rules for obtaining some of the two-generator symmetries are listed in Table 1. The table is not guaranteed to be complete, as the sets of minimal generating sets for the space groups have not been determined systematically. As far as we know, this remains an open problem in group theory.

The design strategy outlined above has been used to generate protein shells. The connection between subunits was made by extending an alpha helix from one protein into the next to create a chimera that is a fusion of two oligomer-forming proteins. The shells are meant to assemble from 12 identical protein subunits into a complex with tetrahedral symmetry (Figure 17). The chimera is a fusion between a dimeric protein, influenza virus matrix protein M1, and a trimeric protein, bromoperoxidase, via an uninterrupted alpha helix between these two domains.

The same strategy has been used to create protein filaments from two dimeric proteins (Figure 18). The helical symmetry of the filaments is the result of combining two 2-fold rotation axes, and this also causes the filament to have 2-fold symmetry perpendicular to the helical axis, making it bipolar (see Figure 5C).

Layers and crystals can, in principle, be designed in the same way. Designed protein layers may feature control over assembly by including proteins that assemble or disassemble in response to the solution conditions.

Tab. 1 Construction rules for some of the 2-generator symmetry groups

Resulting symmetry	***Generating symmetries***	***Angle [°]***	***Intersecting?***
Point groups:			
D_n	2, 2	360/2n	Yes
D_n	2, n	90.0	Yes
T	2, 3	54.7	Yes
T	3, 3	70.5	Yes
O	2, 3	35.3	Yes
O	2, 4	45.0	Yes
O	3, 4	54.7	Yes
O	4, 4	90.0	Yes
I	2, 3	20.9	Yes
I	2, 5	31.7	Yes
I	3, 5	37.4	Yes
I	5, 5	63.4	Yes
Helical:			
helical	2, 2	any	No
Layer:			
p3	3, 3	0.0	No
p321	2, 3	90.0	No
p4	2, 4	0.0	No
$p42_12$	2, 4	90.0	No
p6	2, 3	0.0	No
p6	2, 6	0.0	No
p6	3, 6	0.0	No
Crystal:			
I4	4, 2_1	0.0	No
$I4_1$	2, 4_1	0.0	No
$P4_12_12$	2, 4_1	90.0	No
$P4_32_12$	2, 4_3	90.0	No
$P6_3$	3, 2_1	0.0	No
$P6_122$	2, 6_1	90.0	No
$P6_522$	2, 6_5	90.0	No
R3	3, 3_1	0.0	No
R3	3, 3_2	0.0	No
$P2_13$	3, 3	70.5	No
$I2_13$	2, 3	54.7	No
$P4_132$	2, 3	35.3	No
$P4_332$	2, 3	35.3	No
P432	4, 4	90.0	No
I432	2, 4	45.0	No
F432	3, 4	54.7	No

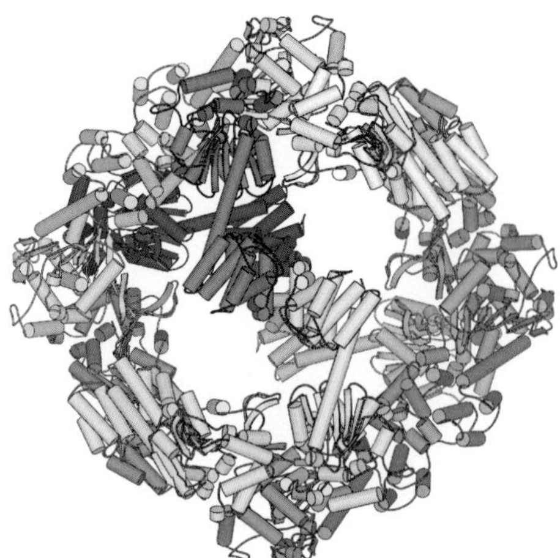

Fig. 17 Model of a designed tetrahedral protein assembly. (From Yeates and Padilla, 2002; reprinted with permission from Elsevier Science.)

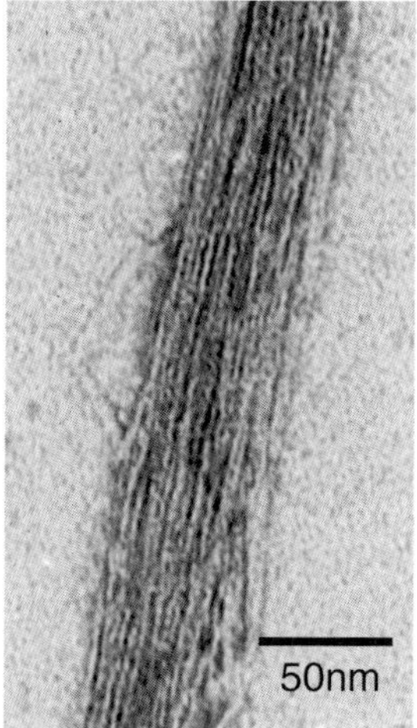

Fig. 18 Negative stain electron micrograph of a designed protein filament. In this image, the filaments have associated laterally into a bundle. (From Padilla et al., 2001.)

6.4 Polyvalency

So far, there has been one example of a designed protein crystal. Dotan et al. (1999) took advantage of the carbohydrate-binding ability of the lectin concanavalin A, together with its natural association as a D_2 tetramer. The tetrameric complex has a roughly tetrahedral shape, and binds carbohydrates at the points of the "tetrahedron". A two-headed carbohydrate was used to join the corners of the tetramers; this causes each protein monomer to be polyvalent, in contact with three or more other molecules (Figure 19). Achieving ordered connectivity via polyvalency promotes crystal growth.

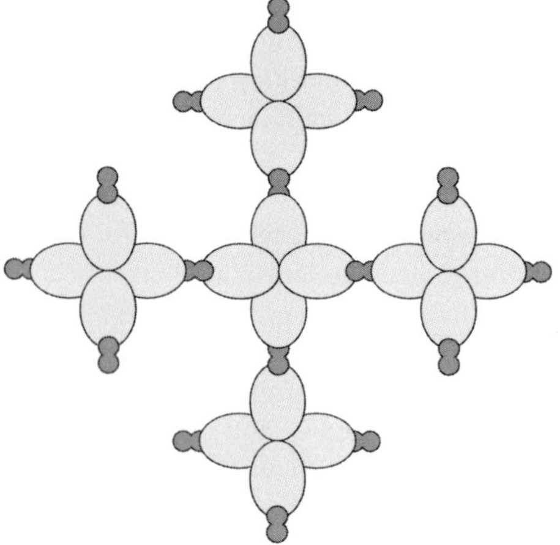

Fig. 19 Schematic drawing of lectin tetramers connected via bivalent carbohydrates. (From Yeates and Padilla, 2002; reprinted with permission from Elsevier Science.)

7 Observing Protein Assemblies

Several methods are available for observing protein complexes of various sizes at resolutions ranging from the atomic to the nanometer scale. These methods are transmission electron microscopy with negative staining, cryoelectron microscopy, atomic force microscopy, electron diffraction, and X-ray diffraction. The differences between these methods include resolution, signal-to-noise ratio, sample damage, and preparation of the sample, making all of these techniques valuable in different ways for the determination of macromolecular structures.

7.1 Transmission Electron Microscopy

Transmission electron microscopy (TEM), which was first developed during the 1930s, is widely used to examine the structures of protein complexes. Materials for TEM are prepared to allow electrons to transmit through the sample, and the internal structure of the sample is then observed by the electrons being diffracted by the sample (Watt, 1985). A negative stain electron micrograph is shown in Figure 18.

There are several ways in which samples can be prepared:

- Negative staining, which was developed during the late 1950s, accounts for many exciting discoveries of supramolecular structures, such as the icosahedral viral capsid, actin filaments, microtubules, and S-layers. In negative staining, the sample is surrounded or embedded in an electron-dense material that provides high contrast and good preservation. Thus, the size, shape and surface structure can be revealed under TEM. The resolution of TEM on negative-stained samples can be 1–2 nm (Harris, 1991).
- Metal shadowing was invented in 1945. In this method, a thin layer of metal is evaporated onto the surface of a freeze-dried specimen. Minimal amounts of metal are able to accumulate on the side of an object which is away from the metal source; thus, any surface features of particles are reflected by the metal coating and, under TEM, the detailed information of the surface is revealed. The resolution of TEM on a metal-shadowed sample is generally poorer than with negative-stained samples (Harris, 1991).

Artifacts may be caused within the sample as a result of chemical procedures such as staining, and unstained specimens are sometimes used to preserve the native state of the sample as much as possible. Wet specimens are examined at room temperature on a hydrated stage, while sugar-embedded samples are prepared just as in negative staining, except that the negative stain solution is replaced by a sugar solution. These techniques are only appropriate for highly ordered samples, however.

7.2 Cryoelectron Microscopy

Cryoelectron microscopy (cryo-EM) differs from negative-staining techniques primarily in sample preparation and image analysis. Both utilize TEM, but in this technique an aqueous sample is flash-frozen in order to preserve all the molecules in an undistorted state. Freezing occurs fast enough to form vitreous ice; no ice crystals have a chance to form. Unfortunately, in the absence of any stain, the contrast between protein and solvent is very low, and image reconstruction is generally necessary.

To reconstruct the 3-D structure of a complex from cryo-EM images requires several stages. In a noncrystalline sample, complexes are usually oriented almost randomly within the vitreous ice; this means that a single image contains several particles seen from different viewpoints. With the aid of computer software, images of the particle from the same view can be averaged to increase the signal-to-noise ratio. Each image is a projection of a 3-D object, and the projection angles must be determined computationally. When the correct orientations are determined, the views are reprojected in 3-D space to create the 3-D structure of the protein complex (van Heel et al., 2000). Figure 10 shows an example of a cryo-EM reconstruction of a viral capsid structure.

The resolution of a cryo-EM reconstruction can reach 1 nm or better, with resolutions being reported up to 5.9 Å. This resolution is limited by the homogeneity of the particles, how accurately the projection angles have been determined, the symmetry of the particle, and how many images have been collected. Symmetry in the particle effectively increases the number of observations. As with other electron microscopy techniques, antibody and gold labeling can be used to identify certain polypeptides within the structure.

7.3 Atomic Force Microscopy

This approach permits the surface structure of a particle or layer to be determined literally by touching it gently. Atomic force microscopy (AFM) features a good signal-to-noise ratio, and permits individual molecules to be seen. A sharp tip, typically made of silicon nitride (Si_3N_4) for biological samples, is scanned across the surface of the sample. There are two modes, contact mode and tapping mode. Tapping mode is used when a particle cannot be immobilized well, and features slightly worse resolution than contact mode, but is suitable for single protein complexes or filaments (Engel et al., 1999).

Contact mode AFM can achieve 1 nm lateral and 0.1 nm vertical resolution. Biological samples are generally viewed in aqueous solution, as dehydration often distorts them. Electrostatic and van der Waals interactions govern the sample–tip interaction, which can also be influenced by the electrolyte composition of the solution.

AFM has been used to image membrane protein oligomers and the variation in flexible regions of their structure. An AFM image of a bacterial S-layer is shown in Figure 16. Conformational changes can also be viewed by AFM as the tip can be attached to the sample or used to manipulate atoms or molecules. It has also been used to remove individual subunits from crystalline S-layer surfaces, or sometimes hexamers or other oligomers are removed, each subunit in succession.

7.4 X-ray and Electron Diffraction

Atomic structures, at better than 1 Å resolution, can be determined using X-ray crystallography. The first stage is to produce a crystal of diffraction quality, after which the X-ray diffraction pattern is recorded. There is no substance that can act as a lens to refocus the X-rays that are scattered by the crystal; thus, X-rays cannot be used in the same way as can visible light be used in a microscope. This leads to the so-called "phase problem" of crystallography, and the lost phase information must be reconstructed using either experimental or computational means in order to form an image of the electron density of the molecules in the crystal. The chemical structure is combined with the electron density to create an atomic model of the molecule, such as those shown in Figures 3, 8, 9, and 14.

The diffraction from single molecules is too weak to reconstruct structural information. The crystal, with millions of copies of the same molecule, acts as an amplifier of the signal – hence the need for crystallization. Crystals unsuitable for X-ray diffraction may be suitable for electron diffraction; this technique requires 2-D or very thin 3-D crystals, and the resolution obtainable is 3.7 Å, or better (Rossmann et al., 2001).

8 Applications

Protein materials that self-assemble based on symmetry are being explored for a variety of uses. Peptide-based linear assemblies appear to show promise as scaffolds for tissue engineering (Holmes, 2002), while peptides that self-assemble into fibers in a β-sheet-like structure have been used to produce a hydrogel material that supports neurite growth and synapse formation (Holmes et al., 2000). A nanofiber formed by the self-assembly of an alkylated peptide was found to direct mineralization of hydroxyapatite with an organization similar to that of the collagen and hydroxyapatite crystals found in bone (Hartgerink et al., 2001). Self-assembling peptides may also be used as antibacterial agents, as cyclic peptides with alternating D- and L-α-amino acids were found to stack into hollow, β-sheet-like tubular structures in lipid membranes and cause rapid cell death, with some of these peptides possessing effective in-vivo antibacterial efficacy (Fernandez-Lopez et al., 2001).

The uses for cage-like protein assemblies have been investigated only minimally. The virion of cowpea chlorotic mottle virus is able to host polyoxometalate mineralization and trap an anionic organic polymer in a pH-dependent manner (Douglas and Young, 1998), which suggests that protein cages might be used for molecular capture and material synthesis. Further information relating to protein cages and their applications may be found in Chapter 14 in Volume 8 of this series, and interested readers should consult Douglas (2002). Other possible uses include molecular delivery vehicles, adjuvants, or selective catalytic units to which only sufficiently small substrates can gain access.

S-layers are excellent representatives of self-assembled protein layers, and many potential uses of these materials have been investigated (Sleytr et al., 2001). The highly symmetric and repetitive nature of S-layers means that they have pores of identical size and morphology, and are also able to self-assemble on solid surfaces. Permeability assays have suggested that S-layers could be used to construct ultrafiltration membranes, and they have also been used as matrices for the covalent attachment of ligands, leading to products such as affinity microparticles, dipsticks in solid-phase immunoassays and enzyme-based biosensors. S-layers are suitable as vaccine adjuvants (because the self-assembly may potentiate immune responses), and as templates for the preparation of regularly arranged nanoparticles. A recent development has been that of designed symmetric protein layers (Padilla et al., 2001), and both naturally occurring, de-novo designed or genetically altered S-layers might prove commercially valuable for a variety of processes in the near future.

9 Patents

Some typical US patents relating to the development and use of self-assembling macromolecular complexes are listed in Table 2. Similar patents are listed together as one item, with the information on the first-mentioned item as being representative.

10 Outlook and Perspectives

It is clear that the future is very bright for nanotechnology, and the role of proteins in this area is continuing to develop. For example, as proteins are able to organize and encapsulate other materials, to present

Tab. 2 US patents relating to the construction and utilization of self-assembling macromolecular complexes

Viral capsids	
5,985,610 (5,871,998) (5,756,284) (5,744,142) (5,716,620) (5,709,996) (5,437,951)	Self-assembling recombinant papillomavirus capsid proteins Lowy; Douglas R. (Washington, DC); Schiller; John T. (Silver Spring, MD); Kirnbauer; Reinhard (Bethesda, MD) "Recombinant papillomavirus capsid proteins ...can be prepared as vaccines to induce a high-titer neutralizing antibody response in vertebrate animals. The self assembling capsid proteins can also be used as elements of diagnostic immunoassay procedures for papillomavirus infection."
5,905,040	Parvovirus empty capsids Mazzara; Gail P. (Winchester, MA); Destree; Antonia T. (Boston, MA); Panicali; Dennis L. (Acton, MA) "Eukaryotic transfectants express self-assembled empty viral capsids which can be used to vaccinate against the virus or antigenically related species of the virus."
5,747,324 (5,736,368) (5,631,154) (5,614,404) (5,420,026)	Self-assembled, defective, non-self-propagating lentivirus particles Mazzara; Gail P. (Winchester, MA); Roberts; Bryan (Cambridge, MA); Panicali; Dennis L. (Acton, MA); Gritz; Linda R. (Somerville, MA); Stallard; Virginia (Sequim, WA); Mahr; Anna (Natick, MA) "The present invention provides recombinant DNA viral vectors which co-express lentivirus genes encoding structural and enzymatic polypeptides capable of assembling into defective nonself-propagating viral particles. The viral DNA vectors as well as the viral particles can be used as immunogens and for targeted delivery of heterologous gene products and genes."
20020081295 (Patent application)	Virus-like particles for the induction of autoantibodies Schiller, John T.; (Silver Spring, MD); Chackerian, Bryce; (Chevy Chase, MD); Lowy, Douglas R.; (Bethesda, MD) "Novel biotechnological tools, pharmaceuticals, therapeutics and prophylactics, which concern chimeric or conjugated virus-like particles, and methods of use of the foregoing are provided for the study of B cell tolerance and the treatment or prevention of human diseases, which involve the onset of B cell tolerance."
Oligomeric proteins	
6,190,886	Trimerizing polypeptides, their manufacture and use Hoppe; Hans-Jurgen (MRC Immunochemistry Unit, Dept. of Biochemistry, University of Oxford,, OX1 3QU, Oxford, GB); Reid; Kenneth B. (MRC Immunochemistry Unit, Dept. of Biochemistry, University of Oxford,, OX1 3QU, Oxford, GB) "Polypeptides comprising a collectin neck region, or variant or derivative thereof or amino acid sequence having the same or a similar amino acid pattern and/or hydrophobicity profile, are able to trimerize." "One use for the polypeptides is in seeding collagen formation."
Protein layers	
4,802,951 (4,728,591)	Method for parallel fabrication of nanometer scale multi-device structures Clark; Noel A. (Boulder, CO); Douglas; Kenneth (Boulder, CO); Rothschild; Kenneth J. (Newton, MA) "Articles exhibiting fabricated structures with nanometer size scale features ... are produced by a method employing a substrate base or coating and a thin layer serving as a lithographic mask or template, consisting of a self-assembled ordered material array, typically a periodic array of molecules such as undenatured proteins ... "

Tab. 2 (cont.)

5,874,267	Expression of surface layer proteins Deblaere; Rolf Y. (Waarschoot, BE); Desomer; Jan (Drongen, BE); Dhaese; Patrick (Drongen, BE) "A host cell which is provided with a *S-layer* comprising a fusion polypeptide consisting essentially of: (a) at least sufficient of a *S-layer* protein for a *S-layer* composed thereof to assemble, and (b) a heterologous polypeptide which is fused to either the carboxy terminus of (a) or the amino terminus of (a) and which is thereby presented on the outer surface of the said cell; can be used as a vaccine, for screening for proteins and antigens and as a support for immobilizing an enzyme, peptide or antigen."
6,296,700	Method of producing a structured layer Sleytr; Uwe B. (Vienna, AT); Pum; Dietmar (Vienna, AT); Loschner; Hans (Vienna, AT) "The present invention relates to a method of producing a structured layer of defined functional molecules on the upper face of a basically substantially flat sheet substrate. For this purpose, the substrate is covered with a monolayer of crystalline cell surface layer (*S-layer*) containing protein deposited by recrystallisation."
5,500,353	Bacterial surface protein expression Smit; John (Richmond, CA); Bingle; Wade H. (Vancouver, CA) "This invention provides a bacterium having an *S-layer* modified such that the bacterium *S-layer* protein gene contains one or more in-frame sequences coding for one or more heterologous polypeptides and, the *S-layer* is a fusion product of the *S-layer* protein and the heterologous polypeptide."
General methods for the construction of nanoscale biomaterials	
5,712,366	Fabrication of nanoscale materials using self-assembling proteins McGrath; Kevin P. (Grafton, MA); Kaplan; David L. (Stow, MA) "The "recognition elements," which are in fact non-naturally occurring amino acid sequences, control the molecular organization and assembly of the materials into which they are incorporated, and can be combined in different patterns and proportions to generate complex molecular architectures and functions."

antigens, and to bind specific targets, it is likely that self-assembling protein materials will be of special use in medical applications, for detection assays, and in the organization of other materials on the nanoscale.

In addition, this is a time of major achievement in the field of structural biology of large protein complexes, using techniques such as X-ray diffraction, cryo-EM, and electron crystallography. It is likely that new structures will emerge as these "molecules of life" are increasingly visualized.

11 References

Adams, M. J., Blundell, T. L., Dodson, E. J., Dodson, G. G., Yijayan, M., Baker, E. N., Harding, M. M., Hodgkin, D. C., Rimmer, B., Sheat, S. (1969) Structure of rhombohedral 2 zinc insulin crystals, *Nature* **224**, 491–495.

Amos, L. A., Klug, A. (1974) Arrangement of subunits in flagellar microtubules, *J. Cell Sci.* **14**, 523–549.

Anderson, W. F., Ohlendorf, D. H., Takeda, Y., Matthews, B. W. (1981) Structure of the cro repressor from bacteriophage lambda and its interaction with DNA, *Nature* **290**, 754–758.

Bennett, M. J., Choe, S., Eisenberg, D. (1994) Domain swapping – entangling alliances between proteins, *Proc. Natl. Acad. Sci. USA* **91**, 3127–3131.

Boisen, M. B., Gibbs, G. V. (1985) Mathematical crystallography: an introduction to the mathematical foundations of crystallography. Washington, DC: Mineralogical Society of America.

Boisvert, D. C., Wang, J. M., Otwinowski, Z., Horwich, A. L., Sigler, P. B. (1996) The 2.4 angstrom crystal structure of the bacterial chaperonin GroEL complexed with ATP gamma S, *Nature Struct. Biol.* **3**, 170–177.

Caspar, D. L. D., Klug, A. (1962) Physical principles in the construction of regular viruses, *Cold Spring Harbor Symp. Quant. Biol.* **27**, 1–24.

Creighton, T. E. (1993) *Proteins: structures and molecular properties.* New York: W. H. Freeman.

Crick, F. H. C., Watson, J. D. (1956) Structure of small viruses, *Nature* **177**, 473–476.

Crowfoot, D. (1935) X-ray single crystal photographs of insulin, *Nature* **135**, 591–592.

Dotan, N., Arad, D., Frolow, F., Freeman, A. (1999) Self-assembly of a tetrahedral lectin into predesigned diamondlike protein crystals, *Angew. Chemie Int. Ed.* **38**, 2363–2366.

Douglas, T., et al. (2002) Self-assembling Protein Cage-Systems and Applications in Nanotechnology, in: *Biopolymers* (Steinbüchel, A., Ed.), Weinheim: Wiley-VCH, Vol. 8.

Douglas, T., Young, M. (1998) Host-guest encapsulation of materials by assembled virus protein cages, *Nature* **393**, 152–155.

Downing, K. H. (2000) Structural basis for the interaction of tubulin with proteins and drugs that affect microtubule dynamics, *Annu. Rev. Cell Dev. Biol.* **16**, 89–111.

Eakins, F., Al-Khayat, H. A., Kensler, R. W., Morris, E. P., Squire, J. M. (2002) 3D structure of fish muscle myosin filaments, *J. Struct. Biol.* **137**, 154–163.

Engel, A., Lyubchenko, Y., Muller, D. (1999) Atomic force microscopy: a powerful tool to observe biomolecules at work, *Trends Cell Biol.* **9**, 77–80.

Fernandez-Lopez, S., Kim, H. S., Choi, E. C., Delgado, M., Granja, J. R., Khasanov, A., Kraehenbuehl, K., Long, G., Weinberger, D. A., Wilcoxen, K. M., Ghadiri, M. R. (2001) Antibacterial agents based on the cyclic D,L-alpha-peptide architecture, *Nature* **412**, 452–455.

Franklin, R. E. (1955) Structure of tobacco mosaic virus, *Nature* **175**, 379–381.

Geeves, M. A., Holmes, K. C. (1999) Structural mechanism of muscle contraction, *Annu. Rev. Biochem.* **68**, 687–728.

Ghadiri, M. R., Granja, J. R., Milligan, R. A., McRee, D. E., Khazanovich, N. (1993) Self-assembling organic nanotubes based on a cyclic peptide architecture, *Nature* **366**, 324–327.

Goodsell, D. S., Olson, A. J. (2000) Structural symmetry and protein function, *Annu. Rev. Biophys. Biomol. Struct.* **29**, 105–153.

Hanson, J., Huxley, H. E. (1953) Structural basis of the cross-striations in muscle, *Nature* **172**, 530–532.

Hanson, J., Lowy, J. (1963) The structure of F-actin and of actin filaments isolated from muscle, *J. Mol. Biol.* **6**, 46–60.

Harris, R. E. (1991) *Electron Microscopy in Biology, a Practical Approach*. Oxford University Press, New York.

Hartgerink, J. D., Beniash, E., Stupp, S. I. (2001) Self-assembly and mineralization of peptide-amphiphile nanofibers, *Science* **294**, 1684–1688.

Hecht, H. J., Sobek, H., Haag, T., Pfeifer, O., Vanpee, K. H. (1994) The metal-ion free oxidoreductase from *Streptomyces aureofaciens* has an alpha/beta hydrolase fold, *Nature Struct. Biol.* **1**, 532–537.

Holmes, T. C. (2002) Novel peptide-based biomaterial scaffolds for tissue engineering, *Trends Biotechnol.* **20**, 16–21.

Holmes, T. C., de Lacalle, S., Su, X., Liu, G., Rich, A., Zhang, S. (2000) Extensive neurite outgrowth and active synapse formation on self-assembling peptide scaffolds, *Proc. Natl. Acad. Sci. USA* **97**, 6728–6733.

Hornung, C. P. (1959) *Handbook of Designs and Devices*. New York: Dover Publications, Inc..

Houwink, A. L. (1953) A macromolecular monolayer in the cell wall of *Spirillum* spec., *Biochim. Biophys. Acta* **10**, 360–366.

Kanaseki, T., Kadota, K. (1969) The "Vesicle in a basket", *J. Cell Biol.* **42**, 202–220.

Kim, C. A., Phillips, M. L., Kim, W., Gingery, M., Tran, H. H., Robinson, M. A., Faham, S., Bowie, J. U. (2001) Polymerization of the SAM domain of TEL in leukemogenesis and transcriptional repression, *EMBO J.* **20**, 4173–4182.

Kim, C. A., Gingery, M., Pilpa, R. M., Bowie, J. U. (2002) The SAM domain of polyhomeotic forms a helical polymer, *Nature Struct. Biol.* **9**, 453–457.

Kim, K. K., Kim, R., Kim, S. H. (1998) Crystal structure of a small heat-shock protein, *Nature* **394**, 595–599.

Klug, A. (1999) The tobacco mosaic virus particle: structure and assembly, *Philos. Trans. R. Soc. Lond. B Biol. Sci.* **354**, 531–535.

Klug, A., Caspar, D. L. D. (1960) The structure of small viruses, in: *Advances in Virus Research*, Volume 7 (Smith, K. M., Lauffer, M. A., Eds.), New York, London: Academic Press Inc., 225.

Kraulis, P. J. (1991) Molscript – a program to produce both detailed and schematic plots of protein structures, *J. Appl. Crystallogr.* **24**, 946–950.

Liu, Y., Eisenberg, D. (2002) 3D domain swapping: as domains continue to swap, *Protein Sci.* **11**, 1285–1299.

Merritt, E. A., Bacon, D. J. (1997) Raster3D: photorealistic molecular graphics, *Macromol. Crystallogr. Pt B* **277**, 505–524.

Ogihara, N. L., Ghirlanda, G., Bryson, J. W., Gingery, M., DeGrado, W. F., Eisenberg, D. (2001) Design of three-dimensional domain-swapped dimers and fibrous oligomers, *Proc. Natl. Acad. Sci. USA* **98**, 1404–1409.

Orlova, A., Galkin, V. E., VanLoock, M. S., Kim, E., Shvetsov, A., Reisler, E., Egelman, E. H. (2001) Probing the structure of F-actin: cross-links constrain atomic models and modify actin dynamics, *J. Mol. Biol.* **312**, 95–106.

Padilla, J. E., Colovos, C., Yeates, T. O. (2001) Nanohedra: using symmetry to design self assembling protein cages, layers, crystals, and filaments, *Proc. Natl. Acad. Sci. USA* **98**, 2217–2221.

Pandya, M. J., Spooner, G. M., Sunde, M., Thorpe, J. R., Rodger, A., Woolfson, D. N. (2000) Sticky-end assembly of a designed peptide fiber provides insight into protein fibrillogenesis, *Biochemistry* **39**, 8728–8734.

Parry, D. A., Steinert, P. M. (1999) Intermediate filaments: molecular architecture, assembly, dynamics and polymorphism, *Q. Rev. Biophys.* **32**, 99–187.

Potekhin, S. A., Melnik, T. N., Popov, V., Lanina, N. F., Vazina, A. A., Rigler, P., Verdini, A. S., Corradin, G., Kajava, A. V. (2001) De novo design of fibrils made of short alpha-helical coiled coil peptides, *Chem. Biol.* **8**, 1025–1032.

Rossmann, M. G., Arnold, E., International Union of Crystallography (2001) International tables for crystallography Volume F. Crystallography of biological macromolecules. Dordrecht, London: Kluwer Academic, published for the International Union of Crystallography.

Scheuring, S., Stahlberg, H., Chami, M., Houssin, C., Rigaud, J. L., Engel, A. (2002) Charting and unzipping the surface layer of *Corynebacterium glutamicum* with the atomic force microscope, *Mol. Microbiol.* **44**, 675–684.

Slautterback, D. B. (1963) Cytoplasmic microtubules, *J. Cell Biol.* **18**, 367–388.

Sleytr, U.B. (2002) Self-assembly Protein Systems: Microbial S-layers, in A. Steinbüchel (ed.) Biopolymers, volume 7, Wiley-VCH, Weinheim, pp. 285–338.

Sleytr, U. B. (1978) Regular arrays of macromolecules on bacterial cell walls: structure, chemistry, assembly, and function, *Int. Rev. Cytol.* **53**, 1–64.

Sleytr, U. B., Beveridge, T. J. (1999) Bacterial S-layers, *Trends Microbiol.* **7**, 253–260.

Sleytr, U. B., Sara, M., Pum, D., Schuster, B. (2001) Characterization and use of crystalline bacterial cell surface layers, *Prog. Surface Sci.* **68**, 231–278.

Smith, C. J., Grigorieff, N., Pearse, B. M. (1998) Clathrin coats at 21 Å resolution: a cellular assembly designed to recycle multiple membrane receptors, *EMBO J.* **17**, 4943–4953.

Squire, J. M. (1971) General model for the structure of all myosin-containing filaments, *Nature* **233**, 457–462.

Stewart, P. L., Cary, R. B., Peterson, S. R., Chiu, C. Y. (2000) Digitally collected cryo-electron micrographs for single particle reconstruction, *Microsc. Res. Technique* **49**, 224–232.

Strauss, J. H., Strauss, E. G. (2002) *Viruses and Human Disease.* San Diego: Academic Press.

van Heel, M., Gowen, B., Matadeen, R., Orlova, E. V., Finn, R., Pape, T., Cohen, D., Stark, H., Schmidt, R., Schatz, M., Patwardhan, A. (2000) Single-particle electron cryo-microscopy: towards atomic resolution, *Q. Rev. Biophys.* **33**, 307–369.

Villeret, V., Clantin, B., Tricot, C., Legrain, C., Roovers, M., Stalon, V., Glansdorff, N., Van-Beeumen, J. (1998) The crystal structure of *Pyrococcus furiosus* ornithine carbamoyltransferase reveals a key role for oligomerization in enzyme stability at extremely high temperatures, *Proc. Natl. Acad. Sci. USA* **95**, 2801–2806.

Wade, R. H., Meurer-Grob, P., Metoz, F., Arnal, I. (1998) Organisation and structure of microtubules and microtubule-motor protein complexes, *Eur. Biophys. J.* **27**, 446–454.

Watt, J. M. (1985) *The Principles and Practice of Electron Microscopy.* Cambridge: Cambridge University Press.

Wilson, H. R. (1966) *Diffraction of X-rays by Proteins, Nucleic Acids and Viruses.* New York: St. Martin's Press.

Wukovitz, S. W., Yeates, T. O. (1995) Why protein crystals favour some space-groups over others, *Nature Struct. Biol. 2*, 1062–1067.

Xu, Z. H., Horwich, A. L., Sigler, P. B. (1997) The crystal structure of the asymmetric GroEL-GroES-(ADP)(7) chaperonin complex, *Nature* **388**, 741–750.

Yeates, T. O., Padilla, J. E. (2002) Designing supramolecular protein assemblies, *Curr. Opin. Struct. Biol.* **12**, 464–470.

11
Self-assembly Protein Systems: Microbial S-layers

Prof. Dr. Uwe B. Sleytr[1], Dr. Margit Sára[2], Dr. Dietmar Pum[3], Dr. Bernhard Schuster[4], Dr. Paul Messner[5], Dr. Christina Schäffer[6]

[1] Center for Ultrastructure Research and Ludwig Boltzmann Institute for Molecular Nanotechnology, University of Agricultural Sciences Vienna, Gregor Mendel Str.33, A-1180 Vienna, Austria; Tel.: +43-1-476542201; Fax: +43-1-4789112; E-mail: sleytr@edv1.boku.ac.at

[2] Center for Ultrastructure Research and Ludwig Boltzmann Institute for Molecular Nanotechnology, University of Agricultural Sciences Vienna, Gregor Mendel Str.33, A-1180 Vienna, Austria; Tel.: +43-1-476542208; Fax: +43-1-4789112; E-mail: sara@edv1.boku.ac.at

[3] Center for Ultrastructure Research and Ludwig Boltzmann Institute for Molecular Nanotechnology, University of Agricultural Sciences Vienna, Gregor Mendel Str.33, A-1180 Vienna, Austria; Tel.: +43-1-476542205; Fax: +43-1-4789112; E-mail: dpum@edv1.boku.ac.at

[4] Center for Ultrastructure Research and Ludwig Boltzmann Institute for Molecular Nanotechnology, University of Agricultural Sciences Vienna, Gregor Mendel Str.33, A-1180 Vienna, Austria; Tel.: +43-1-476542213; Fax: +43-1-4789112; E-mail: bschuste@edv1.boku.ac.at

[5] Center for Ultrastructure Research and Ludwig Boltzmann Institute for Molecular Nanotechnology, University of Agricultural Sciences Vienna, Gregor Mendel Str.33, A-1180 Vienna, Austria; Tel.: +43-1-476542202; Fax: +43-1-4789112; E-mail: pmessner@edv1.boku.ac.at

[6] Center for Ultrastructure Research and Ludwig Boltzmann Institute for Molecular Nanotechnology, University of Agricultural Sciences Vienna, Gregor Mendel Str.33, A-1180 Vienna, Austria; Tel.: +43-1-476542203; Fax: +43-1-4789112; E-mail: crs@edv1.boku.ac.at

Ala	alanine
AMP	affinity microparticles
Asn	asparagine
Asp	aspartic acid
$[AuCl_4]^-$	tetrachloroauric acid
Bet v1	major birch pollen allergen
BIP 1	monoclonal antibody against the major birch pollen allergen
BLM	bilayer lipid membrane
$CdCl_2$	cadmium chloride
CdS	cadmium sulfide
CdSe	cadmium selenide
DNA	deoxyribonucleic acid
EDTA	ethylenediaminetetraacetic acid
FACE	fluorophore-assisted carbohydrate electrophoresis
Fuc3NAc	3-*N*-acetylfucosamine
GalA	galacturonic acid

Gal*f*	galactofuranose
GalNAc	*N*-acetylgalactosamine: GHCl: guanidinium hydrochloride
GlcA	glucuronic acid
GlcNAc	*N*-acetylglucosamine
Glc*p*	glucopyranose: H_2S: hydrogen sulfide
HPAEC/PED	high-performance anion exchange chromatography pulsed electrochemical detection
IdA	iduronic acid
IgE	immunoglobulin E
IgG	immunoglobulin G
IL-8	interleukin 8
$KPtCl_6$	potassium palladium chloride
MALDI-TOF	matrix assisted laser-desorption-ionization mass spectrometry
Man	mannose
ManA	mannuronic acid
$ManA2,3(NAc)_2$	2,3-di-actetamidodideoxymannuronic acid
ManNAc	*N*-acetylmannosamine
MDS	microspheres-based detoxification system
MS	mass spectrometry
$NiSO_4$	nickel sulfate
NMR	nuclear magnetic resonance spectroscopy
*O*Me	*O*-methyl
3-*O*Me-GalA	3-*O*-methylgalacturonic acid
PAS	periodic acid-Schiff staining
$Pb(NO_3)_2$	lead nitrate
$PdCl_2$	palladium chloride
pI	isoelectric point
PO_4^{3-}	phosphate
Pt	platinum: $[PtCl_4]^{2-}$: tetrachloroplatinum acid
Qui3NAc	3-*N*-acetylquinovosamine (3-acetamido-3,6-dideoxyglucose)
RAST	radioallergosorbent test
Rha	rhamnose
SCWP	secondary cell wall polymer
SDS-PAGE	sodiumdodecylsulfate polyacrylamide gel electrophoresis
Ser	serine
S-layer	surface layer
SLH	S-layer homology domain
SO_4^{2-}	sulfate
SsLM	S-layer-supported lipid membrane
SUM	S-layer ultrafiltration membrane
Ta/W	tantalum/tungsten
Thr	threonine
t-PA	tissue type plasminogen activator
Tyr	tyrosine

1 Introduction

The different cell wall structures observed in prokaryotic organisms (archaea and bacteria), particularly the outermost envelope layers exposed to the environment, reflect evolutionary adaptations of the organisms to a broad spectrum of selection criteria (Sleytr and Beveridge, 1999; Sleytr et al., 1993). Crystalline arrays of proteinaceous subunits forming surface layers (S-layers) (Sleytr, 1978; Sleytr et al., 1994) are now recognized as one of the most common outermost cell envelope components of prokaryotic organisms (Sleytr et al., 1996b). Most of the presently known S-layers are composed of a single protein or glycoprotein species endowed with the ability to assemble into monomolecular arrays on the supporting envelope layer.

S-layers, as the most abundant of bacterial cellular proteins, are important model systems for studies of structure, synthesis, genetics, assembly, and functions of proteinaceous supramolecular structures. In fact, S-layers represent the simplest biological membranes developed during evolution. They can provide organisms with a selection advantage by fulfilling various functions (Sleytr et al., 1993) including protective coats, molecular sieves, ion traps, and structures involved in cell surface interactions. In a great variety of archaea, they are involved in determining cell shape, and they were also identified as contributors to virulence when present as a structural component of cell walls of pathogens.

The wealth of information accumulated on the general principles of S-layers led to a broad spectrum of applications in material sciences (Sleytr and Sára, 1997; Sleytr et al., 1999, 2001a,b). Most relevant, S-layers represent very versatile self-assembly systems with unique features as a structural basis for a complete supramolecular construction kit, involving all major species of biological molecules (e.g., lipids, glycans, nucleic acids). The possibility to change the natural properties of S-layer proteins by genetic engineering and to incorporate single or multifunctional domains in S-layer lattices has opened new horizons for the tuning of their structural and functional features. Applications for S-layers have been found (1) in the production of isoporous ultrafiltration membranes, (2) as support for defined immobilization or incorporation of functional molecules (e.g., enzymes, antigens, antibodies, ligands) as required for affinity and enzyme membranes, and (3) in the development of biosensors or in solid-phase immunoassays. S-layers also have been used as support and stabilizing matrices for functional lipid membranes and liposomes. Another line of application concerns the use of S-layers as adjuvants for weakly immunogenic antigens and haptens. More recently, S-layer technologies have provided new approaches for controlled biomineralization, for nanopatterning of surfaces, and for formation of ordered arrays of metal clusters or nanoparticles as required for molecular electronics and non-linear optics.

2 Historical Outline

In 1953 Houwink described the presence of a "macromolecular monolayer" in the cell wall of *Spirillum* sp. (Houwink, 1953) for the first time. Subsequently, it took some time before the relevance of this cell wall structure was appreciated. Regular arrays of macromolecules as part of prokaryotic cell envelopes are now recognized to be widely present in the prokaryotic domains Archaea and Bacteria. Crystalline cell surface layers

(S-layers) have been identified in hundreds of different strains belonging to all major phylogenetic groups of the domain Bacteria and represent an almost universal feature in Archaea (Murray, 1993; Sleytr et al., 1996a; Sleytr and Messner, 1996). Since the beginning of S-layer research, different names have been coined for these regularly arrayed molecules:

- regular surface layer in *Aquaspirillum* sp. (Beveridge and Murray, 1974),
- tetragonal layer in *Bacillus sphaericus* (Aebi et al., 1973),
- S-layer in a number of gram-positive and gram-negative strains (Sleytr, 1978),
- additional layer in *Aeromonas salmonicida* (Udey and Fryer, 1978),
- hexagonally packed intermediate (HPI) layer in *Deinococcus radiodurans* (Baumeister and Kübler, 1978),
- middle and outer wall protein in *Brevibacillus* (formerly *Bacillus*) *brevis* (Yamada et al., 1981),
- regular array in *Clostridium difficile* (Kawata et al., 1984),
- cell surface glycoprotein in *Halobacterium halobium* (Lechner and Sumper, 1987),
- surface-protein antigen in *Rickettsia prowazeki* (Carl and Dasch, 1989), and
- surface-array protein in *Campylobacter fetus* (Blaser and Gotschlich, 1990).

At the Second International Workshop on S-layers, held in Vienna in 1987, it was agreed to use the term "S-layer" for two-dimensional crystalline arrays of proteinaceous subunits that form surface layers on prokaryotic cells (Sleytr et al., 1988a). In the 1960s the early pioniers in S-layer research, in particular R. G. E. Murray, described complex patterns of hexagonally packed subunits in negatively stained preparations of *Micrococcus* (now *Deinococcus*) *radiodurans* (Glauert, 1962; Murray, 1962), *Spirillum* (now *Aquaspirillum*) sp. (Murray, 1963a), and even more complex regular layers in *Lampropedia hyalina* (Murray, 1963b). Different electron microscopy preparation techniques to visualize S-layers have been applied, such as negative staining, staining with auro thioglucose for low dose electron microscopy, cryoelectron microscopy, freeze-drying or freeze-etching and metal-shadowing, and ultrathin sectioning (Sleytr and Glauert, 1975; Sleytr, 1978; Kellenberger and Kistler, 1979; Taylor et al., 1982; Lepault and Pitt, 1984; Sleytr et al., 1988b; Woodcock and Baumeister, 1990). The ultimate goal of all these procedures was imaging S-layer lattices as close as possible to the native structure on the cell surface and with a minimum of preparation artifacts. Freeze-etching preparations in particular provided strong evidence for a dynamic process of assembly of S-layer protein on growing and dividing cells (Sleytr, 1975, 1978; Sleytr and Glauert, 1975; Sleytr and Messner, 1989).

Because of the highly ordered arrangement of their subunits, S-layers have been a widely used subject for structural studies, in particular with sophisticated methods of computer-assisted image analysis and reconstruction. S-layers may be classified on the basis of their symmetries. Thus, powerful tools of modern crystallography have been applied (for review, see Amos et al., 1982), and even distorted S-layer lattices were reconstructed (Crowther and Sleytr, 1977; Saxton and Baumeister, 1982). S-layers were studied mainly by transmission electron microscopy, which yielded a projection of the structure. The projected two-dimensional lattice symmetry of S-layers is a unique feature and was often used for their taxonomical classification. Three-dimensional models of the protein mass distribution obtained from tilt series of S-layer fragments or self-assembly products showed a remarkable difference between the two surfaces of the S-layer. This asymmetry makes sense in biological terms, since the S-layer is facing different environments on the inside and on the outside. Later on, this asymmetry in the topographical features of S-layers could be extended to the physicochemical surface properties, such as surface charge and hydrophobicity (Sára and Sleytr, 1987a; Pum et al., 1993). While the first three-dimensional model of the S-layer lattice from the archaeon *Sulfolobus acidocaldarius* was calculated by Taylor et al. (1982), an atlas of three-dimensional models of S-layers from both archaeal and bacterial origin was made available by Baumeister (Baumeister and Engelhardt, 1987) and Hovmöller et al. (1988). With the introduction of scanning probe microscopy to the biological sciences, the possibility to examine S-layers lattices in the fully hydrated state became reality (Karrasch et al., 1994; Pum and Sleytr, 1995). The expectation was that preparation artifacts resulting from dehydration of the protein, which is required for conventional electron microscopy, could be substantially reduced or completely prevented.

Along with the accumulation of knowledge about the structural organization of S-layer lattices went the chemical and immunological characterization of S-layer proteins. Amino acid analyses of a large number of S-layer proteins have shown a relatively similar overall composition of different S-layer proteins (Sleytr and Messner, 1983; Kandler and König, 1985). Usually, they are slightly acidic proteins with a large portion of hydrophobic amino acids and a very low, if any, content of sulfur-containing amino acids. S-layers were the first prokaryotic proteins that were shown to be glycosylated. This was demonstrated at the same time on archaeal and bacterial S-layer proteins (Mescher and Strominger, 1976; Sleytr and Thorne, 1976). Today, glycosylation has been demonstrated in many archaeal S-layer proteins, but among bacterial species, glycosylation has been proven only for members of the Bacillaceae (for a compilation, see Section 4 of this chapter and Messner and Schäffer, 2002). In this context other glycosylated cell wall constituents should be mentioned, namely secondary cell wall polymers (SCWP), whose complete structures have been identified recently (for review see Section 4 of this chapter). In addition, functional analyses in a number of gram-positive organisms have shown that these polymers connect the S-layer subunits with the underlying peptidoglycan sacchulus (Sára, 2001). The S-layer–SCWP complex of selected organisms plays an important role in specific nanobiotechnological applications (Sleytr et al., 2001a, c; see also Section 12).

From the early 1980s on, the immunological properties of S-layers have been investigated mainly in the fish-pathogen organisms *Aeromonas salmonicida* and *A. hydrophila* (Kay and Trust, 1991; Kokka et al., 1992; Kostrzynska et al., 1992); in *Campylobacter fetus*, which causes ovine abortion (Blaser et al., 1988; Blaser and Pei, 1993); in human pathogens such as *Eubacterium yurii*, *Bacteroides* ssp., and *Wolinella recta* (Lai et al., 1981; Haapasalo et al., 1985; Margaret and Krywolap, 1986; Borinski and Holt, 1990); and in *Clostridium difficile* (Kawata et al., 1984; Cerquetti et al., 2000).

With increasing knowledge about the chemical and immunological properties of S-layers and with the advent of molecular biology and genetics, more efforts were put into cloning, sequencing, and expression of the structural genes of S-layer proteins. Udaka and coworkers published the first bacterial S-layer gene sequence from *Brevibacillus brevis* 47 (Tsukagoshi et al., 1984), and the first archaeal S-layer gene to be published was the *csg* gene from *Halobacte-*

rium halobium (Lechner and Sumper, 1987). Today, more than 60 S-layer genes have been analyzed, and their sequence data are accessible from the major databases (for review, see Section 5 and Sára and Sleytr, 2000). They are the basis for the construction of fusion proteins between S-layers and important target molecules of biotechnological and medical relevance (for review, see Sleytr et al., 2001a and Section 10).

The wealth of structural, chemical, and genetic information on S-layer proteins and S-layer glycoproteins, accumulated over the past 50 years, provides an enormous thrust for the extensive search for applications of these highly ordered cell surface macromolecules, for both biological and nonbiological areas. Today, potential applications reach from the development of ultrafiltration membranes to biosensors, diagnostic kits, vaccine applications, affinity matrices, templates for mineral formation, carrier systems for the functionalization of inorganic surfaces, and many other aspects of molecular nanotechnology, nanobiotechnology, and biomimetics (for reviews see Schultze-Lam et al., 1992; Sleytr et al., 1996a, 1999, 2000, 2001a,b,c; Smit et al., 2000; Simon et al., 2001).

3 Occurrence and Ultrastructure

S-layers are one of the most commonly observed prokaryotic cell envelope structures (for compilation, see Sleytr et al., 1996b). They represent an almost universal feature on archaeal cell envelopes and have been detected in hundreds of different species of nearly every taxonomical group of walled bacteria. S-layer-like structures also have been observed in bacterial sheaths (Beveridge and Graham, 1991) and on the surface of the cell wall of eukaryotic algae (Roberts et al., 1982). Unless abundant glycocalyces are present, the most suitable procedure for identifying the presence of S-layer lattices on intact cells is electron microscopy of freeze-etched preparations (Sleytr, 1978; Sleytr and Messner, 1989; Messner and Sleytr, 1992; Beveridge, 1994). S-layers completely cover the cell surface at all stages of cell growth and division in both archaea and bacteria (Figure 1).

High-resolution studies on the mass distribution of the lattices are generally performed on negatively stained S-layer fragments or on unstained thin, frozen foils (Baumeister et al., 1989; Hovmöller, 1993; Beveridge, 1994). More recently, S-layer lattices also have been studied by scanning tunneling microscopy and atomic force microscopy (Pum and Sleytr, 1999; Müller et al., 1999). The topographical images obtained (Figure 2) strongly resemble the three-dimensional reconstructions of S-layers derived from tilt series of electron microscopical images.

Particularly, atomic force microscopy in water of S-layer proteins recrystallized on a solid support revealed structural resolution down to the 1-nm range. Molecular resolution may be obtained only when the loading

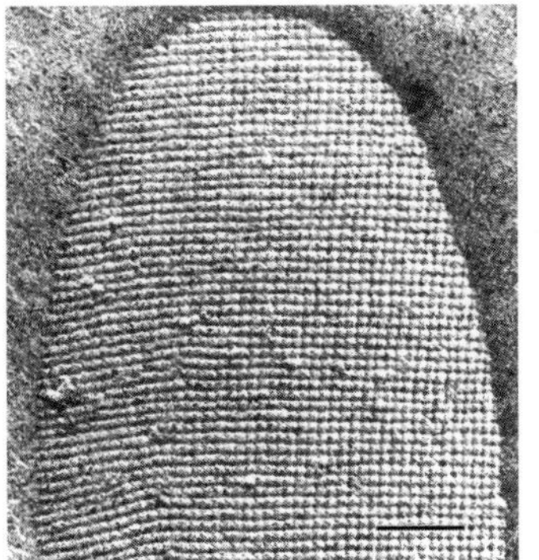

Fig. 1 Electron micrograph of a freeze-etched preparation showing a whole cell exhibiting a square S-layer lattice (Bar 100 nm).

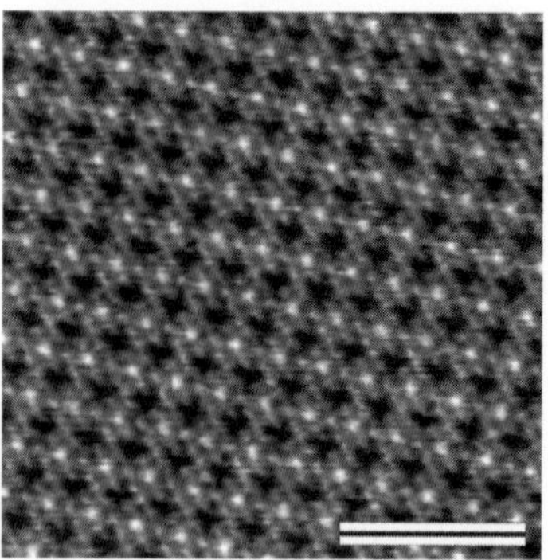

Fig. 2 Scanning force microscopical image of an S-layer with square lattice symmetry recrystallized on a silicone wafer. Image was recorded in contact mode in a liquid cell (Bar 50 nm).

face of the tip is in the range of ~ 100 pN. S-layer protein or glycoprotein subunits can be aligned in lattices with oblique (p1, p2), square (p4), or hexagonal (p3, p6) symmetry (Figure 3), with center-to-center spacings of the morphological units (composed of one, two, three, four, or six identical monomers) of approximately 3.5 – 35 nm.

Among archaea, hexagonal lattices have been shown to be predominant. Most S-layer lattices are 5 – 25 nm thick and have a smooth outer and a more corrugated inner surface. Because S-layers are monomolecular assemblies of identical subunits, they

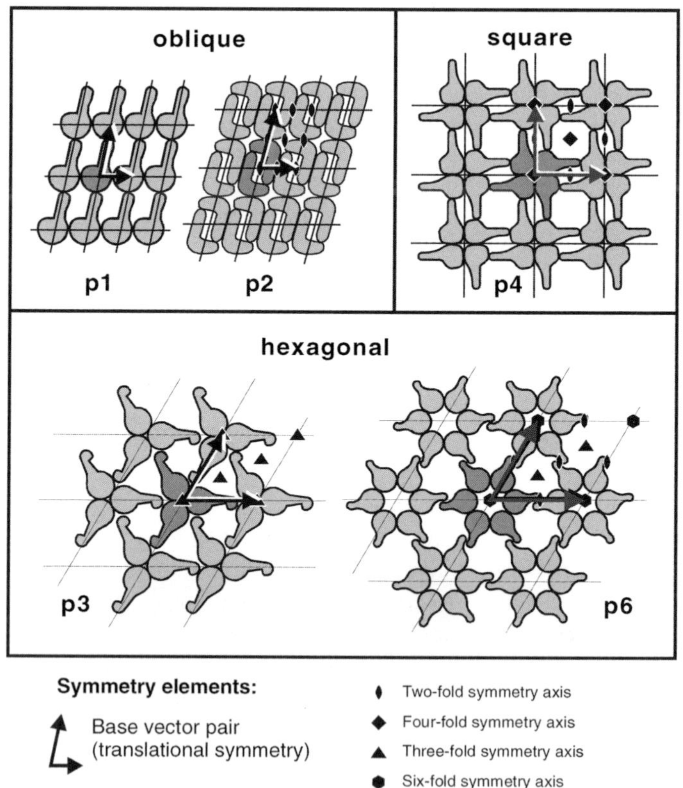

Fig. 3 Schematic representation of S-layer lattice types grouped according to the possible two-dimensional space group symmetries. Morphological units were chosen arbitrarily and are shown in dark gray. Symmetry axes are shown for a single unit cell only. Note: space group symmetries with mirror reflection lines (m) or glide reflection lines (g) (e.g., rectangular space group symmetry [pm]) are not possible in S-layer protein monolayers.

exhibit pores of identical size and morphology (Figure 2). The lattices can display more than one type of pore. High resolution electron microscopy, scanning force microscopy, and permeability studies revealed pore sizes in the lattices ranging from 1.5 to 8 nm with pores occupying up to 70% of the surface area of the protein meshwork. Comparative studies on the distribution and uniformity of S-layers have revealed that in some bacterial species individual strains can show a remarkable diversity regarding lattice symmetry and lattice dimensions. In some organisms two or even more superimposed S-layer lattices have been identified (Sleytr and Messner, 1983, 1988, 1989; Messner and Sleytr, 1992; Sleytr et al., 1996b).

4 Isolation and Chemical Characterization

Although there is considerable variation in the molecular architecture and supramolecular complexity of prokaryotic envelopes, it is reasonable to classify cell wall profiles that contain S-layers into three main groups. In gram-negative archaea, S-layers represent the sole cell wall component and can be closely associated with the plasma membrane so that it is partially integrated in the lipid layer. On both gram-positive bacteria and archaea, S-layers assemble on the surface of a rigid wall matrix (e.g., peptidoglycan or pseudomurein). In the more complex cell envelopes of gram-negative bacteria, S-layers are linked to specific lipopolysaccharide fractions (Sleytr and Beveridge, 1999; Sleytr et al., 1999, 2001a,b; Sára and Sleytr, 2000; Sleytr and Messner, 2000).

Because of the diversity in the supramolecular structures of prokaryotic cell envelopes, different isolation procedures have been developed (Figure 4). Since S-layers usually are not covalently attached to the cell surface, they can be isolated in the presence of dissociating agents such as lithium chloride (Lortal et al., 1992), metal-chelating agents such as ethylenediaminetetraacetic acid (EDTA) and ethylenebis(oxyethylenenitrilo)tetraacetic acid (Thornley et al., 1974), or chaotropic denaturants such as guanidine hydrochloride and urea (Sleytr and Thorne, 1976; Messner and Sleytr, 1988). Extraction and disintegration experiments revealed that the intersubunit bonds in the S-layer are stronger than those binding the subunits to the supporting envelope layer (Sleytr and Glauert, 1975). However, in the halophilic archaea *Halobacterium halobium* and *Haloferax volcanii*, the S-layer protomers are anchored by C-terminally located membrane-spanning domains to the cytoplasmic membrane (Lechner and Sumper, 1987; Trachtenberg et al., 2000). In order to prepare pure S-layer material for analysis, cell envelope preparations were delipidated by repeated extraction with chloroform:methanol = 2:1 at room temperature (Mescher and Strominger, 1976). Recent analyses have further demonstrated that the 200-kDa S-layer glycoprotein of *H. halobium* contains diphytanylglycerol phosphate units at the C-terminus (Kikuchi et al., 1999). This modification seems to be common in halophilic archaea and in part might cause the dramatically reduced electrophoretic mobility of these glycoproteins on SDS-PAGE gels. It is speculated that this hydrophobic modification serves as an additional membrane anchor (Kikuchi et al., 1999).

Chemical and genetic analyses on many bacterial and archaeal S-layers have shown that they are generally composed of a single protein or glycoprotein species with molecular masses ranging from 40 to 170 kDa (Sleytr et al., 1993; Sumper and Wieland, 1995; Sleytr et al., 1999, 2001a; Messner and Schäffer, 2002) (Table 1).

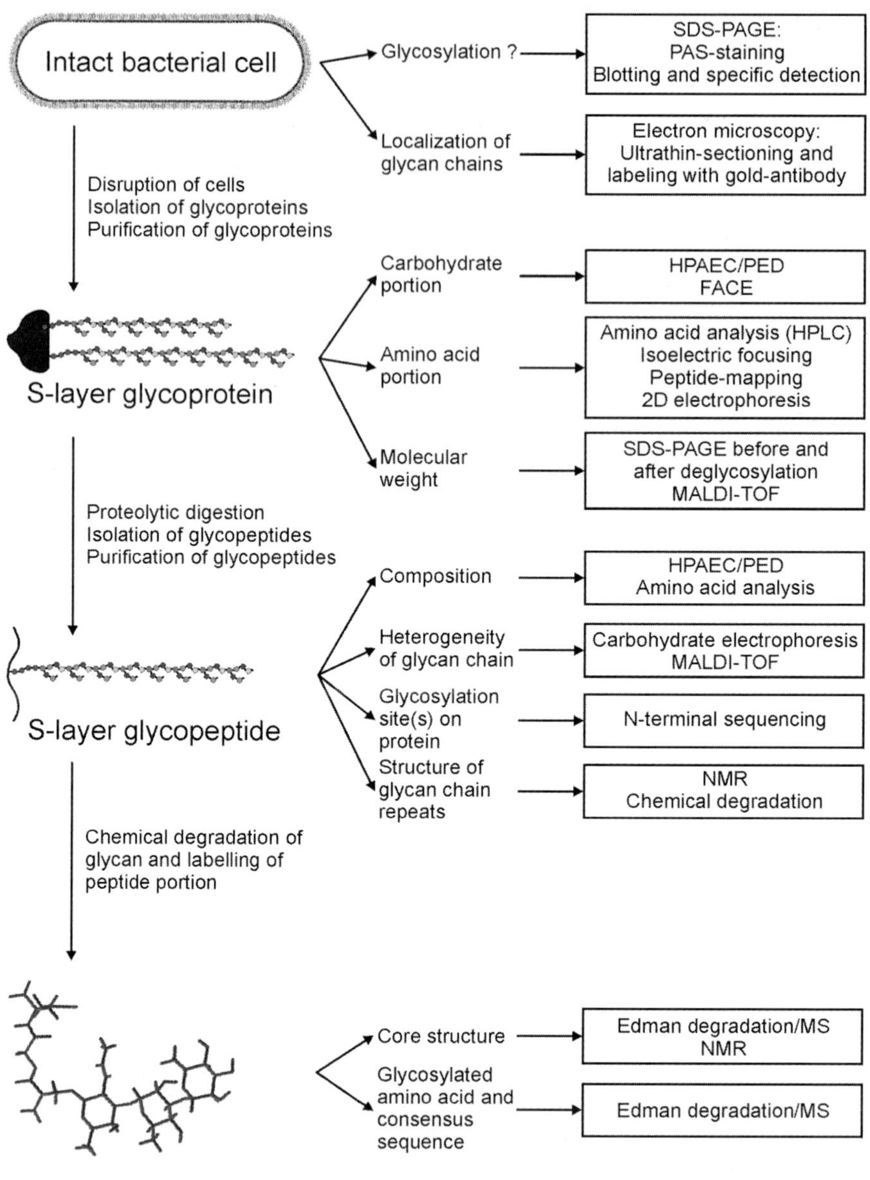

Fig. 4 Protocol for the complete analysis of bacterial S-layer glycoproteins. The protocol starts with the intact bacterial cell and results in the isolation of the linkage unit. Methods applied during different preparation stages are boxed (used with permission from Wiley-VCH Verlag, Weinheim, Germany).

Tab. 1 Chemical and structural properties of S-layers[a]

Property	***Description***
Symmetry	The S-layer lattices can have oblique (p1, p2), square (p4), or hexagonal (p3, p6) symmetry
Composition	S-layers are composed of protein or glycoprotein subunits
	The relative molecular mass of constituent subunits is in the range of 40–200 kDa
	These exhibit a weakly acidic pI between 4 and 6, except for *Methanothermus fervidus* (pI = 8.4) and *lactobacilli* (pI > 9.5)
	Large amounts of glutamic and aspartic acid (about 15 mol%) are present
	There is a high lysine content (about 10 mol%)
	There are large amounts of hydrophobic amino acids (about 40–60 mol%)
Secondary structure	Hydrophobic and hydrophilic amino acids do not form extended clusters
	In most S-layer proteins, about 20% of the amino acids are organized as α-helices and about 40% occur as β-sheets
	Aperiodic foldings and β-turn content may vary between 5 and 45%
Morphology	Outer surface is generally less corrugated than the inner surface
Dimension	Center-to-center spacing of the morphological unit can be 3–35 nm
	The layers are generally 5–20 nm thick (in archaea, up to approximately 70 nm)
Pores	S-layer lattices exhibit pores of identical size and morphology
	In many S-layers two or even more distinct classes of pores are present
	Pore sizes range from approximately 2–8 nm
	Pores occupy 30–70% of the surface area
Modification	Po sttranslational modifications of S-layer proteins include cleavage of N- or C-terminal fragments, glycosylation, and phosphorylation of amino acid residues

[a] Sources: Sleytr and Messner (1983); Baumeister et al. (1990); Beveridge (1994); Sleytr et al. (1996b, 1999, 2001a,b); Sleytr (1997); Sleytr and Beveridge (1999); Sára and Sleytr (2000)

Recently, some interesting observations have been reported from S-layers of the human pathogen *Clostridium difficile*. It is known that *C. difficile* strains possess two immunologically non-cross-reacting S-layers with molecular masses in the range of 32 to 40 kDa (low-molecular-mass [LMM] S-layers) and 45 to 56 kDa (high-molecular-mass [HMM] S-layers) (Kawata et al., 1984; Cerquetti et al., 2000; McCoubrey and Poxton, 2001), respectively. Molecular characterization revealed that both proteins are derived from a single gene product by posttranslational processing. This processing involves removal of the signal peptide and a second cleavage to release two mature S-layer proteins in the described molecular mass ranges (Calabi et al., 2001; Karjalainen et al., 2001). Whereas HMM S-layers were glycosylated in all investigated *C. difficile* strains, glycosylation of LMM S-layers was found in a number of strains (Mauri et al., 1999; Cerquetti et al., 2000). In these S-layer glycoproteins, removal of the glycan chains by trifluoromethanesulfonic acid did not affect the electrophoretic mobility of the specimens in SDS-PAGE gels (Calabi et al., 2001), which is an unexpected effect. New analyses of the cell surface S-layer glycoprotein of *Haloferax volcanii* have shown that, in addition to the known glycan structure (Sumper et al., 1990), other glycosylated protein species can be found in this organism that are neither precursors nor breakdown products of the established S-layer glycan structure (Eichler, 2000). The novel glycoproteins are outwardly oriented and intimately associated with the cytoplasmic membrane. Pulse-chase experiments on *in vitro* synthesized 190-kDa S-layer glycoprotein (Sumper et al., 1990) revealed that this material undergoes a maturation step following translocation across the cytoplasmic membrane (Eichler, 2001). The processing step, detected as an increase in the apparent molecular mass in SDS-PAGE, is unaffected by inhibition of protein synthesis and is obviously unrelated to glycosylation of the protein. Though not yet examined, an additional modification step by isoprenylation, as described for *H. halobium* (Kikuchi et al., 1999), is conceivable (Eichler, 2001).

The most frequent posttranslational modification of S-layer proteins is glycosylation (Messner and Sleytr, 1991; Messner and Schäffer, 2000, 2002; Schäffer and Messner, 2001). For structural analysis of the S-layer glycans, degradation of the S-layer protein moiety is required (Figure 4).

The complete structures of several bacterial and archaeal S-layer glycoprotein glycans and, in addition, SCWPs, which function as anchoring molecules for the S-layer proteins in gram-positive bacteria, are summarized in Tables 2a–c.

While most bacterial S-layer glycoprotein glycans possess a tripartite structure – including an O-antigen-like polysaccharide, a core saccharide, and the linkage region to the S-layer polypeptide (Table 2a, Figure 5) – a few exceptions are known. The tyrosine-linked O-glycan of *Thermoanaerobacterium thermosaccharolyticum* D120-70 is directly linked to the S-layer protein (Schäffer et al., 2000a). In most archaeal S-layer glycoproteins (Table 2b) and, for example, in *T. thermosaccharolyticum* S102-70, only relatively short heterosaccharides are present (Christian et al., 1993). It can, however, be speculated that the short heterosaccharide of *T. thermosaccharolyticum* S102-70 actually represents an incomplete S-layer glycan where, for yet unknown reasons, the O-antigen-like polysaccharide either has not been synthesized or attached to the "preexisting" core structure. Interestingly, in *Aneurinibacillus thermoaerophilus* strains, these cores are of variable length (Schäffer et al., 1999a; Wugeditsch et al., 1999). Recently, the sequence of the structural gene *sgsE* that encodes the glycosylated S-layer protein of *Geobacillus stearothermophilus* NRS 2004/3a has been determined (Schäffer et al., 2002) (Figure 5). Degradation of this material by papain and characterization of the resulting glycopeptides allowed, for

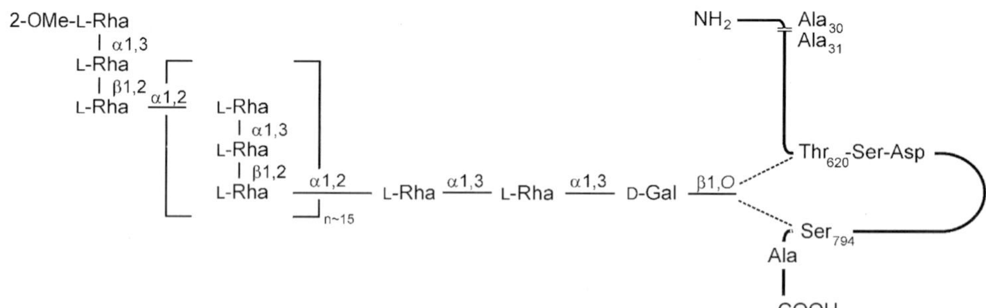

Fig. 5 Schematic drawing of the complete structure of the S-layer glycoprotein of *G. stearothermophilus* NRS 2004/3a. The full curved line symbolizes the S-layer protein SgsE; glycosylation sites are the amino acids Thr 620 and Ser 794. Differences in the glycosylation pattern of the individual S-layer promoters are indicated by dotted lines. The N-terminal signal peptide is cleaved between amino acids Ala 30 and Ala 31 (modified from Schäffer et al. 2002).

Tab. 2 Cell-wall-associated glycan structures

(a) Tripartite S-layer glycoprotein glycans composed of repeating units

Organism[a]	Capping	Repeating unit[c]	n	Core	Linkage
1	2*O*Me	[-2αLRha*p*1-3βLRha*p*1-3αLRha*p*1-]$_n$-3	13–18	αLRha*p*1-3αLRha*p*1-3	βDGal-Ser/Thr
2, 3	3*O*Me	[-3αDRha*p*1-3αDRha*p*1-2αDRha*p*1-2αDRha*p*1-]$_n$-3 2 ↑ 1 αDFuc*p*3NAc (on the 1st αDRha*p*); 2 ↑ 1 αDFuc*p*3NAc (on the 3rd αDRha*p*)	~15	[αDRha*p*]$_{n=0-2}$1-3αDRha*p*1-3	GalNAc-Thr
4	n.d.	[-4αLRhap1-3βD*glycero*D*manno*Hep*p*1-]$_n$	~18	[αLRha*p*]$_{n=0-2}$ 1-3	βDGalNAc-Ser/Thr
5	n.o.	αDGlcp 1 ↓ 6 [-3βDGal*p*1-4βDMan*p*NAc1-]$_n$ –3	~20	αLRha*p*1-3αLRha*p*1-3αLRha*p*1-3 4 ↑ 1 GroA2-OPO$_2$O-4βDMan*p*NAc	βDGal-Tyr
6, 7	3*O*Me	[-3αLRha*p*1-4αDMan*p*1-]$_n$-3-	~27	αLRha*p*1-3αLRha*p*1-3αLRha*p*1-3	βDGal-Tyr
8, 9	n.d.	[-3αDGal*p*NAc1-3αDGal*p*NAc1-]$_n$ 4 ↑ 1 αDGlc*p*NAc1-2βDMan*p*	~25	n.d.	Tyr
10	n.o.	[-4αLRha*p*1-3αDMan*p*1-4βLRha1-3αLGlc*p*1-]$_n$-3 2 ↑ 1 αDGal*p* (on αLRha*p*); 6 ↑ 1 βDGlc*p* (on αDMan*p*)	3–10	"Pseudo-core"[b]	βDGlc-Tyr
11	n.d.	[-4βDGal*p*1-4βDGlc*p*1-4αDMan*p*1-]$_n$ 3 ↑ 1 βDQui*p*3NAc1-6βDGal*f*1-4αLRha*p*	17	n.d.	Tyr
12	n.d.	OSO$_3^-$ (on GlcNAc); OSO$_3^-$ | (on GalNAc) [-4GlcNAc1-4GalA1-3GalNAc1-]$_n$ 6 ↑ 1 3*O*MeGalA (on GlcNAc); 3 ↑ 1 Gal*f* (on GalNAc)	10–15	n.o.	Asn

Tab. 2 (cont.)

(b) Heteropolysaccharide S-layer glycoprotein glycans without a core

Organism[a]	Glycan chain	Linkage amino acid
13	βDGal*f*1-3αDGal*p*1-2αLRha*p*1-3αDMan*p*1-3αLRha*p*1-3βDGlc*p*1-	Tyr
14	αDGlc*p*16)-[αDGlc*p*1-6]$_{n=4-6}$ -αDGlc*p*1-	Ser
12	GlcA1-4GlcA1-4GlcA1-4βDGlc1- 3 3 3 ↑ ↑ ↑ OSO_3^- OSO_3^- OSO_3^- 1/3 of GlcA residues can be replaced by IdA	Asn
12	αDGlc1-2Gal1-	Thr
15	βDGlc1-[4βDGlc1-]$_{n=8}$ 4βDGlc1-	Asn
15	Glc1-2Gal1-	Thr
16	3*O*MeαDMan*p*1-6-3*O*MeαDMan*p*1-[2αDMan*p*1-]$_{n=3}$ 4DGalNAc1-	Asn

(c) Secondary cell wall polymers found in S-layer-carrying organisms

Organism[a]	SCWP	*n*	Type of linkage	Charge
1	[-4βDMan*p*A2,3(NAc)$_2$1-6αDGlc*p*1-4βDMan*p*A2,3(NAc)$_2$1-3αDGlc*p*NAc1-]$_n$	~5	-PP-MurNAc	Negative
5	[(Pyr4,6)-βDMan*p*NAc1-4βDGlc*p*NAc1-)]$_n$-1	11	-PP-MurNAc	Negative
17	[(Pyr4,6)-βDMan*p*NAc1-4βDGlc*p*NAc1-)]$_n$-1	n.d.	n.d.	Negative
4	Biantennary structure without repeats	–	-PP-MurNAc	Neutral
10, 11	[-3βDMan*p*NAc1-4βDGlc*p*NAc1-4βDMan*p*NAc1-3βDGlc*p*NAc1-]$_n$ 4 ↑ 1 αDRib*f*	n.d.	n.d.	Neutral

the first time, the precise determination of glycosylation sites on a bacterial S-layer protein (compare with Section 5).

The chemical characterization of S-layer glycoprotein glycan structures is the basis for subsequent biosynthesis and genetic studies to unravel the pathways of this important posttranslational modification of S-layer proteins.

In gram-positive bacteria, binding of S-layer proteins to the underlying cell wall layer is mediated by SCWP (Araki and Ito, 1989; Archibald et al., 1993; Salton, 1994; Naumova and Shashkov, 1997; Sára, 2001). Recently, the complete structures of several SCWPs from different S-layer-carrying organisms have been elucidated (Table 2c). By mass spectrometry, molecular masses in the range of 4 to 6 kDa have been determined, and ^{31}P-NMR experiments on isolated SCWP-peptidoglycan complexes showed that the glycan portions are linked via pyrophosphate bridges to C-6 of muramic acid units of the peptidoglycan sacculus (Schäffer et al., 1999b, 2000b). Since the linkage to peptidoglycan is not the common phosphodiester bond, these SCWPs are not considered teichoic or teichuronic acids. Some of the analyzed polysaccharide chains are negatively charged by carboxyl groups from hexosamineuronic acid units (Schäffer et al., 1999b) or pyruvic acid residues (Ilk et al., 1999; Schäffer et al., 2000b), whereas others are charge-neutral. The specific functions of these carbohydrate polymers are discussed in Section 10.

5 Molecular Biology, Genetics and Biosynthesis

Comparison of about 60 S-layer genes of organisms from quite different taxonomic affiliations (Table 3) revealed that evolutionary relationship plays a pivotal role for the sequence identity of functionally homologous domains. If present, high sequence identities are found at the N-terminus, which is responsible for the attachment of the S-layer subunits to the underlying cell envelope layer in a large number of organisms from the domain Bacteria. In this context, it is important to note that S-layer homology (SLH) motifs, typically occurring in three repeats consisting of 50 to 60 amino acids each (Lupas et al., 1994), have been identified at the N-terminal part of many S-layer proteins (Sára and Sleytr, 2000). If present, SLH motifs are involved in SCWP-mediated anchoring of the S-layer protein to the peptidoglycan layer (Sára, 2001). In contrast, for the middle and C-terminal parts, which are involved in the self-assembly process of the S-layer protein, sequence identities are below 20% (Sleytr et al., 1999; Sára and Sleytr, 2000).

The biosynthesis of the encoded S-layer protein is a very complex process in which the amount of the protein component, its translocation through the cell wall, and its specific incorporation into the existing S-layer lattice have to be coordinated with the growth rate of the organism and the synthesis of other cell wall components. Such a complex regulation relies on different control mechanisms for protein expression and secretion, acting together to ensure an appropriate level of S-layer protein synthesis. To keep the surface of a bacterium with a generation time of 20 min completely covered with a closed protein lattice, a high synthesis effort of approximately 500 S-layer subunits per second is required (Sleytr and Messner, 1983; Messner and Sleytr, 1992), indicating that S-layer genes must be under the control of strong promoters (Kuen and Lubitz, 1996). The promoter of the S-layer gene of *L. acidophilus* was indeed demonstrated to be twice as strong as that of the gene encoding lactate dehydrogenase, which

Tab. 3 Amino acid sequences of S-layer proteins

Species	***Strain***	***Gene***	***Precursor/mature protein (aa)***	***Lattice type*** [a]	***GenBank accession no.***
Aeromonas hydrophila	TF7	*ahs*	467 / 19	S	L37348
Aeromonas salmonicida	A450	*vapA*	502 / 21	S	M64655
Bacillus anthracis	Sterne derivative, substrain 9131	*sap*	814 / 29	O[b]	Z36946
Bacillus anthracis	Sterne derivative, substrain 9131	eag	862 / 29	O	X99724
Brevibacillus brevis	47	*owp*	1004 / 24	H	M14238
Brevibacillus brevis	47	*mwp*	1053 / 23	H	M19115
Brevibacillus brevis	HPD31	*HWP*	1087 / 23	H	D90050
Bacillus licheniformis	NM105	*olpA*	874 / 29	O	U38842
Bacillus pseudofirmus	OF4	*slpA*	931 / 31	O	AF242295
Bacillus sphaericus	P1	*sequence 8*	1252 / 30	S	A45814
Bacillus sphaericus	2362	*gene 125*	1176 / 30	S	M28361
		gene 80	745 (silent)		
Bacillus sphaericus	CCM 2177	*sbpA*	1,268 / 30	S	AF211170
Bacillus thuringiensis	—	*ctc*	842 / 29	—	AJ012290
Bacillus thuringiensis spp. *galleriae*	NRRL 4045	*slpA*	821 / 29	—	AJ249446
Campylobacter fetus, ssp. fetus	84-32 (23D)	*sapA*	933 / none	H, S [c]	J05577
Campylobacter fetus, ssp. *fetus*	23B	*sapA1*	920 / none	H, S [c]	L15800
Campylobacter fetus, ssp. fetus	82-40LP3	*sapA2*	1,109 / none	H, S [c]	S76860
Campylobacter fetus, ssp. *fetus*	84-91	*sapB*	936 / none	—	U25133
Campylobacter fetus, ssp. *fetus*	CIP 5396T	*sapB2*	1,112 / none	—	AF048699
Campylobacter rectus	314	*crs*	1,361 / none	—	AF010143
Caulobacter crescentus	CB15	*rsaA*	1,026 / none	H	M84760
Clostridium difficile	C253	*slpA*	720 / 24	H, S	AJ291709
Clostridium difficile	R8366	*slpA*	756 / 24	—	AJ300676
Clostridium difficile	R7404	*slpA*	714 / 24	—	AJ300677
Clostridium thermocellum	NCIMB 10682	*slpA*	1,036 / 26	O	U79117
Corynebacterium glutamicum	ATCC 17965	*csp2*	510 / 30	H	X69103
Deinococcus radiodurans	Sark	*HPI gene*	1,036 / 31	H	M17895
Geobacillus stearothermophilus	PV72/p6	*sbsA*	1,228 / 30	H	X71092

Tab. 3 (cont.)

Species	Strain	Gene	Precursor/mature protein (aa)	Lattice type[a]	GenBank accession no.
Geobacillus stearothermophilus	PV72/p2	*sbsB*	920 / 31	O	X98095
Geobacillus stearothermophilus	ATCC 12890	*sbsC*	1,099 / 30	O	AF055578
Geobacillus stearothermophilus [d]	ATCC 12980, mutant	*sbsD*	903 / 30	O	AF228338
Geobacillus stearothermophilus [d]	NRS 2004/3a	*sgsE*	903 / 30	O	AF328862
Haloarcula japonica [d]	TR-1	*csg*	862 / 34	—	D87290
Halobacterium halobium [d]	R_1M_1	*csg*	852 / 34	H	J02767
Haloferax volcanii [d]	DS2	—	827 / 34	H	M62816
Lactobacillus acidophilus	ATCC 4356	*slpA*	444 / 24	O	X89375
Lactobacillus acidophilus	ATCC 4356	*slpB*	456 (silent)	—	X89376
Lactobacillus brevis	ATCC 8287	—	465 / 30	O	Z14250
Lactobacillus crispatus	JCM 5810	*cbsA*	440 / 30	O	AF001313
Lactobacillus crispatus	M247	—	451 / 31	—	AJ007839
Lactobacillus helveticus	CNRZ 892	*slpH1*	439 / 30	O	X91199
Lactobacillus helveticus	CNRZ 1269	*slpH2*	439 / 30	O	X92752
Lactobacillus helveticus	CNRZ 35	—	437 / 30	—	AJ388561
Lactobacillus helveticus	CNRZ 303	—	439 / 30	—	AJ388560
Lactobacillus helveticus	ATCC 15009	—	439 / 30	—	AJ388559
Lactobacillus helveticus	IMPC I60	—	439 / 30	—	AJ388562
Lactobacillus helveticus	IMPC M696	—	439 / 30	—	AJ388563
Lactobacillus helveticus	IMPC HLM1	—	438 / 30	—	AJ388564
Methanococcus voltae	DSM 1537	*sla*	565 / 12	H	M59200
Methanosarcina mazei	S-6	*slgB*	652 / 31	H	X77929
Methanothermus fervidus [d]	DSM 2088	*slgA*	593 / 22	H	X58297
Methanothermus sociabilis [d]	DSM 3496	*slgA*	593 / 22	H	X58296
Rickettsia prowazekii	Brein 1	*ompB*	1,300 / 32	—	M37647
Rickettsia rickettsii	R	*p120*	1,645 / 32	—	X16353
Rickettsia typhii	Wilmington	*slpT*	1,645 / 32	—	L04661
Serratia marcescens	Sr41 / isolate 8000	*slaA*	1,004 / none	—	AB007125
Staphylothermus marinus	F1	—	1,524 / 26	—	U57967

Tab. 3 (cont.)

Species	*Strain*	*Gene*	*Precursor/mature protein (aa)*	*Lattice type*[a]	*GenBank accession no.*
Thermoanaerobacter kivui [d]	DSM 2030	*slp*	762 / 26	H	M31069
Thermus thermophilus	HB8	*slpA*	917 / 27	H, S	X57333

[a] H: hexagonal; S: square; O: oblique [b] Presumably [c] In *Campylobacter fetus* subsp. *fetus*, the lattice type depends on the molecular mass of the S-layer subunits (H: 97 kDa; S: 127 kDa and 149 kDa) [d] S-layer glycoprotein

is considered one of the strongest bacterial promoters (Pouwels et al., 1997). Two, most probably growth stage-dependent, promoters were identified for the S-layer genes of *L. brevis* ATCC 8287 (Kahala et al., 1997), *A. salmonicida* (Chu et al., 1993), and *G. stearothermophilus* ATCC 12980 (Jarosch et al., 2000). For formation of the two superimposed S-layer lattices of *B. brevis* 47, the *cwp* operon is preceded by three tandemly arranged promoters (Adachi et al., 1989). Most of the S-layer protein mRNAs have a long leader sequence and exceptionally high stability (Chu et al., 1991; Boot and Pouwels, 1996), suggesting that S-layer protein export occurs via a common signal sequence secretory pathway. On the other hand, the secretion signal of the *C. crescentus* S-layer protein RsaA was localized to the C-terminal 82 amino acids of the molecule (Bingle et al., 2000; Reichelt et al., 2001), indicating the presence of an alternative type I secretion pathway. The RsaA transport system secretes higher amounts of its substrate protein than known for any other type I secretion system (Awram and Smit, 1998). Signal peptide insertions at the C-terminal end are also present on the S-layer protein of *C. fetus* (Thompson et al., 1998). In the case of the S-layer protein of *Serratia marcescens*, there are indications for a signal sequence-independent pathway involving the *lip* system, which belongs to an ATP-binding cassette exporter (Kawai et al., 1998). In *E. coli*, the expression of the S-layer proteins VapA and AhsA of *A. salmonicida* and *A. hydrophila* which are produced with either a 21- or 19-amino-acid-long signal peptide (Chu et al., 1993; Thomas and Trust, 1995) is affected by the conserved, bifunctional AbcA protein, possessing ATP-binding activity at its N-terminus and a C-terminal leucine zipper functioning as transcriptional activator of the *P2* promoter (Chu and Trust, 1993; Noonan and Trust,

1995a,b). In the case of Thermus thermophilus HB8, *in vitro* evidence suggested that overlapping transcriptional and translational controls act on the 5′ untranslated region of the S-layer protein-encoding gene *slpA* (Fernandez-Herrero et al., 1997; Castán et al., 2001). This 127-bp region could act as binding site for putative transcription factors and be implicated in translational autoregulation. In this model an excess of SlpA would act as a specific repressor of its translation by the direct binding of the 5′ untranslated region of its own transcript. Rep54, which is a C-terminal fragment of SlpA, was identified as the actual molecule responsible for repression. Thus, the 5′ untranslated region stabilizes the mRNA and represses transcription from the *slpA* promoter *in vivo*.

C. difficile is unusual in that it expresses two S-layer proteins that are of varying size (Cerquetti et al., 2000). Both proteins are derived from a common precursor, and processing involves the removal of a signal peptide and a second cleavage to release the two mature S-layer proteins. Investigation of three *C. difficile* strains indicated the higher-molecular-weight protein to be conserved and, potentially, glycosylated (Calabi et al., 2001; Karjalainen et al., 2001). In contrast, in *B. anthracis*, in which the S-layer proteins have been characterized at a molecular level, the two present proteins are encoded by distinct genes (Mesnage et al., 1997; Mignot et al., 2002).

Regulation mechanisms underlying the post-translational modification of S-layer proteins with glycan chains, as well as the coordination of the glycosylation process with protein synthesis, are currently being investigated. Analyses of the S-layer glycosylation of *G. stearothermophilus* NRS 2004/3a and *Aneurinibacillus thermoaerophilus* strains DSM 10155 and L420-91 revealed that genes encoding enzymes for nucleotide sugar biosynthesis, such as GDP-D-*glycero*-β-D-*manno*-heptose, GDP-D-rhamnose, and dTDP-L-rhamnose, are clustered within an S-layer glycan biosynthesis cluster (Kneidinger et al., 2001a,b; Novotny et al., in preparation). Flanking of these clusters with transposons and a decreased G+C content in comparison with the rest of the chromosome indicate that lateral transfer of the whole glycosylation machinery might have taken place. For *G. stearothermophilus* NRS 2004/3a it was shown that the S-layer gene *sgsE*, located immediately upstream of the S-layer biosynthesis cluster, and the cluster are preceded by separate promoters, indicating separate control mechanisms for protein and glycan synthesis (Novotny et al., in preparation).

Interestingly, the occurrence of S-layer glycoproteins seems not necessarily restricted to wild-type strains. With *G. stearothermophilus* ATCC 12980, the development of a glycosylated, non-revertable variant was observed upon shifting the growth temperature from 55°C to 67°C. Thereby, *de novo* generation of the variant-specific sequence *sbsD* as well as disruption of the wild-type-specific sequence *sbsC* might have occurred (Egelseer et al., 2001). Another environmental factor inducing S-layer variation was described for *G. stearothermophilus* PV72/p6, whose hexagonal S-layer lattice formed by SbsA changed into an oblique lattice consisting of SbsB of the variant PV72/p2 when the organism was grown under non-oxygen-limited conditions in continuous culture (Sára et al., 1996). Variant formation was found to be due to integration of *sbsB*, which is located on a natural megaplasmid into the chromosomal expression site. After the switch to *sbsB* expression, *sbsA* is removed from the chromosome but remains are detectable on the plasmid of PV72/p2 (Scholz et al., 2001). S-layer variation, in general, leads either to the expression of different S-layer genes, which are usually smaller than those of the wild-type strain (Egelseer et al., 2001), or to the recombination of partial coding sequences.

S-layer variation as a kind of antigenic variation that might be triggered by the lytic activity of the host immune system is exemplified by the pathogenic *C. fetus* bacteria (Tu et al., 2001a,b). *C. fetus* cells can express any of eight or nine S-layer proteins, each being encoded by a promoterless *sap* homologue (Thompson and Blaser, 2000). Variation of S-layer expression occurs by a mechanism of nested DNA rearrangement that involves the inversion of the 6.2-kb element containing the *sapA* promoter alone or together with one or more flanking *sap* homologues. *Sap* DNA inversion takes place via a high-frequency, RecA-dependent mechanism (Dworkin et al., 1997) or a low-frequency, RecA-independent mechanism (Ray et al., 2000), with the redundancy of mechanisms underlining the importance of *sap* inversion for *C. fetus*.

Insertion sequences (IS) are of importance for the generation of S-layer-deficient phenotypes, by either repressing or activating genes downstream of their integration site into the bacterial chromosome. Variants of *A. salmonicida* without an S-layer were generated by insertion of IS*AS1* and IS*AS2* into different sites of the *vapA* gene and its promoter region (Gustafson et al., 1994). Temperature-dependent transposition, as observed by integration of IS*4712* upon shifting the cultivation temperature from 57°C to 67°C, was demonstrated to inhibit expression of the S-layer gene *sbsA*, leading to the S-layer-deficient phenotype *G. stearothermophilus* PV72/T5 (Scholz et al., 2000). In *G. stearothermophilus* ATCC 12980, ISBst*12*, a novel type of IS element located within the coding region of the *sbsC* gene, was found to cause loss of S-layer expression (Egelseer et al., 2000).

In conclusion, multiple mechanisms leading to S-layer protein variation, modification, or even complete loss of the S-layer indicate the importance of diversification of the surface properties of even closely related organisms for their survival in a competitive habitat.

6 Assembly and Morphogenesis

6.1 Self-assembly *in vivo*

A closed S-layer lattice on an average-sized prokaryotic cell consists of approximately 5×10^5 protein subunits. Thus, at a generation time of approximately 20 min, at least 500 copies of a single polypeptide species with a molecular mass of approximately 100,000 have to be synthesized, translocated to the cell surface, and incorporated into the S-layer lattice per second in order to maintain a closed protein meshwork. With S-layers from various bacteria, it was demonstrated that distinct surface properties of the subunits (charge distribution, hydrophobicity, specific interactions with components of the supporting envelope layer) are important for the proper orientation of the S-layer (glyco)protein subunits during local insertion in the course of lattice growth. Labeling experiments with fluorescent or topographical markers and colloidal gold allowed the localization of incorporation sites of subunits and areas of lattice extension during cell growth (Sleytr and Messner, 1989; Sleytr et al., 2000). Electron microscopical studies on freeze-etched preparations indicated that lattice faults (dislocations and disclinations) in particular serve as incorporation sites for new subunits. Analysis of the distribution of lattice faults in the S-layer of archaea that possess a hexagonal array as the exclusive wall component provided strong evidence that complementary pairs of pentagons and heptagons play an important role in the formation and maintenance of the lobed cell structure and in the cell fission process (Pum et al., 1991). It is now evident that S-layer protein meshworks are "dynamic closed surface crystals" with the intrinsic capability to continuously assume a structure of low free energy during cell growth. Moreover, the morphological potential of S-layer lattices is exclusively determined by the molecular structure of its constituent subunits.

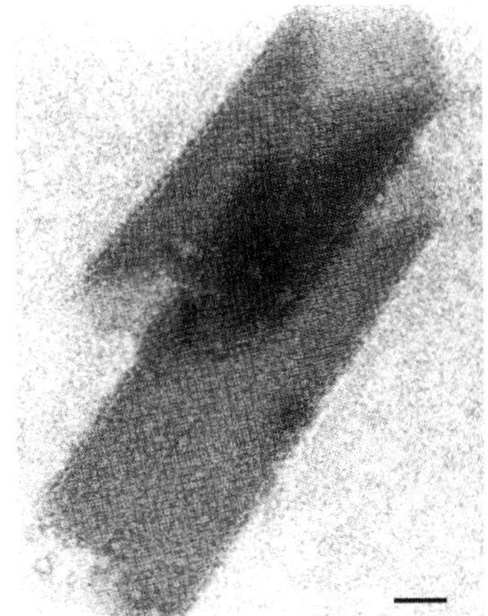

Fig. 6 Electron micrograph of negatively stained S-layer self-assembly products exhibiting a square lattice symmetry (Bar 100 nm).

6.2 Self-assembly *in vitro*

Isolated S-layer subunits from a great variety of bacteria show the ability to reassemble into monomolecular arrays identical to those observed on intact cells (Figures 6 and 7). Monolayer formation occurs after removal of the disrupting agent used in the dissolution process (Sleytr, 1975, 1981; Sleytr and Messner, 1989; Beveridge and Graham,

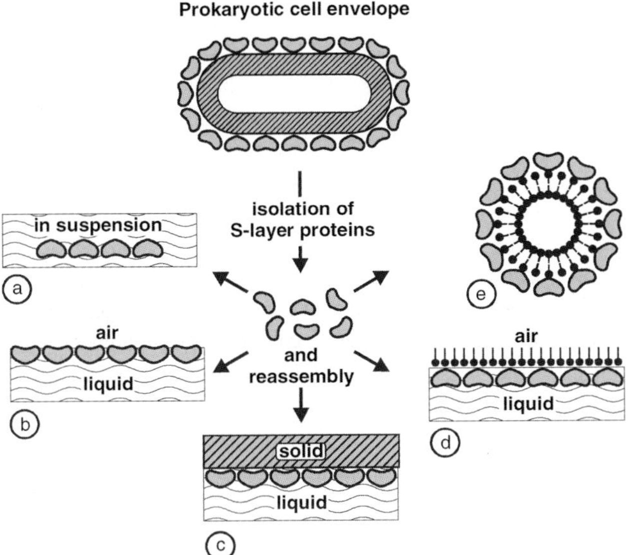

Fig. 7 Schematic illustration of the reassembly of isolated S-layer proteins (a) in suspension, where flat sheets or cylinders are formed; (b) into mono- or double layers at the air–water interface; (c) on solid substrates such as silicon surfaces; (d) at lipid films (phospholipid and tetraether lipid films); and (e) at lipid vesicles (liposomes and nanocapsules).

1991; Beveridge, 1994; Sleytr et al., 2000). Depending on the intrinsic properties of the S-layer (glyco)protein and the reassembly conditions (e.g., temperature, pH value, ionic strength, ion composition), isolated S-layer subunits may crystallize into flat sheets, open-ended cylinders, or closed vesicles (Figure 6). Moreover, protein concentration and polymers associated with S-layers can determine both the rate and extent of association (Ilk et al., 1999; Sára et al., 1998a,b). Most detailed studies regarding the kinetics of *in vitro* self-assembly in suspension, the shape of self-assembly products, the charge distribution, and the topographical properties of the outer and inner surface were carried out with S-layers from different mesophilic and thermophilic Bacillaceae (Messner et al., 1986b; Sleytr et al., 1988b; Sleytr and Messner, 1989). It was shown that the assembly kinetics are multiphasic, with a rapid initial phase leading to oligomeric precursors and slow consecutive rearrangement steps, which finally lead to extended lattices (Jaenicke et al., 1985).

The ability of S-layer proteins to recrystallize into monolayers on solid supports (Pum and Sleytr, 1999; Pum et al., 1997a) (e.g., noble metals, polymers, glass, graphite, silicon wafers), at the liquid–air interface (Wetzer et al., 1997; Pum et al., 1993, 1997a,b; Pum and Sleytr, 1994, 1995), and on Langmuir lipid films and liposomes (Pum et al., 1993; Pum and Sleytr, 1994; Küpcü et al., 1995b; Diederich et al., 1996; Weygand et al., 1999, 2000; Mader et al., 1999) is one of their key features for a broad range of applications in nanotechnology and nanobiotechnology.

Crystal growth at surfaces and interfaces starts at several nucleation points and proceeds inplane until the boundary lines of the growing crystalline domain meet (Figure 8).

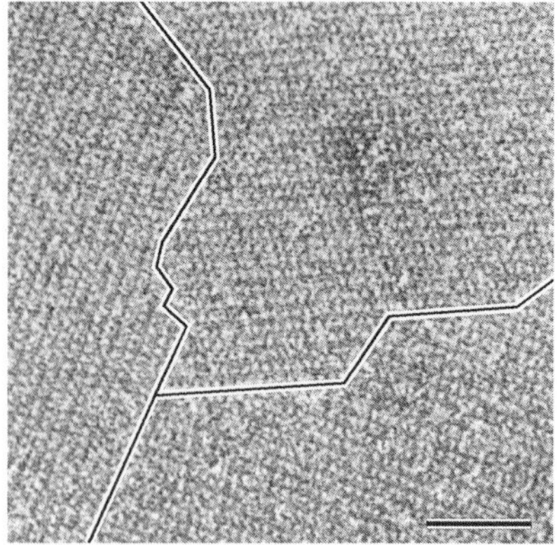

Fig. 8 Transmission election microscopical image of an S-layer recrystallized as a coherent monolayer consisting of crystalline domains at the air–water interface. The S-layer shows oblique lattice symmetry. The grain boundaries are marked by a solid line (Bar 100 nm).

Provided that there is enough protein in solution, a closed mosaic of crystalline areas is formed. Average domain sizes depend on the surface density of the proteins in a very early stage of crystal formation and on their local mobility on the surface, which is determined primarily by the reciprocal influence of the surface properties of the substrate (e.g., hydrophobicity versus hydrophilicity) and the particular protein used (Sleytr et al., 2001a,b). The importance of the environmental conditions, in particular the ionic strength of the subphase, was demonstrated in the recrystallization of the S-layer protein of *B. sphaericus* CCM 2177. Depending on the calcium concentration, a broad range of morphologies varying from tenuous fractal-like structures to large monocrystalline domains was found (Pum and Sleytr, 1994). The orientation of the S-layer with respect to surfaces and interfaces can be controlled by taking advantage of the asymmetry in the physicochemical surface properties of S-layers. This is particularly relevant when S-layer fusion proteins with functional domains exposed on either the inner or outer surface will be recrystallized. Most recently the specific "lectin-type" binding between the N-terminal part of S-layer proteins and SCWP was exploited for recrystallizing S-layer protein in defined orientations on solid supports and lipid films (manuscripts in preparation).

7 Functional Aspects

Considering that S-layer-carrying organisms are ubiquitous in the biosphere, the supramolecular concept of a closed, isoporous, crystalline surface layer could have the potential to fulfill a broad spectrum of functions. When bacteria are no longer subject to natural environmental selection pressures, S-layers are frequently lost, implying that they have importance only in natural habitats. Although no general, all-encompassing natural function has been found so far, it is now recognized that S-layers can have various functions, some of which remain hypothetical (Table 4) (Sleytr and Beveridge, 1999; Sleytr et al., 1999; Sára and Sleytr, 2000).

For assignment of functions, S-layers generally have to be considered as part of the complex bacterial cell envelope rather

Tab. 4 General and specific functions of S-layers

General function	*Specific function*
Determination and maintenance of cell shape	Determination of cell shape and cell division in archaea that possess S-layers as exclusive wall component
Isoporous molecular sieves	Molecular sieves in the ultrafiltration range
	Delineating a compartment (periplasm) in gram-positive bacteria
	Prevention of nonspecific adsorption of macromolecules
	Prevention of molecules reaching the cell wall proper (e.g., lytic enzymes)
Adhesion zones for exoenzymes	High-molecular-weight amylase of *Geobacillus stearothermophilus* wild-type strains
	Pullulanase and glycosyl hydrolases of *Thermoanaerobacter thermosulfurigenes*
Protective coats	Prevention of predation by *Bdellovibrio bacteriovorus* in gram-negative bacteria
	Phage resistance by S-layer variation
	Prevention or promotion of phagocytosis
	Adaptation of *Bacillus pseudofirmus* to alkaline environment
Templates for fine grain mineralization	Induction of precipitation of gypsum and calcite in *Synechococcus* and shedding of mineralized S-layers
Pathogenicity and cell adhesion	Virulence factor in pathogenic organisms
	Important role in invasion and survival within the host
	Specific binding of host molecules
	Protective coat against complement killing
	Ability to associate with macrophages and to resist the effect of proteases
	Production of immunologically noncrossreactive S-layers (S-layer variation)
Surface recognition and cell adhesion to substrates	Physicochemically and morphologically well-defined matrices e
	Masking the net negative charge of peptidoglycan-containing layer in Bacillacea

than as isolated structures. In archaea, which lack a rigid cell wall layer, S-layers are shape-determining or shape-maintaining (Messner et al., 1986a; Hovmöller et al., 1988; Baumeister et al., 1990; Pum et al., 1991; Baumeister and Lembcke, 1992). Because of the presence of pores of identical size and morphology, S-layer lattices work as precise molecular sieves in the ultrafiltration range (Sára and Sleytr, 1987a), determining the exchange of metabolites and nutrients between the bacterium and its environment. S-layer subunits are either tightly associated with or penetrate through the plasma membrane, thereby delineating a kind of periplasm (Graham et al., 1991; Phipps et al., 1991) that is available both for storage of secreted macromolecules involved in nutrient degradation and transport and for folding and export of proteins. S-layers can serve as adhesion zones for exoenzymes possessing SLH domains, such as a high-molecular-weight amylase of *G. stearothermophilus* wild-type strains (Egelseer et al., 1995, 1996) as well as a pullulanase and glycosyl hydrolases from *Thermoanaerobacter thermosulfurigenes* (Matuschek et al., 1994, 1996; Leibovitz et al., 1997). In the case of *Staphylothermus marinus*, a hyperthermostable protease is attached to the glycoproteinaceous stalks of the S-layer (Peters et al., 1995, 1996). S-layers also have been reported to deter predation by bacterial parasites such as *Bdellovibrio bacteriovorus* on several cell types, including *A. salmonicida*, *A. serpens*, *C. fetus*, and *C. crescentus*; these same S-layers, however, were not effective shields against protozoa (Beveridge, 1994; Koval, 1997). S-layers can even act as specific sites for bacteriophage adsorption, and phage resistance is mediated by variation of the relevant adsorption domain on the S-layer protein (Sleytr and Beveridge, 1999). For the S-layer of *B. pseudofirmus* OF4, involvement in the adaptation of the organism to an alkalophilic exterior was proposed, wherein the S-layer supports the Na^+-dependent pH homeostasis (Gilmour et al., 2000). A remarkable type of protective function is attributed to the S-layer of *Synechococcus* GL-24, serving as template for fine grain mineralization of gypsum and calcite from lake water (Schultze-Lam et al., 1992; Schultze-Lam and Beveridge, 1994). Furthermore, S-layers can contribute to virulence when present as a structural component of the cell envelope of pathogens. For the fish pathogen *A. salmonicida*, the S-layer provides resistance to the bactericidal activity of complement in immune and non-immune sera and is involved in uptake of porphyrins and binding of immunoglobulins and extracellular matrix proteins (Doig et al., 1992; Trust et al., 1993; Garduño et al., 1995, 2000). In *Rickettsia prowazekii* and *R. typhii*, which cause endemic and epidemic typhus, S-layer proteins are strongly involved in humoral and cellular immunity (Carl and Dasch, 1989). Interestingly, virulent strains of *B. anthracis* posses two separate S-layer proteins, Sap and EA1, besides its capsule (Fouet et al., 1999). A significant role in the adhesion to host cells was demonstrated for the S-layers of *Campylobacter rectus* (Borinski and Holt, 1990), *C. difficile* (Takeoka et al., 1991), and *L. acidophilus* (Schneitz et al., 1993). For *L. crispatus*, the CbsA (S-layer)-mediated collagen binding represents a true tissue adherence property (Sillanpää et al., 2000). As a special surface structure of *C. fetus* subsp. *fetus* (Dworkin and Blaser, 1997; Grogono-Thomas et al., 2000) and *B. cereus* (Kotiranta et al., 1998, 2000), the S-layer has a significant role not only in adhesion but also in phagocytosis and, in the latter organism, in increased radiation resistance.

8 Biodegradation

Biochemical analyses of isolated S-layers have shown that they are, in most cases, composed of a single polypeptide that may contain covalently linked carbohydrate chains. One problem that becomes obvious as one inspects SDS-PA gels is that many solubilized S-layer proteins reveal more than one band. Several reasons can be considered for an explanation. Some organisms, such as *Aquaspirillum serpens* MW 5 (Kist and Murray, 1984), *Brevibacillus brevis* 47 (Tsuboi et al., 1982), *Campylobacter fetus* (Pei et al., 1988), or *Clostridium difficile* (Kawata et al., 1984; Cerquetti et al., 2000; McCoubrey and Poxton, 2001), possess more than one S-layer protein. Although the reasons for the occurrence of this phenomenon are not yet clear, a plausible explanation has been found in the case of *C. difficile* (Calabi et al., 2001; Karjalainen et al., 2001). In this organism, a specific degradation of the precursor polypeptide takes place, resulting in the release of two mature S-layer proteins (for details see Sections 4 and 5). Other species of bacteria, however, do indeed show multiple protein bands on SDS gels. Radioactive labeling experiments on both gram-positive (Howard et al., 1982) and gram-negative bacteria (Thorne et al., 1976) indicated that once the protomeric units are incorporated into the S-layer lattices, they undergo virtually no turnover. The intact S-layer lattices still attached to the cell surface of the organism are very resistant to solubilization. There is, however, a difference in the resistance to proteases between the outer and the inner surface of an S-layer. After extraction from the underlying peptidoglycan layer, the S-layer proteins become subject to proteolytic cleavage during isolation. With *Aquaspirillum serpens* MW5 and VHA strains, it was demonstrated that degradation of the S-layer proteins was dependent on the chosen isolation procedure (Koval and Murray, 1984), as the low-molecular-mass bands also reacted with specific antisera when tested by Western blotting. Even application of protease inhibitors did not prevent degradation of the S-layer polypeptide (Baumeister et al., 1982; Thompson et al., 1982). However, stabilization of S-layer protein by specific ions such as Ca^{2+} could be achieved in some instances. In other strains such as *Bacillus sphaericus* strain 9662 (Hastie and Brinton, 1979a,b), a 125-kDa polypeptide was obtained either after prolonged storage at 4°C or after proteolytic cleavage of the native protein with pronase. This result indicated that a specific region of the polypeptide was particularly susceptible to cleavage.

On the other hand, some archaeal S-layers showed remarkable chemical stability. In *Thermoproteus tenax*, the intact S-layer could be isolated only by boiling a cell envelope preparation in 2% SDS for 30 min. This material was resistant to treatment with proteinase K, trypsin, and pronase (Kandler and König, 1985), and its stability was explained by cross-linking of the individual S-layer protomers. Another example is the paracrystalline sheath of *Methanospirillum hungatei*. Because of the presence of disulfide bonds, this material proved to be chemically resilient to typical chaotropic and nonionic detergents, denaturants, and hydrolytic enzymes (Beveridge et al., 1985; Southam and Beveridge, 1992).

Although in the past decade almost no attempts have been made to examine the degradation phenomena in greater detail, the conclusions drawn in these early experiments concur well with very recent observations on a domain structure of the S-layer polypeptide in solubilization experiments of S-layers from thermophilic bacilli (Rünzler, Sára, Sleytr, unpublished).

9
Production of S-layer Proteins

For application in molecular biotechnology and nanobiotechnology, wild-type S-layer proteins of Bacillaceae are extracted from the bacterial cell surface, or the recombinant S-layer proteins synthesized in *Escherichia coli* as a heterologous expression system are isolated from the cytoplasm of the host cells. Recombinant S-layer proteins occur either in self-assembled or soluble form, or they form inclusion bodies (Jarosch et al., 2001; Ilk et al., 2002). In the case of Bacillaceae, the S-layer protein may comprise up to 15% of the total cellular protein. The cells are usually grown in continuous culture under steady-state conditions, with a typical productivity of 0.5 g wet pellet.l^{-1}.h^{-1} (Schuster et al., 1995). At a dry mass of 20%, 1.5 g of S-layer protein can be isolated from 50 g of wet biomass pellet. Considering that only 500 ng of S-layer protein are required for generating 1 cm^2 S-layer monolayer, this amount is sufficient for covering an area of 300 m^2.

For isolation of S-layer proteins from Bacillaceae, whole cells are broken by ultrasonication, the plasma membrane is extracted with detergents, DNA and RNA residues are degraded with nucleases, and the remaining S-layer-carrying cell wall fragments are repeatedly washed in buffer and separated by centrifugation (Messner and Sleytr, 1988). During the cell wall preparation procedure, an inner S-layer lattice is formed, which results from the S-layer protein pool being stored within the rigid peptidoglycan-containing cell wall layer of growing cells (Breitwieser et al., 1992). S-layer-carrying cell wall fragments are subsequently treated with guanidine hydrochloride (GHCl) solution (2 to 5 M), which disintegrates the S-layer lattice into the constituent subunits and breaks the bonds between the S-layer subunits and the underlying rigid cell wall layer. After extraction of the S-layer protein, peptidoglycan-containing sacculi are simply removed by centrifugation, and the clear supernatant is subsequently dialyzed against distilled water or buffer. If soluble (monomeric and/or oligomeric) S-layer protein is required for recrystallization on solid supports such as noble metals, silicon wafers, silanized glass, or plastic foils (Sleytr et al., 1999); on lipid layers; or on liposomes (Küpcü et al., 1995a; Mader et al., 1999), the dialysis procedure is stopped at a degree of assembly (% assembled from total S-layer protein) of 20%, and formed self-assembly products are removed by centrifugation (Ilk et al., 1999). The optimum S-layer protein concentration for recrystallization varies between 50 and 100 $\mu g \cdot mL^{-1}$ solution.

In the case of recombinant S-layer proteins produced by *E. coli*, whole cells are lysed in a solution containing Triton X-100, EDTA, and lysozyme (Jarosch et al., 2001; Ilk et al., 2002; Moll et al., 2002). Self-assembled S-layer protein or inclusion bodies are separated by centrifugation and enriched in the pellet, from which the S-layer protein is extracted with GHCl and subsequently purified by gel permeation chromatography. Soluble recombinant S-layer protein can be isolated from lysed cells by precipitation with ammonium sulfate (Jarosch et al., 2001). During dialysis against distilled water, the precipitated S-layer protein is dissolved and subsequently purified by gel permeation chromatography. With the current expression system, about 100 mg of recombinant S-layer protein can be isolated from 5 g wet pellet of biomass (1 g dry weight). This amount is sufficient for generating 20 m^2 of monomolecular S-layer lattice.

10 Application of S-layer Proteins

10.1 S-layer Ultrafiltration Membranes

For production of S-layer ultrafiltration membranes (SUMs), S-layer-carrying cell wall fragments from Bacillaceae were deposited on a microporous support (nylon microfiltration membrane with an average pore size of 0.04 μm) in a pressure-driven process, and the S-layer protein was subsequently cross-linked with glutaraldehyde (Sára and Sleytr, 1987b). To increase the stability properties, Schiff bases were reduced with sodium borohydride. SUMs produced of S-layer-carrying cell wall fragments from Bacillaceae revealed molecular sieving properties identical to those of isolated S-layer lattices. They allowed free passage for proteins with a molecular mass of 30 kDa but rejected proteins with molecular masses > 40 kDa to at least 90% (Sára et al., 1992; Sára and Sleytr, 1987a,b). Because of an excess of free carboxylic acid groups originating from the acidic amino acids of the S-layer protein, SUMs revealed a negative net charge (Weigert and Sára, 1995). Chemical modification of free carboxylic acid groups enabled the preparation of hydrophilic, hydrophobic, positively or strongly negatively charged SUMs. The properties of the nucleophiles linked to the carbodiimide-activated carboxylic acid groups and contact angle measurements correlated well with the membrane adsorption characteristics (Küpcü et al., 1993; Weigert and Sára, 1995, 1996).

10.2 S-layers as Matrix for the Immobilization of Functional Macromolecules

For immobilization of biologically active macromolecules, the free carboxylic acid groups of the S-layer protein were activated with carbodiimide and could subsequently react with amine groups from enzymes, ligands, or antibodies (Sára and Sleytr, 1989; Weiner et al., 1994b; Küpcü et al., 1995a, 1996; Breitwieser et al., 1996). From the amount, molecular mass, and size of the foreign proteins bound to the S-layer lattice, as well as from the molecular mass of the S-layer subunits and the area occupied by one morphological unit, it was derived that most macromolecules formed a monomolecular layer on the surface of the S-layer lattice (Sára and Sleytr, 1996; Sleytr et al., 1997). By immobilization of ferritin, a 12-nm large topographical marker used in electron microscopy, it could be demonstrated that the covalently bound molecules followed the symmetry of the underlying S-layer lattice type (Sára and Sleytr, 1989). In the case of S-layer glycoproteins, the surface-located carbohydrate chains also were exploited for immobilization of macromolecules. Vicinal hydroxyl groups were activated with cyanogen bromide, or the corresponding carbon bonds were cleaved with periodate to form reactive aldehyde groups (Küpcü et al., 1996). Immobilization via spacers such as 6-amino caproic acid was advantageous only when the molecules were small enough to entrapped inside the pores of the S-layer lattice, which, in the case of enzymes, was linked to a significant activity loss (Küpcü et al., 1995a). In general, the activity of enzymes immobilized to S-layer lattices was well preserved (Sleytr et al., 1999).

Affinity microparticles (AMP) represent 1 μm large cup-shaped structures. They are produced from S-layer-carrying cell wall fragments, in which the S-layer protein is cross-linked with glutaraldehyde and Schiff bases are reduced with sodium borohydride (Weiner et al., 1994b). Because of the applied preparation procedure, AMP possess a complete outer and inner S-layer, which can be exploited for immobilization of foreign macromolecules. Protein A as an Fc-binding ligand was linked to the carbodiimide-activated carboxylic acid groups of the S-layer protein of *Thermoanaerobacter thermohydrosulfuricus* L111-69 (Weiner et al., 1994b; Weber et al., 2001). In the case of this hexagonal S-layer lattice, 550 μg of Protein A could be bound per mg S-layer protein. Derived from the area occupied per morphological unit and the molecular mass of the constituent S-layer subunits, this binding capacity corresponded to 320 ng Protein A per cm^2 S-layer surface (Weiner et al., 1994a). Assuming a Stoke's radius of 5 nm for Protein A, the theoretical binding capacity of the S-layer lattice was at least three times above the values calculated with the Stoke's radius. Since Protein A is an extremely long-shaped molecule with a high density of amine groups at one end, the so-called X-region, it became evident that the molecules were bound with their long axis perpendicular to the S-layer lattice. At first, the binding capacity of AMP for IgG from human plasma was investigated in batch experiments. The maximum binding capacity was 40 μg human $IgG.mg^{-1}$ wet pellet of AMP, which corresponded to 4 IgG molecules per hexameric unit cell or to a dense monolayer of uniformly oriented antibody molecules. Bound IgG could be eluted at a pH value of 3.5 (Weiner et al., 1994b; Weber et al., 2001).

AMP were used as escort particles in affinity cross-flow filtration (Weiner et al., 1994a) and as novel immunoadsorbent particles in blood purification (Weber et al., 2001). For both applications, the advantage of AMP can be seen in the cup-shaped structure leading to a high surface-to-volume ratio, as well as in the dense monolayer of Protein A molecules on the outermost surface of both S-layer lattices. When used as escort particles in affinity cross-flow filtration, AMP were circulated in a cross-flow module equipped with an ultrafiltration membrane, which rejected AMP but allowed free passage of IgG. During the first step, the IgG-containing solution was pumped through the cross-flow apparatus, and IgG was bound by AMP. After extensive washing to remove any contaminants, the Protein A–IgG complex was cleaved by decreasing the pH value to 3.5, eluted IgG was collected in the filtrate, and the solution neutralized. Eluted IgG was free of any contaminants (Weiner et al., 1994a; Weber et al., 2001).

Conventional immunoadsorption techniques are based on the perfusion of separated plasma through adsorption columns and reinfusion of the purified plasma together with the concentrated blood cells. In an alternative approach realized in the microspheres-based detoxification system (MDS), the plasma filtrate does not perfuse an adsorption column, but is recirculated into the filtrate compartment of the module. The addition of a suspension of IgG-binding microparticles to the circulating system allows the rapid and specific adsorption of IgG (Weber et al., 2001). Recently, the suitability of AMP for the MDS system was investigated. In comparison with two commercially available immunoadsorbents that could bind either 7 mg human $IgG \cdot g^{-1}$ wet pellet by hydrophobic interactions (Asahi Medical, Tokyo, Japan) or 8 mg human $IgG.g^{-1}$ wet pellet by specific interactions (Protein A–silica affinity matrix, Prosorba, Fresenius), AMP with 40 mg $IgG \cdot g^{-1}$ wet pellet revealed a significantly higher IgG-binding capacity. In biocompatibility tests, supernatants of AMP did not show any limulus amoebocyte lysate activity or cytotoxicity and did not contain any cytokine-inducing substances (Weber et al., 2001).

10.3 S-layer-based Dipsticks

SUMs were used as novel matrices for the development of dipstick-style solid-phase immunoassays. Depending on the test systems, the respective monoclonal antibody was covalently bound to the carbodiimide-activated carboxylic acid groups of the S-layer lattice. The amount that could be immobilized was 370 $ng \cdot cm^{-2}$ S-layer lattice, which corresponded to a monolayer of randomly oriented antibody molecules (Breitwieser et al., 1996, 1998). After immobilization of the monoclonal antibodies, disks 3 mm in diameter were punched out and sandwiched between Teflon foils, leaving the SUM exposed for further binding reactions. Three different types of SUM-based dipsticks were developed for:

- diagnosis of type I allergies (determination of IgE in whole blood or serum against the major birch pollen allergen);
- quantification of tissue-type plasminogen activator (t-PA) in patients' whole blood or plasma for monitoring t-PA levels in the course of thrombolytic therapies after myocardial infarcts;
- determination of interleukin 8 (IL-8) in supernatants of human umbilical vein endothelial cells induced with lipopolysaccharides.

In the case of the IgE dipstick, recombinant birch pollen allergen (rBet v1) was bound to the monoclonal antibody BIP 1. After incubation in whole blood or serum, IgE was quantified via an anti-IgE–alkaline phosphatase conjugate using a substrate that formed an IgE concentration-dependent colored precipitate on the SUM surface (Breitwieser et al., 1998). The intensity was measured with a reflectometer, enabling us to differentiate between RAST classes 1 to 6, which correspond to IgE concentrations of 0.35 to $>$ 100 PrU. The detection limit for IL-8 was below 100 $pg \cdot ml^{-1}$ culture supernatant. The advantages of S-layers as immobilization matrix for producing solid-phase immunoassays may be summarized as follows. Since S-layers are crystalline arrays of identical protein or glycoprotein subunits, they have repetitive surface properties down to the subnanometer scale. S-layer lattices have a high density of functional groups located on the outermost surface of these crystalline arrays.

Because of the molecular size of the catching antibody, immobilization can occur only on the outermost surface of the S-layer lattice, thereby preventing diffusion-limited reactions in further incubation and binding steps. Because the catching antibody is covalently linked to the S-layer lattice, no leakage problems arise during the test procedure. S-layers do not unspecifically adsorb plasma or serum components, which makes the blocking steps required for immunoassays with conventional matrices unnecessary. Stable, concentration-dependent precipitates are formed on the S-layer surface by using appropriate substrates for peroxidase or alkaline phosphatase-conjugated antibodies in the final detection step.

10.4 Supramolecular Structures Generated by Oriented Recrystallization of S-layer Fusion Proteins on Supports Precoated with SCWP

For generating regularly structured functional monomolecular protein lattices, S-layer fusion proteins that incorporated foreign peptide sequences or protein domains were produced by recombinant technologies and used for recrystallization on solid supports, such as gold chips or silicon wafers (Neubauer et al., 2000), on lipid films, or on liposomes (Küpcü et al., 1995b; Mader et al., 1999; Mader et al., 2000). In order to elucidate the structure-function relationship of distinct segments of S-layer proteins, N- or C-terminally truncated forms were produced in a heterologous expression system and their self-assembly and recrystallization properties were investigated (Jarosch et al., 2001; Ilk et al., 2002). Several studies indicated that the N-terminal part is involved in cell wall anchoring by recognizing a distinct type of SCWP that is covalently bound to the peptidoglycan backbone (Ries et al., 1997; Sára et al., 1998a; Egelseer et al., 1998; Ilk et al., 1999, 2002; Jarosch et al., 2000, 2001; Sára and Sleytr, 2000; Sára, 2001). Therefore, the deletion of the N-terminal part was linked to a loss in cell wall binding but had no influence on the self-assembly properties of the S-layer proteins (Howorka et al., 2000; Jarosch et al., 2001; Moll et al., 2002). Most S-layer proteins were also tolerant against deletions at the C-terminal part. For example, the mature S-layer protein SbsC comprises amino acids 31 to 1099, and the deletion of 179 C-terminal amino acids leading to $rSbsC_{31-920}$ influenced neither the size and shape of the self-assembly products nor the formation of the oblique lattice structure (Jarosch et al., 2000, 2001). According to these findings, 200 amino acids could be deleted at the C-terminal end of the S-layer protein SbpA without interfering with its self-assembly properties (Ilk et al., 2002). The obtained C-terminal truncation $rSbpA_{31-1068}$ formed self-assembly products that clearly exhibited the square lattice structure. In contrast, the deletion of even less than 15 C-terminal amino acids in the S-layer protein SbsB led to completely water-soluble forms (Howorka et al., 2000).

Based on the knowledge of the structure-function relationship as derived from the various truncated forms, S-layer fusion proteins were constructed in which the C-terminal part was replaced by a foreign sequence. For example, in rSbsC/Bet v1 (Breitwieser et al., 2002), 179 C-terminal amino acids of the S-layer protein were deleted and replaced with the 161-amino-acid-long Bet v 1 sequence, which is the major birch pollen allergen. In the case of rSbpA, the 9-amino-acid-long Strep-tag I with affinity to streptavidin was exploited for screening amino acid positions that were located on the outer surface of the square S-layer lattice (Ilk et al., 2002). It could be demonstrated that the C-terminal end in the monomers of the full-length form of recombinant rSbpA ($rSbpA_{31-1268}$) was available to only a limited extent but was absolutely not accessible after recrystallization of the S-layer fusion protein on native peptidoglycan-containing sacculi (Ilk et al., 2002). On the other hand, Strep-Tag I was fully accessible in the C-terminally truncated form $rSbpA_{31-1068}$. This was the reason that the C-terminally truncated form was exploited as base form for the construction of further S-layer fusions proteins, incorporating either the Bet v1 sequence (rSbpA/Bet v1) or a camel antibody sequence recognizing lysozyme as an epitope (rSbpA/cAB). In the case of the S-layer protein SbsB, 6 N- and 2 C-terminal fusion proteins comprising the sequence of core streptavidin were constructed (Moll et al., 2002). To obtain functional heterotetramers (1 SbsB–core streptavidin fusion protein molecule plus 3 core streptavidin molecules), the SbsB–core streptavidin fusion protein forms were mixed with core streptavidin in the molar ratio of 1 to 3, and an appropriate refolding procedure was applied (Moll et al., 2002). After purification by gel permeation and affinity chromatography, functional heterotetramers were isolated. The biotin-binding capacity of the heterotetramers was 80% in comparison to free core streptavidin (Moll et al., 2002).

In previous studies, the interactions of the S-layer proteins SbsB, SbsC, and SbpA with the underlying rigid cell envelope layer were investigated (Ries et al., 1997; Sára et al., 1998a; Egelseer et al., 1998; Ilk et al., 1999; Jarosch et al., 2000, 2001; Ilk et al., 2002). The S-layer proteins SbsB and SbpA carry three typical SLH motifs on the N-terminal parts, which recognize a distinct type of secondary cell wall polymer as the proper anchoring structure in the rigid cell envelope layer. To generate a monomolecular S-layer lattice in defined orientation and to investigate location, accessibility, and functionality of introduced foreign sequences, the C-terminal S-layer fusion proteins were recrystallized on native peptidoglycan-containing sacculi, thereby exploiting the specific interactions between the N-terminal part of the S-layer proteins and SCWP (Figure 9).

The surface location of the Bet v1 portion in the S-layer fusion proteins rSbsC/Bet v1 and rSbpA/Bet v1 was demonstrated by labeling the bound Bet v1-specific monoclonal mouse antibody BIP with an anti mouse–colloidal gold conjugate or by labeling bound Bet v1-specific IgE with an anti IgE–colloidal gold conjugate (Breitwieser et al., 2002; Ilk et al., 2002). The surface accessibility of core streptavidin in the oblique S-layer lattice formed by heteroteramers of the C-terminal SbsB–core streptavidin fusion proteins was investigated by its ability to bind biotin or biotinylated ferritin (Moll et al., 2002).

In order to build up functional monomolecular S-layer protein lattices on solid supports, the latent aldehyde group in the polymer chain of the secondary cell wall polymer was modified. In the first reaction step, it was converted into an amine group and finally, by reaction with 2-iminothiolane, into a sulfhydryl group (Mader et al., 2002). Because of the presence of a single thiol group, the polymer chains could attach in

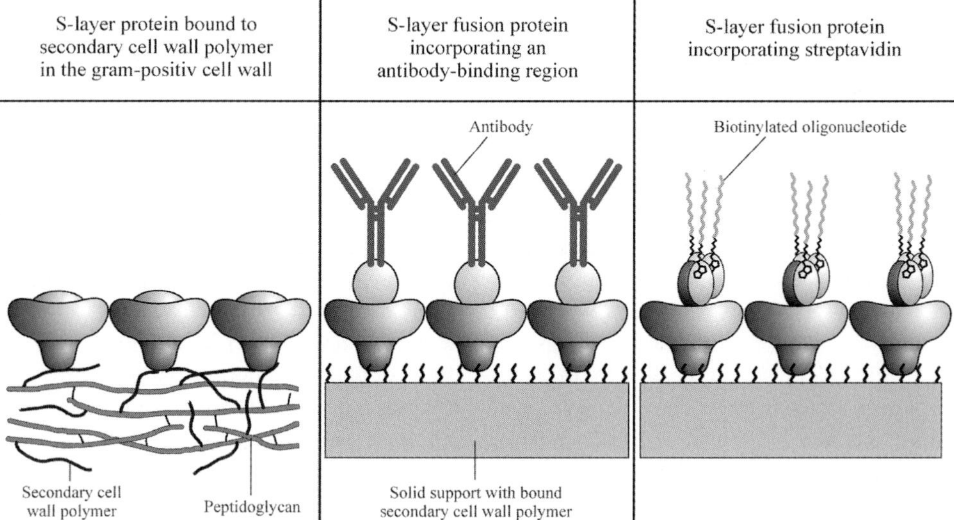

Fig. 9 Schematic drawing illustrating the formation of a regularly structured functional monomolecular protein lattice by oriented recrystallization of an S-layer fusion protein on a solid support precoated with secondary cell wall polymer.

defined orientation to the gold chips. The S-layer fusion proteins bound with the N-terminal end to the modified gold chips, leaving the C-terminal part with the fused functional sequence exposed to the ambient environment. Such supramolecular biomimetic structures, consisting of a solid support, SCWP, and a functional S-layer fusion protein recrystallized in defined orientation, will be exploited for novel developments in gene, protein, and antibody chip developments (Figure 9).

10.5
S-layers as Templates for the Formation of Regularly Arranged Nanoparticles

In the early 1980s, physicists discovered that gaseous metal atoms could spontaneously assemble into small groups, but normally not in large scale. This finding was particularly important since arrays of 1- to 2-nm clusters are ideal for advanced optical and electronic devices. This is because electrons confined in such small spaces show very important new physical properties, such as the production of photons with wavelengths useful for nonlinear optical applications. In addition to these developments in microelectronics, such evenly spaced, identical clusters of nanoparticles provide the basis of nanoscale chemical reactors. A huge amount of knowledge has been accumulated and impressive results have been obtained in the past few decades, but the reproducible formation of nanoparticle arrays in large scale with predefined lattice spacings and symmetries is still a challenge in nanosciences. This is particularly true for self-assembly and bottom-up approaches, since these strategies acquire the highest efficiency in a fabrication process. Thus, biomolecular templating has proven to be very attractive because self-assembly of molecules into monomolecular arrays is an intrinsic feature of many biological systems and has developed into a scientific and engineering discipline that crosses the boundaries of several established fields.

The first approach in using S-layers as lithographic templates in the formation of perfectly ordered nanoparticle arrays was developed by Douglas and coworkers (Douglas and Clark, 1986). It was based essentially on the deposition of a metal vapor onto a nanostructural S-layer. In a three-step process, S-layer fragments of *Sulfolobus acidocaldarius* were deposited on a smooth carbon surface, metal coated by evaporation (1-nm thick Ta/W), and then ion milled. Because this protein–metal heterostructure exhibits differing metal removal under ion milling, a 1-nm thick metal (Ta/W) film with holes 15 nm in diameter was formed. These were hexagonally arranged according to the lattice parameter of the S-layer of 22 nm. In a similar approach, a nanostructured titanium oxide layer was derived after coating S-layer fragments deposited on a smooth graphite surface (Douglas et al., 1992). After oxidation in air and fast-atom beam milling at normal incidence, a thin (~ 3.5 nm) metallic nanoporous mask with pores in the 10-nm range was obtained. Recently, the same group used low-energy, electron-enhanced etching to pattern the surface properties of silicon through the regularly arranged pores of the S-layer (Winningham et al., 1998). After etching and removal of the S-layer the patterned surface was oxidized in an oxygen plasma, leading to a nanometric array of etched holes (18 nm in diameter). In the final step, evaporation of titanium onto the surface led to the formation of an ordered array of nanometric metal clusters. In a similar approach using Ar ion etching in the final step, the S-layer of *Deinococcus radiodurans* was used as a nanometric template for patterning ferromagnetic films (Panhorst et al., 2001). A uniform hexagonal pattern of 10-nm wide dots and lattice constants of 18 nm was fabricated from 2.5-nm thick sputter-coated Co, FeCo, Fe, FeNi, and NiFe films.

Most recently, wet chemical processes were developed for the fabrication of metallic nanoparticle arrays using S-layers as nanometric templates (Shenton et al., 1997; Dieluweit et al., 1998; Mertig et al., 1999; Pompe et al., 1999). First, self-assembled S-layer lattices were exposed to a metal-salt solution (e.g., $[AuCl_4]^-$, $[PtCl_4]^{2-}$), followed by slow reaction with a reducing agent (e.g., hydrogen sulfide [H_2S]). Since the precipitation of the metals was confined to the pores of the S-layer, nanoparticle arrays with prescribed symmetries and lattice geometries could be obtained. The first example exploiting this technique was the precipitation of cadmium sulfide on S-layer lattices of *Bacillus coagulans* E38/V1 and *Bacillus sphaericus* CCM 2177 (Shenton et al., 1997). After incubation of S-layer self-assembly products in a $CdCl_2$ solution for several hours, the hydrated samples were exposed to H_2S for at least one or two days. The CdS nanoparticles were 4–5 nm in size and resembled the oblique lattice symmetry of the S-layer of *B. coagulans* E38/V1 (a= 9.4 nm, b=7.4 nm, γ=80°), or the square lattice symmetry of *B. sphaericus* CCM 2177 (a=b=13.1 nm, γ=90°), respectively. Electron diffraction patterns confirmed the Zinc-Blende crystal structure of CdS. In a similar approach, a superlattice of 4–5 nm gold particles was formed by using the S-layer of *B. sphaericus* CCM 2177 (with pre-induced thiol groups) as a template for the precipitation of a tetrachloroauric (III) acid solution (Dieluweit et al., 1998) (Figure 10).

Reduction of the gold (Au(III)) was performed either by exposing the metallized S-layer to an electron beam in a transmission electron microscope or by slow reaction with H_2S. Transmission electron microscopical studies showed that the metal nanoparticles were formed in the pore region. As determined by electron diffraction, the gold nanoparticles were crystalline (cubic lattice symmetry) but, in the long-range order, were not crystallographically aligned. The wet chemical approach was used in the precipitation of Pd (salt: $PdCl_2$), Ni ($NiSO_4$), Pt ($KPtCl_6$), and Pb ($Pb(NO_3)_2$) nanoparticle arrays (unpublished results). Wet chemistry also was applied for producing platinum nanoparticles on the S-layer of *Sporosarcina ureae* (Mertig et al., 1999; Pompe et al., 1999). One morphological unit of the S-layer lattice of *S. ureae* revealed 7 Pt cluster sites with a diameter of ~ 1.9 nm. Multiple sites on the S-layer of *S. ureae* also could be demonstrated by pulsed laser deposition methods (Pompe et al., 1998). Pulsed laser deposition has already been proven to be a valuable method for generating nanoparticle arrays on S-layer lattices, in particular for the generation of nanometric calibration standards for scanning force microscopy (Neubauer et al., 1997, 1998).

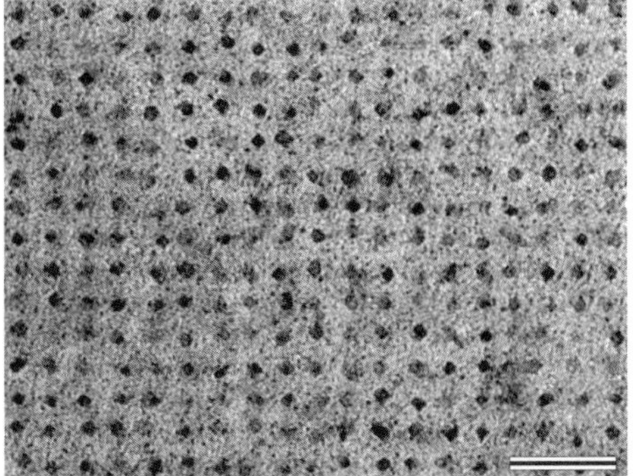

Fig. 10 Transmission electron microscopical image of gold nanoparticles (mean diameter 4.5 nm) obtained by wet chemistry. An S-layer with square lattice symmetry served as a template in the precipitation of the metal salt. The cold nanoparticles were formed in the pore region of the protein meshwork under the electron beam (Bar 50 nm).

Although the wet chemical methods resulted in crystalline arrays of nanoparticles with spacings in register with the underlying S-layer lattice, they do not allow for controlling particle size or composition, such as in core-shell nanoparticles. Thus, the binding of preformed standardized nanoparticles into regular arrays on S-layers will have significant advantages in the development of biomolecule-driven assemblies of nanoscale electronic devices compared to vapor deposition or wet chemical methods. Based on the work on binding biomolecules, such as enzymes or antibodies, onto S-layers in a controlled packing, it has been demonstrated that gold or CdSe nanoparticles can be electrostatically bound in regular arrangements on S-layers (Hall et al., 2001; Györvary, unpublished results) (Figure 11).

The nanoparticles were either negatively charged as a result of surface citrate ions or positively charged as a result of surface capping with poly-L-lysine. These experiments have clearly shown that S-layers are perfectly suited to control the formation of nanoparticle arrays, either by direct precipitation from the vapor or liquid phase or by binding preformed nanoparticles. The S-layer approach provides for the first time a biologically based fabrication technology for the self-assembly of molecular electronic or optic devices.

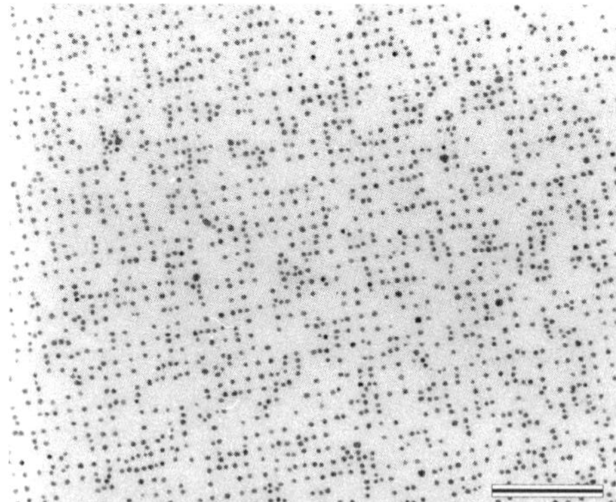

Fig. 11 Transmission electron microscopical image of preformed gold nanoparticles (mean diameter 4 nm) regularly bound on the surface of an S-layer with square lattice symmetry. Electrostatic interactions between the surface of the nanoparticles and functional domains on the S-layer are responsible for the binding (Bar 100 nm).

10.6 S-layers as Supporting Structures for Functional Lipid Membranes (Planar Membranes and Liposomes)

A fascinating feature of S-layer proteins is their ability to recrystallize on liquid/lipid interfaces like Langmuir films and planar or spherical bilayer lipid membranes (BLMs). S-layer-supported lipid membranes (SsLMs; for a compilation, see Schuster and Sleytr, 2000) mimic the supramolecular assembly of archaeal cell envelope structures composed of a cytoplasmic membrane and a closely associated S-layer (Kandler, 1982; König, 1988). In this biomimetic architecture, either a tetraether lipid monolayer or an artificial phospholipid mono- or bilayer replaces the cytoplasmic membrane, and isolated bacterial S-layer proteins are recrystallized on one or both sides of the lipid film (Figure 12).

The first biomimetic SsLMs were generated 10 years ago when isolated bacterial S-layer proteins were recrystallized on a phospholipid Langmuir film (Pum et al., 1993; Pum and Sleytr, 1994). These SsLMs have been characterized by recrystallization studies (Nomellini et al., 1997; Wetzer et al., 1998; Smit et al., 2001) and by various surface-sensitive methods (Diederich et al., 1996; Weygand et al., 1999, 2000). In general, the recrystallization of S-layer proteins on phospholipid films depends on (1) the phase state of the lipid film, (2) the nature of the lipid head group (size, polarity, and charge), and (3) the ionic content and pH of

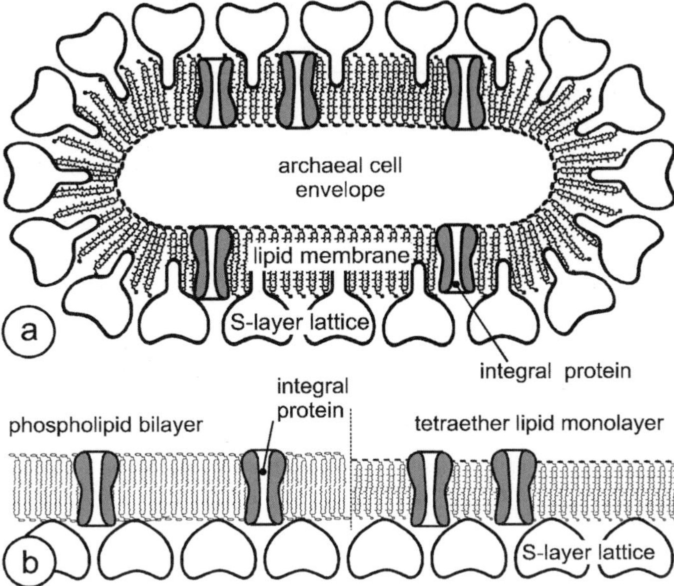

Fig. 12 (a) Schematic illustration of an archaeal cell envelope structure composed of the cytoplasmic membrane with integral membrane proteins and an S-layer lattice, integrated into the cytoplasmic membrane. (b) Using this supramolecular construction, biomimetic membranes can be generated. The cytoplasmic membrane is replaced by a phospholipid bilayer or a tetraether lipid monolayer, and S-layer proteins are recrystallized to form a coherent lattice on the lipid film. Subsequently, integral model membrane proteins can be reconstituted into the S-layer-supported lipid membrane.

the subphase (Table 5). The capability to form a closed lattice differs for various S-layer proteins, as some recrystallize only on lipids with a net neutral charge (Pum et al., 1993; Pum and Sleytr, 1994; Wetzer et al., 1998), others only on net negatively charged phospholipid films (Smit et al., 2001), and others need lipopolysaccharides for recrystallization (Nomellini et al., 1997).

In contrast to solid surfaces, the organic lipid films get slightly modulated during the recrystallization process of S-layer proteins (Table 6). Most important, although peptide side groups of the S-layer protein interpenetrate the phospholipid head groups almost to its entire depth, no impact on the hydrophobic lipid acyl chains has been observed (Schuster et al., 1998b; Weygand et al., 1999, 2000).

Planar phospholipid bilayers (Figure 12b) can be generated by modified Langmuir Blodgett techniques. The BLMs used as relevant models for biological membranes have improved the knowledge on the structure and function of cell envelopes. To utilize the function of cell membrane components for practical applications, stabilization of the BLMs is imperatively necessary, as typically freestanding membranes survive for only a few hours and are very sensitive to vibration and mechanical shocks (Raguse et al., 1998; Tien and Ottova, 2001). Thus, S-layer proteins have been exploited as supporting structures for BLMs. It turned out that BLMs became significantly stabilized (Wetzer et al., 1997; Schuster et al., 1998a, 1999; Schuster and Sleytr, 2002a) and physical features, such as the membrane tension and

Tab. 5 Specific features of S-layer proteins governing interactions with lipid membranes[a]

S-layer proteins adsorb preferentially at lipid films in the liquid-expanded phase [M]
Crystallization is observed only at the liquid-condensed phase [M]
Dynamic crystal growth is initiated at several distant nucleation points [M]
The individual monocrystalline areas grow in all directions until front edges of neighboring crystals meet [M]
Electrostatic interactions are the dominant binding forces [M,L]
"Primary" and "secondary" binding sites with different specificity exist on the S-protein for lipid molecules [M]
The S-layer protein interacts with the head groups of the lipid molecules [M]
Size and charge of the lipid head groups and subphase conditions (e.g., pH value, ionic content) affect the crystallization [M]
The orientation of the S-layer lattice is determined by the subphase conditions [M]
Coating of positively charged liposomes results in inversion of the ζ-potential [L]

[a] S-layer protein was crystallized on [M] lipid monolayers and on [L] liposomes

Tab. 6 Impact of recrystallized S-layer lattices on lipid membranes

A significant change of the lipid head group interactions is observed [L]
Lipid head groups are tilted towards the surface normal of the membrane [B,M]
Peptide material interpenetrates the phospholipid head groups in its entire depth [M]
Head group hydration is reduced by ~ 40% upon peptide interpenetration [M]
S-layer protein does not affect the hydrophobic lipid acyl chains [M,B]
There is no impact on the hydrophobic thickness of the lipid membrane [B]
Fluid lipid film is driven into a state of higher order, especially at low surface pressure [M]
S-layer coating of liposomes leads to an ordering effect of the lipid molecules [L]
Liquid-ordered gel-like state of liposomes is stabilized by S-layer coating [L]
Attachment of S-layers results in a decreased membrane tension [B]
A significant increase of the previously negligible surface viscosity is observed [B]
Highest mobility of lipid probe molecules in S-layer-supported bilayers (compared to silane- or dextran-supported bilayers) [B]
Increase in conformational freedom of the hydrophobic core at temperatures below 29°C [L]
S-layer cover prevents the formation of inhomogeneities in the bilayer [B]

[a] S-layer protein was recrystallized on [M] lipid monolayers, [B] lipid bilayer membranes, and [L] liposomes

fluidity, were kept in good condition (Table 6; Györvary et al., 1999; Hirn et al., 1999).

To assess SsLMs as sensing elements, reconstitution of responsive molecules such as gramicidin, valinomycin, or α-hemolysin has been performed (Schuster et al., 1998a,b). Because the gating of even single α – hemolysin pores can be measured with SsLMs (Schuster and Sleytr, 2002b) and with SsLMs attached on porous supports (Schuster et al., 2001), these composite structures may be used to make rapid and sensitive biosensors. Potential applications range from DNA sequencing and pharmaceutical high throughput screening to lab-on-the-chip technology.

In order to enhance the stability of liposomes and to provide a biocompatible outermost surface structure for controlled immobilization, isolated S-layer proteins have been recrystallized on the outer shell of liposomes (see Figure 13). The recrystallization of S-layer proteins resulted in a completely covered surface and did not affect the morphology of the liposomes (Küpcü et al., 1995b). The high mechanical and thermal stability of S-layer-coated liposomes (Küpcü et al., 1998; Mader et al., 1999; Hianik et al., 1999) and the possibility for immobilizing or entrapping biologically active molecules (Küpcü et al., 1995b; Mader et al., 2000; Krivanek et al., 2002) introduce a broad application potential, particularly as carrier and/or drug-delivery and drug-targeting systems or in gene therapy, e.g., as artificial viruses.

10.7 S-layers for Vaccine Development

Among the different biotechnological and medical application aspects of S-layers, their use in vaccine formulations is obvious because of the surface location. Since surface components frequently mediate specific interactions of a pathogen with its host organism (Williams, 1988), S-layers of pathogenic strains in particular are expected to have an important role in virulence.

S-layer-carrying strains of *Aeromonas salmonicida* and *A. hydrophila* can cause furunculosis in fish in freshwater and marine environments (Cipriano, 1983). The S-layers absolutely are required for virulence, since isogenic mutants are avirulant (Kay and Trust, 1991). Numerous attempts have been undertaken to vaccinate salmon, trout, and catfish against furunculosis using whole cells, cell sonicates, and crude or partially purified cellular preparations (Udey and Fryer, 1978; Evenberg et al., 1988; Thornton et al., 1991; Ford and Thune, 1992). Although there was varying success in the different trials, vaccination is widely used in today's salmon hatcheries to prevent substantial economic damage in case of an infection.

Another severe threat for fish farming is infectious hematopoietic necrosis virus (IHNV), which produces a severe hemorrhagic disease in young salmonid fish

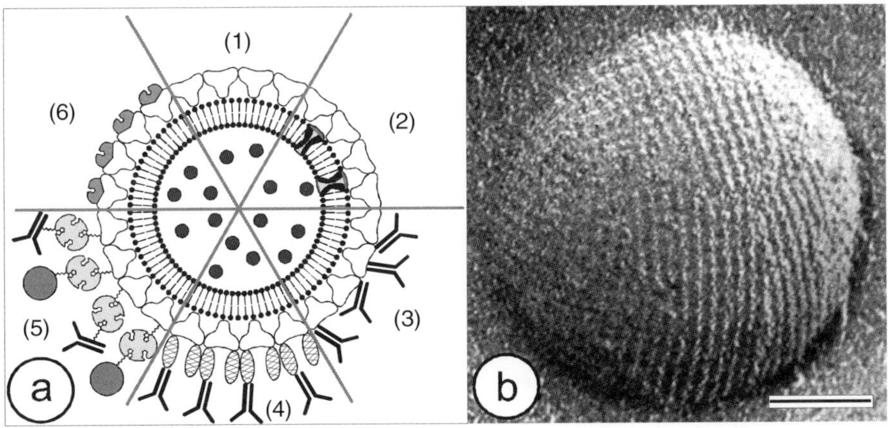

Fig. 13 (a) Schematic drawing of (1) an S-layer-coated liposome with entrapped functional molecules and (2) functionalized by reconstituted integral proteins. S-layer-coated liposomes can be used as immobilization matrix for functional molecules (e.g., human IgG) either by direct binding (3) or by immobilization via the Fc-specific ligand protein A (4), or biotinylated ligands can be bound to the S-layer-coated liposome via the biotin–avidin system (5). Alternatively, liposomes can be coated with genetically modified S-layer subunits incorporating functional domains (6). (b) Electron micrograph of a freeze-etched preparation of an S-layer-coated liposome (Bar 100 nm). Figure courtesy of S. Küpcü.

(Bootland and Leong, 1999). A subunit model vaccine has been developed by fusing a 184-amino-acid segment of the IHNV glycoprotein to the C-terminal portion of the S-layer protein of *Caulobacter crescentus* (Simon et al., 2001). With the IHNV-G/S-layer recombinant-protein vaccine, a relative percentage of survival of 26% to 34% was demonstrated in three laboratory trials. The long-term goal of this project is to develop IHNV vaccines that are safe, inexpensive, and easy to produce and administer.

In the past, several model conjugate vaccines with S-layer (glyco)proteins of thermophilic bacilli and clostridia have been tested in animal models. The antigens were weakly immunogenic carbohydrate antigens (e.g., *Streptococcus pneumoniae*, serotype 8 poly- and oligosaccharides, $[\rightarrow 4\beta Glc1 \rightarrow 4\beta Glc1 \rightarrow 4\alpha\ Gal1 \rightarrow 4\alpha GA1 \rightarrow]_n$) and haptens (e.g., T-disaccharide ($\beta Gal1 \rightarrow 3\alpha GalNAc$), Lewis Y ($Le^y$) tetrasaccharides ($\alpha Fuc1 \rightarrow 2\beta Gal1 \rightarrow 4[\alpha Fuc1 \rightarrow 3]\alpha GalNAc$), or recombinant birch pollen allergens (Jahn-Schmid et al., 1996a). Chemical coupling had been used for covalent binding of the antigens and haptens to the S-layer carrier proteins (Messner et al., 1992; Malcolm et al., 1993a,b; Jahn-Schmid et al., 1996b). Although promising immune responses were elicited in the vaccination trials (Smith et al., 1993; Malcolm et al., 1993a; Jahn-Schmid et al., 1997), even better results will be expected by using recombinant S-layer proteins as carriers for the production of hapten/S-layer fusion proteins. In addition, universal T-cell epitopes that enhance immunogenicity and overcome genetic restrictions of the immune response could be introduced into these second-generation S-layer model vaccines.

11 Outlook and Perspectives

Although self-assembly processes are common throughout nature, there are only a few examples where biomolecules have the intrinsic capability to aggregate into monomolecular arrays. On the other hand, self-assembly is one of the few suitable strategies for generating bottom-up nanostructured materials that can be exploited at the meso- and macroscopic scale. Being composed of a single (glyco)protein species, S-layers represent the simplest self-assembling membranes that developed during the evolution of life. The spontaneous association of identical subunits under equilibrium conditions results in stable, structurally well-defined isoporous protein lattices that can be strengthened by introducing inter- and/or intramolecular covalent bonds.

The repetitive physicochemical properties of S-layer lattices down to the subnanometer scale make them particularly suitable as patterning elements for the defined immobilization of functional molecules or as templates for the fabrication of ordered and precisely located nanometer scale particles, as required in molecular nanotechnology and biomimetics (Sleytr et al., 2000, 2001a,b). Future strategies for exploiting S-layers as unique nanostructured biopolymers will focus on genetic and specific chemical modifications of S-layer proteins. Since S-layers are highly anisotropic structures with regard to the physicochemical properties of their inner and outer surfaces, such modifications will enable the generation of 5- to 10-nm thick, highly porous functional protein lattices with different functions exposed on both surfaces. Moreover, the specific lectin-type binding interactions between S-layer lattices and SCWP open possibilities to build monolayers with defined orientations of functional domains on surfaces and/or interfaces as well as multilayer lattices with the constituent S-layer protein subunits in accurate superposition (Sára, 2001; Sleytr et al., 2001c).

An important line of development is presently directed toward the combination of S-layers with plane lipid membranes and liposomes. This biomimetic approach, copying the supramolecular principle of cell envelopes of archaea that dwell under the most extreme and hostile environmental conditions or envelopes of a great variety of viruses, is expected to lead to new technologies for stabilizing functional lipid membranes at the macroscopic scale (Schuster and Sleytr, 2000; Sleytr et al., 2001a,b). These unique possibilities to improve the stability and lifetime of lipid membranes also can be exploited for a broad spectrum of liposome technologies concerning drug targeting and delivery systems and gene therapy.

S-layer glycoprotein arrays from *Bacillaceae* are promising candidates for their utilization as cell surface display systems of rationally designed glycoslation motifs. This approach, which necessarily requires re-engineering of S-layer glycans by heterologous glycosyltransferases ("carbohydrate-engineering") (Schäffer and Messner, 2001), will eventually produce a versatile tool for applied research, leading to the expression of various antigenic determinants, tissue-targeting signals, or receptor mimics. Regularity and multivalency of S-layer lattices in combination with the general advantages of gram-positive expression systems (Stahl and Uhlen, 1997) make S-layer *neo*glycoprotein lattices worth being further exploited for this type of application.

Although up to now the development of S-layer technologies has focused primarily on life sciences, an important field of future applications concerns non-life sciences. S-layers that have been recrystallized on solid supports can be used as patterning elements for accurate spatial positions of nanometer-scale metal particles or as mask for vapor or wet chemical deposition of metals, as required for molecular electronics and non-linear optics. Such production procedures may even include steps in which the S-layer is sacrificed during one of the production steps (Dieluweit et al., 1998; Shenton et al., 1997).

12 Patents

Details of patents relating to the application of S-layer proteins are listed in Table 7.

Tab. 7 Patents

Patent number	*Holder*	*Inventor*	*Title*	*Date of publication*	*Main claim*
WO8504111	U. B. Sleytr	U. B. Sleytr	Structure with membranes having continuous pores	1985-09-26	A structure that comprises at least one membrane with continuous pores having a diameter within the range of 1 to 8 nm, extending along plane, curved, cylindrical, or vesicular surfaces consisting essentially of at least one layer of continuous identical protein-containing molecules
US4752395				1988-06-21	
US4886604	S. Margit	M. Sára		1989-12-12	
WO8603685	M. Sára	M. Sára	Process for changing the effective pore size of a structure	1986-07-03	Use of a structure with membranes having continuous pores (see WO8504111)
US4886604	U. B. Sleytr	U. B. Sleytr		1989-12-12	
US4728591	Univ .of Boston, Univ. of Colorado Foundation	N. A. Clark, K. Douglas, K. J. Rothschild	Self assembled nanometer lithographic masks and templates and method for parallel fabrication of nanometer scale multi-device structures	1988-03-01	Method of forming a nanometer scale pattern using two-dimensional (protein) layers as lithographic mask
US4802951	Univ. of Boston, Univ. of Colorado Foundation	N. A. Clark, K. Douglas, K. J. Rothschild	Method for parallel fabrication of nanometer scale multi-device structures	1989-02-07	Method of forming a nanometer scale pattern employing parallel fabrication
US5043158	U. B. Sleytr	U. B. Sleytr, P. Messner, W. Mundt	Pharmaceutical structure with protein-supported agents	1989-03-08	In the pharmaceutical structure haptens and/or immunogenic or immunostimulant substances are bound to a protein carrier
WO8909406	U. B. Sleytr, M. Sára	U. B. Sleytr, M. Sára	Process for immobilizing or depositing molecules or substances on a support	1989-10-05	Supporting structure consisting of at least one layer of molecules containing identical proteins extending along a flat, curved, cylindrical or vesicular surface, the molecules being arranged in the form of a crystal lattice with a lattice constant of 1 to 50 nm
WO 9519371	Solvay	R. Y. Deblaere, J. Desomer, P. Dhäse	Expression of surface layer proteins	1995-07-20	A host cell which is provided with an S-layer comprising a fusion polypeptide; can be used as a vaccine, for screening for proteins and antigens and as a support for immobilizing an enzyme, peptide or antigen.

Tab. 7 (cont.)

Patent number	*Holder*	*Inventor*	*Title*	*Date of publication*	*Main claim*
US5500353	Univ. of British Columbia	J. Smit, W. H. Bingle	Bacterial surface protein expression	1996-03-19	Bacterium having an S-layer modified such that the bacterium S-layer protein gene contains one or more in-frame sequences coding for one or more heterologous polypeptides and the S-layer is a fusion product of the S-layer protein and the heterologous polypeptide
WO9728263	W. Lubitz	W. Lubitz	Recombinant expression of S-layer proteins	1997-08-07	Processes for the recombinant preparation of S-layer proteins in gram-negative host cells. Nucleotide se-
DE19603649	U. B. Sleytr	M. Sára, S. Howorka, B. Kuen, S. Resch, G. Schroll, U. B. Sleytr, M. Truppe		1997-08-07	quence of a new S-layer gene and a process for preparation of modified S-layer proteins is disclosed
WO9748837	J. Hofinger	R. Wahl	Metallic nanostructure on the basis of self assembling, geometrically highly	1997-12-24	Metallic nanostructures on the basis of self assembled, geometrically highly ordered proteins, and a process for
DE19603649		R. Kirsch, M. Mertig, W. Pompe, E. Unger	ordered proteins, and process for preparation thereof	1998-01-08	their preparation
WO0181425	U. B. Sleytr, M. Sára	U. B. Sleytr , M. Sára, C. Mader, B. Schuster, F. M. Unger	Use of a secondary cell wall polymer of procaryotic microorganisms	2001-11-01	Use of a secondary cell wall polymer for oriented monomolecular bonding of molecular layers and/or addition of molecules of a carrier

13
References

Adachi, T., Yamagata, H., Tsukagoshi, N., Udaka, S. (1989) Multiple and tandemly arranged promoters of the cell wall protein of *Bacillus brevis* 47, *J. Bacteriol.* **171**, 1010–1016.

Aebi, U., Smith, P. R., Dubochet, J., Henry, C., Kellenberger, E. (1973) A study of the structure of the T-layer of *Bacillus brevis, J. Supramol. Struct.* **1**, 498–522.

Amos, L. A., Henderson, R., Unwin, P. N. T (1982) Three-dimensional structure determination by electron microscopy of two-dimensional crystals, *Prog. Biophys. Mol. Biol.* **39**, 183–231.

Altman, E., Brisson, J.-R., Messner, P., Sleytr, U. B. (1990) Chemical characterization of the regularly arranged surface layer glycoprotein of *Clostridium thermosaccharolyticum* D120-70, *Eur. J. Biochem.* **188**, 73–82.

Altman, E., Brisson, J.-R., Messner, P., Sleytr, U. B. (1991) Chemical characterization of the regularly arranged surface layer glycoprotein of *Bacillus alvei* CCM 2051, *Biochem. Cell Biol.* **69**, 72–78.

Altman, E., Brisson, J.-R., Gagné, S. M., Kolbe, J., Messner, P., Sleytr, U. B. (1992) Structure of the glycan chain from the surface layer glycoprotein of *Clostridium thermohydrosulfuricum, Biochim. Biophys. Acta* **1117**, 71–77.

Altman, E., Schäffer, C., Brisson, J.-R., Messner, P. (1995) Characterization of the glycan structure of a major glycopeptide from the surface layer glycoprotein of *Clostridium thermosaccharolyticum* E207-71, *Eur. J. Biochem.* **229**, 308–315.

Altman, E., Schäffer, C., Brisson, J.-R., Messner, P. (1996) Isolation and characterization of an amino sugar-rich glycopeptide from the surface layer glycoprotein of *Thermoanaerobacterium thermosaccharolyticum* E207-71, *Carbohydr. Res.* **295**, 245–253.

Araki, Y., Ito, E. (1989) Linkage units in the cell walls of gram-positive bacteria, *Crit. Rev. Microbiol.* **17**, 121–135.

Archibald, A. R., Hancock, I. C., Harwood, C. R. (1993) Cell wall structure, synthesis and turnover, in: *Bacillus subtilis and Other Gram-positive Bacteria* (Sonenshein, A. L., Hoch, J. A., Losick, R., Eds.), Washington, DC: ASM Press, 381–410.

Awram, P., Smit, J. (1998) The *Caulobacter crescentus* paracrystalline S-layer protein is secreted by an ABC transporter (type I) secretion apparatus, *J. Bacteriol.* **180**, 3062–3069.

Baumeister, W., Engelhardt, H. (1987) Three-dimensional structure of bacterial surface layers, in: *Electron Microscopy of Proteins* (Harris, J. R., Horne, R. W., Eds.), London: Academic Press, Inc, 109–154, Vol. 6.

Baumeister, W., Kübler, O. (1978) Topographic study of the cell surface of *Micrococcus radiodurans. Proc. Natl. Acad. Sci. USA* **75**, 5525–5528.

Baumeister, W., Lembcke, G. (1992) Structural features of archaebacterial cell envelopes, *J. Bioenerg. Biomembr.* **24**, 567–575.

Baumeister, W., Karrenberg, F., Rachel, R., Engel, A., ten Heggeler, B., Saxton, W. O. (1982) The major cell envelope protein of *Micrococcus radiodurans* (R1). Structural and chemical characterization, *Eur. J. Biochem.* **125**, 535–544.

Baumeister, W., Wildhaber, I., Phipps. B. M. (1989) Principles of organization in eubacterial and archaebacterial surface proteins, *Can. J. Microbiol.* **35**, 215–227.

Baumeister, W., Lembcke, G., Dürr, R., Phipps, B. (1990) Electron crystallography of bacterial surface proteins, in: *Electron Crystallography of Organic Molecules* (Freyer, J. R., Dorset, D. L., Eds.), Dordrecht, The Netherlands: Kluwer, 283–296.

Beveridge, T. J. (1994) Bacterial S-layers, *Curr. Opin. Struct. Biol.* **4**, 204–212.

Beveridge, T. J., Graham, L. L. (1991) Surface layers of bacteria, *Microbiol. Rev.* **55**, 684–705.

Beveridge, T. J., Murray, R. G. E. (1974) Superficial macromolecular arrays on the cell wall of *Spirillum putridiconchylium, J. Bacteriol.* **119**, 1019–1038.

Beveridge, T. J., Stewart, M., Doyle, R. J., Sprott, G. D. (1985) Unusual stability of the *Methanospirillum hungatei* sheath, *J. Bacteriol.* **162**, 728–737.

Bingle, W. H., Nomellini, J. F., Smit, J. (2000) Secretion of the *Caulobacter crescentus* S-layer protein: further localization of the C-terminal secretion signal and its use for secretion of recombinant proteins, *J. Bacteriol.* **182**, 3298–3301.

Blaser, M. J., Gotschlich, E. C. (1990) Surface array protein of *Campylobacter fetus, J. Biol. Chem.* **265**, 14529–14535.

Blaser, M. J., Pei, Z. (1993) Pathogenesis of *Campylobacter fetus* infections: critical role of high-molecular-weight S-layer proteins in virulence, *J. Infect. Dis.* **167**, 372–377.

Blaser, M. J., Smith, P. F., Repine, J. E., Joiner, K. A. (1988) Pathogenesis of *Campylobacter fetus* infections, *J. Clin. Invest.* **81**, 1434–1444.

Bock, K., Schuster-Kolbe, J., Altman, E., Allmaier, G., Stahl, B., Christian, R., Sleytr, U. B., Messner, P. (1994) Primary structure of the O-glycosidically linked glycan chain of the crystalline surface layer glycoprotein of *Thermoanaerobacter thermohydrosulfuricus* L111-69. Galactosyl tyrosine as a novel linkage unit, *J. Biol. Chem.* **269**, 7137–7144.

Boot, H. J., Pouwels, P. H. (1996) Expression, secretion and antigenic variation of bacterial S-layer proteins, *Mol. Microbiol.* **21**, 1117–1123.

Bootland, L. M., Leong, J. C. (1999) Infectious hematopoietic necrosis virus (IHNV), in: *Viral, Bacterial and Fungal infections* (Woo, P. T. K., Bruno, D. W., Eds.), Wallingford, UK: CAB International Publishing, 57–121, Vol. II.

Borinski, R., Holt, S. C. (1990) Studies on the surface of *Wolinella recta* ATCC 33238 and human clinical isolates: correlation of structure with function, *Infect. Immunol.* **58**, 2770–2776.

Breitwieser, A., Gruber, K., Sleytr, U. B. (1992) Evidence for an S-layer protein pool in the peptidoglycan of *Bacillus stearothermophilus, J. Bacteriol.* **174**, 8008–8015.

Breitwieser, A., Küpcü, S., Howorka, S., Weigert, S., Langer, C., Hoffmann-Sommergruber, K., Scheiner, O., Sleytr, U. B., Sára, M. (1996) 2-D protein crystals as an immobilization matrix for producing reaction zones in dipstick-style immunoassays, *BioTechniques* **21**, 918–925.

Breitwieser, A., Mader, C., Schocher, I., Hoffmann-Sommergruber, K., Aberer, W., Scheiner, O., Sleytr, U. B., Sára, M. (1998) A novel dipstick developed for rapid Bet v 1-specific IgE detection: recombinant allergen immobilized via a monoclonal antibody to crystalline bacterial cell surface layers, *Allergy* **53**, 786–793.

Breitwieser, A., Egelseer, E. M., Ilk, N., Moll, D., Hotzy, C., Bohle, B., Ebner, C., Sleytr, U. B., Sára, M. (2002) A recombinant bacterial cell surface (S-layer)-major birch pollen allergen-fusion protein (rSbsC/Bet v1) maintains the ability to self-assemble into regularly structured monomolecular lattices and the functionality of the allergen, *Prot. Eng.* **15**, 243–249.

Calabi, E., Ward, S., Wren, B., Paxton, T., Panico, M., Morris, H., Dell, A., Dougan, G., Fairweather, N. (2001) Molecular characterization of the surface layer proteins from *Clostridium difficile, Mol. Microbiol.* **40**, 1187–1199.

Carl, M., Dasch, G. A. (1989) The importance of the crystalline surface layer protein antigens of rickettsia in T-cell immunity, *J. Autoimmunity* **2**(Suppl.), 81–91.

Castán, P., de Pedro, M. A., Risco, C., Vallés, C., Fernández, L. A., Schwarz, H., Berenguer, J. (2001) Multiple regulatory mechanisms act on the 5 ' untranslated region of the S-layer gene from *Thermus thermophilus* HB8, *J. Bacteriol.* **183**, 1491–1494.

Cerquetti, M., Molinari, A., Sebastianelli, A., Diociaiuti, M., Petruzzelli, R., Capo, C., Mastrantonio, P. (2000) Characterization of surface layer proteins from different *Clostridium difficile* clinical isolates, *Microb. Pathog.* **28**, 363–372.

Christian, R., Schulz, G., Schuster-Kolbe, J., Allmaier, G., Schmid, E. R., Sleytr, U. B., Messner, P. (1993) Complete structure of the tyrosine-linked saccharide moiety from the surface layer glycoprotein of *Clostridium thermohydrosulfuricum* S102-70, *J. Bacteriol.* **175**, 1250–1256.

Chu, S., Trust, T. J. (1993) An *Aeromonas salmonicida* gene which influences A-protein expression in *Escherichia coli* encodes a protein containing an ATP-binding cassette and maps beside the surface array protein gene, *J. Bacteriol.* **175**, 3105–3114.

Chu, S., Cavaignac, S., Feutrier, J., Phipps, B. M., Kostrzynska, M., Kay, W. W., Trust, T. J. (1991) Structure of the tetragonal surface virulence array protein and gene of *Aeromonas salmonicida*, *J. Biol. Chem.* **266**, 15258–15265.

Chu, S., Gustafson, C. E., Feutrier, J., Cavaignac, S., Trust, T. J. (1993) Transcriptional analysis of the *Aeromonas salmonicida* S-layer protein gene *vapA*, *J. Bacteriol.* **175**, 7968–7975.

Cipriano, R. C. (1983) Furunculosis: pathogenicity, mechanism of bacterial virulence and the immunological response of fish to *Aeromonas salmonicida*, in: *Bacterial and Viral Diseases of Fish* (Crosa, J. H., Ed.), Seattle: University of Washington, 41–60.

Crowther, R. A., Sleytr, U. B. (1977) An analysis of the fine structure of the surface layer from two strains of clostridia, including correction for distorted images, *J. Ultrastruct. Res.* **58**, 41–49.

Diederich, A., Sponer, C., Pum, D., Sleytr, U. B., Lösche, M. (1996) Reciprocal influence between the protein and lipid components of a lipid-protein membrane model, *Coll. Surf. B* **6**, 335–346.

Dieluweit, S., Pum, D., Sleytr, U. B. (1998) Formation of a gold superlattice on an S-layer with square lattice symmetry, *Supramol. Sci.* **5**, 15–19.

Doig, P., Emödy, L., Trust, T. J. (1992) Binding of laminin and fibronectin by the trypsin resistant major structural domain of the crystalline virulence surface array protein of *Aeromonas salmonicida*, *J. Biol. Chem.* **267**, 43–49.

Douglas, K., Clark, N. A. (1986) Nanometer molecular lithography, *Appl. Phys. Lett.* **48**, 676–678.

Douglas, K., Devaud, G., Clark, N. A. (1992) Transfer of biologically derived nanometer-scale patterns to smooth substrates, *Science* **257**, 642–644.

Dworkin, J., Blaser, M. J. (1997) Molecular mechanisms of *Campylobacter fetus* surface layer protein expression, *Mol. Microbiol.* **26**, 433–440.

Dworkin, J., Shedd, O. L., Blaser, M. J. (1997) Nested DNA inversion of *Campylobacter fetus* S-layer genes is *recA* dependent, *J. Bacteriol.* **179**, 7523–7529.

Egelseer, E., Schocher, I., Sára, M., Sleytr, U. B. (1995) The S-layer from *Bacillus stearothermophilus* DSM 2358 functions as an adhesion site for a high-molecular weight amylase, *J. Bacteriol.* **177**, 1444–1451.

Egelseer, E., Schocher, I., Sleytr, U. B., Sára, M. (1996) Evidence that an N-terminal S-layer protein fragment triggers the release of a cell-associated high-molecular-weight amylase in *Bacillus stearothermophilus* ATCC 12980, *J. Bacteriol.* **178**, 5602–5609.

Egelseer, E., Leitner, K., Jarosch, M., Hotzy, C., Zayni, S., Sleytr, U. B., Sára, M. (1998) The S-layer proteins of two *Bacillus stearothermophilus* wild-type strains are bound via their N-terminal region to a secondary cell wall polymer of identical chemical composition, *J. Bacteriol.* **180**, 1488–1495.

Egelseer, E. M., Idris, R., Jarosch, M., Danhorn, T., Sleytr, U. B., Sára, M. (2000) ISBst*12*, a novel type of insertion-sequence element causing loss of S-layer-gene expression in *Bacillus stearothermophilus* ATCC 12980, *Microbiology* **146**, 2175–2183.

Egelseer, E. M., Danhorn, T., Pleschberger, M., Hotzy, C., Sleytr, U. B., Sára, M. (2001) Characterization of an S-layer glycoprotein produced in the course of S-layer variation of *Bacillus stearothermophilus* ATCC 12980 and sequencing and cloning of the *sbsD* gene encoding the protein moiety, *Arch. Microbiol.* **177**, 70–80.

Eichler, J. (2000) Novel glycoproteins of the halophilic archaeon *Haloferax volcanii*, *Arch. Microbiol.* **173**, 445–448.

Eichler, J. (2001) Post-translational modification of the S-layer glycoprotein occurs following translocation across the plasma membrane of the haloarchaeon *Haloferax volcanii*, *Eur. J. Biochem.* **268**, 4366–4373.

Evenberg, D., De Graff, P., Lugtenberg, B., van Muiswinkel, W. B. (1988) Vaccine-induced protective immunity against *Aeromonas salmonicida* tested in experimental carp erythrodermatitis, *J. Fish Dis.* **11**, 337–350.

Fernández-Herrero, L. A., Olabarría, G., Berenguer, J. (1997) Surface proteins and a novel transcription factor regulate the expression of the S-layer gene in *Thermus thermophilus*, *Mol. Microbiol.* **24**, 61–72.

Ford, L. A., Thune, R. L. (1992) Immunization of channel catfish with crude, acid-extracted preparation of motile aeromonad S-layer protein, *Biomed. Lett.* **47**, 355–362.

Fouet, A., Mesnage, S., Tosi-Couture, E., Gounon, P., Mock, M. (1999) *Bacillus anthracis* surface: capsule and S-layer, *J. Appl. Microbiol.* **87**, 251–255.

Garduño, R. A., Phipps, B. M., Kay, W. W. (1995) Physical and functional S-layer reconstruction in *Aeromonas salmonicida*, *J. Bacteriol.* **177**, 2684–2694.

Garduño, R., Moore, A, Olivier, G. Lizama, A. L., Garduño, E., Kay, W. W. (2000) Host cell invasion and intracellular residence by *Aeromonas salmo-*

nicida: role of the S-layer, *Can. J. Microbiol.* **46**, 660–668.

Gilmour, R., Messner, P., Guffanti, A.A., Kent, R., Scheberl, A., Kendrich, N., Krulwich, T. A. (2000) Two-dimensional gel electrophoresis analyses of pH-dependent protein expression in facultatively alkaliphilic *Bacillus pseudofirmus* OF4 lead to characterization of an S-layer protein with a role in alkaliphily, *J. Bacteriol.* **182**, 5969–5981.

Glauert, A. M. (1962) Fine structure of bacteria, *Brit. Med. Bull.* **18**, 245–250.

Graham, L., Nanninga, N, Beveridge, T. J. (1991) Periplasmic space and the concept of periplasm, *Trends Biochem. Sci.* **16**, 328–329.

Grogono-Thomas, R., Dworkin, J., Blaser, M. J., Newell, D. G. (2000) Roles of the surface layer proteins of *Campylobacter fetus* subsp. *fetus* in ovine abortion, *Infect. Immunol.* **68**, 1687–1691.

Gustafson, C. E., Chu, S., Trust, T. J. (1994) Mutagenesis of the paracrystalline surface array protein of *Aeromonas salmonicida* by endogenous insertion elements, *J. Mol. Biol.* **237**, 452–463.

Györvary, E., Wetzer, B., Sleytr, U. B., Sinner, A., Offenhäuser, A., Knoll, W. (1999) Lateral diffusion of lipids in silane-, dextrane- and S-layer protein-supported mono- and bilayers, *Langmuir* **15**, 1337–1347.

Haapasalo, M., Lounatmaa, K., Ranta, H., Shah, H., Ranta, K. (1985) Ultrastructure of *Bacteroides capillus, B. buccae, B. pentasaceus, B. oris, B. oralis, B. veroralis* and pentose-sugar-fermenting *Bacteroides* sp. from humans with periapical osteitis: occurrence of external proteinaceous cell wall layer, *Int. J. Syst. Bacteriol.* **35**, 65–72.

Hall, S. R., Shenton, W., Engelhardt, H., Mann, S. (2001) Site-specific organization of gold nanoparticles by biomolecular templating, *Chem. Phys. Chem.* **3**, 184–186.

Hastie, A. T., Brinton, Jr., C. C. (1979a) Isolation, characterization and *in vitro* self-assembly of the tetragonally arranged layer of *Bacillus sphaericus, J. Bacteriol.* **138**, 999–1009.

Hastie, A. T., Brinton, Jr., C. C. (1979b) Specific interaction of the tetragonally arrayed protein layer of *Bacillus sphaericus* with its peptidoglycan sacculus, *J. Bacteriol.* **138**, 1010–1021.

Hianik, T., Küpcü, S., Sleytr, U. B., Rybár, P., Krivánek, R., Kaatze, U. (1999) Interaction of crystalline bacterial cell surface proteins with lipid bilayers in liposomes. A sound velocity study, *Coll. Surf. A* **147**, 331–339.

Hirn, R., Schuster, B., Sleytr, U. B., Bayerl, T. M. (1999) The effect of S-layer protein adsorption and crystallization on the collective motion of a planar lipid bilayer studied by dynamic light scattering, *Biophys. J.* **77**, 2066–2074.

Houwink, A. L. (1953) A macromolecular monolayer in the cell wall of *Spirillum* sp., *Biochim. Biophys. Acta* **10**, 360–366.

Hovmöller, S. (1993) Crystallographic image processing applications for S-layers, in: *Advances in Bacterial Paracrystalline Surface Layers* (Beveridge, T. J., Koval, S. F., Eds.), New York: Plenum Press, 13–21.

Hovmöller, S., Sjögren, A., Wang, D. N. (1988) The structure of crystalline bacterial surface layers, *Prog. Biophys. Mol. Biol.* **51**, 131–163.

Howard, L. V., Dalton, D. D., McCoubrey, Jr., W. K. (1982) Expansion of the tetragonally arrayed cell wall protein layer during growth of *Bacillus sphaericus, J. Bacteriol.* **149**, 748–757.

Howorka, S., Sára, M., Wang, Y. J., Kuen, B., Sleytr, U. B., Lubitz, W., Bayley, H. (2000) Surface-accessible residues in the monomeric and assembled forms of a bacterial surface layer protein, *J. Biol. Chem.* **275**, 37876–37886.

Ilk, N., Kosma, P., Puchberger, M., Egelseer, E. M., Mayer, H. F., Sleytr, U. B., Sára, M. (1999) Structural and functional analyses of the secondary cell wall polymer of *Bacillus sphaericus* CCM 2177 that serves as an S-layer-specific anchor, *J. Bacteriol.* **181**, 7643–7646.

Ilk, N., Völlenkle, C., Egelseer, E. M., Breitwieser, A., Sleytr, U. B., Sára, M. (2002) Molecular characterization of the S-layer gene *sbpA* of *Bacillus sphaericus* CCM 2177 and production of a functional S-layer fusion protein with the ability to recrystallize in defined orientation while presenting the fused allergen, *Appl. Environ. Microbiol.* (accepted)

Jaenicke, R., Welsch, R., Sára, M., Sleytr, U. B. (1985) Stability and self-assembly of the S-layer protein of the cell wall of *Bacillus stearothermophilus, Biol. Chem. Hoppe-Seyler* **366**, 663–670.

Jahn-Schmid, B., Messner, P., Unger, F. M., Sleytr, U. B., Scheiner, O., Kraft, D. (1996a) Towards selective elicitation of Th1-controlled vaccination responses: vaccine applications of bacterial surface layer proteins, *J. Biotechnol.* **44**, 225–231.

Jahn-Schmid, B., Graninger, M., Glozik, M., Küpcü, S., Ebner, C., Unger, F. M., Sleytr, U. B., Messner, P. (1996b) Immunoreactivity of allergen (Bet v 1) conjugated to crystalline bacterial cell surface layers (S-layers), *Immunotechnology* **2**, 103–113.

Jahn-Schmid, B., Siemann, U., Zenker, A., Bohle, B., Messner, P., Unger, F. M., Sleytr, U. B., Scheiner, O., Kraft, D., Ebner, C. (1997) Bet v 1, the major birch pollen allergen, conjugated to crystalline bacterial cell surface proteins, expands allergen-specific T cell clones of the Th1/Th0 phenotype *in vitro* by induction of IL-12, *Int. Immunol.* **9**, 1867–1874.

Jarosch, M., Egelseer, E. M., Mattanovich, D., Sleytr, U. B., Sára, M. (2000) S-layer gene sbsC of *Bacillus stearothermophilus* ATCC 12980: molecular characterization and heterologous expression in *Escherichia coli, Microbiology* **146**, 273–281.

Jarosch, M., Egelseer, E. M., Huber, C., Moll, D., Mattanovich, D., Sleytr, U. B., Sára, M. (2001) Analysis of the structure-function relationship of the S-layer protein SbsC of *Bacillus stearothermophilus* ATCC 12980 by producing truncated forms, *Microbiology* **147**, 1353–1363.

Kahala, M., Savijoki, K., Palva, A. (1997) In vivo expression of the *Lactobacillus brevis* S-layer gene, *J. Bacteriol.* **179**, 284–286.

Kandler, O. (1982) Cell wall structures and their phylogenetic implications, *Zbl. Bakt. Hyg., I. Abt. Orig. C* **3**, 149–160.

Kandler, O., König, H. (1985) Cell envelopes of archaebacteria, in: *The Bacteria* (Woese, C. R., Wolfe, R. S., Eds.), New York: Academic Press, 413–457, Vol. VIII.

Kärcher, U., Schröder, H., Haslinger, E., Allmaier, G., Schreiner, R., Wieland, F., Haselbeck, A., König, H. (1993) Primary structure of the heterosaccharide of the surface glycoprotein of *Methanothermus fervidus, J. Biol. Chem.* **268**, 26821–26826.

Karjalainen, T., Waligora-Dupriet, A. J., Cerquetti, M., Spigaglia, P., Maggioni, A., Mauri, P., Mastrantonio, P. (2001) Molecular and genomic analysis of genes encoding surface-anchored proteins from *Clostridium difficile, Infect. Immunol.* **69**, 3442–3446.

Karrasch, S., Hegerl, R., Hoh, J., Baumeister, W., Engel, A. (1994) Atomic force microscopy produces faithful high-resolution images of protein surfaces in an aqueous environment, *Proc. Natl. Acad. Sci. USA* **91**, 836–838.

Kawai, E., Akatsuka, H., Idei, A., Shibatani, T., Mori, K. (1998) *Serratia marcescens* S-layer protein is secreted extracellularly via an ATP-binding cassette exporter, the Lip system, *Mol. Microbiol.* **27**, 941–952.

Kawata, T., Takeoka, A., Takumi, K., Masuda, K. (1984) Demonstration and preliminary characterization of a regular array in the cell wall of *Clostridium difficile, FEMS Microbiol. Lett.* **24**, 323–328.

Kay, W. W., Trust, T. J. (1991) Form and functions of the regular surface array (S-layer) of *Aeromonas salmonicida, Experientia* **47**, 412–414.

Kellenberger, E., Kistler, J. (1979) The physics of specimen preparation, in: *Unconventional Electron Microscopy for Molecular Structure Determination* (Hoppe, W., Mason, R., Eds.), Braunschweig: F. Vieweg & Sohn, 49–79.

Kikuchi, A., Sagami, H., Ogura, K. (1999) Evidence for covalent attachment of diphytanylglycerol phosphate to the cell-surface glycoprotein of *Halobacterium halobium, J. Biol. Chem.* **274**, 18011–18016.

Kist, M. L., Murray, R. G. E. (1984) Components of the regular surface array of *Aquaspirillum serpens* MW5 and their assembly *in vitro, J. Bacteriol.* **157**, 599–606.

Kneidinger, B., Graninger, M., Adam, G., Puchberger, M., Kosma, P., Zayni, S., Messner, P. (2001a) Identification of two GDP-6-deoxy-D-*lyxo*-4-hexulose reductases synthesizing GDP-D-rhamnose in *Aneurinibacillus thermoaerophilus* L420-91T, *J. Biol. Chem.* **276**, 5577–5583.

Kneidinger, B., Graninger, M., Puchberger, M., Kosma, P., Messner, P. (2001b) Biosynthesis of nucleotide-activated D-*glycero*-D-*manno*-heptose, *J. Biol. Chem.* **276**, 20935–20944.

Kokka, R. P., Vedros, N. A., Janda, J. M. (1992) Immunochemical analysis and possible biological role of an *Aeromonas hydrophila* surface array protein in septicaemia, *J. Gen. Microbiol.* **138**, 1229–1236.

König, H. (1988) Archaebacterial cell envelopes, *Can. J. Microbiol.* **34**, 395–406.

Kosma, P., Neuninger, C., Christian, R., Schulz, G., Messner, P. (1995a) Glycan structure of the S-layer glycoprotein of *Bacillus* sp. L420-91, *Glycoconjugate J.* **12**, 99–107.

Kosma, P., Wugeditsch, T., Christian, R., Zayni, S., Messner, P. (1995b) Glycan structure of a heptose-containing S-layer glycoprotein of *Bacillus thermoaerophilus, Glycobiology* **5**, 791–796. Erratum: *Glycobiology* (1996) **6**, 5.

Kostrzynska, M., Dooley, J. S. G., Shimojo, T., Sakata, T., Trust, T. J. (1992) Antigenic diversity of the S-layer proteins from pathogenic strains of *Aeromonas hydrophila* and *Aeromonas veronii* biotype *sobria, J. Bacteriol.* **174**, 40–47.

Kotiranta, A., Haapasalo, M., Kari, K., Kerosuo, E., Olson, I., Sorsa, T., Meurman, J. H., Lounatmaa, K. (1998) Surface structure, hydrophobicity, phagocytosis, and adherence to matrix proteins of *Bacillus cereus* with and without the crystalline surface protein layer, *Infect. Imm.* **66**, 4895–4902.

Kotiranta, A., Lounatmaa, K., Haapasalo, M. (2000) Epidemiology and pathogenesis of *Bacillus cereus* infections, *Microbes Infect.* **2**, 189–198.

Koval, S. F. (1997) The effect of S-layers and cell surface hydrophobicity on prey selection by bacteriovorus protozoa, *FEMS Microbiol. Rev.* **20**,138–142.

Koval, S. F., Murray, R. G. E. (1984) The isolation of surface array proteins from bacteria, *Can. J. Biochem. Cell Biol.* **62**, 1181–1189.

Krivanek, R., Rybar, P., Küpcü, S., Sleytr, U. B., Hianik, T. (2002) Affinity interactions on a liposome surface detected by ultrasound velocimetry, *Bioelectrochemistry* **55**, 57–59.

Kuen, B., Lubitz, W. (1996) Analysis of S-layer proteins and genes, in: *Crystalline Bacterial Cell Surface Proteins* (Sleytr, U. B., Messner, P., Pum, D., Sára, M., Eds.), Austin, TX: R. G. Landes Comp. and Academic Press, 77–102.

Küpcü, S., Sára, M., Sleytr, U. B. (1993) Influence of covalent attachment of low molecular weight substances on the rejection and adsorption properties of crystalline proteinaceous ultrafiltration membranes, *Desalination* **90**, 65–76.

Küpcü, S., Mader, C., Sára, M. (1995a) The crystalline cell surface layer from *Thermoanaerobacter thermohydrosulfuricus* L111-69 as immobilization matrix: influence of the morphological properties and pore size of the matrix on activity loss of covalently bound enzymes, *Biotechnol. Appl. Biochem.* **21**, 275–286.

Küpcü, S., Sára, M., Sleytr, U. B. (1995b) Liposomes coated with crystalline bacterial cell surface protein (S-layers) as immobilization structures for macromolecules, *Biochim. Biophys. Acta* **1235**, 263–269.

Küpcü, S., Sleytr, U. B., Sára, M. (1996) Two-dimensional paracrystalline glycoprotein S-layers as a novel matrix for the immobilization of human IgG and their use as microparticles in immunoassays, *J. Immunol. Methods* **196**, 73–84.

Küpcü, S., Lohner, K., Mader, C., Sleytr, U. B. (1998) Microcalorimetric study on the phase behaviour of S-layer coated liposomes, *Mol. Membr. Biol.* **15**, 69–74.

Lai, C.-H., Listgarten, M. A., Tanner, A. C. R., Socransky, S. S. (1981) Ultrastructures of *Bacteroides gracilis, Campylobacter concius, Wolinella recta,* and *Eikenella corrodens,* all from humans with periodontal disease, *Int. J. Syst. Bacteriol.* **31**, 465–475.

Lechner, J., Sumper, M. (1987) The primary structure of a procaryotic glycoprotein, *J. Biol. Chem.* **262**, 9724–9729.

Lechner, J., Wieland, F. (1989) Structure and biosynthesis of prokaryotic glycoproteins, *Annu. Rev. Microbiol.* **58**, 173–194.

Leibovitz, E. M., Lemaire, M., Miras, I., Salamitou, S., Beguin, P., Ohayon, H., Gounon, P., Matusckek, M., Sahm, K., Bahl, H. (1997) Occurrence and function of a common domain in S-layer and other exocellular proteins, *FEMS Microbiol. Rev.* **20**, 127–133.

Lepault, J., Pitt, T. (1984) Projected structure of unstained, frozen-hydrated S-layer of *Bacillus brevis, EMBO J.* **3**, 101–105.

Lortal, S., van Heijenoort, J., Gruber, K., Sleytr, U. B. (1992) S-layer of *Lactobacillus helveticus* ATCC 12046: isolation, chemical characterization and re-formation after extraction with lithium chloride, *J. Gen. Microbiol.* **138**, 611–618.

Lupas, A., Engelhardt, H., Peters, J., Santarius, U., Volker, V., Baumeister, W. (1994) Domain structure of the *Acetogenium kivui* surface layer revealed by electron crystallography and sequence analysis, *J. Bacteriol.* **176**, 1224–1233.

Mader, C., Küpcü, S., Sára, M., Sleytr, U. B. (1999) Stabilizing effect of an S-layer on liposomes towards thermal or mechanical stress, *Biochim. Biophys. Acta* **1418**, 106–116.

Mader, C., Küpcü, S., Sleytr, U.B., Sára, M. (2000) S-layer-coated liposomes as a versatile system for entrapping and binding target molecules, *Biochim. Biophys. Acta* **1463**, 142–150.

Mader, C., Moll, D., Huber, C., Sleytr, U. B., Sára, M. (2002) Real-time monitoring of the interaction between the crystalline bacterial cell surface layer protein SbsB and a secondary cell wall polymer functioning as the proper anchoring structure for the S-layer protein in the bacterial cell wall by using surface plasmon resonance biosensor technology (in preparation)

Malcolm, A. J., Messner, P., Sleytr, U. B., Smith, R. H., Unger, F. M. (1993a) Crystalline bacterial cell surface layers (S-layers) as combined carrier/adjuvants for conjugate Vaccines, in: *Immobilized Macromolecules: Application Potentials* (Sleytr, U. B. Messner, P., Pum, D., Sára, M., Eds.), London: Springer-Verlag, 195–207.

Malcolm, A. J., Best, M. W., Szarka, R. J., Mosleh, Z., Unger, F. M., Messner, P., Sleytr, U. B. (1993b) Surface layers from *Bacillus alvei* as a carrier for a *Streptococcus pneumoniae* conjugate Vaccine, in: *Advances in Bacterial Paracrystalline Surface Layers* (Beveridge, T. J., Koval, S. F., Eds.), New York: Plenum, 219–233.

Margaret, B. S., Krywolap, G. N. (1986) *Eubacterium yurii* subsp. *yurii* sp. nov. and *Eubacterium yurii* subsp. *margaretiae* subsp. nov.: test tube brush bacteria from subgingival dental plaque, *Int. J. Syst. Bacteriol.* **36**, 145–149.

Matuschek, M., Burchhardt, G., Sahm, K., Bahl, H. (1994) Pullulanase of *Thermoanaerobacterium thermohydrosulfurigenes* EM1 (*Clostridium thermohydrosulfurigenes*): molecular analysis of the gene, composite structure of the enzyme, and a common model for its attachment to the cell surface, *J. Bacteriol.* **176**, 3295–3302.

Matuschek, M., Sahm, K., Zibat, A., Bahl, H. (1996) Characterization of genes from *Thermoanaerobacterium thermohydrosulfurigenes* EM1 that encode two glycosyl hydrolases with conserved S-layer domains, *Mol. Gen. Genet.* **252**, 493–496.

Mauri, P. L., Pietta, P. G., Maggioni, A., Cerquetti, M., Sebastianelli, A., Mastrantonio, P. (1999) Characterization of surface layer proteins from *Clostridium difficile* by liquid chromatography/electrospray ionization mass spectrometry, *Rapid Comm. Mass Spectrom.* **13**, 695–703.

McCoubrey, J., Poxton, I. R. (2001) Variation in the surface layer proteins of *Clostridium difficile, FEMS Immunol. Med. Microbiol.* **31**, 131–135.

Mertig, M., Kirsch, R., Pompe, W., Engelhardt, H. (1999) Fabrication of highly oriented nanocluster arrays by biomolecular templating, *Europ. Phys. J.* **9**, 45–48.

Mescher, M. F., Strominger, J. L. (1976) Purification and characterization of a prokaryotic glycoprotein from the cell envelope of *Halobacterium salinarium, J. Biol. Chem.* **251**, 2005–2014.

Mesnage, S., Tosi-Couture, E., Mock, M., Gounon, P., Fouet, A. (1997) Molecular characterization of the *Bacillus anthracis* S-layer component: evidence that it is the major cell-associated antigen, *Mol. Microbiol.* **23**, 1147–1155.

Messner, P., Schäffer, C. (2000) Surface layer glycoproteins of Bacteria and Archaea, in: *Glycomicrobiology* (Doyle, R. J., Ed.), New York: Kluwer Academic/Plenum Publishers, 93–125.

Messner, P., Schäffer, C. (2002) Prokaryotic glycoproteins, in: *Progress in the Chemistry of Organic Natural Products* (Herz, W., Falk, H., Kirby, G. W., Moore, R. E., Tamm, C., Eds.). Wien: Springer-Verlag, Vol. 85 (in press)

Messner, P., Sleytr, U. B. (1988) Separation and purification of S-layers from gram-positive and gram-negative bacteria, in: *Bacterial Cell Surface Techniques* (Hancock, I. C., Poxton, I. R., Eds.), Chichester: John Wiley & Sons, 97–104.

Messner, P., Sleytr, U. B. (1991) Bacterial surface layer glycoproteins, *Glycobiology* **1**, 545–551.

Messner, P., Sleytr, U. B. (1992) Crystalline bacterial cell-surface layers, in: *Advances in Microbial Physiology* (Rose, A. H., Ed.), London: Academic Press, Inc, 213–275, Vol. 33.

Messner, P., Pum, D., Sára, M., Stetter, K. O., Sleytr, U. B. (1986a) Ultrastructure of the cell envelope of the archaebacteria *Thermoproteus tenax* and *Thermoproteus neutrophilus, J. Bacteriol.* **166**, 1046–1054.

Messner, P., Pum, D., Sleytr, U. B. (1986b) Characterization of the ultrastructure and the self-assembly of the surface layer of *Bacillus stearothermophilus* NRS 2004/3a, *J. Ultrastruct. Mol. Struct. Res.* **97**, 73–88.

Messner, P., Mazid, M. A., Unger, F. M., Sleytr, U. B. (1992) Artificial antigens. Synthetic carbohydrate haptens immobilized on crystalline bacterial surface layer glycoproteins, *Carbohydr. Res.* **233**, 175–184.

Messner, P., Schuster-Kolbe, J., Schäffer, C., Sleytr, U. B., Christian, R. (1993) Glycoprotein nature of select bacterial S-layers, in: *Advances in Bacterial Paracrystalline Surface Layers* (Beveridge, T. J. Koval, S. F., Eds.), New York: Plenum, 95–107.

Messner, P., Christian, R., Neuninger, C., Schulz, G. (1995) Similarity of "core" structures in two different glycans of tyrosine-linked eubacterial S-layer glycoproteins, *J. Bacteriol.* **177**, 2188–2193.

Mignot, T., Mesnage, S., Couture-Tosi, E., Mock, M., Fouet, A. (2002) Developmental switch of S-layer protein synthesis in *Bacillus anthracis, Mol. Microbiol.* **43**, 1615–1627.

Moll, D., Huber, C., Schlegel, B., Pum, D., Sleytr, U. B., Sára, M. (2002) S-layer-streptavidin fusion proteins as template for nanopatterend molecular arrays, *Proc. Natl. Acad. Sci. USA* (submitted)

Möschl, A., Schäffer, C., Sleytr, U. B., Messner, P., Christian, R., Schulz, G. (1993) Characterization of the S-layer glycoproteins of two lactobacilli, in: *Advances in Bacterial Paracrystalline Surface Layers* (Beveridge, T. J., Koval, S. F., Eds.), New York: Plenum 281–284.

Müller, D. J., Baumeister, W., Engel, A. (1999) Controlled unzipping of a bacterial surface layer with atomic force microscopy, *Proc. Natl. Acad. Sci. USA* **96**, 13170–13174.

Murray, R. G. E. (1962) Fine structure and taxonomy of bacteria, in: *Microbial Classification* (Ainsworth, G. C., Sneath, P. H. A., Eds.), Cambridge: Cambridge University Press, 119–144.

Murray, R. G. E. (1963a) On the cell wall structure of *Spirillum serpens, Can. J. Microbiol.* **9**, 381–392.

Murray, R. G. E. (1963b) Role of superficial structures in the characteristic morphology of

Lampropedia hyalina. Can. J. Microbiol. 9, 593–600.

Murray, R. G. E. (1993) A perspective on S-layer research, in: *Advances in Bacterial Paracrystalline Surface Layers* (Beveridge, T. J., Koval, S. F., Eds.), New York: Plenum, 3–9.

Naumova, I. B, Shashkov, A. S. (1997) Anionic polymers in cell walls of gram-positive bacteria, *Biochemistry (Moscow)* **62**:809–84

Neubauer, A., Kautek, W., Dieluweit, S., Pum, D., Sahre, M., Traher, C., Sleytr, U. B. (1997) A new approach to calibration standards in the ten-nanometer range: metal-coated S-layers, *PTB-Berichte* **F-30**, 188–190.

Neubauer, A., Kautek, W., Pentzien, S., Reetz, S., Sahre, M., Solomun, T., Korntner, R., Pum, D., Sleytr, U. B. (1998) Electrochemical deposition through and electron beam deposition on S-layer templates: a step towards calibration standards in the 10-nm range, *PTB-Berichte* **F-34**, 75–81.

Neubauer, A., Györvary, E., Pum, D., Sára, M., Sleytr, U. B. (2000) Investigation of the orientation of supramolecular protein structures (S-layers) on silicon, gold, and lipid films by scanning force microscopy, *PTB-Berichte* **F-39**, 118–123.

Nomellini, F. J., Küpcü, S., Sleytr, U. B., Smit, J. (1997) Factors controlling *in vitro* recrystallization of the *Caulobacter crescentus* paracrystalline S-layer, *J. Bacteriol.* **179**, 6349–6354.

Noonan, B., Trust, T. J. (1995a) Molecular analysis of an A-protein secretion mutant of *Aeromonas salmonicida* reveals a surface layer-specific protein secretion pathway, *J. Mol. Biol.* **248**, 316–327.

Noonan, B., Trust, T. J. (1995b) The leucine zipper of *Aeromonas salmonicida* AbcA is required for the transcriptional activation of the P2 promoter of the surface-layer structural gene, *vapA*, in *Escherichia coli*, *Mol. Microbiol.* **17**, 379–386.

Novotny, R., Schäffer, C., Messner, P. Organization of the S-layer glycosylation cluster in *Geobacillus stearothermophilus* NRS 20004/3a (in preparation)

Panhorst, M., Brückl, H., Kiefer, B., Reiss, G., Santarius, U., Guckenberger, R. (2001) Formation of metallic surface structures by ion etching using an S-layer template, *J. Vac. Sci. Technol. B* **19**, 722–724.

Pei, Z., Ellison III, R. T., Lewis, R. V., Blaser, M. J. (1988) Purification and characterization of a family of high molecular weight surface-array proteins from *Campylobacter fetus*, *J. Biol. Chem.* **263**, 6414–6420.

Peters, J., Nitsch, M., Kühlmorgen, B., Golbik, R., Lupas, A., Kellermann, J., Engelhardt, H., Pfander, J.-P., Müller, S., Goldie, K., Engel, A., Stetter, K.-O., Baumeister, W. (1995) Tetrabrachion: a filamentous archaebacterial surface protein assembly of unusual structure and extreme stability, *J. Mol. Biol.* **245**, 385–401.

Peters, J., Baumeister, W., Lupas, A. (1996) Hyperthermostable surface layer protein tetrabrachion from the archaebacterium *Staphylothermus marinus*: evidence for the presence of a right-handed coiled coil derived from the primary structure, *J. Mol. Biol.* **257**, 1031–1041.

Phipps, B. M., Huber, R., Baumeister, W. (1991) The cell envelope of the hyperthermophilic archaebacterium *Pyrobaculum organotrophicum* consists of two regularly arrayed protein layers: three-dimensional structure of the outer layer, *Mol. Microbiol.* **5**, 253–265.

Pompe, W., Mertig, M., Kirsch, R., Engelhardt, H., Kronbach, T. (1998) Functionalized biomolecular membranes for microreactors, in: *Proc. 1. Int. Conf. on Microresolution Technol.* (Ehrfeld, W., Ed.), Berlin: Springer, 105–111.

Pompe, W., Mertig, M., Kirsch, R., Wahl, R., Ciachi, L. C., Richter, J., Seidel, R., Vinzelberg, H. (1999) Formation of metallic nanostructures on biomolecular templates, *Z. Metallkd.* **90**, 1085–1091.

Pouwels, P., Kolen, C. P. A. M., Boot, H. J. (1997) S-layer protein genes in *Lactobacillus*, *FEMS Microbiol. Rev.* **20**, 78–82.

Pum, D., Sleytr, U. B. (1994) Large-scale reconstruction of crystalline bacterial surface layer proteins at the air-water interface and on lipids, *Thin Solid Films* **244**, 882–886.

Pum, D., Sleytr, U. B. (1995) Monomolecular reassembly of a crystalline bacterial cell surface layer (S-layer) on untreated and modified silicon surfaces, *Supramol. Sci.* **2**, 193–197.

Pum, D., Sleytr, U. B. (1999) The application of bacterial S-layers in molecular nanotechnology, *Trends Biotechnol.* **17**, 8–12.

Pum, D., Messner, P., Sleytr, U. B. (1991) Role of the S layer in morphogenesis and cell division of the archaebacterium *Methanocorpusculum sinense*, *J. Bacteriol.* **173**, 6865–6873.

Pum, D., Weinhandl, M., Hödl, C., Sleytr, U. B. (1993) Large-scale recrystallization of the S-layer of *Bacillus coagulans* E38-66 at the air/water interface and on lipid films, *J. Bacteriol.* **175**, 2762–2766.

Pum, D., Stangl, G., Sponer, C., Fallmann, W., Sleytr, U. B. (1997a) Deep UV patterning of monolayers of crystalline S layer protein on silicon surfaces, *Coll. Surf. B* **8**, 157–162.

Pum, D., Stangl, G., Sponer, C., Riedling, K., Hudek, P., Fallmann, W., Sleytr, U. B. (1997b) Patterning of monolayers of crystalline S-layer proteins on a silicon surface by deep ultraviolet radiation, *Microelectr. Eng.* **35**, 297–300.

Raguse, B., Braach-Maksvytis, V., Cornell, B. A., King, L. G., Osman, P. D. J., Pace R. J., Wieczorek, L. (1998) Tethered lipid bilayer membranes: Formation and ionic reservoir characterization, *Langmuir* **14**, 648–659.

Ray, K. C., Tu, Z. C., Grogono-Thomas, R., Newell, D. G., Thompson, S. A., Blaser, M. J. (2000) *Campylobacter fetus sap* inversion occurs in the absence of *recA* function, *Infect. Immun.* **68**, 5663–5667.

Reichelt, M., von Specht, B. U., Hahn, H. P. (2001) The *Caulobacter crescentus* outer membrane protein Omp58 (RsaF) is not required for paracrystalline S-layer secretion, *FEMS Microbiol. Lett.* **201**, 277–283.

Ries, W., Hotzy, C., Schocher, I., Sleytr, U. B., Sára, M. (1997) Evidence for the N-terminal part of the S-layer protein from *Bacillus stearothermophilus* PV72/p2 recognizes a secondary cell wall polymer, *J. Bacteriol.* **179**, 3892–3898.

Roberts, K., Hills, G. J., Shaw, P. J. (1982) The structure of algal cell walls, in: *Electron Microscopy of Proteins* (Harris, J. R., Ed.), London: Academic Press, Inc, 1–40, Vol. 3

Salton, M. R. J. (1994) The bacterial cell envelope – a historical perspective, in: *Bacterial Cell Wall* (Ghuysen, J.-M., Hakenbeck, R., Eds.), Amsterdam: Elsevier Science B.V, 1–22.

Sára, M. (2001) Conserved anchoring mechanisms between crystalline cell surface S-layer proteins and secondary cell wall polymers in gram-positive bacteria, *Trends Microbiol.* **9**, 47–49.

Sára, M., Sleytr, U. B. (1987a) Molecular sieving through S-layers of *Bacillus stearothermophilus* strains, *J. Bacteriol.* **169**, 4092–4098.

Sára, M., Sleytr, U. B. (1987b) Production and characteristics of ultrafiltration membranes with uniform pores from two-dimensional arrays of proteins, *J. Membr. Sci.* **33**, 27–49.

Sára, M., Sleytr, U. B. (1989) Use of regularly structured bacterial cell envelope layers as matrix for the immobilization of macromolecules, *Appl. Microbiol. Biotechnol.* **30**, 184–189.

Sára, M., Sleytr, U. B. (1996) Crystalline bacterial cell surface layers (S-layers): from cell structure to biomimetics, *Prog. Biophys. Mol. Biol.* **65**, 83–111.

Sára, M., Sleytr, U. B. (2000) S-layer proteins, *J. Bacteriol.* **182**, 859–868.

Sára, M., Pum, D., Sleytr, U. B. (1992) Permeability and charge-dependent adsorption properties of the S-layer lattice from *Bacillus coagulans* E38-66, *J. Bacteriol.* **174**, 3487–3493.

Sára, M., Kuen, B., Mayer, H. F., Mandl, F., Schuster, K. C., Sleytr, U. B. (1996) Dynamics in oxygen-induced changes in the S-layer protein synthesis from *Bacillus stearothermophilus* PV72 and the S-layer deficient variant T5 in continuous culture and studies of the cell wall composition, *J. Bacteriol.* **178**, 2108–2117.

Sára. M., Dekitsch, C., Mayer, H. F., Egelseer, E.M., Sleytr, U. B. (1998a) Influence of the secondary cell wall polymer on the reassembly, recrystallization, and stability properties of the S-layer protein from *Bacillus stearothermophilus* PV72/p2, *J. Bacteriol.* **180**, 4146–4153.

Sára, M., Egelseer, E. M., Dekitsch, C., Sleytr, U. B. (1998b) Identification of two binding domains, one for peptidoglycan and another for a secondary cell wall polymer, on the N-terminal part of the S-layer protein SbsB from *Bacillus stearothermophilus* PV72/p2, *J. Bacteriol.* **180**, 6780–6783.

Saxton, W. O., Baumeister, W. (1982) The correlation averaging of a regularly arranged bacterial cell envelope protein, *J. Microsc.* **127**, 127–138.

Schäffer, C., Messner, P. (2001) Glycobiology of surface-layer proteins, *Biochimie* **83**, 591–599.

Schäffer, C., Müller, N., Christian, R., Graninger, M., Wugeditsch, T., Scheberl, A., Messner, P. (1999a) Complete glycan structure of the S-layer glycoprotein of *Aneurinibacillus thermoaerophilus* GS4-97, *Glycobiology* **9**, 407–414.

Schäffer, C., Kählig, H., Christian, R., Schulz, G., Zayni, S., Messner, P. (1999b) The diacetamidodideoxyuronic-acid-containing glycan chain of *Bacillus stearothermophilus* NRS 2004/3a represents the secondary cell-wall polymer of wild-type *B. stearothermophilus* strains, *Microbiology* **145**, 1575–1583.

Schäffer, C., Dietrich, K., Unger, B., Scheberl, A., Rainey, F. A., Kählig, H., Messner, P. (2000a) A novel type of carbohydrate-protein linkage region in the tyrosine-bound S-layer glycan of *Thermoanaerobacterium thermosaccharolyticum* D120-70, *Eur. J. Biochem.* **267**, 5482–5492.

Schäffer, C., Müller, N., Mandal, P. K., Christian, R., Zayni, S., Messner, P. (2000b) A pyrophosphate bridge links the pyruvate-containing secondary cell wall polymer of *Paenibacillus alvei* CCM 2051 to muramic acid, *Glycoconjugate J.* **17**, 681–690.

Schäffer, C., Wugeditsch, T., Kählig, H., Scheberl, A., Zayni, S., Messner, P. (2002) The surface layer (S-layer) glycoprotein of *Geobacillus stearothermophilus* NRS 2004/3a. Analysis of its glycosylation, *J. Biol. Chem.* **277**, 6230–6239.

Schäffer C., Kählig, H., Zayni, S., Scheberl, A., Messner, P. (in preparation) Description of neutral secondary cell wall polymers in *Thermoanaerobacterium thermosaccharolyticum* strains E207-71 and D120-70.

Schneitz, C., Nuotio, L., Lounatmaa, K. (1993) Adhesion of *Lactobacillus acidophilus* to avian intestinal epithelial cells mediated by the crystalline bacterial cell surface layer (S-layer), *J. Appl. Microbiol.* **74**, 290–294.

Scholz, H., Hummel, S., Witte, A., Lubitz, W., Kuen, B. (2000) The transposable element IS4712 prevents S-layer gene (*sbsA*) expression in *Bacillus stearothermophilus* and also affects the synthesis of altered surface layer proteins, *Arch. Microbiol.* **174**, 97–103.

Scholz, H. C., Riedmann, E., Witte, A., Lubitz, W., Kuen, B. (2001) S-layer variation in *Bacillus stearothermophilus* PV72 is based on DNA rearrangements between the chromosome and the naturally occurring megaplasmids, *J. Bacteriol.* **183**, 1672–1679.

Schultze-Lam, S., Beveridge, T. J. (1994) Physicochemical characteristics of the mineral-forming S-layer from the cyanobacterium *Synechococcus* strain GL24, *Can. J. Microbiol.* **40**, 216–233.

Schultze-Lam, S., Harauz, G., Beveridge, T. J. (1992) Participation of a cyanobacterial S-layer in fine-grain mineral formation, *J. Bacteriol.* **174**, 7971–7981.

Schuster, B., Sleytr, U. B. (2000) S-layer-supported lipid membranes, *Rev. Mol. Biotechnol.* **74**, 233–254.

Schuster, B., Sleytr, U.B. (2002a) The effect of hydrostatic pressure on S-layer-supported lipid membranes, *Biochim. Biophys. Acta - Biomembr* **1563**, 29–34.

Schuster, B., Sleytr, U. B. (2002b) Single channel recordings of α-hemolysin reconstituted in S-layer-supported lipid bilayers, *Bioelectrochemistry* **55**, 5–7.

Schuster, B., Pum, D., Sleytr, U. B. (1998a) Voltage clamp studies on S-layer supported tetraether lipid membranes, *Biochim. Biophys. Acta* **1369**, 51–60.

Schuster, B., Pum, D., Braha, O., Bayley, H., Sleytr, U. B. (1998b) Self-assembled α-hemolysin pores in an S-layer-supported lipid bilayer, *Biochim. Biophys. Acta* **1370**, 280–288.

Schuster, B., Sleytr, U. B., Diederich, A., Bähr, G., Winterhalter, M. (1999) Probing the stability of S-layer-supported planar lipid membranes, *Eur. Biophys. J.* **28**, 583–590.

Schuster, B., Pum, D., Sára, M., Braha, O., Bayley, H., Sleytr, U. B. (2001) S-layer ultrafiltration membranes: a new support for stabilizing functionalized lipid membranes, *Langmuir* **17**, 499–503.

Schuster, K. C., Mayer, H. F., Kieweg, R., Hampel, W. A., Sára, M. (1995) A synthetic medium for continuous culture of the S-layer carrying *Bacillus stearothermophilus* PV72 and studies on the influence of growth conditions on cell wall properties, *Biotechnol. Bioeng.* **48**, 66–77.

Shenton, W., Pum, D., Sleytr, U. B., Mann, S. (1997) Biocrystal templating of CdS superlattices using self-assembled bacterial S-layers, *Nature* **389**, 585–587.

Sillanpää, J., Martínez, B., Antikainen, J., Toba, T., Kalkkinen, N., Tankka, S., Lounatmaa, K., Keränen, J., Höök, M., Westerlund-Wikström, B., Pouwels, P. H., Korhonen, T. K. (2000) Characterization of the collagen-binding S-layer protein CbsA of *Lactobacillus crispatus*, *J. Bacteriol.* **182**, 6440–6450.

Simon, B., Nomellini, J. F., Chiou, P., Bingle, W., Thornton, J., Smit, J., Leong, J.-A. (2001) Recombinant vaccines against infectious hematopoietic necrosis virus: production by the *Caulobacter crescentus* S-layer protein secretion system and evaluation in laboratory trials, *Dis. Aquat. Org.* **44**, 17–27.

Sleytr, U. B. (1975) Heterologous reattachment of regular arrays of glycoproteins on bacterial surfaces, *Nature* **257**, 400–402.

Sleytr, U. B. (1978) Regular arrays of macromolecules on bacterial cell walls: structure, chemistry, assembly, and function, *Int. Rev. Cytol.* **53**, 1–64.

Sleytr, U. B. (1981) Morphopoietic and functional aspects of regular protein membranes present on prokaryotic cell walls, in: *Cytomorphogenesis in Plants. Cell Biology Monographs* (Kiermayer, O., Ed.), Wien: Springer-Verlag, 3–26, Vol. 8.

Sleytr, U. B. (1997) Basic and applied S-layer research: an overview, *FEMS Microbiol. Rev.* **20**, 5–12.

Sleytr, U. B., Beveridge, T. J. (1999) Bacterial S-layers, *Trends Microbiol.* **7**, 253–260.

Sleytr, U. B., Glauert, A. M. (1975) Analysis of regular arrays of subunits on bacterial surfaces; evidence for a dynamic process of assembly, *J. Ultrastruct. Res.* **50**, 103–116.

Sleytr, U. B., Messner, P. (1983) Crystalline surface layers on bacteria, *Annu. Rev. Microbiol.* **37**, 311–339.

Sleytr, U. B., Messner, P. (1988) Crystalline surface layers on bacteria, in: *Crystalline Bacterial Cell Surface Layers* (Sleytr, U. B., Messner, P., Pum, D., Sára, M., Eds.), Berlin: Springer-Verlag, 160–186.

Sleytr, U. B., Messner, P. (1989) Self-assemblies of crystalline bacterial cell surface layers, in: *Electron Microscopy of Subcellular Dynamics* (Plattner, H., Ed.) Boca Raton, FL: CRC Press, 13–31.

Sleytr, U. B., Messner, P. (1996) Appendix. Crystalline surface layers on eubacteria and archaeobacteria, in: *Crystalline Bacterial Cell Surface Proteins* (Sleytr, U. B., Messner, P., Pum, D., Sára, M., Eds.), Austin, TX: R. G. Landes Comp. and Academic Press, Inc, 211–225.

Sleytr, U. B., Messner, P. (2000) Crystalline bacterial cell surface layers (S layers), in: *Encyclopedia of Microbiology* 2nd ed. (Lederberg, J., Ed.), San Diego: Academic Press, Inc, 899–906, Vol. 1.

Sleytr, U. B., Sára, M. (1997) Bacterial and archaeal S-layer proteins: structure-function relationships and their biotechnological applications, *Trends Biotechnol.* **15**, 20–26.

Sleytr, U. B., Thorne, K. J. I. (1976) Chemical characterization of the regularly arrayed surface layers of *Clostridium thermosaccharolyticum* and *Clostridium thermohydrosulfuricum*, *J. Bacteriol.* **126**, 377–383.

Sleytr, U. B., Messner, P., Pum, D., Sára, M., Eds. (1988a) *Crystalline Bacterial Cell Surface Layers.* Berlin: Springer-Verlag.

Sleytr, U. B., Messner, P., Pum, D. (1988b) Analysis of crystalline bacterial surface layers by freeze-etching, metal shadowing, negative staining and ultrathin sectioning, in: *Methods in Microbiology* (Mayer, F., Ed.), London: Academic Press, Ltd, 29–60, Vol. 20.

Sleytr, U. B., Messner, P., Pum, D., Sára, M. (1993) Crystalline bacterial cell surface layers, *Mol. Microbiol.* **10**, 911–916.

Sleytr, U. B., Sára, M., Messner, P., Pum, D. (1994) Two-dimensional protein crystals (S-layers): fundamentals and applications, *J. Cell. Biochem.* **56**, 171–176.

Sleytr, U. B., Messner, P., Pum, D., Sára, M., Eds. (1996a) *Crystalline Bacterial Cell Surface Proteins.* Austin, TX: R. G. Landes Comp. and Academic Press, Inc.

Sleytr, U. B., Messner, P., Pum, D., Sára, M. (1996b) Occurrence, location, ultrastructure and morphogenesis of S-layers, in: *Crystalline Bacterial Cell Surface Proteins* (Sleytr, U. B., Messner, P., Pum, D., Sára, M., Eds.), Austin, TX: R. G. Landes Comp. and Academic Press, Inc, 5–33.

Sleytr, U. B., Pum, D., Sára, M. (1997) Advances in S-layer nanotechnology and biomimetics, *Adv. Biophys.* **34**, 71–79.

Sleytr, U. B., Messner, P., Pum, D., Sára, M. (1999) Crystalline bacterial cell surface layers (S-layers): From supramolecular cell structure to biomimetics and nanotechnology, *Angew. Chem. International Ed.* **38**, 1035–1054.

Sleytr, U. B., Sára, M., Pum, D. (2000) Crystalline bacterial cell surface layers (S-layers): a versatile self-assembly system, in: *Supramolecular Polymerization* (Ciferri, A., Ed.), New York: Marcel Dekker, 177–213.

Sleytr, U. B., Sára, M., Pum, D., Schuster, B. (2001a) Characterization and use of crystalline bacterial cell surface layers, *Prog. Surf. Sci.* **68**, 231–278.

Sleytr, U. B., Sára, M., Pum, D., Schuster, B. (2001b) Molecular Nanotechnology and nanobiotechnology with two-dimensional protein crystals (S-layers), in: *Nano-surface Chemistry* (Rosoff, M., Ed.), New York: Marcel Dekker, 333–389.

Sleytr, U. B., Sára, M., Mader, C., Schuster, B., Unger, F. M. (2001c) Use of a secondary cell wall polymer of procaryotic microorganisms, International Patent WO0181425, November 1, 2001.

Smit, E., Oling, F., Demel, R., Martinez, B., Pouwels, P. H. (2001) The S-layer protein of *Lactobacillus acidophilus* ATCC 4356: Identification and characterisation of domains responsible for S-protein assembly and cell wall binding, *J. Mol. Biol.* **305**, 245–257.

Smit, J., Sherwood, C. S., Turner, R. F. B. (2000) Characterization of high density monolayers of the biofilm bacterium *Caulobacter crescentus*: evaluating prospects for developing immobilized cell bioreactors, *Can. J. Microbiol.* **46**, 339–349.

Smith, R. H., Messner, P., Lamontagne, L. R., Sleytr, U. B., Unger, F. M. (1993) Induction of T-cell immunity to oligosaccharide antigens immobilized on crystalline bacterial surface layers (S-layers), *Vaccine* **11**, 919–924.

Southam, G., Beveridge, T. J. (1992) Characterization of novel, phenol-soluble polypeptides which confer rigidity to the sheath of *Methanospirillum hungatei* GP1, *J. Bacteriol.* **174**, 935–946.

Stahl, S., Uhlen, M. (1997) Bacterial surface display: trends and progress, *Trends Biotechnol.* **15**, 185–192.

Steindl C., Schäffer C., Wugeditsch T., Graninger M., Matecko I., Müller N., Messner P. (submitted) The first biantennary secondary cell wall polymer of the domain Bacteria, isolated from *Aneurinibacillus thermoaerophilus* DSM 10155, and its influence on the assembly of the S-layer glycoprotein, *Biochem. J.*

Sumper, M., Wieland, F. T. (1995) Bacterial glycoproteins, in: *Glycoproteins* (Montreuil, J., Vliegenthart, J. F. G., Schachter, H., Eds.), Amsterdam: Elsevier, 455–473.

Sumper, M., Berg, E., Mengele, R., Strobl, I. (1990) Primary structure and glycosylation of the S-layer protein of *Haloferax volcanii, J. Bacteriol.* **172**, 7111–7118.

Takeoka, A., Takumi, K., Koga, T., Kawata, T. (1991) Purification and characterization of S layer proteins from *Clostridium difficile* GAI 0714, *J. Gen. Microbiol.* **137**, 261–267.

Taylor, K. A., Deatherage, J. F., Amos, L. A. (1982) Structure of the S-layer of *Sulfolobus acidocaldarius, Nature* **299**, 840–842.

Thomas, S. R., Trust, T. J. (1995) Tyrosine phosphorylation of the tetragonal paracrystalline array of *Aeromonas hydrophila*: molecular cloning and high-level expression of the S-layer protein gene, *J. Mol. Biol.* **245**, 568–581.

Thompson, B. G., Murray, R. G. E., Boyce, J. F. (1982) The association of the surface array and the outer membrane *of Deinococcus radiodurans, Can. J. Microbiol.* **28**, 1081–1088.

Thompson, S. A., and Blaser, M. J. (2000) Pathogenesis of *Campylobacter fetus* infections, in: *Campylobacter*, 2nd ed. (Nachamkin, I., Blaser, M. J., Eds.), Washington, D.C.: ASM Press, 321–347.

Thompson, S. A., Shedd, O. L., Ray, K. C., Beins, M. H., Jorgensen, J. P., Blaser, M. J. (1998) *Campylobacter fetus* surface layer proteins are transported by a type I secretion system, *J. Bacteriol.* **180**, 6450–6458.

Thorne, K. J. I., Oliver, R. C., Glauert, A. M. (1976) Synthesis and turnover of the regularly arranged surface protein of *Acinetobacter* sp. relative to the other components of the cell envelope, *J. Bacteriol.* **127**, 440–450.

Thornley, M. J., Thorne, K. J. I., Glauert A. M. (1974) Detachment and chemical characterization of the regularly arranged subunits from the surface of an *Acinetobacter, J. Bacteriol.* **118**, 654–662.

Thornton, J. C., Garduño, R. A., Newman, S. G., and Kay, W. W. (1991) Surface-disorganized attenuated mutants of *Aeromonas salmonicida* as furunculosis vaccines, *Microb. Pathog.* **11**, 85–99.

Tien, H.T., Ottova A.L. (2001) The lipid bilayer concept and its experimental realization: from soap bubbles, kitchen sink, to bilayer lipid membranes, *J. Membr. Sci.* **189**, 83–117.

Trachtenberg, S., Pinnicki, B., Kessel, M. (2000) The cell surface glycoprotein layer of the extreme halophile *Halobacterium salinarum* and its relation to *Haloferax volcanii*: cryo-electron tomography of freeze-substituted cells and projection studies of negatively stained envelopes, *J. Struct. Biol.* **130**, 10–26.

Trust, T. J., Kostrzynska, M., Emödy, L., Wadström, T. (1993) High affinity binding of the basement membrane protein collagen type IV to the crystalline virulence surface protein array of *Aeromonas salmonicida, Mol. Microbiol.* **7**, 593–600.

Tsuboi, A., Tsukagoshi, N., Udaka, S. (1982) Reassembly *in vitro* of hexagonal surface arrays in a protein-producing bacterium, *Bacillus brevis* 47, *J. Bacteriol.* **151**, 1485–1497.

Tsukagoshi, N., Tabata, R., Takemura, T., Yamagata, H., Udaka, S. (1984) Molecular cloning of a major cell wall protein gene from protein-producing *Bacillus brevis* 47 and its expression in *Escherichia coli* and *Bacillus subtilis, J. Bacteriol.* **158**, 1054–1060.

Tu, Z. C., Dewhirst, F. E., Blaser, M. J. (2001a) Evidence that the *Campylobacter fetus sap* locus is an ancient genomic constituent with origins before mammals and reptiles diverged, *Infect. Immun.* **69**, 2237–2244.

Tu, Z. C., Ray, K. C., Thompson, S. A., Blaser, M. J. (2001b) *Campylobacter fetus* uses multiple loci for DNA inversion within the 5′ conserved regions of *sap* homologs. *J. Bacteriol.* **183**, 6654–6661.

Udey, L. R., Fryer, J. L. (1978) Immunization of fish with bacterins of *Aeromonas salmonicida, Mar. Fish. Rev.* **40**, 12–17.

Weber, V., Weigert, S., Sára, M., Sleytr, U. B., Falkenhagen, D. (2001) Development of affinity microparticles for extracorporeal blood purification based on crystalline bacterial cell surface proteins, *Ther. Apher.* **5**, 433–438.

Weigert, S., Sára, M. (1995) Surface modification of an ultrafiltration membrane with crystalline structure and studies on the interactions with selected protein molecules, *J. Membr. Sci.* **106**, 147–159.

Weigert, S., Sára, M. (1996) Ultrafiltration membranes prepared from bacterial cell surface layers as model systems for studying the influence of surface properties on protein adsorption, *J. Membr. Sci.* **121**, 185–196.

Weiner, C., Sára, M., Dasgupta, G., Sleytr, U. B. (1994a) Affinity cross-flow filtration: purification of IgG with a novel protein A affinity matrix prepared from two-dimensional protein crystals, *Biotechnol. Bioeng.* **44**, 55–65.

Weiner, C., Sára, M., Sleytr, U. B. (1994b) Novel protein A affinity matrix prepared from two-dimensional protein crystals, *Biotechnol. Bioeng.* **43**, 321–330.

Wetzer, B., Pum, D., Sleytr, U. B. (1997) S-layer stabilized solid supported lipid bilayers, *J. Struct. Biol.* **119**, 123–128.

Wetzer, B., Pfandler, A., Györvary, E., Pum, D., Lösche, M., Sleytr, U. B. (1998) S-layer reconstitution at phospholipid monolayers, *Langmuir* **14**, 6899–6906.

Weygand, M., Wetzer, B., Pum, D., Sleytr, U. B., Cuvillier, N., Kjaer, K., Howes, P. B., Lösche, M. (1999) Bacterial S-layer protein coupling to lipids: X-ray reflectivity and grazing incidence diffraction studies, *Biophys. J.* **76**, 458–468.

Weygand, M., Schalke, M., Howes, P. B., Kjaer, K., Friedmann, J., Wetzer, B., Pum, D., Sleytr, U. B., Lösche, M. (2000) Coupling of protein sheet crystals (S-layers) to phospholipid monolayers, *J. Mat. Chem.* **10**, 141–148.

Williams, P. (1988) Role of the cell envelope in bacterial adaptation to growth *in vivo* in infections, *Biochimie* **70**, 987–1011.

Winningham, T. A., Gillis, H. P., Choutov, D. A., Martin, K. P., Moore, J. T., Douglas, K. (1998) Formation of ordered nanocluster arrays by self-assembly on nanopatterned Si(100) surfaces, *Surf. Sci.* **406**, 221–228.

Woodcock, C. L. F., Baumeister, W. (1990) Different representations of a protein structure obtained with different negative stains, *Eur. J. Cell Biol.* **51**, 45–52.

Wugeditsch, T., Zachara, N. E., Puchberger, M., Kosma, P., Gooley, A. A., Messner, P. (1999) Structural heterogeneity in the core oligosaccharide of the S-layer glycoprotein from *Aneurinibacillus thermoaerophilus* DSM 10155, *Glycobiology* **9**, 787–795.

Yamada, H., Tsukagoshi, N., Udaka, S. (1981) Morphological alterations of cell wall concomitant with protein release in a protein-producing bacterium, *Bacillus brevis* 47, *J. Bacteriol.* **148**, 322–332.

12
Cytoskeletons in Eukaryotes and Prokaryotes

Prof. Dr. Frank Mayer
Institut für Mikrobiologie und Genetik, Georg-August-Universität Göttingen, Grisebachstrasse 8, 37077 Göttingen, Germany; Tel.: +49-551-39-3829; Fax: +49-551-39-7081; E-mail: fmayer@gwdg.de

GFAP	glial fibrillary acidic protein
GFP	green fluorescent protein
EF-Tu	elongation factor Tu
IF	intermediate filament
MTOC	microtubule organizing center
WASP	Wiskott–Aldrich syndrome protein

1 Introduction

The preservation of shape of typical multicellular organisms, with all its functional implications, is warranted by skeletons: bones in vertebrates; a cuticula in insects; a carapace in crustaceae; and shells in mollusks, for example. In addition, certain elements also function as skeletons in plants as the cell walls exert shape preservation and provide stability for the entire organism. This is true also for unicellular plants such as algae, and for yeasts and fungi in general.

Any typical multicellular organism, however, passes through a state where no skeleton of this type is present, this being the state of the germ cells. This is a unicellular state, and a cell in this state is not protected by a skeleton. Nonetheless, germ cells are capable of fulfilling all of their functions. The optimization of preparative and imaging techniques for biological samples by light and electron microscopy showed the cytoplasm of cells of eukaryotic organisms in general (not only germ cells) to be not simply a relatively homogeneous "soup" in which the organelles floated, but to contain a highly differentiated and complex intracellular skeleton, the cytoskeleton. The cytoskeleton is composed of a framework of at least three classes of fibers that consist of proteins (Figure 1). One function of the cytoskeleton is to help maintain cell shape by providing mechanical support. In animal cells, the cytoskeleton also assists the function of the extracellular matrix that surrounds the cell and consists of collagen, proteoglycans, and further components that provide rigidity and strength not only to the cell *per se* but also to tissues and organs. In addition, the cytoskeleton provides anchoring sites for other cellular components and serves as a track along which motor molecules are able to move. The cytoskeleton provides a structural framework against which muscles work to elicit movement. However, it is not a static structure, the most obvious dynamic functions being the actions of muscles, the formation and degradation of spindle fibers mediating mitosis and cytokinesis, the movement of amoebae and macrophages, and the streaming of protoplasma in plant cells. Depending on functional necessities or states and types of differentiation of cells, the cytoskeleton can be specifically organized or reorganized. One highly specialized type of organization and dynamic interaction of cytoskeletal elements is that of muscle (Figure 2) (Darnell et al., 1990), though a detailed description of the physiology of muscle is beyond the scope of this chapter.

At the time that sequencing of the human genome was approaching completion, it transpired that over 850 different types of protein could be assumed as structural or functional components of the cytoskeleton (or at least proteins interacting with it), and that over 350 types of protein may belong to the group of motor proteins involved in functions of the cytoskeleton.

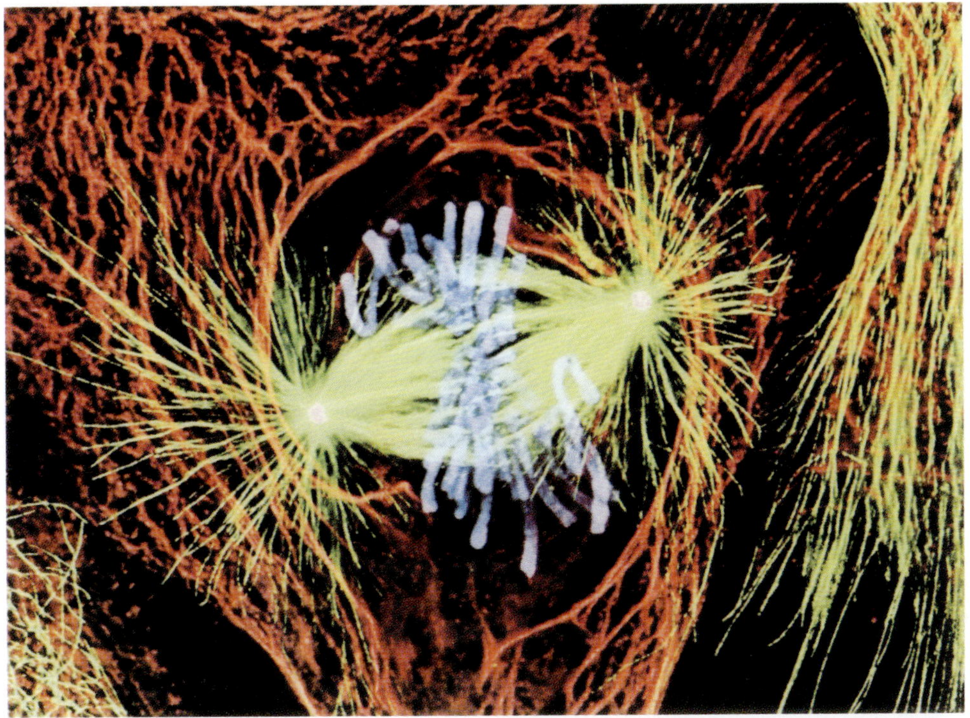

Fig. 1 The cytoskeleton as depicted by application of the immunofluoresence technique. Lung cell of a newt. Metaphase cell stained for centrosomes, microtubules, chromosomes and intermediate filaments. Actin filaments not shown. (Courtesy of Alexey Khodjakov, Albany NY, USA, and Olympus Optical Co. [Europa] GmbH, with permission).

In contrast to eukaryotic cells, the occurrence of a cytoskeleton remained unproven for some time in prokaryotes (Mayer, 1986), the existence of a shape-preserving cell wall being taken as the clear reason for absence of a cytoskeleton in these organisms. Although no bacterial motor proteins have been identified, no need was seen for any other explanation for translocation of daughter chromosomes into daughter cells during cell division besides involvement of the cytoplasmic membrane. The specific sizes and shapes of the many different prokaryotes were assumed to be defined by a genetically determined and regulated interaction of cytoplasmic membrane and cell wall layers. Neither did bacterial locomotion appear to require a cytoskeleton; after all, locomotion is known to be brought about by flagella driven by a 'motor' (consisting of proteins that are unrelated to the known motor proteins of eukaryotic cells) inserted into the cytoplasmic membrane. Cell polarity was thought to be determined by intrinsic properties of the cytoplasmic membrane. Moreover, endocytosis and plasma streaming were shown not to occur in prokaryotes, and fibrillar structures similar to members of the classes of intracytoplasmic fibers known for the eukaryotic cytoskeleton were not detected. In addition, using conventional methods of nucleotide sequencing, sequences similar to those of genes coding for proteins of the eukaryotic cytoskeleton were

Fig. 2 (a) Heart muscle (striated muscle), longitudinal section. Rows (myofibrils) of sarcomeres, with each of the sarcomeres delineated by a Z disk (Z) at each end, can be seen. A striated muscle cell contains numerous myofibrils. The arrow points to a *discus intercalaris*, i.e., the zone where two muscle cells are in contact. Heart muscle cells contain only one cell nucleus, other striated muscles up to 100. M, mitochondria. A striated muscle cell may have a length of up to 40 mm. The distance between two Z disks along the sarcomere measures about 1.5–1.8 μm. (From Lodish et al., 1996, with permission.) (b) Diagram showing arrangement of thick myosin and thin actin filaments in striated muscle in relaxed and in contracted state. The (+) ends of actin filaments are anchored at the Z disks; pivoting of the myosin heads pushes the actin filaments toward the center of the sarcomere, thus reducing sarcomere length. The I band, but not the A band, shortens during contraction. (From Darnell et al., 1990, with permission.)

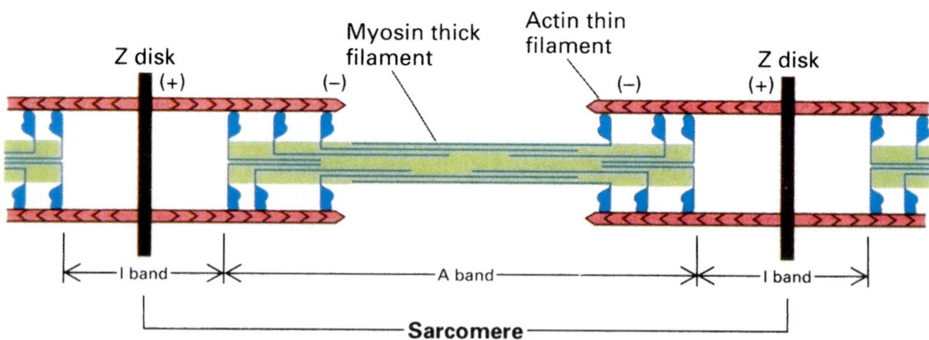

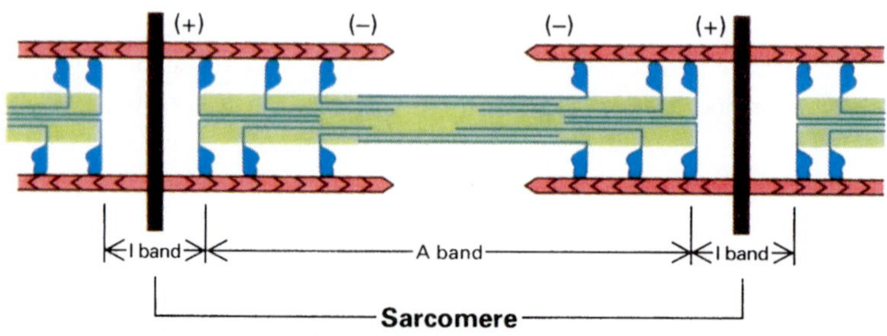

not found in prokaryotes. One exception was the wall-less bacterium *Mycoplasma pneumoniae,* wherein a structure–the rod – was detected that was assumed to be a type of cytoskeletal element. Based on cytological data, the rod was deduced to support the tip structure of the cell–a finger-like extension of the cell body that appeared to be involved in cell adhesion. The presence of filamentous structures in adherent *M. pneumoniae* cells treated with nonionic detergents was also demonstrated (Biberfeld and Biberfeld, 1970; Göbel et al., 1981; Regula et al., 2001).

Besides the mechanical and biochemical properties of cytoskeletons and their constituents, the physico-chemical properties of the cell must also be considered for a better understanding of the function of eukaryotic and prokaryotic cells. Essential cell functions such as contraction, division, transport and communication are thought to be strongly influenced by the gel-like nature of the cell, and a recently proposed working model challenged current views of how a cell "works" by suggesting that the cytoskeleton has a modulatory effect (Pollack, 2001). The model implied that reversible, transient sol–gel transitions, brought about by polymerization/depolymerization/compaction reactions of cytoskeletal elements, influence the state of the cytoplasm. The implications of this would be that modifications of distribution and availability of ions (especially ions involved in cell energetics) would lead to changes in the properties of the immediate environment of biomembranes. An proposed example of this view is the postulated relationship between structural dynamics of the cytoskeleton and the action potential (Figure 3). When the strands of the cytoskeleton are bridged by calcium ions, the cytoskeletal network is first collapsed but then expands as sodium replaces calcium. However, increasing sodium levels eventually neutralize the surface charge, weakening the water structure by water "melting" and allowing the polymer-retractive force to collapse the network, at which stage calcium may easily bridge the strands once again. In this way, both volume and cell potential are restored.

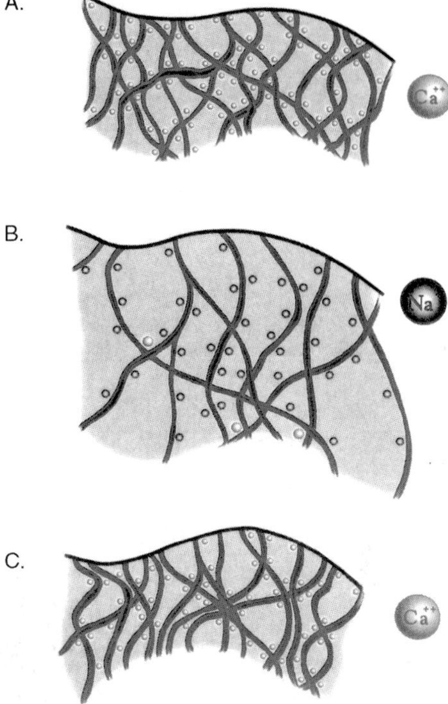

Fig. 3 Structural dynamics and the action potential. Initially (A) the network of the cytoskeleton is collapsed as strands are bridged by calcium. The network expands as sodium replaces calcium (B). The increasing sodium level eventually neutralizes the surface charge, weakens the water structure and allows the polymer-retractive force to collapse the network (C), at which stage calcium may easily bridge the strands once again. (From Pollack, 2001, with permission.)

Investigations of the structural organization of cytoskeletons, interactions of their constituents, interactions of cytoskeletal elements with other cell components, and the effects of introducing structural and functional defaults can be carried out using a variety of imaging techniques. These include light microscopy (either conventional or laser-scanning techniques) in conjunction with immunofluorescence (see Figures 1 and 17a); these techniques are complemented by methods using expression of recombinant proteins (fusion proteins consisting of the cytoskeletal protein proper, fused to green fluorescent protein (GFP) in the analysis of aspects of the cytoskeleton at the cellular level (see Figure 16a). At the macromolecular level, various types of electron microscopic sample preparation and imaging techniques can be applied, including direct visualization of metal-shadowed samples, perhaps in conjunction with freeze-fracturing and deep-etching (Figure 4), of negatively stained samples (see Figure 10a), and of ultrathin sections (see Figure 2a). Immunoelectron microscopy performed at either the cellular (see Figure 20) or molecular level provides access to finer details of structural organization in combination with the identification and localization of defined cytoskeletal elements.

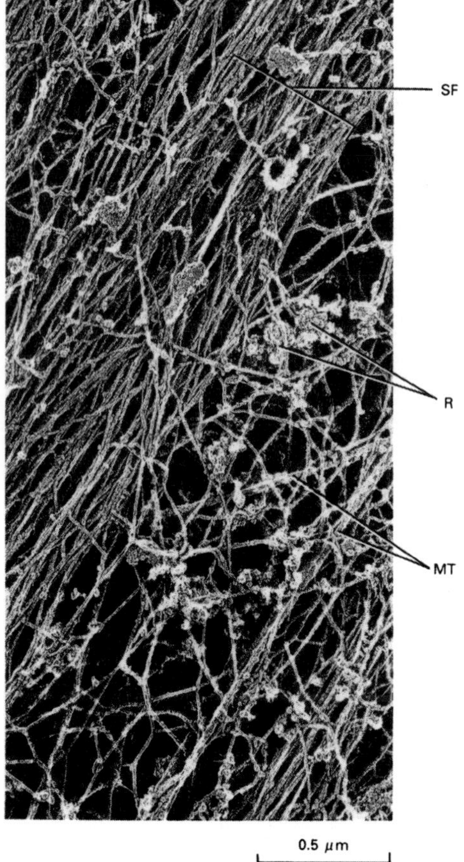

Fig. 4 Electron micrograph of a platinum replica of a cytoskeleton prepared by quick freezing and deep etching from a fibroblast immersed in the detergent Triton X-100 to remove soluble cytoplasmic proteins and membranes. Prominent are bundles of actin microfilaments termed stress fibers (SF), which are thought to connect segments of the plasma membrane and anchor the cell to the substratum. Also visible are two thicker microtubules (MT) and a more diffuse meshwork of filaments studded with grape-like clusters, which are probably polyribosomes (R). (Reproduced from Lodish et al., 1996, with permission.)

2
Historical Outline

The first valuable source of reference material relating to the history of structure–function analyses on muscle, as well as initial investigations into the eukaryotic cytoskeleton, was provided by DuPraw in 1968.

Before the discovery of nonmuscle actin/myosin, tubulin and other filamentous elements now known to be major constituents of the cytoskeleton, research was focused intensely on muscle. These early investigations revealed many aspects of muscle structure and function that formed the basis of studies into the eukaryotic cytoskeleton for several years.

In 1934, the role of ATP in muscle contraction was inferred from the so-called "Lohmann reaction". Dialyzed muscle extracts were shown not to catalyze the hydrolysis of creatine phosphate; instead, catalysis of transphosphorylation to ATP was demonstrated, with the ATP being subsequently hydrolyzed. In 1939, Engelhardt and Ljubimova showed "myosin" itself to be an ATPase, while in 1941, Szent-Györgyi showed that artificial fibers of "actomyosin" contracted *in vitro* after the addition of ATP. By 1942, the term "myosin" was used to describe a complex between myosin and actin, but this was changed when actin was isolated and its properties determined in 1942 by Straub.

The analyses of the macromolecular architecture of the muscle, its constituents, and the interplay of these macromolecules as well as the role of ATP and ATPase were only made possible by major advances in sample preparation and imaging techniques, and by the introduction of identification systems for the structural elements, including immunofluorescence and immunoelectron microscopy. The availability of X-ray diffraction and electron microscopic data for F-actin, actomyosin, and tropomyosin, following the pioneering studies of Huxley, Hansen, and Lowry, led to many of the former observations being explained in detail.

In 1964, Ohnishi discovered "nonmuscle actin/myosin" and isolated it from chloroplasts. This component was thought to be responsible for the long-known phenomenon of chloroplast contraction, and the concept was the first indication that contractile elements related to actin and myosin might exist outside the muscle. In this respect, it is remarkable that, until very recently, the existence of cytoskeletal elements in prokaryotes was not recognized. As early as 1849, "cytoplasmic streaming" had been thought to be due to "ectoplasmic contraction" that was similar to muscle contraction, and to occur by the action of nonmuscle contractile elements. This view was later supported by Wohlfarth-Bottermann (1964) who found 70 Å filaments *in situ* in the peripheral ectoplasma in *Physarum*, possessing ATPase activity ("myxomycin").

An additional component of the cytoskeleton, the cytoplasmic microtubule, was discovered in 1958–1959 by Roth. Before this, microtubules were known to be the basic structures for the "9 + 2 fibrils" scheme in eukaryotic flagella and cilia. Further structural details were revealed after the introduction in 1963 of glutaraldehyde fixation for biological samples. In 1965, DuPraw described an extensive system of cytoplasmic microtubules, while in 1967 Moor speculated on a "contraction of microtubules", at about which time the first evidence appeared that microtubules constitute the spindle fibers. Nonetheless, even by 1968 no clear indication of the role or mechanism of action of microtubules was available.

An outline of the present-day state of research on the known components of eukaryotic as well as prokaryotic cytoskeletons, forms the basis of the following text.

3 Principles of the Architecture of Cytoskeletons at the Cellular Level

In typical eukaryotic cells (animals, plants, and eukaryotic microorganisms), complexes of cytoskeletal elements are located close to the inside of the cytoplasmic membrane and interacting with it, and also as fibrous systems crossing the cytoplasm (see Figure 4). Based on the functions of these cytoskeletal complexes, some of them must be highly variable in terms of their structure and composition, whereas others must be much more stable and persistent. Whatever its function, the cytoskeleton in eukaryotic cells is a highly dynamic structure. Its ability to adapt transiently to different functions and specific needs is based its ability to vary its own composition, with interactions between the cytoskeletal elements and other (accessory) proteins or compounds being modified by changes in conditions both inside and outside the cell.

Analyses of the cellular architecture of prokaryotic cells with regard to the cytoskeleton have been carried out only recently, and only in very few cases (see Section 5). Generally, in these types of cells, specific adaptations of fibrillar cytoskeletal elements, each consisting of a small number of different polypeptides, usually result in structural organizations that are more persistent than in typical nonmuscle eukaryotic cells. Nevertheless, these cytoskeletons exhibit dynamic properties such as shortening, not necessarily caused by dissociation or by a gliding mechanism involving different types of components, but rather by reversible contraction of polypeptides as exemplified in case of a linear motor (see below), or rearrangement or extension brought about by assembly/disassembly reactions (the FtsZ system, see below).

4
Cytoskeletons in Eukaryotes: Their Constituents, Dynamic Properties and Interactions

The cytoskeleton in eukaryotes is composed of at least three classes of fiber: microtubules, microfilaments, and intermediate filaments, though additional types of filament of unknown composition may exist.

4.1
Microtubules

Microtubules, composed of tubulins, are fibers 24 nm in diameter and of variable length (Figure 5a). With few exceptions (mammalian erythrocytes), they appear common to all eukaryotic cells. Microtubules can form both transient and permanent cellular structures, and their tubulins can serve as constituents of different types of microtubules exerting specific functions, dependent upon actual cellular needs. Interphase cells are characterized by a network of microtubules which disappears at the onset of mitosis, the tubulin then being used for formation of the spindle apparatus. Certain protozoa form tentacles to trap prey; these tentacles are highly specific arrays of a number of cross-bridged microtubules. They are transient cellular structures that exhibit a core of microtubules which undergo rearrangements as the tentacles expand or contract. Other microtubular structures, contained as components in a cytoskeletal lattice composed of microtubules and intermediate filaments present along axons and dendrites of neurons, do not regularly disassemble or reassemble. The same holds true for axonemes in cilia and flagella of certain eukaryotic microorganisms and of gametes. These elongated structures exhibit a diameter of 0.25 μm and are circular in cross-section. They are composed of a $9+2$ array of microtubules, with two microtubules located in the center, and nine forming the peripheral ring located inside an enclosing membrane (Figure 5b,c).

Nearly all microtubules show the same principle of construction, based on globular polypeptides 4–5 nm in size. Dimeric subunits formed by the polypeptides are arranged in 13 parallel protofilaments (longitudinal rows) encircling the hollow center. The subunit contains one α- and one β-tubulin, each with a molecular weight of ~50 kDa. Such a dimeric structural subunit is 8 nm in length. The repeating heterodimers are arranged "head-to-tail", thus establishing a defined polarity within the microtubule; as a consequence, all structures with microtubules as components are also polar

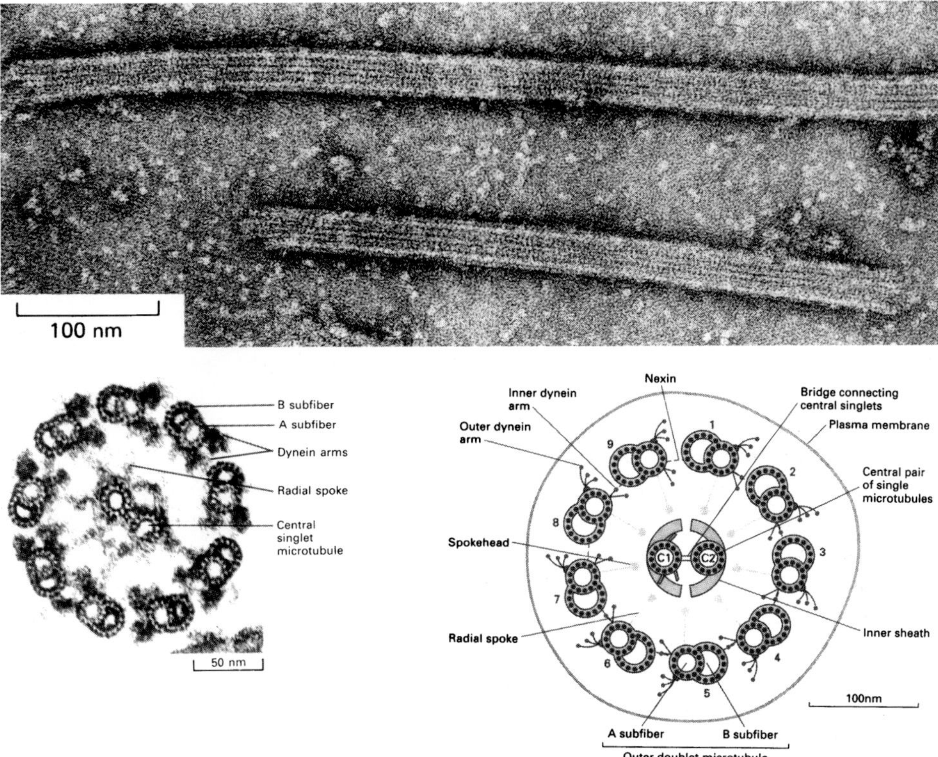

Fig. 5 Microtubules. (a) Electron micrograph of negatively stained isolated microtubules. (From Alberts et al., 1983, with permission.) (b) Micrograph of a transverse section through an isolated demembranated cilium. The two central singlet microtubules are surrounded by nine outer doublets, each composed of an A and B subfiber. (c) Cross-sectional view of an idealized flagellum showing all major structures. The dynein arms and radial spokes with attached heads occur only at intervals along the longitudinal axis. The central microtubules, C1 and C2, are distinguished by fibers bound only to C1. (From Darnell et al., 1990, with permission.)

in nature. Despite the fact that the subunits are organized into 13 protofilaments, i.e., fibrils running parallel to the long axis of the microtubule, the molecular architecture of the microtubule wall could also be envisaged as a helical array of repeating tubulin subunits.

The properties of microtubules reflect the functional differences of the two ends, called the minus (–) end and plus (+) end, respectively, due to the intrinsic polarity. These properties govern the assembly and disassembly of microtubules (Figure 6) (Darnell et al., 1990). Assembly and disassembly both occur preferentially at the (+) end, but they also take place – albeit at a low rate – at the (–) end. In experiments where defined conditions of ion composition were maintained, a free tubulin subunit concentration was observed at which a steady-state situation was reached: the addition of free dimers to some of the microtubules was balanced by loss of dimers from others. Above this critical concentration, microtubules tend to grow; below it they are shortened. However, not all of the free tubulin subunits present in the pool necessarily participate in assembly/disassembly reactions. In *Chlamydomonas*, assembly of the two flagella is highly regulated (Darnell et al., 1990). The large pool of tubulin present during interphase is not used immediately to produce more or longer flagella. A high degree of regulation is also exemplified by observations made after shearing off one of the two flagella. The remaining flagellum is first shortened at the same rate as the new one is regenerated, and only when both flagella have reached half of their normal length do both begin to grow simultaneously. It was proposed that simultaneous occurrence of microtubular assembly and disassembly in one and the same cell may be regulated by different conditions within the two flagella; the concentration of calcium ions might be one factor. This notion would mean that the two flagella in fact constitute two functional compartments. Application of colchicine and other treatments can shift the steady-state situation described above. It is known that tubulin synthesis is autoregulated: unpolymerized tubulin binds to ribosomes synthesizing tubulin and triggers degradation of tubulin mRNA. As treatment of cells with colchicine results in an increase in unpolymerized tubulin subunits, it also inhibits tubulin synthesis. The colchicine effect is reversible (Darnell et al., 1990), as is the disassembly of microtubuli by low temperature and high pressure.

In interphase as well as in mitotic animal cells, microtubules radiate from the centrosome, microtubule organizing center (MTOC) (Figure 7), in which are located the centrioles. Basal bodies–structures similar in function to a MTOC in cells devoid of typical MTOCs – are the centers from which cilia and flagellar microtubules radiate into the cytoplasm. The (–) end of each microtubule points to the MTOC or the basal body.

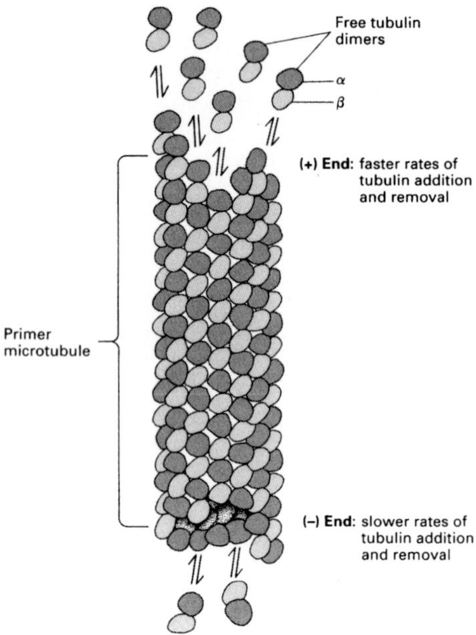

Fig. 6 In-vitro microtubule assembly-disassembly reaction. Disassembly is favored at low temperatures; assembly is favored at 37°C and by the presence of GTP. (From Darnell et al., 1990, with permission.)

In mitotic cells, microtubules radiate from two poles: the microtubular (–) ends point to one of the poles, whereas the (+) ends point to the equator at the center of the cell; hence the (+) ends make contact with the chromosomes. The situation is different in neurons however (Figure 7); in dendrites, there is no obvious MTOC is present and both the (+) and (–) ends of their microtubules can point to the neuron cell body.

In most eukaryotic cells, tubulin proteins are subjected to several types of evolutionarily conserved post-translational modification, comprising detyrosination, acetylation, phosphorylation, palmitoylation, polyglutamylation, and polyglycylation. Recent studies have provided evidence that at least some tubulin post-translational modifications have important functions (Rosen-

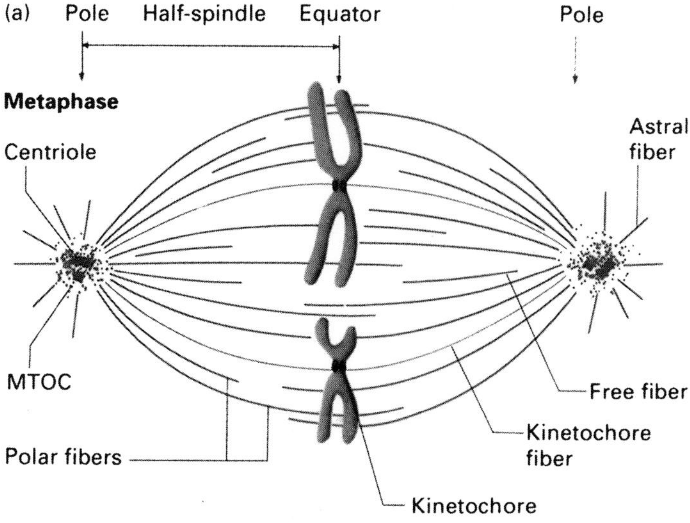
(a) Pole
Half-spindle
Equator
Pole
Metaphase
Centriole
Astral fiber
MTOC
Free fiber
Kinetochore fiber
Polar fibers
Kinetochore

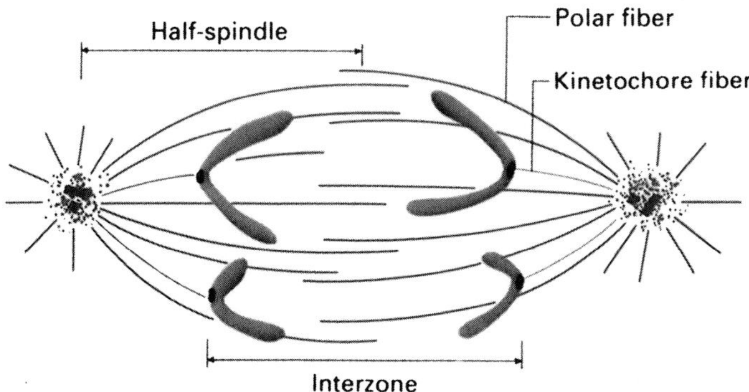
Anaphase
Half-spindle
Polar fiber
Kinetochore fiber
Interzone

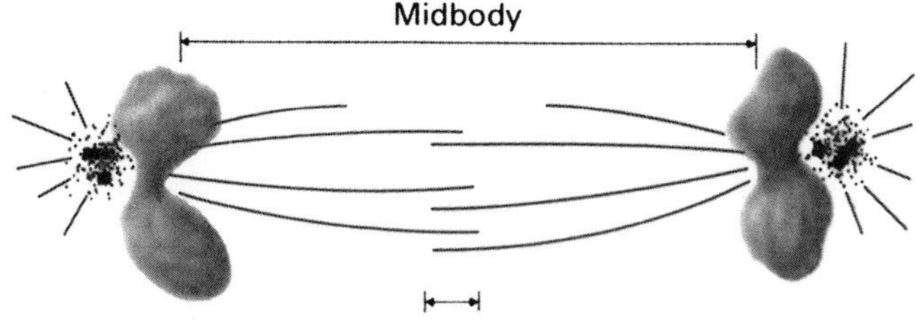
Telophase
Midbody
Zone of interdigitation

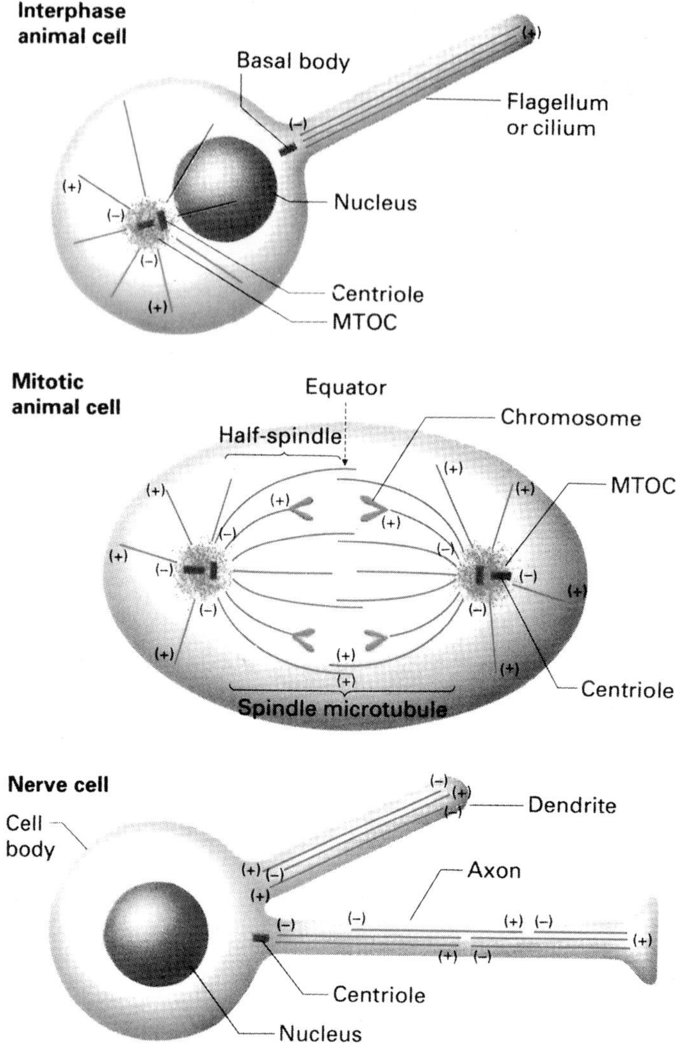

Fig. 7 (a) Schematic diagrams of mitotic spindle in animal cells during phases of mitosis. Each fiber represents many microtubules. (b) Polarity of microtubules. Axonal microtubules in neurons exhibit the usual polarity; dendritic microtubules, however, are unusual in that either their (+) or (–) ends can face the cell body. MTOC, microtubule organizing center. (From Darnell et al., 1990, with permission.)

baum, 2000). An example was found in the ciliate *Tetrahymena* by site-directed mutations of tubulin genes expressed *in vivo* (Xia et al., 2000). The carboxyl termini of both α-tubulin and β-tubulin have multiple sites of polyglycylation. Although the elimination of α-tubulin glycylation did not have a phenotypic effect, a high level of β-tubulin

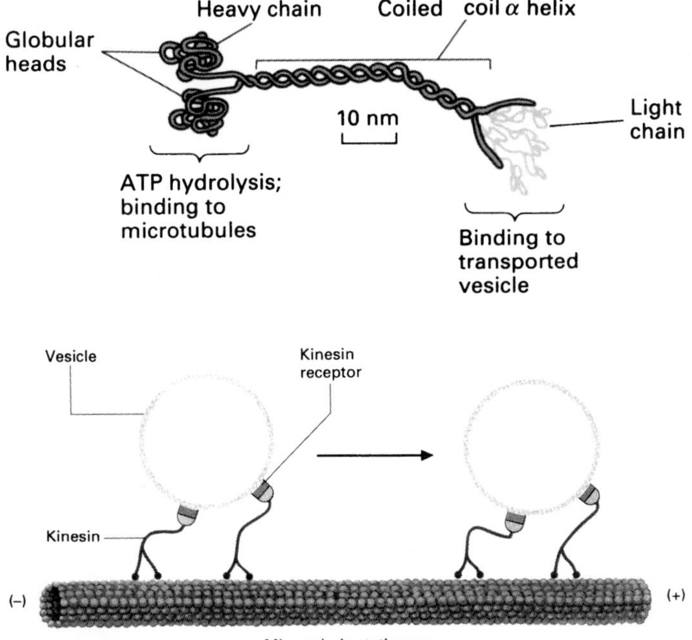

Fig. 8 (a) Schematic model of kinesin. (b) Model of kinesin-catalyzed anterograde transport of vesicles along microtubules. (From Darnell et al., 1990, with permission.)

glycylation was required for cells to survive. By modifying some, but not all, of the β-tubulin glycylation sites, cells were created which moved poorly, suggesting that the β-tubulin glycylation was affecting the function of the ciliary dynein motor. Detyrosination also affects an interaction of microtubules, i.e., that with the motor protein, kinesin. Taking these and other results together, it may be suggested that the post-translational tubulin modifications affect the binding of motors to the microtubule wall.

The molecular motor proteins – kinesin (Figure 8), which mediates movement of vesicles along microtubules, and dynein, which is involved in the movement of cilia and flagella (Figure 9)–interact with microtubules as either tracks or stationary structures.

4.2 Microfilaments

Microfilaments, also called F-actin, are filaments formed by polymerization of a monomeric protein, the G actin (Figure 10). Bundles of actin microfilaments are termed stress fibers (see Figure 4). G actin is a polypeptide that is globular in shape, and has a molecular mass ~43 kDa. Actin filaments can be isolated from both muscle and nonmuscle sources. Within these filaments, the monomers exhibit a defined polarity, and they are polymerized head-to-tail; hence, actin filaments also have a polarity. Under the electron microscope, the monomers making up such a filament can be seen from a variety of angles. A single chain of monomers forms a helical filament in which subunits are connected by a 167°

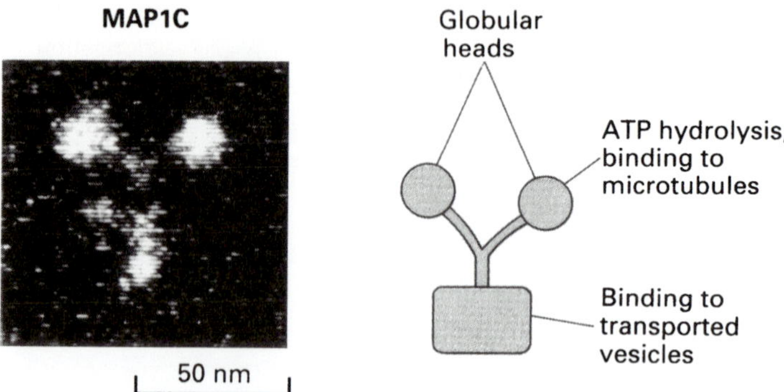

Fig. 9 Electron micrograph and schematic diagram of cytoplasmic dynein (MAP1C), showing the two globular heads that bind to microtubules. (From Darnell et al., 1990, with permission.)

rotation and 2.7 nm axial rise, and two actin protofilaments coil around each other to form a helical filament. Each monomer in the protofilament has major interactions with its two closest neighbors, as well as weaker interactions with other monomers in the neighborhood. Analyses of bacterial proteins MreB and Mbl and filaments formed thereof have been valuable in creating a better understanding of the architecture of filaments consisting of G actin (Erickson, 2001). In fact, it has been shown that these bacterial proteins, constituting types of bacterial cytoskeletons, might share common ancestors with actin (see below).

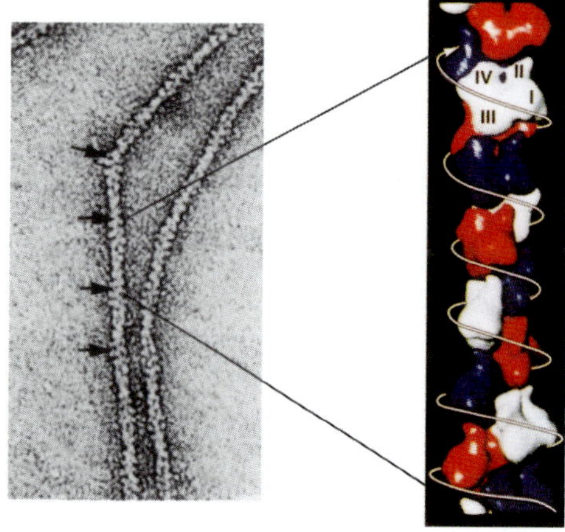

Fig. 10 (Left) Electron micrograph (negative staining) of an F-actin filament; the helical arrangement of the subunits (G-actin) causes variable diameters (9 nm, 7 nm) along the filament (arrows, periodically positioned diameter; distance between the arrows is 36 nm). (Right) Model of F-actin illustrating the arrangement of G-actin molecules along the F-actin filament; the depicted stretch corresponds to the stretch between two arrows in the left-hand illustration. The helical nature of the polymer is indicated. (From Lodish et al., 1996, with permission.)

Similar to the situation with tubulin, actin protofilaments grow by addition of subunits to both ends, with the rate of polymerization at the (+) end being much higher than that at the (–) end. Similar to tubulin, where the monomer carries tightly bound GTP or GDP, actin monomers contain tightly bound ATP or ADP. Filament growth by addition of actin monomers with bound ATP is much faster than that of monomers with bound ADP. However, ATP hydrolysis is not a prerequisite for polymerization. Like microtubules, under steady-state conditions actin filaments are in equilibrium with their monomers. Under conditions simulating the in-vivo ion concentrations, the rate of dissociation of monomers from actin polymers is slow. This is not the case with microtubules, and indicates that actin filaments are much more resistant against unfavorable conditions that cause depolymerization of microtubules. In addition, the critical concentration for actin monomers to assemble into filaments is comparatively low. As a consequence, under steady-state conditions most of the monomers of actin in a cell are polymerized in microfilaments. Actin filaments in muscle were found to be stable for days (stabilization is most likely enhanced by actin-binding proteins in addition to the intrinsic actin filament stability); in comparison, actin filaments in the elongating ends of motile animal cells were observed to be much less stable, and polymerized and depolymerized within minutes. This property, however, appears to be a prerequisite for rapid reactions to occur during cell movement. Other dynamic events involving mechanical forces in non-muscle cells–especially endocytosis–were long believed to be coupled to specific functions of actin, but the actual role of the actin cytoskeleton in endocytosis has been unclear. Recently, three proteins–cortactin, Abp1p, and Pan1p–were identified and characterized and shown to provide a link between the actin cytoskeleton and the endocytic machinery (Jeng and Welch, 2001). Most of the early evidence suggesting a role for actin in endocytic processes was obtained from studies in budding yeast. The disruption of genes encoding actin and many actin-binding proteins was shown to block the uptake of endocytic markers. Now, recent studies have strengthened the association between actin and endocytosis in both yeast and mammalian systems. It has become clear that these three proteins exert their function by activating a complex, the Arp2/3 complex, a conserved ubiquitous complex of seven polypeptides comprising actin-related proteins Arp2 and Arp3, that accelerates the nucleation of actin filaments. Most of the activators known so far are members of the ubiquitous WASP family of proteins (WASP = Wiskott–Aldrich syndrome protein, a scaffolding multimodular protein). An example is the ActA protein from a bacterial pathogen, *Listeria monocytogenes.* This bacterium uses actin polymerization to move within its host – a mode of locomotion also shown by *Shigella flexneri* (Figure 11). A major difference between the activities of these three proteins and that of proteins of the WASP family is that the level of stimulation by the three proteins is weaker than that caused by SCAR/WASP proteins. A possible explanation for this difference is that the former require activation by regulatory factors; another difference is that the three proteins bind to filamentous actin, rather than to actin monomers.

Actin-based motility as described above for bacteria in a host cell or of functionalized microspheres (functionalized by an N-WASP-coat, see above) can be reconstituted *in vitro* from only five pure proteins. Movement results from the regulated site-directed treadmilling of actin filaments, similar to actin dynamics observed in living motile cells. A minimal motility medium should contain ATP as the energy source, the Arp2/3 complex (see above), and actin filaments in the presence of ADF (actin depolymerizing factor, cofilin), profilin, and capping proteins.

An overview of potential states and changes of actin aggregation is provided in Figure 12. One of the most prominent proteins interacting with actin filaments is myosin, which was shown to move along actin filaments, the process being driven by ATP hydrolysis. In fact, myosin is an ATPase, and when pure actin filaments are added to a sample of myosin, measurable ATP hydrolysis increases drastically to a rate similar to that found in contracting muscle. In muscle, myosin is organized into myosin thick filaments which, together with the thin

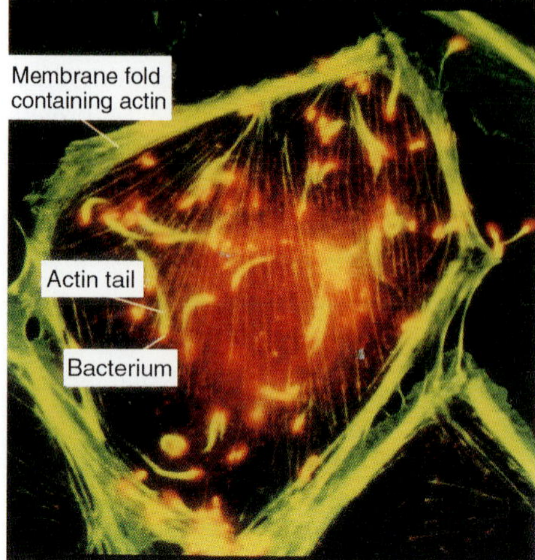

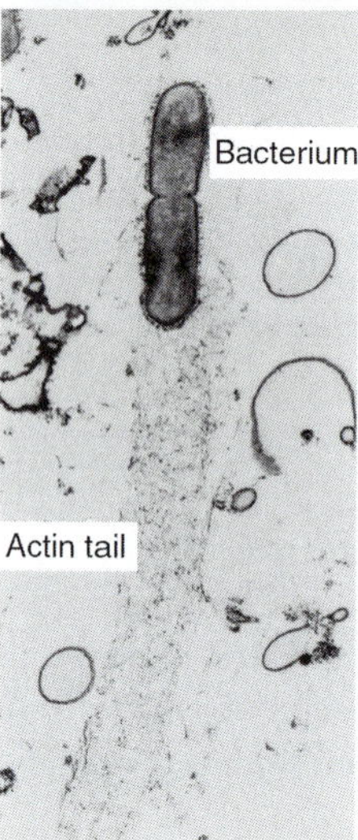

Fig. 11 Locomotion of *Listeria* cells in infected fibroblasts. (a) Individual bacterial cells (recognized by red immunofluorescence obtained by application of an antibody raised against a bacterial membrane protein, ActA) appear to move independently in the cytoplasm, each of them connected to an actin tail (immunolabeled green by an antibody raised against phalloidin, an actin-binding protein). (b) Electron micrograph (ultrathin section) depicting an intracellular *Listeria* cell connected to an actin tail. (From Lodish et al., 1996, with permission.)

filaments formed by actin, slide relative to each other within sarcomeres, thus causing muscle contraction. The interaction of actin filaments and myosin is also important in nonmuscle cells (Darnell et al., 1990), for example for cytokinesis, and also for the transport of vesicles, endoplasmic reticulum and other organelles along actin filaments– the so-called plasma streaming.

Stress fibers, which are bundles of parallel microfilaments in nonmuscle cells (see Figure 4), appear to be primarily involved in the attachment of cells to a substratum (Darnell et al., 1990). In addition, they are thought to generate the stress or tension that determines the shape of fibroblasts. Stress fibers have been shown to contain myosin, α-actinin, and tropomyosin intermittently along their length, and the regulatory protein myosin light-chain kinase. Actin and myosin in stress fibers are believed to form minisarcomeres similar to structures in the mitotic contractile ring. Several studies have suggested that stress fibers are potentially contractile, and a number of additional proteins which interact with actin-based cytoskeletal elements by cross-linking are known (Alberts et al., 1983); these include spectrin (Figure 13), α-actinin, filamin, ABP120, dystrophin, fimbrin, dematin, villin, scruin, fasein, and EF-1a.

4.3
Intermediate Filaments

Intermediate filaments (IF) exhibit diameters between 8 and 12 nm. They are thicker than actin thin filaments and thinner than myosin thick filaments, and their major function seems to be that of the principal structural elements in metazoan cells and tissues (Figure 14) (Darnell et al., 1990). In contrast to microtubules and microfilaments, IF subunit proteins do not exist in dynamic equilibrium with a pool of mono-

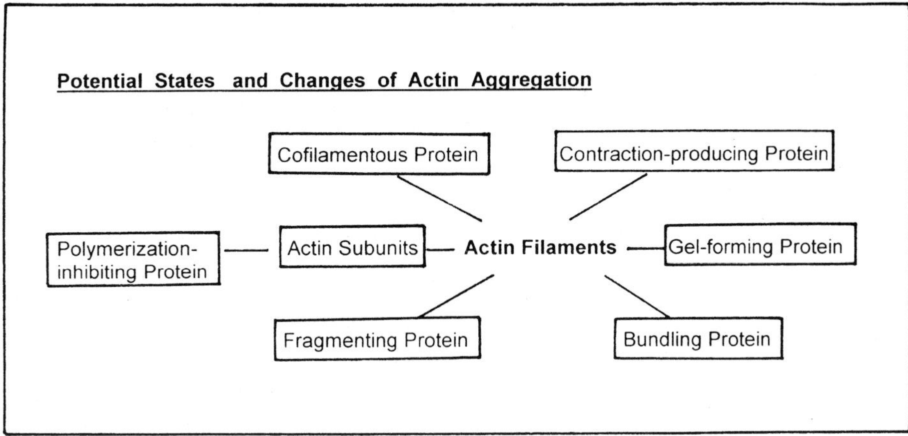

Fig. 12 Potential states and changes of actin aggregation.

meric subunits in the cell. Nearly all of the IF subunits are found in the cell in the polymerized form. The main remnants after extraction of a cell by nonionic detergent are intermediate filaments, whereas microtubuli and microfilaments are solubilized under these conditions. The assembly of intermediate filaments does not require energy, and so is relatively simple. In contrast to actin and myosin that are widely distributed, several cell-specific classes of IF subunit proteins have been identified. In contrast to actin and tubulin (which are almost globular in shape and form filaments that resemble beads on a string), IF subunit proteins are extended molecules that form rope-like polymers (Figure 15a). The denaturation of these polymers with urea or strong detergents is reversible.

Five classes of IF subunit proteins, each characteristically expressed in a specific cell type, have been found among the higher vertebrates (Lodish et al., 1996):

1) Vimentin is expressed in mesenchymal cells, in blood vessel endothelial cells and in certain epithels (layers of epithelial cells). Vimentin fibers are often associated with microtubules and are assumed to keep the nucleus or other organelles in a defined location within the cytoplasm of the cell. Vimentin fibers have also been seen to surround lipid droplets in fat cells.
2) Desmin is expressed in muscle cells, where it forms an interconnecting network across each muscle cell.
3) Neurofilaments are expressed in axons, and are composed of at least three types of neurofilament polypeptides; they are usually found associated with axonal microtubules.
4) Glial fibrillary acidic protein (GFAP) is expressed in glial cells surrounding neurons.
5) Cytokeratins. There are two classes of cytokeratins: acidic (class I), and neutral/basic (class II). Cytokeratins are expressed in epithelial cells, e.g. those that form the nails and wool of sheep, and in other epithelia (Figure 15b). The cells always express multiple cytokeratins, and the filaments formed thereof are always composed of equal numbers of subunits of classes I and II.

Due to the specific occurrence of the various types of IFs, the identification of specific IF subunit proteins in cells can be

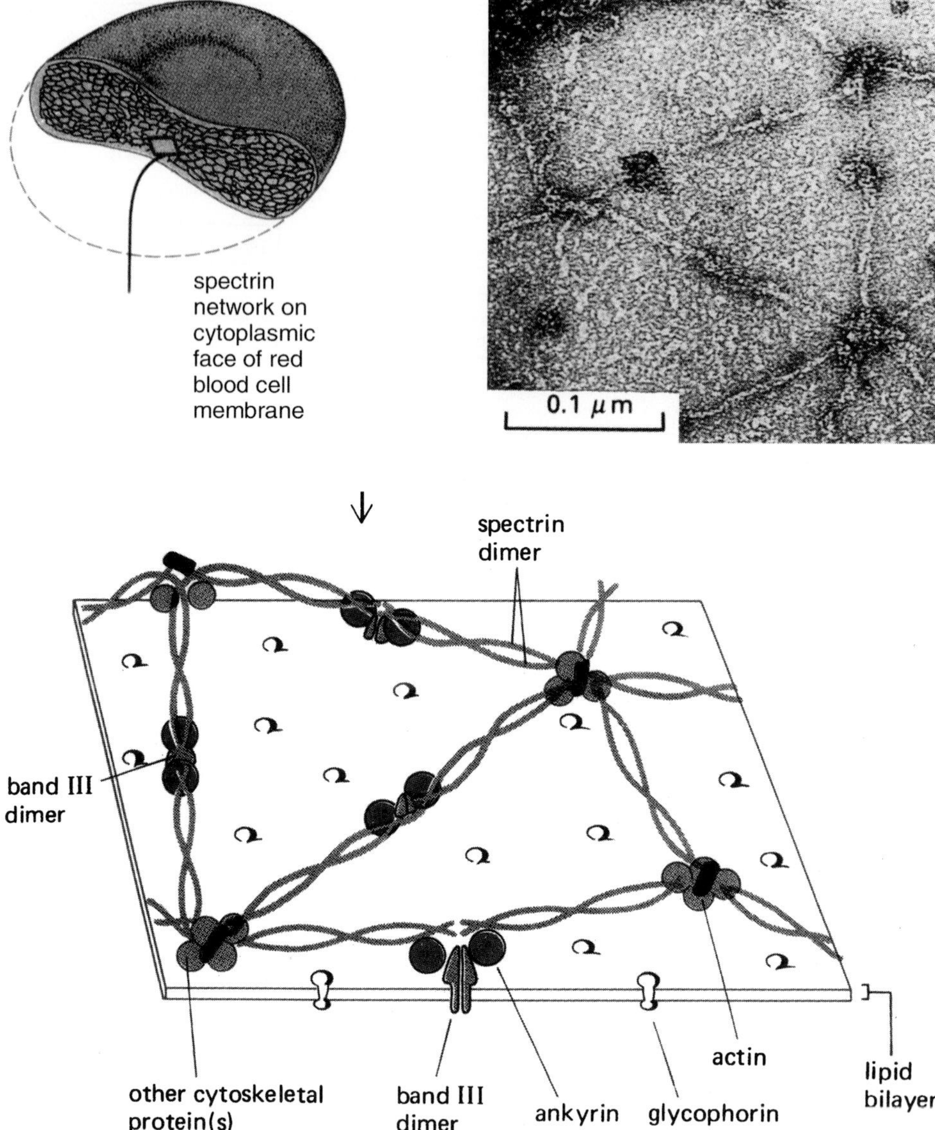

Fig. 13 Human erythrocyte cytoskeleton. (a) Sketch of an erythrocyte. (b) Electron micrograph (negative staining) of a spread cytoskeleton originally attached to the inside of the erythrocyte membrane. (c) Schematic drawing of the probable organization of the constituents of this cytoskeleton. (a and c, from Alberts et al., 1983; b, from Darnell et al., 1990, modified; with permission.)

used for the typing of cells, for example when tumor classification is not possible by conventional histologic means. However, this form of typing can influence the type of treatment undertaken.

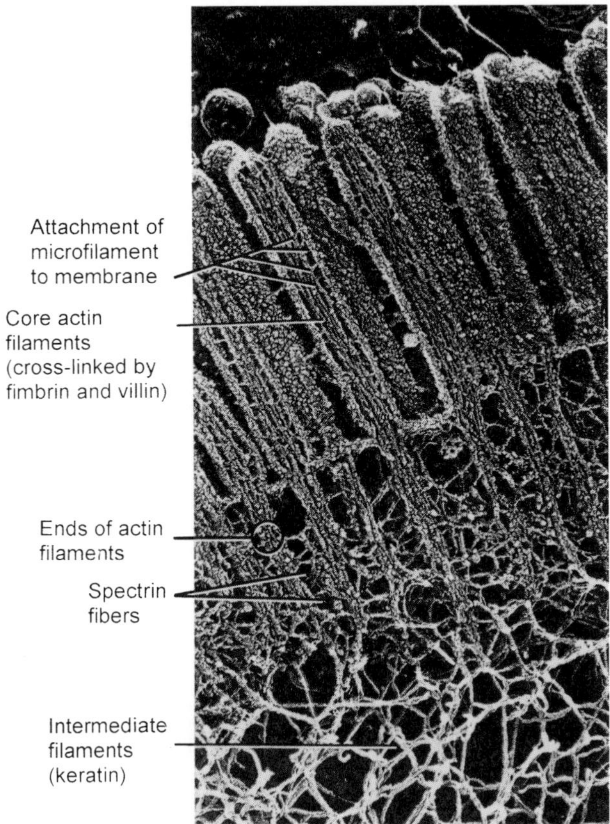

Fig. 14 Ultrastructural organization of microvilli. (From Lodish et al., 1996, with permission.)

5 Cytoskeletons in Prokaryotes

Although, until recently, both the existence of a typical prokaryotic cytoskeleton similar to that in the eukaryotic cell and the occurrence of prokaryotic motor proteins were unknown, indications have been obtained over the past few years for the existence of cytoskeletal elements in bacteria (Erickson, 2001; Nanninga, 2001). Microtubules of the eukaryotic cell had been traced back to the bacterial protein FtsZ, which forms filaments that are involved in bacterial cell division. The origin of actin in the higher cell remained more obscure, though recent findings have supported the view that prokaryotes contain various types of protein that may assemble into cytoskeletons with an architecture specific for prokaryotes, and that relationships of the building elements of these cytoskeletons with those of the eukaryotic cell might be disclosed.

5.1 Cytological Aspects

Before the functions of FtsZ had been elucidated, the simple picture of the architecture of prokaryotic cells in general, i.e., the absence of a cytoskeleton, was not substantially complicated by experimental

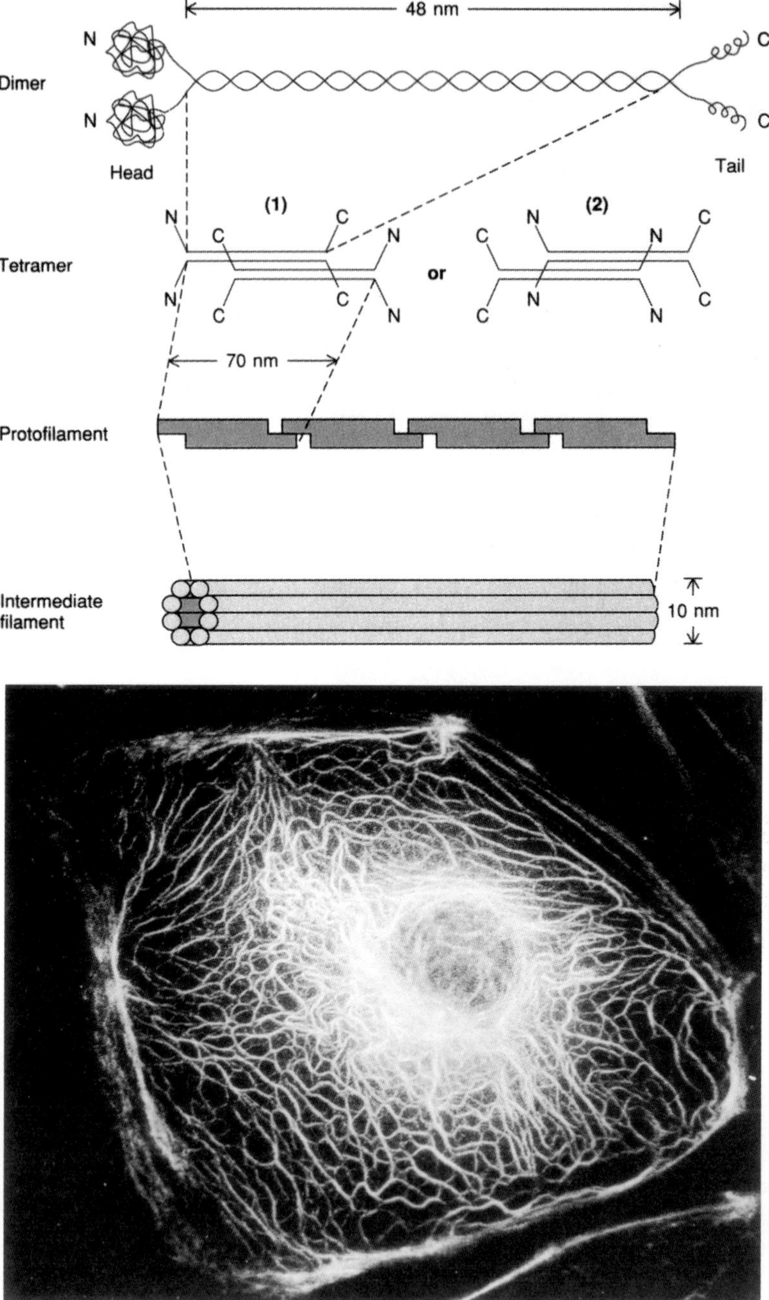

Fig. 15 (a) Level of organization and assembly of intermediate filaments. (b) Network of keratin filaments; the nucleus, located in the center of the cell, is surrounded by keratin filaments; keratin filaments also stabilize the cell periphery. (a, from Darnell et al., 1990; b, from Lodish et al., 1996, with permission.)

findings showing that certain bacterial proteins, especially elongation factor Tu (EF-Tu), after isolation and purification, exhibited fiber formation *in vitro*. As no EF-Tu fibers could be detected *in vivo* even by application of cryo techniques, their existence *in vivo* was denied (Schilstra et al., 1996). Moreover, searches in prokaryotes for typical sequences of genes coding for actin, myosin or tubulin proved negative.

Doubts on the correctness of this simple picture originated not only from the discovery of FtsZ, but also from several other findings. A case was described where an originally elongated typical Gram-positive bacterium, *Thermoanaerobacterium thermosaccharolyticum*, retained its elongated shape and viability even after total removal of the cell wall (S-layer, peptidoglycan) (Antranikian et al., 1987). Immunoelectron microscopy of ultrathin sections of these cells (Mayer et al., 1998), when treated with anti-actin antibodies, showed the presence of cross-reacting proteins along the cell periphery and in the cytoplasm. This suggested the existence of a type of stabilizing structure in this bacterium that consisted of polypeptides which bore some relationship to actin. The same was found to be true for a number of other bacteria, and also for archaea. When wall-less *Th. thermosaccharolyticum* cells were artificially disintegrated on the electron microscopic grid, and the remnants negatively stained, a network of fibers could be seen which contained components (proteins) that cross-reacted with anti-actin antibodies (Mayer et al., 1998). Detailed analyses of the FtsZ system revealed that protein components of this system were the prerequisites for the correct placement of the initial site of cell division and for the formation of a ring-like dynamic structure at the plane of division. It was agreed that such a function could be defined as a cytoskeletal function. In cells not in the state of division, FtsZ was found to be irregularly distributed in the cytoplasm. Recently, a fusion protein consisting of FtsZ and GFP was overexpressed in chloroplasts (Kiessling et al., 2000). Fluorescence microscopy revealed that the interior of the chloroplast was crossed by numerous fibers formed by the fusion protein (Figure 16a). As the GFP part of the fusion protein cannot form fibers, the FtsZ moiety must have been responsible for fiber formation. According to the endosymbiosis theory, chloroplasts should be model systems suitable for the simulation of events occurring in prokaryotes. This finding confirmed results obtained from studies of the in-vitro formation of fibers from isolated FtsZ polypeptides.

Parallel to these investigations, other studies conducted in several laboratories finally led to the description of a variety of new and convincing data, all of which substantiated the occurrence of a cytoskeleton in bacteria.

5.2
Cytoskeletal Elements with Specific Functions, and their Polypeptides

As mentioned above, investigations of the FtsZ system played a pioneering role with regard to analyses of bacterial cytoskeletal elements at the molecular and genetic levels. It was shown *in vitro* that the isolated bacterial cell division protein FtsZ assembles into protofilament sheets and minirings (Figure 16b), which are structural homologues of tubulin polymers. Sequencing and structural studies in fact revealed a relationship between FtsZ and tubulin (see below).

Milestones in the research into bacterial cytoskeletons were findings communicated only recently on a set of proteins (Mbl, MreB) that can be viewed as cytoskeletal polypeptides in bacteria (van den Ent et al., 2001). Helical, actin-like, band-shaped fila-

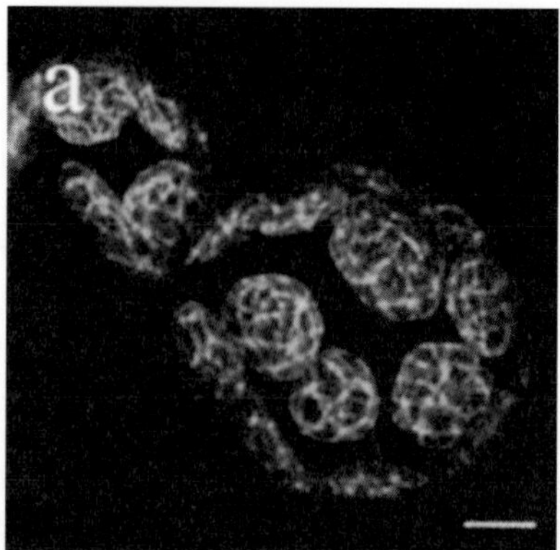

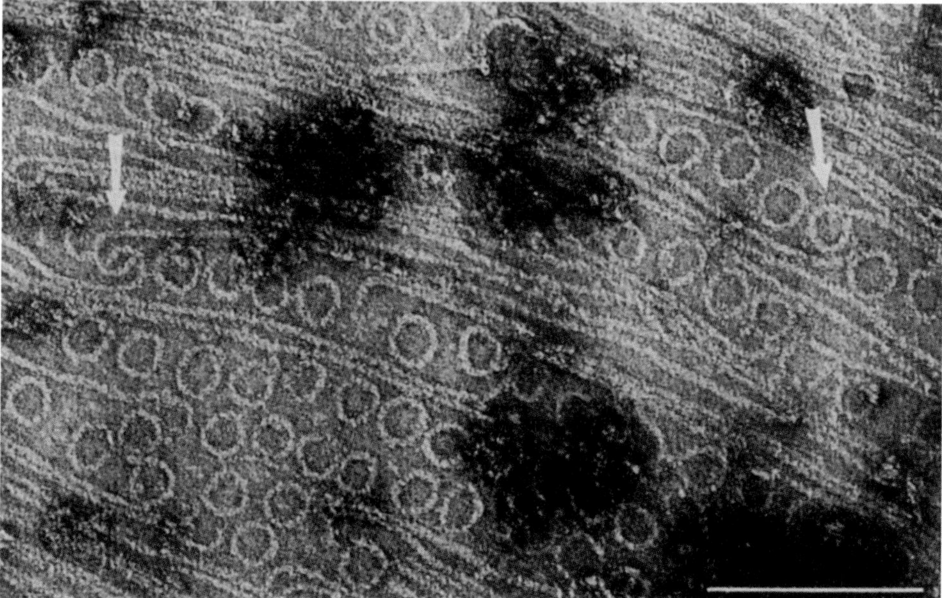

Fig. 16 (a) Visualization of a cytoskeleton-like FtsZ network in chloroplasts of *Physcomitrella*. (b) Protofilaments, small sheets, and minirings formed by isolated FtsZ in GDP (electron micrograph, negative staining). The arrows indicate protofilaments that have separated from a sheet, remaining straight for some length, and then changing abruptly to the curved conformation. (a, from Kiessling et al., 2000; b, from Erickson et al., 1996; with permission.)

ments formed by these proteins were detected in *Bacillus subtilis* (Figure 17) and shown to play distinct, complementary roles in cell shape determination. Fluorescence microscopy was used to depict fibrillar structures within wild-type cells and mutants, and the consequences of mutations for cell shape could be shown. Clearly, the determination of cell length, diameter and shape are distinct parameters that are controlled by these proteins. Numbers of copies of these proteins were determined (an average bacterium contains about 8000 molecules of MreB and 13,000 molecules of Mbl), and it was postulated that the protofilaments (Figure 17) formed by these polypeptides (the building elements of the helical bands supporting cell shape) might provide sufficient mechanical stability. According to the available data, proteins Mbl and MreB are only present in elongated cells (but not only in *B. subtilis*; Table 1), whereas they appear to be missing in coccoid forms of bacteria. It was shown that protein Mbl is, in fact, nonessential for bacterial viability proper, even though cell shape is disturbed in its absence. These observations indicate that proteins Mbl and MreB, though viewed as true cytoskeletal elements with a relation to actin (see below), do not really correspond to actin of the eukaryotic cell.

The situation might be similar in a case that was also reported recently in which a bacterial linear motor was described (Trachtenberg and Gilad, 2001). The helical bacterium *Spiroplasma melliferum* contains a contractile helical cytoskeleton composed of complexes of protein molecules that can vary their diameter (Figure 18). The helix is composed of parallel helical rows of these protein molecules, such that both the pitch of the helix of the cytoskeleton and of the cell body proper can be reproducibly varied by the cell, clearly according to needs. Besides shape variation, this change in pitch also influences locomotion of the cell. Although the proteins forming the cytoskeletal helix appear to be essential components of the *S. melliferum* cell, they seem to be specific forms of a cytoskeletal protein. Similar proteins do not appear to be generally present in prokaryotes; therefore, they might not truly correspond to actin or other substantial cytoskeletal proteins in eukaryotes.

In gliding bacteria, the inherent features of a motility apparatus detected by electron microscopy of frozen samples seem to point

Tab. 1 Distribution of mreB-like genes in the bacterial subkingdoms

Organism/group	***Shape***
mreB present	
Bacillus subtilis (3 genes); *B. cereus, E. coli, Rickettsia prowazekii, Pseudomonas fluorescens, Haemophilus influenzae, Aquifex aeolicus, Thermotoga maritima* (2 genes), *Methanobacterium thermoautotrophicum**	Rod
Vibrio cholerae, Caulobacter crescentus, Pasteurella multocida	Curved rod
Streptomyces coelicolor (3 genes)	Filamentous
Campylobacter jejuni, Helicobacter pylori (3 isolates), *Wolinella succinogenes, Treponema pallidum, Borrelia burgdorferi*	Helical
Chlamydia (2 species)	Pleiomorphic
mreB absent	
*Streptococcus pneumoniae, Enterococcus faecalis, Staphylococcus aureus, Deinococcus radiodurans, Methanococcus jannaschii**, *Neisseria* (2 species), *Archaeoglobus fulgidus**, *Pyrococcus horikoshii**, *Synechocystis* sp.	Round
Mycoplasma (2 species), *Mycobacterium tuberculosus*	Pleiomorphic

BLAST searches were run on EMBnet-CH or The Institute for Genomic Research databases using the predicted amino acid sequences of *B. subtilis* MreB protein. The family of MreB homologues formed a homogeneous group distinct from the DnaK-like proteins (also members of the actin superfamily).*Archaebacterial organism.

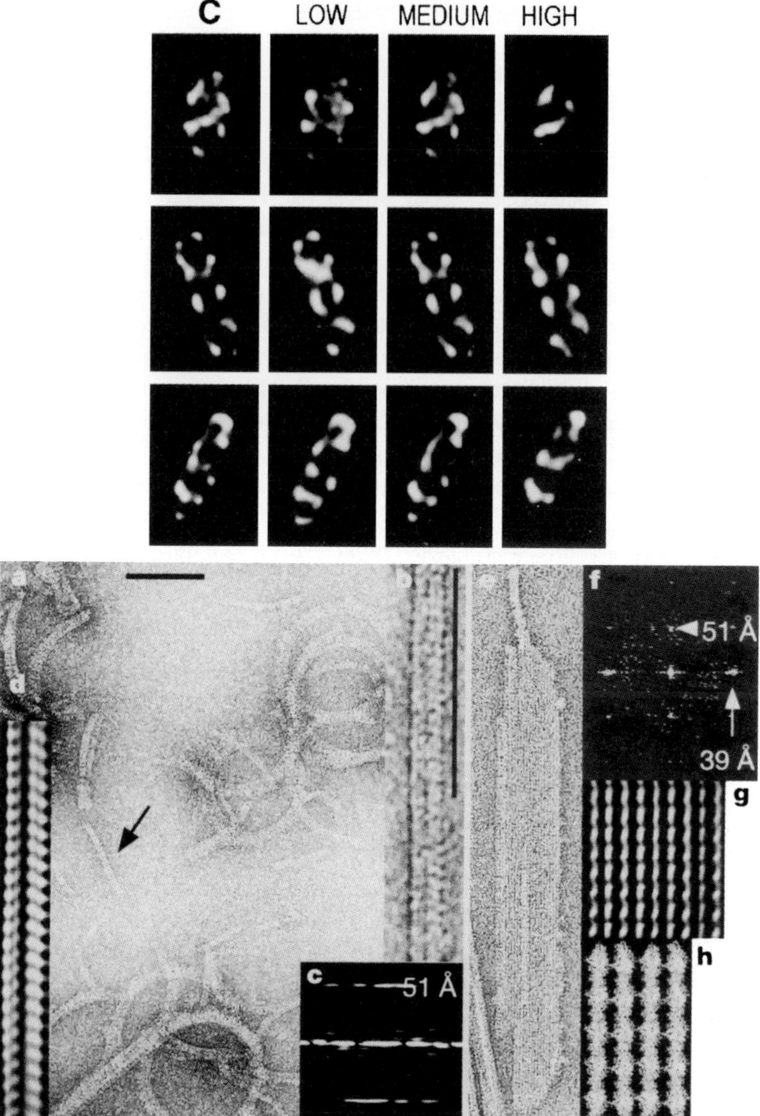

Fig. 17 (a) Subcellular localization of protein Mbl, a relative to MreB, in *Bacillus* cells, depicted by immunofluorescence. C shows cell images after deconvolution of an image stack; in addition, three separate optical sections taken at low, medium, and high positions in the z axis of the cells are presented. (b) Electron micrographs of negatively stained MreB filaments; a, typical view when salt and neutral pH are used; b, double filament formed at high pH; c, diffraction image of polymer in b; d, filtered image of b; e, MreB sheet; f, diffraction image of sheet in e; g, filtered image of sheet in e; h, the protofilaments found in the crystal of MreB fit well with g. The lateral spacing in the sheets suggests that the protofilaments interact with their flat sides. (A, from Jones et al., 2001; B, from van den Ent et al., 2001; with permission.)

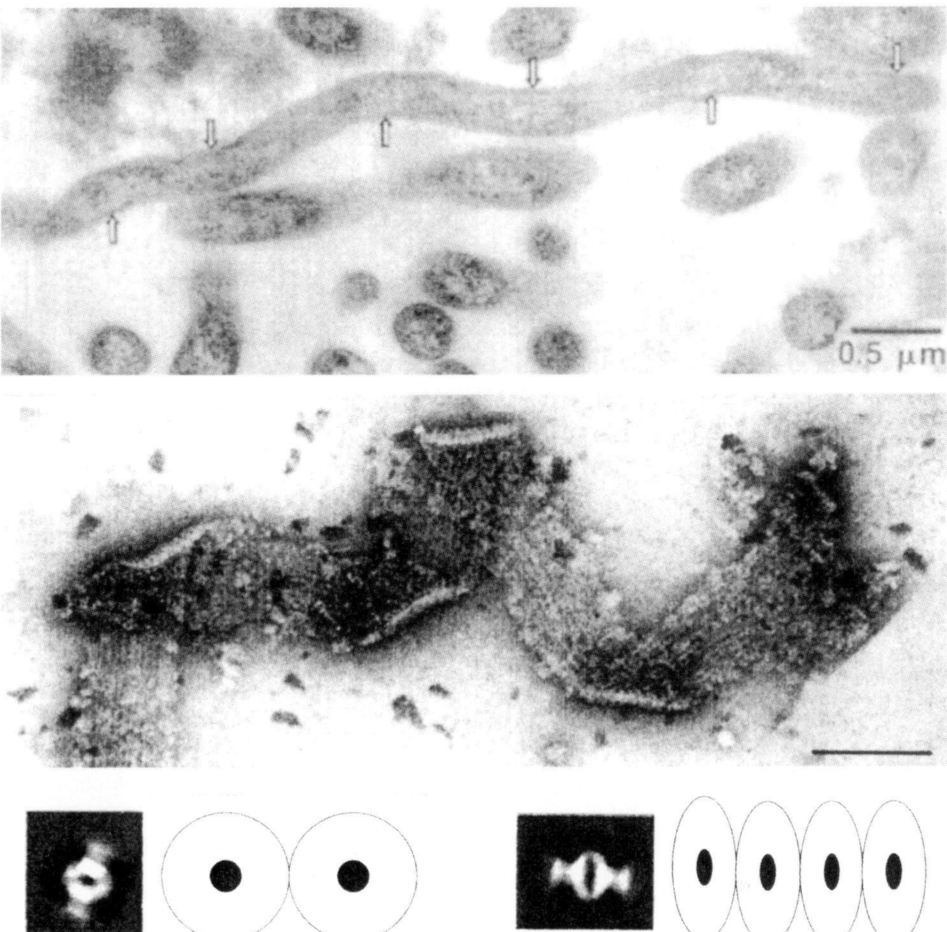

Fig. 18 A bacterial linear motor in *Spiroplasma melliferum*. (a) In the longitudinally sectioned, long, stretched cells the cytoskeleton follows the innermost helical line in the coiled cell (arrows). (b) Negatively stained ribbon-like cytoskeletons obtained by on-grid lysis and detergent extraction of whole cells. (c) Subunit images; left, circular molecule; right, elliptical molecule. Transition of subunits from circular to elliptical causes a fibril composed of subunits to shorten. (From Trachtenberg and Gilad, 2001, with permission.)

to the existence of cytoskeletal elements and motor proteins being involved in cellular locomotion (Lünsdorf and Schairer, 2001). This system awaits further detailed analyses, and only then might the relationships with cytoskeletal elements in eukaryotes be detectable.

5.3
A "Basic" Cytoskeleton?

According to our present knowledge, the cytoskeletal structures and their proteins described above (Mbl and MreB, FtsZ, the contractile proteins of the linear motor) cannot be viewed as true equivalents of a typical cytoskeleton in eukaryotic cells, though the elements making up these cytoskeletons may be structurally related to typical cytoskeletal elements in eukaryotes. Mbl, MreB and FtsZ turned out to be either species-specific, or occasionally missing or nonessential, or they do not form explicit cytoskeletal elements during all states of the bacterial cell life. Hence, the question might be asked, "Do cytoskeletal proteins exist in prokaryotes that form a "basic" cytoskeleton, i.e., a permanent, albeit dynamic, cytoskeleton, with functions of a very basic nature?" Such a basic cytoskeleton would be characterized by a number of properties. It should be ubiquitously present and common in all prokaryotes; its constituting proteins should be present in a number of copies which is sufficient for the formation of a cytoskeleton enclosing the entire cytoplasm and exhibiting numerous fibers crossing the cytoplasm (the number of copies [Jones et al., 2001] would be too low in case of proteins Mbl and MreB; a number around 100,000 would be more realistic); its absence (by mutation of the genes coding for the constituting proteins), or disturbance of its assembly should inhibit cell viability; it should provide the function of a scaffold for the proper placement/assembly of associated cytoskeletal elements; it should be able to act as a site of attachment for many other functional macromolecules in the cell; it should be composed of units that allow dynamic alterations of states by reversible assembly and disassembly reactions; and it should provide means for establishment of cell polarity. In addition, as described above for the cytoskeleton proposed to exist in *Th. thermosaccharolyticum*, a main component of such a cytoskeleton should be a protein that exhibits cross reactivity with anti-actin antibodies.

As mentioned above, some decades ago the bacterial protein EF-Tu was thought to have not only a function in translation (synthesis of polypeptides), but also in other aspects (Beck et al., 1978; Helms and Jameson, 1995), including the formation of cytoskeletal elements. This idea was abandoned when immunoelectron microscopy of cryo sections seemed to provide evidence for a near-homogeneous distribution of the respective label in the cytoplasm (Schilstra et al., 1996). Data on the formation of fibrillar structures (protofilaments, sheets) by EF-Tu *in vitro* (Figure 19), of sequencing experiments, and of high-resolution X-ray studies on crystallized EF-Tu (Song et al., 1999) can now be combined. The fibers formed by EF-Tu *in vitro* exhibited properties which indicated that their presence in bacterial cells may have remained undetected in earlier experiments because the sample preparation procedures may have caused disassembly of the fibers *in vivo*, perhaps due to low temperature and/or inadequate concentrations of specific ions. This may have led to a homogeneous distribution of the remnants of the fibers in the cytoplasm.

Immunoelectron microscopy of ultrathin sections labeled with anti-EF-Tu antibodies, performed on *Th thermosaccharolyticum* and *Mycoplasma pneumoniae*, and of negatively stained samples of glutaraldehyde-fixed *M. pneumoniae* cells from which the cytoplasmic membrane had been removed by Triton X-100 treatment, and prepared by the whole-mount technique (unpublished results; Figure 20), revealed a strong positive reaction of antigens located both within the cytoplasm and close to the cell periphery. However, care must be taken regarding these data as long as it has not been demonstrated that the EF-Tu has not escaped the interior of the cells. The pattern formed by the colloidal gold label within the cytoplasm of ultrathin-sectioned *Th. thermosaccharolyticum* cells – rows of gold label, rather than a homogeneous distribution in the cytoplasm – did not

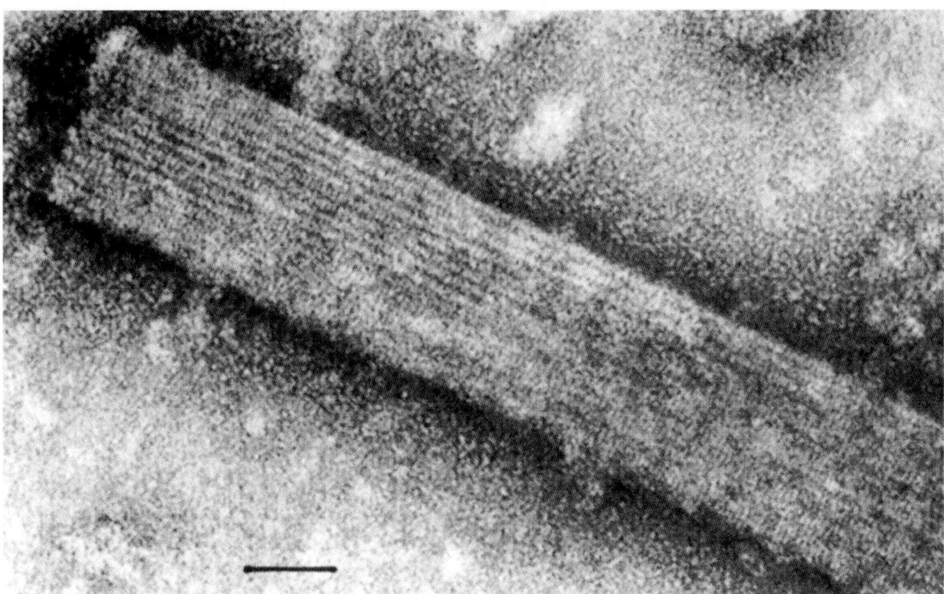

Fig. 19 Paracrystalline sheet obtained, *in vivo*, from a mixture of elongation factor EF-Tu and EF-Ts from *Escherichia coli*. Preparation for electron microscopy by negative staining. (From Beck et al., 1978, with permission.)

confirm the distribution of the respective antigens (EF-Tu) as expected from earlier studies by cryoelectron microscopy (Schilstra et al., 1996), though they did demonstrate the exclusive organization of EF-Tu molecules into fibrillar structures. Although rows of label may be expected in these experiments (EF-Tu might be attached to polysomes), the absence of a homogeneous distribution of the label also indicates that the huge surplus of EF-Tu present in a bacterial cell (only about 25% of the cellular EF-Tu is known to be involved in translation; for the remaining 75% no function could be envisaged) is exclusively organized into fibers. It was therefore concluded that a bacterial cytoskeleton exists that contains, as a major constituent, protofilaments made up of EF-Tu as a main constituent. This notion was supported by investigations with *M. pneumoniae*, where a number of specific cytoskeletal elements (the rod with its linkers, the supports for the rod, fibrils attached to one of the supports) could be seen to be in close contact with a peripheral network of fibers containing polypeptides reacting positively with anti-EF-Tu antibodies (unpublished results). In this context it should be mentioned that analyses of the entire nucleotide sequence of *M. pneumoniae* did not reveal the occurrence of proteins with sequence similarity to Mbl and MreB–the bacterial proteins assumed to be the shape-determining proteins in bacteria in general. Nevertheless, this bacterium, though wall-less, exhibits a well-differentiated tip that consists not only of a rod but also of a cytoplasmic membrane held in place. Hence, the data collected so far may make the revival of the old notion–that EF-Tu is a constituent of a bacterial cytoskeleton–reasonable. After all, EF-Tu is common and essential in any bacterial cell, as it occurs in an amount sufficient to form the peripheral part of a cytoskeleton enclosing the entire cytoplasm as well as filaments crossing the cytoplasm. As a member of the actin superfamily, it cross-reacts with anti-actin antibodies.

Recent unpublished data produced in the author's laboratory revealed that induced expression of a truncated EF-Tu gene – coding just for *Escherichia coli* EF-Tu domain 3, without domains 2 and 1 (for structural details of EF-Tu, see Section 6) – in addition to the EF-Tu gene present in a wild-type *E. coli* cell, resulted in collapse of the cells, with subsequent lysis. This occurred despite the presence of the native full-size EF-Tu gene coding for a sufficient amount of EF-Tu copies for an intact cytoskeleton. The events leading to cell death could be further analyzed. A first feature observable by electron microscopy was the loss of the cell envelope (cytoplasmic membrane, cell wall). At this stage, the original content of the cell (polysomes, DNA with binding proteins) appeared to be well preserved as far as the gross architecture was concerned. The peripheral polysomes were arranged, in parallel, in a helical fashion, indicating that they had also been arranged in this manner in the living cell. One might speculate that they had been in contact with the helical cytoskeleton (consisting of EF-Tu?).The next step was the entire disintegration of the cell content. An explanation for this observation may be that integration of domain 3 in the existing or growing "basic" cytoskeleton (located close to the cell periphery and spanning the cell cytoplasm) consisting of EF-Tu, leads to destabilization and disintegration of this cytoskeleton, with consequent cell death.

One further aspect should be mentioned: In *M. pneumoniae*, it was shown that the cytoskeleton interacts with the cytoplasmic membrane by stalks which extend from the surface of the network of the peripheral cytoskeleton fibrils (unpublished data). *In*

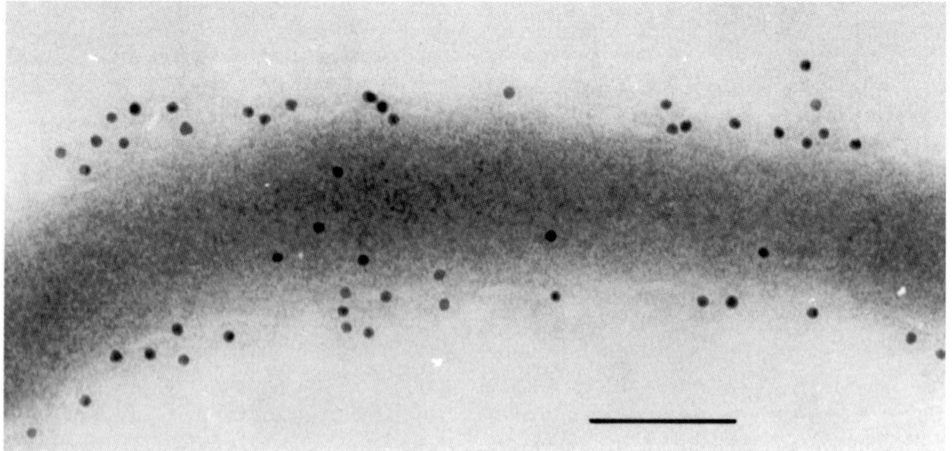

Fig. 20 Immunolabeling of the peripheral cytoskeleton of *Mycoplasma pneumoniae* with gold-conjugated anti-EF-Tu antibodies. Negatively stained whole-mount sample. Prior to immunolabeling, the cells were chemically fixed, in suspension, with glutaraldehyde, the cytoplasmic membrane was partially removed by Triton X-100 treatment, and the cells were washed several times with buffer. Label can be seen in areas where the membrane had been removed and where, hence, the cytoskeleton was exposed to the environment. No label was present in areas from which the membrane had not been removed. (Courtesy of Jan Hegermann.)

vivo, the ends of these stalks are in contact with the cytoplasmic membrane; the membrane, a "fluid mosaic" without sufficient stability to withstand mechanical forces (note that, in the absence of a wall, osmotic pressure would unavoidably cause rupture of the cell), is supported and held in place by the stalks. This is similar to an umbrella where the frame supports the covering fabric. In fact, with this layout of the cell periphery, the cell manages to place a hydrophobic phase, the cytoplasmic membrane, as an interface between two hydrophilic phases (the cytoplasm, the outside environment). This enables the cell to concentrate hydrophobic and amphiphilic components in a defined area or volume. Under these conditions, reactions are favored which are necessary for the assembly of complexes of proteins in an interface between two hydrophilic phases with different concentrations of ions, whereby these complexes are involved in cell energetics by utilizing ion gradients. The cell is allowed to reach this goal with only relatively few copies of the functional components of the complexes due to the restricted volume of the membrane. The high probability of encounters is brought about by the very restricted volume of the cytoplasmic membrane.

6 Molecular Characteristics of Cytoskeletal Polypeptides, and Comparison between Cytoskeletal Polypeptides of Eukaryotes and Prokaryotes

Tubulin and actin sequences from a multitude of sources have been determined at the nucleotide and amino acid levels. It is established that both tubulin genes and actin genes form families, each of which contains many members that exhibit close relationships. Until recently, it was not known whether bacteria possessed proteins that are closely related to the tubulin or actin of eukaryotic cells.

A close structural relationship has now been found for eukaryotic tubulin and bacterial FtsZ protein (Burns, 1998; Löwe and Amos, 1998). Although at first glance this relationship was not evident at the nucleotide sequence level (only ~15% sequence identity was determined), at the macromolecular architecture level the two proteins showed an astonishing similarity (Figure 21). This finding was somewhat surprising, though it was acknowledged that both types of protein need to accomplish specific functions in the cytoskeleton, namely the fulfillment of highly specific prerequisites. Seemingly, nature has managed to do this with different amino acid sequences making up the respective protein. Recently, a closer look into sequence similarities indicated that both proteins shared considerable amino acid homology at several key sequences surrounding the site responsible for GTP binding and cleaving.

A comparison of actin sequences with sequences of the bacterial protein MreB, and of the macromolecular architecture of actin and protofilaments formed thereof with the architecture of MreB and protofilaments formed from this polypeptide, resulted in a further surprise (Jones et al., 2001). Similarities at the nucleotide and amino acid sequence levels, though clearly existing, were not striking (Figure 22) and resembled more a patchwork. It became clear that, as in the comparison between tubulin and FtsZ, sequence homologies were evident around the nucleotide (ATP) binding site that is situated deep in a cleft between the two halves of the folded polypeptide chain.

Again, the two polypeptides exhibited extensive structural similarities (Figure 23), and specific projections of the two proteins could indeed be superimposed. MreB poly-

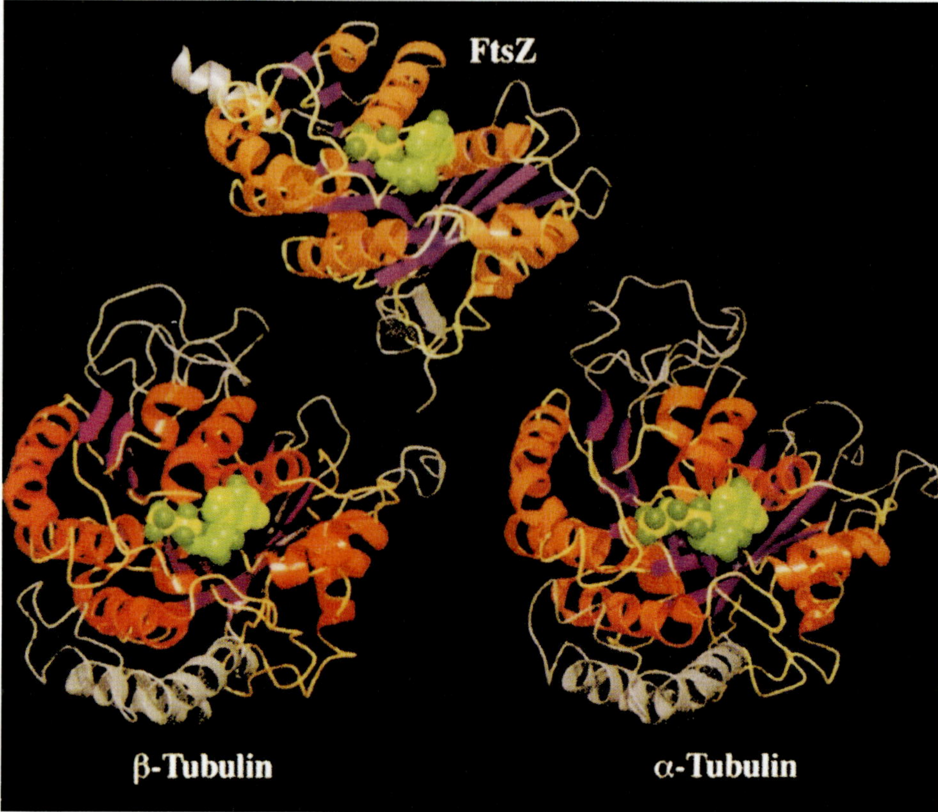

Fig. 21 Comparison of FtsZ and tubulin. The orientation of the polypeptides was chosen such that similarities and differences in the features and projections of the two kinds of protein become evident. (From Burns, 1998, with permission.)

merizes into protofilaments that pair lengthwise, but MreB double filaments are not nearly as helical as actin double filaments (Pantaloni et al., 2001).

On the basis of these findings, discussion has now commenced regarding the evolutionary aspects of cytoskeleton proteins (Doolittle, 1995; Hartman and Fedorov, 2002).

It was assumed that the geometry of a protein subunit that can assemble into a straight protofilament sheet is so complex and precise that this assembly could not exist *in vitro* unless it were constantly selected for and, hence, functionally important *in vivo*. To form a straight protofilament, as observed *in vitro* not only for tubulin but also for EF-Tu, longitudinal bonding interfaces must be exactly 180° apart on the subunit. Because all subunits make lateral bonds in the same plane, rotation of subunits about the protofilament axis must be precisely 0°. This precise geometry would not be preserved in evolution unless it were functionally important *in vivo*. The meaning of this geometry could very well also be envisaged for EF-Tu. The introduction of a helical shape into a protofilament (e.g., in actin) may be an additional feature not simply explainable by the rules for bond formation

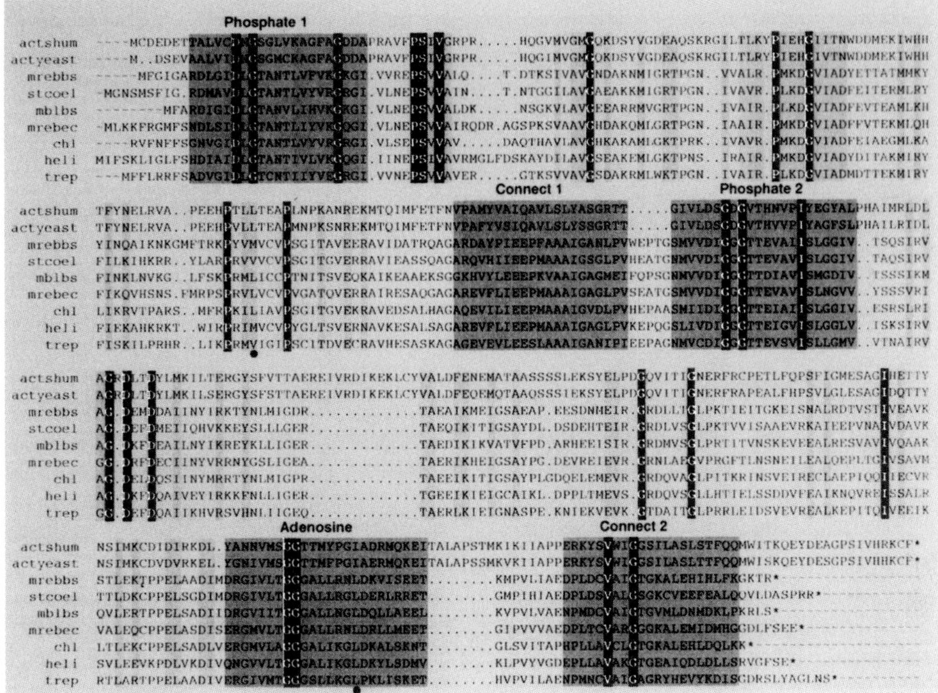

Fig. 22 Sequence alignment of MreB proteins from various bacteria with eukaryotic actins: actshum, human ActS; actyeast, *Schizosaccharomyces pombe*; mrebbs, *Bacillus subtilis* MreB; mblbs, *B. subtilis* Mbl; stcoel, *Streptomyces coelicolor*; mrebec, *Escherichia coli*; chl, *Chlamydia trachomatis*; heli, *Helicobacter pylori*; trep, *Treponema pallidum*. (From Jones et al., 2001, with permission.)

mentioned above (see Biopolymers, Vol. 7, pp. 261–284).

X-ray data revealed the architecture of the EF-Tu polypeptide in much detail (Song et al., 1999). The existence and detailed layout of three domains within the polypeptide was confirmed. Domain 1, responsible for the known function of EF-Tu in translation, is the only domain of the protein for which a defined function is known. A closer inspection of domains 2 and 3 revealed that their exposed surfaces can be arranged, by modeling, in such a way that they fit nicely into each other, forming contact areas (interfaces) as a basis for fibril (protofilament) formation (Figure 24). As units in such a fibril, EF-Tu polypeptides could very well exert their function in translation without disturbance because domain 1 would not be involved in protofilament formation.

The calculated diameter of such a fiber and its macromolecular architecture are close to that derived from the electron microscopy investigations of fibrillar in-vitro polymerization products (unpublished data). The fibrils observed in remnants of negatively stained *Th. thermosaccharolyticum* cells (see above) exhibited a very similar ultrastructure, and they reacted positively with anti-EF-Tu antibodies.

Comparison of DNA sequences of EF-Tu-coding genes in various bacteria revealed that the presence of three domains is a

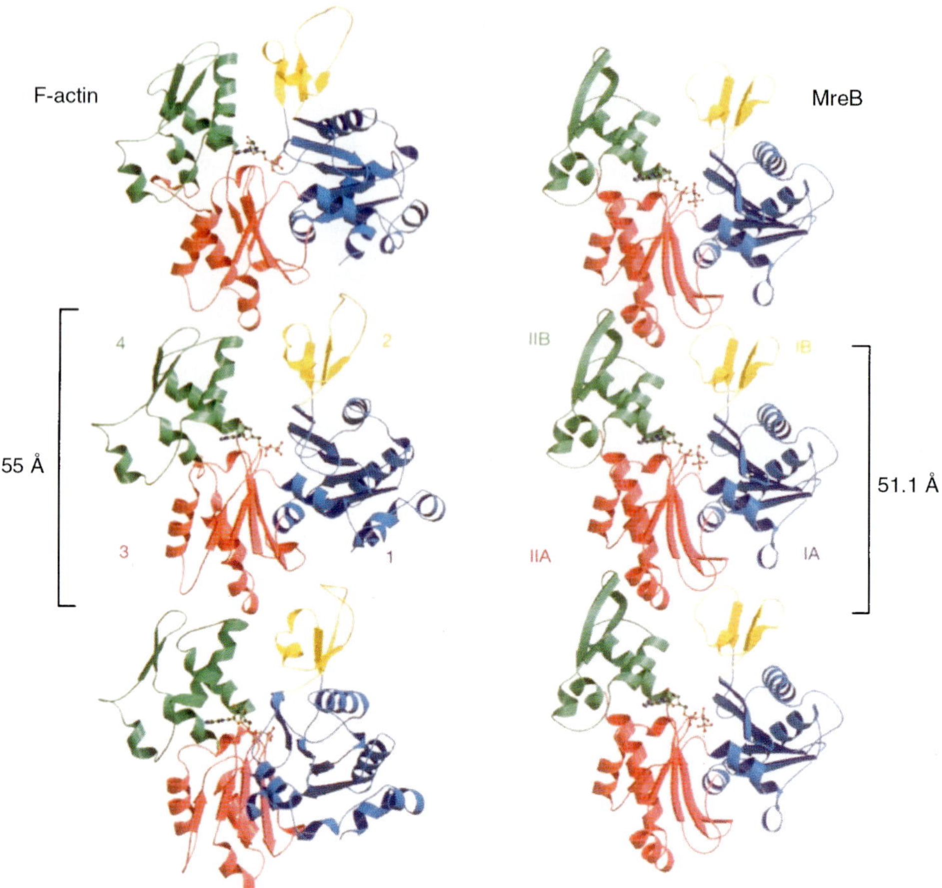

Fig. 23 Comparison of F-actin and MreB protofilament. Residues at the bottom of subdomains IA and IIA (1 and 3 in actin) insert into the cleft formed by subdomains IB and IIB (2 and 4 in actin). This is basically the same interaction that has been proposed for the longitudinal interaction in the two strands of F-actin, shown in the left panel. (From van den Ent et al., 2001, with permission.)

highly conserved property (unpublished data). In addition, the distance between the proposed sites of contact (interfaces) in the assumed formation of protofilaments from EF-Tu polypeptides, measured as the number of amino acids, was constant (126 amino acids). An analysis and comparison of the local amino acid sequences–with lengths of oligopeptides–at the exposed sites of the interfaces was performed. It was found that at these sites the amino acid sequences were also highly conserved, whereas variations in other stretches of the EF-Tu polypeptide were common.

EF-Tu shares, along with actin, tubulin and the bacterial protein MreB, the ability to carry a nucleotide binding site. Though it is accepted that, in EF-Tu, the bound nucleotide, GTP, is needed for one of the steps of translation (protein synthesis), involvement of bound GTP in other functions, especially in a role of EF-Tu as a component of a

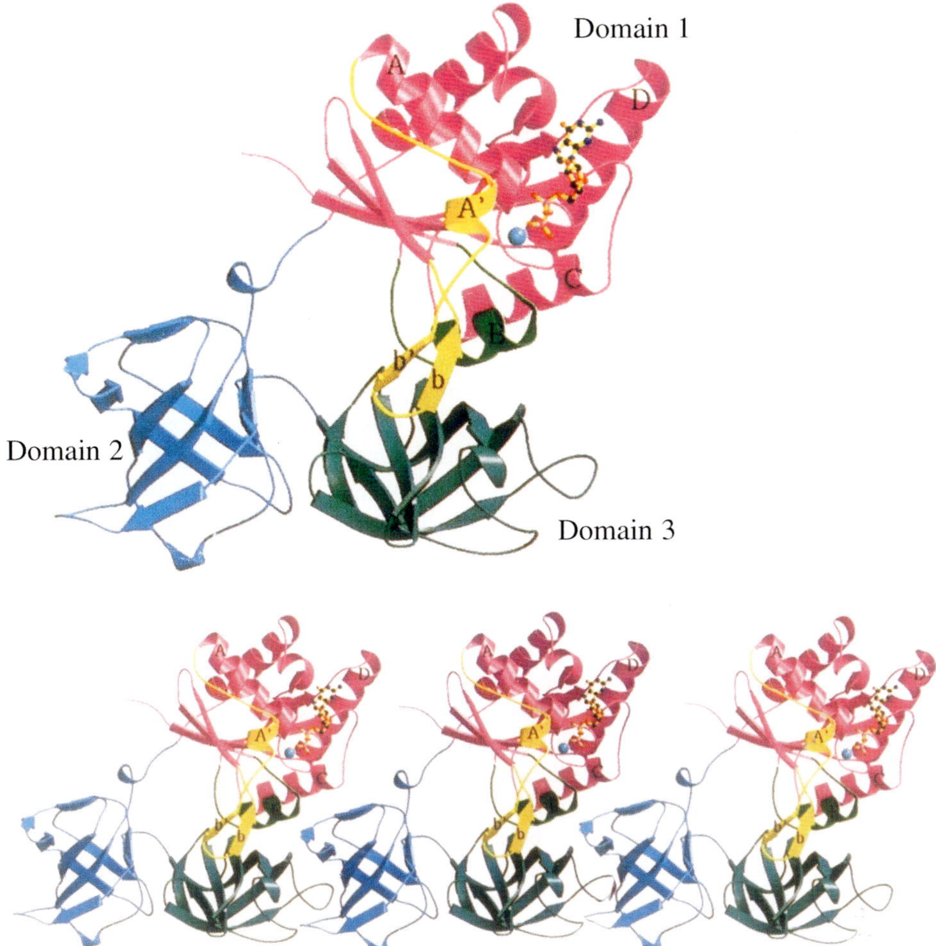

Fig. 24 Elongation factor EF-Tu from *Escherichia coli*, GDP form. (a) Ribbon diagram. (b) Protofilament as proposed by modeling; formation of a filament by interaction of the exposed domains 2 and 3. (a, from Song et al., 1999, with permission.)

dynamic bacterial cytoskeleton, could be envisaged. As a consequence, interaction with other cytoskeleton-related proteins, for example proteins carrying GTPase activity, would be necessary.

Taking these observations and conclusions together, EF-Tu cannot be considered as being closely evolutionarily related to actin, tubulin, and MreB. However, EF-Tu exhibits a number of properties that could make it a candidate protein for the formation of an early, "basic" type of cytoskeleton common in all bacteria, a cytoskeleton carrying out a function as the site of attachment of ribosomes/polysomes to a support, and a function as scaffold for positioning of more specific cytoskeletal elements evolved later in a variety of bacteria. Polypeptides similar in sequence to typical bacterial EF-Tu could also be found in archaea, and proteins related to bacterial EF-Tu are known to occur in mitochondria, chloroplasts, and in the cytoplasm of eukaryotic cells.

7 Aspects of Evolution of Cytoskeletal Polypeptides

Within this chapter, it is not of prime interest to discuss conclusions and assumptions regarding evolutionary aspects within the classes of the three or four types of cytoskeletal proteins (Doolittle, 1995; Erickson, 2001; Palevitz, 2001; Hartman and Fedorov, 2002). More important appears to be that a paradigm shift took place when data became available for proteins making up cytoskeletons in bacteria. Though these data are still scarce, speculative proposals towards a description of the evolution of cytoskeletal proteins should be allowed. It was proposed to assume that the pairs FtsZ and tubulin, and MreB and actin each share a common ancestor, and that FtsZ/tubulin were first on stage. One of the key steps, in going from prokaryotes to eukaryotes, may have been the development of a mechanism to form the double-helical filament of actin from the single MreB protofilament structure.

One more aspect is discussed: It is known that actins can function along with the motor enzyme myosin to produce cellular motion, whereas microtubules utilize dynein and kinesin instead. The question remains, "Are there motor proteins in bacteria that interact with FtsZ and MreB?" Currently, their existence appears not very likely as judged from sequence studies, but one guess is that dynein, kinesin and myosin evolved in eukaryotes, after the evolution of microtubules and actin filaments. However, the general notion that bacteria lack motor proteins is no longer valid: as described above, *Spiroplasma melliferum* contains a bacterial linear motor (Trachtenberg and Gilad, 2001).

8 Patents

Patents regarding a target function of structural elements of the eukaryotic cytoskeleton, or of a target function of cellular molecules interacting with these elements, or regarding drugs functioning by interference with the eukaryotic cytoskeleton have not been sought for this chapter. Now, as the human genome has been sequenced, the field is very much in progress, and discoveries may be expected that indicate involvement of numerous so far unidentified proteins in some type of interaction with cytoskeletal elements, giving rise to the development of new drugs. The field is expected to expand drastically, especially with respect to macromolecules, as targets for new drugs, introducing defaults into the cytoskeletal architecture.

A preliminary search for patents and patent applications related to prokaryotic cytoskeletal elements revealed that the fields of FtsZ/FtsA and of EF-Tu might be interesting for research and development with respect to antibacterial agents.

Examples for patent applications include:

- FtsZ:FtsA multimeric proteins and their uses. Publication date: 2000–05-04. (SmithKline Beecham Corp. (US)). Application no: WO1999US24653 19991020. Abstract: The invention provides multimeric FtsZ:FtsA polypeptides and polynucleotides encoding multimeric FtsZ:FtsA polypeptides and methods for producing such polypeptides by recombinant techniques. Also provided are methods for utilizing multimeric FtsZ:FtsA polypeptides to screen for antibacterial compounds

- FtsZ polypeptides from *Streptococcus pneumoniae*. Publication date: 1999–03-03. (SmithKline Beecham Corp. (US)). Application no. 98306077.3. Abstract: The invention provides ftsZ polypeptides and polynucleotides encoding ftsZ polypeptides and methods for producing such polypeptides by recombinant techniques. Also provided are methods for utilizing ftsZ polypeptides to screen for antibacterial compounds.
- Antibakterielles Mittel. Application date: 2001–04-30. (IBIS GmbH Bio-Innovationen). Application no. DE 101 21 145.7.
- Antibakterielles Mittel. Application date: 2001–06-21. (IBIS GmbH Bio-Innovationen). Application no. DE 101 29 870.6.
- Antibakterielles Mittel. Application date: 2002–04-22. (IBIS GmbH Bio-Innovationen). Application no. PCT/EP02/04410.

All three Antibakterielles Mittel applications relate to EF-Tu as a target for antibacterial agents, and specify the formula of such agents and the mode of their interaction with the target.

9 Outlook and Perspectives

Future research into cytoskeletons will concentrate on the acquisition of more detailed data on types of interactions and regulation of cytoskeletal elements with cytoskeleton-binding proteins and the implications thereof, in conjunction with the consequences of malfunction. A better understanding of these interactions and malfunctions will be important for treatment of diseases, especially cancer. Currently available drugs such as paclitaxel and derivatives thereof, and vincristine might be complemented by new compounds. These efforts may be supported by new data on properties of cytoskeletons in bacteria. Basic research in the field will provide further insight into the evolution of cytoskeletal elements. Clearly, knowledge of these structures will promote a better understanding of the very basic processes of function in cells, organs, and organisms.

The structural organization of bacterial cytoskeletons may allow the design and application of a new class of antibiotic drug, aimed at destabilizing the cytoskeleton and, hence, interfering with cell survival. First attempts have been made with FtsZ as a candidate target protein (see above). The name "Nanocillin®" has been introduced for drugs under design that are expected to interact with components of the basic bacterial cytoskeleton.

10
References

Alberts, B., Bray, D., Lewis, L., Raff, M., Roberts, K., Watson, J. D. (1983) *Molecular Biology of the Cell*, New York, London: Garland Publishing.

Antranikian, G., Herzberg, C., Mayer, F., Gottschalk, G. (1987) Changes in the cell envelope structure of *Clostridium* sp. strain EM1 during massive production of alpha-amylase and pullulanase, *FEMS Microbiol. Lett.* **41**, 193–197.

Beck, B. D., Arscott, P. G., Jacobson, A. (1978) Novel properties of bacterial elongation factor Tu, *Proc. Natl. Acad. Sci. USA* **75**, 1250–1254.

Biberfeld, G., Biberfeld, P. (1970) Ultrastructural features of *Mycoplasma pneumoniae*, *J. Bacteriol.* **102**, 855–861.

Burns, R. (1998) Synchronized division proteins, *Nature* **413**, 121–123.

Darnell, J., Lodish, H., Baltimore, D. (1990) *Molecular Cell Biology*, New York: Scientific American Books.

Doolittle, R. F. (1995) The origins and evolution of eukaryotic proteins, *Philos. Trans. R. Soc. Lond. Biol. Sci.* **349**, 235–240.

DuPraw, E. J. (1968) *Cell and Molecular Biology*, New York, London: Academic Press.

Erickson, H. P. (2001) Evolution in bacteria, *Nature* **413**, 30.

Erickson, H. P., Taylor, D. W., Taylor, K. A., Bramhill, D. (1996) Bacterial cell division protein FtsZ assembles into protofilament sheets and minirings, structural homologs of tubulin dimers, *Proc. Natl. Acad. Sci. USA* **93**, 519–523.

Göbel, U., Speth, V., Bredt, W. (1981) Filamentous structures in adherent *Mycoplasma pneumoniae* cells treated with nonionic detergents, *J. Cell Biol.* **91**, 537–543.

Hartman, H., Fedorov, A. (2002) The origin of the eukaryotic cell: a genomic investigation, *Proc. Natl. Acad. Sci. USA* **99**, 1420–1425.

Helms, M. K., Jameson, D. M. (1995) Polymerization of an *Escherichia coli* elongation factor Tu, *Arch. Biochem. Biophys.* **321**, 303–310.

Jeng, R. L., Welch, M. D. (2001) Cytoskeleton: actin and endocytosis – no longer the weakest link, *Curr. Biol.* **11**, R691–R694.

Jones, L. J. F., Carballido-Lopez, R., Errington, J. (2001) Control of cell shape in bacteria: helical, actin-like filaments in *Bacillus subtilis*, *Cell* **104**, 913–922.

Kiessling, J., Kruse, S., Rensing, S. A., Harter, K., Decker, E. L., Reski, R. (2000) Visualization of a cytoskeleton-like FtsZ network in chloroplasts, *J. Cell Biol.* **151**, 945–950.

Lodish, H., Baltimore, D., Berk, A., Zipurski, S. L., Masudaira, P., Darnell, J. (1996) *Molekulare Zellbiologie*, Berlin, New York: Walter de Gruyter.

Löwe, J., Amos, L. A. (1998) Crystal structure of the bacterial cell division protein FtsZ, *Nature* **391**, 203–206.

Lünsdorf, H., Schairer, H. U. (2001) Frozen motion of gliding bacteria outlines inherent features of the motility apparatus, *Microbiology* **147**, 939–947.

Mayer, F. (1986) *Cytology and Morphogenesis of Bacteria*, Berlin, Stuttgart: Gebrüder Bornträger.

Mayer, F., Vogt, B., Poc, C. (1998) Immunoelectron microscopic studies indicate the existence of a cell shape preserving cytoskeleton in prokaryotes, *Naturwissenschaften* **85**, 278–282.

Nanninga, N. (2001) Cytokinesis in prokaryotes and eukaryotes: common principles and different solutions, *Microbiol. Molec. Biol. Rev.* **65**, 319–333.

Palevitz, B. A. (2001) Deciphering protein evolution. *The Scientist* **15** [23]:18, November 26: e-mail article: http://www.the-scientist.com/yr2001/nov/palevitz p 18 011126.html.

Pantaloni, D., Le Clainche, C., Carlier, M.-F. (2001) Mechanism of actin-based motility, *Science* **292**, 1502–1506.

Pollack, G. H. (2001) *Cells, Gels and the Engines of Life,* Seattle: Ebner & Sons.

Regula, J. T., Boguth, G., Görg, A., Hegermann, J., Mayer, F., Frank, R., Herrmann, R. (2001) Defining the mycoplasma 'cytoskeleton': the protein composition of the Triton X-100 insoluble fraction of the bacterium *Mycoplasma pneumoniae* determined by 2-D gel electrophoresis and mass spectroscopy, *Microbiology* **147**, 1045–1057.

Rosenbaum, J. (2000) Cytoskeleton: functions of tubulin modifications at last, *Curr. Biol.* **10**, R801–R803.

Schilstra, M. J., Slot, J. W., van der Meide, P. H., Posthuma, G., Cremers, A. F., Bosch, L. (1996) Immunocytochemical localization of the elongation factor Tu in *E. coli* cells, *Biochim. Biophys. Acta* **1291**, 122–130.

Song, H., Parsons, M. R., Rowsell, S., Leonard, G., Phillips, S. E. V. (1999) Crystal structure of intact elongation factor EF-Tu from *Escherichia coli* in GDP conformation at 2.05 Å resolution, *J. Mol. Biol.* **285**, 1245–1256.

Trachtenberg, S., Gilad, R. (2001) A bacterial linear motor: cellular and molecular organization of the contractile cytoskeleton of the helical bacterium *Spiroplasma melliferum* BC3, *Molec. Microbiol.* **41**, 827–848.

van den Ent, F., Amos, L. A., Löwe, J. (2001). Prokaryotic origin of the actin skeleton, *Nature* **413**, 39–44.

Xia, L., Hai, B., Gao, Y., Burnette, D., Thazhath, R., Duan, J., Bre, M.-H., Levilliers, N., Gorovsky, M. A., Gaertig, J. (2000) Polyglycylation of tubulin is essential and affects cell motility and division in *Tetrahymena thermophila, J. Cell Biol.* **149**, 1097–1106.

13 Enzymes for Technical Applications

Thomas Schäfer[1], Ole Kirk[2], Torben Vedel Borchert[3], Claus Crone Fuglsang[4], Sven Pedersen[5], Sonja Salmon[6], Hans Sejr Olsen[7], Randy Deinhammer[8], Henrik Lund[9]

[1] Novozymes A/S,Krogshojvej 36, 2880 Bagsvaerd, Denmark; Tel.: +45-4442-6444; Fax: +45-4442-7828; E-mail: TSch@novozymes.com

[2] Novozymes A/S,Krogshojvej 36, 2880 Bagsvaerd, Denmark; Tel.: +45-4442-3206; Fax: +45-4442-1999; E-mail: Oki@novozymes.com

[3] Novozymes A/S,Krogshojvej 36, 2880 Bagsvaerd, Denmark; Tel.: +45-4442-6977; Fax: +45-4444-0246; E-mail: TVB@novozymes.com

[4] Novozymes A/S,Krogshojvej 36, 2880 Bagsvaerd, Denmark; Tel.: +45-4442-1406; Fax: +45-4444-4096; E-mail: CCF@novozymes.com

[5] Novozymes A/S,Krogshojvej 36, 2880 Bagsvaerd, Denmark; Tel.: +45-4442-2239; Fax: +45-4442-1237; E-mail: SvP@novozymes.com

[6] Novozymes North America, Inc., 77, Perry Chapel Church Road, Franklinton, NC 27525, USA; Tel.: +919-494-3000; E-mail: SiSa@novozymes.com

[7] Novozymes A/S,Krogshojvej 36, 2880 Bagsvaerd, Denmark; Tel.: +45-4442-2045; Fax: +45-4442-1237; E-mail: HSO@novozymes.com

[8] Novozymes A/S,Krogshojvej 36, 2880 Bagsvaerd, Denmark; Tel.: +45-4442-1810; Fax: +45-4442-7181; E-mail: RDe@novozymes.com

[9] Novozymes A/S,Krogshojvej 36, 2880 Bagsvaerd, Denmark; Tel.: +45-444-8502; Fax: +45-4442-6645; E-mail: HLu@novozymes.com

6-APA	6-aminopenicillin acid
CBD	cellulose binding domain
CBH	cellobiohydrolase
CGTase	cyclodextrin glycosyltransferase
CMC	carboxymethyl cellulose
EPA	Environmental Protection Agency
ETBE	ethyl tertiary butyl ether
HFCS	high-fructose corn syrup
L-DOPA	L-dioxyphenylalanine
MOW	mixed office waste
MTBE	methyl tertiary butyl ether
NMR	nuclear magnetic resonance
NREL	National Renewable Energy Laboratory

OCC	old corrugated containers
ONP	old newsprint
PAA	polyacrylic acid
PVA	polyvinyl alcohol
SGW	stone groundwood
WM	waste magazines

1 Introduction

Enzymes are major contributors to clean industrial products and processes (Bull et al., 1999). They show a variety of advantages over chemicals, e.g., their specificity, their high efficiency and their compatibility with the environment. Enzymes can be produced from renewable resources and are in turn degraded by microbes in nature. Various industries have replaced old processes using chemicals that cause detrimental effects on the environment and equipment with new processes that use biodegradable enzymes under less corrosive conditions.

Currently, industrial enzymes are manufactured by three major suppliers, Novozymes A/S (headquartered in Denmark), Genencor International Inc. (headquartered in the U.S.), and DSM N.V (headquartered in the Netherlands). Their main market segments are food (e.g., dairy, baking, brewing, beverage), animal feed, and technical applications. Novozymes A/S is the largest supplier in each of these three sectors, with an estimated market share between 41% and 44% of the industrial enzyme market in 1999. Genencor International Inc., which operates in the technical and feed segments, and DSM N.V., which focuses on food and feed, had, according to estimates by Novozymes A/S, market shares of around 21% and 8%, respectively, that year. The rest of the market is divided among a few smaller enzyme producers, some of which produce enzymes for their own use, in the U.S., Canada, Europe, and Japan, as well as a number of small local producers in China.

The applications in which enzymes are used are many and diverse. So far, technical enzymes represent the largest part of the market, with a value of approximately 1 billion USD in 1999. Enzymes for detergent are the largest single market for enzymes, with a value of around USD 0.5 billion. The other dominating markets are baking, beverage, and dairy; as well as feed and pulp and paper applications. All of these industries are traditional users of enzymes. Overall, the estimated value of the worldwide use of industrial enzymes has grown from 1 billion USD (Godfrey and West, 1996) in 1995 to 1.5 billion USD in 2000 (McCoy, 2000a).

In the following sections, we review enzymes for technical applications and briefly describe key technologies for discovery and optimization of industrial enzymes.

2 Historical Outline

Enzymes have been exploited by humans for centuries. Classical foods and beverages like cheese, yogurt and kefir, bread, beer, vinegar, wine, and other fermented drinks, as well as paper and textiles, were produced with the help of enzymes as early as 6000 B.C. in China, Sumer, and Egypt. The epoch of classical biotechnology was marked by the landmark discoveries of Leeuwenhook, who observed and described microbes; Pasteur, who defined fermentation as a biological process; Buchner, who discovered that enzymes are proteins that function as catalysts; and Sumner, who crystallized the first enzyme. The modern era of industrial enzymology began in 1913 when Otto Röhm was granted a patent for the use of a crude protease mixture isolated from pancreases in laundry detergents. In the following years, an increasing number of enzymes were found in microorganisms, and these microbes were cultured in large-scale fermentations to produce enzymes. However, the number of enzymes that could be produced in this fashion was limited, because not all microbes are amenable to large-scale fermentation. The pioneering work of Avery and MacLeod, Hershey and Chase, Watson and Crick, Cohen and Boyer, and many others who introduced the era of recombinant biotechnology revolutionized industrial enzyme production. With the advent of genetic engineering, genes encoding interesting enzymes could be transferred to and expressed in selected host microbes for industrial-scale production. Today, gene technology plays a major role in both the discovery of novel enzymes and the optimization of existing proteins and is the basis for production of the majority of industrial enzymes.

3 Enzymes for the Detergent Industry

The use of enzymes as performance enhancers in detergents is arguably the biggest innovation of the detergent industry during the past 20 years. Enzymes have found broad-based application in a variety of commercially relevant areas.

3.1 Introduction

One of the most important and profitable applications for enzymes is in detergents, where the total global market size was ~ 0.6 billion USD in 2000 (Novozymes data). The breadth of commercial detergent products is quite large, and they are used in such diverse applications as laundering, dishwashing, and in industrial and institutional cleaning (Eriksen, 1996; Olsen and Falholt, 1998). There, they have shown great utility in providing noticeable improvements in the appearance of a particular item of value to the consumer, such as clothes or dishes. Penetration of enzymes into world markets is high, with nearly complete penetration into North American, European, and Japanese markets and somewhat lower, but increasing, penetration into developing markets such as China and Latin America (Showell, 1999). The main enzyme manufacturers are Novozymes A/S, headquartered in Denmark, and Genencor International, headquartered in the U.S.

To provide desirable benefits, enzymes must be stable and function well in the presence of a variety of potentially “enzyme unfriendly” detergent ingredients (e.g., anionic/nonionic/cationic surfactants, chelants, builders, polymers, bleaches) and in various forms of detergent products (i.e., liquids and powders). In addition, many detergent manufacturers, to simplify production processes and to lower costs, have been working toward producing “globally applicable” detergents from which a single enzyme would be expected to provide its benefits across a range of global wash conditions (i.e., water hardness, presence of transition metals, chlorine) and wash practices (i.e., type of washing machine, wash temperature).

It seems clear from the growth of the detergent enzyme business that enzymes have been found or engineered to meet these challenges. In many parts of the world, particularly in Europe and in North America, the environmental impact of detergents has been disputed (Ho Tan Tai and Rataj, 2001). This concern has influenced many detergent makers to improve the environmental compatibility of their detergents. In this vein, enzymes can be considered "ideal ingredients" in detergents of the future, since their catalytic nature allows them to provide good benefits at much lower dosage levels compared to conventional detergent ingredients such as surfactants, bleaches, and polymers. In addition, enzymes are completely biodegradable and therefore do not accumulate in the environment.

3.2 History

The first use of enzymes in detergents occurred in 1913 when Rohm & Haas introduced crude trypsin into their detergent Burnus® based on a German patent issued to Otto Röhm (1913). Issues with the performance and stability of this enzyme in their detergent, which was designed for use in a laundry presoak context to remove biological stains such as blood, did not excite consumers at the time about the potential of enzymes. It was not until Novo Industri A/S in Denmark introduced the protease Alcalase® in 1963, together with small detergent producers in Switzerland and the Netherlands, that the benefits of using enzymes in detergents became noticed. Until the 1980s, proteases were considered to be the only commercially relevant enzymes. Amylases, lipases, and cellulases were then developed, and the market began to grow substantially.

Today, many laundry-detergent products contain at least a protease, and many contain cocktails of enzymes including proteases, amylases, cellulases, and lipases. With new innovations in the areas of fermentation-process optimization and genetic modification of host organisms continually being made, as well as an increased emphasis on the environmental impact of detergents, the enzyme market will likely continue to grow in coming years.

3.3 Overview of Enzymes

Essentially all of the enzymes found in today's detergents are hydrolytic in that they catalyze the hydrolysis of chemical bonds present within a polymeric substrate. Most commonly, these enzymes act in an "endo-" manner, meaning that the hydrolysis occurs randomly in the interior of the polymer. General descriptions of the commonly used hydrolases such as proteases, amylases, lipases, and cellulases are given below. Because of space considerations, mention is not made of methods for assessing the activity and performance of these enzymes. Discussions of these items can be found elsewhere (Eriksen, 1996; Showell, 1999; Olsen and Folholt, 1998). As discussed in the "Latest Innovations" section, mannanases, which were introduced into the detergent market in 2000 based on a collaboration between Procter & Gamble and Novozymes, have been shown to provide exciting new types of claimable benefits such as the prevention of "reappearing stains" during washing. Finally, in the "Future Perspectives" section, oxidoreductases for stain bleaching and approaches for enzymatically generated fabric-care benefits are discussed as new and exciting frontiers for detergent enzymes.

3.3.1 Proteases

Proteases are hydrolases that catalyze the hydrolysis of amide bonds within proteinaceous substrates that are present in soils. Stains such as blood, grass, spinach, and keratin from collar and cuff soil are most relevant for laundry applications, whereas baked-on egg soils are of interest for dishwashing applications. Proteases also are used for cleaning membranes and endoscopes in the industrial and institutional area (Eriksen, 1996). These enzymes are by far the most commonly used types in detergents. Hydrolysis breaks the proteinaceous substrates down into smaller fragments (i.e., amino acids or oligopeptides), thereby increasing the ease with which the soils can be solubilized in the wash liquor by surfactants and the like. Proteases also help to prevent the redeposition of proteins on fabrics, particularly hydrophobic ones present in soils, such as blood, thereby also providing a whiteness benefit (Venegas, 1997). Figure 1 illustrates an example of the whiteness benefit that can be provided by proteases, as visualized through the redeposition of particulate soil following initial washing in detergent both with and without protease.

Nearly all commercial proteases are so-called serine proteases originating from the *Bacillus* family and contain the catalytic triad of amino acids (i.e., aspartic acid – histidine – serine) in their active sites. They are also unspecific endoproteases, meaning that they cleave peptide bonds within proteins in an unspecific way, leading to complex reaction product mixtures of oligopeptides. Given these similarities, commercial detergent proteases mainly differ in their temperature and pH optima, bleach sensitivity, and dependence on Ca and Mg ion concentration for stability. Tuning of these parameters through structural changes in the enzymes has resulted in a range of proteases that are suited to different types of tasks, i.e., improved stability in bleach-containing granulated detergents or egg-stain removal in an auto-dishwashing detergent.

Given that the function of proteases is to break down proteins, autoproteolysis and compatibility with other detergent enzymes are of concern during the formulation and storage of detergents, particularly those in the liquid form. This issue has been addressed by adding reversible protease inhibitors such as boric acid and propylene glycol to the detergent (Showell, 1999). Upon dilution of the detergent in the wash, the

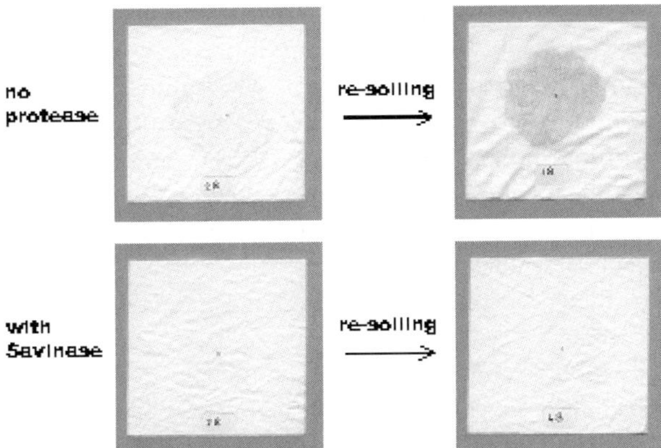

Fig. 1 Images of blood spots taken after washing in detergent with and without Savinase® and after additional particulate soil treatment.

inhibitors are released from the enzyme active site, thereby allowing the enzyme to perform its function.

3.3.2 Amylases

Amylases are also hydrolases that catalyze the hydrolysis of glucosidic linkages in gelatinized starch polymers. Starch polymers are commonly found in foods such as pasta, fruit, chocolate, baby food, barbeque sauce, and gravy. As colored stains, their removal is of interest in both detergent and dishwashing contexts. Removal of starch from surfaces is also important in providing a whiteness benefit, since it is known that starch can be an attractant for many types of particulate soils (Ryom and Gibson, 2001). The most common class of detergent amylases is the α-amylases, which hydrolyze the 1,4-α-glucosidic bonds in starch. On some food stains, such as those from cocoa and barbeque sauce, which contain both starch and proteinaceous soils, synergistic benefits can be seen between amylases and proteases (Showell, 1999; Gormsen, 1997). Most commercial amylases derive from either the *Bacillus* or *Aspergillus* genera and typically consist of three different domains, with the active site existing between two of the domains.

3.3.3 Lipases

Lipases are a fairly new addition to the commercial detergent market. The first commercially successful lipase was introduced by Novo Nordisk A/S in 1988 under the trade name of Lipolase®, which originated from the fungus *Humicola lanuginosa*. Lipases break down triglycerides into their component glycerol and fatty acid units, thereby increasing their water solubility, particularly at a pH > 8 (Aaslyng et al., 1991). Owing to their hydrophobicity, these stains are among the most difficult to remove via the conventional surfactant technology commonly found in detergents (Showell, 1999). As a result, stain-removal benefits are observed on greasy/fatty stains such as lard, butter, and lipstick and on body soils such as sebum. Enzymatic removal of these soils also can generate a whiteness benefit on hydrophobic fabrics such as polyester and polyester/cotton blends, which tend to bind strongly to the unhydrolyzed soils. On cotton, oily soils can also penetrate into the lumen, and lipases can greatly aid in the removal of these soils as well (Obendorf et al., 2001).

In addition to their tenacity as stains, greasy/oily stains can, over time, undergo oxidation to form unsaturated compounds, which can impart a rancid odor and a yellowish color to fabrics (Showell, 1999). Lipases can help to remove these soils and improve the odor and appearance of fabrics.

Binding to the water-substrate interface is critical for lipase action. As one might envision, other surface-active detergent components such as surfactants can show an inhibitory effect on lipases through a blocking mechanism (Svendsen, 1997). This problem can be reduced through judicious choice of a surfactant system that contains the optimal ratio of anionic to nonionic surfactants.

Traditionally, lipase benefits have not been observed until after multiple wash cycles, and a drying step was required to show activity (Eriksen, 1996). Recent work has shown that lipases can be developed that show good benefits in the first wash cycle, as shown in Figure 2 (Callisen and Damhus, 2000). When using these lipases, a high degree of soil removal is obtained in the first wash cycle. This removal can facilitate the removal of other soil substances from the fabric by surfactants, enzymes, etc., thereby improving the overall fabric appearance.

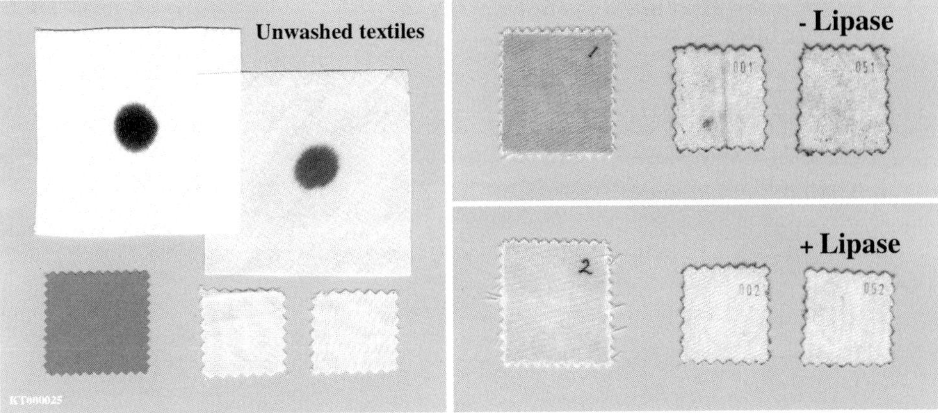

Fig. 2 An example of in-the-wash benefits from new first-wash lipases: high stain removal of a lipstick (reddish swatches) and reduced redeposition of carbon black from soiled swatches containing dirty motor oil (large swatches in left picture) onto polyester tracer swatches (small whitish swatches). Tergitometer wash under EU conditions with or without a first-wash lipase.

3.3.4
Cellulases

The use of cellulases in detergents began in the late 1980s when Kao introduced an alkaline cellulase into their Attack® detergent. In contrast to the enzyme classes discussed above, the main function of cellulases in detergents is to hydrolyze the cellulose in cotton and polycotton fabrics to provide cleaning and fabric-care benefits. Cleaning effects arise from the ability of cellulases to remove particulate and oily soils. In contrast, fabric-care effects arise from the ability of cellulases to provide anti-pilling, softness, whiteness, and color clarification benefits (Gormsen, 1997). Cellulases can be either endo- or exo- and function by catalyzing the cleavage of β-1,4-glycosidic bonds in cellulose. Commercial cellulases come from both bacterial and fungal sources. They are composed of up to three different domains, including the catalytic domain, the linker, which is usually a short peptide, and a cellulose-binding domain (CBD). The CBD typically contains a hydrophobic region that affords cellulase affinity to cotton. This binding is critical for observing fabric-care benefits from these enzymes (Klyosov, 1990). As with lipases, the enzymatic mechanism requires the enzyme to first bind to the cellulose surface, followed by glycosidic bond hydrolysis.

To observe fabric-care benefits from cellulases, washing tests typically consist of multiple washing and drying cycles, since fabric decolorization and pilling increase as the garments are mechanically worn. Typically, multi-cycle tests consisting of 10 or more wash/dry cycles are needed to observe benefits, as illustrated in Figure 3.

Since cellulases act to degrade cellulose, the dosage of enzymes in the detergent must be chosen carefully to allow for fabric-care benefits to be observed without causing structural damage to the garments over their lifetime. To address this, tensile-strength-loss tests are often conducted on sensitive fabrics as a check for fabric damage. In addition to modifying dosage, tensile strength loss can be minimized by choosing a cellulase that preferentially degrades amorphous versus crystalline cel-

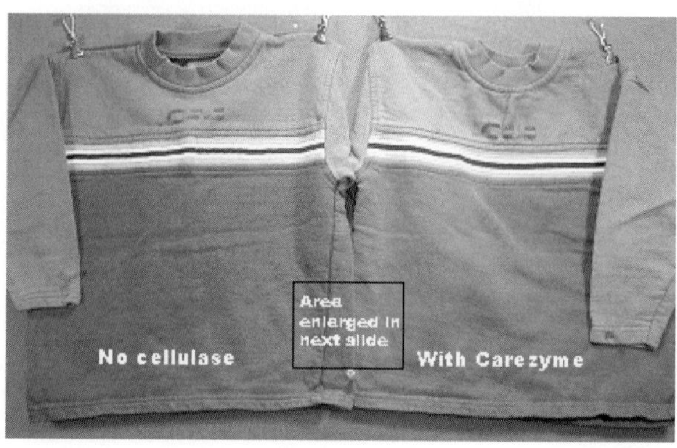

Fig. 3 Image showing fabric-care anti-pilling effects after multi-cycle washing in European detergent, with and without the cellulase Carezyme®.

lulose (Lenting and Warmoeskerken, 2001). Weak, surface-based fibers are thought to consist of amorphous cellulose, and their presence is thought to give rise to much of the decolorization and pilling observed on worn cotton. In contrast, crystalline cellulose resides in the interior of the cotton, and damage to it is thought to be largely responsible for tensile strength losses.

3.4 Latest Innovations

Sometimes, the true benefit of a new detergent enzyme to the consumer may not be apparent during the early stages of its development, possibly because observation of the benefit relies on a complex set of factors that are sometimes not part of standard laboratory wash-testing protocols. This "true" benefit may be realized only after testing the effect of the new enzyme in consumers' homes as part of a consumer test. The example below illustrates such an enzyme and shows how capitalizing on such performance benefits can lead to marketplace success.

3.4.1 Mannanases

A particularly exciting new development in the area of detergent enzymes was the introduction of Mannaway® in Procter & Gamble's Tide Deep Clean® liquid detergent formula in 2000. This enzyme, which was developed in collaboration with Novozymes, addresses a key cleaning problem commonly seen in consumers' laundry. This problem revolves around the tendency of certain polysaccharides such as guar gum and starch to strongly bind particulate soils. Guar gum is used commonly as a thickener in foods, and its cationic variant is also used as a softener and as a secondary active delivery agent in many personal care products. Owing to its very high molecular weight, it exhibits a strong adherence to fabric. Stains containing colorless guar gum can appear removed in the wash, but often the guar gum is left behind. Particulates arising from wear of the garment or coming from a second wash can bind to the gum and show up as stains. Mannaway®, which originates from *Bacillus*, was shown to effectively cleave the β-1,4-linkages between mannose units in guar, thereby dramatically reducing the "reappearing stain" phenomenon.

Figure 4 shows an example of the effectiveness of Mannaway® at removing guar gum from fabric. The images were taken by confocal fluorescence microscopy by flowing a solution of liquid Tide® containing Mannaway® over a piece of cotton textile stained with fluorescently labeled guar gum. The bright spots correspond to the guar gum. After the wash, the guar gum is almost completely removed.

In the laundry context, Mannaway® has been shown to provide a broad stain-removal profile directly on stains containing guar gum, as well as whiteness benefits on other items, as a result of its ability to effectively degrade guar gum and prevent its redeposition. As alluded to above, the broad benefit profile is likely due to the pervasiveness of guar gum use in the food and personal care industries. Figure 5 shows an example of how Mannaway® can significantly improve the removal of a range of stains when using a liquid laundry detergent.

3.5 Future Perspectives

As detergent manufacturers continually work to grow sustainable market share, they often look to develop new products or product forms that provide clear, claimable benefits to consumers and that cannot be easily copied by competitors, i.e., a liquid detergent with bleach or improved fabric

New Liquid Tide (w/Mannanase)
Guar Gum
25X
Initial
Final

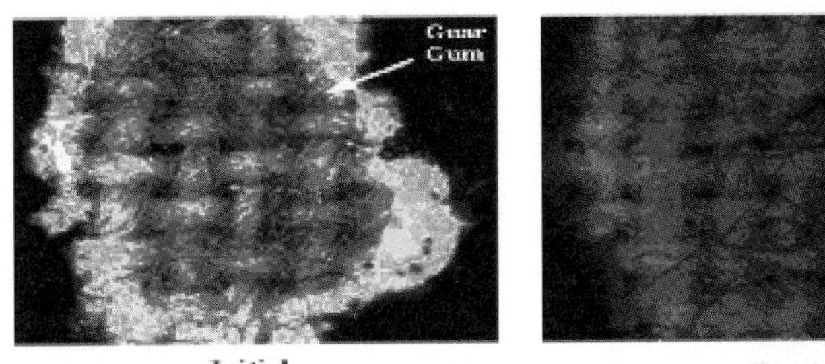

Fig. 4 Confocal fluorescence microscopy images showing the removal of fluorescently-labeled guar from cotton by Liquid Tide® containing Mannaway®. Figure kindly provided by Proctor & Gamble.

Performance of Mannanase in Laundry Application
% Stain Removal
100
95
90
85
80
75
70
Control
w/Mannanase
Ice Cream
BBQ Sauce
Fudge Bars
Teriyaki BBQ
Chocolate

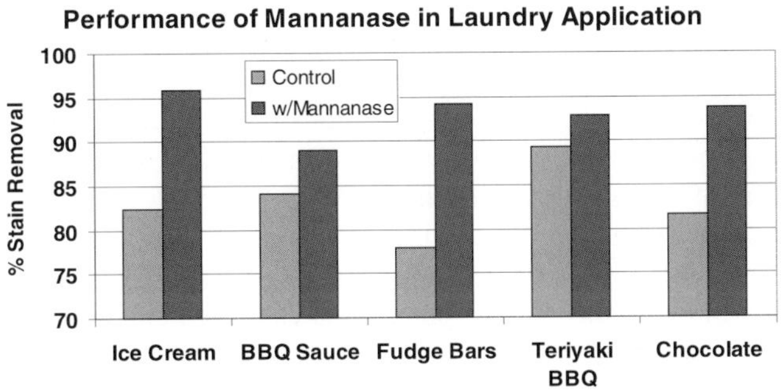

Fig. 5 Stain-removal performance of Mannaway® on a range of different guar-containing stains. Figure kindly provided by Proctor & Gamble.

care in the wash cycle. The two sections below briefly discuss two such groups of enzymes.

3.5.1 Oxidoreductases

Investigating the potential use of oxidoreductases in detergents is a relatively new and exciting area of research. Such enzymes catalyze oxidation/reduction reactions and could conceivably provide benefits in detergents such as bleaching of stains on fabrics or tableware, reduction of general dinginess on fabrics, bleaching of dyes removed from fabrics for color-care benefits, and disinfection (Gormsen, 1997). There has been a proliferation of patents appearing in the literature in recent years that describe the identification of enzymes for these applications (Davis et al., 2001; Antheunisse et al., 2001; Bodie et al., 2000). In the stain- and dinginess-bleaching areas, for example, oxidoreductases have potential to be a more attractive ingredient than conventional non-enzymatic bleaching technology owing to (1) the possibility of formulating them into liquid detergents that, as of now, cannot be formulated with traditional bleach chemistry because of incompatibility issues; and (2) their possible better performance at lower wash temperatures (Gormsen, 1997). Unfortunately, no such enzymes have yet appeared in the marketplace.

3.5.2 Fabric-care Enzymes

One area that has recently gained much attention is that of enhanced fabric care, i.e., providing benefits beyond those provided by the cellulases of today, such as anti-wrinkling benefits; dye-transfer inhibition; odor removal; anti-static, anti-wear, and anti-shrinkage benefits; and enhanced softness. Enzymes may be a particularly attractive option for providing these benefits because of their ability to carry out chemical reactions directly on the fabric surface or in solution to effect desirable changes.

Several documents have appeared recently in the patent literature that discuss enzymes such as transferases, which can chemically bind sugar moieties to cellulose fabrics. The inventors claim that such reactions lead to anti-wrinkling and fabric anti-wear benefits on cotton fabrics (Smets et al., 1999a). Other patents in the field discuss linking various enzymes such as proteases, cellulases, and amylases to CBDs to enhance their delivery to the cotton fabric surface, thereby enhancing their stain-removal and fabric-care benefits (Smets et al., 1999b; Busch et al., 1999). As detergent manufacturers strive to become involved in every stage of the garment-production, wearing, and cleaning cycle, the industry will likely continue to work toward finding uses for such enzymes in the fabric-care arena.

4 Enzymes for the Starch Industry

Enzymes for this industry cover a variety of hydrolases that act mainly on starch or starch-derived saccharides, transferases (CGTases), and isomerases. Starches are converted to a variety of products, high-fructose corn syrup being the most important product.

4.1 Introduction

Starch is the second most available carbohydrate source (after cellulose) on earth and is found in crops such as rice, wheat, potatoes, corn, and tapioca. The total world production in 1998 was 37 million tonnes, of which more than 70% came from corn.

Starch consists of chains of glucose molecules, which are linked together by $\alpha(1>4)$ and $\alpha(1>6)$ glycosi-dic bonds. Industrial starch can be used as native starch, as modified starch, or as starch hydrolysates. Examples of important enzymatically modified starch are shown in Figure 6.

By far, the most important industrial enzyme process is the conversion of starch to high-fructose corn syrup (HFCS). After a steeping process in which the corn kernel is separated into its components (starch, gluten, germ oil, and fiber), the production of HFCS from the starch fraction comprises three enzymatic process steps: liquefaction (α-amylase), saccharification (glucoamylase and a debranching enzyme), and isomerization (glucose isomerase) (Pedersen, 1993).

Depending on the maltose content, maltose syrups are made (after liquefaction to 10 dextrose equivalent [DE] with α-amylase) by fungal α-amylase (50–55% maltose); by fungal α-amylase, maltogenic amylase (Outtrup and Norman, 1984), and pullulanase (55–65%); by β-amylase and pullulanase (70–80%); or by β-amylase, maltogenic amylase, and pullulanase (> 80%), all at 30% DS. Properties and production methods of high maltose syrups, maltodextrins, and oligosaccharides are described in Okada and Nakakuki (1992).

Confectionary syrups and specialty syrups (brewing syrups) are made from liquefied starch (thinned by acid treatment or α-amylase), followed by treatment with fungal amylase or combinations of fungal amylase and glucoamylase.

Cyclodextrins are oligosaccharides with a closed ring structure in which the glucose units are joined together by α-1,4 linkages. Cyclodextrins containing 6, 7, or 8 glucose units are most common and are known as α-, β-, and γ-CD, respectively (Pedersen et al., 1995; Hedges, 1992). Cyclodextrins are made from starch by the enzyme cyclodextrin glycosyltransferase (CGTase). When acceptors such as sucrose are present in the reaction mixture in addition to starch, the CGTase will catalyze transfer of glucosyl residues from starch to sucrose. This product is known as "coupling sugar" (Okada, 1988).

Trehalose is a non-reducing oligosaccharide with two glucose molecules linked by an α-1,1 linkage. It is produced from starch by two enzymes: maltooligosyl trehalose synthase (MTSase) and maltooligosyl trehalose trehalohydrolase (Mukai et al., 1997).

A number of other enzymatic reactions on starch are described in the literature but so far have no or limited industrial use. Examples are debranching (heat-stable pullulanase) of starch at high temperature for preparation of modified starches (e.g., fat replacers) (Rüdiger et al., 1995; Chiu, 1991b); the use of branching enzymes for

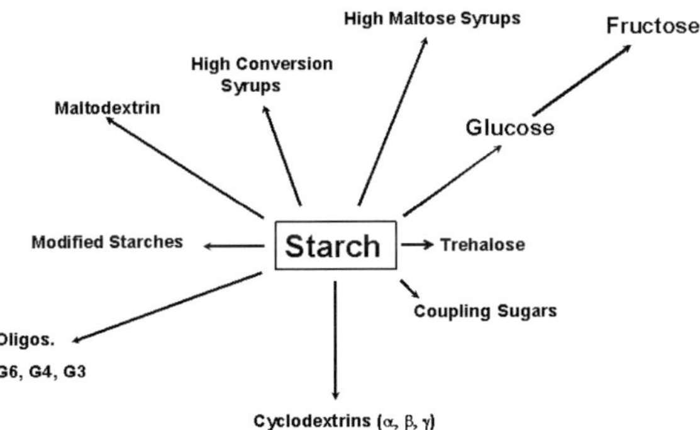

Fig. 6 Examples of enzymatic conversions of starch.

preparation of specialty starches with high solubility (Yanase et al., 1995; Bruinenberg et al., 1995); production of thermoreversible gels by amylomaltase (Euvering and Binnema, 1997); and preparation of starch emulsifiers (Chiu, 1991a).

4.2 History

The use of hydrochloric acid for hydrolysis of starch has been known since the Napoleon era (discovered by the Russian chemist Kirchhoff). Acid hydrolysis became a well-known and proven technology for the starch industry, and it was not before the end of the 1950s that the American starch industry started to look into the use of enzymes, despite the fact that the use of enzymes for starch hydrolysis had been known for more than 100 years at the time (Buchholz and Poulsen, 2000).

Bacterial amylases were introduced commercially in the early 1950s (from *B. subtilis*), and the American company Miles introduced the first glucoamylase, Diazyme, in the late 1950s. It was, however, not before the development of a bacterial amylase that could work at temperatures above 100°C that the industrial application of enzymes in the starch industry really took off. The introduction of such a heat-stable enzyme in the starch industry was a revolution and allowed the industry to carry out liquefaction in a one-step process (see Table 1).

Market development reflects the gradual acceptance of HFCS as a substitute for sucrose by soft drink producers in the United States (see Table 2).

Tab. 1 Highlights in enzyme developments in the starch industry

Event	*Reference*
The first heat-stable bacterial amylase is introduced	Madsen et al., 1973
Sweetzyme – A new immobilized glucose isomerase is introduced	Zittan et al., 1975
A novel debranching enzyme for application in the glucose syrup industry is introduced	Norman, 1982

Tab. 2 Highlights in the development of the glucose isomerization process

Event	*Reference/year*
A glucose isomerase is described for the first time in the literature	Marshall and Kooi, 1957
A partially reusable immobilized glucose isomerase is developed in Japan	Takasaki et al., 1969
Clinton Corn Orocessing Company, a division of Standard Branch, Inc., produces the first commercial fructose corn syrup in the United States by the immobilized enzyme developed by Takasaki	Takasaki et al., 1969
The first continuous isomerization process is developed at Clinton Corn Processing Company	Lloyd et al., 1972
An immobilized glucose isomerase that can be used in industrial fixed-bed reactors is developed	Zittan et al., 1975
A 55% enriched HFCS (EFCS) becomes available on an industrial scale as a result of the development of a new fractionation technology involving chromatographic separation of fructose from glucose by the use of a strong acid cation-exchange resin	Biezer and de Ressett, 1977
Coca Cola and Pepsi Cola approve 100% substitution of sucrose by EFCS in the United States	1984

4.3
Enzymes

Only enzymes used in the HFCS process or in the production of high maltose syrups will be described here.

4.3.1
Bacterial α-amylases

α-Amylases catalyze the hydrolysis of internal α-1,4 glycosidic linkages by which the starch molecule is cut into smaller fragments. This reduces the viscosity, which is necessary for further processing. The degree of hydrolysis is measured by the dextrose equivalent (DE), a measure of the reducing power relative to glucose. DE after liquefaction is typically 8–12 (8–12% of the reducing power of glucose). Industrially important bacterial α-amylases for starch processing are derived from *B. licheniformis*, *B. stearothermophilus*, and *B. amyloliquefaciens* (see Table 3, which includes fungal α-amylase).

4.3.2
Fungal α-amylases

This enzyme hydrolyses α-1,4 glycosidic linkages. Gelatinized starch is broken down to dextrins and oligosaccharides. Prolonged reaction results in formation of maltotriose and large amounts of maltose. The enzyme is therefore used in the production of high maltose syrups.

4.3.3
Glucoamylase

Industrial production of this exo-amylase is based on fermentation of *A. niger*. During the hydrolysis of starch, glucose units are removed in a stepwise manner from the non-reducing end of the molecule. The rate of hydrolysis depends on the chain length, maltotriose and, in particular, maltose being hydrolyzed at a lower rate than higher oligosaccharides. Furthermore, α-1,6 linkages are broken down more slowly than α-1,4 linkages, but, eventually, almost complete conversion of starch into glucose is possible. Application parameters are 60–65°C and pH 4–5.

4.3.4
Pullulanase

Industrially used pullulanases are heat-stable enzymes that act simultaneously with glucoamylase during saccharification (typically 60°C, pH 4.3–4.5). Pullulanases catalyze the hydrolysis of the α-1,6 linkages in amylopectin, especially in partially hydrolyzed amylopectin. This enables the enzyme to act specifically on the branched oligosaccharides obtained from starch, provided there are at least two D-glucose units in the side chain.

4.3.5
β-amylase

The prefix β- does not refer to the substrate, but to the reaction product maltose, which is released in its β- form. β-Amylase is an exo-amylase that catalyzes the release of maltose

Tab. 3 Industrially important α-amylases

Enzyme	*Microorganism*	*Application temperature (°C)*	*Application pH*
Bacterial mesophilic α-amylase	*B. amyloliquefaciens*	80–85	6–7
Bacterial thermophilic α-amylase	*B. licheniformis*	95–105	6–7
Bacterial thermophilic α-amylase	*B. stearothermophilus*	95–105	5.8–7.0
Fungal α-amylase	*A. oryzae*	55–60	4–5

units from the non-reducing end of the starch molecule. The enzyme is not able to hydrolyze maltotriose and therefore produces maltose syrups with a very low glucose content. β-Amylase has no α-1,6-hydrolyzing activity, and when it acts on amylopectin without pullulanase present, a limit dextrin will be produced. Industrially used β-amylase is extracted from barley.

4.3.6
Glucose Isomerase

Glucose isomerase is used as an immobilized enzyme. The following are reasons for immobilization:

- Immobilization allows for a minimization of reaction time in order to prevent degradation, especially of fructose, to organic acids and carbonyl compounds.
- Glucose isomerase is an expensive enzyme because of its high K_M (it is actually a xylose isomerase with only low activity towards glucose).
- Glucose isomerase is an intracellular enzyme, giving a natural limit to the amount of enzyme that can be formed within the cell.

Glucose isomerase converts glucose to fructose. The equilibrium conversion under industrial conditions is 50%, making chromatographic separation necessary in order to obtain the industrial product 55% fructose, which matches the sweetness of sucrose.

A description of the most commonly used glucose isomerases is given in Table 4. The description is based on literature and information from the manufacturers and is not necessarily a description of the exact methods used.

4.4
Latest Innovations

As seen in Table 5, pH varies considerably during the HFCS process (in addition, the steeping process is carried out at pH 4.0–4.5), and the direction in research and development has been to lower pH during liquefaction. Because of modern protein-engineering techniques, this reduction has been highly successful: during the past five years, the pH value during the liquefaction process has been reduced by a whole pH unit (pH 6.0 to 5.0) (Bisgaard-Frantzen et al., 1999; Nielsen and Borchert, 2000; Novozymes A/S, 2001a).

Previously, the α-amylase had to be inactivated before saccharification (heat treatment at pH 3.8–3.9 at 95°C for 5–10 minutes). This was necessary in order to prevent formation of large amounts of panose. Again, through modern protein-

Tab. 4 Examples of commercially immobilized glucose isomerases

Manufacturer	*Trade name*	*Enzyme source*	*Immobilization method*
Novozymes	Sweetzyme IT	*S. murinus*	Cross-linking of cell material with glutaraldehyde, extruded
Genencor International	GENSWEET	*S. rubigonosus*	The enzyme is cross- linked with or without cellular debris using PEI and glutaraldehyde Granular particles are formed by extrusion/ marumerization
Godo Shusei	AGI-S-600	*S. griseofuseus*	Chitosan-treated glutaraldehyde cross-linked cells, granulated
UOP	Ketomax 100	*S. olivochromogenes*	PEI (polyethylenimine)-treated ceramic alumina with glutaraldehyde cross-linked, purified GI

Tab. 5 Typical process conditions for production of HFCS from starch

Process	*Temperature (°C)*	*Dry substance content (%)*	*pH*	*Process time (h)*
Jet cooking/ dextrinization	105/95	30–35	5.1–5.8	0.1/1–2
Saccharification	60	30–35	4.3–4.5	40–60
Isomerization	50–60	40–50	7–8	0.3–3.0

engineering techniques, it has been possible to change the specificity of the bacterial α-amylase so that it does not form panose when active during saccharification (Novozymes A/S, 2001a).

In saccharification, glucoamylase products with a higher content of acid fungal α-amylase have been introduced. *A. niger* produces this enzyme together with glucoamylase, but the ratio of the two enzymes has been changed by genetic engineering techniques. The improved enzyme blend gives a faster initial saccharification rate as well as a higher glucose content of the final syrup (Novozymes A/S, 2001b).

Considerable efforts have been used in developing enzymatic steeping methods with the purpose of reducing both steeping time and use of SO_2 (Johnston and Singh, 2001). These methods appear promising but have not yet been applied in industrial practice.

4.5 Perspectives

Because of the increased use of fermentation processes for the production of industrial chemicals, the perspectives of increased use of starch are very good. In addition to fuel alcohol, the fermentation of organic acids (especially lactic acid) is expected to increase rapidly. Whether the increased use of starch will be based on dry milling, as in the case of fuel alcohol, or wet milling is difficult to predict and will be dependent on the future value of the byproducts and on the developments in the recovery processes after fermentation.

Increased use of starch for specialty starches with new functional properties and for oligosaccharides with nutritional properties will continue with the methods described in the introduction.

5 Enzymes for Biofuel

The category of biofuels includes biodiesel, bioethanol, and biomethanol. Main starting materials are starch- and cellulose-containing materials, indicating that amylolytic and cellulytic enzymes play a major role in the corresponding processes that are described in the following sections.

5.1 Introduction

Biodiesel is a methyl or ethyl ester derived from vegetable oils, e.g., rapeseed. It can replace diesel entirely or it can be mixed with diesel in different proportions for running diesel engines that require little or no modification.

Bioethanol and its derivative, ethyl tertiary butyl ether (ETBE), are oxygenated products that are produced from a range of agricultural feedstocks, e.g., starch, grains, and sugar crops. It can be used in existing, slightly modified petrol engines, although cold starting requires the addition of a small amount of a volatile fuel component – usually petrol. ETBE is used as an additive to enhance the octane rating and as a replacement for the fossil oxygenate methyl tertiary butyl ether (MTBE).

Bioethanol, also called fuel alcohol, can be produced from nearly any readily available sugar- or starch-based crop, though globally, the dominant substrates are sugar cane (Brazil) and corn (U.S., Canada, and China). Wheat and sugar beets are used in Northern Europe, and sweet sorghum is used in Southern Europe for both ethanol and ETBE. In western Canada, wheat and barley are used. Research on the conversion of cellulose to ethanol has intensified in recent years, with the first large-scale biomass-to-ethanol production facility scheduled to begin commercial operation in 2003 (BAFF, 2001). The economic use of cellulosic substrates for ethanol production has posed a great challenge because of the difficulty of breaking down these complex substrates into simple sugars.

Biomethanol can be produced from wood and used in existing petrol engines in the same way as bioethanol. It also can be used as a feedstock in the production of biodiesel.

Global production of oil from conventional sources is likely to peak and decline permanently during the next decade, according to the most thoughtful analyses. Forecasts about the abundance of oil are usually warped by inconsistent definitions of "reserves". In truth, every year for the past two decades, the industry has pumped more oil than it has discovered, and production will soon be unable to keep up with rising demand (Campbell and Laherrère, 1998).

The world's production of bioethanol (beverage ethanol and fuel ethanol) in 2000 was approximately 42 million m^3, and it was distributed as shown in Figure 7 (Licht, 2001).

Biofuel ethanol is mainly produced in the U.S., Brazil, Canada, Spain, and France in amounts as shown in Figure 8. Because bioethanol is the most significant biofuel,

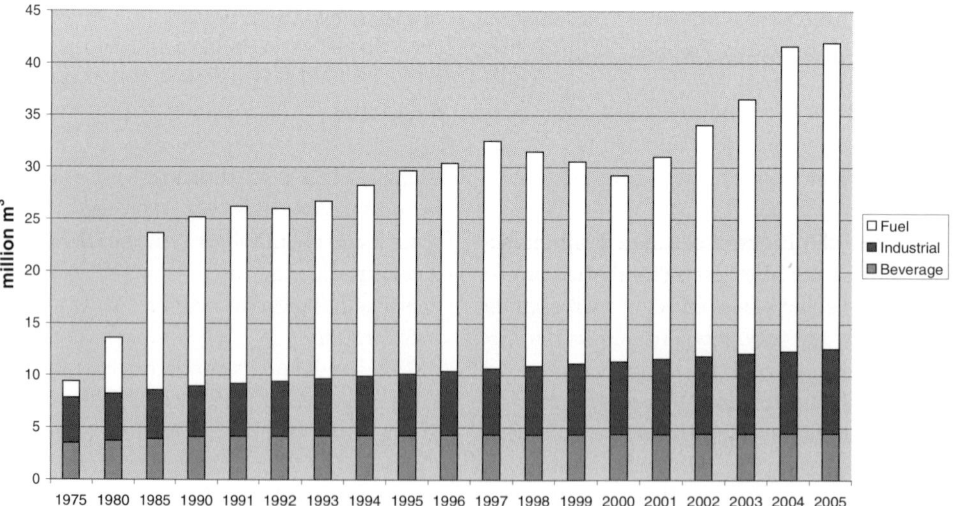

Fig. 7 World ethanol production according to Licht, 2001.

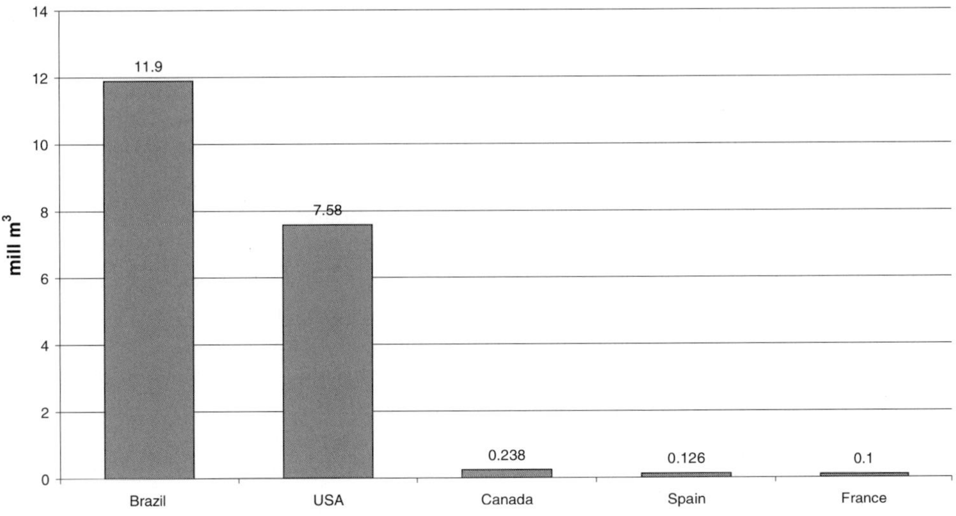

Fig. 8 Production forecast of bioethanol for fuel 2001.

the majority of the rest of this section will be concerned with it.

5.2 History

In 1908, Henry Ford designed his Model T to run on alcohol. He continued to champion its use into the 1920s because he thought it would benefit the farm industry. "There's enough alcohol in one year's yield of an acre of potatoes to drive the machinery necessary to cultivate the fields for one hundred years," said Ford.

During the 1930s, more than 2000 service stations in the Midwest sold ethanol (made from corn) in blends of 6–12% with gasoline. This "gasohol" worked well in combustion engines but was replaced in the 1940s with cheaper petroleum.

The history of ethyl alcohol fuel has been explored partially by Giebelhaus (1979) and Bernton et al. (1982) but focuses on the U.S. Farm Chemurgy Movement in the 1930s.

American farmers embraced the vision of new markets for farm products, especially alcohol fuel, three times in the 20th century: first, around 1906, again in the 1930s with Ford's blessing, and most recently, during the oil crisis of the 1970s. By the mid-1980s, more than 100 corn alcohol production plants had been built and more than a billion gallons of ethyl alcohol were sold per year in the fuel market. In the late 1980s and 1990s, with an apparently permanent world oil glut and rock-bottom fuel prices, most of the alcohol plants shut down. Some observers joked that ethyl alcohol was the fuel of the future – and always would be. "Gasohol" had become passé.

The production of ethanol from crops rich in starch has been practiced for centuries. The enzymatic degradation of starch to fermentable sugars was achieved by adding malt or koji.

The use of industrial enzymes for alcohol production from starch was reviewed early on by Aschengreen (1969). A presentation of various enzyme-based cooking procedures was published by Hagen (1981). A review of production of ethanol from whole grain was given by Lyons (1983) and later on by Lewis (1996).

Degradation of biomass using cellulases has been a major research area for over 30 years; however, it has not been economically feasible, in part because of the complex substrates, cost of enzymes, and overall lack of efficiency. Sheehan and Himmel (1999) remark that, since Pringsheim published about enzymatic degradation of cellulose in 1912, we still are not aware of all the details about how a successful application of random mutagenesis, coupled with high-throughput microbial screening, can result in enzymes that can be industrially used for complete saccharification of cellulose to fermentable sugar.

5.3 Enzymes and Latest Innovations

The most important enzymes in the biofuel industry are α-amylases; pullulanase; glucoamylases, which are known from the starch industry, as well as proteases; and β-glucanases. Their main purpose is to generate sugars that can be fermented to ethanol by yeast.

5.3.1 Ethanol Production Using Starch-based Raw Materials

Today, koji has been replaced with industrial enzymes in distilling operations. The advantages of this are many. A few liters of enzyme preparation can be used to replace 100 kg of koji. Enzymes are therefore easier to handle and store. When switching to commercial enzymes, savings of 20 – 30% of raw material costs can be expected. Furthermore, because industrial enzymes have a uniform, standardized activity, distilling becomes more predictable and there is a better chance of obtaining a good yield from the fermentations. The actual enzyme activity level of koji, on the other hand, can vary from year to year and from batch to batch, as can malt. Microbial amylases are available that have activities covering a broad pH and temperature range and therefore are suitable for the low pH values found in the mash. In view of these advantages, it is hardly surprising that commercial enzymes have replaced malt in all but the most conservative parts of the alcohol industry.

5.3.1.1 The Liquefaction Amylases

Starch liquefaction is usually carried out by pressure-cooking. Potatoes, for example, are heated to 150°C at a pressure of 5 atmospheres. When the pressure is released suddenly, the cell walls of the potatoes literally explode and release the starch. In this case, the enzymes are added to the mash tun after cooking, but in other cases, a highly heat-stable enzyme can be used in the cooker itself. Grain-based raw material also can be treated like this. Recently, the old non-pressure-cooking method has gained popularity, especially in smaller distilleries. Instead of temperatures of around 150°C, the maximum temperature ranges from 60 – 95°C. This results in obvious energy savings and there is no need to invest in pressure vessels.

The main process stages in alcohol production from starch-containing crops are summarized in Figure 9. First, the raw material is gelatinized with steam and liquefied with α-amylase to dissolve and dextrinize the starch carbohydrate. This treatment is referred to as cooking. Then, the resulting crude mash is saccharified with glucoamylase and fermented with ordinary yeast. Finally, the fermented mash is separated by distillation into alcohol and stillage.

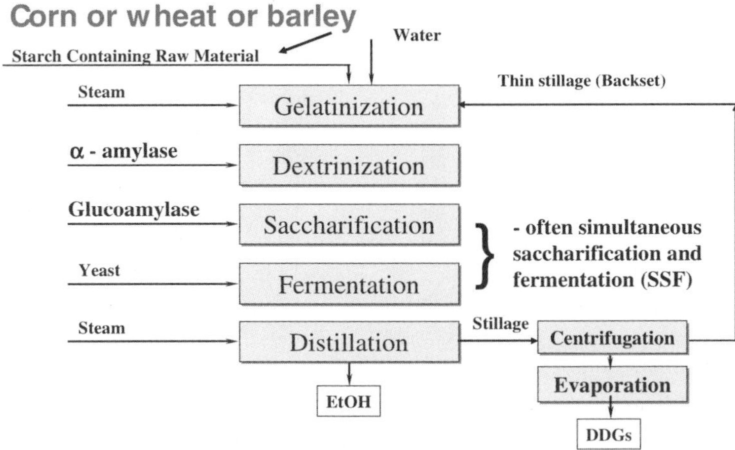

Fig. 9 Alcohol production: Main process stages.

The starch may be liquefied and pre-saccharified using first α-amylase and then glucoamylase. The resulting sugar is cooled and transferred to the fermenter, where yeast is added. The fermentation processes are usually performed continuously, using holding times during the fermentation step of around 24–30 hours.

After fermentation, the resulting beer is transferred to a separation process where the beer and brewer's yeast are separated. The beer stream is then transferred to the distillation process where the ethanol is separated from the remaining stillage. The ethanol is concentrated by using conventional distillation and then is dehydrated. The anhydrous ethanol is blended with denaturant and is ready for shipment into the fuel market.

In the so-called dry-milling process, a set of hammer mills grind the corn to a fine powder. The resulting meal is mixed with water to form a mash. In the principal process shown in Figure 10, the pre-liquefaction consumes a minimum of steam for mash cooking. This raw material slurring and two-step liquefaction process was introduced as the "Hot Slurry Pre-liquefaction Process" by Hagen (1981).

The most economical effect is judged to be when the same plant volume is applied to treat more corn per hour. The intake of corn is increased without altering the investment. The effect of the enzyme treatment here would influence the rest of the unit operations of the complete plant. Higher plant productivity versus invested capital results in lower production cost. In recent years, the dry-milling process has become automated and highly controlled. This has highlighted the importance of thorough liquefaction to the efficiency of the process. Some of the concerns of the dry-milling industry for a liquefaction amylase include consistent conversion at decreased calcium ion levels and at lower pH values. Further, rapid viscosity reduction in the mash, energy cost reductions, and efficient utilization of recycle streams are desired.

Lower viscosity of the slurry in the downstream processes provides energy savings, better utilization of the heat exchangers, and

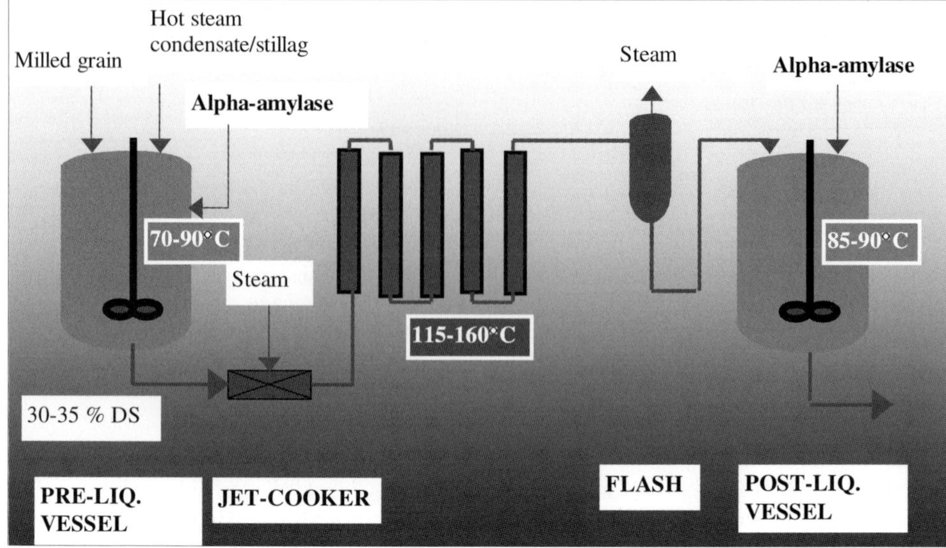

Fig. 10 "Warm or Hot Slurry Pre-liquefaction Processes".

improved process conditions in general. The newly developed α-amylase Termamyl®SC from Novozymes was introduced by Peckous at the 1999 Starch Convention in Detmold, Germany. Termamyl®SC is a protein-engineered *B. stearothermophilus* for which no additional calcium is required, and its broad pH tolerance (5.0–6.0) does not require any pH adjustment by lime addition in the industrial plants.

5.3.1.2
The Saccharification Glucoamylase

The second step in ethanol processes is saccharification before fermentation or saccharification carried out simultaneously with the fermentation process. Types of commercial saccharification enzymes used for alcohol production are mentioned in Table 6.

5.3.2
Fermentation

The fermentation processes for distilling alcohols are usually carried out through use

Tab. 6 Saccharification enzymes used for fermentation

Enzyme type	*Effect for production of fermentable sugars*
Amyloglucosidase	Break down starch molecules and dextrins to glucose
Amyloglucosidase and acid protease	Break down starch molecules and dextrins to glucose and produce amino acids and small peptides for the yeast's nutrition
Fungal α-amylase	Break down starch molecules and dextrins to mainly maltose at 55–65°C, pH 4.8–5.5
Balanced mixture of amyloglucosidase and pullulanase	Dextrozyme gives higher glucose yield (~ 2 DX) than AMG (glucoamylase) as well as correspondingly higher alcohol yield

of *Saccharomyces cerevisia*, where glucose or maltose is converted into alcohol by the yeast cell metabolism. Today, efficient glucoamylase produced from genetically modified strains of *Aspergillus niger*, with high amounts of acid α-amylase side activity, secures that residual starch is broken down efficiently and that a more rapid generation of fermentable sugars is achieved. For facilitating the fermentation and distillation processes, it may be necessary to reduce the viscosity of the fermented broth by using β-glucanase/pentosanase preparations (Adler-Nissen et al., 1984). Thus, better yeast nutrition results in improved fermentation. Increased utilization of released glucose was measured, and this resulted in a shorter fermentation time.

The bottleneck in an alcohol plant is often the fermentation tanks. The effect of adding enzymes, which improve the nutritional status of the yeast, may result in higher production capacity of the other unit operations. Cereals tend to be low in soluble nitrogen compounds. This deficiency results in poor yeast growth and increased fermentation time, which can be overcome by adding a small amount of protein-degrading enzyme to the mash. This method was mentioned by Aschengreen (1969), who described that amyloglucosidase from *Aspergillus niger* contained some α-amylase and proteinase.

Improving yeast nutrition by addition of enzymes has been shown to be able to secure that a higher intake of corn per hour in the plant can be made without extra investments in tanks, distillation towers, etc. It is thus assumed that a capacity increase based on corn of up to 20%–30% may be implemented without extra investments that change the investment costs. A way to do this may be through reduction of yeast flocculation effects or through increase of the content of free amino acids and yeast nutrition compounds like minerals and vitamins.

5.3.3
Energy Reduction

Typical values for energy consumption in the most demanding process stages are listed in Table 7. Obviously, substantial savings can be obtained by replacing traditional batch pressure-cooking with continuous processes. Additional savings can be obtained in the distillation and stillage evaporation stages – mainly by improving heat recovery. Recently, various engineering companies have quoted even lower energy consumption figures.

5.4
Future Perspectives

The main perspective for the use of enzymes in the biofuel industry is the generation of fermentable sugars from cellulose-containing raw materials. The focus will be to develop cellulases that are improved with respect to thermostability, cellulose-binding domain, active site, and reduced non-specific binding.

Tab. 7 Process energy consumption[a]

Process	*Energy consumption MJ/L ethanol*		
	Cooking	*Distillation*	*Stillage evaporation*
Traditional batch	7–8	10–16	0–18
Modern continuous	1–2	5–7	0–12

[a] Source: Hagen and Nielsen (1982)

5.4.1
Ethanol Production Using Lignocellulose-based Raw Materials

Cellulose is the most abundant organic material on earth. In the future, it could be an unlimited source of energy. The process used in most economical analysis can be briefly described as using co-current dilute acid prehydrolysis of the lignocellulosic biomass with simultaneous enzymatic saccharification of the remaining cellulose and co-fermentation of the resulting glucose and xylose to ethanol. In addition to these unit operations, the process involves feedstock handling and storage, product purification, wastewater treatment, enzyme production, lignin combustion, product storage, and other utilities (McAloon et al., 2000).

The most promising techniques to release cellulose fibers have been adaptation of the pulping techniques used for preparation of cellulose fibers for paper production. Among those, continuous steam explosion pulping at high pressures and low residence times has been shown to produce pulps with sufficient release of cellulose for further processing to ethanol. Batch steam explosion was commercially introduced in 1931, along with the Masonite process for the production of fiberboard. Beginning in the 1970s, the Canadian company Iogen and several others began conducting research within applications of steam explosion for ethanol processes.

The wet oxidation process is also promising for the pretreatment of biomass into fermentable sugars. Olsson et al. (1999) have investigated different types of biomass as substrates for fermentation of ethanol. For wheat straw, it was found that alkaline wet oxidation was efficient for fractionating of the polysaccharides into fermentable sugars. The reaction time needed for efficient fractionation was typically 10 minutes with a reaction temperature of 195°C, 10 min., under alkaline conditions and 12 bar O_2.

The National Renewable Energy Laboratory (NREL) has undertaken a complete review and update of the process design and economic model for the biomass-to-ethanol process based on co-current dilute acid prehydrolysis, along with simultaneous saccharification (enzymatic) and co-fermentation (Wooley et al., 1999a).

Wooley et al. (1999b) concluded that the cost of cellulases for industrial alcohol must be significantly reduced and that cellulases must be improved with regard to thermostability, cellulose-binding domain, active site, and reduced non-specific binding.

Research has been initiated thanks to the U.S. Department of Energy Grant that Novozymes has obtained. The research project is aimed at reducing the cost of cellulase enzymes used in converting plant waste material (biomass) into fermentable sugars. Currently, the enzyme costs per gallon of ethanol produced are far too high to make this process commercially relevant. Cost estimates could be made based on different assumptions, e.g., on location enzyme production. For the most optimistic scenarios, the direct enzyme-related costs/performance must be reduced by one-tenth for the process to be economical in the U.S. Therefore, at least a 10-fold increase in the specific activity or production efficiency is needed. Thus, bioethanol may be produced in the future by using the cob, the stalk, and the husks of corn or other waste material that contains cellulose.

6
Enzymes for the Textile Industry

During the year 2000, in the United States alone, sales of textile enzymes accounted for about 40 million USD (BCC, 2001). Today, enzymes are being used or anticipated for use in all process steps of the textile industry, comprising hydrolases, oxidoreductases, and lyases.

6.1 Introduction

Global textile production handles about 50 million metric tons of fiber annually, where, on average, 275 USD worth of process chemicals are used per ton of fiber (Freedonia, 2000). Approximately 40% of the value of applied chemicals is in the dyes and colorants. The remaining value is in chemicals including emulsion polymers, surfactants, silicones, gums, starches, solvents, oils, waxes, and enzymes.

Enzyme treatments are used during the textile "wet process" to (1) facilitate removal of (natural or applied) impurities and (2) help modify the physical properties of the textile. Mild enzyme treatments lead to improved process quality, efficiency, and effectiveness and reduced environmental impact.

6.2 History

Even before the special catalytic role of enzymes was understood, textiles were treated with microorganisms to improve properties. Flax retting – the separation of flax fiber from non-fiber tissues in the stems – by the action of enzymes from indigenous soil fungi is the oldest of these microbial processes and is still in use today (Akin et al., 2001). At the turn of the 20th century, crude amylase preparations derived from molds, pancreas, or malt were used to "desize" (remove) starch from woven fabric, thereby overcoming the fiber-damaging effects of conventional acid processes. Based on a patent awarded in 1917, the French company Société Rapidase spearheaded the modern use of fermented enzymes for textile applications with bacterial amylases (Aunstrup, 2001). Following this came a range of commercial amylases with desirable features, particularly, high heat stability (Nielsen et al., 1994). Also during this time, the use of protease enzymes for silk and wool processing was explored, resulting in commercial products, but small markets.

Introduction of cellulases for treatment of denim during the 1980s truly revolutionized this segment of textile manufacture. Although the degradation effect of fungi (which secrete cellulase enzymes) on cotton and rayon textiles was well known – and had been studied from the point of view of preventing fiber strength loss (Blum and Stahl, 1952) – a major commercial market was opened by the discovery that cellulases could provide a "stone-washed" effect on warp-dyed denim fabrics that previously was achieved only by using pumice stones (Olson, 1989). Using cellulases significantly reduced process time, labor, energy, and strain on equipment, sometimes totally eliminating the need for stones and the added expense of removing residual stones by hand.

During the 1990s, two broad new enzyme classes were added to the textile portfolio: (1) oxidoreductases (represented by catalase, laccase, and peroxidase) and (2) lyases (represented by a special pectate lyase-type of pectinase). The oxidoreductases, in particular, prompted a new mode of thinking in the way enzymes and mediator-type chemicals could work together to create significant process advantages.

6.3 Enzymes

Hydrolytic enzymes (amylase, cellulase, lipase, and protease) catalyze reactions in which a chemical bond is broken by addition of the elements of water. Oxidoreductase enzymes (catalase, laccase, and peroxidase) catalyze the electron transfer reaction that can lead to chemical bond cleavage. Lyase enzymes (pectate lyase) catalyze reactions involving chemical group elimination to form double bonds, which also results in a breakdown of polymeric molecules. To date, all commercial textile enzymes catalyze reactions that degrade large, often polymeric, molecules into smaller ones. For many years, amylases were the only enzymes used in textile processing. Now enzymes are used, or are anticipated for use, in almost all process steps (Table 8, Figure 11).

During cotton "preparation" – desizing, scouring, and bleaching – raw fabric is converted to a material that is ready for

Tab. 8 Enzymes for commercial textile applications

Enzyme	*Commercial application*	*Benefits*
Amylase	Removes starch-based size during cotton preparation	Removes starch without causing fiber damage
Catalase	Removes hydrogen peroxide after bleaching	Eliminates excessive rinses to save water, energy, and time No need for chemical reducing agents Does not interfere with dyeing steps
Cellulase	Modifies the surface of cellulosic fibers or fabrics to achieve a desired hand or surface effect	Eliminates/reduces need for stones during denim "stone-washing" Saves time, energy, and labor Softens and improves surface and prevents pilling of cotton fabric Provides defibrillation of lyocell
Laccase	Selective decolorization of indigo during denim processing	Easy process control avoids risk of over-bleaching Eliminates/reduces indigo "back-staining" Mild process with minimum strength loss for lightweight and sensitive fabrics Saves time and water Provides broad range of fashion effects
Lipase	Removes triglyceride-based size lubricants during cotton preparation	Aids removal of natural size components Increases fabric absorbency for improved levelness of dyeing
Pectinase (pectate lyase)	Removes pectin from raw cotton during preparation	Aids in natural wax removal Provides good fabric wettability Significantly reduces the need for caustic substances Saves water, heat energy, and electricity Minimizes fiber damage Process conditions are mild to mill, operator, and environment
Peroxidase	Removes hydrolyzed reactive dyes from the rinse after dyeing	Significant savings in water, heat energy, and time Used to obtain highly wash-fast fabrics
Protease	Removes gum layer from raw silk during preparation. Modifies the surface of wool or silk fabrics to achieve a desired hand or surface effect	Mild degumming of silk Stiffness-reduction or "peach-skin" effect on silkSoftens wool Enhances dye uptake

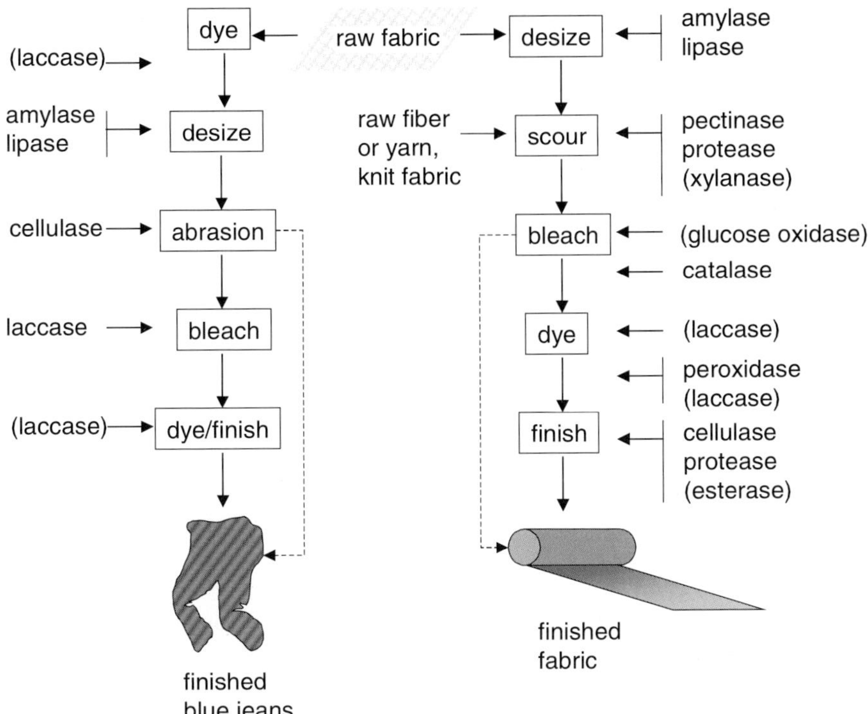

Fig. 11 Commercial (and emerging/potential) enzyme applications throughout textile wet processing.

dyeing and finishing. The material must be uniform and wettable, with minimal imperfections and the required degree of whiteness. Woven fabrics must be desized to remove sizing agents that were applied prior to weaving. Starch-based sizing agents are selectively degraded by amylase without damage to the textile fibers. Amylases are typically applied at pH 5–8 and, depending on the product, can operate at temperatures ranging from 20–115°C (Novozymes, 2001). Although not a major application, lipases can be used to improve the efficiency of size removal when vegetable oils or animal fats are present. The use of synthetic sizes, based on polyvinyl alcohol (PVA), polyacrylic acid (PAA), or carboxymethylcellulose (CMC), and starch-synthetic blends can create problems in an amylase-only desize but also offers opportunities for new enzyme activities or chemo-enzymatic systems – both areas of active research.

"Scouring" follows desizing for woven fabrics and is the first wet-processing step for raw cotton fibers, yarns, and knit fabrics. The main objective is to improve fiber wettability. During scouring, hemicelluloses and hydrophobic waxes in the primary cell wall of cotton are removed. Conventional chemical scouring uses a "brute force" alkaline treatment. Enzymatic scouring selectively targets pectin scaffold substances in the primary cell wall with enzymes commonly called pectinases (Husain et al., 1999; Etters et al., 1999). Pectin is a complex polymeric substrate, containing many different sugar subunits. An individual pectinase may have one of several activities, including pectin lyase, galactanase, arabinanase, pectin esterase, mannanase, polygalacturonase, and pectate lyase. Pectinase degradation of the primary cell wall components causes breaks and allows release of the waxy outer layer, leading to excellent fiber wettability (Etters, 1999; Li and Hardin, 1997; Waddell, 2000). A commercial enzyme product based on bacterial pectate lyase offers significant cost savings compared to the conventional alkaline process (Table 9, figure 12).

During cotton bleaching, fabric whiteness is increased, and "motes" (dark seed coat fragments) are decolorized or degraded. Although positive results have been shown when using glucose oxidase to mildly generate hydrogen peroxide *in situ* as the bleaching agent (Tzanov et al., 2002), it is not clear that the enzyme process can yet compete with conventional hydrogen peroxide bleaching. Immediately after bleaching, the first oxidoreductase for textile applications, catalase, can be used to efficiently decompose residual hydrogen peroxide to (1) stop the bleaching reaction before fiber damage can occur, (2) prevent hydrogen peroxide interference with subsequent dyeing, and (3) save water and time by reducing the number of rinses needed after bleaching and before dyeing (Schmidt, 1995).

Tab. 9 Process savings during enzymatic scouring of cotton yarn[a]

Process variable	*Savings (USD/ton yarn)*
Run time	66
Water	17
Heat energy	6
Electricity	1
Total	90

[a] Source: Durden et al. (2001)

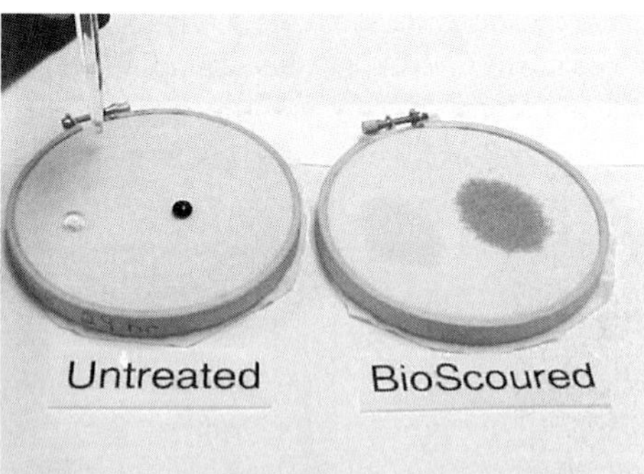

Fig. 12 Pectinase for cotton scouring. The enzyme provides a softer hand, improved dye levelness, full fabric strength retention and improved dyeing efficiency ass illustrated in this figure. Bioscouring with pectate lyase allows water to rapidly penetrate cotton textiles during the preparation step of "wet processing".

Like the cotton process, natural protein fibers, such as silk and wool, also go through preparation steps. Protease enzymes mildly degrade the protective surface protein (sericin) during silk degumming. Lipase enzymes may be used to clean natural oils from wool, but competing surfactant technologies may offer more advantages. During finishing, proteases are used to soften wool fibers or to produce a "sandwash" finish on silk fabrics (Novo Nordisk, 1999a). Proteases also may have some limited ability to improve shrink resistance of wool fabrics by blunting the edges of wool scales, but usually the corresponding strength loss is high. A technique using alcohol solvents to block enzyme penetration into the fiber and to localize protease effects to the surface scales has been able to overcome the strength loss issue while providing significantly improved shrink resistance (McDevitt and Shi, 2000).

The "cellulase revolution" in denim blue jean processing (Stewart, 1996; Kumar et al., 1997) remains the largest value-added market for textile enzymes, followed by a range of finishing applications including cotton softening and depilling (Liu et al., 1998; Snyder, 1997) and lyocell defibrillation (Kumar and Harnden, 1999) (Figure 13).

Cellulases are also used, to a lesser extent, to degrade CMC found in some sizing agents and print paste thickeners. The combined action of three major classes of cellulase enzymes contributes to the complete degradation of cellulose to glucose (Béguin and Aubert, 1994). Endoglucanases randomly cleave cellulose chains at internal glycosidic bonds. The (non-reducing) cellulose chain ends are cleaved by cellobiohydrolase (CBH) to soluble cellobiose dimers, followed by β-glucosidase hydrolysis of cellobiose to two glucose monomers. Commercial enzyme products may contain one enzyme activity – a monocomponent product – or a mixture of enzyme activities – a multicomponent product. The catalytic stereochemistry (Davies and Henrissat, 1995; Claeyssens and Henrissat, 1992) and protein family classification (Henrissat and Bairoch, 1993) of many cellulases are known in detail.

The latest revolution in denim finishing came with the introduction of laccase for indigo decolorization. Compared to other enzymes, laccase has relatively low substrate specificity, acting on a range of substituted aromatic compounds. Laccase-catalyzed oxidation of these compounds produces transient intermediates that can degrade upon further oxidation, condense to form oligomers or polymers, or transfer the oxidation effect to other molecules. This transfer of oxidizing power via a so-called "mediator" compounds makes it possible for laccase to catalyze oxidation reactions outside its direct substrate specificity or steric accessibility. A commercial product based on a laccase-mediator system selectively decolorizes indigo to give fabric-safe enhanced abrasion (Kierulff, 1997; Novo Nordisk, 1999b).

Peroxidases, like laccases, catalyze electron transfer reactions. They have been studied in both microbial (Heinfling et al., 1997) and free enzyme (Morita et al., 1996; Conrad et al., 1997) systems for their role in dye decolorization. As with laccase, peroxidase catalytic activity can be enhanced or adjusted to target specific substrates by using a mediator compound (Morita et al., 1997). This feature has been used in developing a peroxidase system that can selectively remove unfixed dyestuff and reduce the number of rinse cycles needed after reactive dyeing (Damhus and Vogt, 2000).

Enzymes alone can offer great process advantages; however, combinations of enzymes together with chemicals, "chemo-enzymatic" systems, afford practical benefits as well. Successful examples of these approaches are the laccase-mediator system

Fig. 13 Cellulase enzymes remove surface fuzz and fibrils from cotton (and other cellulosic) fibers to improve appearance and surface texture in the finishing process.

for denim bleaching and the peroxidase-mediator system for reactive dye post-wash. Auxiliaries, like surfactants, can often "boost" enzyme performance by improving substrate accessibility or helping to suspend and remove reaction products and residues. Other chemistries, such as mild bleach-boosting agents (Cai et al., 2000) may also find applications together with enzymes when the target substrates are too complex or inaccessible for enzymes alone.

6.4 Latest Innovations

The past 20 years have seen significant growth in the number of new textile applications for enzymes. The textile industry is becoming increasingly familiar with related process advantages. This growth includes creative new uses for existing commercial enzymes and the introduction of entirely new enzyme activities (Table 10). Increasing global textile trade creates opportunities for new uses of enzymes, even in traditionally old applications, such as desizing. Possibilities include new amylases that are more robust to different starting substrates, xyloglucanases to degrade natural gum-based sizes, cellulases to degrade CMC size agents (or printing thickeners), or even entirely new enzyme activities that can target synthetic sizes such as PVA and PAA.

Although the use of enzymes for processing natural fibers is well established, and existing commercial enzymes can be used to provide special benefits during processing of sensitive blended fabrics (especially, avoiding the need for harsh, fabric-damaging chemicals), the next major milestone will be to anchor the use of enzymes for direct processing of synthetic textiles. These new opportunities are made possible only through corresponding rapid advances in the sciences of enzyme biotechnology and engineering.

7 Enzymes for the Pulp and Paper Industry

All major hydrolytic enzymes – cellulases, xylanases, amylases, lipases, and amylases – are used in the pulp and paper industry in a variety of applications. A crude idea of current enzyme use in the pulp and paper industry can be gained by the volume of enzyme sales. An estimated worldwide value of enzyme sales to all markets has been placed at about 1.5 billion USD. Since 1985,

Tab. 10 Future opportunities for enzymatic processing of textiles

Enzyme	*Application*
Amidase, nitrile hydratase	Modification of acrylic substrates (Tauber et al., 2000)
Amidase, manganese peroxidase	Modification of nylon substrates (Deguchi et al., 1998)
Esterase (cutinase)	Modification of polyester substrates for oligomer removal (Riegels et al., 2001) and surface modification effects (Sato, 1983)
Glucose oxidase	Mild, *in situ* generation of hydrogen peroxide for bleaching (Tzanov et al., 2002)
Laccase, peroxidase	Mild reoxidation of vat and sulfur dyes (Xu et al., 1999), dye discharge printing (Hall et al., 1999), and (*in situ*) dye synthesis (Barfoed and Kirk, 1999)
PVA oxidase	Degradation of polyvinyl alcohol-based sizing agents (Mori et al., 1997)
Xylanase	Bast fiber finishing (Sreenath et al., 1996) and enhancement of bleaching (Kundu et al., 1993)

the sales volume of industrial enzymes into the industry appears to have grown by more than 1%, but less than 2%, of the total figure provided above.

7.1 Introduction

Over the last two decades, the use of enzymes in production of pulp and paper products has increased dramatically. Beginning with the early use of amylases in starch-coating applications, the use of enzymes in the pulp and paper industry now involves eight different enzyme classes and even more application areas. Furthermore, research and development efforts have led to the introduction of enzyme products that are customized for special processes and process conditions, e.g., cellulases optimized for such diverse applications as drainage improvement, energy reduction during pulping, deinking, or fiber modification.

Still, the industry's acceptance of enzyme technology has been slow in coming, but there are positive signs that the pace of adoption is accelerating. This seems to be particularly true in the area of paper-machine operations, where increased efficiencies have provided significant value to the end user.

7.2 Overview of Selected Applications

When discussing the pulp and paper industry it can be worthwhile to first make a few distinctions. Pulp manufacture denotes the process of defibration of lignocellulosic raw material – notably wood. This process is done in a pulp mill, and the means for obtaining the defibration can be mechanical (i.e., stone groundwood pulp, thermomechanical pulp, etc.) or chemical (i.e., kraft pulp, where wood chips are cooked in the presence of $NaOH/Na_2S$). Furthermore, the pulp will often be bleached (e.g., with ClO_2, oxygen, peroxide, etc.) or brightened (e.g., with H_2O_2 or hydrosulfite). Alternatively, the pulp can be produced from recycled paper, where the material to be recycled undergoes such steps as repulping, deinking, and bleaching/brightening. The pulp can then be converted into a variety of end uses – i.e., writing paper, newsprint, boards, or towel and tissue products. This is done in a paper mill (or tissue mill, board mill, etc).

In the paper mill, the pulp is usually brought into a dilute suspension that can then be converted into paper by homogeneously spraying the pulp suspension onto a forming fabric that allows water to drain from the pulp.

7.3 Enzymes

Table 11 summarizes most of the enzyme-based processes that are in use commercially or have been described in mill-scale testing.

7.3.1 Xylanases for Bleach Boosting of Kraft Pulp

Xylanases hydrolyze the 1,4-glycosidic linkages in the hemicellulose xylan that makes up 20–30% of hardwoods and about 10% of softwoods. Xylanases received major attention as a bleaching pretreatment for unbleached kraft pulps (Viikari et al., 1987) because they could be used to remove re-precipitated xylan from so-called kraft brownstock and thereby improve the pulp's bleachability. Use of this enzyme can offer bleaching chemical reduction in attaining a given brightness target, and there are now a number of mills in the Americas, as well as in Europe and the Far East, where xylanases are in regular use (Tolan and Thibault, 2000).

Tab. 11 Enzyme-based processes for pulp and paper applications

Segment	*Subsegment*	*Application*	*Main enzyme*
Pulp manufacturing	Chemical pulp	Bleach boosting	Xylanase
	All pulps	Refining	Cellulase Xylanase
	Recycled pulp	Deinking	Amylase Cellulase Lipase
	Recycled pulp	Repulping	Cellulase Amylase
	Sulfite/Mechanical pulp	Pitch control	Lipase
	Recycled pulp	Stickies control	Esterase Lipase
Papermaking		Starch coating	Amylase
		Fiber modification (softening)	Cellulase
		Cleaning processes	Protease Lipase Amylase

The main focus for developments in this area has been the identification of enzymes that can withstand the high temperatures and high alkalinity found at the relevant treatment points. Enzyme suppliers have developed products for this use that are both more alkali- and heat-tolerant, which better meet the industry's needs, and today xylanase treatment steps can usually be run without having to make major process changes at the mills.

During recent years, the implementation of the EPA Cluster Rules in North America has caused a renewed interest in the use of xylanases. Xylanase use has been documented to significantly decrease chlorine dioxide demand in pulp bleaching systems; hence, their use offers a capital cost savings for additional capacity of ClO_2 in mills which are limited by their bleaching capacity (and effluent levels of adsorbable organic halides).

7.3.2
Cellulases and Other Enzymes for Refining of Pulp

Ensuring that a pulp is separated into individual fibers is not only one of the most important aspects of high-quality papermaking but also one of the most energy-requiring processes, as the refining process usually relies on high-impact mechanical shear to provide the result.

Enzymes can play an important role in terms of hydrolyzing the cellulose and hemicellulose that takes part in "gluing" the fibers together. So far most enzyme products in this area are based on multi-component cellulase preparations produced by *Trichoderma*. Meanwhile, the latter type of products generally comprises a broad array of cellulolytic and hemicellulolytic activities, and it is still not fully clarified how this array of enzymes affects the refining process. Some publications have documented a positive impact of cellobiohydrolase (CBH)-type enzymes as well as of mannanases to be important (Pere et al., 1999, 2000), while others have described a positive impact resulting from xylanase treatment of pulp (Dickson et al., 1999).

Recent work suggests that enzymes other than those acting on cell wall polysaccharides also may have a potential role in enzymatic refining of wood – amongst others, Wong et al. (1999) have described positive effects resulting from treatment with both protease and laccase.

With ever-increasing energy prices and constant attention on reduction and conservation of energy in industry, it is expected that enzymes for facilitating refining processes will be a significant growth area in the coming decade.

7.3.3
Cellulase, Amylases and Lipases for Deinking and Repulping of Waste Paper

The large volumes of waste paper subjected to recycling have led to developments in which enzymes are replacing conventional chemistries, such as caustics and surfactants. The deinking area can be divided into two major segments – mixed office waste (MOW) and old newsprint and waste magazines (ONP/WM). The two different grades differ substantially both in terms of the predominant nature of the print/ink that needs to be removed and in terms of the grades of pulp originally used for their manufacture.

Most work has focused on the enzymatic deinking of MOW, which is now a commercial reality (Knudsen et al., 1997). The advantages have been shown to include improved toner ink removal, brighter pulps, and improved pulp drainage. These advantages appear to come with lower water treatment costs in the recycling operation.

Most early work in enzymatic deinking of MOW assumed that cellulases would serve as the key active agents with this waste paper grade (Figure 14) (Heise et al., 1996). However, since the manufacture of paper for this grade employs starch surface sizing, amylases are – alone as well as in combination with other enzymes (cellulases and lipases) – gaining more and more importance for deinking use (Sakaguchi et al., 2000; Gelhoff, 1998).

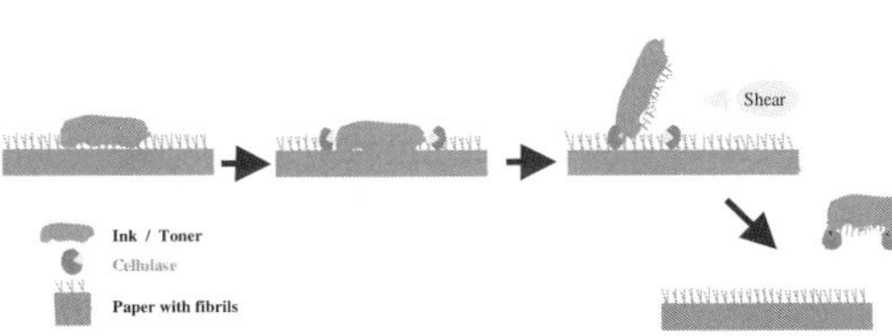

Fig. 14 A model for enzymatic deinking with cellulases.

A preferred mode of recycling waste paper is the so-called flotation deinking process, where enzymes are added during the initial repulping of the waste paper, whereby the enzymes along with the mechanical treatment release the ink particles from the fibers. Subsequently, the ink is removed in a flotation cell in which air is blown though the pulp slurry after the addition of a surfactant.

Since most white papers contain a significant amount of calcium carbonate, the neat pH of a MOW furnish is often in the neutral to alkaline range; hence, the preferred enzymes for use in this application are those that have performance optimum in this range as well. The use of cellulases should always be balanced in order not to significantly reduce the yield in the process or the strength of the fibers. A development in this area that has been particularly important has been the introduction of recombinant (monocomponent) enzymes, as it allows ink removal to take place with little or no negative impact on fiber quality.

The application of enzymes in the deinking of ONP has been discussed in the literature, but it has been less successful in terms of commercial implementation as compared to the office waste segment (Putz et al., 1994). A major problem with enzymatic deinking of ONP relates to particle size, type of ink, and the development of surfactant systems to deal with the liberated inks. A commonly used process of ONP deinking is to follow the flotation stage with an alkaline peroxide-bleaching step. One of the major drivers for the continuous search for improved enzymes for ONP deinking is the possibility to obtain a neutral deinking process where the targeted brightness can be obtained with hydrosulfite rather than with the more expensive peroxide-bleaching chemicals.

A third major segment for recycling is termed old corrugated containers (OCC). The use of enzymes in the recycling of OCC is less focused on the visual impact of the treatment (i.e., brightness) but rather focuses on the drainage properties of the recycled pulp, i.e., the productivity of the mills that recycle OCC. The reason for the improvement in mill output is the finding that both cellulases and amylases can improve pulp drainage (Lascaris et al., 1997), which in turn has a significant impact on the maximum speed of the paper machine. The effect of cellulases can be ascribed to the hydrolytic action causing a reduction in the amount of small fiber fragments or "fines". The benefits of amylases relate to the significant amounts of starch that are present on the containers and that increase viscosity during repulping. While amylases can counteract this viscosity effect, the benefits may come with the price of lower strength of the resulting product. The reason for the latter is that starch acts as an interfiber bonding aid, a characteristic that is lost upon too-extensive degradation.

7.3.4
Lipases for Pitch Control

The ability of lipases to rapidly break down triglycerides in fatty acids and glycerol has found widespread use in the area of mechanical pulping – mainly for newsprint manufacture. The segment where this application started out was the so-called stone groundwood (SGW) segment, and the effort was pioneered commercially by the former Jujo Paper in Japan (now Nippon Paper).

In kraft pulping, the triglycerides present in the wood are usually hydrolyzed effectively during the process; this is usually not the case with mechanical pulp or with sulfite pulp (Fischer and Messner, 1992). In the latter cases, the triglyceride component present in "pitchy" woods, primarily pines, represents a prime cause of deposits and operating problems (Fujita et al., 1992) on the paper machine.

With the addition of a lipase to the pulp, the triglycerides are hydrolyzed to free fatty acids, which in turn can either be removed by washing or be fixed on the fibers with alum or cationic polymers.

The prevention or reduction of deposit problems have substantial merit in themselves, but with the expanding use of enzymatic pitch control, several other benefits have been observed. In particular, it has been found that lipase treatment of mechanical pulp can yield significant strengthening benefits (Mustranta et al., 2001) and also improve the wear life of the expensive felts and forming fabrics on the paper machine.

The use of lipases also has been shown to bring some value in the manufacture of fluff pulp (Malmgren et al., 2002) used for the manufacture of diapers and other personal hygiene products. In the case of the former, an increased rate of water absorption results from the lipase treatment.

7.3.5
Esterases for Stickies Control

When paper is recycled for use in production of new paper materials, a very common problem is "stickies". Stickies are agglomerates of different glue materials with fiber bundles or other non-fiber material. Stickies deposit on conveying felts, machinery, and the paper sheet; the result is increased downtime and lost production. Stickies consist of many water-insoluble materials, mainly glues and adhesives, and one of the big contributors is polyvinyl acetate.

Some very recent work (Fitzhenry et al., 2000) describes the use of a system that includes the esterase class of hydrolases to significantly effect stickies reduction in recycled applications. The advantages include fewer sheet breaks and improved machine cleanliness and efficiency.

7.3.6 Amylases for Starch-coating Applications

Coating and surface sizing of paper will usually employ a modified starch as part of the formulation applied to the paper. Starch as a film-forming material can help to improve the surface graphic properties of lightweight coated papers and also improve the surface properties of papers used in non-contact printing applications. In some paper products, starch penetration into the paper also can result in increased strength.

The modified starch is produced from native starch by either chemical or enzymatic treatment with amylase, typically by a chemical supplier. Amylase modification of native starch directly at the paper mill is an alternative that provides advantages from an economical perspective; it also enables the operator to produce a range of different starches, for different applications, from the same raw material.

All natural starch materials can be used – maize, potato, wheat, tapioca etc. – and the conversions can run either as batch processing or continuous. The important part is to control temperature, pH, and time to obtain the desired level of viscosity reduction as well as to have an effective inactivation procedure to stop the amylase action.

7.3.7 Cellulases for Fiber Modification

One of the most promising growth areas for using enzymes in the pulp and paper industry is the use of highly selective cellulases for fiber modification. The area in which the most attention is currently focused is on the softening of tissue and towel products with cellulases (Franks et al., 1997).

When going over literature in this area, one easily gets the sense that cellulases generally have a negative impact on fiber quality; however, this does not necessarily have to be the case. In particular, it has been found that certain cellulases can actually increase strength while at the same time improving softness of sanitary products (Franks et al., 1997).

The mechanisms underlying these findings are not fully understood. One proposed mechanism for the ability of certain cellulases to improve wet strength is that the enzymes' action on the fiber surface increases the number of reducing ends, which in turn can lead to an increased formation of interfiber, hemiacetal bonds.

7.3.8 Proteases, Lipases, and Amylases for Cleaning Applications

With some similarity to other hard-surface cleaning applications in industry, enzymes find use in formulations for applications such as felt cleaning and for the cleaning of starch deposits in the coating section of paper mills. Also there is now the commercial application of certain classes of hydrolases used as adjuncts to standard biocide systems. The hybrid systems can increase the efficiency of standard biocides in white-water systems so that the load of biocides can be reduced or eliminated. In this application, the enzymes act to prevent the microorganisms, "slime", from adhering to the surface by hydrolyzing the exo-polysaccharides and proteins that anchor the microorganisms to the solid surface.

7.4 Future Perspectives

So far, most commercial developments have focused on the countless possibilities with hydrolytic enzymes. No doubt the coming years will add more applications and more optimized enzymes to this list (Cooper, 1998; McCoy, 2000b) (see Figure 15). Other enzyme classes are also in the late stages of development or early stage of commercialization, namely the oxidoreductases and transglycosylases (transferase).

A major area of future opportunity is the availability of the oxidoreductase enzyme family. This family includes, among others, catalases, laccases, and peroxidases. Both of the latter enzymes have been shown to be able to modify lignin in pulp fibers. The early potential uses for this class have included enzymatic delignification of kraft pulps (Chakar et al., 2000), improved fiber-fiber bonding in high-yield pulps (Felby at al., 2001), fiber board manufacture, and wastewater treatment. While the laccases and peroxidases still have not gained any significant commercial use, catalases are today used as a means for elimination of residual peroxide after pulp bleaching.

Transglycosylases, and in particular the enzymes referred to as xylogucan-endo transferases, are enzymes that form rather than degrade glycosidic linkages. Early work in this area has shown that these enzymes may have the potential to strengthen lignocellulosic materials significantly (Kofoed and Lund, 2000). However, at this point in time, it is not yet clear whether these enzymes can be produced in sufficient levels for commercial use.

Fig. 15 Use of enzymes for gluing fiberboards may be one of the future growth areas.

8 Enzymes for Organic Synthesis

Hydrolytic enzymes comprising lipases, nitrilases, esterases, amidases, peptidases, and hydantoinases are the most common enzymes in many diverse applications. Oxidoreductases requiring cofactor-regenerating systems are of increasing importance.

8.1 Introduction

The huge diversity and complexity of organic molecules in nature is a reflection of the remarkable power of enzyme catalysis. The phenomenal rate acceleration together with the unique selectivity and mild reaction conditions offered by enzymes make them highly attractive as catalysts for organic synthesis. The chemical industry has, however, been somewhat slow to take advantage of the obvious potential offered by enzymes. Several myths around enzymatic catalysis have, unfortunately, initially limited its use. These myths have been focused around enzymes being regarded as highly costly and fragile catalysts. A number of large-scale, enzyme-based processes have, however, now been established, and these success stories have helped dispel the myths once and for all (Rozzell, 1999). Accumulated experience and confidence in the chemical industry in enzyme catalysis have now finally paved the way for much broader use. Enzymes are now about to gain the position within the chemical industry that they deserve.

8.2 History

The ability of enzymes to catalyze synthetic useful reactions and their ability to work in organic solvents, which is a requirement in many synthetic reactions because of the poor aqueous solubility of the substrates, are often regarded as very recent discoveries. It is therefore a surprise to many chemists that the first enzyme-catalyzed reactions performed in organic solvents were, in fact, reported very early in the last century. At that point in time, only very few enzymes were available in well-characterized form, and even fewer, if any, were available in industrially relevant quantities. This picture was changed during the 1960s when many different enzymes became available in industrial scale as a result of the evolution of modern biotechnology. The development of immobilization as a tool to stabilize enzymes and to enable easy recovery and recycling made it possible to introduce enzyme catalysis into the first industrial-scale bio-processes in the 1970s. Initially, enzymes were used in aqueous processes such as resolution of racemic amino acids using aminoacylases, production of 6-APA by penicillin acylase, and production of high-fructose corn syrup by glucose isomerase. The two latter reactions are still among the biotransformations being performed in the largest scale. In the mid-1970s, Unilever made a highly valuable contribution when Coleman and Macrae (1977) described the use of an immobilized lipase in non-aqueous media. This work led to the current thinking of enzymes as potential chemical catalysts. Further work pioneered by Klibanov and coworkers in the 1980s (Klibanov, 1986) showed that many enzymes were active in organic media and able to catalyze a number of synthetic useful reactions. During the 1990s the chemical industry started to consider enzyme catalysis as a serious alternative to classical organic chemistry. In particular, the unique enantioselectivity of enzymes has been valued by the industry, fuelled by the growing demand for enantiopure pharmaceuticals and agrochemicals. Today, a number of enzyme-catalyzed processes are, as outlined below, established in industrial scale and, because many chemical companies are still at an early stage in their exploitation of enzyme catalysis, the number will surely be expanded significantly in the coming decade (Schmidt et al., 2001).

8.3 Enzymes

A variety of different enzymatic activities have been explored within organic synthesis. However, most enzymes used, in particular in true large-scale processes, are various hydrolases. Lipases have, in particular, found use in a number of different applications. Because these enzymes are designed in nature to operate at an oil-water inter-phase they are generally very compatible with organic solvents. Furthermore, lipases catalyze a variety of different reactions that are very different from the reaction they are designed for in nature, i.e., the hydrolysis of triglycerides. One particular lipase, the B-component from the yeast *Candida antarctica*, has in numerous applications been shown to be a particularly efficient enzyme in catalyzing a great number of different reactions including both regio- and enantioselective syntheses and even reactions involving both sulfur- and nitrogen-based nucleophiles (Andersson et al., 1998). Other hydrolases of significant importance within organic synthesis are nitrilases, esterases, amidases, peptidases, and hydantoinases. A few oxidoreductases also have gained industrial importance, but this potential of this enzyme class is, as outlined below, still in the early phase of exploitation.

A selection of commercially established products produced by enzyme-based processes is presented in Table 12.

8.4
Latest Innovations

Oxidoreductases catalyze a number of reactions that are highly relevant to organic synthesis (Deveaux-Basseguy et al., 1997). Some of the most useful oxidoreductases are, however, co-factor requiring. Nicotinamide-based co-factors such as NADH and HADPH are both highly costly and unstable. The use of co-factor-requiring oxidoreductases has accordingly been inhibited by some of the same myths that, in general, initially limited the use of enzyme catalysis within the chemical industry. Over the past decade, several schemes for effective co-factor regeneration involving coupled enzyme reactions have been developed (Adlercreutz, 1996), and several processes have been scaled up based on formate dehydrogenase as the regenerating enzyme (Liese et al., 1998; Tishkov et al., 1999).

Directed evolution is the latest powerful addition to the toolbox used to develop new enzymes (Tobin et al., 2000). Naturally, this new tool also has been applied to the development of enzymes for organic synthesis (Bornscheuer, 2001; Powell et al., 2001). In particular, if an enzyme that exhibits the desired selectivity or selectivity stability is not available, directed evolution offers techniques that can be used to develop a tailor-made catalyst fulfilling the demands of the chemical process.

8.5
Perspectives

As outlined above, it has taken some time for enzymatic catalysis to be widely accepted by the chemical industry, and the full potential of enzymes is still far from being fully explored. However, many success stories have helped to pave the way for the development of the next generation of large-scale, biocatalytic processes, and the tools to facilitate development of the enzymes needed are at hand. The next decade is, therefore, expected to be an exciting area for biocatalysis.

9
Enzymes for Processing of Fats and Oils

Lipases are the most important enzymes for the processing of fats and oils and they cover a variety of applications. Next to classical

Tab. 12 Examples of products produced by enzymatic catalysis on an industrial scale

Product	*Enzyme*	*Approximate scale (tons/year)*
Acrylamide	Nitrile hydratase	> 10,000
Aspartame	Thermolysin	> 1000
Enantiopure alcohols	Lipases	> 1000
6-Aminopenicillanic acid (6-APA)	Penicillin acylase	> 1000
L-Aspartic acid	Ammonia lyase	> 1000
D-phenylglycine	Hydantoinase	> 1000
Enantiopure amines	Lipases	> 100
L-DOPA	Lipase	> 100
L-Phenylalanine	Tyrosinase	> 100
(S)-2-Chloropropionic acid	Dehalogenase	> 100

hydrolysis, lipases in immobilized form are used for synthesis and interesterification of triglycerides for the production of, for example, cocoa butter equivalents and human milk fat replacers. Phospholipases are used in the degumming process, i.e., the removal of phospholipids, in vegetable oils.

9.1 Introduction

Fats and oils play an important role in human diet, and a great variety of different fats and oils and derivatives thereof have profound industrial importance. Well-known products comprise bulk fats and oils such as margarine and cooking oils, more specialized lipids such as cocoa butter, and derivatives such as mono- and diglycerides that are the most widely used emulsifiers in the food, pharmaceutical, and cosmetic industries. Many of these products are processed in different ways to provide specific properties, most important, a well-defined melting point. Lipolytic enzymes offer obvious potential as a catalyst in some of these processes.

9.2 History

The development of enzymes for the processing of fats and oils follows very much the path of development of enzymes for organic synthesis. As outlined above, a great number of enzymes became available during the 1960s and mid-1970s. Unilever made a pioneering contribution by developing a process for enzyme-catalyzed interesterification of fats and oils that uses an immobilized lipase (Coleman and Macrae, 1977). Interesterification is a process that is used to modify the properties of triglyceride mixtures by changing the distribution of fatty acyl residues among the triglycerides (Sreenivasan, 1978). By exploiting the specificity of the lipases, Unilever showed that it is possible to produce useful triglycerides, which cannot be obtained by chemical catalysis (Macrae, 1983). The mild reaction conditions, the high catalytic efficiency, and the unique selectivity of various lipolytic enzymes are now widely recognized by the fats and oils industry, and a number of different applications have, as outlined below, been established.

9.3 Enzymes

Lipases are the dominating enzymes used within the fats and oils industry. Various lipases are used as catalysts in several different processes. Immobilized lipases are now widely used as catalysts for the interesterification of triglycerides (Xu, 2000). Depending on the specificity of the immobilized lipase used, very well defined, structured lipids can be produced. These lipids are utilized on an industrial scale for the production of specialized lipids such as cocoa butter equivalents and human milk fat substitutes. There is strong interest in the industry to expand their use to the production of bulk fats and oils such as margarine and cooking oils. The recent focus on eliminating *trans*-fatty acids from these products calls for the use of new processes, and a lipase-based process would be ideal. However, the cost of the available immobilized products so far has been prohibitive. As outline below, new technological developments may be about to change this picture. Immobilized lipases are also used to catalyze the synthesis of homogeneous triglycerides when very mild reaction conditions are required, e.g., in the preparation of long-chain, ω-3-type polyunsaturated fatty acid triglycerides that are characteristic of marine fat (Haraldsson et al., 1995).

Lipases are also used to catalyze hydrolysis of triglycerides. Non-specific lipases are used in the full hydrolysis of triglycerides, also known as "fat splitting", in particular when hydrolyzing triglycerides that contain sensitive fatty acyl residues. Lipases also can be used for the partial hydrolysis of triglycerides to afford monoglycerides. Lipase catalysis, however, offers no obvious benefits in this process, and few, if any, processes have been implemented on an industrial scale (Bornscheuer, 1995).

Another important class of lipolytic enzymes used within the fats and oils industry is the phospholipases. Depending on their specificity, these enzymes are used for different purposes. Most importantly, phospholipases are used on an industrial scale for the removal of phospholipids in vegetable oils ("degumming") (Clausen, 2001). Compared to chemical alternatives, the phospholipase-based process offers savings of both energy and water, to the benefit of both the industry and the environment. Furthermore, specific phospholipases, such as phospholipase A_2, are used industrially for the production of lyso-lecithins.

9.4 Latest Innovations

So far, lipase-catalyzed interesterification has not been sufficiently cost-effective to be introduced into true large-scale applications such as the production of margarine. Even though enzyme production has become greatly more efficient, the cost of immobilization has remained an obstacle. Recent developments have, however, changed this picture. A new process for immobilization of lipases based on granulation of silica has dramatically lowered process cost, and processes based on this new material are now being implemented for the production of commodity fats and oils with no content of *trans*-fatty acids (Christensen et al., 2001).

Another recent development is the use of an immobilized 1,3-specific lipase for the selective synthesis of 1,3-diglycerides. These lipids have been found to posses highly desired properties as low-calorie fat replacements (Nagao et al., 2000). So far, they have been introduced with great success in a number of products on the Japanese market, and they are currently being introduced in the U.S.

9.5 Perspectives

The potential of enzyme catalysis is widely recognized by the fats and oils industry. However, lipolytic enzymes so far have been used primarily for the production of relatively high-priced niche products. We hope that the availability of more cost-effective enzyme products will pave the way for wider use in the not-too-distant future.

10 Key Technologies for the Discovery of Industrial Enzymes

In the following sections, we will briefly describe main technologies for the discovery of novel enzyme product candidates. The focus is on the screening of enzymes from natural environments and on improvement of enzymes by protein engineering and directed evolution to develop products that show high performance in their target application. Accordingly, this section is divided into two sections: (1) Exploring Nature's Diversity and (2) Protein Optimization

10.1 Exploring Nature's Diversity

The traditional route of screening for novel enzymes is based on microbes that can be found in natural ecological niches and that produce the enzyme of interest. Latest developments include the so-called non-cultivable microbes by taking advantage of the genes in the environment. Both approaches are discussed below.

10.1.1 Evolutionary Diversification as Basis for Enzyme Screening

One important basis for screening of novel biocatalysts is the natural diversity of microorganisms. Prokaryotes have been evolving on earth for over 3.5 billion years (Woese, 1987). The amount of diversification and adaptation within this group is vast, and the largest fraction of the existing diversity has not yet been described (DeLong, 1997; Tiedje and Stein, 1999). The extent of bacterial diversity is unknown and cannot be extrapolated in a reasonable manner (Tiedje and Stein, 1999). However, the number of known fungi species is about 74,000, and the total number of fungal species is conservatively estimated to be 1.5 million (Hawksworth, 2001). Microorganisms can be found in all ecological niches on earth, in both natural and manmade habitats, including hydrothermal vents in the deep sea (Prieur, 1997), geothermal areas such as acidic solfataric fields (Albuquerque et al., 2000), soda and saline lakes (Kristjansson and Hreggvidsson, 1995), and in and on plant leaves (Hallmann et al., 1997). The diversity of microbial habitats reflects a fascinating diversity of physiological adaptations covering, for example, organisms growing at high or low pH and high or low temperature. The enzymes show activity profiles that normally resemble the growth physiology of the corresponding microbe, meaning that enzymes from psychrophiles are active at 0–30°C (Gerday et al., 1997), while enzymes from extreme thermophiles are active at temperatures up to and above the boiling point of water (Adams et al., 1995; Antranikian and Vorgias, 2000).

Today, it is generally accepted that only minor parts of the natural diversity have been cultured or even can be cultured in the laboratory, thereby leaving an enormous potential of yet-undiscovered physiological and biochemical traits as well as enzyme genes in the so-called metagenome (Rondon et al. 1999). Accordingly, we present the following methods: (1) the traditional screening approach, which has its roots in cultivation of microbes, and (2) the metagenomics approach, which directly exploits gene diversity without prior cultivation (Short, 1997).

10.1.2 Traditional Enzyme Screening

Industrial enzymes on the market today are, in the vast majority, extracellular enzymes that derive mainly from two major groups of microorganisms, i.e., bacteria and fungi. All these products originate from traditional screening, i.e., from microbes that have been cultured. A screening program for industrial enzymes starts with a problem to be solved by enzymes and a business justification (Cheetham, 1998). Based on insight into the industrial process where the enzyme will be used, criteria are defined that include the substrate to be converted, temperature, pH, and presence or absence of ions, and from which a sensitive and specific functional screening assay is developed for detection of the enzyme from natural microbial isolates (Figure 16). It is then decided which physiological or taxonomic groups most probably produce the desired activity (Bull et al. 2000, Ogawa and Shimizu, 1999). Generally,

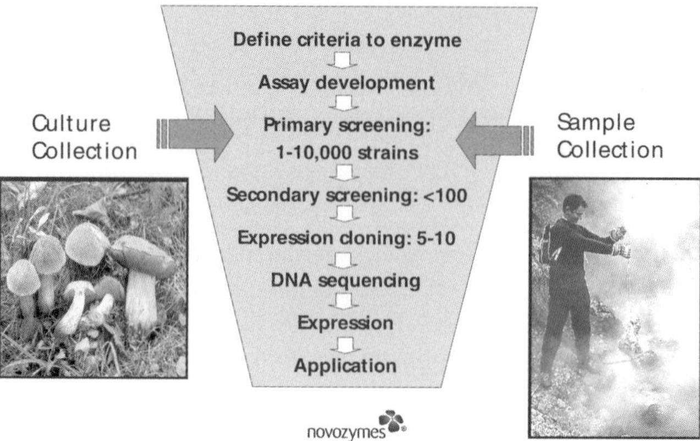

Fig. 16 Overview on traditional screening route including primary and secondary screening rounds.

microbial screening programs are iterative processes divided into two main phases. In a broad primary screening, all microbes that are positive in the applied assay are selected. In the next round, these hits are subjected to a secondary screening, where a highly selective enzyme assay is applied that allows ranking and selection of the best-performing enzyme candidates. Obtained hits that perform well are further characterized, often including purification and testing in small-scale models of the real application. Finally, the remaining enzymes are compared with respect to performance and production feasibility to form a short list of top candidates. As a consequence of the above-described screening route, large and highly diverse sample and culture collections have been established and continue to grow. The aim of these collections is to have representatives of as many taxonomical, physiological, ecological, and geographical groups of microorganisms and as many samples as possible and to be optimally equipped when the screening of any type of enzyme is initiated.

10.1.2.1
Cloning as an Important Step to Obtain Monocomponent Enzymes for Application Testing

For further selection among the remaining top candidates from secondary screening, the encoding genes are cloned and expressed by a variety of different methods (Sambrock and Russel, 2001). This enables production of quantities of pure enzymes for thorough characterization and application testing. In most cases, a technique called expression cloning is used, which is an effective means of isolating a gene from a gene library based on its encoded activity (Dalbøge and Heldt-Hansen, 1994). This technology combines the ability of the host organism, normally baker's yeast (*Saccharomyces cerevisiae*), and of *E. coli* to express heterologous enzyme genes with the utilization of sensitive, simple, and reliable enzyme assays. The method allows simultaneous screening of the cDNA and genomic libraries for many different enzyme activities simply by making several replicas of the corresponding agar- or microtiter plates and applying different screening assays.

10.1.2.2
Importance of the Screening Assay and Screening Technology

The screening assay has a central role during the whole selection/deselection program, as the quality of the assay determines the quality of the resulting candidate. The strength of activity-based screening is the possible discovery of totally novel enzymatic activities and corresponding enzyme genes. These assays are used to isolate microbes from environmental samples (Apitz and Van Pee, 2001; Horikoshi, 1995), specific clones in gene libraries actively expressing the gene of interest, and variants from artificial evolution projects (see below). Much effort is invested to develop screening assays and novel screening technologies that give enhanced freedom in assay design. Accordingly, a variety of publications and patent applications cover this field, of which only selected examples are referred to here (Joo et al., 1999; Ruijssenaars and Hartmans 2001; Meeuwsen et al. 2000; Preisig and Byng 2001; Kongsbak et al., 1999; Short and Keller 2001; Schellenberger, 1997). As a routine part of the screening process, the cloned genes are sequenced. This process yields valuable information for understanding the structures and specific functions of the enzymes and allows classification and comparison to similar enzymes.

10.1.2.3
Homology-based Screening

Functional screening is often supplemented with screening techniques that are based on similarities between enzyme-encoding gene sequences (Dalbøge and Lange, 1998). Sequence information from a set of related enzyme genes is used to screen for similar genes from other organisms. An alignment of either the nucleotide sequences or the corresponding amino acid sequences followed by identification of evolutionarily conserved regions makes it possible to design degenerate PCR primers containing the DNA sequence corresponding to the homologous region. These primers can be used to amplify and sequence a fragment of a homologous gene from a sample of genomic DNA. By using this method, a number of genes homologous to the initial gene sequences can be identified quickly. Using this method, Schülein et al. (1996) have described screening and cloning of 32 cellulase genes from cellulose family 45 (Henrissat and Romeu, 1995), representing all four major fungal taxa, i.e., ascomycota, zygomycota, chytridiomycota, and basidiomycota, based on only four homologous gene sequences known at the start of the project.

10.1.3
Bioinformatics as a Basis for Enzyme Discovery

It is apparent that the above-described screening approaches require tools to handle and compare sequence data obtained from cloning projects. Additionally, the expanding inventory of cloned enzyme genes in the public domain, which is rapidly increasing the available sequence information for a large selection of enzymes and enzyme families, has to be retrieved. The situation became more complex and challenging when whole bacterial, fungal, and archaeal genomes were sequenced as exemplified for *Bacillus subtilis* (Kunst et al., 1997), *Aspergillus nidulans* (Dunn-Coleman and Prade, 1998), and *Methanococcus jannaschii* (Bult et al., 1996). The current status on established genomes and those underway can be obtained by visiting the homepage of the TIGR institute (http://www.tigr.org/). The importance of bioinformatics is reviewed in Bull et al., (2000) and cannot be described here in further detail because of space limitations.

10.1.4 Metagenomics or Cloning from Non-cultivable Microorganisms

Metagenomics (Handelsman et al., 1998; Lorenz and Schleper, 2002; Rondon et al., 1999) is an alternative, complementing approach to screening for a diversity of enzymes. DNA or mRNA is isolated directly from an environmental sample and then is purified, digested, and cloned into suitable cloning vectors to construct complex "environmental libraries". These gene libraries need to be screened as described above by using either sequence-based techniques (Dalbøge and Lange, 1998) or activity assays (Rondon et al., 2000). Ideally, cultivation-independent approaches enable microbiologists to fully exploit the biological potential of a microbial community in its totality. The fundamentals of this fascinating, but challenging, approach were described by microbial ecologists who applied novel molecular techniques to study microbial diversity of communities in the beginning of the 1980s:

- There is a huge microbial diversity in soil and sediment samples (Torsvik, 1980; Torsvik et al., 1998; Torsvik et al., 1990).
- Only minor parts of the microbes from a given sample can be cultivated (Jannasch and Jones, 1959; Gonzalez et al., 1996; Kogure et al., 1979; Amann et al., 1995; Hugenholtz and Pace, 1996).
- The total microbial diversity comprises phyologenetic groups of organisms not related to any organism described so far, as revealed by analysis of the rRNA sequences (Hugenholtz and Pace, 1996; Pace et al., 1986; Schmidt et al., 1991; Amann et al.,1995).
- DNA can be isolated directly from these samples (Olsen et al., 1986; Steffan et al., 1988; Torsvik, 1980).
- PCR reactions can be carried out on environmental DNA templates (Olsen et al., 1986; Stahl et al., 1985; Steffan et al., 1988).
- DNA can be digested and cloned in suitable cloning vectors. Genes can be screened and cloned directly from environmental libraries (Schmidt et al., 1991; Stein et al., 1996).

It is tempting to exploit this huge diversity of microbes for the discovery of industrial enzymes (Short, 1997), though there are also some important constraints and limitations discussed in the literature (Lorenz and Schleper, 2002). The following are the most noteworthy:

- quality of the environmental DNA (Akkermans et al., 1994, 1995; Lorenz and Schleper, 2002; Torsvik, 1980; Steffan et al., 1988);
- normalization of the library in order to avoid repetitive screening of highly abundant genes (Johnson et al., 1986; Short and Mathur, 1999; Bonaldo et al., 1996; Li et al., 1999);
- size of the metagenomics libraries, which for many screenings presupposes high-throughput screening equipment and investment in that as well as alternative cloning methods, such as cosmid (Entcheva et al., 2001), fosmid (Schleper et al., 1998; Stein et al., 1996), or BAC libraries (Rondon et al., 2000; Lorenz and Schleper, 2002);
- applicability of the approach for eucaryotic microbes undoubtedly present in the environment;
- product and production approval that in many countries might require a defined source, i.e., a bacterial or fungal species;
- production potential of the detected gene in a production organism (Lorenz and Schleper, 2002).

Published results to date show that environmental libraries can be screened for functional activities, though it can be stated that screening efficiency seems surprisingly low. It is apparent from Cottrell et al. (1999) that screening efficiency is dependent on the DNA sample and the screening method (microtiter plate assays versus plaques). While 9 chitinase clones were found from 75,000 screened from an estuarian sample, only 2 (identical) clones were detected from 750,000 in the coastal sample. No cellulose hits were detected, although 750,000 clones were screened.

Henne et al. (2000) describe screening of lipase/esterase activity in three different environmental libraries constructed according to the same procedure. The hit rate is remarkably low. In library 1 for example, no hit was found among 430,000 clones on triolein agar, and in total only four clones were found from ca. 1 million clones screened.

An even lower hit rate was described for a Na^+/H^+ antiporter (Majernik et al., 2001): from about 1.5 million clones screened only two were found positive, indicating that only two bacteria in the whole diverse consortium show this activity, even though this activity is important for Na^+ transport in all living cells.

The BAC libraries made by Rondon et al. (2000) contained DNAse, amylase, antibacterial activity, and lipase, while cellulase, chitinase, esterase, keratinase, protease, and hemolytic activity were not detected in the same library.

In conclusion, metagenomics has yielded evidence for new organisms and new genetic resources containing millions of uncharacterized genes. However, construction and screening of environmental libraries is still in its infancy, and screening efficiency is low compared to costs and technical complexity. The initial results are promising, but methodological problems need to be solved before the virtually boundless diversity of genes hidden in the non-cultivable world of microbes can be fully accessed (Lorenz and Schleper, 2002; Rondon et al., 1999; Tiedje and Stein, 1999; Ogram, 2000).

10.2 Protein Optimization

Industrial applications are often carried out under conditions that are far from those that exist in nature. Therefore, it cannot be expected that enzymatic solutions to an industrial process can necessarily be identified in nature. The ability to optimize natural enzymes for properties relevant to specific industrial application has enhanced our capabilities for providing the enzymes necessary to address a wider range of industrial processes.

10.2.1 Rational Protein Engineering

After relying solely on nature as the source of diversity for new industrial enzymes for decades, developments within molecular biology in the 1980s promoted a new discipline: rational protein engineering.

The dual capabilities of heterologous protein expression and directed DNA mutagenesis provide the possibility of inserting, deleting, or changing any amino acid in a protein and evaluating the consequences of the alteration (Smith, 1985). Thus, detailed knowledge of the three-dimensional structure and the biochemical and biophysical properties of an enzyme should, in theory, guide a trained protein engineer in selecting the residues to be altered in order to change the properties of this enzyme in the desired direction. Naturally, rational engineering has been a process of trial and error, as the understanding of structure/function relationships is a difficult science. We are not yet at the stage where inspection of the protein structure enables exact prediction of the amino acids to be changed in order to end up with the desired properties in the resulting enzyme. The reasons for this limitation reside primarily in the delicate balance observed in most enzymes between stability and efficient catalysis, as well as in our limited understanding of the catalytic event.

However, in many cases we are able to pinpoint regions, positions, or even specific substitutions that are more likely to have the desired effect on the molecule than will changes in other regions/positions/substitutions. Often enzymes are obtained that are indeed improved for a particular property after construction of only a limited number of variants.

Despite the limitations of rational engineering, the trial-and-error process has led to a vast amount of literature on improved proteins, and a number of commercial products are currently on the market. The majority of commercial success stories are presented only in patents. The literature on the protein engineering of various enzymes is too large to cover here, but Table 13 provides references to reviews on using protein engineering to alter properties for a number of the most commercially interesting classes of enzymes.

Tab. 13 Recent reviews on protein engineering of industrially relevant enzymes

Enzyme	*Reference*
Proteases	Bryan, 2000
α-Amylases	Nielsen and Borchert, 2000
Cellulases	Schülein, 2000
Lipases	Svendsen, 2000
Glucose isomerase	Hartley et al., 2000
Glucoamylase	Sauer et al., 2000
Cyclodextrin glycosyl transferase	Van der Veen et al., 2000

Engineered versions of a number of hydrolytic enzymes such as proteases, amylases, lipases, and cellulases are currently available. An example of rational engineering to reach commercial goals is presented by Bisgaard-Frantzen et al. (1999). They describe the successful development of new α-amylases for two very diverse applications, i.e., as an additive to detergent formulations to aid the removal of starch-containing stains in laundry and as the catalyst in the starch-liquefaction step in the high-fructose corn syrup process. These applications are very different with respect to demands on the enzyme catalyst, as washing is typically carried out at alkaline conditions and starch liquefaction is carried out at extreme temperature (initially above 100°C) and acidic pH.

10.2.1.1 Idea Generation

The first step in rational protein engineering has to be a hypothesis on the limitations of the present protein molecule with respect to the desired property. Stability issues tend to be the easiest to address, as there often are many solutions to a particular problem. Addressing properties such as activity or general application performance is less trivial.

The methods normally used to suggest mutagenesis strategies include visual inspection of protein structures obtained from x-ray crystallography or NMR studies and computational approaches, such as studies of the dynamic properties of the enzymes by molecular dynamics simulations. Protein electrostatics calculations are used as guiding tools when investigating properties involving charges, i.e., pH dependency of catalysis or pH-induced inactivation of the protein (Honig and Nicholls, 1995). Docking studies are used to gain knowledge about protein interactions with ligands, i.e., enzyme interactions with substrates, inhibitors, and cofactors in the case where the primary interest is within catalysis.

Successes obtained through years of trial and error have resulted in a constantly improved understanding of the molecular basis for functional properties of enzymes and have provided an essential basis for carrying out rational engineering.

Most stability issues can now be addressed by rational means, and even more complicated properties such as catalysis, substrate and product specificities, and so forth have been successfully addressed by rational means (Cedrone, et al., 2000; Beier et al., 2000).

10.2.1.2
The Natural Limitation

Nature has been able to solve nearly any catalytic challenge by the combinatorial use of only 20 amino acids, but the protein engineer rapidly realizes that the use of "non-natural" amino acids could increase the chances of utilizing the natural scaffolds in a more optimal way. If a greater diversity of functional groups can be tried in crucial positions, the chances for moving protein function in new directions is higher. Ibba and Hennecke (1994) have reviewed the possibilities of introducing non-natural amino acids into protein by utilizing *in vivo* systems. Of course, the number and variety of amino acid analogues that can be introduced are still limited, and a low efficiency of incorporation also poses problems for the commercial utilization of interesting semi-natural enzymes.

An alternative way of introducing new functional groups by posttranslational chemical modification has been described by Kenyon and Bruice (1977) and Khumtaveeporn et al. (2001). This technique is based on the introduction of a unique cysteine residue into a particular exposed position in a protein. The thiol side chain of such a cysteine can be chemically modified by reaction with various methanethiosulfonate reagents and new functional groups are introduced in this position. A large variety of new functional groups can be introduced by this method, but the availability of proteins lacking additional cysteine residues limits its use.

10.2.2
Directed Molecular Evolution

The 1990s placed another tool in our toolbox – directed molecular evolution – fulfilling a long-held desire to mimic nature in its methods of evolving new properties and traits, i.e., mutagenesis and recombination.

The laboratory process consists of repeated rounds of three steps (Figure 17). First, diversity is generated by carrying out random mutagenesis of a gene. In the second step, the connection between the generated gene diversity and the encoded functional diversity is established. In most cases, the pool of mutagenized genes is transformed into a cell where the protein is expressed, thereby generating a so-called library of variants. The last step is the identification and isolation of the fittest individuals in the library, which are subsequently utilized as starting material in the next round of directed molecular evolution. This process is typically based on a screening, selection, or a panning procedure.

10.2.2.1
Step 1: Initial Diversity Generation

Mutations can be introduced randomly by utilizing chemical or physical means or by using DNA polymerases with relatively high error frequencies. Such polymerases can additionally be used under non-optimal conditions, such that misincorporation happens at an even higher frequency in the synthesized DNA (Leung et al., 1989). Any

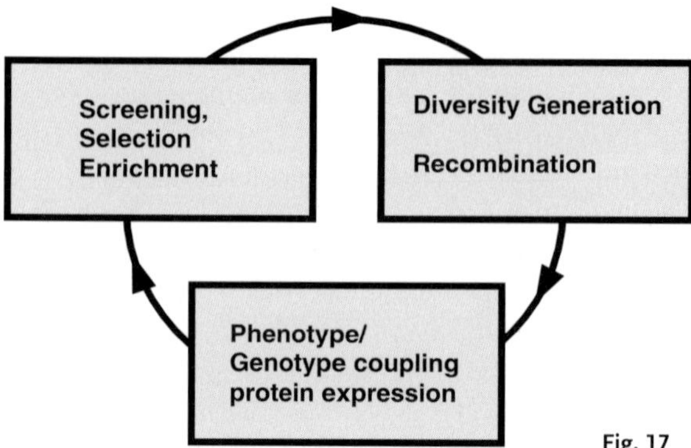

Fig. 17 The directed molecular evolution cycle.

kind of random mutagenesis method will display an intrinsic bias towards specific kinds of changes. With the modest mutagenesis frequency used in most directed evolution cases, the probability of changing more than one base in a codon is rather small (Steipe, 1999). Consequently, it is very unlikely to observe more than five or six different amino acid substitutions for a particular position in the protein, and the belief that completely random mutagenesis methods explore all possible substitutions remains an illusion.

In other methods, small degenerate oligonucleotides can be utilized to direct a higher frequency of mutations towards small, defined regions of the protein. In those cases, the design of the mutagenic oligonucleotide determines the randomization of the designated region of the gene.

Alternatively, the initial diversity can be generated *in vivo* in a living cell, e.g., in a mutator strain, thereby combining step 1 and step 2 of the directed molecular evolution procedure (Greener et al., 1996; Fabret et al., 2000).

10.2.2.2
Step 2: Coupling of Sequence and Functional Diversity

The most frequently used method of coupling the diversity seen in the gene with the expressed variant protein is to express the encoded protein by the use of a cellular system, e.g., bacteria, yeast, fungus, or insect cells. The obtainable size of the library is limited by the transformation frequency of the selected host. If the most popular molecular biology host, *E. coli*, is utilized, libraries will typically be limited to around 10^7 individuals. Even smaller libraries can be expected if more complicated hosts are used.

Variations on establishing the connection between genotype and phenotype exist. In some cases, the protein can be generated by *in vitro* translation, thereby eliminating the transformation step and rendering the cells superfluous. Such a procedure is typically applied if a coupling between the genetic diversity and the translated protein can be made, e.g., by the elegant mRNA display technique (Hanes and Plückthun, 1997). If the subsequent enrichment procedure is sufficiently efficient that single protein molecules are the basis for the isolation of the improved candidates, this method has the advantage of generating and screening extremely large libraries of the size of 10^{15} individuals.

10.2.2.3
Step 3: Screening/Selection or Enrichment

The generated library of protein variants is subjected to a procedure in which the variant proteins are distinguished based on their functional properties and typically compared with the wild-type enzyme. Such procedures can be based on screening methods that typically are downscaled to small volumes in microtiter plates in order to enhance the throughput. It is a very demanding task to set up a meaningful assay that addresses all relevant properties for a particular industrial application. Some of the important parameters have been described above in the section on screening of natural diversity. The demands are, however, stricter when looking at laboratory-generated diversity, as we typically search for marginal improvements in a well-functioning wild-type enzyme rather than an "on/off" answer when dealing with wild-type screening.

Various attempts to look at large libraries in screening procedures have been carried out (Wölcke and Ullmann, 2001).

If expression of a particular property can provide a growth advantage to the host, selection procedures can be applied. Naki et al. (1998) demonstrated that efficient production of a subtilisin results in a growth advantage to the host cells during growth in a defined medium with serum albumin as the sole nitrogen source.

The coupling of phenotype and genotype also can be carried out by transformation into a cellular system that promotes the display of the protein (phenotype) on the surface of the cell or phage, thus establishing the connection to the genotype. The typical example is phage display (O'Neil and Hoess, 1995), but cells also have been utilized for this purpose (Ståhl and Uhlen, 1997). The same demand for selection at the single-molecule level applies to these systems as that described above for the enrichment procedures used for mRNA display

10.2.2.4
The Iterative Element

The simplest approach is to utilize sequential rounds of random mutagenesis. After the first round of directed molecular evolution, the fittest candidate is exposed to another round of random mutagenesis, library generation, and isolation of the best variant.

Two major disadvantages exists for such methods: 1) there is no easy way of getting rid of harmful mutations, as both beneficial and detrimental changes tend to accumulate over a number of cycles; and 2) only the best candidate from a mutagenesis round is typically utilized; thus, beneficial mutations might be discarded in each round. The invention of DNA shuffling by W.P. Stemmer solved these problems (Stemmer, 1994a,b). The genes encoding all the improved clones from the first round are randomly fragmented, and these fragments are used in a PCR reaction where they act as both primers and templates of each other. After sufficient rounds of annealing and extension, the full-length gene is reassembled and typically reamplified in another PCR. A library is generated and screened, and the complete process is repeated until the desired properties have been obtained.

This method provides clear advantages, as improvements are not discarded from a round of evolution, and it allows for removal of detrimental diversity (Figure 18).

Variations on this theme have been reported covering both alternative *in vitro* techniques (Zhao et al., 1998; Coco et al., 2001) and formats utilizing the recombination potential of living cells (Cherry et al., 1999).

Rather than using synthetic variants of a gene, the starting material can be a number of homologous genes from different species, resulting in a technology called "family shuffling" (Crameri et al., 1998). The initial

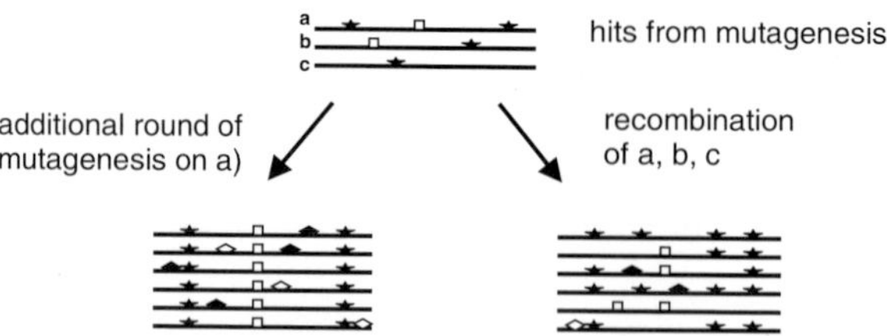

Fig. 18 Schematic illustration demonstrating the differences between sequential rounds of mutagenesis and recombination based directed molecular evolution. a, b and c represent isolated positives from the first round of mutagenesis. Filled stars and open squares are beneficial and detrimental mutations, respectively. To the left the fittest individual, denoted a), is subjected to a second round of mutagenesis and new beneficial or detrimental mutations are introduced (filled or open diamonds) in the progeny. Beneficial and detrimental mutations accumulate. Alternatively (to the right) all three positive hits are recombined in a reaction where additional mutations can be introduced. The possibility exists of combining the beneficial mutations without accumulating detrimental mutations in the progeny.

genes represent unique points in sequence space that all represent unique answers to a particular catalytic property. The shuffling reaction creates random hybrids between the starting genes, thus generating new spots in sequence space that address the same question. This technology results in a surprisingly high fraction of active hybrids, which might reflect the fact that the complete starting diversity has been preselected by nature. In other words, in each position, the amino acids found in the progeny have already fulfilled the requirements for function in one of the parents. The evolutionary steps are thus larger than observed in the case where the starting material is point mutations in one starting gene.

Ness et al. (1999) have carried out a family shuffling experiment with a very large family of 26 subtilisin fragments. The resulting library was screened for five different properties, and candidates that were improved for each of the screened-for properties could be identified from a library of less than 600 active clones. Even new subtilisins improved simultaneously for two or more properties were identified from the screening.

Alternative formats of artificial hybrid formation have been reported (Ostermeier et al., 1999; Sieber et al., 2001)

10.2.2.5
Understanding the Evolutionary Experiment

Recently, hypotheses for explaining and understanding the molecular evolutionary process have been presented. Voigt et al. (2001) have tried to explain the outcome of directed molecular evolution experiments from structural and computational considerations, and they propose the hypothesis that beneficial mutagenic or recombinogenic events cause the least disturbance of the structure. It will be interesting to see whether such theories result in the future design of better and more successful directed evolution techniques.

10.2.3
Protein Optimization Outlook

All technologies used for comparing the properties of proteins are limited in throughput capacity. Thus, the number of clones that can be screened from a library is limited, and it is preferable to screen "quality diversity" rather than any diversity.

The computational methods used in rational engineering also can be applied to designing mutagenesis strategies for directed evolution experiments, and as much knowledge as possible should be utilized.

A recent example of the elegant combination of rational engineering and directed evolution was published by Altamirano et al. (2000), who introduced a new phosporibosylanthranilate isomerase activity into an enzyme scaffold with an active site able to carry out a completely different reaction, indole-3-glycerol phosphate synthetase.

The efficient development of future biocatalysts relies on combining our current technologies. We foresee a situation where directed molecular evolution and rational engineering as complementing technologies will enable the future development of an increasingly broad range of enzyme-based industrial catalysis.

10.3
Conclusion

Discovery of industrial enzymes is a multidisciplinary effort involving a wide array of different technologies. In the sections above, we have outlined some major routes for the discovery of enzymes for industrial applications. Nature holds a wonderful diversity of organisms and a corresponding wealth of enzymes, but even nature's assortment faces some limitations. It is the challenge of scientists to optimize the natural enzymes and to generate manmade diversity to custom make enzymes for a given application. Finally, many more enzymes have already been made available as products for several applications. For all approaches it is important to stress that it is not the *broadest* possible diversity but rather the highest possible *quality* of diversity that is the ultimate goal. In this respect selection/deselection via perfectly designed assays are of utmost importance, indicating the significance of linking process understanding to biochemistry. A key technology toolbox combined with perfect understanding of the target application, in other words, the link between research and development, is the key to successful development of industrial enzymes. Within discovery we consider natural diversity approaches and optimizing strategies as complementing routes, and both are equally important to develop an unlimited diversity for enzymes. As Cheetham points out, it needs to be highlighted that "people with many different skills are needed to create successful biocatalysts, and with the contribution of each specialism absolutely vital for the success" (Cheetham, 1998). It will be the combination of skills, knowledge, new technologies, and novel enzymes that will ensure future benefits by the development of truly eco-friendly processes.

Acknowledgments

The authors gratefully acknowledge the help of many colleagues at Novozymes. Special thanks from Randy Deinhammer to Eric Gormsen, Ture Damhus, Thomas Callisen, Niels Munk, and Keith Gibson at Novozymes and to Mike Showell at Procter & Gamble for their help in reviewing this article and for use of figures; from Henrik Lund to Neal Franks, Hui Xy, Hanne Høst Pedersen, and Gerda Jensen; from Thomas Schäfer to Fiona Becker and Mary Stringer for their patience reading and correcting my "German-English," as well as to Gerda Jensen for the help with Reference Manager; and from Torben Vedel Borchert to Mary Stringer, Allan Svendsen, and Steffen Danielsen for critically reading the manuscript.

11
References

Aaslyng, D., Gormsen, E., Malmos, H. (1991) Mechanistic studies of proteases and lipases for the detergent industry, *J. Chem. Technol. Biotechnol.* **50**, 321–330.

Adams, M. W. W., Perler, F. B., Kelly, R. M. (1995) Extremozymes: Expanding the limits of biocatalysis, *Biotechnology* (NY), **13(7)**, 662–668.

Adlercreutz, P. (1996) Cofactor regeneration in biocatalysis in organic media, *Biocatal. Biotransform.* **14**, 1–30.

Adler-Nissen, J., Gürtler, H., Jensen, G.W., Olsen, H.S., Riisgaard, S., Schülein, M. (1984) Method for treating plant polysaccharides using SPS-ase for decomposing soluble polysaccharides. US Patent 4,478,854.

Akin, D. E., Foulk, J. A., Dodd, R. B., McAlister III, D. D. (2001) Enzyme-retting of flax and characterization of processed fibers, *J. Biotechnol* **89(2–3)**, 193–203.

Akkermans, A. D. L., Mirza, M. S., Harmsen, H. J. M., Blok, H. J., Herron, P. R., Sessitsch, A., Akkermans, W. M. (1994) Molecular ecology of microbes: A review of promises, pitfalls and true progress, *FEMS Microbiol Rev.* **15(2–3)**, 185–194.

Akkermans, A. D. L., Van Elsas, J. D., De Bruijn, F. J. E. (1995) *Molecular Microbial Ecology Manual*, Dordrecht, Kleuwer.

Albuquerque, L., Rainey, F. A., Chung, A. P., Sunna, A., Nobre, M. F., Grote, R., Antranikian, G., da Costa, M. S. (2000) Alicyclobacillus hesperidum sp nov and a related genomic species from solfataric soils of Sao Miguel in the Azores, *Int J Syst Evol Microbiol* **50**, 451–457.

Altamirano, M.M., Blackburn, J.M., Aguayo, C, Fersht, A.R. (2000) Directed evolution of a new catalytic activity using the alpha/beta-barrel scaffold, *Nature* **403**, 617–622.

Amann, R. I., Ludwig, W., Schleifer, K. (1995) Phylogenetic identification and *in situ* detection of individual microbial cells without cultivation, *Microbiol Rev.* **59(1)**, 143–169.

Andersson, E. M., Larsson, K. M., Kirk, O. (1998) One biocatalyst - many applications: The use of *Candida antarctica* B-lipase in organic synthesis, *Biocatal. Biotransform.* **16**, 181–204.

Antheunisse, W., Van Der Logt, C.P.E., Parry, N.J., Swarthoff, T (2001) Bleaching detergent compositions. Patent WO200148135-A1.

Antranikian, G., Vorgias, C. E. (2000) Glycosyl hydrolases from extremophiles, *Glycomicrobiology* **11**, 313–340.

Apitz, A., Van Pee, K. H. (2001) Isolation and characterization of a thermostable intracellular enzyme with peroxidase activity from *Bacillus sphaericus*, *Arch Microbiol*, **175(6)**, 405–412.

Aschengreen N.H. (1969). Microbial enzymes for alcohol production. *Process Biochemistry*, A5648. Bagsvaerd, Denmark: Novo Terapeutisk Laboratorium (our old name).

Aunstrup, K. (2001) *Growing Novozymes*. Bagsvaerd, Denmark: Novozymes A/S.

Barfoed, M., Kirk, O. (1999) Method for dyeing a material with a dyeing system which contains an enzymatic oxidizing agent. US Patent 5,972,042.

BAFF (2001) I drift 2003, Nyhetsbrev BAFF Nr 1 December 2001, (page 1), Örnsköldsvik, Sweden: BioAlcohol Fuel Foundation (BAFF) (in Swedish).

BCC (2001) Enzymes for industrial applications, Section 5.1, June 2001 Market Report. Business Communications Company, Inc.

Béguin, P., Aubert, J-P. (1994) The biological degradation of cellulose, *FEMS Microbiol Rev*, **13**, 25–58.

Beier, L., Svendsen, A., Andersen, C., Frandsen, T. P., Borchert, T. V., Cherry J. R. (2000)

Conversion of the maltogenic alpha-amylase novamyl into a CGTase, *Protein Eng.* **13(7)**, 509–513.

Bernton, H. Kovarik, B. Sklar, S. (1982). *The Forbidden Fuel: Power Alcohol in the 20th Century*, New York: Griffin.

Bieser, H.J., de Ressett, A. J. (1977). Continuous counter current separation of saccharides with inorganic adsorbents, *Stärke* **29**, 392–397.

Bisgaard-Frantzen, H., Svendsen, A., Norman, B., Pedersen, S., Kjærulff, S., Outtrup, H., Borchert T.V. (1999). Development of industrially important α-amylases, *J. Appl. Glycosci.* **46**, 199–206

Blum, R., Stahl, W. H. (1952) Enzymic degradation of cellulose fibers, *Textile Res. J.* **22**, 178–192.

Bodie, E.A., De Vries, C.H., Wang, H. (2000) Detergent compositions comprising phenol oxidase enzymes. Patent No. WO200039306-A2.

Bonaldo, M. F., Lennon, G., Soares, M. B. (1996) Normalization and subtraction: two approaches to facilitate gene discovery, *Genome Research*, **6(9)**, 791–806.

Bornscheuer, U. T. (1995) Lipase-catalyzed synthesis of monoacylglycerols, *Enzyme Microb. Technol.* **17**, 578–586.

Bornscheuer, U. T. (2001) Directed evolution of enzymes for biocatalytic applications, *Biocatal. Biotransform.* **19**, 85–97.

Bruinenberg, P. M., Hulst, A. C., Faber, A., Voogd, R. H. (1995). A process for surface sizing or coating paper. EP 0 690 170 A1.

Bryan, P. (2000) Protein engineering of subtilisins, *Biochem. Biophys. Acta - Protein Structure and Molecular Enzymology* **1553**, 203–222.

Buchholz, K. B., Poulsen, P. B. (2000). Introduction, in: *Applied Biocatalysis*, 2nd edition, (A. J .J. Straathof, P. Adlerkreutz Eds.), New York: J Harwood Academic Publishers, 1–15.

Bull, A. T., Bunch, A. W., Robinson, G. K. (1999) Biocatalysts for clean industrial products and processes, *Curr. Opin. Microbiol.* **2(3)**, 246–251.

Bull, A. T., Ward, A. C., Goodfellow, M. (2000) Search and discovery strategies for biotechnology: The paradigm shift, *Microbiol. Mol. Biol. Rev.* **64(3)**, 573–606.

Bult, C. J., White, O., Olsen, G. J., Zhou, L. X., Fleischmann, R. D., Sutton, G. G., Blake, J. A., Fitzgerald, L. M., Clayton, R. A., Gocayne, J. D., Kerlavage, A. R., Dougherty, B. A., Tomb, J. F., Adams, M. D., Reich, C. I., Overbeek, R., Kirkness, E. F., Weinstock, K. G., Merrick, J. M., Glodek, A., Scott, J. L., Geoghagen, N. S. M., Weidman, J. F., Fuhrmann, J. L., Nguyen, D., Utterback, T. R., Kelley, J. M., Peterson, J. D., Sadow, P. W., Hanna, M. C., Cotton, M. D., Roberts, K. M., Hurst, M. A., Kaine, B. P., Borodovsky, M., Klenk, H. P., Fraser, C. M., Smith, H. O., Woese, C. R., Venter, J. C. (1996) Complete genome sequence of the methanogenic archaeon, methanococcus-jannaschii, *Science* **273(5278)**, 1058–1073.

Busch, A., Bettiol, J.P., Smets, J., Boyer, S.L. (1999) Laundry detergent and/or fabric care compositions comprising a modified cellulase. WO9957260-A1.

Cai, J. Y., Evans, D. J., Smith, S. M. (2000) Bleaching of cotton and cotton/wool blends with TAED and NOBS activated peroxide systems, in: *AATCC International Conference & Exhibition*, CD-ROM. Research Triangle Park, NC: AATCC.

Callisen, T.H., Damhus, T. (2000) Anti-redeposition effects by lipases in relation to fat-associated build-up of diffuse soiling/dinginess, *AOCS Conference*, USA, April 25–28 (paper available from authors).

Campbell, C. J., Laherrère, J. H. (1998). The end of cheap oil. *The Global Hubbert Peak in Scientific American*, **278(3)**, 78–83.

Cedrone, F., Menez, A., Quemeneur, E. (2000) Tailoring new enzyme functions by rational redesign, *Curr. Opin. Struct. Biol.* **10**, 405–410.

Chakar, F., Allison, L., Kim, D. H., Ragauskas, A. J., Elder, T. J. (2000) The path forward to practical nascent laccase biobleaching technologies, *Proceedings Tappi Pulp./Process & Prod. Quality Conf. 2000.*

Cheetham, P. S. J. (1998) What makes a good biocatalyst? *J. Biotechnol.* **66(1)**, 3–10.

Cherry, J.R., Lamsa, M.H., Schneider, P., Vind, J., Svendsen, A., Jones, A., Pedersen, A.H. (1999) Directed evolution of a fungal peroxidase, *Nat. Biotechnol.* **17**, 379–334.

Chiu, C.-W. (1991a). Modified starch emulsifier characterised by shelf stability. US Patent 4,977,252.

Chiu, C.-W. (1991b). Short chain amylose as a replacement for fats in foods. EP 0 486 936 A1.

Christensen, M.W., Andersen, L., Kirk, O., Holm, H.C. (2001) Enzymatic interestification of commodity oils and fats: approaching the tonnes scale, *Lipid Technol. News* **2001**, 33–37.

Claeyssens, M., Henrissat, B. (1992) Specificity mapping of cellulolytic enzymes: Classification into families of structurally related proteins confirmed by biochemical analysis, *Protein Science* **1**, 1293–1297.

Clausen, K. (2001) Enzymatic oil-degumming by a novel microbial phospholipase, *Eur. J. Lipid Sci. Technol.* **103**, 333–340.

Coco W. M., Levinson, W. E., Crist, M. J., Hektor, H. J., Darzins, A., Pienkos, P. T., Squires, C. H., Monticello, D. J. (2001) DNA shuffling method for generating highly recombined genes and evolved enzymes, *Nat. Biotechnol.* **19**, 354–359.

Coleman, M. H., Macrae, A. R. (1977) Catalytic rearrangement of fatty acid groups in glyceride fats or oils by contacting with an enzyme transesterification catalyst activated with water. Dutch Patent No. NL 7,701,449.

Conrad, L. S., Damhus, T., Kirk, O., Schneider, P. (1997) Enzymatic inhibition of dye transfer, *Inform* **8(9)**, 950–957.

Cooper, E. W. (1998) Treatment of hardwood pulp with a mixture of cellulase(s) and xylanase(s) – to reduce vessel element picking in subsequent printing on paper made from the pulp. US Patent 5,725,732-A, Glatfelter Co, P. H., (patent).

Cottrell, M. T., Moore, J. A., Kirchman, D. L. (1999) Chitinases from uncultured marine microorganisms, *Appl. Environ. Microbiol.* **65**, 2553–2557.

Crameri, A., Raillard, S. A., Bermudez, E., Stemmer, W. P. (1998) DNA shuffling of a family of genes from diverse species accelerates directed evolution, *Nature* **391**, 288–291.

Dalbøge, H. Heldt-Hansen, H. P. (1994) A novel method for efficient expression cloning of fungal enzyme genes, *Mol. Gen. Genet.* **243(3)**, 253–260.

Dalbøge, H., Lange, L. (1998) Using molecular techniques to identify new microbial biocatalysts, *Trends Biotechnol.* **16(6)**, 265–272.

Damhus, T., Vogt, U. (2000) Process for removal of excess dye from printed or dyed fabric or yarn. US Patent 6,048,367.

Davies, G., Henrissat, B. (1995) Structures and mechanisms of glycosyl hydrolases, *Structure* **3(9)**, 853–859.

Davis, P.J., Van Der Logt, C.P.E., Parry, N.J. (2001) Bleaching detergent compositions. WO200107555-A1.

Deguchi, T., Kitaoka, Y., Kakezawa, M., Nishida, T. (1998) Purification and characterization of a nylon-degrading enzyme, *Appl. Environ. Microbiol.* **64(4)**, 1366–1371.

DeLong, E. F. (1997) Marine microbial diversity: the tip of the iceberg, *Trends Biotechnol.* **15(6)**, 203–207.

Deveaux-Basseguy, R., Bergel, A., Comtat, M. (1997) Potential applications of NAD(P)-dependent oxidoreductases in synthesis: A survey, *Enzyme Microb. Technol.* **20**, 248–258.

Dickson, A. R., Wong, K. K. Y., Mansfield, S. D. (1999) Escher-Wyss and PFI refining of kraft pulp treated with a commercial xylanase, *Proceedings 10th Int. Symp. Wood and Pulping Chem.*, **3**, 252–255.

Dunn-Coleman, N., Prade, R. (1998) Toward a global filamentous fungus genome sequencing effort, *Nat. Biotechnol.* **16(1)**, 5.

Durden, D. K., Etters, J. N., Sarkar, A. K., Henderson, L. A., Hill, J. E. (2001) Advances in commercial Biopreparation of cotton with alkaline pectinase, *AATCC Review*, **1(8)**, 28–31.

Entcheva, P., Liebl, W., Johann, A., Hartsch, T., Streit, W. R. (2001) Direct cloning from enrichment cultures, a reliable strategy for isolation of complete operons and genes from microbial consortia, *Appl. Environ. Microbiol.* **67(1)**, 89–99.

Eriksen, N. (1996) Detergents, in: *Industrial Enzymology* (Godfrey, T. West, S., Eds.), New York: Stockton Press, 187–200.

Etters, J. N. (1999) Cotton preparation with alkaline pectinase: an environmental advance, *Textile Chem. Color. & Am. Dyestuff Rep.* **1(3)**, 33–36.

Etters, J. N., Condon, B. D., Husain, P. A., Lange, N. K. (1999) Alkaline pectinase: Key to cost-effective, environmentally friendly preparation, *Am. Dyestuff Rep.* **88(6)**, 19–23.

Euvering, G. J., Binnema, D. J. (1997). Use of modified starch as an agent for forming a thermoreversible gel, WO 98/15347.

Fabret, C., Poncet, S., Danielsen, S., Borchert, T., Ehrlich, S.D., Janniere, L. (2000) Efficient gene targeted random mutagenesis in genetically stable *Eschericia coli* strains, *Nucleic Acids Res.* **28**, 21–95.

Felby, C., Pedersen, L. S., Pedersen, L. (2001) Manufacture of lignocellulose-based product for fibre board, plywood, paper etc. mfr. - comprises treating lignocellulosic material and phenolic polysaccharide having substits. contg. phenolic hydroxy gp. with enzyme in presence of oxidising agent. ES2158951-T3, Novo-Nordisk, A. S. And Novozymes, A. S., (Patent).

Fischer, K., Messner, K. (1992) Reducing troublesome pitch in pulp mills by lipolytic enzymes, *Tappi J.* **75(2)**, 130–134.

Fitzhenry, J. W., Hoekstra, P. M., Glover, D. (2000) Enzymatic stickies control in MOW, OCC and ONP furnishes, *Proceedings Tappi Pulping Conf. 2000.*

Franks, N. E., Kida, K., Onishi, M., Sakaguchi, H., Schulein, M., Takahashi, M., Tamagawa, H.,

Ohishi, M., Schuelein, M., Sharyo, M. (1997) Producing improved sanitary paper products - by treating paper pulp with cellulase in absence of cellulose binding domain, WO9727363-A1, Novo-Nordisk, A. S., Novozymes, A. S., and Crecia Corp. (patent).

Freedonia (2000) *World Textile Chemicals – Industry Study 1328*. Cleveland, OH: The Freedonia Group, Inc.

Fujita, Y., Awaji, H., Taneda, H., Mausukura, M., Hata, K., Shimoto, H., Sharyo, M., Sakaguchi, H., Gibson, K. (1992) Recent advances in enzymatic pitch control, *Tappi J.* **75(4)**, 117–122.

Gelhoff, W. S. (1998) The benefits of using enzymes to improve flotation deinking of office paper furnishes, *Proceedings Tappi Recycling Symp 1998*, 277.

Gerday, C., Aittaleb, M., Arpigny, J. L., Baise, E., Chessa, J. P., Garsoux, G., Petrescu, I., Feller, G. (1997) Psychrophilic enzymes: A thermodynamic challenge, *Biochem. Biophys. Acta - Protein Structure and Molecular Enzymology* **1342(2)**, 119–131.

Giebelhaus, A.W. (1979) Resistance to long-term energy transition: The case of power alcohol in the 1930s, Paper to the American Association for the Advancement of Science, 4 January 1979.

Godfrey, T., West, S. I. (1996) Introduction to industrial enzymes, in: *Industrial Enzymology* (Godfrey, T., West, S. I., Eds.), New York: Macmillan Press, 1–8.

Gonzalez, J. M., Whitman, W. B., Hodson, R. E., Moran, M. A. (1996) Identifying numerically abundant culturable bacteria from complex communities: an example from a lignin enrichment culture, *Appl. Environ. Microbiol.* **62**, 4433–4440.

Gormsen, E. (1997) Enzymes, in: *Powdered Detergents, Surfactant Science Series* (Showell, M. S., Ed.), Marcel Dekker: New York, 137–163.

Greener, A., Callahan, M., Jerpseth, B. (1996) An efficient random mutagenesis technique using an *E. coli* mutator strain, *Methods Mol. Biol.* **57**, 375–385.

Hagen, H. A. (1981) Production of ethanol from starch-containing crops - various cooking procedures. Paper given at a Meeting on bio-fuels in Bologna, June 1981. Available as Article A-5762a GB from Novozymes A/S.

Hagen H. A., Helwiig-Nielsen B. (1982) Ethanol from starch-containing crops. – energy-saving cooking processes. Paper presented at the 5th Int. Fuel Alc. Symp. in Auckland, N.Z., May 1982. Available as Article A-5783a GB from Novozymes A/S.

Hall, G. K., Stewart, C. W., Screws, G. A. (1999) Enzymatic discharge printing of dyed textiles. US Patent 5,951,714.

Hallmann, J., Quadt-Hallmann, A., Mahaffee, W. F., Kloepper, J. W. (1997) Bacterial endophytes in agricultural crops, *Can. J. Microbiol.*, **43**, 895–914.

Handelsman, J., Rondon, M. R., Brady, S. F., Clardy, J., Goodman, R. M. (1998) Molecular biological access to the chemistry of unknown soil microbes: A new frontier for natural products, *Chem Biol (London)*, **5**, R245–R249.

Hanes, J., Plückthun, A. S. (1997) *In vitro* selection and evolution of functional proteins by using ribosome display, *Proc. Natl. Acad. Sci.* **94**, 4937–4942.

Haraldsson, G. G., Gudmundsson, B. Ö., Almarsson, Ö. (1995) The synthesis of homogeneous triglycerides of eicosapentaenoic acid and docosahexaenoic acid by lipase, *Tetrahedron* **51**, 941–952.

Hartley, B. S., Hanlon, N., Jackson, R. J., Rangarajan, M. (2000) Glucose isomerase: insights into protein engineering for increased thermostability, *Biochem. Biophys. Acta - Protein Structure and Molecular Enzymology* **1553**, 294–335.

Hawksworth, D. L. (2001) The magnitude of fungal diversity: the 1.5 million species estimate revisited, *Mycol. Res.* **105**, 1422–1432.

Hedges, A. R. (1992) Cyclodextrins: production, properties, and applications, in: . *Starch Hydrolysis Products. Worldwide Technology, Production, and Applications*, (Schenck F. W., Hebeda R. E., Eds.). New York: VCH, 319–333.

Heinfling, A., Bergbauer, M., Szewzyk, U. (1997) Biodegradation of reactive dyes by white-rot fungi – perspectives for an application in wastewater treatment, in: *Scriftenreihe des Sonderforschungsbereiches 193 der Technischen Universität Berlin "Biologische Behandlung industrieller und gewerblicher Abwässer,"* Berlin: DTU, 171–185.

Heise, O.U., Unwin, J.P., Klungness, J.H., Fineran, W.G., Sykes, M., Abubakr, S. (1996) Industrial scaleup of enzyme-enhanced deinking of nonimpact printed toners, *Tappi J.* **79**, 207–212.

Henne, A., Schmitz, R.A., Bomeke, M., Gottschalk, G., Daniel, R. (2000) Screening of environmental DNA libraries for the presence of genes conferring lipolytic activity on *Escherichia coli*, *Appl. Environ. Microbiol.* **66**, 3113–3116.

Henrissat, B., Bairoch, A. (1993) New families in the classification of glycosyl hydrolases based on amino acid sequence similarities, *Biochem. J.* **293**, 781–788.

Henrissat, B., Romeu, A. (1995) Families, superfamilies and subfamilies of glycosyl hydrolases, *Biochem. J.* **311**, 350–351.

Ho Tan Tai, L., Rataj, N. (2001) Detergents of the 21st century, *OCL* **8(2)**, 155–159.

Honig, B., Nicholls, A. (1995) Classical electrostatics in biology and chemistry, *Science* **268**, 1144–1149.

Horikoshi, K. (1995) Discovering novel bacteria, with an eye to biotechnological applications, *Curr. Opin. Biotechnol.* **6**, 292–297.

Hugenholtz, P., Pace, N. R. (1996) Identifying microbial diversity in the natural environment: a molecular phylogenetic approach, *Trends Biotechnol* **14(6)**, 190–197.

Husain, P. A., Lange, N.K., Henderson, L., Liu, J., Condon, B.D. (1999) Biopreparation – a new industrial enzyme process, in: *Book of Papers, AATCC International Conference & Exhibition*, 170–182.

Ibba, M., Hennecke, H. (1994) Towards engineering proteins by site-directed incoporation *in vivo* of non-natural amino acids. *Biotechnology* **12**, 678–682.

Jannasch, H. W., Jones, G. E. (1959) Bacterial populations in sea water as determined by different methods of enumeration, *Limnol. Oceanogr.* **4**, 128–139.

Johnson, A. M., Dubey, J. P., Dame, J. B. (1986) Purification and characterization of Toxoplasma gondii tachyzoite DNA, *Aust. J. Exp. Biol. Med. Sci.* **64(Pt 4)**, 351–355.

Johnston, D. B., Sing, V. (2001). The use of proteases to reduce steep time and SO_2 requirements in a corn wet-milling process. *Cereal Chemistry* **78**, 405–411.

Joo, H., Arisawa, A., Lin, Z. L., Arnold, F. H. (1999) A high-throughput digital imaging screen for the discovery and directed evolution of oxygenases, *Chem. Biol.* **6**, 699–706.

Kenyon, G. L., Bruice, T.W. (1977) Novel sulfhydryl reagents, *Methods Enzymology* **47**, 407–430.

Khumtaveeporn, K., Ullmann, A., Matsumoto, K., Davis, B. G., Jones, J. B. (2001) Expanding the utility of proteases in synthesis: Broadening the substrate acceptance in non-coded amide bond formation using chemically modified mutants of subtilisin, *Tetrahedron Asymmetry*, **12(2)**, 249–261.

Kierulff, J. V. (1997) Denim bleaching, *Textile Horizons*, Oct/Nov, 33–36.

Klibanov, A. M. (1986) Enzymes that work in organic solvents, *Chemtech* **16**, 354–359.

Klyosov, A.A. (1990) Trends in biochemistry and enzymology of cellulose degradation, *Biochemistry* **29**, 10577.

Knudsen, O., Young, J. D., Yang, J. L. (1997) Long-term use of enzymatic deinking at Stora Dalum plant, *Proceedings 7th Int. Conf. Biotech. P&P Industry*, Volume A, 17.

Kofod, L. V., Lund, H. (2000) Environmentally friendly preparation of cellulosic material - comprises contacting cellulosic material with xyloglucan endo:transglycosylase in aqueous medium, and imparts improved strength and/or shape-retention and/or anti-wrinkling properties, JP2000502410-W, Novo-Nordisk, A. S., (patent).

Kogure, K., Simidu, U., Taga, N. (1979) A tentative direct microscopic method for counting living marine bacteria, *Can. J. Microbiol.* **25**, 415–420.

Kongsbak, L., Jørgensen, K.S., Jørgensen, C.T., Husum, T.L., Ernst, S., Møller, S. (1999) A fluorescence polarisation screening method. Novo Nordisk A/S, (patent).

Kristjansson, J.K., Hreggvidsson, G.O. (1995) Ecology and habitats of extremophiles, *World Journal of Microbiology and Biotechnology*, **11(1)**, 17–25.

Kumar, A., Harnden, A. (1999) Cellulase enzymes in wet processing of lyocell and its blends, *Textile Chem. Color. & Am. Dyestuff Rep.*, **1(1)**, 37–41.

Kumar, A., Yoon, M-Y., Purtell, C. (1997) Optimizing the use of cellulase enzymes in finishing cellulosic fabrics, *Textile Chem. Color.*, **29(4)**, 37–42.

Kundu, A. B., Ghosh, B. S., Chakrabarti, S. K. (1993) Enhanced bleaching and softening of jute pretreated with polysaccharide degrading enzymes, *Textile Res. J.* **63(8)**, 451–454.

Kunst, F., Ogasawara, N., Moszer, I., Albertini, A. M., Alloni, G., Azevedo, V., Bertero, M. G., Bessires, P., Bolotin, A., Borchert, S., Borriss, R., Boursier, L., Brans, A., Braun, M., Brignell, S. C., Bron, S., Brouillet, S., Bruschi, C. V., Caldwell, B., Capuano, V., Carter, N. M., Choi, S. K., Codani, J. J., Connerton, I. F., Danchin, A. (1997) The complete genome sequence of the gram-positive bacterium *Bacillus subtilis*, *Nature* **390**, 249–256.

Lascaris, E., Lonergan, G., Forbes, L. (1997) Drainage improvement using a starch degrading enzyme blend in a recycling paper mill, *Proceedings Tappi Biol. Sci. Symp. 1997*, 271.

Lenting, H. B. M., Warmoeskerken, M. M. C. G. (2001) Guidelines to come to minimized tensile

strength loss upon cellulase application, *J. Biotech.* **89**, 227–232.
Leung, D. W., Chen, E., Goeddel, D. V. (1989) A method for random mutagenesis of a defined DNA segment using a modified polymerase chain reaction, *Technique* **1**, 11–15.
Lewis, S. M. (1996). Fermentation alcohol, in: *Industrial Enzymology.* (Godfrey T., West S.. Eds.), London: Macmillan Publishers.
Li, W. B., Jessee, J., Nisson, P. E. (1999) Normalized nucelic acid libraries and methods for production thereof, WO 99/15702, Life Technologies Inc, (patent).
Li, Y., Hardin, I.R. (1997) Enzymatic scouring of cotton: effects on structure and properties, *Textile Chem. Color* **29(8)**, 71–76.
Lichts, F. O. (2001). *International Molasses and Alcohol Report.* Vol. 38 No. 22, November 16.
Liese, A., Zelinsky, T., Kula, M.-R., Kierkels, H., Karutz, M., Kragl, U., Wandrey, C. (1998) A novel reactor concept for the enzymatic recuction of poorly soluble ketones, *J. Mol. Catal. B: Enzyme* **4**, 91–99.
Liu, J., Otto, E., Lange, N. K., Husain, P., Condon, B. (1998) Biopolishing of cotton knit: from multicomponent cellulase complex to mono-component, in: *Book of Papers, AATCC International Conference & Exhibition*, 445–454.
Lloyd, N. E., Lewis, L. T., Logan, R. M., Patel, D. N. (1972). US Patent 3,694,314.
Lorenz, P., Schleper, C. (2002) Metagenome - a challenging source of enzyme discovery. *Journal Of Molecular Catalysis B: Enzymatic* (submitted).
Lyons, T. P (1983). Alcohol – Power/fuel, in: *Industrial Enzymology.* (Godfrey T., Reichelt J., Eds.), London: Macmillan Publishers.
Macrae, A. R. (1983) Lipase-catalyzed interesterification of oils and fats, *J. Am Oil Chem Soc.* **60**, 291–294.
Madsen, G. B., Norman, B.E., Slott, S. (1973). A new, heat stable bacterial amylase and its use in high temperature liquefaction, *Stärke* **25**, 304–308.
Majernik, A., Gottschalk, G., Daniel, R. (2001) Screening of environmental DNA libraries for the presence of genes conferring Na^+ $(Li^+)/H^+$ antiporter activity on *Escherichia coli*: Characterization of the recovered genes and the corresponding gene products, *J. Bacteriol.* **183**, 6645–6653.
Malmgren, K., Saaby-Pedersen, L., Skjold-Jorgensen, S., Pedersen, L. (2002) Hydrolysis of resin in pulp - using enzyme present during peroxy bleaching, CZ9800143-A3, Novo-Nordisk, A. S., SCA Graphic Paper, A. B., SCA Wifsta Oestrand, A. B., SCA Pulp, A. B.(patent).
Marshall, R. O., Kooi, E. R. (1957). Enzymatic conversion of D-glucose to D-fructose, *Science* **125**, 648–649
McAloon, A., Taylor, F., Yee, W., Ibsen, K., Wooley, R. (2000). Determining the cost of producing ethanol from corn starch and lignocellulosic feedstocks. NREL/TP-580-28893, National Renewable Energy Laboratory, Golden, CO 80401, USA.
McCoy, M. (2000a) Novozymes emerges. *Chem. Eng. News.* **19**, 23–25.
McCoy, M. (2000b) Additives: Where all the magic is, *Chem. Eng. News.* **19**, 26.
McDevitt, J. P., Shi, X. (2000) Method for treatment of wool. US Patent 6,099,588.
Meeuwsen, P.J.A., Vincken, J.P., Beldman, G., Voragen, A.G.J. (2000) A universal assay for screening expression libraries for carbohydrases, *J. Biosci. Bioeng.* **89**, 107–109.
Mori, T., Sakimoto, M., Kagi, T. (1997) Enzymatic desizing of polyvinyl alcohol from cotton fabrics, *J. Chem. Tech. Biotechnol.* **68**, 151–156.
Morita, M., Ito, R., Kamidate, T., Watanabe, H. (1996) Kinetics of peroxidase catalyzed decolorization of Orange II with hydrogen peroxide, *Textile Res. J.* **66(7)**, 470–473.
Morita, M., Kamidate, T., Shibata, T., Watanabe, H. (1997) Activators for decolorization of azo dyes catalyzed by peroxidase, *J. Jpn. Oil Chem. Soc.* **46(7)**, 67–69.
Mukai, K., Tabuchi, A., Nakada, T., Shibuya, T., Chaen, H., Fukuda, S., Kurimoto, M., Tsujisaka, Y. (1997). Production of trehalose from starch by thermostable enzymes from *Sulfolobus acidocaldarius*, *Stärke* **49**, 26 – 30.
Mustranta, A., Buchert, J., Spetz, P., Holmbom, B. (2001) Treatment of mechanical pulp and process waters with lipase, *Nord. Pulp Paper Res. J.* **16**, 125–129.
Nagao, T., Watanabe, H., Goto, N., Onizawa, K., Taguchi, H., Matsuo, N., Yasukawa, T., Tsushima, R., Shimasaki, H., Itakura, H. (2000) Dietary diacylglycerol suppresses accumulation of body fat compared to triacylglycerol in men in a double-blind controlled trial, *J. Nutr.* **130**, 792–797.
Naki D., Paech, C., Ganshaw, G., Schellenberger, V. (1998) Selection of a subtilisin-hyperproducing *Bacillus* in a highly structured environment, *Appl. Microbiol. Biotechnol.* **49**, 290–294.
Ness, J. E., Welch, M., Giver, L., Bueno, M., Cherry, J. R., Borchert, T. V., Stemmer, W. P. C., Minshull,

J. (1999) DNA shuffling of subgenomic sequences of subtilisin, *Nat. Biotechnol.* **17**, 893–896.

Nielsen, J. E., Borchert, T. V. (2000), Protein engineering of bacterial alpha-amylases, *Biochim Biophys Acta - Protein Structure and Molecular Enzymology* **1553**, 253–274.

Nielsen, P., Malmos, H., Damhus, T., Diderichsen, B., Nielsen, H., Simonsen, M., Schiff, H., Oestergaard, A., Olsen, H., Eigtved, P., Nielsen, T. (1994) Enzyme applications (Industrial), in: *Kirk-Othmer Encyclopaedia of Chemical Technology*, 4th editon, New York: John Wiley & Sons, 567–620, Vol. 9.

Norman, B. E. (1982). A novel debranching enzyme for application in the glucose syrup industry, *Stärke* **34**, 340–346.

Novo Nordisk (1999a) Application Sheet for Wool and Silk Finishing, B1165a-GB. Bagsvaerd, Denmark: Novo Nordisk A/S.

Novo Nordisk (1999b) The time is right for DeniLite II, *BioTimes* **1**, 10–11.

Novozymes (2001) Desizing at any temperature, DDB No. 2001–04329-01. Bagsvaerd, Denmark: Novozymes A/S.

Novozymes A/S (2001a). Application sheet for Liquezyme X

Novozymes A/S (2001b). Application sheet for Dextrozyme GA

Obendorf, S. K., Varanasi, A., Mejldal, R., Thellersen, M. (2001) Function of lipase in lipid soil removal as studied using fabrics with different chemical accessibility, *J. Surf. Det.* **4(3)**, 233–245.

Ogawa, J., Shimizu, S. (1999) Microbial enzymes: New industrial applications from traditional screening methods, *Trends Biotechnol.* **17**, 13–21.

Ogram, A. (2000) Soil molecular microbial ecology at age 20: methodological challenges for the future, *Soil Biol. Biochem.* **32**, 1499–1504.

Okada, M., Nakakuki, T. (1992) Oligosaccharides: Production, properties, and applications, in: *Starch Hydrolysis Products. Worldwide Technology, Production, and Applications*, (Schenck F. W., Hebeda R. E., Eds.), New York: VCH, 335–366.

Okada, S. (1988) Maltooligosylsucrose syrup, in: *Handbook of Amylases and Related Enzymes. Their sources, Isolation Methods, Properties and Applications*, (Amylase Research Society of Japan, Ed.), New York: Pergamon Press, 226–228.

Olsen, G. J., Lane, D. J., Giovannoni, S. J., Pace, N. R., Stahl, D. A. (1986) Microbial ecology and evolution: a ribosomal RNA approach, *Annu. Rev. Microbiol.* **40**, 337–365.

Olsen, H. S., Falholt, P. (1998) The role of enzymes in modern detergency, *J. Surfact. Deterg.* **1(4)**, 555–567.

Olson, L. (1989) Compositions and methods that introduce variations in the color density into cellulosic fabrics, particularly indigo dyed denim. US Patent 4,832,864.

Olsson, L., Thomsen, A. B., Ahring, B. K., Klinke H. B. (1999) Influence of inhibitors from wet-oxidized wheat straw on *Saccharomyces cerevisiae. IEA Workshop - Biotechnology for the conversion of Lignocellulose.* Itala Game Reserve, South Africa, 22–26 August 1999.

O'Neil, K. T., Hoess, R. H. (1995) Phage display: protein engineering by directed evolution, *Curr. Opin. Struct. Biol.* **5**, 443–449.

Ostermeier, M., Shim, J. H., Benkovic, S. J. (1999), A combinatorial approach to hybrid enzymes independent of DNA homology, *Nat. Biotechnol.* **17**, 1205–1209.

Otto Röhm (1913), German Patent DE28923.

Outtrup, H., Norman, B. E. (1984). Properties and application of a thermostable maltogenic amylase produced by a strain of *Bacillus* modified by recombinant-DNA techniques, *Stärke* **36**, 405–411.

Pace, N. R., Stahl, D. A., Lane, D. J., Olsen, G. J. (1986) The analysis of natural microbial populations by ribosomal RNA sequences. *Adv. Microb. Ecol.* **9**, 1–55.

Pedersen, S. (1993). Industrial aspects of immobilised glucose isomerase, in: *Industrial Applications of Immobilised Biocatalysts*, Tanaka A., Tosa T., Kobayashi, T., Eds.), New York: Marcel Dekker, 185–208

Pedersen, S., Dijkhuizen, L., Dijkstra, B. W., Jensen, B. F., Jørgensen, S. T. (1995). A better enzyme for cyclodextrins, *Chemtech* **25**, 19–25.

Pere, J., Siika-aho, M., Viikari, L. (1999) Mechanical pulp prodn. to reduce crystallinity of cellulose - including treatment with cellobiohydrolase and mannanase, US Patent 5,865,949-A, Valtion Teknillinen, T.U.T.K., (patent).

Pere, J., Siika-aho, M., Viikari, L. (2000) Biomechanical pulping with enzymes: Response of coarse mechanical pulp to enzymatic modification and secondary refining, *Tappi J.* **83(5)**, 85.

Powell, K. A., Ramer, S. W., del Cardayre, S. B., Stemmer, W. P. C., Tobin, M. B., Longchamp, P. F., Huisman, G. W. (2001) Directed evolution and biocatalysis, *Angew. Chem. Int. Edn.* **40**, 3948–3959.

Preisig, C., Byng, G. (2001) Applications of mass spectrometry in screening for new biocatalysts, *J. Mol. Catal. B Enzymatic*, **11(4–6)**, 733–741.

Prieur, D. (1997) Microbiology of deep-sea hydrothermal vents, *Trends Biotechnol.* **15(7)**, 242–244.

Putz, H.-J., Gottsching, L., Renner, K., Jokinen, O. (1994) Enzymatic deinking in comparison with conventional deinking of offset news, *Proceedings Tappi pulping Conf. 1994*, 877.

Riegels, M., Koch, R., Pedersen, L. S., Lund, H. (2001) Enzymatic hydrolysis of cyclic oligomers, US Patent 6,184,010.

Rondon, M. R., Goodman, R. M., Handelsman, J. (1999) The Earth's bounty: Assessing and accessing soil microbial diversity, *Trends Biotechnol.* **17**, 403–409.

Rondon, M. R., August, P. R., Bettermann, A. D., Brady, S. F., Grossman, T. H., Liles, M. R., Loiacono, K. A., Lynch, B. A., MacNeil, I. A., Minor, C., Tiong, C. L., Gilman, M., Osburne, M. S., Clardy, J., Handelsman, J., Goodman, R. M. (2000) Cloning the soil metagenome: a strategy for accessing the genetic and functional diversity of uncultured microorganisms, *Appl. Environ. Microbiol.* **66**, 2541–2547.

Rozzell, J. D. (1999) Commercial scale biocatalysis: Myths and realities, *Bioorg. Med. Chem.* **7**, 2253–2261.

Rüdiger, A., Jorgensen, P. L., Antranikian, G. (1995). Isolation and characterisation of a heat-stable pullulanase from the hyperthermophilic archaeon *Pyrococcus woesei* after cloning and expression of its gene in *Escherichia coli, Appl. Environ. Microbiol.* **61**, 567–575.

Ruijssenaars, H. J., Hartmans, S. (2001) Plate screening methods for the detection of polysaccharase- producing microorganisms, *Appl. Microbiol. Biotechnol.* **55(2)**, 143–149.

Ryom, N. M., Gibson, K. W., (2001) Multiple enzymes in household detergents, *La Rivista Italiana Delle Sostanze Grasse* **78**, 425–430.

Sakaguchi, H., Sharyo, M., Shimoto, H. (2000) Paper-making pulp prodn. from starch-coated printed paper - by disintegrating paper to give pulp, treating with starch- degrading enzyme before, during or after disintegration and sepg. ink particles, DE69423364-E, Novo-Nordisk, A. S., (patent).

Sambrock, J., Russel, D. W. (2001) *Molecular Cloning.* Cold Spring Harbor, NY: Cold Spring Harbor Laboratory Press.

Sato, M. (1983) Deterioration of filaments and films of polyethylene terephthalate with enzyme of *Cladosporium cladosporioides. FERM J-8, Sen-I Gakkaishi*, **39(5)**, 67–77.

Sauer, J., Sigurdskjold, B. W., Christensen, U., Frandsen, T. P., Mirgorodskaya, E., Harrison, M., Roepstorff, P. Svensson, B. (2000) Glucoamylase: structure/function relationsships, and protein engineering. *Biochem. Biophys. Acta - Protein Structure and Molecular Enzymology* **1543**, 275–293.

Schellenberger, V. (1997) Compartmentalization method for screening microorganisms, Genencore Intl.Inc, (patent WO 87/37036).

Schleper, C., DeLong, E. F., Preston, C. M., Feldman, R. A., Wu, K. Y., Swanson, R. V. (1998) Genomic analysis reveals chromosomal variation in natural populations of the uncultured psychrophilic archaeon *Cenarchaeum symbiosum, J. Bacteriol.* **180**, 5003–5009.

Schmidt, A., Dordick, J. S., Hauer, B., Kiener, A., Wubbolts, M., Witholt, B. (2001) Industrial biocatalysis today and tomorrow, *Nature* **409**, 258–268.

Schmidt, M. G. (1995) Bleach cleanup with catalase, in: *Book of Papers, AATCC International Conference & Exhibition*, RTP, NC: AATCC, 248–255.

Schmidt, T. M., DeLong, E. F., Pace, N. R. (1991) Analysis of a marine picoplankton community by 16S rRNA gene cloning and sequencing, *J. Bacteriol.* **173**, 4371–4378.

Schülein, M. (2000) Protein engineering of cellulases *Biochem. Biophys. Acta -Protein Structure and Molecular Enzymology* **1553**, 239–252.

Schülein, M., Andersen, L. N., Lassen, S. F., Lange, L., Kauppinen, S., Nielsen, R., Ihara, Takagi, S. (1996) Novel endoglucanases. US Patent 6,001,639, Novo Nordisk A/S, (patent).

Sheehan, J., Himmel, M. (1999). Enzymes, energy, and the environment: A strategic perspective on the U.S. Department of Energy's research and development activities, *Bioethanol. Biotechnol. Prog.* **15**, 817–827

Short, J. M. (1997) Recombinant approaches for accessing biodiversity, *Nat. Biotechnol.* **15**, 1322–1323.

Short, J. M., Keller, M. (2001) High-throughput screening for novel enzymes. US Patent 6,174, 673 B1, Diversa Corporation, (patent).

Short, J. M., Mathur, E. J. (1999) Production of normalised DNA libraries, US Patent 6,001,574, Diversa Corporation, (patent).

Showell, M. S. (1999) Enzymes, detergent, in: *Encyclopedia of Bioprocess Technology-Fermentation, Biocatalysis, and Bioseparation* (Flickinger,

M. C., Drew, S. W., Eds.), New York: Wiley, 958–971, Vol. 2.
Sieber, V., Martinez, C. A., Arnold, F. A. (2001) Libraries of hybrid proteins from distantly related sequences, *Nat. Biotechnol.* **19**, 456–460.
Smets, J., Barnabas, M. V., Showell, M. S., Boyer, S. L., Convents, A. C. (1999a) Laundry detergent and/or fabric care compositions comprising a modified transferase, WO9957258-A1.
Smets, J., Bettiol, J. P., Boyer, S. L., Busch, A. (1999b) Laundry detergent and/or fabric care compositions comprising a modified enzyme, WO9957252-A1.
Smith, M. (1985) *In vitro* mutagenesis, *Ann. Rev. Genet.* **19**, 423–462
Snyder L. G. (1997) Improving the quality of 100% cotton knit fabrics by defuzzing with singeing and cellulase enzymes, *Textile Chem. Color* **29(6)**, 27–31.
Sreenath, H. K., Shan, A. B., Yang, V. W., Gharia, M. M., Jeffries, T. W. (1996) Enzymatic polishing of jute/cotton blended fabrics, *J. Ferment. Bioeng.* **81(1)**, 18–20.
Sreenivasan, B. (1978) Interesterification of fats, *J. Am. Oil Chem. Soc.* **55**, 796–805.
Stahl, D. A., Lane, D. J., Olsen, G. J., Pace, N. R. (1985) Characterization of a Yellowstone, USA, hot spring microbial community by 5s ribosomal RNA sequences, *Appl. Environ. Microbiol.* **49**, 1379–1384.
Ståhl, S., Uhlen, M. (1997) Bacterial surface display, *Tibtech* **15**, 185–192.
Steffan, R. J., Goksoyr, J., Bej, A. K., Atlas, R. M. (1988) Recovery of DNA from soils and sediments, *Appl. Environ. Microbiol.* **54**, 2908–2915.
Stein, J. L., Marsh, T. L., Wu, K. Y., Shizuya, H., DeLong, E. F. (1996) Characterization of uncultivated prokaryotes: isolation and analysis of a 40-kilobase-pair genome fragment from a planktonic marine archaeon, *J. Bacteriol.* **178**, 591–599.
Steipe, B. (1999), Evolutionary approaches to protein engineering, *Curr. Top. Microbiol. Immunol.* **243**, 55–86.
Stemmer, W. P. (1994a) DNA shuffling by random fragmentation and reassembly: *In vitro* recombination for molecular evolution, *Proc. Natl. Acad. Sci.* **91**, 10747–10751.
Stemmer, W. P. (1994b) Rapid evolution of a protein by *in vitro* DNA shuffling, *Nature* **370**, 389–391.
Stewart, C. W. (1996) Enzyme finishing technology, in: *Book of Papers, AATCC International Conference & Exhibition*, Research Triangle Park, NC: AATCC, 212–217.
Svendsen, A., Clausen, I. G., Patkar, S. A., Borch, K., Thellersen, M. (1997) Protein engineering of microbial lipases of industrial interest, *Methods in Enzymology* **284**, 317–340.
Svendsen, A. (2000) Lipase protein engineering, *Biochem. Biophys. Acta - Protein Structure and Molecular Enzymology* **1553**, 223–238.
Takasaki, Y., Kosugi, Y., Kanbuyashi, A. (1969). Streptomyces glucose isomerase, in: *Fermentation Advances* (Perlman, D., ed.) New York: American Press, 561–589.
Tauber, M., Gubitz, G., Cavaco-Paulo, A. (2000) Enzymatic treatment of acrylic fibres and granulates, in: *AATCC International Conference & Exhibition*, CD-ROM, Research Triangle Park, NC: AATCC.
Tiedje, J. M., Stein, J. L. (1999) Microbial diversity: Strategies for its recovery, , *Manual of Industrial Microbiology and Biotechnology* (Demain, A. L., Davies, J. E., Eds.), Washington, DC: American Society for Microbiology, 682–692.
Tishkov, V. I., Galkin, A. G., Fedorchuck, V. V., Savitsky, P. A., Rojkova, A. M., Gieren, H., Kula, M-R. (1999) Pilot scale production and isolation of recombinanat NAD^+ and $NADP^+$-specific formate dehydrogenases, *Biotechnol. Bioeng.* **64**, 187–193.
Tobin, M. B., Gustafsson, C., Huisman G. W. (2000) Evolution: The "rational" basis for "irrational" design, *Curr. Opin. Struct. Biol.* **10**, 421–427.
Tolan, J., Thibault, L. (2000) Mill scale implementation of enzymes in pulp bleaching, *Proceedings Tappi Pulping Conf. 2000.*
Torsvik, V. L. (1980) Isolation of bacterial DNA from soil, *Soil Biol Biochem* **12(1)**, 15–22.
Torsvik, V., Goksoyr, J., Daae, F. L. (1990) High diversity in DNA of soil bacteria, *Appl. Environ. Microbiol.* **56**, 782–787.
Torsvik, V., Daae, F. L., Sandaa, R., Ovreas, L. (1998) Novel techniques for analysing microbial diversity in natural and perturbed environments, *J. Biotechnol.* **64**, 53–62.
Tzanov, T., Costa, S. A., Gübitz, G. M., Cavaco-Paulo, A. (2002) Hydrogen peroxide generation with immobilized glucose oxidase for textile bleaching, *J. Biotechnol.* **93**, 87–94.
van der Veen, B. A., Uitdehaag, J. C. M., Dijkstra, B. W., Dijkhuizen, L. (2000) Engineering of cyclodextrin glycosyltransferase reaction and product specificity *Biochemica et Biophysica Acta* **1543**, 336–360.
Venegas, M. G. (1997) Detergency mechanisms, in: *Powdered Detergents, Surfactant Science Series*

(Showell, M. S., Ed.), New York: Marcel Dekker, 285–312.

Viikari, L., Ranua, M., Kanteelinen, A., Linko, M., Sundquist, J. (1987) Application of enzymes in bleaching, *Proceedings 4th International Symp.Wood Pulp.Chem.* 151, Vol. 1.

Voigt, C. A., Kauffman S., Wang Z. G. (2001) Rational evolutionary design: The theory of *in vitro* protein evolution, *Adv. Protein Chem.* **55**, 79–160.

Waddell, R. (2000) Bioscouring of cotton fabric – commercial applications of alkaline stable pectinase, in: *AATCC International Conference & Exhibition*, CD-ROM. Research Triangle Park, NC: AATCC.

Woese, C. R. (1987), Bacterial evolution, *Microbiol. Rev.* **51**, 221–271.

Wölke, J., Ullmann, D. (2001) Miniturized HTS technologies –uHTS, *Drug Discovery Today* **6**, 637–646.

Wong, K. K. Y., Mansfield, S. D. (1999) Enzymatic processing for pulp and paper manufacture - a review, *Chem. Eng. News*, **52(6)**, 409–418.

Wooley, R. Ruth, M., Sheehan, J., Ibsen, K., Majdeski, H., Galvez, A. (1999a). Lignocellulosic biomass to ethanol process design and economics utilizing co-current dilute acid prehydrolysis and enzymatic hydrolysis current and futuristic scenarios. NREL/TP-580-26157, National Renewable Energy Laboratory, Golden, CO 80401, USA.

Wooley, R., Ruth, M., Glassner, D., Sheehan, J. (1999b). Process design and costing of bioethanol technology, *Biotechnol. Prog.* **15**, 794–803.

Xu, F., Salmon, S., Deussen, H-J.W. (1999) Enzymatic methods for dyeing with reduced vat and sulfur dyes. US Patent 5,948,122.

Xu, X. (2000) Production of specific-structures triacylglycerols by lipase-catalyzed reactions: a review, *Eur. J. Lipid Sci. Technol* **2000**, 287–303.

Yanase, M., Okada, S., Takata, H., Nakamura, H., Fujii, K. (1995). Glucans having cyclic structure and method for producing the same. EP 0 710 674 A3

Zhao, H., Giver, L., Shao, Z., Affholter, J. A., Arnold, F. H. (1998) Molecular evolution by staggered extension process (StEP) *in vitro* recombination, *Nat. Biotechnol.* **16**, 258–261.

Zittan, L., Poulsen, P. B., Hemmingsen, S.H. (1975). Sweetzyme: a new immobilized glucose isomerase. *Stärke* **27**, 236–241.

14
Proteasomes: Molecular Machines for Protein Degradation

Dr. Susanne Witt[1], **Prof. Dr. Wolfgang Baumeister**[2]
[1] Department of Molecular Structural Biology, Max-Planck-Institute of Biochemistry, Am Klopferspitz 18a, 82152 Martinsried, Germany; Tel.: +49-89-8578-2648; Fax: +49-89-8578-2641; E-mail: witt@biochem.mpg.de
[2] Department of Molecular Structural Biology, Max-Planck-Institute of Biochemistry, Am Klopferspitz 18a, 82152 Martinsried, Germany; Tel.: +49-89-8578-2652; Fax: +49-89-8578-2641; E-mail: baumeist@biochem.mpg.de

AAA	ATPases associated with a variety of activities
AC	antechamber
AMP	adenosine monophosphate
ARC	ATPase forming ring-shaped complex
ATP	adenosine triphosphate
CC	central cavity
eIF3	eukaryotic initiation factor 3
MHC	major histocompatibility complex
MPN	found in Mpr1 and Pad1 in the N-terminus
Ntn	N-terminal nucleophile
ODC	ornithine decarboxylase
PAN	proteasome activating nucleotidase
PCI	proteasome-COP9 complex initiation
PINT	proteasome-Int6-Nip1-Trip15
RC	regulatory complex
Rpn	regulatory particle non triple-A protein
Rpt	regulatory particle triple-A protein
STEM	scanning-transmission electron microscopy
TAP	transporter associated with antigen processing
TPPII	tripeptidyl peptidase II
Ub	ubiquitin

1 Introduction: Controlling Intracellular Proteolysis

Intracellular protein degradation is a necessity for many reasons. First, homeostasis must be maintained while cellular structures are continually rebuilt, in particular during development or in response to external stimuli. Second, proteins, which are misfolded as a consequence of errors in transcription or translation, or because of thermal or oxidative stress, must be scavenged and removed as they are prone to aggregation. Beyond these more mundane "housekeeping" functions, protein degradation provides a means to terminate the lifespan of many regulatory proteins; amongst them are the cyclins, transcription factors, and many components of signal transduction pathways (for reviews, see Coux et al., 1996; Hilt and Wolf, 1996; Varshavsky, 1997). Moreover, the immune system relies on the availability of immunocompetent peptides generated by the degradation of foreign antigens (for reviews, see Goldberg et al., 1995; Heemels and Ploegh, 1995).

To avoid havoc, protein degradation must be subject to spatial and temporal control in order to prevent the destruction of proteins not destined for degradation. A basic stratagem in controlling protein degradation is compartmentalization; that is, the confinement of the proteolytic action to sites that can only be accessed by proteins displaying degradation signals. Such a compartment can be an organelle, delimited by a membrane, as in the case of the lysosome. Proteins to be degraded must be imported into the lysosome via specific pathways, and the hydrolases carrying out the task must also be sorted from other proteins and are translocated by means of specific transport vesicles (Harter and Wieland, 1996).

Prokaryotic cells, possessing neither membrane-bound compartments nor vesicular transport systems, have developed a different form of compartmentalization, namely self-compartmentalization (Lupas et al., 1997). This principle is seen at work in several unrelated proteases that have all converged toward a common architecture in which proteolytic subunits self-assemble to form barrel-shaped complexes. These enclose inner cavities, which are several nanometers in diameter and harbor the active sites. Access to these inner compartments is usually restricted to unfolded polypeptides, which can pass through the narrow channels guarding the entrance. The target proteins thus require interaction with a machinery capable of binding and presenting them in an unfolded form to the proteolytic core complexes; these interactions may be of either a transient or a continuous nature. Since protein folding and unfolding are closely related mechanistically, it is not surprising that this task is performed by ATPase complexes, which bear some resemblance to the chaperonins and have been referred to as "reverse chaperones"' or "unfoldases" (Lupas et al., 1993). Since their action requires the hydrolysis of ATP, protein degradation becomes energy-dependent, although the hydrolysis of the polypeptide chain itself is an exergonic process.

Self-compartmentalizing proteases are common in all three domains of life: archaea, bacteria, and eukarya. This bears testimony to an old evolutionary principle. In fact, in contrast to organelles such as the lysosome, self-compartmentalizing molecular devices offer far greater flexibility: when equipped with the appropriate localization signals, they can be deployed to different cellular locations in the cytosol or in the nucleus, wherever their action is needed. The advances made in recent years in understanding the structure of the proteasome and its mechanism of action has helped to shape the concept of self-compartmentalization, and the proteasome became the paradigm of this form of regulation (Baumeister et al., 1998).

In eukaryotic cells, the vast majority of proteins in the cytosol and in the nucleus is degraded via the proteasome-ubiquity pathway (Figure 1) (for reviews, see Coux et al., 1996; Hilt and Wolf, 1996). The 26S proteasome is a huge protein degradation machine of 2.5 MDa, built of approximately 35 subunits. It comprises two subcomplexes, the 20S core particle and one or two regulatory complexes, the 19S caps (for reviews, see Tanaka, 1998; DeMartino and Slaughter, 1999; Voges et al., 1999; Gorbea and Rechsteiner, 2000). The proteolytic action takes place in the 20S core complex, sequestered from the cellular environment. The 19S regulatory complexes recruit substrates marked for degradation and prepare them for translocation into the 20S core complex (Lupas et al., 1993; Larsen and Finley, 1997). The sequence of events encountered by a substrate until it is eventually degraded is reflected by a linear arrangement of functional modules within the 45 nm-long supramolecular assembly (Figure 2).

Whereas the structure and function of the 20S proteasome have been elucidated in great detail (for reviews, see Baumeister et al., 1998; Voges et al., 1999), the 19S regulator is understood only dimly at present. Structural studies are hampered by the low intrinsic stability of this assembly, which makes it notoriously difficult to obtain homogeneous preparations. Nevertheless, analyses of yeast, *Drosophila*, and human 26S proteasomes have revealed a common set of 17–18 subunits, although a few species-specific differences have been found (Udvardy, 1993; Tanaka, 1998; Glickman et al., 1998b; Ferrell et al., 2000; Hölzl

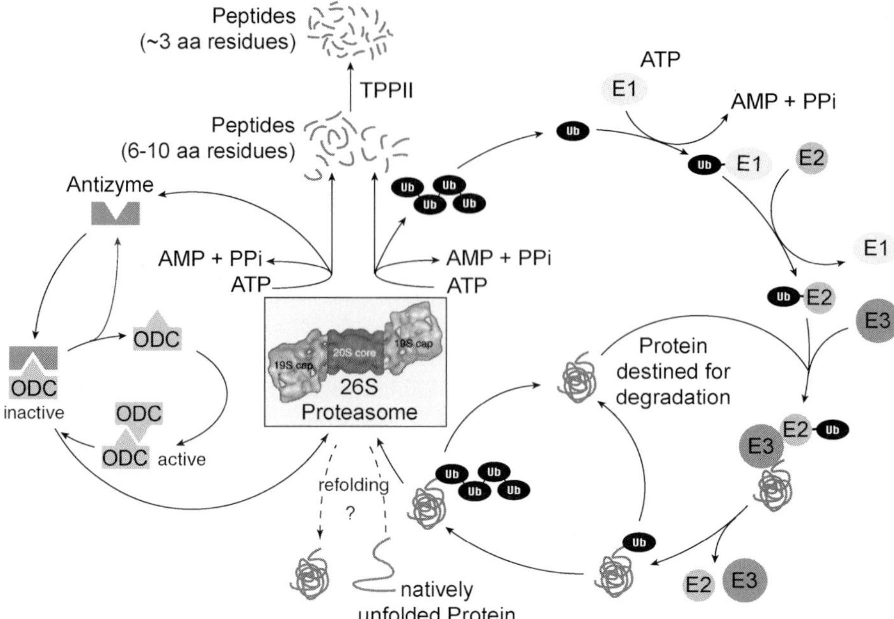

Fig. 1 Schematic representation of the ubiquitin-proteasome pathway of protein degradation and the degradation of ODC by the 26S proteasome. Proteins destined for degradation (misfolded proteins, short-lived regulatory proteins or foreign proteins which are degraded to make antigenic peptides) receive a multiubiquitin chain which serves as a "death" signal that is recognized by the 26S proteasome. Initially, in an ATP-dependent step, ubiquitin (Ub), a highly conserved 76 amino-acid protein, is activated by an enzyme (E1), from which it is transferred to the ubiquitin-carrier protein (E2) and finally ligated to the target protein an E3. The E3 proteins have a key role in selecting the substrates. Once bound to the regulatory 19S complexes of the 26S proteasome, the multiubiquitin chains are cleaved off and ubiquitin is recycled. Although the precise route a substrate takes until it reaches the central cavity inside the 20S core particle, where proteolysis takes place, is understood only dimly at present, it is clear that the regulatory 19S complex must unfold the substrate in an ATP-dependent manner and assists in translocating it through the entrance gate of the 20S core complex. Degradation steps further downstream convert the six- to ten-residues long peptides released by the proteasome into free amino acids. This appears to involve an intermediate cleavage step into tripeptides by another large complex, tripeptidyl peptidase II (TPPII). There are alternative routes of protein degradation by the 26S proteasome: Ornithine decarboxylase (ODC) is the prototype substrate which can be degraded in its inactive, antizyme-bound form by the 26S proteasome without ubiquitinylation in a process which requires ATP. The putative role of the 26S proteasome as a refolding machine is yet not firmly established.

et al., 2000; Verma et al., 2000). These subunits can be assigned to two subcomplexes (Glickman et al., 1998a): the "base" part, which comprises an array of six paralogous AAA ATPases and is sufficient to assist in the degradation of (partially) unfolded proteins; and the "lid" part, which provides the link to the ubiquity system and, in turn, confers selectivity (Pickart, 2000; Wilkinson, 2000). Consistent with the lack of a ubiquitin system in archaea and bacteria (Ruepp et al., 2000), minimal homohexameric AAA ATPase complexes are sufficient for regulating prokaryotic 20S proteasomes (Horwich et al., 1999; Wickner et al., 1999; Schmidt et al., 1999; Zwickl et al., 2000a).

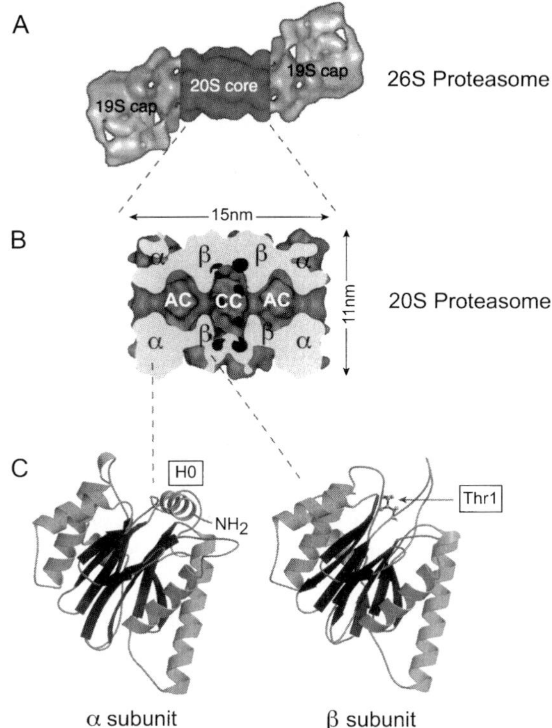

Fig. 2 (A) Composite model of the three-dimensional (3-D) structure of the 26S proteasome combining a three-dimensional reconstruction from electron micrographs of *Drosophila* 26S complexes and the crystal structure of the *Thermoplasma* 20S proteasome. (B) The structure of the 20S proteasome is cut along the seven-fold axis for display of the three inner compartments. The central cavity (CC) is formed by the two seven-membered rings of β subunits, and the antechambers (AC) are formed jointly by one α and one β subunit ring. In the central cavity the positions of eight of the 14 active sites are marked in black. (C) Folds of the α- and β-subunits of the *Thermoplasma* proteasome. A sandwich of five-stranded β sheets is flanked on both sides by α helices. The active-site threonine (Thr1) in the cleft on top of the β-subunit is shown in a ball-and-stick representation. In the α subunit which has basically the same fold, this cleft is occupied by an α-helix (H0), which in the assembled complex sits on top of the cylinder radiating outwards from the central channel.

Current knowledge of the structure, assembly, and function of the 20S proteasome and its regulators in prokaryotic and eukaryotic cells will be reviewed in the following sections.

2 Historical Outline

The earliest report of a protein complex with proteasome-like features dates back to 1968 when an abundant cylinder-shaped structure was observed on electron micrographs of cell lysate from human erythrocytes (Harris, 1968). The observation went largely unnoticed and the cylinder-shaped particle or "cylindrin" (Harris, 1988) remained a structure without a recognized function for quite some time. During the following two decades the particle was rediscovered several times, and the plethora of names it accumulated is a reflection of the problems encountered in defining its biochemical and cellular functions.

A remarkable report describing a tube-shaped "alkaline protease" complex of about 600 kDa, 15 nm in length and 10 nm in diameter and composed of four rings, received surprisingly little attention (Hase et al., 1980). At about the same time, a large multisubunit protease had been isolated from bovine pituitary gland (Wilk and Orlowski, 1980; Orlowski and Wilk, 1981). The complex of approximately 700 kDa was initially believed to be composed of three to five subunits ranging from 24 to 28 kDa in size. It displayed three distinct proteolytic activities (trypsin-like, chymotrypsin-like, and peptidylglutamyl-peptide hydrolyzing) when assays were performed with small synthetic peptides. During the following years, the concept of a multicatalytic protease complex put forward by Wilk and Orlowski was confirmed and further developed by others (Wilk and Orlowski, 1983; Dahlmann et al., 1985; Tanaka et al., 1986; Hough et al., 1987; Rivett, 1989). It was noted that the integrity of the 20S complex was essential for all proteolytic activities, and attempts were made to assign the specific activities to distinct subunits.

Along a different line, a particle called the "prosome" was under intensive investigation in the mid-1980s (Schmidt et al., 1984; Martins de Sa et al., 1986). Reminiscent in size and subunit composition of the multicatalytic protease complex, it appeared to be associated with RNA and it was suggested to have a role in the regulation of gene expression. In 1988, it was established beyond doubt that the prosome and the multicatalytic protease complex were one and the same particle (Arrigo et al., 1988; Falkenburg et al., 1988). More recently, it has been shown that the RNA content of the particle is sub-stoichiometric and probably originates from a cellular pool of abundant tRNAs and 5S rRNAs which bind nonspecifically to several large protein complexes (Pamnani et al., 1994). In 1988, the name proteasome was coined (Arrigo et al., 1988), in recognition of its only established function, the proteolytic one, and in view of its character as a molecular machine.

3 Occurrence and Subunit Composition of 20S Proteasomes

Over the first 20 years the proteasome was structurally rather featureless – it remained the cylinder-shaped particle – and its subunit composition stoichiometry was ill-defined. This led to a search for proteasomes of hopefully simpler subunit composition in prokaryotes. While initial attempts to find proteasomes in eubacteria were unsuccessful (Zwickl et al., 1990), they were found in the archeon *Thermoplasma acidophilum* (Dahlmann et al., 1989). The archaeal proteosomes turned out to be similar in size and shape to proteasomes from eukaryotic sources, but much simpler in subunit composition. The archaebacterial proteasome is built of only two subunits, designated α (25.8 kDa) and β (22.3 kDa). The two subunits have significant sequence similarities (26% identity), suggesting that their genes arose from a common ancestor through a gene duplication event (Zwickl et al., 1991). Because of its relative simplicity, the *Thermoplasma* proteasome has taken a pivotal role in elucidating the structure and enzymatic mechanism of this intriguing macromolecular assembly.

As the number of sequenced genomes grows, a progressively clearer picture of the species distribution of proteasomes is emerging. In general, it appears that proteasomes are ubiquitous and essential in eukaryotes (Heinemeyer, 2000), ubiquitous but not essential in archaea (Ruepp et al., 1998), and rare and nonessential in bacteria, where other energy-dependent proteases abound (Knipfer and Shrader, 1997).

While only one α- and one β-subunit gene have been found in most archaeal genomes, a few species contain a second α or a second β gene (Maupin-Furlow et al., 2000; Zwickl et al., 2000b). To what extent these additional genes are expressed and the subunits incorporated into proteasome particles remains to be investigated. In bacteria, genuine proteasomes of the α + β-type have hitherto only been found in species belonging to the order Actinomycetales. With the exception of *Rhodococcus erythropolis*, where the 20S proteasomes are built of two α- and two β-type subunits (Tamura et al., 1995), proteasomes from all other species, are composed of a single α- and a single β-type subunit (De Mot et al., 1999; Zwickl et al., 2000b). Many other bacteria contain a gene, called hslV in *E. coli*; it encodes a protein that is closely related to the proteasomal β-type subunits (Lupas et al., 1994) and forms a ring-shaped dodecamer (Rohrwild et al., 1997). The occurrence of proteasomes and the HslV protease seems to be mutually exclusive; either one or the other, or none of them is present in the same species (De Mot et al., 1999; Zwickl et al., 2000b).

In eukaryotes such as *Saccharomyces cerevisiae* and *Drosophila melanogaster*, seven different α and seven different β genes have been found (Hughes, 1997). In higher eukaryotes with an adaptive immune system, γ-interferon stimulates expression of three additional β-type subunits, which can replace the closely related, constitutively expressed active β-type subunits (Rock and Goldberg, 1999; Rechsteiner et al., 2000). This results in a modulation of proteolytic specificity. Interestingly, a superfamily of 23 proteasomal genes has been identified in *Arabidopsis thaliana*, and these can be grouped into 14 distinct subfamilies encoding isoforms (Parmentier et al., 1997; Fu et al., 1998a). The functional significance of these isoforms is unclear at present. Alignment of all the available sequences of proteasome subunits shows that the two families, α and β, originate from a gene duplication event, which must have taken place early in evolution (Zwickl et al., 1992a, 2000b; Hughes and Yeager, 1997; Bouzat et al., 2000).

4
Structural Features of the 20S Proteasome

In spite of the differences in subunit complexity, the quaternary structure is highly conserved in 20S proteasomes from all three domains of life – archaea, bacteria, and eukaryotes (Figure 3). The 28 subunits, 14 of the α-type and 14 of the β-type, constitute four seven-membered rings (two α-type and two β-type), which collectively form a barrel-shaped complex with a length of 15 nm and a diameter of 11 nm (Figure 2). The two adjacent β-subunit rings enclose the central cavity (CC); this has a diameter of approximately 5 nm and harbors the active sites. It is connected via two narrow constrictions with two slightly smaller outer cavities, the "antechambers" (AC), which are formed jointly by one α and one β ring. An axial pore in the rings gives access to the antechambers (Figure 2).

In prokaryotic proteasomes, built of 14 α- and 14 β-subunits, the rings are homomeric and the complex is described by an $\alpha_7\beta_7\beta_7\alpha_7$

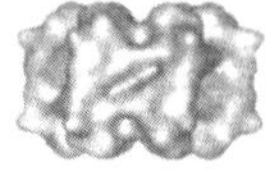

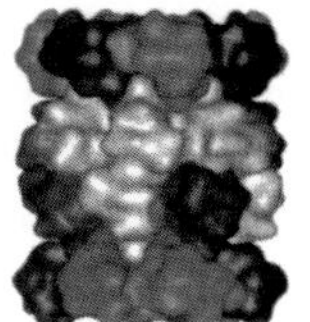

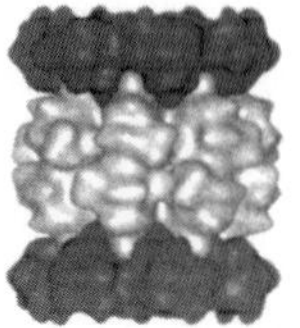

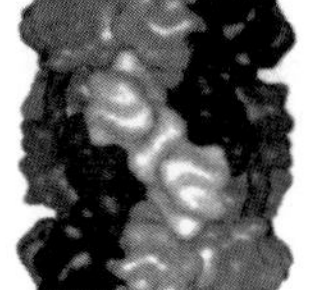

Fig. 3 Comparison of the basic architectures of prokaryotic and eukaryotic 20S proteasomes. Surface models of the *Thermoplasma acidophilum* and *Saccharomyces cerevisiae* proteasome and the *E. coli* HslV protease were obtained by low-pass filtering (1.2-nm cut-off) of the crystal structures. The *Thermoplasma* model was used to represent the *Rhodococcus* proteasome, for which a crystal structure is as yet not available. The (outer ring) α subunits of the proteasome are shown in a darker tone than the inner ring β subunits. HslV is a homologue of the proteasome lacking α-type subunits and has a different symmetry (six-fold versus seven-fold in genuine α-β-type proteasomes). The different gray levels indicate that between one and seven distinct α-type and β-type subunits, respectively, form the α and the β rings in 20S proteasomes. Each of the seven different subunits has a well-defined position in the *Saccharomyces* proteasome, whereas the positions of the two different α-type and β-type subunits in the seven-membered rings of the *Rhodococcus* proteasome are most likely random.

stoichiometry. Correspondingly, the stoichiometry of eukaryotic 20S proteasomes is $\alpha_{1\text{-}7}\beta_{1\text{-}7}\beta_{1\text{-}7}\beta_{1\text{-}7}$. Each of the 14 different subunits is present in two copies within one complex and occupies a precisely defined position. Thus, the multiple axes of symmetry of the *Thermoplasma* proteasome are reduced to a C2 symmetry in the eukaryotic proteasome (Schauer et al., 1993; Löwe et al., 1995; Groll et al., 1997; Kopp et al., 1997). To indicate subunit positions within the rings, a systematic nomenclature for proteasome subunits has been proposed: Subunits are numbered α1 to α7 and β1 to β7, and those related by C2 symmetry are distinguished by the prime symbol (α1′ to α7′ and β1′ to β7′). In mammals, the complexity of proteasomes is further enhanced by the fact that, after γ-interferon stimulation, three of the constitutive subunits (β1, β2, and β5) can be replaced by closely related subunits to form immunoproteasomes; according to the new nomenclature the inducible subunits are termed β1i, β2i, and β5i (Groll et al., 1997). Since only three subunits (β1 or β1i, β2 or β2i, β5 or β5i), of the β-type subunits present in a single proteasome particle, are proteolytically active, eukaryotic proteasomes contain a total of six active sites, whereas prokaryotic proteasomes, built of identical copies of β subunits, have 14 active sites.

Another feature that appeared to distinguish eukaryotic from prokaryotic proteasomes relates to the path of substrate entry into the inner cavities. Although the crystal structure of the *Thermoplasma* proteasome has revealed the existence of an approximately 1.3 nm wide axial channel in the α rings (Löwe, et al., 1995), and electron microscopy of 20S proteasomes incubated with gold-labeled substrates had demonstrated that it is in fact the entry site (Wenzel and Baumeister, 1995), there was no obvious passageway for substrate entry or exit in the structure of the yeast 20S proteasome (Groll et al., 1997). Meanwhile, it has become clear that the yeast particle had been crystallized in a closed form, in which the N-terminal tails of the α-subunits interdigitate to form a plug, whereas these tails are disordered in the archaeal proteasome and, therefore the gate appears to be open.

In vivo, associations with regulatory complexes, e.g., the 19S regulatory complex or the PA28 activator, are known to activate 20S proteasomes. All these regulatory complexes interact with the terminal α rings and are likely to function by mechanisms that open the gate for substrate uptake (or product release) (Pickart and VanDemark, 2000; Rechsteiner et al., 2000; Whitby et al., 2000).

As anticipated from their sequence similarity, the (non-catalytic) α- and the (catalytic) β-type subunits have the same fold (Löwe et al., 1995; Groll et al., 1997): α four-layer α+β structure with two antiparallel five-stranded β sheets, flanked on one side by two, on the other side by three α helices (Fig. 2). In the β-type subunits, the β-sheet sandwich is closed at one end by four hairpin loops and open at the opposite end to form the active-site cleft; the cleft is oriented towards the central cavity. In the α-type subunits an additional helix formed by an N-terminal extension crosses the top of the β-sheet sandwich and fills this cleft. Initially, the proteasome fold was believed to be unique; however, it turned out to be prototypical of a new superfamily of proteins referred to as Ntn (N-terminal nucleophile) hydrolases (Brannigan et al., 1995). Beyond the common fold, members of these family share the mechanisms of the nucleophilic attack and self-processing (Dodson and Wlodawer, 1998; Oinonen and Rouvinen, 2000).

5 Catalytic Mechanism of the 20S Proteasome

Site-directed mutagenesis and the crystal structure analysis of a proteasome–inhibitor complex were used to identify the amino-terminal threonine (Thr1) of *Thermoplasma* β subunits as both the catalytic nucleophile and primary proton acceptor (Löwe et al., 1995; Seemüller et al., 1995). The hydroxyl group of Thr1 attacks the carbonyl carbon of the scissile peptide bond, while its amino group serves as the primary proton acceptor, which enhances the nucleophilicity by stripping the proton from the side-chain hydroxyl; for steric reasons a water molecule is likely to mediate the proton transfer (Figure 4). Such a "single residue" active site is a characteristic feature of Ntn hydrolases, as is the autocatalytic removal of a prosequence, which is necessary to expose the nucleophilic group. For this autocatalytic cleavage, the hydroxyl oxygen of Thr1 of the proteasome β-subunit precursor (Thr1O) adds to the carbonyl carbon of Gly1, and a water molecule is supposed to fulfil the base function of the Thr1 α-amino group in the mature enzyme (Figure 4) (Ditzel et al., 1998). Both, Gly1 and Thr1 are invariant residues of all "active" proteasome subunits. In other Ntn hydrolases, the nucleophile is provided either by a threonine, cysteine or serine residue, and the residues at the (–1) position are variable. In proteasomes, replacement of the active-site threonine by serine does not alter the rates of hydrolysis of small fluorogenic peptide substrates, but the mutant is significantly slower in degrading larger peptides and proteins (Seemüller et al., 1995; Maupin-Furlow et al., 1998; Kisselev et al., 2000). Self-processing of β-subunit precursors is inefficient with serine at the +1 position (Chen and Hochstrasser, 1996; Seemüller et al., 1996; Heinemeyer et al., 1997). Conversely, substitution of the

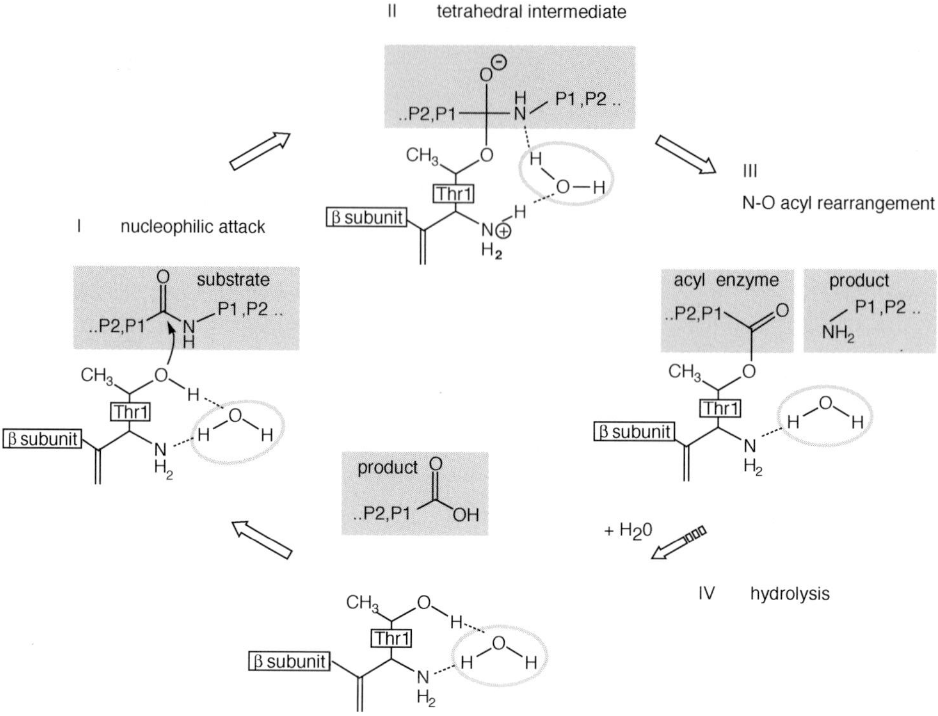

Fig. 4 Putative mechanism of substrate hydrolysis by 20S proteasomes. A proton transfer from the hydroxyl group of Thr1 of β subunits to its own terminal amino group initiates the nucleophilic attack (I). As a result of the nucleophilic addition to the carbonyl carbon of the scissile peptide bond, a tetrahedral intermediate is formed (II). By an N–O rearrangement, an ester is formed (the acyl enzyme); and the amino-terminal cleavage product is released (III). Finally, hydrolysis of the acyl enzyme yields the carboxyl-terminal cleavage product and frees the enzyme for another reaction cycle (IV).

threonine by a cysteine yields proteasomes that are unable to cleave substrates, but the self-cleavage reaction remains unaffected by this mutation (Seemüller et al., 1996). Besides the N-terminal threonine, several other residues of proteasome β subunits (Glu17, Lys33, Asp166) are required for the proteolytic activity, although their precise contribution to the catalytic mechanism awaits further clarification (Löwe et al., 1995; Chen and Hochstrasser, 1996; Seemüller et al., 1996; Heinemeyer et al., 1997).

The catalytic role of the N-terminal threonine of β subunits was soon extended to eukaryotic proteasomes by the observation that binding of lactacystin, a *Streptomyces* metabolite, to the mammalian β-type subunit β5/X irreversibly inhibits its proteolytic activity (Fenteany et al., 1995). Meanwhile, a large number of mutational studies with other archaeal, bacterial, yeast and mammalian β-type subunits have confirmed a common proteolytic mechanism for all types of proteasomes. It remains enigmatic, however, why in eukaryotic proteasomes four out of seven β subunits lack the N-terminal active site threonine, and therefore are presumably inactive.

In recent years, a number of proteasome inhibitors, both natural and synthetic, have become available which are invaluable tools for dissecting the role of the proteasome in intracellular proteolysis (Lee and Goldberg, 1998; Bogyo and Wang, 2002).

6
Processing and Assembly of 20S Proteasomes

The mature active state of the 20S proteasome is reached via a folding and assembly pathway which in the case of the eukaryotic proteasome must be able to orchestrate the correct positioning of two copies each of 14 different but related subunits (Figure 5A). Moreover, in the course of the assembly, β-type subunits are processed by an autocatalytic mechanism that removes the propeptide and exposes the catalytic nucleophile, i.e., the N-terminal threonine (Figure 6). Not surprisingly, in view of the increased subunit complexity, archaeal, bacterial and eukaryotic proteasomes do not follow exactly the same assembly pathways.

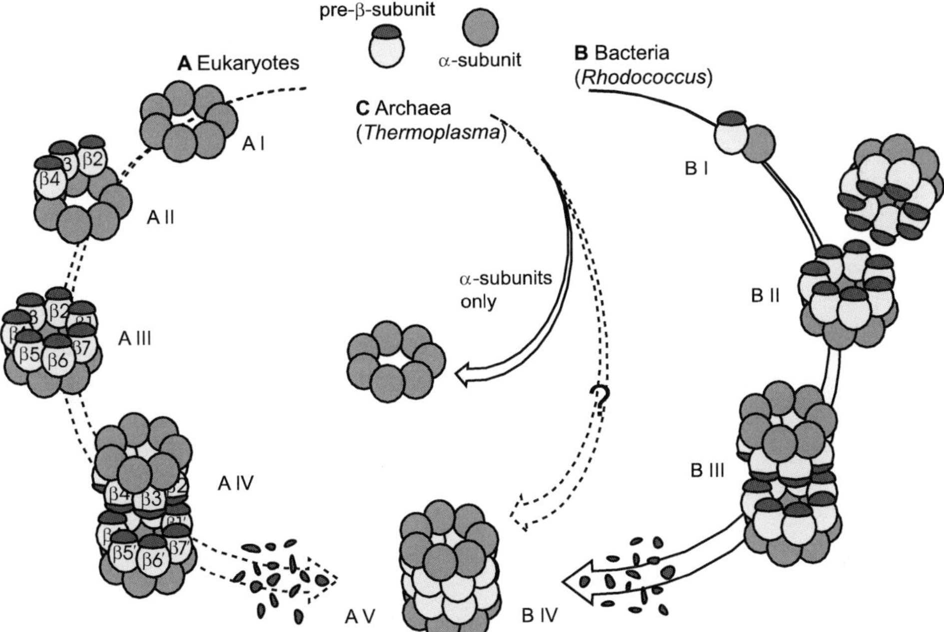

Fig. 5 Assembly pathways of (A) eukaryotic, (B) bacterial, and (C) archaeal 20S proteasomes. The current model for early events of proteasome assembly in mammals proposes that proteasomal α subunits self-assemble into seven-membered rings (AI) that serve as a template onto which the β subunits assemble. Three of the seven β-type subunits, β2, β3, and β4 have been identified as part of a 13S precursor complex (AII). The other four β-type subunits, β5, β6, β7, and β1 were identified as part of a 16S precursor complex (AIII). In a final step these precursor complexes, the "half-proteasomes", assemble to form preholoproteosomes (AIV), which are eventually converted into proteolytically active holoproteasomes (AV). The key step in this transition is the autocatalytic removal of the propeptides. The assembly of the *Rhodococcus* proteasome is initiated by the formation α/β heterodimers (BI), which assemble further to half proteasomes (BII). As in the eukaryotic system, dimerization of the half proteasomes leads to preholoproteosomes (BIII), which are processed to form active holoproteasomes (BIV). Autocatalytic removal of the propeptide is, most likely, triggered by the docking of two half proteasomes.
In archaea, α subunits alone are capable of self-assembling into seven-membered rings which, as in eukaryotes, seem to serve as a template for the assembly of the β subunits. Half-proteasomes have not yet been identified in archaea.

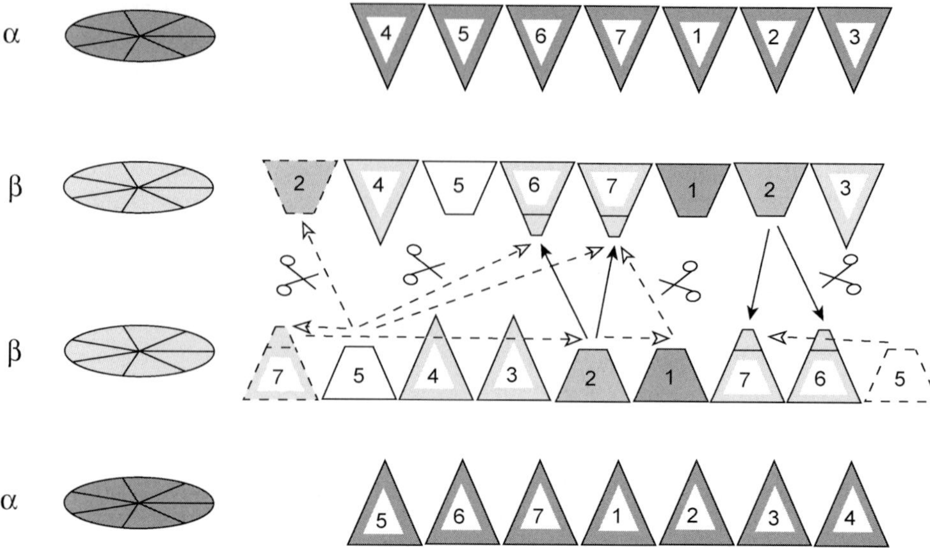

Fig. 6 Current model for intersubunit processing in eukaryotic 20S proteasomes. The open triangles represent inactive β subunits, the filled ones active subunits. The specificities of the three active β-type subunits β1 (peptidylglutamyl-hydrolyzing activity), β2 (trypsin-like activity) and β5 (chymotrypsin-like activity) are represented by different tinges. Cut-off tips of the triangles symbolize the processing of the propeptide. The propeptides of the subunits β5, β1, β2 (or β5′, β1′, and β2′) are removed and the active-site Thr1 is exposed. The propeptides of the subunits β6, β7 (or β6′ and β7′) are only truncated and therefore the Thr1, or the residue replacing it, remain capped and no active site is formed. In wild-type proteasomes, inactive precursor subunits β6 and β7 are processed by the active subunit β2 of the opposite ring (cleavages are indicated by scissors and bold arrows). Mutations of β2 (and other active subunits) induce alternate cleavages exerted by the nearest active neighbor, respectively (cleavages are indicated by scissors and dashed arrows). For the sake of clarity, alternate cleavages are shown in one direction only.

The *Thermoplasma* β-subunit is synthesized as a precursor with a N-terminal propeptide of eight amino acid residues (Zwickl et al., 1994), that is removed by autocatalytic cleavage during proteasome assembly (Seemüller et al., 1996). When expressed alone, the α-subunits of *Thermoplasma* assemble into seven-membered rings (Figure 5C), whereas β-precursors and processed β-subunits, lacking propeptides, do not form an ordered structure (Zwickl et al., 1994). Co-expression of α and β genes yields fully assembled and proteolytically active proteasomes, independent of the presence or absence of the propeptide of the β-precursors (Zwickl et al., 1992b). *In vitro, Thermoplasma* proteasomes can be reassembled after complete dissociation, re-emphasizing the fact that the propeptide is not essential for assembly (Grziwa et al., 1994).

The assembly of *Rhodococcus* proteasomes has been particularly well studied (Figure 5B). The four different subunits (α1, α2, β1, β2) of isolated *Rhodococcus* proteasomes assemble into proteolytically active particles both *in vivo* and *in vitro* with any combination of α- and β-type subunits (Zühl et al., 1997a). Contrary to *Thermoplasma* α-subunits, the *Rhodococcus* α-type subunits alone do not form rings and remain monomeric. The *Rhodococcus* β1- and β2-subunits are translated as precursor proteins with relatively long propeptides of 65 and 59 residues, respectively. The β-type precursors alone do not form complexes and remain unprocessed and thus, inactive (Zühl et al., 1997b). When *Rhodococcus* α- and β-type subunits are mixed *in vitro*, it is quite likely that they form α/β-precursor heterodimers, which quickly assemble into proteolytically inactive half-proteasomes (Zühl et al., 1997a). After dimerization of half-proteasomes, and concomitant with the processing of β-type precursors, the resulting fully assembled proteasomes become proteolytically active. The propeptides of the *Rhodococcus* β1- and β2-subunit precursors are not essential for the incorporation of these subunits into proteasomes, but in their absence the rate of proteasome complex formation is strongly retarded. The propeptides of the β1- and β2-subunits are suggested to act in a chaperone-like manner in the folding of β-subunits and in the assembly of 20S proteasomes. The propeptides can exert their function either in *cis* (covalently linked) or in *trans* (added as separate peptide) (Zühl et al., 1997a). When α- and β-type subunits, which are rendered incompetent for self-processing by mutation of Lys33 of the β-type subunits, are co-expressed, they assemble into pre-holoproteasomes, which are fully assembled particles accommodating the uncleaved propeptides in their central cavities and antechambers. Thus, it appears that pre-holoproteasome formation triggers the autocatalytic cleavage of the propeptides and their degradation (Mayr et al., 1998). Such a self-triggering mechanism is fundamentally different from the activation mechanism of other proteases where inhibitory propeptides are either cleaved off by means of enzymatic cofactors, or where external signals, such as pH changes, trigger the autocatalytic conversion of zymogens to active proteases (Khan and James, 1998). The timing of the self-triggering is dictated by the proteasome assembly pathway and is sufficient to ensure that activation follows sequestration, thus avoiding the risk of uncontrolled proteolysis (Seemüller et al., 2001).

For eukaryotic proteasomes it has been reported that assembly requires additional factors that are only transiently associated with the nascent complex and probably act as chaperones of some kind (Schmidtke et al., 1997; Ramos et al., 1998). The first detected intermediates of proteasome assembly contain all α-type subunits and a subset of β-type subunits (β2, β3 and β4) in a 300 kDa or 13S complex (Figure 5A) (Nandi et al., 1997; Schmidtke et al., 1997). The subsequent incorporation of the residual β-type subunits (β1, β5–β7) triggers rapid dimerization of these precursor complexes into processing competent complexes (Nandi et al., 1997). In the next step, propeptides of β-type subunits are cleaved off in autocatalytic reactions to complete the assembly of proteasomes. Again, as observed in *Rhodococcus*, the propeptides of the β-type subunits – some of which are quite long but are not conserved in sequence – appear to have a chaperone-like function and promote the efficient incorporation of β-type subunits, as has been shown with the yeast β5-subunit (Doa3) (Chen and Hochstrasser, 1996).

7 Size Distribution of 20S Proteasome Products

It is intriguing to observe that, in spite of cleaving protein substrates in an apparently nonspecific manner, peptide products fall into a relatively narrow size range, averaging around seven to nine residues. It is this feature which predisposed the proteasome for a role it assumed in the course of evolution, namely the generation of immunocompetent peptides (Goldberg et al., 1995; Heemels and Ploegh, 1995). The observation that peptide products have a restricted range of sizes led to the proposal that proteasomes may possess an intrinsic molecular ruler. One of the options considered at the time was that the distance between active sites, acting in concert, could provide the mechanistic basis for such a rule (Wenzel et al., 1994). Indeed, the crystal structure of the *Thermoplasma* proteasome revealed a distance of 2.8 nm between neighboring active sites, which corresponds to a hepta- or octapeptide in an extended conformation. Thus, the crystal structure seemed to provide strong evidence in support of this hypothesis (Löwe et al., 1995).

On the other hand, recent more quantitative analyses of product length, although in agreement with an average length of 8 ± 1 residues, showed larger size variations, which are difficult to reconcile with a purely geometry-based rule which should yield products more focused in length (Kisselev et al., 1998). Moreover, reduction in the number of active sites to four or two in mutant yeast proteasomes has had little effect upon the size of generated peptides (Dick et al., 1998). Therefore, it is unlikely that the distance between active sites is a major determinant of the product size.

Recent studies with synthetic peptides which vary in length, but display the same pattern of cleavage sites, indicate that below a certain threshold in length ($<12-14$ residues) degradation products have a high probability of exiting the proteolytic nanocompartment formed by the proteasome (Dolenc et al., 1998). Although they may reenter and be degraded further, this appears to be a slow and inefficient process and thus products smaller than 12–14 residues accumulate. It should be noted here, that the lower "affinity" of the proteasome for shorter peptides was considered to be another option when the molecular rule hypothesis was originally put forward (Wenzel et al., 1994). There is evidence that in *Thermoplasma* (and likewise several other archaea and bacteria), the tricorn protease and its interacting aminopeptidases further degrade products released by the proteasome to amino acids in order to complete the turnover of cellular proteins (Tamura et al., 1998). It has been proposed that, in eukaryotes, this role is assumed by tripeptidyl peptidase II (TPPII) (Tomkinson, 1999).

8
The 19S Regulatory Complex

In eukaryotic cells, the 20S proteasome assembles with one or two 19S regulatory complexes, the "caps", to form the 26S proteasome (for review, see Voges et al., 1999). This is a huge protein degradation machine of 2.5 MDa (Hölzl et al., 2000). On electron micrographs, the elongated complex (which is 45 nm long) shows a characteristic shape – the (double) "dragon head motif" (Peters et al., 1993; Rechsteiner, 1998).

The 26S holoenzyme is the most downstream element of the ubiquitin–proteasome pathway. It is an abundant complex, both in the nucleus and in the cytoplasm, and studies with proteasome inhibitors have indicated that the majority of cellular proteins (80–90%) is degraded via this pathway (Rock et al., 1994; Craiu et al., 1997). It is not only used as a disposal machinery for misfolded proteins, but the lifespan of many short-lived regulatory proteins is also controlled via this pathway (Hershko and Ciechanover, 1998; Bonifacino and Weissman, 1998; Hicke, 1999; Plemper and Wolf, 2000). Interestingly, the 26S proteasome does not only degrade proteins completely, but also activates certain transcription factors by limited proteolysis of precursors.

Probably the best-studied examples of this type are human NFκB and NFκB2 (Palombella et al., 1998; Heusch et al., 1999). In addition to its function in ubiquitin-dependent proteolysis, the 26S proteasome was also found to mediate ubiquitin-independent degradation of several proteins, e.g., ornithine decarboxylase (ODC) in association with antizyme (Murakami et al., 2000). Moreover, a not yet fully understood nonproteolytic role of proteasomes in nucleotide excision repair has been described recently (Russell et al., 1999; Jelinsky et al., 2000; Ortolan et al., 2000).

9 Subcomplexes and Subunits of the 19S Regulator

Slightly differing sets of regulatory complex (RC) subunits have been reported for different organisms (DeMartino et al., 1994; Dubiel et al., 1995; Glickman et al., 1998b; Hölzl et al., 2000) and the abundance of factors that have been described to associate to the RC in a tissue and development-specific manner and with variable affinities, makes it difficult to draw the line between interacting factors, transiently bound subunits or integral components of the RC. However, as the mass of the *Drosophila* RC (~890 kDa), determined by scanning-transmission electron microscopy (STEM) measurements is in good agreement with the summed masses of the 18 individual subunits (932 kDa) identified on two-dimensional gels (Hölzl et al., 2000), it seems likely that the catalogue of integral subunits is now complete.

The 19S regulator of the yeast proteasome can be dissociated into two subcomplexes, the "base" and the "lid", which are located proximally and distally with respect to the 20S core (Figure 7) (Glickman et al., 1998a). The complex formed by the base and the 20S proteasome is sufficient for the degradation of nonubiquitylated protein substrates, most of which are probably natively unfolded proteins (Uversky, 2002). It is not sufficient for the degradation of ubiquitylated substrate proteins. From the location of the base in the 26S complex it had been inferred to have a role in substrate unfolding, acting as a "reverse chaperone", and in controlling the gate in the coaxially apposed α-rings of the 20S particle (Lupas et al., 1993). Recently, a chaperone-like activity was demonstrated for the base (Braun et al., 1999; Strickland et al., 2000), and, likewise for its ancestral precursor proteasome activating nucleotidase (PAN) (Benaroudj and Goldberg, 2000). Both have the ability to recognize proteins destined for degradation, to unfold them in an ATP-dependent manner, and to assist in their translocation into the proteolytic core complex. The degradation signals remain to be defined. Possibly some local unfolding and an exposure of hydrophobic segments is sufficient for targeting substrates. As only the 26S holoenzyme, but not the 20S-base complex, degrades ubiquitylated proteins (Glickman et al., 1998a; Thrower et al., 2000), it appears that recognition and binding of ubiquitin-tagged substrates is mediated by the subunits of the lid subcomplex. In fact, very little is known about the molecular mechanisms of substrate shuttling from the E3 ubiquitin–ligase complexes, where the final step of ubiquitylation takes place, to the lid subcomplex of the 26S proteasome. Free diffusion of ubiquitylated substrates to the 26S proteasomes would be consistent with the rather high binding constants observed *in vitro* (Baboshina and Haas, 1996). On the other hand, subcomplexes or subunits of the RC, such as the lid or S5a/Rpn10, could serve as substrate carriers. Indeed, free lid complexes exist and can be isolated from human erythro-

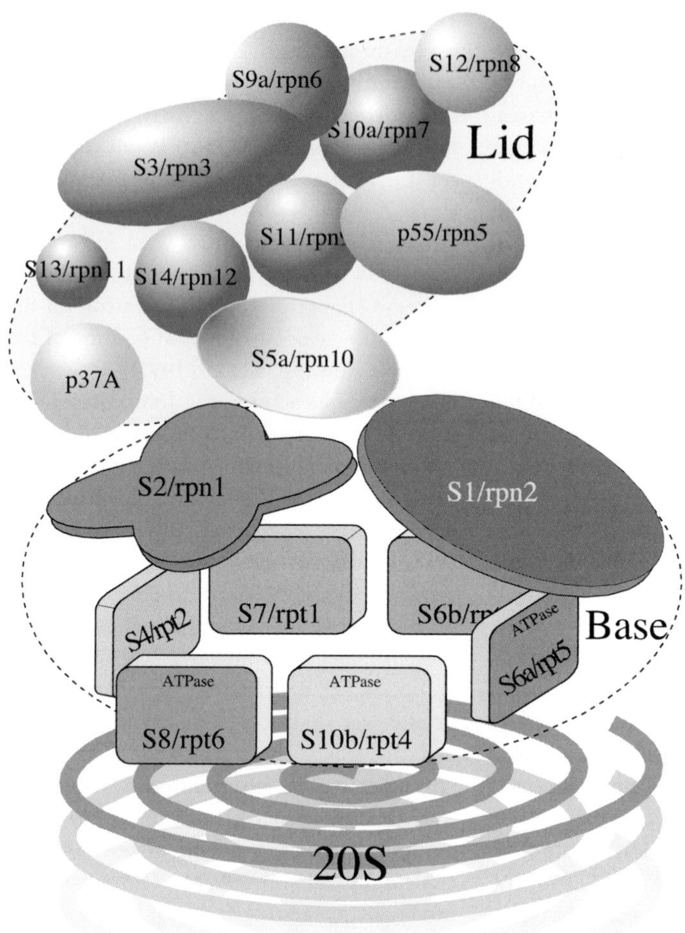

Fig. 7 Model for the subunit organization in the 19S regulatory particle of the eukaryotic 26S proteasome. The regulatory complex is attached to either one or both ends of the 20S complex (bottom). It comprises two subcomplexes – the "base", which is proximal to the 20S core article, and the "lid", which is distal. Six ATPases, the so-called Rpt proteins (regulatory particle AAA ATPases) and the two large Rpn (regulatory particle non AAA ATPase) subunits, Rpn1 and Rpn2, form the base. Rpn10 is at the interface of the two subcomplexes. The remaining subunits comprise the lid. The detailed arrangement of the subunits within the lid and the base complex is largely undefined.

cytes as stable particles (Henke et al., 1999; Kapelari et al., 2000).

The base complex contains all six ATPases, S7/Rpt1, S4/Rpt2, S6/Rpt3, S10b/Rpt4, S6′/Rpt5, and S8/Rpt6, the two largest subunits, S1/Rpn2 and S2/Rpn1, as well as S5a/Rpn10 (Figure 7). The six distinct ATPases in the 19S complex contain an AAA

(ATPases Associated with a variety of Activities) domain, and are all essential and confer ATP dependence on protein degradation (Rubin et al., 1998). Although prokaryotes do not contain 19S complexes, most likely their 20S proteasomes also interact with ATPases of the AAA superfamily (Zwickl et al., 2000b). The recombinant *Methanococcus* protein PAN (P. Zwickl, unpublished results) forms a high-molecular weight complex and stimulates the degradation of protein substrates by *Thermoplasma* 20S proteasomes in a nucleotide-dependent manner. The related AAA ATPase ARC (ATPase-forming Ring-shaped Complexes) from *Rhodococcus erythropolis* also forms a hexameric complex (Wolf et al., 1998). However, the formation of a distinct complex of the 20S proteasome with PAN or ARC has not yet been shown. In both archaea and bacteria, which lack ubiquitin as a degradation signal, ATPases may not only unfold and translocate, but also recognize substrate proteins (Gottesman, 1996). Specific interactions between different 19S ATPases suggest that the six ATPases assemble into a ring resembling other AAA ATPase complexes. The ATPases also interact with 20S α-type subunits and form the interface for the interaction of the 19S complex with the 20S core (Figure 7) (Baumeister et al., 1998; Voges et al., 1999), similar to other protease–ATPase complexes (Gottesman et al., 1997). It is possible that the N-terminal regions of the 19S ATPases, which are predicted to form coiled coils (Rechsteiner et al., 1993; Russell et al., 1996), are involved in the binding of substrate proteins (Wang et al., 1996). The two largest subunits (S1/Rpn2, S2/Rpn1) of the base complex are related in sequence and contain leucine-rich repeats. Their role is still an enigma, though it is possible that they serve as tethering sites for partially unfolded protein substrates en route to the 20S core or recognizing and binding natively unfolded proteins.

S5a/Rpn10 is the only subunit in the 19S complex yet identified which binds Lys-linked ubiquitin (Ub) chains *in vitro* (Deveraux et al., 1994). Ub binding sites have been localized in the C-terminal part of the protein and involve the regions GVDP and LAL/MALRV/LSM, that have alternating large and small hydrophobic residues (Haracska and Udvardy, 1997; Fu et al., 1998b; Young et al., 1998). S5a interacts with the hydrophobic stretch UbLeu8, UbIle44, UbVal70 present on the surface of Ub chains (Beal et al., 1996, 1998). It remains controversial as to whether the Ub binding properties of free S5a and the 26S proteasome are identical. Furthermore, S5a is not essential in yeast (van Nocker et al., 1996; Rubin et al., 1997), suggesting that there are other ubiquitin recognition components in the 19S complex. The N-terminal part of S5a/Rpn10 seems to be involved in interactions between the lid and the base complexes (Glickman et al., 1998a).

For the degradation of ubiquilated target proteins the entire 26S complex, including the lid complex, is essential. The lid is composed of eight subunits, S3/Rpn3, Rpn5, S9/Rpn6, S10a/Rpn7, S11/Rpn9, S12/Rpn8, S13/Rpn11, and S14/Rpn12. Five of these subunits, S3/Rpn3, Rpn5, S9/Rpn6, S10a/Rpn7, and S11/Rpn9 contain a PINT/PCI domain in their C-terminal region and S12/Rpn8 and S13/Rpn11 contain a MPN domain in their N-terminal regions (Aravind and Ponting, 1998; Hofmann and Bucher, 1998). These motifs are found in subunits of other large protein complexes, such as the COP9/signalosome complex (Seeger et al., 1998; Wei et al., 1998) and the eukaryotic initiation factor 3 (eIF3) complex (Asano et al., 1997). There is possibly a one-to-one correspondence between 19S and COP9 signalosome subunits that make a common evolutionary origin likely for the complexes. The functions of lid subunits is largely unknown, but some may be involved in the recognition of Ub conjugates. It is possible that the lid subcomplex and COP9 are exchangeable in the 26S complex and thus allow the modulation of substrate recognition. For a more comprehensive discussion of the 19S subunits and the associated phenotypes, the reader is referred to Voges et al. (1999).

10 The PA28 Activator

The PA28 activator or 11S regulator, is an ATP-independent activator of the 20S proteasome, which stimulates the hydrolysis of small peptides, but not of denatured or ubiquitylated proteins (Dubiel et al., 1992). PA28 is a predominantly cytosolic 200 kDa complex formed by two different but related 28 kDa subunits, PA28α and PA28β, which are believed to form a heteroheptamer (Zhang et al., 1999). The PA28 complex can bind to both ends of the 20S proteasome, as shown by electron microscopy (Gray et al., 1994; Koster et al., 1995). Immunoprecipitation experiments indicate that 20S proteasomes, PA28 and the 19S regulator can also form "hybrid proteasomes" (Hendil et al., 1998), which are able to degrade ODC in an ATP-dependent, but ubiquitin-independent reaction (Tanahashi et al., 2000). Recently the existence of "hybrid proteasomes" has been shown directly by electron microscopy (Cascio et al., 2002). When expressed in *E. coli*, in the absence of PA28 β-subunits, PA28 α-subunits assemble into a heptameric particle, the crystal structure of which has been determined. The PA28 heptamer is a cone-shaped particle formed by bundles of α helices; it is traversed by a central channel, that has a 3-nm opening on the side proximal and a 2-nm opening on the side distal to the proteasome (Knowlton et al., 1997).

In vitro studies have shown that PA28 can enhance the generation of antigenic peptides by inducing dual substrate cleavages by the 20S proteasome, which suggests an *in vivo* role for PA28 in antigen processing (Dick et al., 1996; Groettrup et al., 1996). Consistent with these data, genes for PA28α and PA28β are found only in organisms with adaptive immune systems (Tanaka and Kasahara, 1998). The cytokine γ-interferon, induces expression of both PA28 subunits, as well as both subunits of the transporter associated with antigen processing (TAP1 and TAP2), and the immunoproteasome subunits β1i, β2i, and β5i (Früh and Yang, 1999).

11 Outlook and Perspectives

Over the past decade, great strides have been made in revealing the structure, assembly and enzymatic mechanism of the 20S proteasome which has become the paradigm of multimeric self-compartimentalizing proteases. Nevertheless, several questions remain unanswered, including details of the precise mechanism of substrate translocation into the inner proteolytic compartment.

In prokaryotes, the 20S proteasome cooperates with relatively simple homooligomeric ATPases, which appear to recognize substrates targeted for degradation, unfold them and assist in opening the channel that controls access to the interior of the 20S particle. However, the signals which mark proteins for degradation in archaea and bacteria remain to be identified.

In eukaryotes, ubiquitin serves as the most common degradation signal and the 19S regulatory complexes, which associate with the ends of the barrel-shaped 20S complex mediate interaction with ubiquitylated substrates. The functional organization of the 26S proteasome is understood only vaguely at present. The constituent subunits of the 19S complex have been identified, but specific functions have only been assigned to a few of them. The molecular details of the various steps from the initial recognition of the substrate to the translocation into the 20S core complex where degradation takes place, have remained elusive so far.

More evidence for interactions between chaperones and the 26S proteasome has emerged recently, lending support to the notion that protein folding and protein degradation "machines" cooperate in the quality control of cellular proteins.

12 References

Aravind. L., Ponting, C.P. (1998) Homologues of 26S proteasome subunits are regulators of transcription and translation, *Protein Sci.* **7**, 1250–1254.

Arrigo, A. P., Tanaka, K., Goldberg, A. L., Welch, W. J. (1988) Identity of the 19S 'prosome' particle with the multifunctional protease complex of mammalian cells (the proteasome). *Nature* **331**, 192–194.

Asano, K., Vornlocher, H. P., Richtercook, N. J., Merrick, W. C., Hinnebusch A. G., Hershey, J. W. B. (1997) Structure of cDNAs encoding human eukaryotic initiation factor 3 subunits–possible roles in RNA binding and macromolecular assembly, *J. Biol. Chem.* **272**, 27042–27052.

Baboshina, O. V., Haas, A. L. (1996) Identification of a novel family of ubiquitin-conjugating enzymes with distinct amino-terminal extensions. *J. Biol. Chem.* **271**, 2789–2794.

Baumeister, W., Walz, J., Zühl, F., Seemüller, E. (1998) The proteasome–paradigm of a self-compartimentalizing protease, *Cell* **92**, 367–380.

Beal, R., Deveraux, Q., Xia, G., Rechsteiner, M., Pickart, C. (1996) Surface hydrophobic residues of multiubiquitin chains essential for proteolytic targeting, *Proc. Natl. Acad. Sci. USA* **93**, 861–866.

Beal, R. E., Toscanocantaffa, D., Young, P., Rechsteiner, M., Pickart, C. M. (1998) The hydrophobic effect contributes to polyubiquitin chain recognition, *Biochemistry* **37**, 2925–2934.

Benaroudj, N., Goldberg, A. L. (2000) PAN, the Proteasome-Activating Nucleotidase from Archaebacteria, is a protein-unfolding molecular chaperone, *Nature Cell Biol.* **2**, 833–839.

Bogyo, M., Wang, E. W. (2002) Proteasome inhibitors: complex tools for a complex enzyme, in: *The Proteasome-Ubiquitin Protein Degradation Pathway* (Zwickl, P., Baumeister, W., Eds.), Berlin, Heidelberg, New York: Springer-Verlag, 185–208.

Bonifacino, J. S., Weissman, A. M. (1998) Ubiquitin and the control of protein fate in the secretory and endocytic pathways, *Annu. Rev. Cell Dev. Biol.* **14**, 19–57.

Bouzat, J. L., McNeil, L. K., Robertson, H. M., Solter, L. F., Nixon, J. E., Beever, J. E., Gaskins, H. R., Olsen, G., Subramaniam, S., Sogin, M. L., Lewin, H. A. (2000) Phylogenomic analysis of the a proteasome gene family from early-diverging eukaryotes, *J. Mol. Evolution* **51**, 532–543.

Brannigan, J. A., Dodson, G., Duggleby, H. J., Moody, P. C. E., Smith, J. L., Tomchick, D. R., Murzin, A. G. (1995) A protein catalytic framework with an N-terminal nucleophile is capable of self-activation, *Nature* **378**, 416–419.

Braun, B. C., Glickman, M., Kraft, R., Dahlmann, B., Kloetzel, P. M., Finley, D., Schmidt, M. (1999) The base of the proteasome regulatory particle exhibits chaperone-like activity, *Nature Cell Biol.* **1**, 221–226.

Cascio, P., Call, M., Petre, B. M., Walz, T., Goldberg, A. L. (2002) Properties of the hybrid form of the 26S proteasome containing both 19S and PA28 complexes, *EMBO J* **21**, 2636–2645.

Chen, P., Hochstrasser, M. (1996) Autocatalytic subunit processing couples active site formation in the 20S proteasome to completion of assembly, *Cell* **86**, 961–972.

Coux, O., Tanaka, K., Goldberg, A. L. (1996) Structure and function of the 20S and 26S proteasomes, *Annu. Rev. Biochem.* **65**, 801–847.

Craiu, A., Gaczynska, M., Akopian, T., Gramm, C. F., Fenteany, G., Goldberg, A. L., Rock, K. L. (1997) Lactacystin and clasto-lactacystin beta-lactone modify multiple proteasome beta-subunits and inhibit intracellular protein degradation and major histocompatibility complex Class I antigen presentation, *J. Biol. Chem.* **272**, 13437–13445.

Dahlmann, B., Kuehn, L., Rutschmann, L., Reinauer, H. (1985) Purification and characterization of a multicatalytic high molecular mass proteinase from rat skeletal muscle, *Biochem. J.* **228**, 161–170.

Dahlmann, B., Kopp, F., Kuehn, L., Niedel, B., Pfeifer, G., Hegerl, R., Baumeister, W. (1989) The multicatalytic proteinase (proteasome) is ubiquitinous from Eukaryotes to Archaeabacteria, *FEBS Lett.* **251**, 125–131.

De Mot, R., Nagy, I., Walz, J., Baumeister, W. (1999) Proteasomes and other self-compartmentalizing proteases in procaryotes, *Trends Microbiol.* **7**, 88–92.

DeMartino G. N. and Slaughter C. A. (1999) The proteasome, a novel protease regulated by multiple mechanisms, *J. Biol. Chem.* **274**, 22123–22126.

DeMartino, G. N., Moomaw, C. R., Zagnitko, O. P., Proske, R. J., Chu-Ping, M., Afendis, S. J., Swaffield, J. C., Slaughter, C. A. (1994) PA700, an ATP-dependent activator of the 20 S proteasome, is an ATPase containing multiple member of a nucleotide-binding protein family, *J. Biol. Chem.* **269**, 20878–20884.

Deveraux, Q., Ustrell, V., Pickart, C., Rechsteiner, M. (1994) A 26 S protease subunit that binds ubiquitin conjugates, *J. Biol. Chem.* **269**, 7059–7061.

Dick, T. P., Ruppert, T., Groettrup, M., Kloetzel, P. M., Kuehn, L., Koszinowski, U. H., Stevanovic, S., Schild, H., Rammensee, H. G. (1996) Antigen presentation–recent developments, *Int. Arch. Allergy Immunol.* **110**, 299–307.

Dick, T. P., Nussbaum, A. K., Deeg, M., Heinemeyer, W., Groll, M., Schirle, M., Keilholz, W., Stevanovic, S., Wolf, D. H., Huber, R., Rammensee, H. G., Schild, H. (1998) Contribution of proteasomal β-subunits to the cleavage of peptide substrates analyzed with yeast mutants, *J. Biol. Chem.* **273**, 25637–25646.

Ditzel, L., Huber, R., Mann, K., Heinemeyer, W., Wolf, D. H., Groll, M. (1998) Conformational constraints for protein self-cleavage in the proteasome, *J. Mol. Biol.* **279**, 1187–1191.

Dodson, G., Wlodawer, A. (1998) Catalytic triads and their relatives, Trends Biochem. Sci. **23**, 347–352.

Dolenc, I., Seemueller, E., Baumeister, W. (1998) Decelerated degradation of short peptides by the 20s proteasome, *FEBS Lett.* **424**, 357–361.

Dubiel, W., Pratt, G., Ferrell, K., Rechsteiner, M. (1992) Purification of an 11 S regulator of the multicatalytic protease, *J. Biol. Chem.* **267**, 22369–22377.

Dubiel, W., Ferrell, K., Rechsteiner, M. (1995) Subunits of the regulatory complex of the 26S protease, *Mol. Biol. Reports* **21**, 27–34.

Falkenburg, P. E., Haass, C., Kloetzel, P. M., Niedel, B., Kopp, F., Kuehn, L., Dahlmann, B. (1988) *Drosophila* small cytoplasmic 19S ribonucleoprotein is homologous to the rat multicatalytic proteinase, *Nature* **331**, 190–192.

Fenteany, G., Standaert, R. F., Lane, W. S., Choi, S., Corey, E. J., Schreiber, S. L. (1995) Inhibition of proteasome activities and subunit-specific amino-terminal threonine modification by lactacystin, *Science* **268**, 726–731.

Ferrell, K., Wilkinson, C. R. M., Dubiel, W., Gordon, C. (2000) Regulatory subunit interactions of the 26S proteasome, a complex problem, *Trends Biochem. Sci.* **25**, 83–88.

Früh, K., Yang, Y. (1999) Antigen presentation by MHC Class I and its regulation by interferon gamma, *Curr. Opin. Immunol.* **11**, 76–81.

Fu, H. Y., Doelling, J. H., Arendt, C. S., Hochstrasser, M., Vierstra, R. D. (1998a) Molecular organization of the 20s proteasome gene family from *Arabidopsis thaliana*, *Genetics* **149**, 677–692.

Fu, H. Y., Sadis, S., Rubin, D. M., Glickman, M., Vannocker, S., Finley, D., Vierstra, R. D. (1998b) Multiubiquitin chain binding and protein degradation are mediated by distinct domains within the 26 S proteasome subunit Mcb1, *J. Biol. Chem.* **273**, 1970–1981.

Glickman, M. H., Rubin, D. M., Coux, O., Wefes, I., Pfeifer, G., Cjeka, Z., Baumeister, W., Fried, V. A., Finley, D. (1998a) A subcomplex of the proteasome regulatory particle required for ubiquitin-conjugate degradation and related to the Cop9-signalosome and Eif3, *Cell* **94**, 615–623.

Glickman, M. H., Rubin, D. M., Fried, V. A., Finley, D. (1998b) The regulatory particle of the *Saccharomyces cerevisiae* proteasome, *Mol. Cell. Biol.* **18**, 3149–3162.

Goldberg, A., Gaczynska, M., Grant, E., Michalek, H., Rock, K. L. (1995) Functions of the proteasome in antigen presentation, *Cold Spring Harbor Symp. Quant. Biol.* **60**, 479–490.

Gorbea, C., Rechsteiner, M. (2000) The mammalian regulatory complex of the 26S proteasome, in: *Proteasomes: The World of Regulatory Proteolysis* (Hilt, W., Wolf, D. H., Eds.), George Town: Eurekah.com/Landes Biosciences, 91–128.

Gottesman, S. (1996) Proteases and their targets in *Escherichia coli*, *Annu. Rev. Genet.* **30**, 465–506.

Gottesman, S., Maurizi, M.R., Wickner, S. (1997) Regulatory subunits of energy-dependent proteases, *Cell* **91**, 435–438.

Gray, C. W., Slaughter, C. A., DeMartino, G. N. (1994) PA28 activator protein forms regulatory caps on proteasome stacked rings, *J. Mol. Biol.* **236**, 7–15.

Groettrup, M., Soza, A., Eggers, M., Kuehn, L., Dick, T. P., Schild, H., Rammensee, H. G., Koszinowski, U. H., Kloetzel, P. M. (1996) A role for the proteasome regulator PA28a in antigen presentation, *Nature* **381**, 166–168.

Groll, M., Ditzel, L., Löwe, J., Stock, D., Bochtler, M., Bartunik, H. D., Huber, R. (1997) Structure of 20s proteasome from yeast at 2.4-Angstrom resolution, *Nature* **386**, 463–471.

Grziwa, A., Maack, S., Pühler, G., Wiegand, G., Baumeister, W., Jaenicke, R. (1994) Dissociation and reconstitution of *Thermoplasma* proteasome, *Eur. J. Biochem.* **223**, 1061–1067.

Haracska, L., Udvardy, A. (1997) Mapping the ubiquitin-binding domains in the P54 regulatory complex subunit of the *Drosophila* 26s protease, *FEBS Lett.* **412**, 331–336.

Harris, A. L. (1988) Erythrocyte cylindrin: possible identity with the ubiquitous 20S high molecular weight protease complex and the prosome particle, *Indian J. Biochem. Biophys.* **25**, 459–466.

Harris, J. R. (1968) Release of a macromolecular protein component from erythrocyte ghosts, *Biochim. Biophys. Acta* **150**, 534–537.

Harter, C., Wieland, F. (1996) The secretory pathway: mechanisms of protein sorting and transport, *Biochim. Biophys. Acta* **1286**, 75–93.

Hase, J., Kobashi, K., Nakai, N., Mitsiui, K., Iwata, K., Takadera, T. (1980) The quaternary structure of carp muscle alkaline protease, *Biochim. Biophys. Acta* **611**, 205–213.

Heemels, M. T., Ploegh, H. (1995) Generation, translocation, and presentation of MHC Class I-restricted peptides, *Annu. Rev. Biochem.* **64**, 463–491.

Heinemeyer, W. (2000) Active sites and assembly of the 20S proteasome, in: *Proteasomes: The World of Regulatory Proteolysis* (Hilt, W., Wolf, D., Eds.), Georgetown: Eurekah.com/Landes Biosciences, 48–70.

Heinemeyer, W., Fischer, M., Krimmer, T., Stachon, U., Wolf, D. H. (1997) The active sites of the eukaryotic 20 S proteasome and their involvement in subunit precursor processing, *J. Biol. Chem.* **272**, 25200–25209.

Hendil, K. B., Khan, S., Tanaka, K. (1998) Simultaneous binding of Pa28 and Pa700 activators to 20 S proteasomes, *Biochem. J.* **332**, 661–665.

Henke, W., Ferrell, K., Bech-Otschir, D., Seeger, M., Schade, R., Jungblut, P., Naumann, M., Dubiel, W. (1999) Comparison of human COP9 signalosome and 26S proteasome 'lid', *Mol. Biol. Reports* **26**, 29–34.

Hershko, A., Ciechanover, A. (1998) The ubiquitin system, *Annu. Rev. Biochem.* **67**, 425–479.

Heusch, M., Lin, L., Geleziunas, R., Greene, W. C. (1999) The generation of nfkb2 P52: mechanism and efficiency, *Oncogene* **18**, 6201–6208.

Hicke, L. (1999) Gettin' down with ubiquitin: turning off cell-surface receptors, transporters and channels, *Trends Cell Biol.* **9**, 107–112.

Hilt, W., Wolf, D. H. (1996) Proteasomes: destruction as a programme, *Trends Biochem. Sci.* **21**, 96–102.

Hofmann, K., Bucher, P. (1998) The PCI domain: a common theme in three multiprotein complexes, *Trends Biochem. Sci.* **23**, 204–205.

Hölzl, H. Kapelari, B. Kellermann, J. Seemüller, E. Sumegi, M. Udvardy, A. Medalia, O. Sperling, J. Muller, S. A. Engel A. and Baumeister W. (2000) The regulatory complex of *Drosophila melanogaster* 26S proteasomes: subunit composition and localization of a deubiquitylating enzyme, *J. Cell Biol.* **150**, 119–129.

Horwich, A. L., Weber-Ban, E. U., Finley, D. (1999) Chaperone rings in protein folding and degradation, *Proc. Natl. Acad. Sci. USA* **96**, 11033–11040.

Hough, R., Pratt, G., Rechsteiner, M. (1987) Purification of two high molecular weight proteases from rabbit reticulocyte lysate, *J. Biol. Chem.* **262**, 8303–8313.

Hughes, A. L. (1997) Evolution of the proteasome components, *Immunogenetics* **46**, 82–92.

Hughes, A. L., Yeager, M. (1997) Molecular evolution of the vertebrate immune system, *BioEssays* **19**, 777–786.

Jelinsky, S. A., Estep, P., Church, G. M., Samson, L. D. (2000) Regulatory networks revealed by transcriptional profiling of damaged *Saccharomyces cerevisiae* cells: Rpn4 links base excision repair with proteasomes, *Mol. Cell. Biol.* **20**, 8157–8167.

Kapelari, B., Bech-Otschir, D., Hegerl, R., Schade, R., Dumdey, R., Dubiel, W. (2000) Electron microscopy and subunit-subunit interaction studies reveal a first architecture of COP9 signalosome, *J. Mol. Biol.* **300**, 1169–1178.

Khan, A. R., James, M. N. (1998) Molecular mechanisms for the conversion of zymogens to active proteolytic enzymes, *Protein Sci.* **7**, 815–836.

Kisselev, A. F., Akopian, T. N., Goldberg, A. L. (1998) Range of sizes of peptide products generated during degradation of different proteins by archaeal proteasomes, *J. Biol. Chem.* **273**, 1982–1989.

Kisselev, A. F., Songyang, Z., Goldberg, A. L. (2000) Why does threonine, and not serine, function as the active site nucleophile in proteasomes? *J. Biol. Chem.* **275**, 14831–14837.

Knipfer, N., Shrader, T. E. (1997) Inactivation of the 20S proteasome in *Mycobacterium smegmatis*, *Mol. Microbiol.* **25**, 375–383.

Knowlton, J. R., Johnston, S. C., Whitby, F. G., Realini, C., Zhang, Z. G., Rechsteiner, M., Hill, C. P. (1997) Structure of the proteasome activator Reg-a (Pa28-a), *Nature* **390**, 639–643.

Kopp, F., Hendil, K. B., Dahlmann, B., Kristensen, P., Sobek, A., Uerkvitz, W. (1997) Subunit arrangement in the human 20S proteasome, *Proc. Natl. Acad. Sci. USA* **94**, 2939–2944.

Koster, A. J., Walz, J., Lupas, A., Baumeister, W. (1995) Proteasomes of the yeast *S. cerevisiae* – genes, structure and functions, *Mol. Biol. Reports* **21**, 3–10.

Larsen, C. N., Finley, D. (1997) Protein translocation channels in the proteasome and other proteases, *Cell* **91**, 431–434.

Lee, D. H., Goldberg, A. L. (1998) Proteasome inhibitors–valuable new tools for cell biologists, *Trends Cell Biol.* **8**, 397–403.

Löwe, J., Stock, D., Jap, B., Zwickl, P., Baumeister, W., Huber, R. (1995) Crystal structure of the 20S proteasome from the archaeon *T. acidophilum* at 3.4 Angstrom resolution, *Science* **268**, 533–539.

Lupas, A., Flanagan, J. M., Tamura T., Baumeister, W. (1997) Self-compartmentalizing proteases, *Trends Biochem. Sci.* **22**, 399–404.

Lupas, A., Koster, A. J., Baumeister, W. (1993) Structural features of 26S and 20S proteasomes, *Enzyme Protein* **47**, 252–273.

Lupas, A., Zwickl, P., Baumeister, W. (1994) Proteasome sequences in Eubacteria, *Trends Biochem. Sci.* **19**, 533–534.

Martins de Sa, C., Grossi de Sa, M. F., Akhayat, O., Broders, F., Scherrer, K., Horsch, A., Schmid, H. P. (1986) Prosomes: ubiquity and inter species structural variations, *J. Mol. Biol.* **187**, 479–493.

Maupin-Furlow, J. A., Aldrich, H. C., Ferry, J. G. (1998) Biochemical characterization of the 20S proteasome from the Methanoarchaeon *Methanosarcina thermophila*, *J. Bacteriol.* **180**, 1480–1487.

Maupin-Furlow, J. A., Wilson, H. L., Kaczowka, S. J., Ou, M. S. (2000) Proteasomes in the Archaea: from structure to function, *Front. Biosci.* **5**, D837–D866A.

Mayr, J., Seemüller, E., Müller, S. A., Engel, A., Baumeister, W. (1998) Late events in the assembly of 20S proteasomes, *J. Struct. Biol.* **124**, 179–188.

Murakami, Y., Matsufuji, S., Hayashi, S., Tanahashi, N., Tanaka, K. (2000) Degradation of ornithine decarboxylase by the 26S proteasome, *Biochem. Biophys. Res. Commun.* **267**, 1–6.

Nandi, D., Woodward, E., Ginsburg, D. B., Monaco, J. J. (1997) Intermediates in the formation of mouse 20S proteasomes–implications for the assembly of precursor beta subunits, *EMBO J.* **16**, 5363–5375.

Oinonen, C., Rouvinen, J. (2000) Structural comparison of Ntn-hydrolases, *Protein Sci.* **9**, 2329–2337.

Orlowski, M., Wilk, S. (1981) A multicatalytic protease complex from pituitary that forms enkephalin and enkephalin containing peptides, *Biochem. Biophys. Res. Commun.* **101**, 814–822.

Ortolan, T. G., Tongaonkar, P., Lambertson, D., Chen, L., Schauber, C., Madura, K. (2000) The DNA repair protein rad23 is a negative regulator of multi-ubiquitin chain assembly, *Nature Cell Biol.* **2**, 601–609.

Palombella, V. J., Conner, E. M., Fuseler, J. W., Destree, A., Davis, J. M., Laroux, F. S., Wolf, R. E., Huang, J. Q., Brand, S., Elliott, P. J., Lazarus, D., McCormack, T., Parent, L., Stein, R., Adams, J., Grisham, R. B. (1998) Role of the proteasome and NF-kappa B in streptococcal cell wall-induced polyarthritis, *Proc. Natl. Acad. Sci. USA* **95**, 15671–15676.

Pamnani, V., Haas, A. L., Pühler, G., Sanger, H. L., Baumeister, W. (1994) Proteasome associated RNAs are non specific, *Eur. J. Biochem.* **225**, 511–519.

Parmentier, Y., Bouchez, D., Fleck, J., Genschik, P. (1997) The 20S proteasome gene family in *Arabidopsis thaliana*, *FEBS Lett.* **416**, 281–285.

Peters, J. M., Cejka, Z., Harris, J. R., Kleinschmidt, J. A., Baumeister, W. (1993) Structural features of the 26S proteasome complex, *J. Mol. Biol.* **234**, 932–937.

Pickart, C. M. (2000) Ubiquitin in chains, *Trends Biochem. Sci.* **25**, 544–548.

Pickart, C. M., VanDemark, A. P. (2000) Opening doors into the proteasome, *Nature Struct. Biol.* **7**, 999–1001.

Plemper, R. K., Wolf, D. H. (2000) Function of the proteasome in the protein quality control process of the endoplasmic reticulum, in: *Proteasomes: The World of Regulatory Proteolysis* (Hilt, W., Wolf, D. H., Eds.), Georgetown: Eurekah.com/Landes Bioscience, 264–301.

Ramos, P. C., Hockendorff, J., Johnson, E. S., Varshavsky, A., Dohmen, R. J. (1998) Ump1p is required for proper maturation of the 20s proteasome and becomes its substrate upon completion of the assembly, *Cell* **92**, 489–499.

Rechsteiner, M. (1998) Ump1 is required for proper maturation of the 20S proteasomes and becomes its substrate upon completion of the assembly, in: *Ubiquitin and the Biology of the Cell* (Peters, J. M., Harris, J. R., Finley, D., Eds.), New York: Plenum Press, 147–189.

Rechsteiner, M., Hoffmann, L., Dubiel, W. (1993) The multi-catalytic and 26S proteases, *J. Biol. Chem.* **268**, 6065–6068.

Rechsteiner, M., Realini, C., Ustrell, V. (2000) The proteasome activator 11S REG (PA28) and Class I antigen presentation, *Biochem. J.* **345**, 1–15.

Rivett, A. J. (1989) The multicatalytic proteinase of mammalian cells, *Arch. Biochem. Biophys.* **268**, 1–8.

Rock, K. L., Goldberg, A. L. (1999) Degradation of cell proteins and the generation of MHC Class I-presented peptides, *Annu. Rev. Immunol.* **17**, 739–779.

Rock, K. L., Gramm, C., Rothstein, L., Clark, K., Stein, R., Dick, L., Hwang, D., Goldberg, A. L. (1994) Inhibitors of the proteasome block the degradation of most cell proteins and the generation of peptides presented on MHC Class I molecules, *Cell* **78**, 761–771.

Rohrwild, M., Pfeifer, G., Santarius, U., Muller, S. A., Huang, H. C., Engel, A., Baumeister, W., Goldberg, A. L. (1997) The ATP-dependent HslVU protease from *Escherichia coli* is a four-ring structure resembling the proteasome, *Nature Struct. Biol.* **4**, 133–139.

Rubin, D. M., Vannocker, S., Glickman, M., Coux, O., Wefes, I., Sadis, S., Fu, H. Y., Goldberg, A., Vierstra, R., Finley, D. (1997) ATPase and ubiquitin-binding proteins of the yeast proteasome, *Mol. Biol. Reports* **24**, 17–26.

Rubin, D. M., Glickman, M. H., Larsen, C. N., Dhruvakumar, S., Finley, D. (1998) Active site mutants in the six regulatory particle ATPases reveal multiple roles for ATP in the proteasome, *EMBO J.* **17**, 4909–4919.

Ruepp, A., Eckerskorn, C., Bogyo, M., Baumeister, W. (1998) Proteasome function is dispensable under normal but not under heat shock conditions in *Thermoplasma acidophilum*, *FEBS Lett.* **425**, 87–90.

Ruepp, A., Graml, W., Santoz-Martinez, M. L., Koretke, K. K., Volker, C., Mewes, H. W., Frishman, D., Stocker, S., Lupas, A. N., Baumeister, W. (2000) The genome sequence of the thermoacidophilic scavenger *Thermoplasma acidophilum*, *Nature* **407**, 508–513.

Russell, S. J. Sathyanarayana, U. G., Johnston, S. A. (1996) Isolation and characterization of Sug2–a novel ATPase family component of the yeast 26 S proteasome, *J. Biol. Chem.* **271**, 32810–32817.

Russell, S. J., Reed, S. H., Huang, W. Y., Friedberg, E. C., Johnston, S. A. (1999) The 19S regulatory complex of the proteasome functions independently of proteolysis in nucleotide excision repair, *Mol. Cell* **3**, 687–695.

Schauer, T. M., Nesper, M., Kehl, M., Lottspeich, F., Muller, T. A., Gerisch, G., Baumeister, W. (1993) Proteasomes from *Dictiostelium discoideum*: characterization of structure and function, *J. Struct. Biol.* **111**, 135–147.

Schmidt, H. P., Akhayat, O., Martins de Sa, C., Koehler, K., Scherrer, K. (1984) The prosome: a ubiquitous morphologically distinct RNP particle associated with repressed mRNPs and containing ScRNA and a characteristic set of proteins, *EMBO J.* **3**, 29–34.

Schmidt, M., Zantopf, D., Kraft, R., Kostka, S., Preissner, R., Kloetzel, P. M. (1999) Sequence information within proteasomal prosequences mediates efficient integration of β-subunits into the 20S proteasome complex, *J. Mol. Biol.* **288**, 117–128.

Schmidtke, G., Schmidt, M., Kloetzel, P. M. (1997) Maturation of mammalian 20S proteasome–purification and characterization of 13S and 16S proteasome precursor complexes, *J. Mol. Biol.* **268**, 95–106.

Seeger, M., Kraft, R., Ferrell, K., Bechotschir, D., Dumdey, R., Schade, R., Gordon, C., Naumann, M., Dubiel, W. (1998) A novel protein complex involved in signal transduction possessing similarities to 26S proteasome subunits, *FASEB J.* **12**, 469–478.

Seemüller, E., Lupas, A., Stock, D., Loewe, J., Huber, R., Baumeister, W. (1995) Proteasome of *Thermoplasma acidophilum*: a threonine protease, *Science* **268**, 579–582.

Seemüller, E., Lupas, A., Baumeister, W. (1996) Autocatalytic processing of the 20S proteasome, *Nature* **382**, 468–470.

Seemüller, E., Zwickl, P., Baumeister, W. (2001) Self-processing of subunits of the proteasome, in:

The Enzymes, Co- and Posttranslational Proteolysis of Proteins, 3rd edition (Dalbey, R. E., Sigman, D. S., Eds.), San Diego: Academic Press, 335–371, Vol. XXII.

Strickland, E., Hakala, K., Thomas, P. J., DeMartino, G. N. (2000) Recognition of misfolding proteins by PA700, the regulatory subcomplex of the 26S proteasome, *J. Biol. Chem.* **275**, 5565–5572.

Tamura, N., Lottspeich, F., Baumeister, W., Tamura, T. (1998) The role of tricorn protease and its aminopeptidase-interacting factors in cellular protein degradation, *Cell*, **95**, 1–20.

Tamura, T., Nagy, I., Lupas, A., Lottspeich, F., Cejka, Z., Schoofs, G., Tanaka, K., De Mot R., Baumeister, W. (1995) The first characterization of a eubacterial proteasome: the 20S complex of *Rhodococcus*, *Curr. Biol.* **5**, 766–774.

Tanahashi, N., Murakami, Y., Minami, Y., Shimbara, N., Hendil, K. B., Tanaka, K. (2000) Hybrid proteasomes - induction by interferon-γ and contribution to ATP-dependent proteolysis, *J. Biol. Chem.* **275**, 14336–14345.

Tanaka, E., Ji, K., Ichichara, A., Waxman, L., Goldberg, A. (1986) A high molecular weight protease in the cytosol of rat liver, *J. Biol. Chem.* **261**, 15197–15203.

Tanaka, K. (1998) Molecular biology of the proteasome, *Biochem. Biophys. Res. Commun.* **247**, 537–541.

Tanaka, K., Kasahara, M. (1998) The MHC Class I ligand-generating system–roles of immunoproteasomes and the interferon-γ-inducible proteasome activator PA28, *Immunol. Rev.* **163**, 161–176.

Thrower, J. S., Hoffman, L., Rechsteiner, M., Pickart, C. M. (2000) Recognition of the polyubiquitin proteolytic signal, *EMBO J.* **19**, 94–104.

Tomkinson, B. (1999) Tripeptidyl peptidases: enzymes that count, *Trends Biochem. Sci.* **24**, 355–359.

Udvardy, A. (1993) Purification and characterization of a multiprotein protein of the *Drosophila* 26S (1500 kDa) proteolytic complex, *J. Biol. Chem.* **268**, 9055–9062.

Uversky, V. N. (2002) Natively unfolded proteins: a point where biology waits for physics, *Protein Sci.* **11**, 739–756.

van Nocker, S., Sadis, S., Rubin, D. M., Glickman, M., Fu, H., Coux, O., Wefes, I., Finley, D., Vierstra, R. D. (1996) The multiubiquitin-chain-binding protein Mcb1 is a component of the 26S proteasome in *Saccharomyces cerevisiae* and plays a nonessential, substrate-specific role in protein turnover, *Mol. Cell. Biol.* **16**, 6020–6028.

Varshavsky, A. (1997) The ubiquitin system, *Trends Biochem. Sci.* **22**, 383–387.

Verma, R., Chen, S., Feldman, R., Schieltz, D., Yates, J., Dohmen, T., Deshaies, R. J. (2000) Proteasomal proteomics: identification of nucleotide-sensitive proteasome-interacting proteins by mass spectrometric analysis of affinity-purified proteasomes, *Mol. Biol. Cell* **11**, 3425–3439.

Voges, D., Zwickl, P., Baumeister, W. (1999) The 26S proteasome: a molecular machine designed for controlled proteolysis, *Annu. Rev. Biochem.* **68**, 1015–1068.

Wang, W., Chevray, P. M., Nathans, D. (1996) Mammalian SUC1 and c-FOS in the nuclear 26S proteasome, *Proc. Natl. Acad. Sci. USA* **93**, 8236–8240.

Wei, N., Tsuge, T., Serino, G., Dohmae, N., Takio, K., Matsui, M., Deng, X. W. (1998) The Cop9 complex is conserved between plants and mammals and is related to the 26S proteasome regulatory complex, *Curr. Biol.* **8**, 919–922.

Wenzel, T., Baumeister,W. (1995) Conformational constraints in protein degradation by the 20S proteasome, *Nature Struct. Biol.* **2**, 199–204.

Wenzel, T., Eckerskorn, C., Lottspeich, F., Baumeister, W. (1994) Existence of a molecular ruler in proteasomes suggested by analysis of degradation products, *FEBS Lett.* **349**, 205–209.

Whitby, F. G., Masters, E. I., Kramer, L., Knowlton, J. R., Yao, Y., Wang, C. C., Hill, C. P. (2000) Structural basis for the activation of 20S proteasomes by 11S regulators, *Nature* **408**, 115–120.

Wickner, S., Maurizi, M. R., Gottesman, S. (1999) Posttranslational quality control: folding, refolding, and degrading proteins, *Science* **286**, 1888–1893.

Wilk, S., Orlowski, M. (1980) Cation-sensitive neutral endopeptidase: isolation and specificity of the bovine pituitary enzyme, *J. Neurochem.* **35**, 1172–1182.

Wilk, S., Orlowski, M. (1983) Evidence that pituitary cation-sensitive neutral endopeptidase is a multicatalytic protease complex, *J. Neurochem.* **40**, 842–849.

Wilkinson, K. D. (2000) Ubiquitination and deubiquitination: targeting of proteins for degradation by the proteasome, *Semin. Cell Dev. Biol.* **11**, 141–148.

Wolf, S., Nagy, I., Lupas, A., Pfeifer, G., Cejka, Z., Muller, S. A., Engel, A., Demot, R., Baumeister, W. (1998) Characterization of Arc, a divergent member of the Aaa ATPase family from *Rhodococcus erythropolis*, *J. Mol. Biol.* **277**, 13–25.

Young, P., Deveraux, Q., Beal, R. E., Pickart, C. M., Rechsteiner, M. (1998) Characterization of two polyubiquitin binding sites in the 26 S protease subunit 5a, *J. Biol. Chem.* **273**, 5461–5467.

Zhang, Z. G., Krutchinsky, A., Endicott, S., Realini, C., Rechsteiner, M., Standing, K. G. (1999) Proteasome activator 11S REG or PA28: recombinant REG a/REG b hetero-oligomers are heptamers, *Biochemistry* **38**, 5651–5658.

Zühl, F., Seemüller, E., Golbik, R., Baumeister, W. (1997a) Dissecting the assembly pathway of the 20S proteasome, *FEBS Lett.* **418**, 189–194.

Zühl, F., Tamura, T., Dolenc, I., Cejka, Z., Nagy, I., De Mot, R., Baumeister, W. (1997b) Subunit topology of the *Rhodococcus* proteasome, *FEBS Lett.* **400**, 83–90.

Zwickl, P., Pfeifer, G., Lottspeich, F., Kopp, F., Dahlmann, B., Baumeister, W. (1990) Electron microscopy and image analysis reveals common principles of organization in two large protein complexes: GroEL-type and proteasomes, *J. Struct. Biol.* **103**, 197–203.

Zwickl, P., Lottspeich, F., Dahlmann, B., Baumeister, W. (1991) Cloning and Sequencing of the gene encoding the large α-subunit of the proteasome from *Thermoplasma acidophilum*, *FEBS Lett.* **278**, 217–221.

Zwickl, P., Grziwa, A., Pühler, G., Dahlmann, B., Lottspeich, F., Baumeister, W. (1992a) Primary structure of the *Thermoplasma* proteasome and its implication for the structure, function, and evolution of the multicatalytic proteinase, *Biochemistry* **31**, 964–972.

Zwickl, P., Lottspeich, F., Baumeister, W. (1992b) Expression of functional *Thermoplasma acidophilum* proteasomes in *Escherichia coli*, *FEBS Lett.* **312**, 157–160.

Zwickl, P., Kleinz, J., Baumeister, W. (1994) Critical elements in proteasome assembly, *Nature Struct. Biol.* **1**, 765–770.

Zwickl, P., Baumeister, W., Steven, A. (2000a) Disassembly lines: the proteasome and related ATPase-assisted proteases, *Curr. Opin. Struct. Biol.* **10**, 242–250.

Zwickl, P., Goldberg, A. L., Baumeister, W. (2000b) Proteasomes in prokaryotes, in: *Proteasomes: The World of Regulatory Proteolysis* (Wolf, D. H., Hilt, W., Eds.), Georgetown: Eurekah.com/Landes Bioscience, 8–20.

15
Modifications of Proteins and Poly(amino acids) by Enzymatic and Chemical Methods

Dr. Kousaku Ohkawa[1], **Prof. Dr. Hiroyuki Yamamoto**[2]

[1] Institute of High Polymer Research, Faculty of Textile Science and Technology, Shinshu University, Ueda 386-8567, Japan; Tel.: +81-268-215573; Fax: +81-268-215571; E-mail: kohkawa@giptc.shinshu-u.ac.jp

[2] Institute of High Polymer Research, Faculty of Textile Science and Technology, Shinshu University, Ueda 386-8567, Japan; Tel.: +81-268-215572; Fax: +81-268-215571; E-mail: hyihpr2@giptc.shinshu-u.ac.jp

GA	glutaraldehyde
GTP	guanosine 5′-triphosphate
NBS	*N*-bromo-succinimide
NMR	nuclear magnetic resonance
PIC	polyion complex
PLG	poly(α,L-glutamic acid)
PLL	poly(α,L-lysine)
PMLG	poly(γ-methyl-α, L-glutamate)
TGc	tissue transglutaminase type II

1 Introduction

The problems of protein cross-linking have been a central subject in protein chemistry. In the 1950s, Brown (1950) reviewed the protein-stabilizing forces and mentioned the importance of covalent cross-links. Starting with the study on the disulfide bond of keratin, which is an important class of structural proteins of natural occurrence, accelerated research on protein chemistry has been done in close cooperation with material science of natural products of industrial importance, such as wool and feathers. In the 1960s, the identification of cross-links in elastin (Partridge et al., 1963) and resilin (Anderson, 1963) was made. The polymer chemistry of natural rubbers has contributed to our understanding of the elastic properties of elastin (Hoeve and Flory, 1958).

Development of the chemistry for polypeptide synthesis has had a great impact on the study of structural proteins. Synthetic homo- and copolypeptides provide significant information on the secondary structures of the proteins. Studies on the tensile properties of collagen have been advanced, and in the 1970s, stable cross-links were detected in collagen that may be permanent constituents in that protein (Siegel et al., 1970; Housley et al., 1975). Because collagen and elastin are essential components in the connective tissues of mammals, biochemical studies of these proteins' cross-linking have become significant in the study of pathology in tissue diseases. In the 1980s and 1990s, molecular biology and gene technology became the mainstream for investigation of the relationship between structure and function of these proteins. Researchers were able to recognize the domain, which is the amino acid sequence underlying the protein functions and mechanical properties.

At the present time, in addition to progress in the technology of precision analysis, the natural products, many of which have complex chemical structures, have become revealed in detail. The accumulating knowledge about the structural proteins of organisms, of which the performances occasionally are superior to synthetic polymeric materials, encourages biopolymer chemists to develop bio-inspired technology for diverse areas in material science.

This chapter contains a review of the modifications of proteins and polypeptides by enzymatic and chemical methods. Because the literature in this field is indeed vast – encompassing biology, biochemistry, polymer chemistry, and material science – it would be impossible to present here of all the reactions reported in such an interdisciplinary field of investigation. Instead, the present goal is to survey and systematically summarize the various types of biopolymer formulations through cross-linking reactions between the side chains of proteins or poly(amino acids) from the standpoints of biopolymer chemistry and applied material science.

2 Naturally Occurring Cross-linked Materials

Knowledge about the chemical characteristics of structural materials in organisms has been an interplay of polymer chemistry, biology, and materials science. Terpenoid polymers (natural rubber), polysaccharides (cotton fiber), polyphenols (lacquer and lignin), and proteins (silk and wool) are of scientific interest and industrial importance. The structural proteins occurring in vertebrate tissues, such as keratin, collagen, and elastin, provide the clues for understanding an important system for protein cross-linking in life.

2.1 Keratin, Collagen, and Elastin

The structural proteins of animals, such as keratin, collagen, elastin, and other matrix proteins, themselves have a long history of investigation over the last 50 years. Their occurrences, fine structures, mechanical properties, and chemical nature of cross-linkages have been reviewed by Borysko (1963) (collagen), Lundgren and Ward (1963) (keratin), Tanzer (1976) (collagen cross-linking), and Franzblau (1970) (elastin cross-linking). The detailed chemical structures and mechanism of cross-bridge formulation are more recently revealed from the standpoint of biochemistry. Furthermore, the biochemical and physicochemical properties of enzymes, which are involved in the modification of the protein side chains for cross-linking, confer the mechanism associated with the cross-linking formation. Scheme 1 summarizes the protein cross-linking structure found in nature.

The disulfide bond **(1)** between cysteine side-chain sulfurs has been described as a force for stabilizing the protein structure for a prolonged period. Among many structural proteins composing the vertebrate body, keratin is the most abundant protein and is stabilized by the disulfide bonds between the α-helical polypeptide chains. Keratins are tough, elastic, and insoluble in water and are elaborated into structures adapted for mechanical protection, warmth, abrasion, and resistance, e.g., horns, hooves, scales, hair, and feathers.

Collagen and elastin are the important classes of structural proteins in both vertebrates and invertebrates. Knowledge of the abundance and types of cross-links in collagen and elastin should provide important insight into these proteins' function and structure in organisms (Tanzer and Waite, 1982). Cross-links have evolved in collagen to provide sufficient resistance to tensile forces, while retaining the extensibility of the protein.

The first purification of elastin from elastic tissues in the 1840s was described by Richards and Gies (1902). Elastin is an insoluble protein found in most connective tissues in conjunction with other structural elements such as collagen and mucopolysaccharide. Burton (1954) described that the

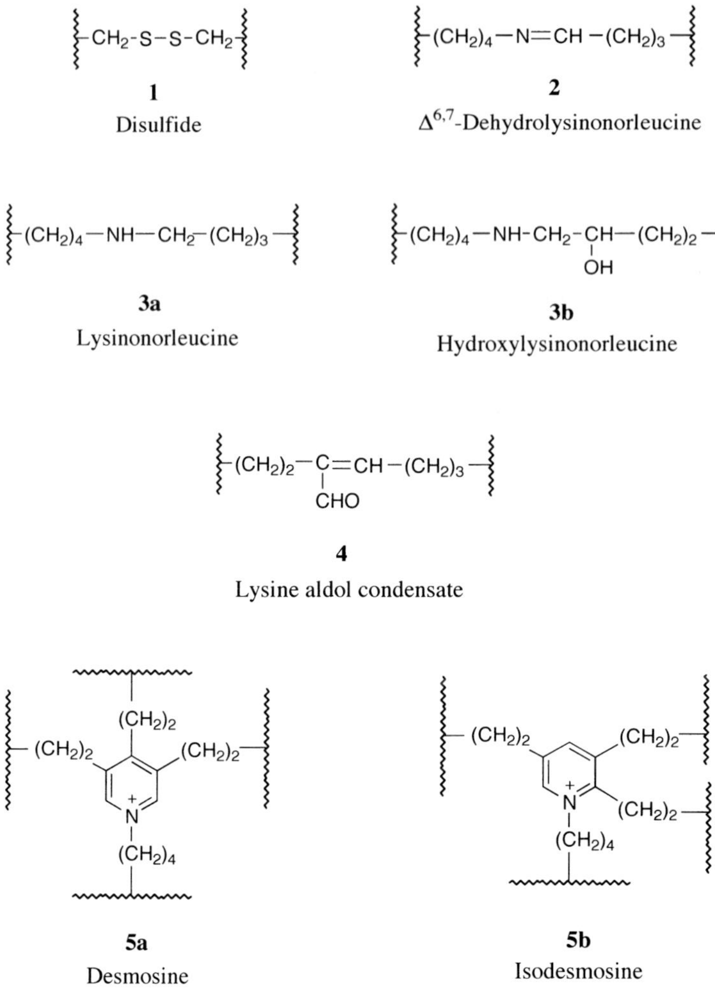

$-CH_2-S-S-CH_2-$

1
Disulfide

$-(CH_2)_4-N{=}CH-(CH_2)_3-$

2
$\Delta^{6,7}$-Dehydrolysinonorleucine

$-(CH_2)_4-NH-CH_2-(CH_2)_3-$

3a
Lysinonorleucine

$-(CH_2)_4-NH-CH_2-CH(OH)-(CH_2)_2-$

3b
Hydroxylysinonorleucine

$-(CH_2)_2-C(CHO){=}CH-(CH_2)_3-$

4
Lysine aldol condensate

5a
Desmosine

5b
Isodesmosine

Scheme 1 Cross-linking structures found in natural proteins.

elastic constants of elastin (elastic fibers) are quite extensible compared to most substances, being more extensible than rubber. Elastin possesses a high extensibility combined with a low modulus, which is very similar to the elastic properties of rubber.

The repetitive triplet amino acid sequence -Pro-Hyp-Gly- and the occurrence of δ-hydroxylysine are inherent characteristics of collagen. Elastin contains enriched Ala and Lys residues in its C-terminal region, which are responsible for intermolecular cross-linking. Although the amino acid composition, chemical reactivity, and physical properties of elastin are quite different from those of collagen, their cross-link chemistry is strikingly similar. All of the cross-links in elastin and collagen are derived from a combination of lysyl residues incorporated into the backbone polypeptide chains. Franzblau (1970) concisely reviewed the cross-linking structures occurring in

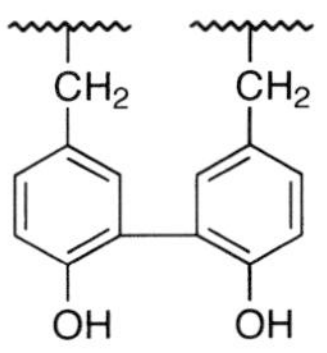

6
Dityrosine

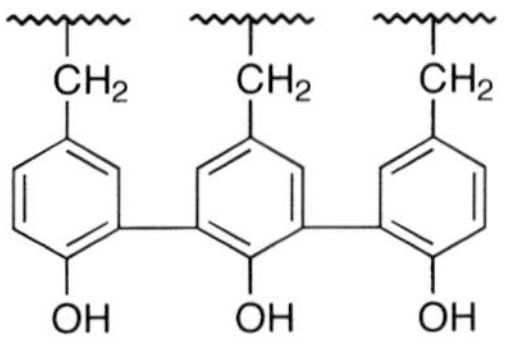

7
Trityrosine

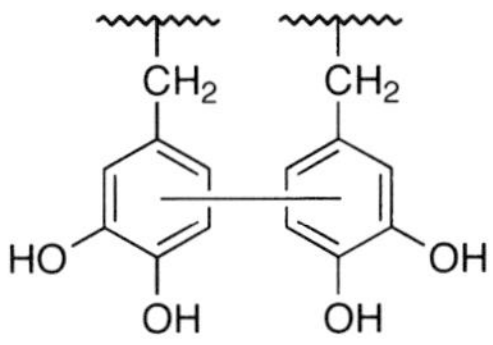

8
Diphenol adduct

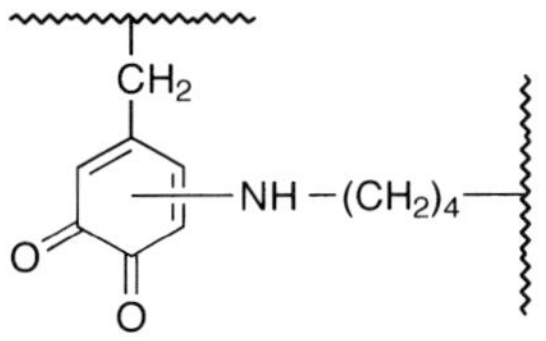

9
Lysyl-quinone adduct

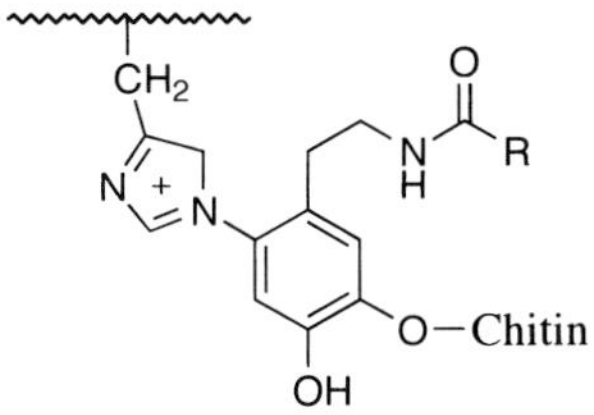

10
Aryl-histidine adduct

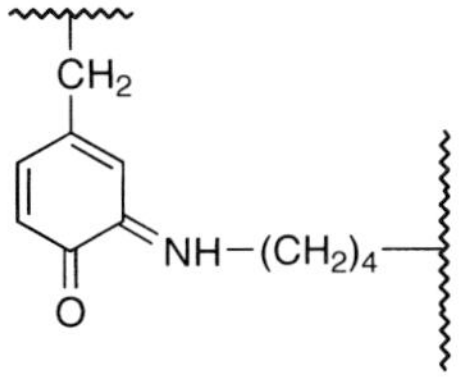

11
Imine-type adduct

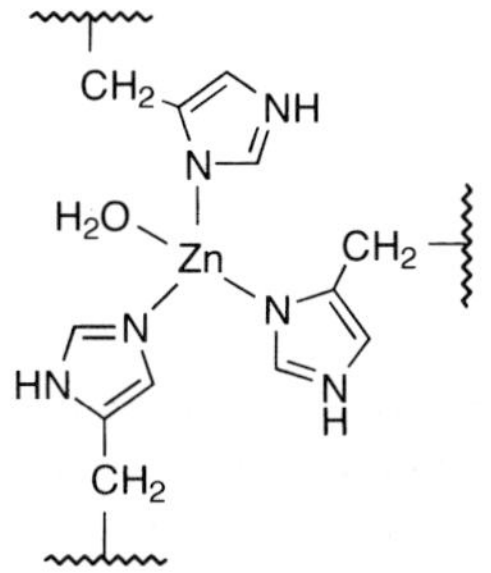

12
Multiple complexation of histidine

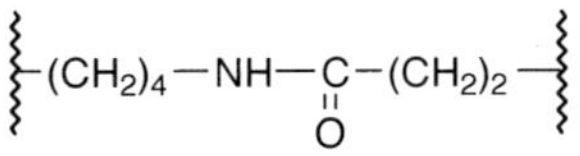

13
ε-(γ-Glutamyl)-lysine

elastin. The common cross-linked structures are presented as $\Delta^{6,7}$-dehydrolysinonorleucine **(2)**, lysinonorleucine **(3a)**, and lysine aldol condensate **(4)** (Franzblau et al., 1965). Bailey and Peach (1968) isolated hydroxylysinonorleucine **(3b)** from rat-tail tendon collagen. Its structure suggests that **3b** is formed via a **2**-like Schiff base between α-aminoadipic acid δ-semialdehyde (see Section 3.1 for details) and δ-hydroxylysine residues. The hydroxylysinonorleucine **(3b)** in collagen differs from the formation of lysinonorleusine **(3a)** in elastin because the latter utilizes a residue of lysine instead of δ-hydroxylysine.

Desmosine **(5a)** and isodesmosine **(5b)** are well-characterized cross-linking structures that were found in elastin by Thomas et al. (1963) and Partridge et al. (1963). They demonstrated that desmosine and isodesmosine are involved in the cross-linking of elastin. Francis et al. (1973) described the mechanism of these two cross-bridge formations that consist of four lysine residues. Structures **(5a)** and **(5b)** potentially join four chains of polypeptides of elastin; thus, two, three, or four chains could be cross-linked by one desmosine or isodesmosine molecule. The elasticity of the elastin is ascribed to these complex cross-linkages. The mechanism of elastin cross-linking remains one of the major subjects in the biochemistry of the structural and matrix proteins in connective tissues.

2.2 Structural Proteins in Invertebrate Bodies

Invertebrates organize both rigid cuticular and flexible tendon tissues for making their bodies. Interestingly, the invertebrate tissues having such a variety of mechanical and physicochemical properties are produced from chemically analogous compounds, i.e., monophenols, *o*-diphenols, and *o*-quinones. The cross-linking reaction, in which a monophenolic side chain of Tyr residue and its oxidation derivatives are involved, is referred to as sclerotization or quinone tanning. These are key terms for understanding living strategy, especially of insects and mollusks.

2.2.1 Insect Cuticles and Eggshells

A review described by Sugumaran (1988) is the most helpful reference to understand systematically the sclerotization in insects from the standpoint of chemistry. Mainly for body protection and to resist microbial infections, several organs of invertebrates have adopted a sclerotization strategy. The biochemical components that participate in the sclerotization process are proteins, chitin, enzymes, and phenols. The cross-linking reaction in the sclerotization process is quite complex and forms a major part of the structural biology of insects and mollusks. The strategy of sclerotization in nature affords significant insight into the material science of manmade polymers.

Anderson (1963) isolated and characterized two unusual amino acids, dityrosine **(6)** and trityrosine **(7)**, from the acid hydrolysates of resilin (Weis–Fogh, 1960), which is a rubber-like protein identified in the elastic tendons of insects. Gross and Sizer (1959) suggested that **(6)** and **(7)** are produced by the peroxidase-catalyzed oxidative coupling of protein-bound tyrosine. Cross-linkages **(6)** and **(7)** provide the rubber-like nature of the cross-linked tendon proteins.

One of the most important discoveries in the chemistry of protein cross-linking is quinone tanning. Quinones are known for their redox reactivities for oxidative coupling. As early as 1940, Pryor (1940) proposed the quinone-tanning hypothesis from an observation that protocatechuate quinone was produced by the interaction of a sclerotizing precursor with peroxidase in cockroach ootheca formation. Quinones are generated by the actions of phenoloxidases on low molecular catechols, such as *N*-acetyl dopamine. Protein-bound quinone is produced by the catalysis of an oxidase, tyrosinase, on tyrosine residue to yield a L-β-3,4-dihydroxy-α-alanine (Dopa) residue, followed by Dopa residue to Dopa quinone residue. The quinones react spontaneously with cuticular proteins, resulting in the formation of cross-linked proteins. The detailed properties of the enzyme involved in the production of quinone are described in Section 3.2. Once quinone is produced, the pathway for the coupling reactions is considerably diverse; for example, the possible coupling reactions are the bimolecular coupling of quinone **(8)** or the addition of amine **(9)** (Lys residue), imidazol (His), guanide (Arg), and sulfur groups (Cys and Met). Hackman and Todd (1953) observed the facile nucleophilic addition of amines to quinones. Schaefer et al. (1987) reported a solid-state nuclear magnetic resonance (NMR) study of ^{15}N- and ^{13}C-labeled *Manduca sexta* cuticle and found the presence of quinoid-histidine adducts **(10)**. The formation of **(10)** was presumed to occur by the reaction of quinone of *N*-β-alanyldopamine with histidine imidazol ring nitrogens. Cys reacts readily with dopaquinone, forming cysteinyl Dopa as well as dicysteinyl Dopas (Ito and Fujita, 1981). Met also reacts with quinones in a similar fashion (Vithayathil and Murthy, 1972). Cys and Met are found only in low levels in cuticular proteins, but from the standpoint of the material science, the reactivity of Cys and Met suggests a novel strategy for cross-linking formations in synthetic poly(amino acids), as described in Section 4.

2.2.2
Byssus of Mussels

Nature's powerful adhesives, secreted by marine mollusks such as mussels, oysters, and barnacles, which must routinely cope with the force of surf and tides, have attracted the researchers in biology, biochemistry, and recently biopolymer chemistry and material science. Among them, byssus of the marine mussel is the most well characterized cross-linked material on a molecular basis. As early as 1970, Pujol (1970) observed that byssus consists of collagen-like protein and phenolic substances. Waite and Tanzer (1980) first identified the phenolic substances as four kinds of proteins, all of which contain Dopa residues. Based on this property, the byssus precursor proteins are designated as polyphenolic proteins. Waite and co-workers have made successive contributions to the biochemistry involved in the adhesion mechanism of mussel species, and detailed information on adhesive proteins of the byssus of mussels is available in another chapter of this book series (Lucas et al., 2003). The protein-bound Dopa is currently considered to play a central role in both surface adhesion (Ooka and Garrell, 2000) and cross-linking reactions of the polyphenolic proteins.

Earlier histochemical analysis of byssus revealed that dityrosine **(6)** (DeVore and Gruebel, 1978) and diphenol **(8)** cross-linking (Ravindranath and Ramalingam, 1972) occurred in the hardening process of byssus. Lindner and Dooley (1973) first proposed the quinone cross-linking mechanism of byssus precursor proteins. They observed that the absorption spectrum of a low-molecular-weight catechol derivative (280 nm) changes during the tyrosinase-catalyzed oxidation to produce quinone (360 nm), followed by an addition reaction of the amino group. Later, Yamamoto (Yamamoto, 1987a; Nagai and Yamamoto, 1989) proved that tyrosinase-catalyzed cross-linking actually caused the insolubilization of the synthetic copolypeptides containing Tyr and Lys.

As an example, Figure 1 shows a spectral change occurring upon the tyrosinase-catalyzed oxidation of a synthetic sequential polydecapeptide (Yamamoto, 1996b), poly(Ala-Lys-Pro-Ser-Tyr-Pro-Pro-Thr-Tyr-Lys),

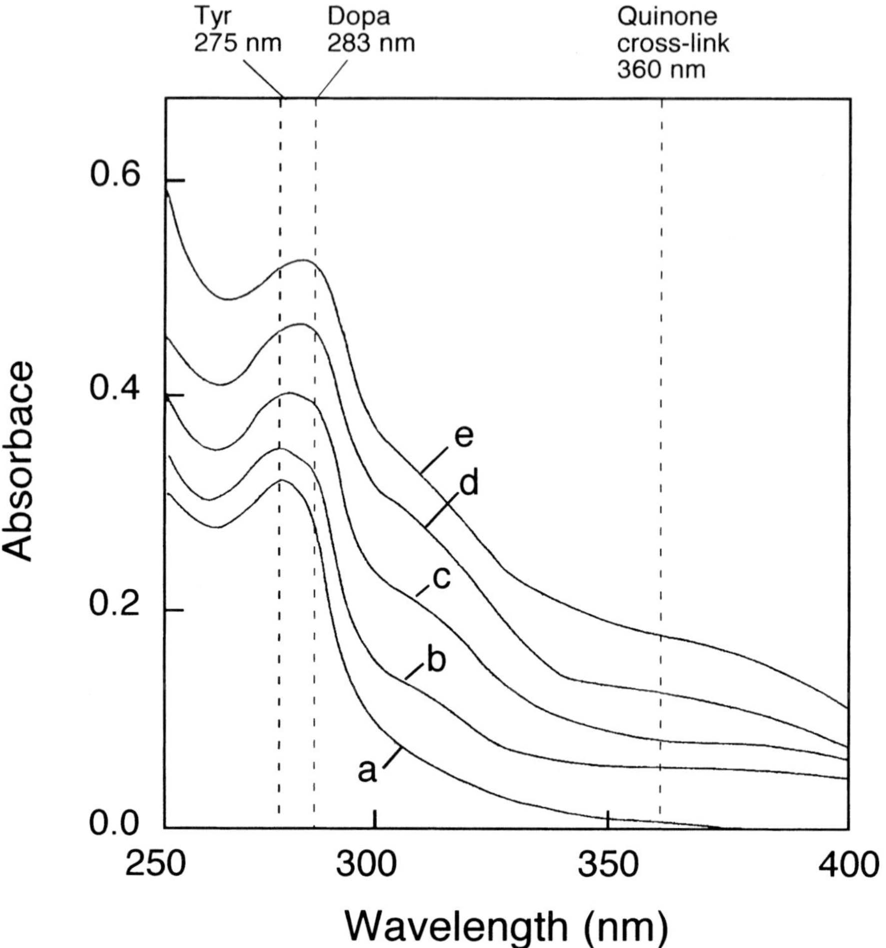

Fig. 1 Time-course of the absorption spectra of the synthetic adhesive protein poly(Ala-Lys-Pro-Ser-Tyr-Pro-Pro-Thr-Tyr-Lys) upon tyrosinase-catalyzed oxidation: (A) 0 min; (B) 40 min; (C) 60 min; (D) 90 min; (E) 180 min.

which is inspired from a byssus precursor protein of the blue mussel, *Mytilus edulis* (Waite and Benedict, 1984). The synthetic polydecapeptide was dissolved in buffer solution and treated with tyrosinase at pH 8.0. The characteristic absorption band of Tyr (278 nm) shifted to 283 nm, to which Dopa has been assigned. After 5 h, the Dopa residues in the polydecapeptide were further oxidized to Dopa quinone, judging from the new broad absorption band at around 350 nm. An insolubilization reaction proceeds, and after 8 h precipitates could be observed by the naked eye. Using a synthetic polynanopeptide (Yamamoto, 1996b), poly(Gly-Gly-Gly-Tyr-Gly-Gly-Tyr-Gly-Lys), which was derived from the viteline protein of a liver fluke, *Fasciola hepatica* (Waite and Benedict, 1984), a similar insolubilization reaction also was observed when the Lys

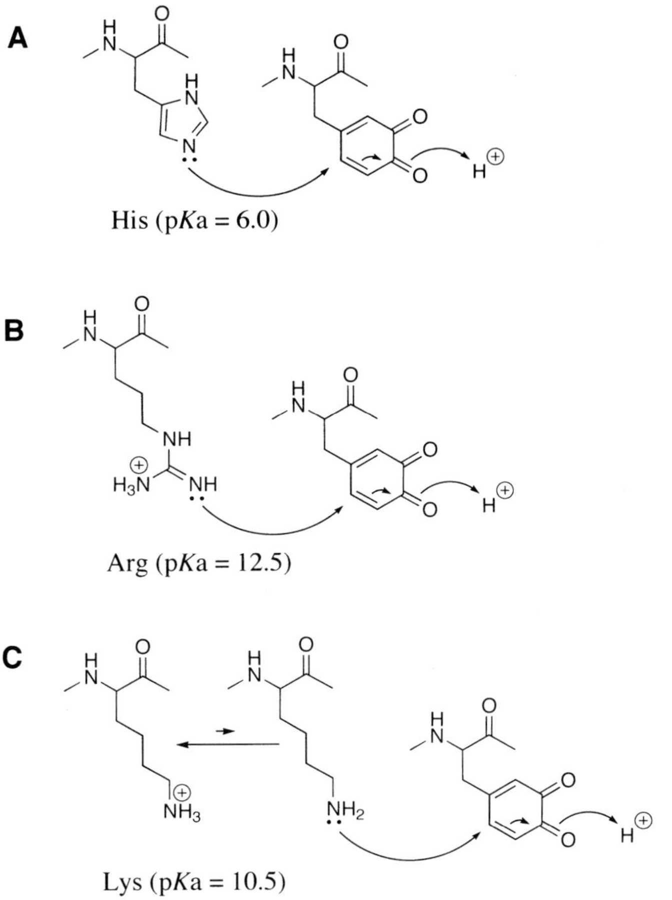

A

His (p*K*a = 6.0)

B

Arg (p*K*a = 12.5)

C

Lys (p*K*a = 10.5)

Fig. 2 Putative mechanisms of the nucleophilic attack of (A) His, (B) Lys , and (C)Arg on Dopa quinone.

residue was replaced by His or Arg, suggesting that the imidazol and guanidino groups are potent nucleophiles that possibly participate in the cross-linking reaction with Dopa quinone (Figure 2).

Recently, more detailed results on the cross-linking reaction of tyrosinase-catalyzed insolubilization of byssal proteins have been reported by Burzio and Waite (2000). They found that the 5,5′-bimolecular adduct of the Dopa residues is the major product isolated from the hydrolysates of cross-linked products of the decapeptide, Ala-Lys-Pro-Ser-Dopa-Hyp-Hyp-Thr-Dopa-Lys. The imine derivative **(11)** is also suggested as a minor product based on the data from mass spectroscopy. Their results first revealed that the complex chemistry of the quinone-mediated coupling reaction of the polyphenolic protein produces unexpectedly simple products. As for the collagen-like protein found in the byssus, multiple complexation of histidines **(12)** is proposed by Waite et al. (1998).

2.3
Isopeptide Cross-linking of Cellular Proteins

Cellular proteins involved in cell response to the foreign signals were functionally organized at the inner and outer surfaces of the plasma membrane. Ligand acceptor, transducer, effector, and amplifier proteins have to be correctly localized in the lipid bilayer of the plasma membrane, where several kinds of cytoskeletal proteins are cooperatively arranged for the action of transducing proteins, resulting in formation of the signal transduction apparatus of the cell. Successful cellular response is achieved not only by systematic localization of these proteins but also by stabilization of the extracellular matrix proteins correctly organized beside the blood circulating system.

Several proteins of the signal transduction apparatus and extracellular matrix are cross-linked by an enzyme, transglutaminase, for the regulation of spatial localization. The cross-bridge structure of transglutaminase-catalyzed cross-linking is commonly γ-glutaminyl-ε-lysyl cross-linkage **(13)**. The proteins cross-linked by the enzyme are the guanosine 5′-triphosphate (GTP)-binding protein, cell-surface-associated fibronectin, collagens type I and V, and Ca^{2+}-binding proteins, such as osteonectin and osteopontin, which play a role in the nucleation and growth of hydroxyapatite crystals for tissue mineralization. In the blood circulation system, clotting of polymerized fibrin is stabilized by formation of the cross-linkage **(13)**, and this reaction is catalyzed by Factor XIII, which has transglutaminase activity.

Isopeptide cross-linkage like **(13)** is also found in the homeostatic regulation of the cellular proteins mediated by the proteasome-ubiquitin system (Ciechanover, 1994). Proteins damaged by foreign stress are hazardous to many cellular functions. The damaged proteins are rapidly cross-linked with a small protein, ubiquitin, through the cross-linkage between the ε-amino group of Lys residue in the damaged protein and the C-terminal carboxyl group of ubiquitin. The repeating addition of ubiquitin forms the protein-grafted polyubiquitin chain, which is the signal for the degradation by proteasome, and the polyubiquitinated protein is finally removed by proteasome-mediated protein degradation. Thus, isopeptide bond **(13)** is utilized not only for stabilization of the cellular structural proteins but also for degradation of the damaged proteins through the cross-linking with ubiquitin.

3
Enzymes Involved in the Protein Cross-links

The enzymatic cross-linking methods of polypeptides have two advantages when compared to the chemical methods described in Section 4. First, the enzymes have high substrate specificities, and this property allows for a much finer regulation when forming the cross-bridge structures in the polymer networks. Second, the enzymatic methods can possibly be applied to the *in vivo* utilization of cross-linked materials, because an enzyme catalyzes the cross-linking formations under the given physiological conditions. Three kinds of enzymes responsible for the protein and/or polysaccharide cross-linking are currently known, i.e., tyrosinase, lysyl oxidase, and transglutaminase.

3.1
Lysyl Oxidase

Since the earlier works on this enzyme by Pinnel and Martin (1968), lysyl oxidase has had about a 30-year history of investigation. Kagan and co-workers have made significant contributions to the biochemistry of lysyl oxidase, especially in the cross-linking mechanisms mediated by this enzyme. The physiological roles of lysyl oxidase, including substrate preferences, oxidation kinetics, and active site structure, have been investigated in detail (Smith-Mungo and Kagan, 1998).

Lysyl oxidase is a copper-containing amine oxidase, which initiates the biosynthesis of lysine-derived cross-links in collagen and elastin by catalyzing the oxidative deamination of lysyl residues to yield the α-amino-adipic-δ-semialdehyde. The reactive aldehyde then can condense to form covalent cross-links such as the aldol condensation product and anhydrolysinonorleucine, the latter occurring as a Schiff base, as described in Section 2.2.1.

The method for purification of lysyl oxidase was described in a review by Kagan and Sullivan (1982). Thoracic aorta of 2- to 6-week calves were an appropriate source of the large-scale preparations. Four kinds of isoforms were found in the preparation from the calf aorta. Molecular weights of the isoforms were estimated to be 32,000 $\pm$ 1000. Two kinds of assay methods for the lysyl oxidase activity were reported. One is an isotopic method using a defined system consisting of highly purified lysyl oxidase and chick aorta labeled by L-[4,5-^{3}H]lysine (Pinnell and Martin, 1968). The resulting tritiated water was quantified using a liquid scintillation counter. Another is originally referred to as the peroxidase-coupled assay (Trackman et al., 1981). The H_2O_2 release was stoichiometrically coupled with lysyl oxidase-catalyzed aldehyde formation; thus the released amount of H_2O_2 is determined by fluorescence measurement using the peroxidase-catalyzed oxidation of homovanilates.

At present, there is no report on the purifications of lysyl oxidase from microorganisms or invertebrate tissues. The lysinonorleucine derivatives, however, have been found in invertebrate tissues (Tanzer, 1976), suggesting the occurrence of lysyl oxidase in invertebrate tissues. Young vertebrate tissues, such as bovine aorta or chick embryo, are the only source of the lysyl oxidase. Unfortunately, lysyl oxidase is not commercially available at the present time; therefore, lysyl oxidase-associated material science, e.g., tissue engineering, still remains to be developed, although lysyl oxidase is one of the most important enzymes for vertebrate tissue development and has an investigation history as long as that of the two kinds of enzymes described below, tyrosinase and transglutaminase.

3.2 Tyrosinase

In 1967, a concise review on the organic chemical and biochemical aspects of the oxidative coupling of phenols was published, and Brown (1967) described enzyme-catalyzed phenol coupling reactions in a chapter on enzymes. The enzymatic oxidation and coupling of phenol are subjects of great importance in biochemistry. Biosynthesis pathways to a wide range of natural products – e.g., tannins, lignins, melanins, pigments, antibiotics, and alkaloids – involve oxidation and coupling of phenols as key actions.

Tyrosinase occurs widely in plants, fungi, bacteria, and animal tissues. A review by Kertesz and Zito (1962) is a helpful reference for the preparation, physicochemical properties, and oxidation kinetics of tyrosinase. Mushroom tyrosinase has a molecular weight of 128,000–133,000 and is commercially available at present. Tyrosinase catalyzes two kinds of oxidation reaction; one is monophenol hydroxylation (Tyr to Dopa) and the other is polyphenol oxidation (Dopa to Dopa quinone).

Monophenol hydroxylase activity of tyrosinase was assayed by the quantification of the polyphenolic substances, which are produced during the tyrosinase-catalyzed oxidation of the monophenolic substances.

Spectrophotometric methods are conventional for many purposes, such as the purification and kinetic studies of tyrosinase. Waite and Benedict (1984) described the optimization of Arnow's (1937) nitration reaction for the detection of Dopa residues in the marine adhesive proteins. Ohkawa et al. (1999) described a fluorometric method, which was originally reported by Yagi and Nagatsu (1960), for assaying the Dopa residues involved in the adhesive protein from the foot of the Asian freshwater mussel. The redox reaction between Dopa and nitroblue tetrazolium was utilized for *in situ* staining of Dopa- and quinone-containing proteins electrophoretically separated on polyacrylamide gel (Paz et al., 1991). Recently, a new method for assaying polyphenol oxidase activity was reported by Winder and Harris (1991). They applied Besthorn's hydrazone (3-methyl-2-benzothiazolinone-hydrazone) for entrapping Dopa-quinone to produce pink pigments, and the resulting pigments were monitored by the absorbance.

In vertebrates, tyrosinase occurs in skin melanoma cells (Pomerantz, 1964) and plays an important role in melanin synthesis. Albinism is a result of defeat of the tyrosinase gene. The biosynthesis of melanin pigment is involved in the protection of cellular DNA from hazardous ultraviolet rays. The utilization of the tyrosinase-mediated hardening reaction of the structural proteins is not found in vertebrates.

3.3 Transglutaminase

Transglutaminase (EC 2.3.2.13) catalyzes the post-translational modification of proteins, i.e., the amine-γ-glutamyl transferase reaction, which forms either an intra- or intermolecular γ-glutamyl-ε-lysine isopeptide bond. Most commonly, the γ-carboxamide group of a peptide-bound glutamine residue and the ε-amino group of Lys are involved in the cross-bridge structure. Such bonds stabilize the structural proteins of many tissues. Since the early work on this enzyme by Folk and Cole (1966), the roles of transglutaminase in tissue homeostasis in many biological systems have been well established (Lorand, 1984). Recently, Aeschlimann and Thomazy (2000) described a review on the cellular roles of the transglutaminases. Their review is also helpful to understand the pathology that is associated with transglutaminase and provides several insights for medical applications of this enzyme as a tissue glue for wound healing and for the formation of bioartificial materials for tissue engineering. Transglutaminase is commercially available, and polymer chemists and material scientists could utilize the enzyme for their ideas for investigations.

There are three types of transglutaminase – I, II, and III. The physiological function of transglutaminase type II (TGc, tissue transglutaminase) is diverse and it is referred to as a multifunctional enzyme. The TGc is thought to participate in the process of stabilization of extracellular matrices in development and wound healing, GTP-binding protein-mediated cell signaling, and programmed cell death. Factor XIII is the last zymogen to become activated in the blood coagulation cascade and is also one member of the TGc gene family. Factor XIII has more restricted functions that are highly specialized for the rapid cross-linking of fibrin γ-chains.

Transglutaminases have been purified as a calcium- and cysteine-dependent enzyme with a molecular weight of 55,000 from bovine snout epidermis. Localization was confirmed by using the fluorescein-conjugated antibody labeling of the enzyme *in situ* or a fluorescent lysine analogue (dansyl cadaverine) and calcium. The fluorescence localization revealed that the epidermal transglutaminase cross-links epidermal proteins during the final stages of keratinization.

4 Chemical Cross-linking

The chemical cross-linking of polymeric materials is effective for mechanical reinforcement, insolubilization, and resistance to chemical degradation. Thus, there are many applications of chemical cross-linking methods for material formulations as fibers, films, gels, microparticles, and adhesives from polymer solutions. Chemical cross-linking methodology is divided into four reaction types, i.e., bifunctional reagents, oxidation, photochemical coupling, and electrostatic force interaction.

4.1 Bifunctional Cross-linking Reagents

Reactive side chains of polymers, especially their nucleophilic character, could be utilized in the cross-linkage formation by using bifunctional cross-linking reagents. As for proteins and polypeptides, the side-chain amino group of Lys and the hydroxyl group of Ser or Thr could be the available nucleophiles to attack the cross-linking reagents.

The most important feature of utilization of the chemical cross-linking reagent is the rapid cross-linking reaction. Bifunctional cross-linking reagents are especially useful for a situation in which rapid phase transition is required, i.e., gelation, fiber spinning, or adhesive hardening. Bifunctional cross-linking reagents commonly have two functional groups at both ends of the molecule. Several practical cross-linking reagents are summarized in Scheme 2. Bifunctional compounds, such as glutaraldehyde (GA) **(14)**, glyoxal **(15)**, phthalaldehydes **(16–18)**, 2,5-hexanedione **(19)**, 2,4-pentanedione **(20)**, ethyleneglycol diglycidyl ether **(21)**, and hexamethylene diisocyanate **(22)**, are important for cross-linking not only of synthetic polymers but also of polypeptides and polysaccharides

In many cases, because protein cross-linking reactions are performed in aqueous solutions, the essentially required properties for the protein cross-linking reagents are solubility and stability in water. Acid halides are the most reactive towards the nucleophiles; however, they are labile to hydrolysis in a watery environment and thus are not appropriate for polypeptide cross-linking reactions in aqueous solutions. Aldehyde is one of the most practical functional groups for cross-linking, and there have been many reports on the use of GA for protein cross-linking since the earlier studies. The resulting cross-bridge structures, however, still are not clearly determined. The nucleophilic attack of the amino group of the Lys residue yields the Schiff base-type condensates, while the equilibrium state between aldol condensation and dissociation of the aldehyde group interferes with determination of the cross-linking distance. Apart from these complexities in the cross-linking reaction, dialdehydes produce the cross-linked materials of poly(amino acids), such as adhesives, hydrogels, and fibers, with promising performance.

Bifunctional cross-linking reagents are used not only for the material formulation of polypeptides but also for the analysis of the interaction of polypeptides in the multisubunit enzymes. Dimethylsuberimidate · 2HCl **(23)** and disuccinimidyl suberate **(24)** also have several applications for fixing proteins on column resins for affinity chromatographies.

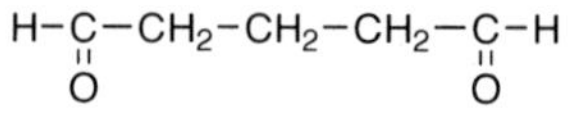

$H-C(=O)-CH_2-CH_2-CH_2-C(=O)-H$

14
Glutaraldehyde

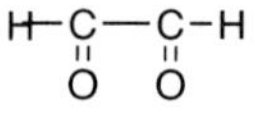

$H-C(=O)-C(=O)-H$

15
Glyoxal

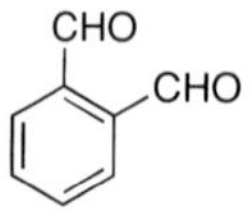

16
o-Phthalaldehyde

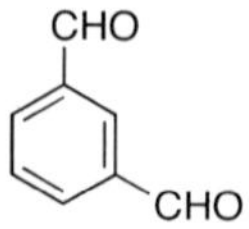

17
m-Phthalaldehyde

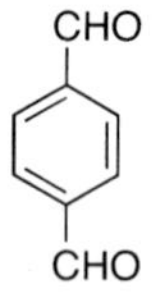

18
p-Phthalaldehyde

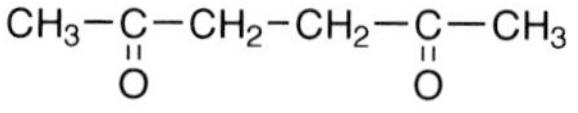

$CH_3-C(=O)-CH_2-CH_2-C(=O)-CH_3$

19
2,5-Hexanedione

$CH_3-C(=O)-CH_2-C(=O)-CH_3$

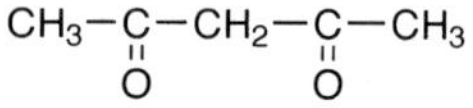

20
2,4-Pentanedione

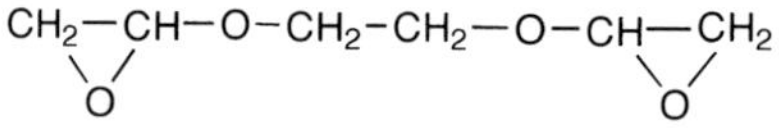

21
Ethyleneglycoldiglycidylether

$O=C=N-(CH_2)_6-N=C=O$

22
Hexamethylenediisocyanate

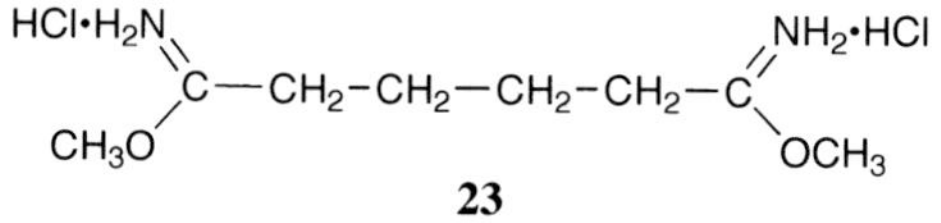

23
Dimethylsuberimidate dihydrochloride

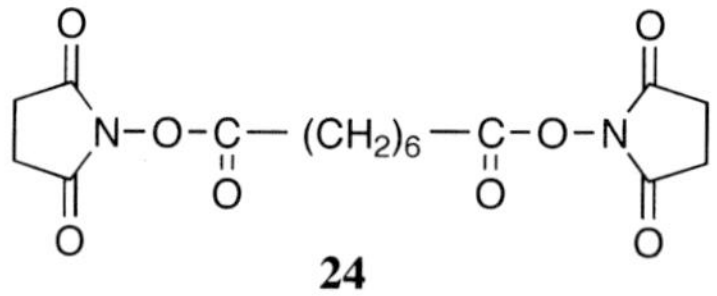

24
Disuccinimidyl suberate

Scheme 2 Bifunctional cross-linking agents.

4.2 Oxidants

Chemical oxidants also could be utilized for protein cross-linking reactions. As an example from our report (Ohkawa, 2001a), *N*-bromo-succinimide (NBS) oxidizes the side-chain amino groups of Lys residues to yield aldehyde via oxidative deamination, and this reaction product, α-aminoadipic-δ-semialdehyde, is same as that in the case of lysyl oxidase-catalyzed oxidation of Lys residues, as mentioned in Section 3.1. The oxidation reaction kinetics of NBS is considerably faster than that of lysyl oxidase.

Figure 3 shows the NBS-mediated insolubilization and oxidation time courses of poly(Lys) and copoly(Lys Ala)s. The insolubilization rate is represented by transmittance (%T) at 650 nm, and the rate of %T decrease was in the following order: copoly(Lys^1 Ala^3) > copoly(Lys^1 Ala^1) > copoly(Lys^3 Ala^1) > poly(Lys). Poly(Lys) exhibited the highest aldehyde concentration peak at 30 S. The peak aldehyde concentration value increased with an increase in the Lys content of the polypeptides. This result indicates that the chemical oxidation rate was governed by the stoichiometry between the amino group and NBS. Thus, the difference between the chemical and enzymatic cross-linking reactions is that the former kinetics depends purely on the stoichiometry between the substrate and the cross-linking reagents, while latter is ruled by the affinity of substrates for the enzyme, namely the Michaelis-Menten kinetics constants, K_m and V_{max}.

Another typical use of oxidants for the cross-linking reaction of polypeptides is found in a report by Yu and Deming (1998). They synthesized copoly(Dopa Lys)

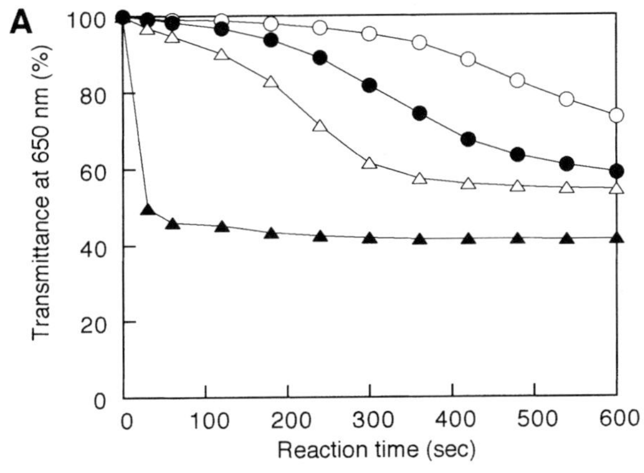

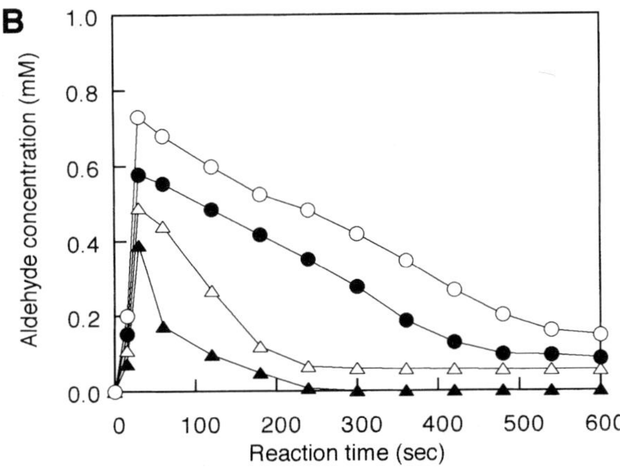

Fig. 3 Time-dependent changes in (A) transmittances and (B) aldehyde concentrations during NBS-induced insolubilization reactions: poly(L-Lys); copoly(Lys^3 Ala^1); copoly(Lys^1 Ala^1); copoly(Lys^1 Ala^3). Reprinted with permission from Ohkawa et al. (2001a). Copyright 2001 American Chemical Society.

and copoly(Dopa Glu) as mimics of the byssus precursor proteins of blue mussel and compared the effects of chemical oxidants, such as hydrogen peroxide and iodine, with tyrosinase on the oxidation of Dopa to Dopa quinone. The chemical oxidation of Dopa to Dopa quinone causes a cross-linking reaction similar to that of tyrosinse, and cross-linking by the chemical oxidants results in a faster and stronger adhesion than by tyrosinase.

4.3 Photo-Induced Cross-linking – Photodimerization of Polypeptide Side Chain

Photosensitive biopolymers are interesting systems because of their relevance to the molecular mechanism of photochemistry in biological materials and processes. They have received increased attention because of their broad applications as new formulation systems. Our group has been studying photoresponsive peptide and polypeptide systems. The major objective in this research field is the design of biopolymer systems reversibly responsible for photo-induced events, including new methods for synthesis of photoresponsive polymers (Imai et al., 2000; Reihmann and Ritter, 2000), the dimerization mechanisms of the side-chain photoresponsive groups (Coqueret, 1999), and photochromic or chiroptical switches (Carlini et al., 2000; Frank et al., 2000).

The formation of dimers and cycloaddition products upon irradiation of compounds containing olefinic constituents is one of the first types of photochemical reactions but is still a useful strategy for the formation of a cross-linking polymer network. Early workers mainly reported on irradiation in sunlight (Mustafa, 1952). At the present, separation and purification of the stereochemical isomers from the photodimerization products are more easily performed by means of high-performance liquid chromatographic procedures. In addition, NMR and the mass-spectrometric techniques can determine the absolute structure and composition of the stereochemistry in the photodimerization. They are also powerful tools for the photodimerization of protein-bound photosensitive groups.

When a photosensitive group is introduced into the polypeptide side chains, the photodimerization upon irradiation causes interchain cross-linking, resulting in the insolubilization or gelation of the polymer. Cinnamoyl, cinnamidyl, thymine, and coumarin have been used as the polypeptide-bound photoresponsive groups capable of inducing photo-cross-linking. The photodimerization of low-molecular-weight coumarin has been rather extensively investigated in both solid and solution states (Anet, 1962). Irradiation of coumarin in solution yields a mixture of four stereoisomeric dimers. Irradiation in polar solvents, such as acetonitrile, dimethylformamide, and methanol, leads to mixtures of syn and *trans* head-to-head, with the former predominating (Krauch et al., 1966; Morrison et al., 1966). Use of nonpolar solvents, e.g., dioxane, benzene, and ethyl acetate, or the presence of benzophenone as a sensitizer results in a predominance of the *trans* (anti) head-to-head isomers (Hammond et al., 1964; Morrison et al., 1966). Yamamoto et al. (1999) described a facile synthetic method to prepare *cis* (syn) head-to-head cyclocoumarin using dichloromethane and ethanol. In addition, according to our experiments on photodimerization of ε-coumaryloxyacetyl-L-lysine [Lys(Cou)] derivatives, the dimerization of monomeric coumarin moieties upon irradiation can be followed by a decrease in the absorption bands at 311 nm and 273 nm.

Dimerization of the coumarine moieties in copolymers of Orn and δ-coumaryloxyacetyl-L-ornthine [Orn(Cou)] was observed upon irradiation at 360 nm in dilute aqueous solution (Ohkawa et al., 2001b). The absorption bands at 311 nm and 273 nm decreased with the irradiation time, exhibiting cross-linking (dimerization) between the coumarin moieties in the side chains (Figure 4). This spectrometric change copoly[Orn Orn(Cou)] is similar to the change in copoly[Lys Lys(Cou)] (also refer the later Section 5).

4.4
Noncovalent Cross-linking – Electrostatic Force Interaction

Polyion complexes (PICs) are formed by the reaction of a polyelectrolyte with an oppositely charged polyelectrolyte in aqueous solution. Considering that almost all biopolymers are polyelectrolytes, the studies on PIC as models of complicated biological systems and as biodegradable material formulations are very important. Polyion complexes have long been investigated from the standpoints of polyacid-polybase interaction, stoichiometry, and self-assembly (Bronich et al., 1998; Mita et al., 1977) and also have numerous applications such as membranes, antistatic coatings, surfactants, and microcapsules.

The most promising system developed is the encapsulation of biological constituents, in particular for the immune and proteolytic protection of enzymes (Margolin et al., 1985) and for biotransport of DNAs (Kabanov and Kabanov, 1995). Polysaccharide polyion complexes also have been widely studied, using chitosan ([1,4]-2-amino-2-deoxy-β-D-glucan) (Muzzarelli, 1977) with amino functional groups and carboxymethyl cellulose. Yamamoto (Yamamoto and Senoo, 2000; Yamamoto et al., 2001) and Ohkawa et al. (2001c) reported that some of the interactions between cationic chitosan and poly(α,L-lysine) (PLL) and anionic gellan ([3]-β-D-glucose-[1,4]-β-D-glucuronic acid-[1,4]-β-D-glucose-[1,4]-α-L-rhamnose-[1]) (Gibson, 1992) and poly(α,L-glutamic acid) (PLG) can produce characteristic structures, including capsule and fiber formation, at the interface in aqueous solutions. PICs are formed by the interaction between polyanionic and polycationic biopolymers. Cross-linking between polyelectrolytes is achieved by charge interaction, and thus interchain non-covalent bonds are formed. The resulting PIC materials, however, are insoluble in a given aqueous environment for retaining association by the charge interaction, e.g., under physiological conditions of salinity and pH.

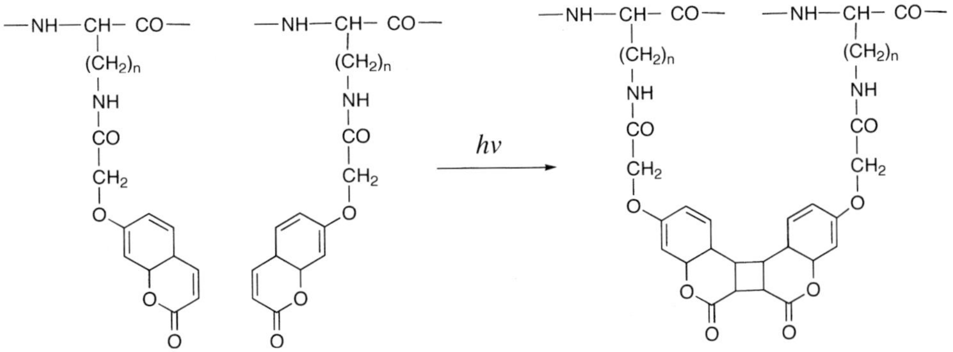

Fig. 4 Cross-linking reaction of polypeptides by photo-induced dimerization of side-chain coumarin.

5 Material Science of Cross-linked Proteins and Poly(amino acids)

Cross-linked biological polymers in watery systems have long been an important class of materials and are used in a diverse assortment of applications (Dikie et al., 1988; Labana and Dickie, 1984). Progress continues in developing approaches to describe the molecular structure of cross-linked polymers. We have investigated four methods for the cross-linking of biological polymers in a watery system: (1) chemical cross-linking using the bifunctional cross-linking reagents; (2) photo-induced cross-linking through the photochemical dimerization of coumarin; (3) enzymatic cross-linking caused by an oxidase, tyrosinase: and (4) electrostatic cross-linking via polyion complex formation between cationic and anionic biological polyelectrolytes. This section mainly describes our investigations on poly(amino acid) and protein formulations of industrial importance, namely, hydrogels, capsules, fibers, and adhesives, by means of the above-mentioned methods for cross-linking (for the authors' patents, see Table 1).

The synthesis of poly(amino acids) has a history of over 60 years, and some monographs are still helpful for considerations on the synthesis and biological properties of poly(amino acids) (Bamford et al., 1956; Stahmann, 1962). Synthetic poly(amino acids) have received a great deal of attention even now, mainly in their capacity as protein models. The earlier concepts for utilization of poly(amino acids) in material science were summarized by Silman and Sela (1967). Polycondensation reactions of *N*-carboxy anhydrides, which are derived from amino acids to be polymerized, remain a practical method for the synthesis of homo- and copoly(amino acids). They have yielded important information on problems of the structure of the polypeptide chain, its electrochemical behavior, and its interaction with living systems (Fasman, 1967).

The polycondensation of oligo-peptide *p*-nitrophenyl active ester was applied for the preparation of sequential polypeptides. Us-

Tab. 1 List of the authors' patents

Publication No.	*Applicants*	*Inventors*	*Title*	*Publication date*
JP04-334396	Hitachi Chem. Co., Ltd.	H. Yamamoto	Insolubilizing protein containing lysine and tyrosine and insolubilized protein	20 November 1992
JP05-017499	Hitachi Chem. Co., Ltd.	H. Yamamoto	New heptapeptide and insolubilization thereof	26 January 1993
JP10-279604	Nippon Soda Co., Ltd.	H. Yamamoto	Spinnable composition comprising chitosan and anionic polysaccharide and fiber produced there from	20 October 1998
JP2000-004837	Chiba Flour Milling Co., Ltd.	H. Yamamoto, T. Yamauchi, M. Sakurai	Bonding agent for raw meat piece and bonding of raw meat piece	11 January 2000
JP2000-281699	Ueda-Seni-Kagaku-Shinkokai, Hiroyuki Yamamoto	H. Yamamoto, K. Ohkawa, A. Nishida	Adhesive protein having good biocompatibility and its amino acid sequence	10 October 2000

ing this method, Yamamoto (1987b) first synthesized a Dopa-containing polydecapeptide, poly(Ala-Lys-Pro-Ser-Tyr-Hyp-Hyp-Thr-Dopa-Lys), the sequence of which was derived from the adhesive protein of the blue mussel *Mytilus edulis*. The details of recent synthetic methodology for sequential marine adhesive polypeptides are reviewed by Yamamoto (1996a,b). The following sections describe the material formulation of the cross-linked random and sequential polypeptides, which were synthesized by the above methods.

5.1 Cross-linked Hydrogels

Cross-linked hydrogel itself is one of most important materials that have numerous applications for medical, food industrial, sanitary, and desert-greening purposes. Chemical, enzymatic, and photo-induced cross-linking methods can be used for the production of protein and poly(amino acid) hydrogels. Organic cross-linking agents, such as dialdehydes and diketones, insolubilized water-soluble cationic Lys- and Orn-containing polypeptides, resulting in rapid precipitation or gel formation (Yamamoto and Tanisho, 1993). The cross-linking reaction using dialdehyde is much faster than that using diketones (Yamamoto and Tanisho, 1993). The PLL and poly(L-ornithine) hydrogels exhibited the selective adsorption of anionic amino acids. Detailed information on the preparation and characteristics of the chemically cross-linked polypeptide hydrogels is available in a review by Yamamoto and Ohkawa (2002).

Photo-induced dimerization of coumarin also induces the gelation of polypeptide solution (Yamamoto et al., 1999). Figure 5 shows the photographs of photo-cross-linked copoly[$Orn^{89}Orn(Cou)^{11}$] gels, which were prepared from the 20% (w/v) solutions upon being irradiated for 24 h in water. The photo-cross-linked copoly[Orn Orn(Cou)] hydrogels expand when they are immersed in water and shrink when they are immersed in ethanol, and the photo-cross-linked hydrogels exhibited ultimate biodegradability (Ohkawa et al., 2001b). Based on these characteristics, the photo-cross-linked hydrogels are expected to be useful as desert-greening materials. Cross-linked poly(γ-glutamic acid), which is a fermentation product of a *Bacillus* species, is also a practical candidate for desert-greening purposes. The poly(γ-glutamic acid) hydrogels are prepared by γ-ray irradiation or chemical cross-linking using diglycidyl ethers.

An enzymatic approach for hydrogel formation was described by Payne and coworkers, using tyrosinase, chitosan, and low-

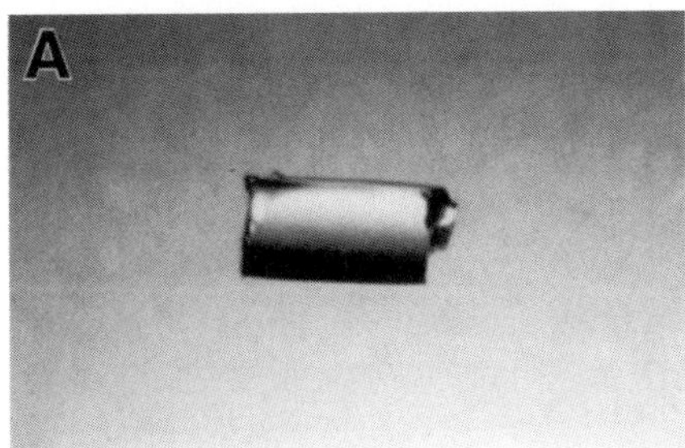

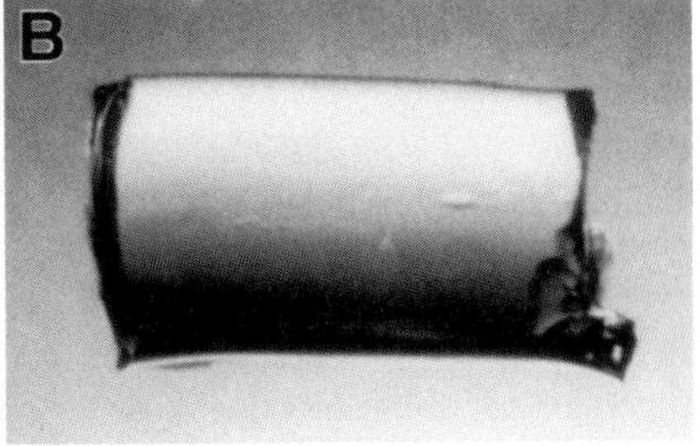

Fig. 5 Photographs of the photo-cross-linked copoly[$Orn^{89}Orn(Cou)^{11}$] cylindrical gel prepared from 20% (w/v) aqueous solution upon irradiation for 24 h: (A) before swelling; (B) swelling in water. Reprinted with permission from Ohkawa et al. (2001b). Copyright 2001 Wiley-VCH.

molecular-weight phenol (Kumer et al., 2000). The resulting chitosan hydrogel is a biomimetic material for byssus adhesion (Yamada et al., 2000) or insect cuticle sclerotization (Kerwin et al., 1999) and is applicable for underwater adhesives, as described in Section 5.3. Sperinde and Griffith (1997) reported the transglutaminase-mediated gelation of a glutaminyl- and lysyl-peptide-bound polyethyleneglycol. More recently, we have reported the lysyl oxidase-catalyzed insolubilization of synthetic polypeptides, such as copoly(Lys Ala) (Ohkawa et al., 2001a). The resulting insolubilized precipitate was a cluster of white turbid particles that were highly hydrated in aqueous solution, suggesting hydrogel formation.

5.2 Polyion Complex Material Formulations

The recently focused applications of PIC, i.e., the non-covalent cross-linked formulations, are for medical and environmental uses. PLG, as a polyanion, is used as a drug carrier system, a microfiltration membrane for heavy metal sorption (Bhattacharyya et al., 1998), an antifreeze agent for foodstuff (Mitsuki et al., 1998), an electrode (Yu and Chen, 1997), and a surgical adhesive (Sekine et al., 2001). PLL, as a polycation, was recently reported for its usefulness as a DNA vector for gene therapies (Ramsay et al., 2000; Zhang et al., 2001), a vehicle for local drug delivery (Sakharov et al., 2001), a medical adhesive for tissue repair (Hwang and Stupp, 2000), a carrier material that binds biologically active substances (Arnold et al., 1979; Shen et al., 1985), and a stabilizer of lipid bilayers (Krylov et al., 2001). The major advantage of PIC material formulation is its preparation method, i.e., simply mixing a polyelectrolyte aqueous solution with a countercharged polyelectrolyte solution. This allows a broad spectrum of applications for PIC materials, such as films, capsules, and fibers.

5.2.1 Capsules

The benefits of encapsulating and releasing a therapeutic agent from a polymer matrix are drug protection and sustained release. In particular, microspheres and microcapsules are very promising, designed small-sized particles that enable repetitive administration as the therapeutic bolus via either injection or oral route. Polymeric delivery systems using microcapsules and microspheres have been a focused subject in pharmaceutical science. Fabrication of the capsules and spheres from a biodegradable polymer eliminates the need for surgical removal of the carrier materials. Polysaccharides such as chitosan, alginate, and xanthane and synthetic biodegradable polymers such as poly(lactic-*co*-glycolic acid) and poly(lactic-*co*-amide) have been thoroughly investigated for this purpose.

Water-soluble poly(amino acids) are also very suitable for this purpose, because of their adaptability to the mild and fully watery process of drug encapsulation and because of their high biodegradability. PLG and PLL are well suited for material formulations via the PIC in aqueous environments. Two kinds of poly(amino acid)–polysaccharide hybrid capsules, PLL-gellan (Yamamoto et al., 2001) and PLG-chitosan (Ohkawa et al., 2001c) capsules, have been reported from our laboratory. As an example, PLG-chitosan hybrid capsules were prepared by the dropwise addition of chitosan solution into the PLG solution (Figure 6). This procedure gives true spherical droplet structures with various diameters, and the droplets thus formed are stable when washed with water, finger pinching, or

Fig. 6 PLG-chitosan PIC capsule with sphere shapes.

magnetic stirring in distilled water, acid, alkaline, or boiling water.

A major subject in the regulation of drug release is how to control the permeability of the capsule membrane. Charge density, molecular weight, and stoichiometry of the polyelectrolytes affect the pore size of the capsule membrane and the chemical affinity of encapsulants and thus could be the factors for regulation of the sustained release or adsorption of the encapsulants. The external pH and temperature and the mechanical strength of the capsule are also factors in designing the releasing system.

5.2.2
Fibers

Regenerated protein fibers were among the earliest efforts for the use of protein cross-linking reactions in the material science of fibers (Wormell, 1954). Milk casein, groundnut proteins, and alcohol-soluble corn proteins have been spun by extruding their alkaline solutions into an acidic coagulation bath, followed by a hardening process using formaldehyde or cyanate, which renders the fibers insoluble in water (Carroll-Porczynski, 1961). In the 1920s to 1940s, commercial development of the regenerated protein fibers succeeded and came onto the market as carpet and felt. Parallel with the development of synthetic fiber technology such as nylon and rayon in 1930s, regenerated protein fibers have failed to achieve commercial success. The production of the regenerated protein fibers was discontinued in the1950s to 1960s.

Silk, one of the naturally occurring polymers of α-amino acids, has attracted the attention of organic chemists for many years, and the synthetic fiber industry, such as that for rayon and nylon, grew out of this interest (Ballard, 1968). Synthetic fibers were considered to be substitutes for natural wool, cotton, and silk fibers. Poly(α-amino acid) fibers were expected to become a new class of animal-like fibers, having properties similar to those of silk, wool, and the intermediates (Noguchi et al., 1972). Many kinds of organic solvent-soluble polypeptides were spun as fibers from the polypeptide solutions in dichloroacetic acid, benzene, and methylene dichloride from 1945 to the 1970s. In particular, poly(γ-methyl-α,L-glutamate) (PMLG) fiber had been considered to be the most promising for industrial production. In this connection, the PLG fiber was prepared only by the saponification of PMLG, but the resulting PLG fiber prepared is water-soluble. The expensive costs and strong acidity of the halogen-containing solvents for spinning impaired the commercial success of poly(amino acid) fibers. In principle, water-soluble polypeptides in aqueous solutions cannot be spun as water-insoluble fibers.

A fabric method via PIC formation is a unique strategy that realizes the formulation of poly(amino acid) fibers from their aqueous solutions. Our attempt at the development of new-type poly(amino acid) PIC fibers is a kind of reactive spinning (Yamamoto and Senoo, 2000). PLG-chitosan (Ohkawa et al., 2001c) and PLL-gellan (Yamamoto et al., 2001) fibers can be made by self-assemblies to form PICs at the aqueous solution interface (Figure 7). The resulting PIC films were drawn out from the interface to spin the PIC, as in the case of interfacial polycondensation. The PIC fibers can be spun continuously by using a simple wet spinning apparatus accompanied by a roller, and the produced fibers have a silk-like luster (Figure 8). The characteristic features of the present PLG-chitosan, PLL-gellan, and PLG-PLL PIC fibers are (1) a simple operation for spinning, (2) low-cost media (water, acetic acid, and alcohols), (3) enhanced tensile and knot strengths that are due to cross-linking agents, (4) dyeing with many colors, and (5) environmental biodegradability (Ohkawa et al., 2000). As for applications, the adsorption of dyes, anionic medicines, and endocrine disruptors into the capsule and onto the PLG-chitosan fiber have been described (Takahashi et al., 2001). Although the details of the adsorption mechanism still remain to be resolved, it will be possible to design combinations of bio-related polyelectrolytes that are more suitable for environmental or medical uses.

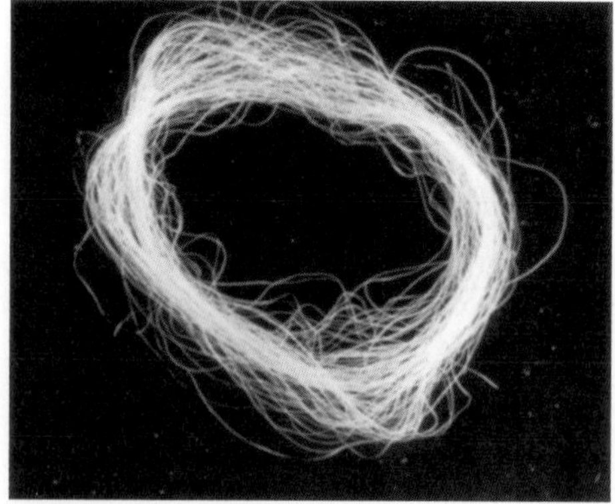

Fig. 8 PIC fabric gives a silk-like fiber from PLG and chitosan.

Fig. 7 Self-assembling PLG-chiosan PIC fiber formed at an aqueous solution interface.

5.3
Medical Adhesives

Medical adhesives have been used in surgery as hemostatic and wound-healing agents. Progress has continued in developing strategies to solve the most difficult issue from the viewpoint of adhesion science, i.e., moisture-resistant and biocompatible adhesion. The classical use of artificial polymers, i.e., the derivatives of poly(cyanoacrylate) (Musierowicz, 1976) and the gelatin-resorcinol-formaldehyde glue (Koehnlein and Lemperle, 1969), has been limited because of the release of cytotoxic chemicals during their biodegradation processes. Improvements in these two glues are still continuing (Sung et al., 1999).

More recently, several composite adhesives of water-soluble natural and synthetic polymeric materials, such as the gelatin-*N*-

hydroxysuccinimide-activated PLG (Iwata et al., 1998), the azide-lactose-modified chitosan (Ono et al., 2000), and the ethylenediamine-modified gelatin-starch systems, have been developed (Mo et al., 2000). The hardening reactions of these adhesives are based on the chemical and photochemical cross-linking of the polymers in aqueous environments. Fibrin glues, such as Tisseel®, also have been used to good effect in a wide variety of surgical and endoscopic procedures. However, the components of fibrin glue, fibrinogen, and thrombin obtained from human blood are not completely free from the risk of virual infection.

Protein-based medical adhesives from the blue mussel have been described (Burzio et al., 1990), suggesting applications such as biocompatibility and adherence and curing in moist environments. The cross-reaction of the adhesive proteins purified from three Chilean mussels with an antiserum produced against the protein of the species was studied. From an immunological examination, the adhesive proteins were poor antigens and exhibited an advantageous property for the potential use of these adhesive proteins in medicine.

As a polypeptide bioadhesive, the simple sequential polypeptide poly(X-Tyr-Lys) (X= Gly, Ala, Pro, Ser, Leu, Ile, Phe), which contains the essential amino acids Tyr and Lys for a quinone cross-link, was inspired by marine mussels (Tatehata et al., 2000). Tissue adhesion properties of poly(X-Tyr-Lys) were compared to those of a commercial fibrin glue, Tisseel® (Tatehata et al., 2001). Buffer solutions of the poly(X-Tyr-Lys)-tyrosinase formulation and the Tisseel® solution were poured into a pigskin incision site. As an example, representative photographs of the subcutaneous tissues of the incision site are shown in Figure 9. When the macrophage development was quantitatively measured using image analysis, the degree of macrophage development in the poly(Gly-Tyr-Lys)-tyrosinase system was 58% ±9% of the Tisseel® system, suggesting that the poly(Gly-Tyr-Lys)-tyrosinase had less immune reactivity than did Tisseel®. Because the totally synthetic adhesive proteins are essentially free from potential virus hazards, they will be an important class of medical glues.

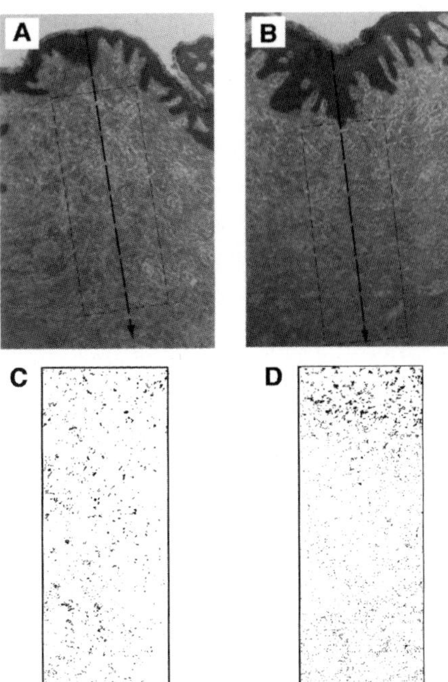

Fig. 9 Optical photographs of histological sections of the subcutaneous tissues of incised pigskin after 1 week (magnification, ×20). The tissue sections were stained with hematoxyline and eosin: (A) poly(Gly-Tyr-Lys) with tyrosinase; (B) Tisseel®. The broken lines with arrowheads indicate the directions and depths of incisions. The boxes rounded with broken lines indicate the microscopic fields that were subjected to the image analysis. The analyzed images of macrophage development: (C) poly(Gly-Tyr-Lys) with tyrosinase; (D) Tisseel®. Reprinted with permission from Tatehata et al. (2001). Copyright 2001 VSP V. B. International Publishers.

6
Outlook and Perspectives

The side-chain amino group of the Lys residue is one of the most important nucleophiles involved in the cross-linking reaction in proteins. In all of the cross-linking reactions catalyzed by lysyl oxidase, tyrosinase, and transglutaminase, the ε-amino group of Lys is an essential constituent. Low-molecular-weight phenolic substances and protein-bound Dopa residues are universal chemicals in invertebrate protein cross-linking. The pathways of catechol-mediated cross-linking (quinone tanning) are complicated because of the numerous possible coupling reactions of the catechol and because of the heterogeneity in the production of adducts., i.e., cross-linking structures. However, complementarily, the resulting cross-linked materials have high mechanical strength and anti-infection, anti-corrosion and anti-erosion characteristics, attracting researchers in interdisciplinary fields, i.e., biology, polymer chemistry, and material science.

The phenomenon of protein modification and cross-linking is full of interesting suggestions for designs of new artificial biological materials, which involve the formation of composite interfaces between different macromolecules in the presence of water. Considerable efforts for discovering, revealing, and engineering the protein modification are still being carried out, and contributions from different disciplines will unify the mutual resources, techniques, and knowledge for exploring a new realm of science.

Acknowledgments
This work was partly supported by Grants-in-Aid for COE Research (10CE2003) and for Scientific Research (No 12450330, No 13555178, No 13556057) by the Ministry of Education, Culture, Sports, Science, and Technology of Japan. The authors are grateful to the American Chemical Society (Figure 3), Wiley-VCH (Figure 5), and VSP V. B. International Publishers (Figure 9).

7 References

Aeschlimann, D., Thomazy, V. (2000) Protein cross-linking in assembly and remodelling of extra-cellular matrices: The role of transglutaminases, *Connect. Tissue Res.* **41**, 1–27.

Anderson, S. O. (1963) Characterization of a new type of cross-linkage in resilin, a rubber like protein, *Biochim. Biophys. Acta* **69**, 249–262.

Anet, R. (1962) The photodimers of coumarin and related compounds, *Can. J. Chem.* **40**, 1249–1257.

Arnold, L. J., Dagan, A., Gutheil, J., Kaplan, N. O. (1979) Antineoplastic activity of poly(L-lysine) with some ascites tumor cells, *Proc. Natl. Acad. Sci. USA.* **76**, 3246–3250.

Arnow, L. E. (1937) Colorimetric determination of the components of 3,4-dihydroxyphenylalanine-tyrosine mixtures, *J. Biol. Chem.* **118**, 351–357.

Bailey, A. J., Peach, C. M. (1968) Isolation and structural identification of a labile intermolecular cross-link in collagen, *Biochem. Biophys. Res. Commun.* **33**, 812–819.

Ballard, D. G. H. (1968) Synthetic polypeptides, in: *Man-Made Fibers: Science and Technology*, (Mark, H. F., Atlas, S. M., Cernia, E., Eds.), New York: John Wiley & Sons, 401–433, Vol. 2.

Bamford, C. H., Elliott, A., Hanby, W. E. (1956) *Synthetic Polypeptides.* New York: Academic Press.

Bhattacharyya, D., Hestekin, J. A., Brushaber, P., Cullen, L., Bachas, L. G., Skidar, S. K. (1998) Novel poly-glutamic acid functionalized micro-filtration membranes for sorption of heavy metals at high capacity, *J. Membrane Sci.* **141**, 121–135.

Borysko, E. (1963) Collagen, in: *Ultrastructure of Protein Fibers* (Borasky, R., Ed.), New York: Academic Press, 19–37.

Bronich, T. K., Cherry, T., Vinogradov, S. V., Eisenberg, A., Kabanov, V. A., Kabanov, A. V. (1998) Self-assembly in mixtures of poly(ethyleneoxide)-*graft*-poly(ethyleneimine) and alkyl sulfates, *Langmuir* **14**, 6101–6106.

Brown, B. R. (1967) Biochemical aspects of oxidative coupling of phenols, in: *Oxidative Coupling of Phenols* (Taylor, W. I., Battersby, A. R., Eds.), New York: Marcel Dekker, 167–201, Vol. 1.

Brown, C. H. (1950) A review of the methods available for the determination of the types of forces stabilizing structural proteins in animals, *Q. J. Microsc. Sci.* **91**, 331–339.

Burton, A. C. (1954) Relation of structure to function of the tissues of the wall of blood vessels, *Physiol. Rev.* **34**, 619–642.

Burzio, L. A., Waite, J. H. (2000) Cross-linking in adhesive quinoproteins: Studies with model decapeptides, *Biochemistry* **39**, 11147–11153.

Burzio, L. O., Gutierrez, E., Pardo, J., de la Fuente, E., Brito, M., Saez, C. (1990) Bioadhesives: A biotechnological opportunity, *Arch. Biol. Med. Exp.* **23**, 173–178.

Carlini, C., Fissi, A., Galletti, A. M. R., Sbrana, G. (2000) Optically active polymers bearing side-chain photochromic moieties: Synthesis and chiroptical properties of methacrylic homopoly-mers with pendant *trans*-azobenzene chromo-phores bound through L-leucine, L-valine and L-proline amino acid spacers, *Macromol. Chem. Phys.* **201**, 1540–1551.

Carroll-Porczynski, C. Z. (1961) *Natural Polymer Man-Made Fibres.* London: National Trade Press.

Ciechanover, A. (1994) The ubiquitin-proteasome proteolytic pathway, *Cell* **79**, 13–21.

Coqueret, X. (1999) Photoreactivity of polymers with dimerizable side-groups: Kinetic analysis for probing morphology and molecular organ-ization, *Macromol. Chem. Phys.* **200**, 1567–1579.

DeVore, D. P., Gruebel, R. J. (1978) Dityrosine in adhesive formed by the sea mussel, *Mytilus edulis*, *Biochem. Biophys. Res. Commun.* **80**, 993–999.

Dikie, R., Labana, S., Bauer, R. (Eds.) (1988) *Cross-linked Polymers.* Washington, DC: American Chemical Society.

Fasman, G. D. (Ed.) (1967) *Poly-α-Amino Acids: Protein Models for Conformational Studies.* New York: Marcel Dekker.

Folk, J. E., Cole, P. W. (1966) Mechanism of action of guinea pig liver transglutaminase. I. Purification and properties of the enzyme: Identification of a functional cysteine essential for activity, *J. Biol. Chem.* **241**, 5518–5525.

Francis, G., John, R., Thomas, J. (1973) Biosynthesis pathway of desmosines in elastin, *Biochem. J.* **136**, 45–55.

Frank, R., Jakob, M., Thunecke, F., Fischer, G., Schutowski, M. (2000) Thioxylation as one-atom-substitution generates a photoswitchable element within the peptide backbone, *Angew. Chem. Int. Edn.* **39**, 1120–1122.

Franzblau, C., Sinex, F. M., Faris, B., Lampidis, R. (1965) Identification of a new cross-linking amino acid in elastin, *Biochem. Biophys. Res. Commun.* **21**, 575–581.

Franzblau, C. (1970) Elastin, in *Comprehensive Biochemistry* (Florkin, M., Stotz, E. H., Eds.), Amsterdam: Elsevier, 659–712, Vol. 26, Part C.

Gibson, W. (1992) Gellan gum, in: *Thickening and Gelling Agents for Food* (Imeson, A., Ed.), London: Chapman & Hall, 227–249.

Gross, A. J., Sizer, I. W. (1959) The oxidation of tyramine, tyrosine, and related compounds by peroxidase, *J. Biol. Chem.* **234**, 1611–1614.

Hackman, R. H., Todd, A. R. (1953) Some observations on the reaction of catechol derivatives with amines and amino acids in presence of oxidizing agents, *Biochem. J.* **55**, 631–637.

Hammond, G. S., Stout, C. A., Lamola, A. A. (1964) Mechanisms of photochemical reactions in solution. XXV. The photodimerization of coumarin, *J. Am. Chem. Soc.* **86**, 3103–3106.

Hoeve, C. A. J., Flory, P. J. (1958) The elastic properties of elastin, *J. Am. Chem. Soc.* **80**, 6523–6526.

Housley, T., Tanzer, M. L., Henson, E., Gallop, P. M. (1975) Collagen cross-linking: Isolations of hydroxyaldol-histidine, a naturally-occurring cross-link, *Biochem. Biophys. Res. Commun.* **67**, 824–830.

Hwang, J. J., Stupp, S. I. (2000) Poly(amino acid) bioadhesives for tissue repair, *J. Biomater. Sci. Polym. Ed.* **11**, 1023–1038.

Imai, Y., Ogoshi, T., Naka, K., Chujo, Y. (2000) Formation of IPN organic-inorganic polymer hybrids utilizing the photodimerization of thymine, *Polym. Bull.* **45**, 9–16.

Ito, S., Fujita, K. (1981) Formation of cysteine conjugates from dihydroxyphenylalanine and its *S*-cysteinyl derivatives by peroxidase-catalyzed oxidation, *Biochim. Biophys. Acta* **672**, 151–157.

Iwata, H., Matsuda, S., Mitsuhashi, K., Itoh, E., Ikada, Y. (1998) A novel surgical glue composed of gelatin and *N*-hydroxysuccinimide activated poly(L-glutamic acid): Part 1. Synthesis of activated poly(L-glutamic acid) and its gelation with gelatin, *Biomaterials* **19**, 1869–1876.

Kabanov, A. V., Kabanov, V. A. (1995) DNA complexes with polycations for the delivery of genetic material into cells, *Bioconjug. Chem.* **6**, 7–20.

Kagan, H. M., Sullivan, K. A. (1982) Lysyl oxidase: Preparation and role in elastin biosynthesis, *Methods. Enzymol.* **82**, 637–650.

Kertesz, D., Zito, R. (1962) Phenolase, in: *Oxygenases* (Hayashi, O., Ed.), New York: Academic Press, 307–354.

Kerwin, J. L., Whitney, D. L., Sheikh, A. (1999) Mass spectrometric profiling of glucosamine, glucosamine polymers and their catecholamine adducts. Model reactions and cuticular hydrolysates of *Toxorhynchites amboinensis* (Culicidae) pupae, *Insect. Biochem. Mol. Biol.* **29**, 599–607.

Koehnlein, H. E., Lemperle, G. (1969) Experimental studies with a new gelatin-resorcin-formaldehyde glue, *Surgery* **66**, 377–382.

Krauch, C. H., Farid, S., Schenck, G. O. (1966) Photo-C_4-cyclodimersation von Cumarin, *Chem. Ber.* **99**, 625–633.

Krylov, A. V., Kotova, E. A., Yaroslavov, A. A., Antonenko, Y. N. (2001) Stabilization of *O*-pyromellitylgramicidin channels in bilayer lipid membranes through electrostatic interaction with polylysines of different chain lengths, *Biochim. Biophys. Acta* **1509**, 373–385.

Kumer, G., Bristow, J. F., Smith, P. J., Payne, G. F. (2000) Enzymatic gelation of the natural polymer chitosan, *Polymer* **41**, 2157–2168.

Labana, S., Dickie, R. (Eds.) (1984) *Characterization of Highly Cross-linked Polymers.* Washington, DC: American Chemical Society.

Lindner, E., Dooley, C. (1973) Chemical bonding in cirriped adhesive, in: *Proceedings 3rd International Congress on Marine Corrosion and Fouling,* Evanston, IL: Northwestern University Press, 653–673.

Lorand, L., Conrad, S. M. (1984) Transglutaminases, *Mol. Cell. Biochem.* **58**, 9–35.

Lucas, J. M., Vaccaro, E., Waite J. H. (2003) Extra-organismic adhesive proteins, in: *Biopolymers* (Fahnestock, S., Steinbüchel, A., Eds.), Weinheim: Wiley-VCH, Vol. 7/8.

Lundgren, H. P., Ward, W. H. (1963) The keratins, in: *Ultrastructure of Protein Fibers* (Borasky, R., Ed.), New York: Academic Press, 39–122.

Margolin, A. L., Sherstyuk, S. F., Izumrudov, V. A., Zezin, A. B., Kabanov, V. A. (1985) Enzymes in polyelectrolyte complexes. The effect of phase transition on thermal stability, *Eur. J. Biochem.* **146,** 625–632.

Mita, K., Zama, M., Ichimura, S. (1977) Effect of charge density of cationic polyelectrolytes on complex formation with DNA, *Biopolymers* **16,** 1993–2004.

Mitsuki, M., Mizuno, A., Tanimoto, H., Motoki, M. (1998) Relationships between the antifreeze activities and the chemical structures of oligo- and poly(glutamic acid)s, *J. Agricul. Food Chem.* **46,** 891–895.

Mo, X., Iwata, H., Matsuda, S., Ikada, Y. (2000) Soft tissue adhesive composed of modified gelatin and polysaccharides, *J. Biomater. Sci. Polym. Ed.* **11,** 341–351.

Morrison, H., Curtis, H., McDowell, T. (1966) Solvent effects on the photodimerization of coumarin, *J. Am. Chem. Soc.* **88,** 5415–5419.

Muzzarelli, R. A. A. (1977) *Chitin.* Oxford: Pergamon.

Musierowicz, A. (1976) Use of tissue glue (butyl-2-cyanacrylate) for creation of loop ureterostomy, *Pol. Przegl. Chir.* **48,** 203–204.

Mustafa, A. (1952) Dimerization reactions in sunlight, *Chem. Rev.* **51,** 1–23.

Nagai, A., Yamamoto, H. (1989) Insolubilizing studies of water-soluble poly(Lys Tyr) by tyrosinase, *Bull. Chem. Soc. Jpn.* **62,** 2410–2412.

Noguchi, J., Tokura, S., Nishi, N. (1972) Poly-α-amino acid fibers, *Angew. Makromol. Chem.* **22,** 107–131.

Ohkawa, K., Nishida, A., Ichimiya, K., Matsui, Y., Nagaya, K., Yuasa, A., Yamamoto, H. (1999) Purification and characterization of a dopa-containing protein from foot of the Asian freshwater mussel, *Limnoperna fortunei, Biofouling* **14,** 181–188.

Ohkawa, K., Yamada, M., Nishida, A., Nishi, N., Yamamoto, H. (2000) Biodegradation of chitosan-gellan and poly(L-lysine)-gellan polyion complex fibers by pure-culture of soil filamentous fungi, *J. Polym. Environ.* **8,** 59–66.

Ohkawa, K., Fujii, K., Nishida, A., Yamauchi, T., Ishibashi, H., Yamamoto, H. (2001a) Lysyl oxidase-catalyzed cross-linking and insolubilization reactions of Lys-containing polypeptides and synthetic adhesive proteins, *Biomacromolecules* **2,** 773–779.

Ohkawa, K., Shoumura, K., Yamada, M., Nishida, A., Shirai, H., Yamamoto, H. (2001b) Photoresponsive peptide and polypeptide systems. 14: Biodegradation of photocross-linkable copolypeptide hydrogels containing L-ornithine and δ-7-coumaryloxuacetyl-L-ornithine residues, *Macromol. Biosci.* **1,** 149–156.

Ohkawa, K., Takahashi, Y., Yamada, M., Yamamoto, H. (2001c) Polyion complex fiber and capsule formed by self-assembly of chitosan and poly(α,L-glutamic acid) at solution interface, *Macromol. Mater. Eng.* **286,** 168–175.

Ono, K., Saito, Y., Yura, H., Ishikawa, K., Kurita, A., Akaike, T., Ishihara, M. (2000) Photocross-linkable chitosan as a biological adhesive, *J. Biomed. Mater. Res.* **49,** 289–295.

Ooka, A. A., Garrell, R. L. (2000) Surface-enhanced Raman spectroscopy of DOPA-containing peptides related to adhesive protein of marine mussel, *Mytilus edulis, Biopolymers* **57,** 92–102.

Partridge, S. M., Elsden, D. F., Thomas, J. (1963) Constitution of the cross-linkages in elastin, *Nature* **197,** 1297–1298.

Paz, M. A., Flückiger, R., Boak, A., Kagan, H. M., Gallop, P. M. (1991) Specific detection of quinoproteins by redox-cycling staining, *J. Biol. Chem.* **266,** 689–692.

Pinnell, S. R., Martin, G. R. (1968) The cross-linking of collagen and elastin: Enzymatic conversion of lysine in peptide linkage to α-aminoadipic-δ-semialdehyde (allysine) by an extract from bone, *Proc. Natl. Acad. Sci. USA.* **61,** 708–716.

Pomerantz, S. H. (1964) Tyrosine hydroxylation catalyzed by mammalian tyrosinase: An improved method of assay, *Biochem. Biophys. Res. Commun.* **16,** 188–194.

Pryor, M. G. M. (1940) On the handling of the ootheca of *Blatta orientalis, Proc. Royal Soc. London B.* **128,** 378–393.

Pujol, J. P. (1970) The collagen of the byssus in *Mytilus edulis* L. II. Autoradiographic study on the incorporation of ^{3}H-proline, *Z. Zellforsch Mikrosk Anat.* **104,** 358–374.

Ramsay, E., Hadgraft, J., Birchall, J., Gumbleton, M. (2000) Examination of the biophysical interaction between plasmid DNA and the polycations, polylysine and polyornithine, as a basis for their differential gene transfection in-vitro, *Int. J. Pharmaceutics* **210,** 97–108.

Ravindranath, M. H., Ramalingam, K. (1972) Histochemical identification of dopa, dopamine and catechol in the phenol gland and the mode of tanning of byssus threads of *Mytilus edulis*, *Acta Histochem.* **42**, 87–94.

Reihmann, M. H., Ritter, H. (2000) Enzymatically catalyzed synthesis of photocross-linkable oligophenols, *Macromol. Chem. Phys.* **201**, 1593–1598.

Richards, A. N., Gies, W. J. (1902) Chemical studies of elastin, mucoid, and other proteids in elastic tissue, with some notes on ligament extractives, *Am. J. Physiol.* **7**, 93–134.

Sakharov, D. V., Jie, A. F. H., Bekkers, M. E. A., Emeis, J. J., Rijken, D. C. (2001) Polylysine as a vehicle for extracellular matrix-targeted local drug delivery, providing high accumulation and long-term retention within the vascular wall, *Arteiosclerosis Thrombosis and Vascular Biology* **21**, 943–948.

Schaefer, J., Kramer, K. J., Garbow, J. R., Jacobs, G. S., Stejskal, E. O., Hopkins, T. L., Speirs, R. D. (1987) Aromatic cross-links in insect cuticle: Detection by solid state ^{13}C and ^{15}N NMR, *Science* **235**, 1200–1204.

Sekine, T., Nakamura, T., Shimizu, Y., Ueda, H., Matsumoto, K., Takimoto, Y., Kiyotani, T. (2001) A new type of surgical adhesive made from porcine collagen and polyglutamic acid, *J. Biomed. Mater. Res.* **54**, 305–310.

Shen, W. C., Ryser, H. J., LaManna, L. (1985) Disulfide spacer between methotrexate and poly(D-lysine). A probe for exploring the reductive process in endocytosis, *J. Biol. Chem.* **260**, 10905–10908.

Siegel, R. C., Pinnell, S. R., Martin, G. R. (1970) Cross-linking of collagen and elastin. Properties of lysyl oxidase, *Biochemistry* **9**, 4486–4492.

Silman, H. I., Sela, M. (1967) Biological properties of poly-α-amino acids, in: *Poly-α-Amino Acids: Protein Models for Conformational Studies* (Fasman, G. D., Ed.), New York: Marcek Dekker, 605–673.

Smith-Mungo, L. I., Kagan, H. M. (1998) Lysyl oxidase: Properties, regulation and multiple functions in biology, *Matrix Biol.* **16**, 387–398.

Sperinde, J. J., Griffith, L. G. (1997) Synthesis and characterization of enzymatically-cross-linked poly(ethylene glycol) hydrogel, *Macromolecules* **30**, 5255–5264.

Stahman, M. A. (Ed.) (1962) *Polyamino Acids: Peptides and Proteins*, Madison: The University of Wisconsin Press.

Sugumaran, M. (1988) Molecular mechanisms for cuticular sclerotizaion, in: *Advances in Insect Physiology* (Evans, P., Wigglesworth, V., Eds.), London: Academic Press, 179–231, Vol. 21.

Sung, H. W., Huang, D. M., Chang, W. H., Huang, L. L., Tsai, C. C., Liang, I. L. (1999) Gelatin-derived bioadhesives for closing skin wounds: An *in vivo* study, *J. Biomater. Sci. Polym. Ed.* **10**, 751–771.

Takahashi, Y., Ohkawa, K., Ando, M., Yamada, M., Yamamoto, H. (2001) Adsorption of endocrine disruptors and related compounds using natural polymer composite fibers formed by polyion complex, *Macromol. Mater. Eng.* **286**, 733–736.

Tanzer, M. L. (1976) Cross-linking, in: *Biochemistry of Collagen* (Ramachandran, G. N., Reddi, A. H., Eds.), New York: Plenum, 137–162.

Tanzer, M. L., Waite, J. H. (1982) Collagen cross-linking, *Collagen Rel. Res.* **2**, 177–180.

Tatehata, H., Mochizuki, A., Kawashima, T., Yamashita, S., Yamamoto, H. (2000) Model polypeptide of mussel adhesive protein I. Synthesis and adhesive studies of sequential polypeptides (X-Tyr-Lys)n and (Y-Lys)n, *J. Appl. Polym. Sci.* **76**, 929–937.

Tatehata, H., Mochizuki, A., Ohkawa, K., Yamada, M., Yamamoto, H. (2001) Tissue adhesive using synthetic model adhesive proteins inspired by the marine mussel, *J. Adhesion Sci. Technol.* **15**, 1003–1013.

Thomas, J., Elsden, D. F., Partridge, S. M. (1963) Partial structure of two major degradation products from the cross-linkages in elastin, *Nature* **200**, 651–652.

Trackman, P. C., Zoski, C. G., Kagan, H. M. (1981) Development of a peroxidase-coupled fluorometric assay for lysyl oxidase, *Anal. Biochem.* **113**, 336–342.

Vithayathil, P. J., Murthy, G. S. (1972) New reaction of *o*-benzoquinone at the thioether group of methionine, *Nature* **236**, 101–103.

Waite, J. H., Benedict, C. V. (1984) Assay of dihydroxyphenylalanine (Dopa) in invertebrate structural proteins, *Method Enzymol.* **107**, 397–413.

Waite, J. H., Tanzer, M. L. (1980) The bioadhesive of Mytilus byssus: A protein containing L-dopa, *Biochem. Biophys. Res. Commun.* **96**, 1554–1561.

Waite, J. H., Qin, X. X., Coyne, K. J. (1998) The peculiar collagens of mussel byssus, *Matrix Biol.* **17**, 93–106.

Weis-Fogh, T. (1960) A rubber-like protein in insect cuticle, *J. Exp. Biol.* **37**, 889–906.

Winder, A. J., Harris, H. (1991) New assays for the tyrosine hydroxylase and dopa oxidase activities of tyrosinase, *Eur. J. Biochem.* **198**, 317–326.

Wormell, R. L. (1954) *New Fibres from Proteins.* London: Butterworths Scientific Publications.

Yagi, K., Nagatsu, T. (1960) Condensation products of ethylenediamine with catechol derivatives, *J. Biochem.* **48**, 439–452.

Yamada, K., Chen, T., Kumar, G., Vesnovsky, O., Topoleski, L. D., Payne, G. F. (2000) Chitosan based water-resistant adhesive. Analogy to mussel glue, *Biomacromolecules* **1**, 252–258.

Yamamoto, H. (1987a) Adhesion studies of synthetic polypeptides: a model for marine adhesive proteins, *J. Adhesion Sci. Technol.* **1**, 177–183.

Yamamoto, H. (1987b) Synthesis and adhesive studies of marine polypeptides, *J. Chem. Soc. Parkin Trans I*, 613–618.

Yamamoto, H. (1996a) Marine adhesive protein, Synthetic, in: *The Polymeric Materials Encyclopedia* (Salamone, J., Ed.), Boca Raton, Florida: CRC Press, 4025–4032, Vol. 6.

Yamamoto, H. (1996b) Marine adhesive proteins and some biotechnological applications, in: *Biotechnology & Genetic Engineering Reviews* (Tombs, M., Ed.), Andover: Intercept, 133–165, Vol. 13.

Yamamoto, H., Ohkawa, K. (2002) Polypeptide behavior at interfaces, in: *Encyclopedia of Surface and Colloid Science* (Hubbard, A., Ed.), New York: Marcel Dekker, 4242–4254.

Yamamoto, H., Senoo, Y. (2000) Polyion complex fiber and capsule formed by self-assembly of chitosan and gellan at solution interfaces, *Macromol. Chem. Phys.* **201**, 84–92.

Yamamoto, H., Tanisho, H. (1993) Gel formation and its properties as hydrogel of cross-linked lysine polypeptides using organic cross-linking agents, *Mater. Sci. Eng.* **C1**, 45–51.

Yamamoto, H., Kitsuki, T., Nishida, A., Asada, K., Ohkawa, K. (1999) Photoresponsive peptide and polypeptide systems. 13. Photo-induced cross-linked gel and biodegradation properties of copoly(L-lysine) containing ε-7-coumaryloxyacetyl-L-lysine residues, *Macromolecules* **32**, 1055–1061.

Yamamoto, H., Horita, C., Senoo, Y., Nishida, A., Ohkawa, K. (2001) Polyion complex fiber and capsule formed by self-assembly of poly-L-lysine and gellan at solution interfaces, *J. Appl. Polym. Sci.* **79**, 437–446.

Yu, A.-M., Chen, H.-T. (1997) Electrocatalytic oxidation and determination of ascorbic acid at poly(glutamic acid) chemically modified electrode, *Anal. Chim. Acta* **344**, 181–187.

Yu, M., Deming, T. J. (1998) Synthetic polypeptide mimics of marine adhesives, *Macromolecules* **31**, 4739–4745.

Zhang, X., Collins, L., Sawyer, G. J., Dong, X., Qiu, Y., Fabre, J. W. (2001) *In vivo* gene delivery via portal vein and bile duct to individual lobes of rat liver using a polylysine-based nonviral DNA vector in combination with chloroquine, *Human Gene Therapy* **12**, 3346–3363.

16
Index

b

d

n